Understanding Quantum Physics

A User's Manual

Understanding Quantum Physics

A User's Manual

Michael A. Morrison

University of Oklahoma

Prentice Hall, Upper Saddle River, New Jersey 07458

Library of Congress Cataloging-in-Publication Data

Morrison, Michael A., (date)
Understanding quantum physics : a user's manual / by Michael A. Morrison.
p. cm.
Includes bibliographical references.
Contents: v. 1. Fundamentals
ISBN 0-13-747908-5 (v. 1)
1. Quantum theory. I. Title.
QC174.12.M69 1990 530.1'2–dc20 89-48295
CIP

Editorial/production supervision: Debra Wechsler
Cover design: Ben Santora
Manufacturing buyer: Paula Massenaro

Upper Saddle River, New Jersey 07458

Printed in the United States of America
10

ISBN 0-13-747908-5

PRENTICE-HALL INTERNATIONAL (UK) LIMITED, *LONDON*
PRENTICE-HALL OF AUSTRALIA PTY. LIMITED, *SYDNEY*
PRENTICE-HALL CANADA INC., *TORONTO*
PRENTICE-HALL HISPANOAMERICANA, S.A., *MEXICO*
PRENTICE-HALL OF INDIA PRIVATE LIMITED, *NEW DELHI*
PRENTICE-HALL OF JAPAN, INC., *TOKYO*
PEARSON EDUCATION ASIA PTE. LTD., *SINGAPORE*
EDITORA PRENTICE-HALL DO BRASIL, LTDA., *RIO DE JANEIRO*

To my wife, Mary
who made it possible

Contents

Preface

"I can't stand it, you know," he said,
"your paradoxes are too monstrous..."
"You are quite wrong," said Ambrose.
"I never make paradoxes; I wish I could."

—*The White People*
Arthur Machen

This is a book about physics, physics as natural philosophy. Its themes are embedded in the myriad beautiful connections among the physical concepts that govern the atomic and sub-atomic domain of matter. It is also a book about how these concepts manifest themselves in the mathematical machinery of quantum mechanics. In practice, these formulas are tools for solving problems involving microscopic systems; by applying them to a host of such problems you can begin to gain "quantum insight" into the nature of reality as understood in modern physics. Such understanding amounts to nothing less than a new way of thinking about the universe. My goal in this book is to provide an in-depth introduction to the concepts of quantum physics and their practical implementation.

Although definitely a physics text, this book contains several dollops of history, biography, and philosophy. I want you to appreciate that quantum physics is more than a collection of abstract laws and equations; it is a paradigm of reality at its most fundamental level, a paradigm with profound, provocative implications. And I want you to meet a few of the men and women who, early in the 20th century, created this towering intellectual achievement. So in these pages you'll have a chance to read some of their words.

I have tried to write a "user-friendly" textbook and an honest one. I have not dodged or hidden the difficulties of quantum mechanics or its applications. This is not *Quantum Physics Made Simple*—I don't believe quantum physics can be made *simple*. But I have tried to make it clear, sacrificing where necessary the elegance of brevity and rigor on the altar of clarity.

Nor is this book *Quantum Mechanics for Everybody*. You'll need a firm foundation in freshman and sophomore physics (mechanics, electricity and magnetism, and thermodynamics), a solid working knowledge of calculus and algebra, and at least a passing acquaintance with the physics of waves.

This book contains more material than one can cover in a single semester, so be selective in what you study. I and others who have used the book have found that by omitting some of the optional (starred) sections, undergraduates (at the junior level) can cover Volume I in about 15 weeks.

Since I began writing this book in 1981, many colleagues have asked, "Why write *another* quantum mechanics text?" A fair question. Neither the selection of topics nor their sequence in this book is particularly unconventional. But what I hope distinguishes it is its persistent emphasis on comprehension of the fundamentals of quantum physics and on how various aspects of this subject relate to one another, the sense in which quantum mechanics is a coherent intellectual structure. I care at least as much that you *understand* quantum mechanics as that you know how to *use* it to solve problems.

Consequently Parts I and II focus on explaining and motivating the postulates that underlie the machinery of quantum mechanics. You can cover this material rather quickly, but do take time to think about it. The postulates are the heart and soul of quantum physics. Our first applications, in Part III, are simple single-particle systems in one dimension. Such problems highlight the essentials of quantum physics but avoid the mathematical clutter and complexity of three-dimensional systems. These are not mere pedantic exercises; research physicists often use one-dimensional models to approximate real-world systems that are too complicated to solve or understand. So at the end of most chapters in Part III we'll take a look at such applications.

In Volume II we'll apply these fundamentals to the "real world" of three-dimensional, many-particle microscopic systems and will learn how to solve problems that are too difficult to admit exact solution. There we'll take up approximation methods such as perturbation theory and the variational method, three-dimensional properties such as orbital angular momentum, a new kind of angular momentum called spin, and many-particle systems, including atoms and molecules.

In both volumes I have provided considerable structure to help you organize your study. In addition to the customary chapters, sections, and subsections, they contain all sorts of asides, footnotes, boxed results, tables of formulae, etc. Think of this book as a layer cake with several levels of flavor. The most important material is in the top layer, the text. In the asides and footnotes you'll find all sorts of tasty morsels: supplementary information, extensions of material, suggested readings and references—things that would interrupt your train of thought were they in the text proper.

The best source of advice about how to study a book is not the author. Several students who used this manuscript in preliminary form, suggest the following: *Plan to go through each chapter several times. On your first reading, skip sections or subsections that seem difficult, planning to return to them later. The first time through, don't even read the asides and footnotes; come back to them also on subsequent readings. Above all, work the Questions, Exercises and Problems.*

The best advice I have ever read about how to study was written by the mathematician Paul R. Halmos in his "automathography" *I Want to Be A Mathematician* (New York: Springer-Verlag, 1985), "[S]tudy actively. Don't just read [the text]; fight it! Ask your own questions, look for your own examples, discover your own proofs." Halmos was writing about math, but his advice pertains as well to physics. If you follow it, you will come away from this book with a solid, working knowledge of quantum physics; if you ignore it, the prospects are grim. You can no more learn quantum mechanics without practice than you could learn how to ride a bike, play a piano, or swim without riding, playing, or swimming.

That's why you'll find in each chapter a host of Questions that will provoke you to think about the material and, at the end of each chapter, a wide selection of Exercises and Problems. Some of these anticipate material in subsequent chapters; others extend the content of the chapter you just read; still others provide opportunities to become familiar with new ideas through simple practice. I hope you will at least think through *all* the Questions and work *all* the Exercises. The Problems are a bit more challenging, but I've provided lots of help: most contain detailed directions, suggestions, helpful hints, etc. I cannot overemphasize the importance of your working as many of these as possible.

You'll also find throughout the book a host of very detailed examples. Rather than simply work these examples, I've tried to show you how to approach them, how to figure out how to work them. Please don't treat these as recipes in a cookbook. Rather, consider them as paradigms; each illustrates one of many possible ways to apply the principles you have studied. You will learn more and have more fun if instead of parroting what I do, you discover your own approaches to problem solving

Of course, presenting examples in great detail takes a lot of space, and I've been able to include only a handful of the many useful illustrations of introductory quantum mechanics. So seek out others. Dig around in the many fine quantum books you'll find in your library (see the Bibliography at the end of this book). Better yet, make up your own examples. Again, Halmos's advice is invaluable: "A good stock of examples, as large as possible, is indispensable for a thorough understanding of any concept, and when I want to learn something new, I make it my first job to build one."

Each chapter also contains recommended readings, annotated lists of additional sources that will enhance your understanding and enjoyment of quantum physics. Please don't ignore these recommendations. Reading several explanations of important topics is good strategy for the study of any subject, but it's particularly important for quantum mechanics, which is so counter-intuitive that you need to see it from several vantage points.

This book grew out of my desire to convey a sense of the fascination and beauty of quantum physics. This won't be the only time you study this subject, nor should it be. Quantum mechanics is like a many-faceted jewel: every time you look at it you discover amazing new things. But I have the privilege of introducing the subject to you, so I particularly hope that this book communicates some of my enthusiasm for it. I believe that learning quantum physics can and should be a joyous activity. Although undeniably hard work, the study of quantum mechanics and its applications to atoms, molecules, solids, nuclei, and the whole rest of the microcosm can be a tremendously exciting intellectual adventure.

Finally, I want to know what you think. I hope that, if you feel so inclined, you'll write to me at the Department of Physics and Astronomy, University of Oklahoma, Norman, Oklahoma 73019 with your opinions of this book, your reaction to quantum physics, any errors you find in the text, or any suggestions you have for improvement should there be a second edition. But for now, settle down in a comfortable chair, get out some scratch paper, and dig in. You are about to be plunged into the Marvelous.

Michael A. Morrison
Norman, Oklahoma

Acknowledgments

One of the great pleasures of finishing a book is thanking those who helped in its creation. First and foremost I owe a great intellectual debt to the mentors who have helped me learn how to think about quantum physics: my professors at Rice University, G. King Walters, Neal F. Lane, and Thomas L. Estle, and my long-time friend and colleague Dr. Lee A. Collins. My other mentors I know only through their works. Like most later-generation textbooks, mine is in part a mosaic of variations on the best of what I have found in many sources, primarily in the vast library of quantum mechanics texts that preceded this one. I have acknowledged these sources collectively in the Bibliography at the end of this book and, where memory served, in footnotes to the text.

Closer to home, I am grateful to my fellow faculty members in the Department of Physics and Astronomy at the University of Oklahoma. During the many years since I began work on this book, they have consistently tolerated, encouraged, and supported me. In particular, Drs. Robert Petry, Thomas Miller, Deborah Watson, and Gregory Parker used various incarnations of this manuscript in classes and provided valuable feedback and brickbats. And Drs. Miller, Parker, Jack Cohn, Ryan Doezema, and Suzanne Willis read portions of the manuscript for technical accuracy.

Students played a crucial role in the latter stages of work on this book. I have tested all or part of the manuscript in five courses. Although space precludes my thanking all of the long-suffering students in those classes, I must single out a few—Thomas Haynes, Brian Argrow, José Rodriguez, Bok-hyun Yoon, and Bryan Biegel—for their extraordinary efforts on behalf of this project. Three graduate students, Wayne Trail, Bill Isaacs, and Brian Elza, carefully read the manuscript in proof and found errors I would not have believed were still in it. Finally, I'm indebted to three colleagues, Drs. Ron Olowin, Richard Henry, and Sheridan Simon, who provided solace and suggestions during our eleventh-hour search for a new title.

I and this book also benefited from the advice of a number of outside reviewers, including Drs. Robert Hallock, Sheridan Simon, John L. Stevens, Sanford Kern, H. A. Gersh, Stephan Estreicher, and Robert Silbey. Thanks folks: I didn't adopt all your suggestions, but I considered each one carefully.

Preparing this manuscript for production would have been impossible had it not been for superb technical help from two sources. Ms. Johnette Ellis provided assistance beyond the call of duty during the trying end-game months of the summer of 1987. And my wife Mary, who does not share my fondness for physics, nevertheless proofread this entire, vast manuscript.

Finally, I must acknowledge the staff at Prentice Hall and the physics editors with whom I worked on this project: Logan Campbell, who initiated it; Doug Humphrey, who served as midwife and father confessor through the long years of its evolution; and Holly Hodder, who saw the final manuscript into production. Most of all, I'm grateful to my production editor, Debra Wechsler, whose care, concern, and oversight has helped immeasurably to make this the book I envisioned when I started.

It is characteristic of all deep human problems
that they are not to be approached without
some humor and some bewilderment—
science is no exception.

—Freeman Dyson
Disturbing the Universe

In trying to achieve success
No envy racks our hearts,
For all we know and all we guess,
We mutually impart.

—Gilbert and Sullivan
Princess Ida

We shall go on seeking [truth] to the end,
so long as there are men on the earth.
We shall seek it in all manner of strange ways;
some of them wise,
and some of them utterly foolish.
But the search will never end.

—Arthur Machen
"With the Gods in Spring"

CHAPTER 1

Introductory

Why Quantum Physics?

I wish to guide you on an excursion
which is long and rather difficult
but worth the trouble,
so I am going to climb ahead of you,
slowly.

—Golem's Inaugural Lecture
in *Imaginary Magnitude*
by Stanislaw Lem

How does the universe work? What are the principles that underlie natural phenomena? Philosophers have pondered such questions for millennia; their answers reflect the *world view* of Western civilization—how we perceive ourselves in relation to the universe.

In the early years of this century the dominant world view was *determinism*, a tenet that had its origins in the philosophical musings of Descartes. Descartes' metaphor for the universe was a gigantic clockwork machine, a Rube Goldberg device that grinds relentlessly onward, towards eternity. According to this interpretation, the future is preordained and fully predictable. All of existence is described precisely by physical laws, and, in principle, can be known by man.

Nature, however, had a surprise in store for philosophers. In the first decades of the twentieth century, they were to see this world view subverted, uprooted, and toppled by a young upstart physical theory called *quantum mechanics*.

Theoretical physics has come a long way since those fateful days in the 1920's and 30's when physicists were taking the first tentative steps towards a radical new definition of reality. So far have we come that in 1981 the physicist Steven Hawking could reasonably predict that[1]

> ... by the end of the century ... we might have a complete, consistent, and unified theory of the physical interactions which would describe all possible observations.

The foundation of that unified theory is quantum mechanics, the subject matter of this book.

What is quantum mechanics? You probably know that if you want to understand how a transistor works or the principles behind the laser in your compact disc player you'd better not use classical physics. Newton's mechanics and Maxwell's electromagnetic theory can explain *macroscopic* phenomena, such as the motion of billiard balls or rockets, but fail spectacularly when applied to *microscopic* phenomena, such as proton-atom scattering or the flow of electrons in a semiconductor. An understanding of such processes requires a better mousetrap: quantum mechanics.

More precisely, **quantum mechanics** is a collection of *postulates* based on a huge number of experimental observations, and the tools derived from those postulates. We use these tools to analyze, predict, and understand microscopic phenomena. Hence quantum theory forms the bedrock of the modern physics of matter—atoms, molecules, and solids—and is at the core of such diverse sciences as astrophysics and biochemistry. In its more pragmatic guises, quantum mechanics plays a vital role in a vast array of technological applications, some of which we use every day—*e.g.*, the microchips in the computer on which I am writing this chapter.

[1] In an article optimistically entitled "Is the End in Sight for Theoretical Physics?" [*Phys. Bull.* **32**, 15 (1981)].

Thus, in one form or another, quantum physics pervades the life of everyone in our high-tech world. Discussing the importance of quantum mechanics to non-physicists, Heinz R. Pagels has written[2]

> When the history of this century is written, we shall see that political events—in spite of their immense cost in human lives and money—will not be the most influential events. Instead the main event will be the first human contact with the invisible quantum world and the subsequent biological and computer revolutions.

You are about to make contact with the quantum world.

In this chapter, we'll ease into this bizarre world, beginning in § 1.1 with a short review of the structure of classical mechanics. We'll also see the way in which that structure supported the deterministic ideas of philosophers of the early 1900's. Then in § 1.2 we'll take a whirlwind tour of the world according to quantum physics, focusing on some of the new ideas that so devastated the Newtonian view.

This chapter and the next provide a bird's eye view of quantum theory; in them, we'll examine qualitatively the central ideas of this theory. Then, beginning in Chap. 3, we'll set off on our voyage of discovery, formulating these notions mathematically and applying them to the denizens of the microscopic universe.

1.1 THE CLASSICAL POINT OF VIEW

Consider Newtonian mechanics. In their simplest form, the laws of classical mechanics are written in terms of particle trajectories. So the *trajectory* underlies the structure of classical physics, and the *particle* underlies its model of physical reality. We start with these basic elements.

Particles and Their Trajectories

Physicists, classical or otherwise, study the universe by studying isolated fragments of it. These little chunks of existence are called *systems*: classical physics pertains to macroscopic systems, quantum physics to microscopic systems.[3] A **system**, then, is just a collection of particles that interact among themselves via internal forces and that may interact with the world outside via external fields.

To a classical physicist, a **particle** is an indivisible mass point possessing a variety of physical properties that can be measured. In physical theory, measurable quantities are called **observables**. By listing the values at any time of the observables of a particle, we can specify its **state**. (The trajectory is an equivalent, more convenient way to specify a particle's state.) The *state of the system* is just the collection of the states of the particles comprising it.

Aside: Extrinsic versus Intrinsic Properties. Classical physicists often characterize properties of a particle as *intrinsic* or *extrinsic*. **Intrinsic properties** don't depend on

[2] *The Cosmic Code: Quantum Physics as the Language of Nature* (New York: Simon and Schuster, 1982), page 98.

[3] The validity of partitioning the universe into systems for the convenience of scientists is discussed in Chap. 4 of *The Origin of Knowledge and the Imagination* by J. Bronowski (New Haven: Yale University Press, 1978).

the particle's location, don't evolve with time, and aren't influenced by its physical environment; rest mass and charge are intrinsic properties. **Extrinsic properties**, on the other hand, evolve with time in response to the forces on the particle; position and momentum are extrinsic properties.

According to classical physics, all properties, intrinsic and extrinsic, of a particle *could* be known to infinite precision. For example, we could measure the precise values of the position and momentum of a particle at the same time. Of course, precise knowledge of everything is a chimera—in the real world, neither measuring apparatus nor experimental technique is perfect, and experimental errors bedevil physicists as they do all scientists. But *in principle* both are perfectible: that is, our knowledge of the physical universe is limited only by ourselves, not by nature.

How does a classical theorist predict the outcome of a measurement? He uses trajectories.[4] The **trajectory** of a single particle consists of the values of its position and momentum at all times after some (arbitrary) initial time t_0:[5]

$$\{\mathbf{r}(t), \mathbf{p}(t)\,;\, t \geq t_0\}, \text{trajectory} \tag{1.1}$$

where the (linear) momentum is, by definition,

$$\mathbf{p}(t) \equiv m\frac{d}{dt}\mathbf{r}(t) = m\mathbf{v}(t), \tag{1.2}$$

with m the mass of the particle.

Trajectories are the "state descriptors" of Newtonian physics. To study the evolution of the state represented by the trajectory (1.1), we use Newton's Second Law,

$$m\frac{d^2}{dt^2}\mathbf{r}(t) = -\nabla V(\mathbf{r}, t), \tag{1.3}$$

where $V(\mathbf{r}, t)$ is the potential energy of the particle.[6] To obtain the trajectory for $t > t_0$, we need only know $V(\mathbf{r}, t)$ and the **initial conditions**, the values of $\mathbf{r}$ and $\mathbf{p}$ at the initial time t_0. With the trajectory in hand, we can study various properties of the particle in the state that the trajectory describes, *e.g.*, its energy or orbital angular momentum.

Notice that classical physics tacitly assumes that we can measure the initial conditions without altering the motion of the particle—*e.g.*, that in measuring the momentum of the particle, we transfer to it only a negligible amount of momentum.[7] Thus *the scheme of classical physics*—which is illustrated in Fig. 1.1—*is based on precise specification of the position and momentum of a particle.*

I do not mean to imply by this cursory description that classical physics is easy. Solving the equations of motion for a system with lots of particles interacting among

[4]This is, of course, a gross oversimplification. For a closer look at Newtonian physics, see Herbert Goldstein's classic text *Classical Mechanics*, now in its second edition (Reading, Mass.: Addison-Wesley, 1983).

[5]Conventionally, we think of a trajectory as a path in geometrical space. In classical physics, however, a trajectory is usually taken to mean a path in "phase space" defined by the position *and* momentum at all times $t \geq t_0$.

[6]As written, Eq. (1.3) is valid for conservative forces, *i.e.*, ones that do not depend on the velocity $\mathbf{v}$. Recall that the force is related to the potential energy by $\mathbf{F} = -\nabla V(\mathbf{r}, t)$.

[7]To determine $\mathbf{p}(t_0)$ we must measure $\mathbf{v}(t_0)$. This quantity is the rate of change of position, so to measure it we must observe $\mathbf{r}(t)$ at two nearby times—say, t_0 and $t_0 + \Delta t$, making the (known) time increment Δt as small as our apparatus will allow. Such position measurements necessarily entail a change in the particle's momentum, but this effect is insignificant for a macroscopic particle.

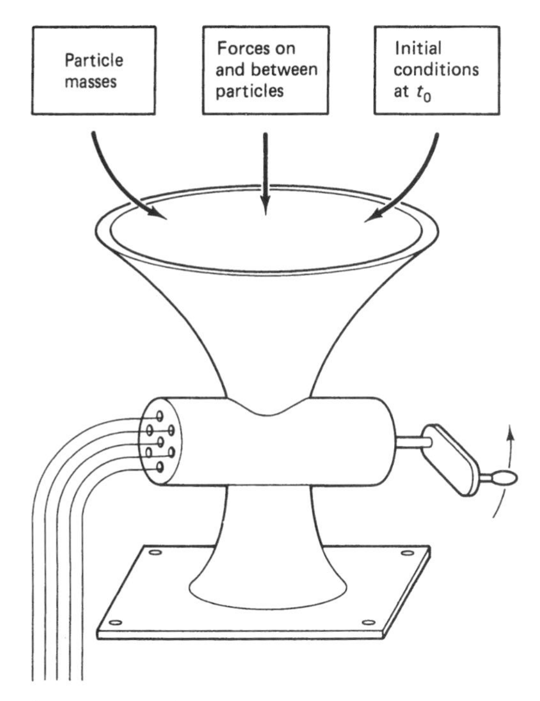

Figure 1.1 The Newtonian Universe Grinder. Fed the proper information, the great machine of classical mechanics can predict the future of the universe. This machine is now obsolete.

themselves or exposed to complicated external fields is a formidable task. Contemporary applications of Newtonian mechanics to problems like celestial mechanics (the motion of the planets in our solar system) require massive computers and budgets to match. At worst, one may be unable, using present technology, to solve these equations. The important point is that *in principle*, given sufficient time and scratch paper, we could predict from the initial positions and momenta of all particles the *precise* future behavior of any system—indeed, of the entire universe. That's the classical view.

The Determinate Universe

The underlying assumptions and philosophical implications of classical physics are so familiar that you may have never given them a second thought. For example, classical physics ascribes to the universe an **objective reality**, an existence external to and independent of human observers. This assumption reassures us that when we do research in physics we are studying what is actually "out there," beyond and distinct from our consciousness, and that we can design experiments that do not affect in any significant way the systems they measure. That is, in an objective universe we can control the interaction between the *observer* and the *observed* and, if clever enough, can make this interaction negligibly weak.

Our other central assumption about the nature of the classical universe is that it's predictable: knowing the initial conditions of the constituents of any system, however complicated, we can use Newton's Laws to predict its future. Precisely, without ambiguity or uncertainty. This notion is the essence of **determinism**, which, supported by Newtonian mechanics, dominated philosophical thinking until the advent of quantum theory.

If the universe is determinate, then for every effect there must be a cause. After all, if *all* events, *all* phenomena can be predicted and explained precisely by physics, then it must be possible to backtrack from any event to find, somewhere in the dim recesses of history, its cause. We may have to look long and hard—the bird that just fell off a tree limb outside my window may have been influenced by an event during the Peloponnesian wars—but somewhere there *is* a cause. Or so says the principle of **causality**.

Causality was an important element of nineteenth century philosophy. It had important implications for physicists of that era, for it guaranteed the reproducibility of experiments—*i.e.*, that two *identical* systems with the *same* initial conditions (*i.e.*, in the same state) subject to the *same* measurement will yield *identical* results. Same causes: same effects. Very neat and tidy.

I suppose the deterministic world view implied by Newtonian mechanics could, even today, seem comforting, for it implies that the universe can be understood fully and dealt with rationally. But it has a dark side. For example, if the universe is ruled by causality, then free will is a meaningless concept, for the universe is a vast, mindless machine that controls our every action, our every decision. Everything that happens happens because of something that happened in the past, not because we chose to make it happen. According to this view, if you "decide" to get married or to take a job or to collect water buffalo, you do so not by choice but because of past events. In such a universe, our loves, hopes, and dreams are but delusions, hiding the grim reality that we are but cogs in Descartes' clockwork machine. Goals are irrelevant; human aspirations pointless.

Thus determinism is a dehumanizing philosophy. It describes a universe that is infinitely predictable, holding few surprises. All can be known; nothing is unexpected. Determinism is boring. It is also wrong.

1.2 THROUGH THE LOOKING GLASS: THE QUANTUM POINT OF VIEW

> We shall be plunged normally into the marvelous.
>
> —Noël Arnaud

We have seen that Newtonian mechanics is constructed from particle trajectories. One reason classical physics fails when applied to *microscopic* systems is that the constituents of such systems are not particles in the classical sense of the word. Moreover, they do not have trajectories. Thus to understand the physics of the microworld—that is, to understand the results of experiments on microscopic particles—we must radically revise our conceptual description of a particle.

When Is a Particle Not a Particle?

In describing the quantum world, we must abandon the classical notion of a *particle* as an indivisible mass point with an independent existence and well-defined, measurable extrinsic properties. Whatever microscopic entities *are*, they are certainly not indivisible mass points. We can verify this assertion in the laboratory. In suitably designed experiments quantum "particles" act like classical particles—appearing, for example, as spots on a screen. But in other experiments their transit through space is like that of a wave, manifesting behavior such as diffraction and interference, as though they were diffuse wave fronts (see §2.5). This apparently contradictory behavior is a manifestation of the **wave-particle duality** that characterizes the domain where quanta dwell.

In our study of duality in Chap. 2, we'll see why it poses such a challenge to the student of quantum physics: it renders useless the conventional mental images with which we visualize physical phenomena. Thus, an atomic electron is *not* a tiny "planet" orbiting a "nuclear sun," as in the Rutherford and Bohr models of the atom. But neither is it a "fuzzy" thing, smeared out over a region of space, as it is portrayed in many introductory physics texts. The electron is *something else*, neither particle nor wave but eerily reminiscent of both.[8]

The dual nature of subatomic particles subverts the classical concept of a particle. But the true nature of microscopic entities is even more nebulous than is implied by wave-particle duality, for *the properties of quantum particles are not, in general, well-defined until they are measured.* In a sense, the physical properties of electrons, protons, and the like are "potential" or "latent" properties until an experimenter—a *macroscopic* being—performs a measurement (see Chap. 5).

You will encounter this disquieting aspect of quantum mechanics if you ask a quantum physicist to predict the value you would obtain were you to measure, say, the position of an electron in a metal. He cannot give you a definite answer, even if he knows fully the state of the electron just prior to the proposed measurement. The inability of quantum theory to provide precise answers to such simple questions is not a deficiency of the theory; rather it is a reflection of its essential nature. We can see this if we look at how quantum mechanics specifies the state of a particle.

Unlike a classical state, a quantum state is a conglomeration of several *possible* outcomes of measurement of physical properties. At most, quantum physicists can tell you only the *possible* outcomes and the *probability* that you will obtain one or another of them. Quantum mechanics is expressed in the language of *probabilities*, not certainties. (A grammar of this new language appears in Chap. 3.) It is inherently *statistical* in nature, describing not definite results of a measurement on an individual system, but rather possible results of measurements on a large number of identical systems. What, then, controls what actually happens in a particular measurement—*e.g.*, which of the possible values of position a particle exhibits when we measure this quantity? *Random chance.*

Of course, were we to carry out a position measurement on a single particle, we would get a single value. So immediately *after* the measurement, we can meaningfully talk about *the* position of the particle—its position is the number we got in the measurement. But what about immediately *before* the measurement? According to quantum mechanics, *the particle does not then have a position.* Rather, its position prior to measurement is latent—a mere possibility, waiting to be made actual.

Thus by the act of measurement, we change the state of the particle from one in which it is characterized by a plethora of possible positions to one in which it has a single, well-defined position. Clearly, measurement will play a more vital role in quantum

[8]Nonetheless, we persist in referring to subatomic whatevers as "particles." (To avoid an excrescence of punctuation, we leave off the quote marks.) The distinguished physicist Sir Arthur Eddington once suggested the term *wavicles* for quantum particles. Perhaps fortunately, the word never caught on.

physics than in classical physics. When we study the microworld, experimentation is not just a way of discovering the nature of external reality, but rather is a way of creating certain aspects of reality! In contrast to the assumptions of classical physics, *an observer cannot observe a microscopic system without altering some of its properties.*

Intrinsically indeterminate interactions between the observer and the observed are an inevitable feature of the quantum universe, one that applies not just to position, but to all properties of microscopic particles. Some physicists believe that we macroscopic observers "create" the microscopic building blocks of the universe by actualizing via measurement their various physical properties.

This interaction is *unavoidable*: the effect of the observer on the observed *cannot be reduced to zero*, in principle or in practice. This fact, which is reflected in the mathematical structure of quantum theory and has been verified by countless experiments, demolishes the philosophical notion of an objective universe, the idea that what we study in physics is necessarily a "real world" external to and independent of our perceptions.[9]

The observer-observed interaction is also *uncontrollable*. Although we can be sure that when we measure the properties of a microscopic particle we will change its state, we can neither predict nor influence precisely *how* the state will be changed. Random chance determines that.

In fact, randomness governs microscopic phenomena even when man is not studying them. A famous example of this seeming perversity of nature is radioactive decay of an atom. Quantum mechanics prohibits us from predicting precisely when a particular atom will decay, emitting radiation. At most, we can determine the probability that it will decay in a given time interval. The role of randomness in microscopic physical processes shatters the illusion that the universe is deterministic.

The Decline and Fall of the Trajectory

So much for the idea of using the classical model of a particle to describe a microscopic entity. But can we salvage the basic state descriptor of classical physics, the trajectory?

The first step in determining the trajectory of a particle is measuring its initial conditions, $x(t_0)$ and $p_x(t_0)$.[10] To determine the accuracy of our results, we would perform such a measurement not on just one particle, but on a large number of *identical* particles, all in the same state.[11]

Each *individual* measurement yields a value for x and a value for p_x (subject to experimental uncertainties). But the results of *different* measurements are not the same, even though the systems are identical. If graphed, these results are seen to fluctuate about a central peak, as illustrated in Fig. 1.2. (A similar spread of results characterizes measurement of the y- or z- components of position and momentum.)

At first, the fact that the results of many identical experiments are not the same doesn't worry us; we think it's just a manifestation of the experimental error that bedevils

[9]It also gives rise to a host of fascinating metaphysical questions: Do particles exist if their properties are not being measured? Is what we study in physics reality or merely our perceptions of reality? What are the implications of quantum theory for the existence of God? With some regret, I'll not discuss these questions in this text.

[10]Actually, we have to measure all three Cartesian components of position and momentum: $x(t_0)$, $y(t_0)$, $z(t_0)$, and $p_x(t_0)$, $p_y(t_0)$, $p_z(t_0)$. But I can make my point by considering only the x components.

[11]Such an aggregate of identical systems in the same state is called an *ensemble*, a concept you'll meet formally in Chap. 3.

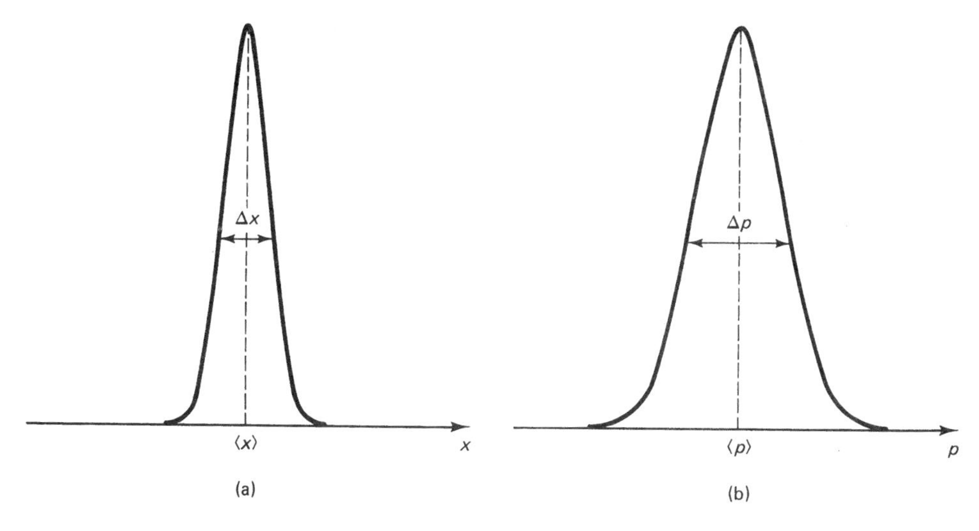

Figure 1.2 The results of measurement of the x components of the position and momentum of a large number of identical quantum particles. Each plot shows the number of experiments that yield the values on the abscissa. Results for each component are seen to fluctuate about a central peak, the mean value.

all measurements. According to classical physics, we can reduce the errors in x and p_x to zero and thereby determine precisely the initial conditions.

But *we cannot unambiguously specify the values of these observables for a microscopic particle.* This peculiarity of nature in the microscopic realm is mirrored in the *Heisenberg Uncertainty Principle* (HUP). In its simplest form, the HUP shows that any attempt to simultaneously measure $x(t_0)$ and $p_x(t_0)$ necessarily introduces an imprecision in each observable.[12] No matter how the experiment is designed, the results are *inevitably* uncertain, and the *uncertainties* $\Delta x(t_0)$ and $\Delta p_x(t_0)$, which are measures of fluctuations like those in Fig. 1.2, cannot be reduced to zero. Instead, their product must satisfy the condition (see Chap. 4)

$$\Delta x(t_0)\,\Delta p_x(t_0) \geq \frac{1}{2}\frac{h}{2\pi}, \tag{1.4}$$

where h is **Planck's Constant** (See Chap. 2)

$$h \equiv 6.626 \times 10^{-34}\,\text{J-sec}. \tag{1.5}$$

Not a very big number, but not zero either. [Similar constraints apply to the pairs of uncertainties $\Delta y(t_0)$, $\Delta p_y(t_0)$ and $\Delta z(t_0)$, $\Delta p_z(t_0)$.] Position and momentum are *fundamentally incompatible observables*, in the sense that knowing the precise value of one precludes knowing anything about the other.

But there is a deeper dimension to the Heisenberg Uncertainty Principle. Quantum theory reveals that the limitation reflected by Eq. (1.4) on our ability to simultaneously

[12] You are probably familiar with one illustration of this principle: the gamma-ray microscope. This experiment is discussed in almost all lower-level modern physics texts as a *gedanken* (thought) experiment. But today, with the advent of electron microscopes, it is a reality. See B. G. William, *Amer. Jour. Phys.*, **52**, 435 (1984).

measure x and p_x is *implicit in nature*. It has nothing to do with a particular apparatus or with experimental technique. Quantum mechanics proves that a particle *cannot* simultaneously have a precise value of x and a precise value of p_x. Indeed, in Chap. 4 we'll derive the HUP from first principles.

Similar uncertainty principles constrain our ability to measure other pairs of incompatible observables. But uncertainty relations such as (1.4) are not veiled criticisms of experimental physicists. They are *fundamental limitations on knowledge: the universe is inherently uncertain.* Man cannot know all of existence. We might think of uncertainty relations as nature's way of restraining our ambitions.

The Heisenberg Uncertainty Principle strikes at the very heart of classical physics: the trajectory. Obviously, if we cannot know the position and momentum of a particle at t_0 we cannot specify the initial conditions of the particle and hence cannot calculate its trajectory. Indeed, since the HUP applies *at any time*, it makes no sense to ascribe a trajectory to a microscopic particle. We are forced, however reluctantly, to conclude that *microscopic particles do not have trajectories.* But once we throw out trajectories, we must also jettison Newton's Laws. And there goes the ball game: stripped of its basic elements and fundamental laws, the whole structure of classical physics collapses.

The demise of classical physics occurred around 1930. Sitting amidst the rubble, physicists of that era realized that their only alternative (other than to change careers) was to rebuild—to construct from scratch a new physical theory, one based on elements other than trajectories and on laws other than those of Newton and Maxwell. Thus began the quantum revolution, whose reverberations are still being felt.

Classical and Quantum Physics Get Together

To say that classical physics died around 1930 is a bit too dramatic. The physics of Newton and Maxwell still accurately and beautifully describes the macroscopic world. Knowing this, physicists developing quantum mechanics demanded that when applied to macroscopic systems, the new physics must reduce to the old physics. Thus, as the size of the system being studied increases, the quantum laws of motion (which we'll discover in Chap. 6) must go over smoothly into those of Newtonian mechanics, and non-classical phenomena such as uncertainty and duality must become undetectable. Neils Bohr codified this requirement into his *Correspondence Principle*, the essential elements of which are sketched in Fig. 1.3. We shall return to this principle repeatedly, since the ability of quantum mechanics to satisfy it is an important theoretical justification of the new physics.

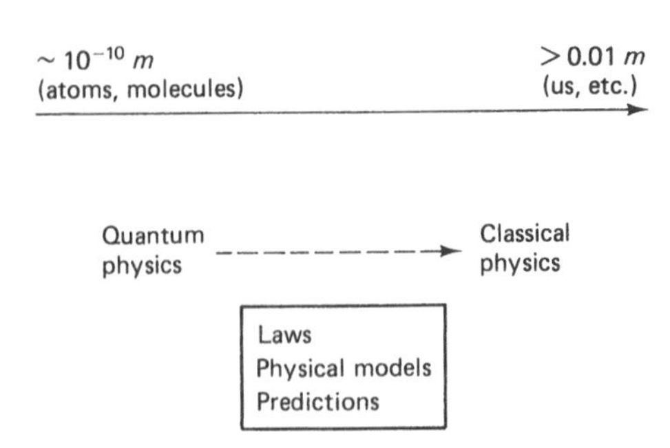

Figure 1.3 The rudiments of the Correspondence Principle, the theoretical link between the microworld and the macroworld.

The *experimental* justification of quantum theory is that it works. The predictions, qualitative and quantitative, of quantum mechanics have been verified by a host of experiments on a diverse collection of systems. These verifications are of paramount importance to physics, for in some ways quantum theory is a fragile creation. It is an *inferential* theory, in which we devise a handful of postulates (see Part II) and from them derive equations, which can then be put to the test of experimental justification. If the equations pass this test, then we have confidence that the postulates, which cannot be directly tested, are correct. But it would take *only one repeatable experiment* whose results confounded the equations of quantum mechanics to bring the whole structure tumbling down. To date, the edifice stands. Quantum mechanics remains the only theory we have to explain how matter is constructed and how physical processes at the atomic level work.

The Indeterminate Universe

As this whirlwind overview may suggest, the new physics, quantum physics, differs from classical physics in nearly every way imaginable—from its elements and laws to the kind of knowledge it provides. It is in this last respect that the implications of the quantum revolution are most profound.

Classical physics yields *precise* information about the properties and behavior of individual, independent systems. It holds out to the scientist the promise that, in principle at least, he can know everything. Quantum physics says that this is just not so, that nature imposes fundamental limitations on knowledge, constraints such as the Heisenberg Uncertainty Principle.

These limitations derive from the *probabilistic nature of quantum mechanics*. The new physics yields only *statistical* information about *aggregates of identical systems*. It is mute about the behavior of individual systems. Moreover, the statistical information provided by quantum theory is limited to the results of measurements.[13] That is, quantum physicists are not allowed to infer facts about a system unless these facts could be verified by experiment. This is a severe limitation—for example, it prohibits us from ascribing an orbit (a path) to a particle, because measurements of position are necessarily performed at discrete times (see § 2.5). Even if we measure the position of a particle at two very nearby times, t_1 and t_2, we cannot even make the simple inference that it traveled from $\mathbf{r}(t_1)$ to $\mathbf{r}(t_2)$ via a path in geometric space; we can only say that we found it at $\mathbf{r}(t_1)$ at t_1 and then at $\mathbf{r}(t_2)$ at t_2.

This limitation, that physics describes only observed phenomena, forces us to think about reality in a strange, new way. At its most fundamental level, reality seems to be discontinuous. Here is Erwin Schrödinger, the father of quantum mechanics, on this most non-classical feature of his theory:

> ... it is better to regard a particle not as a permanent entity but as an instantaneous event. Sometimes these events form chains that give the illusion of permanent beings—but only in particular circumstance and only for an extremely short period of time in every single case.

Quantum physics is a harsh taskmaster.

[13] In the words of one textbook author: *Thou shalt not make any statement that can never be verified.* [R. G. Winter, *Quantum Physics* (Belmont, Ca.: Wadsworth, 1979)].

In closing, I should note that all this has not set well with many physicists. Even today, some find the world view implied by quantum mechanics—a subjective universe, acausal and indeterminate, where the knowledge accessible to man is limited—to be intolerably offensive. And the battle over the interpretation of quantum mechanics goes on. (You'll find accounts of minor skirmishes in Parts II and IV of this book.) But all agree on the usefulness of the theory for predicting and understanding the physics of matter. It is, after all, the only game in town.

1.3 FINAL THOUGHTS: WHAT'S IT ALL ABOUT?

From the overview in § 1.2, you may have concluded that quantum mechanics is not a particularly easy subject. 'Tis true; to pretend otherwise would be deceitful. But *why* is learning quantum mechanics hard?

There are, I think, three reasons. First, we grew up in and now inhabit a macroscopic world. Our intuition—our sense of how things *ought to behave*—has been reinforced every waking minute by experiences in a world that abides by classical physics. Moreover, as students of physics, our intuition has been deepened and further reinforced by our study of classical physics. Quantum physics is an affront to that intuition. To understand it, to use it effectively, we must develop a new way of thinking about the way things work, because, as Richard P. Feynman has written,[14]

> Things on a very small scale behave like nothing you have any direct experience about. They do not behave like waves, they do not behave like particles, they do not behave like clouds, or billiard balls, or weights on springs, or like anything that you have ever seen.

Second, quantum physics operates at a level that is one step removed from reality. It is more abstract than classical physics. While classical physicists can deal with well-defined trajectories that describe particles they can visualize, quantum physicists must forever wander through a haze of uncertainty, probabilities, and indeterminacy, trying to understand a universe they cannot directly observe. The microcosm can be understood, but it cannot be seen.

Third, quantum mechanics is inherently mathematical. There is an important distinction here between classical and quantum physics. In classical physics we use mathematical methods to implement the ideas of the theory, but we can discuss those ideas without recourse to mathematics. This is not possible with quantum mechanics. Notwithstanding the achievements of writers who have recently tried to popularize quantum theory—several of whose books I have included in this chapter's Selected Reading list—I believe it is impossible to fully grasp the principles of quantum theory without seeing them expressed mathematically. Math is more than a tool for solving quantum mechanics problems: *mathematics is the language of quantum physics*. As we move deeper into quantum theory, we'll see physics and math become inextricably intertwined. Thus quantum physics demands that we think in a new, abstract, inherently mathematical way—no easy task.

[14]In Volume III of *The Feynman Lectures on Physics* (Reading, Mass.: Addison-Wesley, 1965). Only the third volume of Feynman's lecture series is devoted to quantum mechanics, but I strongly recommend the whole series to every serious student of physics.

Finally, we must re-think and redefine many familiar words and concepts, and when we use these words to discuss quantum concepts, we must do so with great care and precision. Familiar terms such as "particle" and "motion" assume in the quantum domain subtle overtones and shades of meaning that we ignore at our peril. We have already seen an example of this in our discussion in § 1.2 of the "position" of a microscopic particle. The "position" of a quantum particle prior to its measurement is not a single, well-defined number, such as 10.7 m. Rather it must be specified as a collection of several—maybe an infinity of—values, none of which represent *the position* of the particle. Instead, each value represents a possibility, a location at which the particle *might* be found.

In spite of these roadblocks—our classical intuition, the inherently abstract, mathematical nature of quantum mechanics, and the need to use old words in new ways—I believe that you can understand quantum physics. Even if you cannot visualize what goes on in the microworld of atoms and molecules, you can grasp the beautiful and powerful physical laws that govern that world. After all, countless physicists have done it.

This is our world. And welcome to it.

ANNOTATED SELECTED READINGS

> ... if you were acquainted with certain books on the subject,
> I could recall to your mind phrases
> which might explain a good deal
> in the manuscript that you have been reading.
>
> —from *The White People*
> by Arthur Machen

Quantum Physics for Beginners

If you like to read and are interested in quantum physics, you are in luck. Several excellent books on quantum physics for lay readers have recently hit the stands. I strongly recommend that you acquire at least one of these to supplement your study of this book:

1. Hoffmann, Banesh, *The Strange Story of the Quantum* (Magnolia, Mass.: Peter Smith, 1963). Originally published in 1947, this whimsical, often hilarious book was one of the first attempts to explain quantum physics to non-specialists. It is now available in a Dover paperback.
2. Zukav, Gary, *The Dancing Wu-Li Masters: An Overview of the New Physics* (New York: Bantam, 1980). In spite of its title, this is not a book about mysticism. Zukav is not a physicist, but he has gone to considerable lengths to ensure accuracy in this lively account. His book covers a lot of territory, including relativity and elementary particle physics, with admirable enthusiasm. It is available in both mass-market (cheap) and trade (long-lasting) paperback editions.
3. Pagels, Heinz R., *The Cosmic Code: Quantum Physics as the Language of Nature* (New York: Simon & Schuster, 1982). (Available in paperback.)
4. Gribbin, John, *In Search of Schrödinger's Cat: Quantum Physics and Reality* (New York: Bantam, 1984). (Available in paperback.)
5. Wolf, Fred Alan, *Taking the Quantum Leap: The New Physics for Non-scientists* (New York: Harper & Row, 1981). A lot of people like this whimsical introduction to quantum mechanics; in 1982 it won the American Book Award for the best original science paperback.

6. Davies, Paul, *Other Worlds: A Portrait of Nature in Rebellion* (New York: Simon & Schuster, 1980). A fine introduction by a reliable British writer of popularizations.
7. Davies, R. C. W., and J. R. Brown, eds., *The Ghost in the Atom* (New York: Cambridge University Press, 1986). This collection of interviews with distinguished modern physicists, addressed to lay readers, offers provocative overviews of quantum physics and some of its more bizarre aspects.

At a more advanced level than the above books, yet still accessible to persistent lay readers is

8. Popper, Karl R., *Quantum Theory and the Schism in Physics* (Totowa, N. J.: Rowman and Littlefield, 1982).

Physics and Human Consciousness

Some of the most provocative recent writings on the implications of quantum theory address the interface between physics and the human mind. Although the views expressed in these books are controversial, they are worth your serious consideration:

9. Harth, Erich, *Windows on the Mind: Reflections on the Physical Basis of Consciousness* (New York: Quill, 1983).
10. Wolf, Fred Alan, *Star Wave: Mind, Consciousness, and Quantum Physics* (New York: Macmillan, 1984).

Finally, several books have appeared that probe what the authors perceive as parallels between modern physics and Eastern mysticism. *Beware of these books!* Most are incorrect and irresponsible in their claims about modern physics. The most interesting of the bunch, in my opinion, is

11. Capra, Fritjof, *The Tao of Physics*, 2nd ed. (Boulder, Co.: Shambhala, 1983).

EXERCISES AND PROBLEMS

> "It sounds a dreadful prospect,"
> he commented. "Completely insane.
> As you say, though,
> it may be the only way out."
>
> —from "Thirteen to Centaurus"
> by J. G. Ballard

Exercises

1.1 The Uncertainty Principle Strikes Back.
The diameter of an atom in its ground state is roughly $1\,\text{Å}$. Suppose we want to locate an electron in an atom; to do so using a microscope would require us to use radiation of wavelength $\lambda \approx 1\,\text{Å}$.

(a) From the wavelength of a photon, we can calculate its momentum using the de Broglie relation [see Eq. (2.16)] $p = h/\lambda$. **Calculate** the momentum of a photon corresponding to a wavelength of $1\,\text{Å}$.

(b) Suppose a photon with momentum p collides with the atomic electron. The amount of momentum transferred from the photon to the electron lies somewhere in the range from 0 to p. Consequently, the electron is left with a momentum uncertainty Δp. What is Δp for the electron in this problem?

(c) Using the Heisenberg Uncertainty Principle, **calculate** how accurately you could determine the position of the electron in the atom. What does this calculation tell you about the advisability of using a microscope to try to "locate" an electron in an atom?

1.2 Time Evolution of Uncertainty.

The Heisenberg Uncertainty Principle Eq. (1.4) relates the uncertainties in the position and momentum of quantum particles *at a particular time.* But what happens to these uncertainties as time passes?

To answer this question rigorously, we would need quantum mechanical machinery that we won't learn until Part IV, but with our present knowledge we can get a rough idea based on a *semi-classical analysis.*[15]

Consider a beam of identical microscopic particles of mass m. The particles are free, *i.e.*, they move (at a non-relativistic velocity) in the absence of external forces. For simplicity's sake, we'll consider only motion in one dimension, $-\infty < x < +\infty$. Suppose the position of a particle at an initial time $t_0 = 0$ is known to within an uncertainty Δx_0 about x_0, and the initial momentum to within Δp_0 about p_0.

(a) How accurately can we determine the velocity at t_0?

(b) **Write down** a classical expression for $x(t)$, the position at time $t \geq t_0$, assuming that the position at t_0 is x_0 and the momentum at this time is p_0. Express your answer in terms of x_0, the initial velocity v_0, and t.

(c) Of course, the position at t_0 is unknown to within the uncertainty Δx_0, and the momentum to within Δp_0. From your answer to [b], **determine** upper and lower bounds on $x(t)$.

(d) Now, **derive** an expression for $\Delta x(t)$ in terms of Δx_0, h, m, and t.

(e) Suppose we work very hard to determine the position at t_0 to great precision, so that $\Delta x_0 \ll 1$. What are the consequences of all this effort for our knowledge of the position at some time $t > t_0$?

1.3 The Energy-Time Uncertainty Relation.

Consider a beam of microscopic free particles, each of mass m, traveling along the x axis. In this chapter we noted the uncertainty relation (1.4)

$$\Delta x \, \Delta p \geq \frac{1}{2}\hbar. \tag{1.3.1}$$

This relation expresses the limitation nature imposes on our ability to measure simultaneously the position and momentum of the particle. There is another uncertainty relation (which we shall study in §11.8), the **energy-time uncertainty relation**:

$$\Delta E \, \Delta t \geq \frac{1}{2}\hbar. \tag{1.3.2}$$

In the energy-time relation, Δt is the time required to measure the energy E of a system, and ΔE is the uncertainty in the resulting value. In effect, this relation says that to measure an energy *exactly* would take forever.

Since the energy and momentum of a free particle are related, we should expect these uncertainty relations to be related. **Derive** the energy-time uncertainty relation (1.3.2) from the position-momentum uncertainty relation (1.3.1). Be sure to explain or justify each step of your derivation. (Consider very small uncertainties ΔE and Δp.)

Hint: If the particle moves through a distance Δx in a time interval Δt, then $\Delta x = (p/m)\Delta t$.

[15] A semi-classical analysis is one in which classical physics is discretely and carefully used for certain aspects of a quantum problem. Applied correctly, semi-classical techniques can yield accurate approximations to a full quantal analysis. Applied indiscriminately or without thought, they can yield utter garbage. *Caveat emptor.*

1.4 Revenge of the Heisenberg Uncertainty Principle.

A locomotive engine is at rest in a train yard. The locomotive is macroscopic, so we can see that it is at rest and where it is—*i.e.*, we know its position with certainty. Suppose we use the Heisenberg Uncertainty Principle to estimate the position uncertainty Δx. Consider the following argument:

> Since the engine is at rest, its momentum p is zero. Moreover, since $\Delta p = 0$, the uncertainty in its position is infinite. Therefore *the position of the locomotive is completely unknown.*

It would seem that the HUP has let us down, for it has led to a conclusion that directly contradicts the experience of our own eyes. Find the flaw in the above argument.

Problems

1.1 Testing the Heisenberg Uncertainty Principle.

An ambitious undergraduate decides to check out the uncertainty principle for macroscopic systems. He goes to the top of the University clock tower and drops a marble (of mass m) to the ground, trying to hit one of the (many) cracks in the pavement beneath him. To aim his marble, he teeters precariously directly over the desired crack and uses a very sophisticated apparatus of the highest possible precision, which he has borrowed from one of the superbly equipped freshman physics laboratories.

(a) Alas, try as he might, the student cannot hit the crack. **Prove** that the marble will inevitably miss the crack by an average distance of the order of $[h/(2\pi m)]^{\frac{1}{2}}(H/g)^{\frac{1}{4}}$, where g is the acceleration due to gravity and H is the height of the tower.

(b) Assuming reasonable values for H and m, **evaluate** the order of magnitude of the distance in part (a) in MKS units.

(c) **Discuss** your result in terms of the Correspondence Principle.

1.2 No Rest for the Weary (in the Microworld).

One of the most extraordinary features of microscopic systems is that if they are bound (*i.e.*, confined by forces) they can never be at rest. The explanation for this bizarre fact, as for so many bizarre features of the microworld, is related to the Heisenberg Uncertainty Principle. The state of minimum total energy of a system is its **ground state**. The ground-state energy of a system is its **zero-point energy**. Consider a beam of particles of mass m moving in one dimension. Suppose their potential energy is that of a simple harmonic oscillator of natural frequency ω_0,

$$V(x) = \frac{1}{2}m\omega_0^2 x^2.$$

Use the Heisenberg Uncertainty Principle to **derive** an approximate expression for the ground-state energy of the particles in the beam.

Hint: The average momentum of the particles as they undergo simple harmonic motion is zero, so the *value* of their momentum is of the order $h/\Delta x$.

1.3 Quantum Ideas Incarnate: The Particle in a Box.

Consider an ensemble that consists of a huge number of identical systems. Each system is a particle of mass m confined to a one-dimensional square well of width L. The potential energy of such a particle is

$$V(x) = \begin{cases} \infty & x < -\frac{L}{2} \\ 0 & -\frac{L}{2} \le x \le \frac{L}{2} \\ \infty & x > \frac{L}{2} \end{cases}$$

(a) From arguments based on the Heisenberg Uncertainty Principle **derive** an expression for the minimum kinetic energy T_{min} that the particle can have. **Provide a clear, complete explanation of your reasoning.**

(b) **Calculate** the minimum kinetic energy of an electron that is confined to a nucleus. Take as the diameter of the nucleus the typical value 10 fermi. *Express your answer in MeV. Hint:* The dimensions of a nucleus correspond to a *very small* infinite square well. From your answer to part (a) what qualitative conclusion can you draw about the magnitude of the kinetic energy and hence of the velocity of the particle?

(c) In early studies of beta decay, nuclei were observed to emit electrons with energies of several hundred keV. Initially, these observations were taken as evidence that electrons are the building blocks of nuclei. **Comment briefly** on this hypothesis, basing your remarks on your answer to part (b).

(d) Suppose the mass of the particle is $m = 1$ milligram and the width of the box is $L = 0.1$ mm. **Evaluate** the minimum energy *in eV*. Can the fact that $E_{\text{min}} \neq 0$ for this system be observed? Why or why not?

1.4 Yet Another Uncertainty Relation.

Consider a microscopic particle of mass m that moves in a circle of radius r. Let ΔL denote the uncertainty in the angular momentum of the particle and $\Delta\theta$ the uncertainty in its angular displacement. **Derive** the following uncertainty principle:

$$\Delta L \, \Delta\theta \geq \frac{1}{2}\hbar.$$

Be sure to explain and justify each step of your derivation.

1.5 A Not Particularly Realistic Application of the HUP.

Use the Heisenberg Uncertainty Principle to **estimate** how long an ordinary lead pencil can be balanced upright on its point.

CHAPTER 2

Into the Microworld

Duality and the Double Slit

I have no data yet.
It is a capital mistake to theorize
before one has data. Insensibly
one begins to twist facts to suit theories,
instead of theories to suit facts.

—Sherlock Holmes to Dr. Watson
in *A Scandal in Bohemia*
by Sir Arthur Conan Doyle

Quantum physics is chock full of bizarre, non-classical ideas; weird concepts haunt nearly every section of this book. Three of these ideas—*quantization*, *uncertainty*, and *duality*—deserve special treatment, for they are the bedrock of quantum theory.

Quantization is probably familiar to you, but you wouldn't have encountered it in a classical physics book. The observables of *macroscopic* systems are not quantized. The energy of a ball on a string, the linear momentum of a hockey puck, and the angular momentum of an astronaut in a centrifuge—all can take on any of a *continuum* of values, subject only to whatever constraints are imposed by forces acting on the system. In the macroscopic world nature is continuous.

Not so in the microworld, where there is abundant evidence that nature is inherently discrete. Physicists have been aware of this fact since pre-quantum days. By 1900, for example, spectroscopists knew that radiation was emitted by atoms and molecules at *discrete frequencies*. But the first direct evidence of *quantization of energy* appeared in collision experiments performed in the early 1900's by James Franck (1882–1964) and Gustav Hertz (1887-1975).[1]

For several years, Franck and Hertz had been measuring the ionization potentials of atoms, and this experience influenced their interpretation of the results of their collision experiments.[2] In these experiments, they scattered electrons from various atoms; for example, in their most famous experiment, Franck and Hertz accelerated electrons emerging from a heated platinum wire, and sent the resulting beam through a gas of mercury vapor. By measuring the current in the *scattered* electron beam as a function of the energy of the incident electrons, Franck and Hertz could study the energy loss suffered by the electrons in their collisions with atoms of the gas. They found that the scattered current exhibited a series of sharp drops, each occurring at an incident energy equal to an integral multiple of 4.9 eV.

Franck and Hertz were initially led astray by their prior experience studying ionization; they concluded that 4.9 eV is the ionization potential of mercury. But Neils Bohr, on learning of Franck and Hertz's results, realized that the electrons were not *ionizing* the mercury atoms but rather were *exciting* them—losing energy to the atoms through inelastic collisions. Drawing on his model of the atom, Bohr further deduced that the energy spacing at which the current drops occur, 4.9 eV, is the separation of two *discrete* energy levels of the mercury atom.[3]

In 1914, when Franck and Hertz reported their work, physicists thought that it supported the Bohr theory of the atom—which has since been superseded by quantum

[1]Gustav was the nephew of Heinrich Hertz, whom we'll meet in § 2.1.

[2]For a detailed account of the Franck-Hertz experiment, see § 3.5 of *Introduction to Modern Physics*, 2nd ed. by J. D. McGervey (New York: Academic Press, 1983).

[3]Recent precise measurements of the energy separation between the ground state and the first excited state of mercury give the value 4.86 eV. The ionization potential of mercury, by the way, is 10.43 eV.

theory.[4] Nevertheless, the Franck-Hertz experiment stands as one of the first demonstrations of energy quantization in atoms, a striking example of non-classical shenanigans in the microworld.

Another underlying concept of this world is *uncertainty*. In Chap. 1, I showed you an example of this principle: Nature imposes a peculiar indeterminacy on any attempt to simultaneously measure the position and momentum of a quantum particle. The mathematical manifestation of this indeterminacy is the *Heisenberg Uncertainty Principle*, which we'll return to (and *derive*) in Chap. 4.

Duality, the third great quantum idea, is the focus of this chapter. *Wave-particle duality* is one of the more confounding ideas of quantum physics, for it most directly challenges our classical intuition about how the world behaves. But we must face the challenge of duality; before we can begin to understand the ways of the microworld, we must disabuse ourselves of the idea that quantum particles can be understood as either classical particles or classical waves. The essence of wave-particle duality is beautifully encapsulated in a single experiment, the famous *double-slit experiment*. After laying some background in § 2.1–2.3, we'll investigate this provocative experiment in § 2.4.

2.1 WAVES VERSUS PARTICLES

The behavior of a microscopic "particle" differs in subtle ways from that of a classical particle; in some respects, it resembles the behavior of a classical wave. So before we plunge into the microworld, let's remind ourselves of the *classical* pictures of particles and waves.

In Chap. 1 we reviewed the characteristics of a classical particle and the analysis of its motion using Newtonian mechanics. For example, a macroscopic particle has a well-defined position and linear momentum at any time, and from these observables we can calculate its energy. Essential to this description is the notion of *spatial localization*. The idea that a particle is a localized thing is implicit, for example, in the description of transport of energy through space by a particle in a localized lump. This quality of the classical description of a particle is, in turn, reflected in the "state descriptor" of Newtonian theory: the trajectory.

The characteristics of a *wave* are quite different from those of a particle. A wave is not spatially localized—this quality is reflected in the properties by which we characterize a wave, such as the *wavelength*. Like a particle, a wave carries energy and momentum, but it does so in a *non-localized* manner, distributed over a *wave front*. And a wave exhibits distinctive, non-particle-like behavior such as diffraction, interference, and polarization. Not surprisingly, then, the theory classical physicists use to describe the propagation of a wave and its interaction with matter—the electromagnetic theory of James Clerk Maxwell (1831–1879)—is quite different from that of Newtonian particle dynamics. So at the macroscopic level, classical physics elegantly and cleanly separates into wave mechanics and particle dynamics.

Alas, this division does not apply to microscopic particles, which adamantly refuse to adhere to either the classical wave model or the classical particle model. In some circumstances microscopic particles behave according to the laws of classical mechanics.

[4] Although the Bohr model of the atom, the heart of what is called "the old quantum theory," is of enormous historical importance, we shall not study it, because this model is almost 100% wrong. For a good treatment, see Chap. 4 of *Quantum Physics of Atoms, Molecules, Solids, Nuclei and Particles*, 3rd ed. by R. Eisberg and R. Resnick (New York: Wiley, 1985).

For example, some collisions of highly energetic atoms with molecules can be explained by using classical collision theory. Yet, in other circumstances quantum particles behave like waves: *e.g.*, electrons that have been scattered by a crystal exhibit a diffraction pattern when they are detected. Analysis of this pattern reveals that in this experiment the electrons propagate precisely as though they had a well-defined wavelength and frequency ... as though they are waves.[5]

"This situation," you may be thinking, "is a mess. You tell me that light isn't a wave. And electrons aren't particles. Well, then, *what are they?* And how can I understand the behavior of electrons if not with Newtonian mechanics, and of waves if not with Maxwell's theory? *What is going on here?*"

*2.2 ON THE NATURE OF LIGHT

> Remember,
> light and shadow never stand still.
>
> —John Constable,
> 19th century landscape artist.

To probe deeper into wave-particle duality, we're going to trip lightly through the history of man's evolving ideas about the nature of electromagnetic radiation. This overview need not be comprehensive; you can find several excellent, more complete historical accounts in the Suggested Readings list for this chapter.

Corpuscular theories of light, which treated light as though it were composed of particles, have been batting around since the days of Newton and Laplace. But light has an irritating tendency to exhibit distinctively non-particle-like behavior, such as diffraction and interference, which corpuscular theories could not explain. Once Maxwell introduced his wave theory of electromagnetic radiation (in 1870) and it became clear that this theory could beautifully explain such phenomena, most physicists abandoned corpuscular theories of light.

According to Maxwell, all electromagnetic radiation—light included—consists of real waves propagating through space, waves that carry energy distributed over continuous, non-localized spherical wave fronts. In the late 19th century, it seemed that Maxwell's theory could explain even the most complicated electromagnetic phenomena; so convincing were its successes that in 1886, the distinguished physicist Heinrich Hertz (1857–1894)—uncle of Gustav—wrote: "the wave theory of light is, from the point of view of human beings, a certainty." Hertz, as we shall see, was wrong.[6]

Walking the Planck

It was Albert Einstein (1879–1955) who, in 1905, resurrected the notion that electromagnetic radiation is particle-like rather than wave-like in nature. But where did this idea originate? Einstein has written that the seeds of his 1905 theory were planted by research

[5]Never let it be said that physics is without irony. The fact that electrons propagate like waves was demonstrated in 1927 by G. P. Thompson (1892–1975). By scattering electrons through a thin gold foil, he conclusively verified the *wave nature of the electron*, and, in 1937, received the Nobel prize. Several years before, in 1906, G. P.'s father, J. J. Thompson (1856–1940), won the Nobel prize for discovering the electron (1897) and for measuring its charge to mass ratio. J. J.'s experiment was interpreted as conclusive proof of the *particle nature of the electron*.

[6]"When a distinguished but elderly scientist states that something is possible, he is almost certainly right. When he states that something is impossible, he is very probably wrong."—Arthur C. Clarke, in *Profiles of the Future*.

carried out at the turn of the century by the German physicist Max Planck (1858–1947). Although encouraged by his physics teacher to pursue a career as a musician, Planck persevered in physics. His teacher's advice was lousy; Planck's studies of radiation inside a heated cavity led, via a line of reasoning he himself described as an "act of desperation," to the concept of quantization of energy and thence to the birth of quantum physics.

Planck did not set out to revolutionize physics. Instead, following in the footsteps of his teacher G. R. Kirchhoff (1824–1887), he sought to understand why hot bodies glow. This phenomenon, which is called *black-body radiation*, may be familiar to you if you sculpt. Suppose you have crafted a clay pig. To harden the clay, you *fire* the pig—*i.e.*, put it in a kiln (an oven) and heat to roughly 2000° F for about 10 to 12 hours. Suppose there is a tiny hole in the oven, too small to admit light but large enough to see through. At first, of course, you see darkness. But as the pig gets hotter and hotter, it begins to glow. As the temperature of the kiln further increases, this glow becomes orange, then yellow, then white, and fills the oven, obliterating all detail of the pig. Why?

Planck formulated this question in slightly more abstract terms, asking: *what is the spectrum of electromagnetic radiation inside a heated cavity*? More specifically: how does this spectrum depend on the temperature T of the cavity, on its shape, size, and chemical makeup, and on the frequency ν of the electromagnetic radiation in it? By the time Planck got into the game, part of the answer was known. Kirchhoff and others had shown that once the radiation in the cavity attains equilibrium with the walls, the energy in the field depends on ν and T but, surprisingly, is independent of physical characteristics of the cavity such as its size, shape, or chemical composition.

The cavity, of course, encompasses a finite volume. Planck was interested in the radiative energy inside the cavity, not on effects that depend on its volume, so he worked in terms of an energy *density*. In particular, he sought an expression for the *radiative energy density per unit volume* $\rho(\nu, T)$. If we multiply this quantity by an infinitesimal element of frequency, we obtain $\rho(\nu, T)d\nu$, the energy *per unit volume* in the radiation field with frequencies between ν and $\nu + d\nu$ at cavity temperature T.

Rather than confront the distracting complexities of a real heated cavity, Planck based his work on a model originally introduced by his mentor Kirchhoff. Kirchhoff called his model of a heated cavity in thermal equilibrium a "black-body radiator." A **black body** is simply anything that absorbs *all* radiation incident on it. Thus, a black-body radiator neither reflects nor transmits; it just absorbs or emits.

From the work of W. Wien (1864–1928), Planck knew that the radiative energy density $\rho(\nu, T)$ for a black body is proportional to ν^3 and, from the work of J. Stefan (1835–1893), that the *integrated energy density* $\int_0^\infty \rho(\nu, T)d\nu$ is proportional to T^4. But this information did not fully describe the dependence of $\rho(\nu, T)$ on ν and T; experimental evidence implied a further, unknown dependence on ν/T.

Wien had actually proposed an equation for the energy density of a black-body radiator, but the theoretical foundation of his theory was shaky. Moreover, his equation worked only in certain circumstances; it correctly explained the ν and T dependence of $\rho(\nu, T)$ for *low temperatures and high frequencies*. But it predicted that heated black bodies should emit a *blue* glow at all temperatures—a prediction confounded by our pig. Planck knew of this defect in Wien's theory, for experiments published in 1901 by H. Rubens and F. Kurlbaum conclusively showed it to fail for high temperatures and low frequencies.

In his research, Planck focused on the exchange of energy between the radiation field and the walls. He developed a simple model of this process by imagining that the molecules of the cavity walls are "resonators"—electrical charges undergoing simple harmonic motion. As a consequence of their oscillations, these charges emit electromagnetic radiation at their oscillation frequency, which, at thermal equilibrium, equals the frequency ν of the radiation field. According to classical electromagnetic theory, energy exchange between the resonators and the walls is a continuous process—*i.e.*, the oscillators can exchange any amount of energy with the field, provided, of course, that energy is conserved in the process.

By judiciously applying classical thermodynamics and a healthy dose of physical intuition to his model, Planck deduced an empirical formula for the radiative energy density:[7]

$$\rho(\nu, T) = \frac{A\nu^3}{e^{B\nu/T} - 1}. \tag{2.1}$$

In this equation, A and B are constants to be determined by fitting to experimental data; that's why it is called "empirical."

Equation (2.1) agreed beautifully with experimental data for a wide range of frequencies and temperatures. And, in the limit $\nu \longrightarrow \infty$ and $T \longrightarrow 0$, it reduced properly to Wien's law. But Planck was concerned that he could not rigorously justify his formula. In a letter dated 1931, he wrote of his dissatisfaction: "a theoretical interpretation *had* to be supplied, at all costs, no matter how high ... I was ready to sacrifice every one of my previous convictions about physical laws." Planck had to do just that.

For his second assault on the energy density of a black body, Planck adopted a statistical method based on the concept of entropy as interpreted probabilistically by Ludvig Boltzmann (1844–1906). At one point in his derivation, Planck introduced a simplifying assumption (*i.e.*, a trick to facilitate his mathematics). His ploy was conventional in mathematical analysis, and Planck expected to remove his assumption at the end. Planck was in for a rude shock.

His assumption is both simple and radical. He knew that a classical black body could exchange any amount of energy with the walls of the cavity; nevertheless, for purposes of his derivation, Planck assumed that *only discrete amounts of energy can be absorbed or emitted by the resonators that comprise the walls of the black body.* He called these discrete amounts of energy "quanta." To each quantum Planck assigned an energy equal to an integral multiple of $h\nu$, where h is the constant $h = 6.63 \times 10^{-34}$ J-sec.

Having made this assumption, Planck easily derived the radiation law

$$\rho(\nu, T) = \frac{8\pi\nu^2}{c^3} \frac{h\nu}{e^{h\nu/k_B T} - 1}, \tag{2.2}$$

where $c = 3.0 \times 10^8$ m–sec is the speed of light and $k_B = 1.38 \times 10^{-23}$ J-K^{-1} is Boltzmann's constant. Comparing this result with Eq. (2.1), we see that in his new equation Planck *derived* expressions for the constants A and B that appeared in his empirical form.

[7] You can find a very nice account of Planck's analysis in Chap. 1 of *The Quantum Physicists and an Introduction to Their Physics* by William H. Cropper (New York: Oxford University Press, 1970).

This derivation was the theoretical justification Planck sought for the distribution of radiation in a black body. But in laying the foundation for his theory, Planck paid an enormous price, for try as he might, he could not get rid of his artificial constant h. Setting $h = 0$ inevitably led to a result that disagreed with a huge body of experimental data. Yet, if h *is non-zero*, then Planck's theory is seriously at odds with physicists' understanding of energy exchange as a continuous process.

Planck's rash assumption heralded the strange new physics of the quantum, which dominated the physics community for the next three decades. But physicists, who were then and are today a rather conservative lot, did not take well to being told that their understanding of so basic a process as energy exchange was fundamentally incorrect. Attempts to derive Planck's result without making his drastic assumption failed, and for several years Planck's quanta languished in obscurity, largely ignored. Yet, Planck's assumption was nothing compared to the surprise Einstein had in store.

Particles of Light

Einstein thought he saw an inconsistency in the way Planck used Maxwell's wave theory of electromagnetic radiation in his derivation. With characteristic intellectual courage, Einstein decided that this inconsistency implied a flaw not in Planck's theory but in Maxwell's. This radical contention shifted the emphasis in research on black bodies from the resonators that comprise the walls to the radiation field itself. Ultimately, it completely altered the way we think about light.[8]

In 1905 Einstein proposed that the energy in an electromagnetic field is *not* spread out over a spherical wave front, as Maxwell would have it, but instead is localized in indivisible clumps—in *quanta*. Each quantum of frequency ν, Einstein averred, travels through space at the speed of light, $c = 3.0 \times 10^8$ m–sec, carrying a discrete amount of energy $h\nu$ and momentum $h\nu/c$. Thus, in Einstein's model, light transports energy in the same way particles do. G. N. Lewis subsequently dubbed Einstein's quantum of radiation energy a *photon*, the name we use today.

The photon model cast Planck's theory of black-body radiation in, so to speak, a new light. Planck thought that energy exchange between the resonators of the cavity and the field occurs in units of $h\nu$ because of some strange property of the resonators. To this Einstein said: No; the explanation is that *the radiation field itself is quantized.* Planck's result is consistent with this extraordinary notion; if the energy in the field is contained in photons—quanta of magnitude $h\nu$—then of course only integral multiples of the photon energy can be exchanged with the walls.

The photon model explained more than just black-body radiation. One of Einstein's greatest achievements was using it to understand the *photoelectric effect*—the ejection of electrons from a metal, such as sodium, when light impinges on it. (You see the photoelectric effect in action every time an electric eye opens an elevator door for you or closes one on your foot.) Yet, the photon was too radical for physicists of the early 1900's, and Einstein's model of the electromagnetic field encountered strong opposition. Even as late as 1913—years after the publication of Einstein's work on the photoelectric effect—four distinguished German physicists (including Max Planck) wrote in a petition recommending Einstein's appointment to the Prussian Academy of Science:

[8]Einstein's radical theory of light grew out of his study of the entropy of high-frequency radiation in a low-temperature cavity. For a readable account of this background research, see *The Quantum Physicists and an Introduction to their Physics* by William H. Cropper (London: Oxford University Press, 1970).

> ... That he may sometimes have missed the target in his speculations, as, for example, in his hypothesis of light quanta, cannot really be held too much against him, for it is not possible to introduce fundamentally new ideas, even in the most exact sciences, without occasionally taking a risk.

One year later the American experimentalist R. A. Millikan (1868–1953) reported a precise verification of Einstein's equation for the energy of a photon, $E = h\nu$, and the first measurement of Planck's constant. Yet physicists still resisted abandoning their long-cherished idea that light was a wave.

Then, in 1923, Arthur H. Compton (1892–1962) published the results of his x-ray scattering experiments, and drove the last nail into the coffin of the wave theory of light. Compton scattered x-rays—electromagnetic radiation with a wavelength around 10^{-10} m and a frequency around 10^{18}sec^{-1}—from a thin layer of a light element such as carbon and measured *the shift in the wavelength of the x-rays due to the scattering.* His results were seriously at odds with the predictions of Maxwell's beleaguered theory.

Compton's data clearly showed that the wavelength of the scattered radiation is larger than that of the incident radiation. After several foredoomed attempts to explain this result with classical theory, Compton tried the photon idea. If x-rays carry energy in localized clumps, he reasoned, then we can apply *classical collision theory* to their interaction with the electrons of the target. Compton used the classical laws of conservation of energy and linear momentum—as though he were analyzing a game of billiards—and was able to derive the correct expression for the wavelength shift. This analysis vindicated Einstein's idea. It was too much even for diehards: the photon was accepted.

What is Light?

Einstein's brilliant notion *still* does not fully illuminate the nature of light. The photon is a particle-like model, but clearly light does not actually consist of classical particles. Particles do not diffract. They don't interfere. But light does. Yet, this model demolishes the idea that light is a classical wave. Indeed, the nature of electromagnetic radiation after Einstein seemed more ephemeral than ever; depending on the physical conditions, light seemed to behave either like a wave or like a particle. J. J. Thompson wrote that the conflict between these two models was like "a struggle between a tiger and a shark; each is supreme in his own element, but helpless in that of the other."

The more you think about wave-particle duality, the more befuddling it seems. If you are new to this idea, you can probably imagine the shock, frustration, and confusion that the interpretive problems of duality induced during the early years of quantum theory. Banesh Hoffmann has written of the impact of this puzzle on physicists of the time:[9]

> They could but make the best of it, and went around with woebegone faces sadly complaining that on Mondays, Wednesdays, and Fridays they must look on light as a wave; on Tuesdays, Thursdays, and Saturdays as a particle. On Sundays, they simply prayed.

[9]See his delightful and often hilarious book *The Strange Story of the Quantum*, (New York: Dover, 1959).

*2.3 UNDERSTANDING DIFFRACTION THE CLASSICAL WAY

Diffraction, which was first observed by Leonardo da Vinci, is often considered to be the signature of a wave. Diffraction occurs when ripples in a pond encounter a pair of logs that are close together, when light passes through a narrow slit in a window shade, or when x-rays scatter from a crystal. In each case, we can explain the distinctive pattern that forms using classical wave theory. Understand what happens to a wave when it passes through a small aperture—that is, understand diffraction—and you are well on your way to understanding the behavior of waves in more complicated situations. In this section, I'll refresh your memory about this characteristic wave phenomena.[10]

A schematic of a single-slit diffraction experiment with light is shown in Fig. 2.1. Monochromatic light of frequency ν is incident on a diaphragm in which there is a single slit of width w. We'll assume that the light source is far to the left of the diaphragm, so the incident radiation can be represented by a plane wave. The width of the slit must be small compared to the wavelength $\lambda = c/\nu$ of the radiation if the slit is to appreciably diffract the light; for example, to diffract visible light enough to observe this phenomenon, we require a slit of width $w \approx 10^{-4}$ cm. Light scattered by the diaphragm falls on a detector, such as a photographic plate or a photocell, located at a distance D far to the right of the slit. The detector measures the energy delivered by the scattered wave as a function of the distance x in Fig. 2.1.

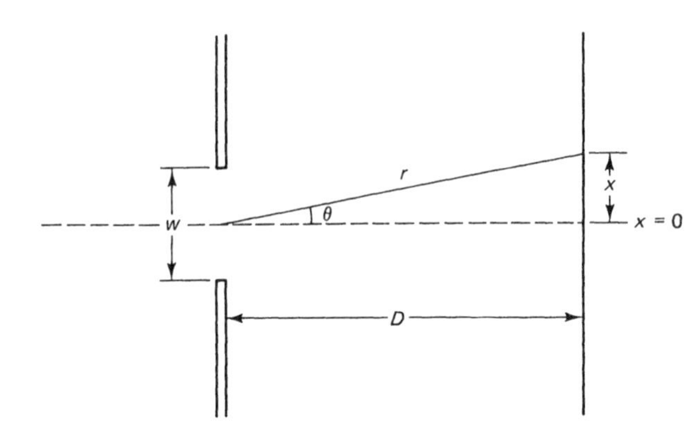

Figure 2.1 A highly simplified single-slit diffraction apparatus. At the detector, the distance x is measured from a point on a perpendicular from the midpoint of the slit. The analysis of this experiment is performed in terms of the radial distance r from the midpoint of the slit to the point x and the corresponding angle θ. (In this figure, the ratio of the slit width w to the distance D from the diaphragm to the detector has been exaggerated; in an actual double-slit experiment, $D \gg w$.)

Light scattered by the single-slit diaphragm forms a beautiful *diffraction pattern* at the detector. This pattern is characterized by a very bright central band located directly opposite to the slit, surrounded by a series of alternating light and dark regions. The light regions on either side of the central band are called *secondary bands*, because they are much less intense than the central band. Indeed, the intensity of the secondary bands drops off so dramatically on either side of the central band that only one pair of secondary bands is visible. Additional weak secondary bands exist, though, as you can see in Fig. 2.2, which is a graph of the intensity measured by the detector. If we play around with the frequency control on the light source and study the resulting diffraction

[10]If you recently studied classical wave mechanics, you needn't read this review or that in §2.4. Just glance at Eqs. (2.4) and (2.8) and press on to §2.6. If you need more review than is provided here, seek out nearly any undergraduate physics textbook. Several useful references can be found in this chapter's Selected Readings list.

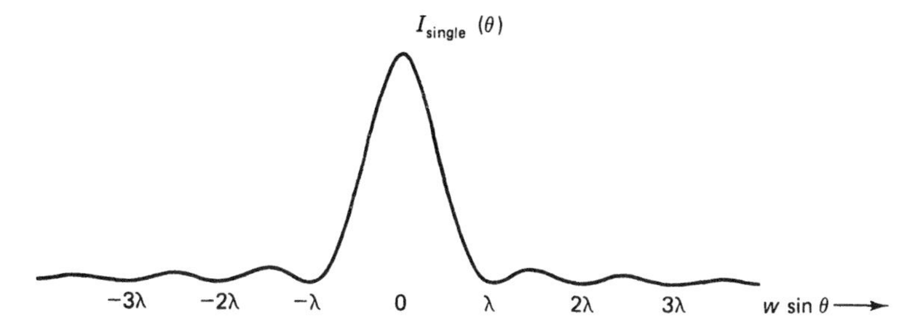

Figure 2.2 The diffraction intensity (2.4) for a single-slit (Fraunhofer) pattern plotted as a function of $w \sin\theta$.

patterns, we discover that the separation between adjacent bright bands is proportional to the wavelength λ of the incident radiation.

A classical physicist would call upon Maxwell's electromagnetic wave theory to explain the pattern in Fig. 2.2. To understand qualitatively what happens when a plane wave passes through a diaphragm we invoke **Huygens's Principle**, which lets us replace the plane wave and the slit (in our mind's eye) by a large number of closely-spaced, discrete radiating charges that fill the region of space where the slit is located.[11] "Scattered waves" radiated by different oscillators—*i.e.*, waves that emerge from different locations in the slit region—have different phases, so the superposition of these scattered waves exhibits regions of high and low intensity, as in Fig. 2.2.

To put this reasoning on a quantitative footing, we would first use Maxwell's theory to derive an expression for the electric field at a point (r, θ) on the detector (see Fig. 2.1). But the quantity measured in this experiment is not the electric field; it is the *intensity*—the rate at which the scattered radiation delivers energy to the detector. This quantity is proportional to the *time-averaged energy flux*—*i.e.*, to the average over one period of the square of the modulus of the electric field. I'll denote by $I_{\text{single}}(\theta)$ the intensity at a fixed value of r due to radiation scattered by a single slit.

Omitting the details of the derivation of $I_{\text{single}}(\theta)$, I'll just quote the result.[12] For convenience, I'll write the intensity in terms of its value I_0 at the principal maximum—*i.e.*, at the point $r = D$, $\theta = 0$, in the central peak in Fig. 2.2—and the handy intermediate quantity

$$\alpha \equiv \pi \frac{w \sin\theta}{\lambda}, \tag{2.3}$$

where, of course, λ is the wavelength of the incident radiation. With these definitions, the single-slit intensity at fixed r is

$$I_{\text{single}}(\theta) = I_0 \left(\frac{\sin\alpha}{\alpha} \right)^2 . \tag{2.4}$$

[11]For more on Huygens's Principle, see Chap. 9 of *Waves* by Frank S. Crawford (New York: McGraw-Hill, 1968).

[12]See, for example, § 12.8 and 12.9 of *Classical Electromagnetic Radiation* by Jerry B. Marion (New York: Academic Press, 1965).

The intensity (2.4) is graphed in Fig. 2.2 as a function of $w \sin\theta$; plotted this way, $I_{\text{single}}(\theta)$ exhibits a characteristic pattern of equally-spaced nodes [where $I_{\text{single}}(\theta) = 0$], which occur at

$$w \sin\theta = \pm n\lambda \qquad \text{for } n = 1, 2, 3, \ldots \qquad [\text{nodes in } I_{\text{single}}(\theta)]. \tag{2.5a}$$

The principal (zeroth-order) maximum of the intensity pattern occurs at $\theta = 0$, and higher order maxima occur (approximately) at

$$w \sin\theta = \pm\frac{2n+1}{2}\lambda \qquad \text{for } n = 1, 2, 3, \ldots \qquad [\text{maxima in } I_{\text{single}}(\theta)]. \tag{2.5b}$$

Equation (2.4) fully accounts for the properties of patterns such as the one in Fig. 2.2. Thus does classical electromagnetic theory rend the veil of mystery from the phenomenon of diffraction.

*2.4 UNDERSTANDING INTERFERENCE

The theory of diffraction is the foundation for study of the double-slit interference experiment. When first performed in 1800 by Thomas Young (1773–1829), this experiment was considered definitive proof of the wave nature of light. We can modify the single-slit apparatus of Fig. 2.1 to suit Young's experiment by simply cutting a second slit in the diaphragm.

As shown in Fig. 2.3, the width of the second slit is the same as that of the first, w. We must position the second slit so that the two slits are close together—but not too close: for observable interference to occur, the slit separation s must be greater than w. Again, we shine a beam of monochromatic light of frequency ν on the diaphragm and see what happens at the detector, which is far to the right of the diaphragm.

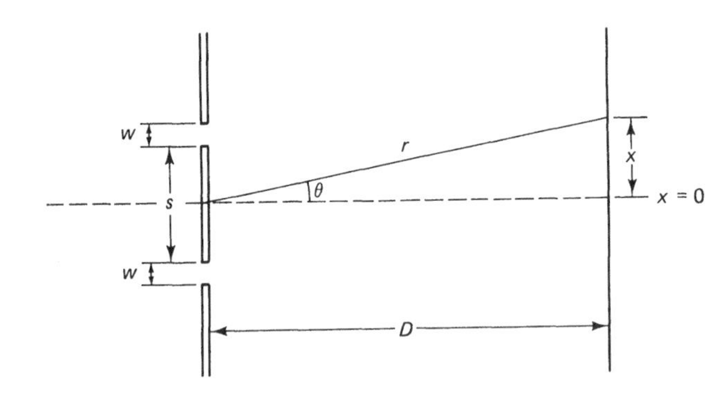

Figure 2.3 Schematic of a double-slit apparatus for Young's interference experiment. The distance r to an observation point on the detector is measured from the midpoint of the part of the diaphragm separating the two slits.

This time, our detector shows an *interference pattern* like the one in Fig. 2.4. At first glance, this figure may look similar to the diffraction pattern of Fig. 2.2, but on closer examination we see striking differences. For one thing, the interference pattern exhibits more bright bands than the diffraction pattern. This means that the energy of the radiation scattered by the double-slit diaphragm is more evenly distributed than that scattered by a single slit. (In the diffraction pattern about 90% of the energy appears in the central band.) Finally, the individual bands in the interference pattern, which are called *interference fringes*, are narrower than those of the diffraction pattern.

If we study how the interference pattern changes as we fiddle with the incident wavelength λ, the slit separation s, and the slit width w, we discover that the separation between the bright bands increases with increasing λ—just as it did in the diffraction pattern, for which the separation is proportional to λ/w. But also we find a difference: the separation in Fig. 2.4 is independent of the slit width w but is inversely proportional to the slit *separation* s.

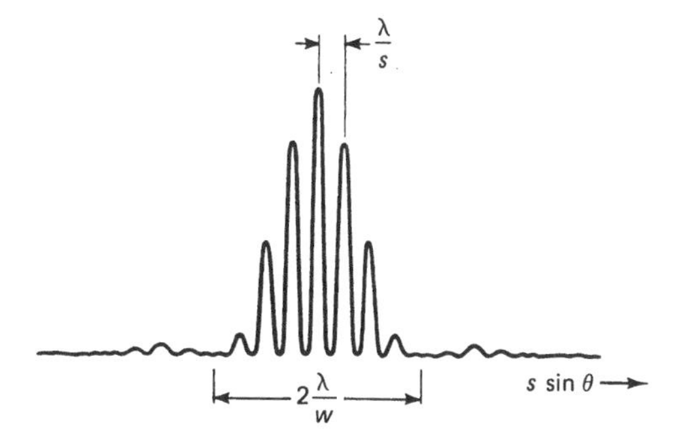

Figure 2.4 The double-slit interference intensity function $I_{\text{double}}(\theta)$ [Eq. (2.6)] for $s \gg \lambda$. (This condition produces several maxima in a small region of θ.) The angular interval between successive maxima is λ/s.

The key to understanding interference is *superposition.* When the incident plane wave encounters the double-slit diaphragm it "splits," and a diffracted wave emerges from each slit. (In a sense, each slit becomes a source of radiation that travels to the detector.) These waves add in the region between the slit and the detector, and this device measures the intensity of their superposition. Now, at the detector the amplitudes of the electric fields of the diffracted waves are equal, but their phases are not. Consequently, their superposition manifests regions of constructive and destructive interference, as seen in Fig. 2.4.

The trick to deriving an equation for the intensity measured in the double-slit experiment—which we'll call $I_{\text{double}}(\theta)$—is therefore to add the electric fields of the waves diffracted by each slit. (This step is legitimate because Maxwell's equation for these fields is linear.) Having done so, we could calculate the aforementioned phase difference and would find the field from the lower slit *lags* the field from the upper slit by an amount $(2\pi s/\lambda)\sin\theta$. The last step is to average the squared modulus of the total electric field at a point (r, θ) on the detector over one period, which yields

$$I_{\text{double}}(\theta) = 4I_0 \left[\frac{\sin\left(\pi\frac{w\sin\theta}{\lambda}\right)}{\left(\pi\frac{w\sin\theta}{\lambda}\right)}\right]^2 \cos^2\left(\pi\frac{s\sin\theta}{\lambda}\right). \tag{2.6}$$

Now this, I would argue, is a curious result. For one thing, I_0 in Eq. (2.6) is the maximum intensity we would obtain in the *diffraction pattern of a single slit of width* w—*i.e.*, the very quantity I_0 we used in Eq. (2.4). Looking deeper, we find buried in Eq. (2.6) the single-slit intensity $I_{\text{single}}(\theta)$. To unearth this function, I'll use the definition (2.4) of α and introduce yet another intermediate quantity

$$\Delta\varphi \equiv 2\pi\frac{s\sin\theta}{\lambda}. \tag{2.7}$$

With these definitions the double-slit intensity (2.6) stands revealed as

$$I_{\text{double}}(\theta) = 4I_{\text{single}}(\theta)\cos^2\left(\frac{1}{2}\Delta\varphi\right). \tag{2.8}$$

This analysis explains the complicated oscillations we see in $I_{\text{double}}(\theta)$ as θ is varied (Fig. 2.4). This structure is compounded of two separate oscillations, one due to the single-slit intensity $I_{\text{single}}(\theta)$—which depends on the slit width w but not, of course, on the separation—the other due to factor $\cos^2\left(\pi\frac{s}{\lambda}\sin\theta\right)$. The separation s of the two slits is greater than their width w, so the latter oscillation is the more rapid of the two. In fact, the slower oscillation of the single-slit intensity forms the *envelope* of the double-slit intensity, as you can see in Fig. 2.5.

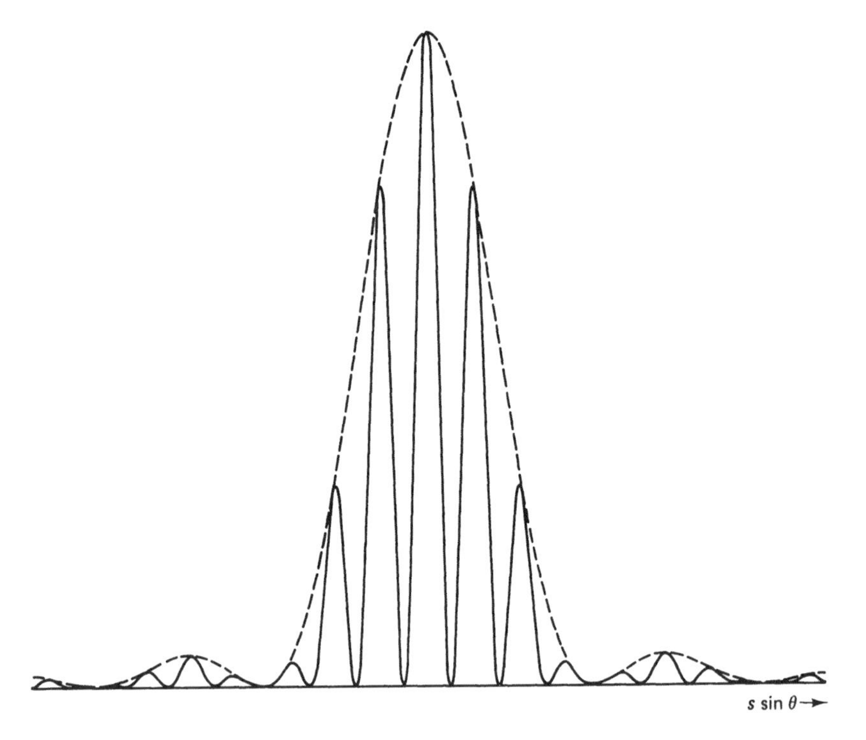

Figure 2.5 The double-slit intensity (solid curve) and its envelope, the single-slit intensity (dashed curve). For this graph, we have chosen $s = 3w/2$.

The double-slit pattern is more complicated than the single-slit pattern. And, as you might expect, even more elaborate interference patterns can be generated by using diaphragms with more than two slits. But no matter how baroque the resulting patterns, they can be explained by Maxwell's electromagnetic wave theory.

But wait! According to Einstein, a light beam consists of *photons* and therefore transports energy in *spatially-localized* clumps with particle-like properties. But no model based on particle dynamics can explain interference—or, for that matter, diffraction.[13] The implications of our double-slit experiment appear to contradict Einstein's theory of light.

[13]We might, for example, argue that a diffraction pattern could result from particle-like scattering of photons from the edges of a single slit. This argument sounds good at first, but when translated into a quantitative prediction, it leads to an intensity that does not agree with experiment. For a delightful account of this argument see Chap. 6 of *The Character of Physical Law* by R. P. Feynman (Cambridge, Mass.: M. I. T. Press, 1965).

Yet, a model based purely on classical wave theory is also inadequate, for it cannot explain phenomena such as the photoelectric effect and x-ray (Compton) scattering. These experiments support a model in which light interacts with other subatomic particles (*e.g.*, electrons) according to the laws of classical particle dynamics.

Maybe we can resolve this conundrum by performing the double-slit experiment using a beam of *particles*—say, electrons. The behavior of these particles in the experiment should differ strikingly from that of light. This is a good idea, but, as we'll see in the next section, the experiment deepens rather than solves the mystery.

2.5 DOUBLE TROUBLE

> Now is not this ridiculous—
> and is not this preposterous?
> A thorough-paced absurdity—
> explain it if you can.
>
> —Chorus of Dragoons,
> in *Patience*
> by Gilbert and Sullivan

Our strategy in the double-slit experiment is to send electrons through a double-slit diaphragm and see how the intensity measured by the detector differs from the interference pattern formed by light. To implement this strategy, we must make a few modifications in the apparatus Young used when he performed this experiment with light.

First we replace the light source with an *electron gun*—a device that produces a (nearly monoenergetic) beam of electrons of energy E. A heated tungsten wire, for example, produces a stream of electrons that we can accelerate to the desired velocity. Second, we replace the photographic plate with an electron detector: a device that counts the number of electrons that arrive in each square meter of unit area per sec. (Like the photographic plate used in Young's experiment, our electron detector measures the rate at which energy arrives at each point on the detector.) A screen covered with phosphor will do; when an electron arrives at the screen, it produces a spot.

What would we expect to see at the detector if the electrons were particles, subject to the same physical laws as, say, marbles? Imagine for a moment that we block one slit—say, the *lower slit* in Fig. 2.3—so that all electrons must come through the other, open slit. Most electrons that make it through the diaphragm will go straight through this slit, "piling up" at the detector directly opposite it. We therefore expect to see a maximum in the measured intensity opposite the upper slit. But some particles will scatter from the edges of this slit, so we expect some amount of spread in the pattern. A reasonable guess for the intensity for a beam of particles passing through this apparatus with only the upper slit open is the curve $I_{\text{u}}(\theta)$ sketched in Fig. 2.6a. The curve $I_{\text{l}}(\theta)$ should be obtained if only the *lower* slit is open.

What *should* happen when both slits are open? Well, if the electrons are indeed particles, then the measured intensity should be simply $I_{\text{u}}(\theta)+I_{\text{l}}(\theta)$. This rather featureless curve is sketched in Fig. 2.6b. (Were you to scale the apparatus to macroscopic size and send through it a uniform beam of marbles—with, of course, a suitable detector—this is what you would see.) But what we *actually* see when we run the experiment with electrons is altogether different.

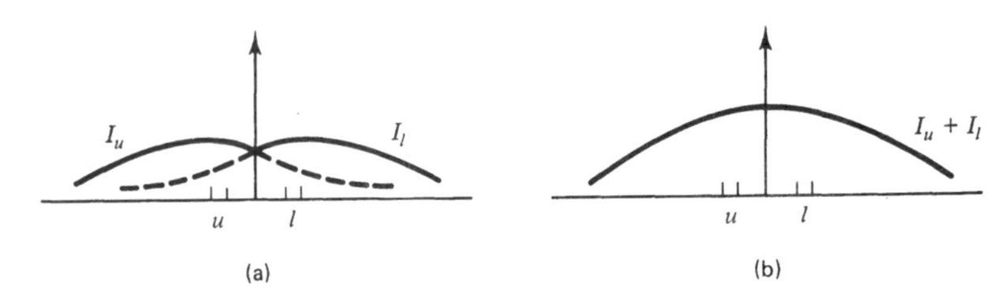

Figure 2.6 (a). Intensity patterns formed by blocking one of the slits in the diaphragm in Fig. 2.3 and sending particles through the apparatus. (b). What we *expect* to get when both slits are open.

The measured intensities in Fig. 2.7 clearly exhibit bands of alternating high and low intensity: an interference pattern, like the one formed by light (see Fig. 2.4). This observation seems to imply that the electrons are *diffracted* by the slits at the diaphragm and then *interfere* in the region between the diaphragm and the detector. We can even fit the measured intensity of the scattered electrons to the double-slit function $I_{\text{double}}(\theta)$ of Eq. (2.6) provided we assign to each electron (of mass m and energy E) a wavelength

$$\lambda = \frac{h}{\sqrt{2mE}}. \tag{2.9}$$

But to a classical physicist, steeped in the idea that electrons are particles, Eq. (2.9) is nonsense!

Figure 2.7 An electron interference pattern formed in a double-slit experiment. [From P. G. Merli, G. F. Missiroli, and G. Pozzi, *Amer. Jour. Phys.*, **44**, 306 (1976). Used with permission.]

Aside: The Real Double-Slit Experiment. The double-slit experiment is no mere thought experiment. For several years double-slit experiments with electrons have been performed in the laboratory. The first such experiment was reported in 1954 by G. Möllenstedt and H. Düker, who observed an electron interference pattern in an apparatus analogous to a Fresnel biprism.[14] Nowadays double-slit experiments are usually performed with an electron microscope.[15] In a typical experiment of this kind, electrons with energies around 60 keV—*i.e.*, with a wavelength of roughly

[15] Their article appeared in *Naturwiss.*, **42**, 41 (1954). See also C. Jönsson, *Zeitschrift für Physik*, **161**, 454 (1961); a translation of this article was published in *Amer. Jour. Phys.*, **42**, 4 (1974).

[15] See O. Donati, G. F. Missiroli, and G. Pozzi, *Amer. Jour. Phys.*, **41**, 639 (1973) and P. G. Merli, G. F. Missiroli, and G. Pozzi, op. cit., **44**, 306 (1976).

0.05 Å—are scattered by $0.5\,\mu$ slits ($1\,\mu = 10^{-6}$ m) that are separated by about $2.0\,\mu$. The distance from the diaphragm to the detector is typically about 10^6 times the slit separation s. The interference pattern shown in Fig. 2.7 was obtained in such an experiment.

Maybe They're Waves?

At first, the results of the electron double-slit experiment seem consistent with the (deviant) notion that electrons are waves, not particles.[16] But as we investigate further, we discover a problem with this hypothesis.

Suppose we turn down the intensity of the *incident* electron beam so that *very few electrons per second pass through the diaphragm.* Were electrons waves, the interference pattern that forms at the detector after a few seconds would be very weak. As time passes and more and more electrons are scattered by the double-slit diaphragm, this pattern would intensify, ultimately looking like Fig. 2.7. But this isn't what happens.

Shortly after we switch on the electron gun, we see at the detector not a diffraction pattern, but *spots*—just what we'd expect to see were the electrons particles! If you think I'm making this up, look at the actual data in Fig. 2.8a. This result is incompatible with the notion that electrons are waves, since waves transport energy via non localized wave fronts, not concentrated lumps.[17]

Curiously, the spots in Fig. 2.8a show no particular pattern, neither interference as in Fig. 2.4, nor a double-hump, as in Fig. 2.6. It looks like the electrons are particles that arrive *randomly* at the diaphragm.

But they aren't. With the passage of time, the electrons begin to form the interference pattern of Fig. 2.7. The photographs in Fig. 2.8(b) and (c) show this pattern forming, as if by magic. These pictures are eerie—it's as though each electron knows precisely where it should go to contribute to the interference pattern. This gives rise to the question: *If the electrons are particles, what guides them to just the right place on the detector?*

We can summarize our observations thusly: when each electron arrives at the detector, its interaction with the material of the detector produces a spot, a spatially-localized point—behavior we associate with a classical particle. Yet, when a number of electrons *propagate* from the diaphragm to the detector, they somehow form the distinctive pattern in Fig. 2.7—behavior we associate with classical waves. Perhaps strangest of all, *this pattern appears only after sufficient time has elapsed that a statistically significant number of electrons has arrived at the detector.*

'Tis a situation worthy of Gilbert and Sullivan. In the same apparatus we can see electrons behaving like waves or like particles, depending on how long we let the experiment run. The short-term behavior of the electrons is incompatible with the wave model, but their long-term behavior is incompatible with the particle model. So what *are* they?

[16]This idea is quite radical—a wealth of experimental evidence to the contrary exists. Early experiments on electrons, such as Thompson's 1907 measurements of e/m, Millikan's 1917 measurements of the electron charge e, and Rutherford's 1911 scattering experiments, strongly support the model of the electron as a discrete, indivisible unit of electric charge—definitely not a wave.

[17]You can see similar behavior—spots at the detector—if you perform Young's interference experiment (the double-slit experiment with light) at low enough intensity that only a few photons per second arrive at the photographic plate.

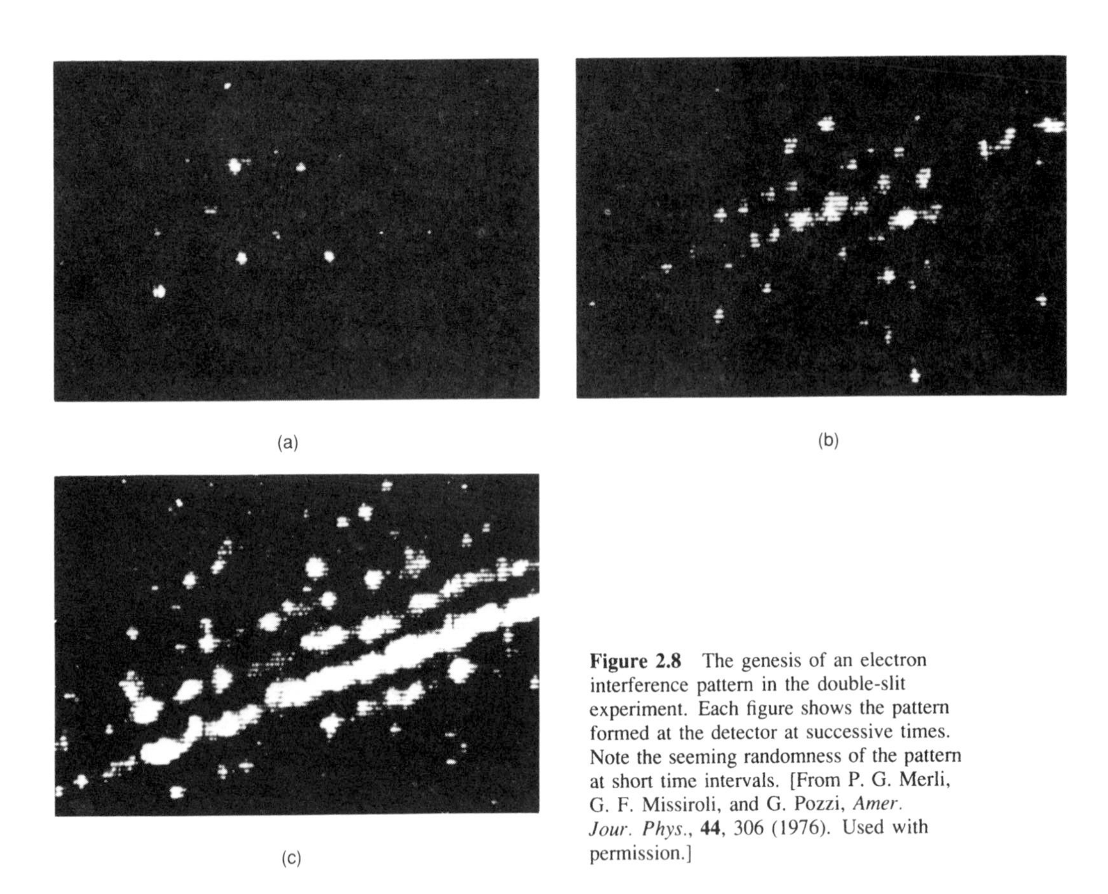

Figure 2.8 The genesis of an electron interference pattern in the double-slit experiment. Each figure shows the pattern formed at the detector at successive times. Note the seeming randomness of the pattern at short time intervals. [From P. G. Merli, G. F. Missiroli, and G. Pozzi, *Amer. Jour. Phys.*, **44**, 306 (1976). Used with permission.]

The Mystery Deepens

Whatever theory we dream up to make sense of these observations must preserve the spatially-localized property of electrons and yet explain the pattern formed when many electrons arrive at the detector. Let's try the hypothesis that an electron is a new kind of particle, one that somehow propagates according to wave dynamics. This hypothesis, at least, would let us explain the observed pattern as due to the interference of two electrons, one from the upper slit and one from the lower slit. To test this idea, we devise another variation on the double-slit experiment.

If this model of the electron is correct, we should be able to get rid of the interference pattern by turning the incident beam density *so low that only one electron goes through the diaphragm at a time*. (Such an experiment is technologically feasible using an electron microscope.) A single electron, after all, has nothing to interfere with. But the results of this experiment are not what we expect. *Even at the lowest possible incident density, an interference pattern forms.* This behavior is totally beyond the power of a particle model to explain. Any such model implicitly assumes that the electron has a trajectory, *i.e.*, that each electron goes through one slit or the other. Yet, in this experiment each electron acts as though it were interfering with itself![18] Such results force us to accept that in no sense can an electron be thought of as a classical particle or a classical wave.

The double-slit experiment beautifully illustrates one of the central mysteries of quantum physics: the dual nature of microscopic entities. I believe that the difficulties many people have understanding the results of this experiment arise because we are hidebound by our classical training and intuition. We persist in trying to force electrons and photons to fit *either* the classical particle model *or* the classical wave model. It is perhaps unfortunate that electrons do behave *like* classical particles in some experiments and *like* waves in others, for this behavior can seduce us into thinking that there must be some way to retain one of these classical models. But there isn't; neither model by itself is correct, for neither model conforms to the observed behavior of electrons. To explain the double-slit results, we need a wholly new theory: quantum mechanics.[19]

More Trouble with Trajectories

In Chap 1, I argued that the very idea of ascribing a trajectory to a microscopic particle is rendered nonsensical by the Heisenberg Uncertainty Principle—on the grounds that if we can't know precisely a particle's position and momentum at any time, then we cannot even define its trajectory. The implications of this argument are nothing if not far-reaching: it requires us to jettison the entire machinery of Newtonian physics! The double-slit experiment provides further insight into this problem by showing us how misleading is the idea that microscopic particles follow paths in geometrical space.

If the incident electrons did follow paths through the diaphragm to the detector, then each would have to pass through *either* the upper slit *or* the lower slit. Yet, the results of the double-slit experiment imply that the electrons go through *both* slits, for an interference pattern forms at the detector even if the incident intensity is so low that electrons traverse the region from source to detector singly. But interference is understood (in classical physics) as arising from the superposition of two (or more) waves, and such an explanation is inconsistent with the notion that the electrons follow classical paths.

Aside: Failed Attempts to Trick the Electrons. You may be wondering if the issue of whether trajectories can be ascribed to electrons could be settled by a variation on this experiment—one that would "trick" the electrons into revealing which slit they went through. Suppose, for example, that we put an electron detector immediately behind each slit. By seeing which detector registers, we could determine once and for all which slit each electron went through.

This gambit works! If we carry out such an experiment, we find that one or the other device (not both) registers for each electron that arrives at the phosphor. But after many electrons have undergone the journey through the diaphragm, if we look for an interference pattern at the phosphor, we find instead that the measured

[18]There are still more subtleties to the double-slit experiment. For a fascinating discussion of this experiment, see Volume III of *The Feynman Lectures on Physics*, by R. P. Feynman, R. B. Leighton, and M. Sands, (Reading, Mass.: Addison-Wesley, 1965).

[19]Regrettably, in this book we won't get quite far enough to work out the quantum mechanics of the double-slit experiment. It is, however, important to note that theory can fully explain the results of this experiment. The argument uses a method from the quantum theory of scattering called the *eikonal approximation* to describe the elastic scattering of electrons by the two slits and the formation of the observed interference pattern. If you're interested in the details, see G. Anido and D. G. Miller's article in the *Amer. Jour of Phys.*, **52**, 49 (1984).

intensity forms the "double hump" pattern characteristic of particles (see Fig. 2.6). So this experiment in no way explains the *wave-like* behavior of the propagation of electrons in the original apparatus, for in this experiment the behavior of the electrons is consistently particle-like.

What went wrong was that we interfered. By measuring position immediately behind the diaphragm, we altered the state of the incident electrons, changing their subsequent behavior and therefore the pattern they formed on the phosphor. In particular, our observation of which slit the electrons went through imparted momentum to them. This transfer of momentum destroyed the interference pattern. This experiment illustrates the interaction between the observer and the observed that we discussed in Chap. 1: a measurement on a microscopic system inevitably alters its state in an uncontrollable way.

The Heisenberg Uncertainty Principle affords some insight into this interaction. To be sure which slit the electrons pass through, we must measure position (just behind the diaphragm) accurately to within an uncertainty Δx that is less than the spacing between the slits. But since the uncertainty product $\Delta x \, \Delta p_x$ must be positive, a non-zero position uncertainty implies a non-zero momentum uncertainty. A detailed quantitative analysis along these lines reveals that the resulting Δp_x is just large enough to wash out the interference pattern at the detector.[20]

Duality characterizes all the whole panoply of quantum particles: protons, neutrons, atoms, molecules, etc. Invariable, microscopic entities seem to propagate according to wave dynamics but to interact with each other according to particle dynamics.

An explanation of wave-particle duality was not to appear until the advent of the quantum mechanics of Schrödinger, Heisenberg, and other physicists that will come forward. But a significant step on the road to this theory was taken in 1923 by Prince Louis Victor Pierre Raymond de Broglie, whom we'll meet in the next section.

2.6 IT ALL STARTED WITH DE BROGLIE

The experiments discussed in § 2.3–2.5 raise a host of questions. If electromagnetic radiation consists of photons—localized clumps of energy—how can we explain phenomena such as diffraction and interference? If not, then why did Compton have to use classical collision theory to explain the scattering of *x rays* by metals? On the other hand, if electrons are particles, why do they produce an interference pattern at the detector in the double-slit experiment? The behavior of electrons and photons in these experiments seems provocatively similar—crazy, to be sure, but crazy in the same way. Are electrons and photons in some sense the same?

Einstein was deeply puzzled by this question until he noticed a possible answer in the doctoral thesis of a young French physicist. In 1924, Einstein wrote in a letter to his Dutch colleague Hendrik Lorentz (1853–1928) that the research of Prince Louis de Broglie (1892–1975) "...is the first feeble ray of light to illuminate this, the worst of our physical riddles." De Broglie's achievement was to synthesize the wave-like and particle-like aspects of microscopic matter. Although de Broglie seems to have only dimly understood the nature of quantum particles, and his rather nebulous physical models of quanta have since been superseded, the importance of his contribution has not diminished. It initiated the development of modern quantum mechanics.

[20]For details, see J. S. Marsh, *Amer. Jour. Phys.*, **43**, 97 (1975).

Sometimes a Great Notion

In 1910 de Broglie began studying history at the University of Paris; soon, however, he switched to physics. His studies were interrupted in 1913 by a six-year stint in the French army, during which he and his brother Maurice worked on wireless telegraphy. Then in 1919 he returned to Paris for his doctoral research.

From work on the x-ray spectra of heavy elements, de Broglie knew of photons and the Bohr model of atomic structure. And he was particularly intrigued by "Planck's mysterious quanta." So he set himself the task of "[uniting] the corpuscular and undulatory points of view and thus [penetrating] a bit into the real nature of quanta."

In 1923, lightning struck. As de Broglie tells it:

> As in my conversations with my brother we always arrived at the conclusion that in the case of x-rays one had [both] waves and corpuscles, thus suddenly—I cannot give the exact date when it happened, but it was certainly in the course of summer 1923—I got the idea that one had to extend this duality to the material particles, especially to electrons.

Thus did de Broglie come up with the idea of *matter waves*. This idea led him to the important notion that *all microscopic material particles are characterized by a wavelength and a frequency, just like photons.*

Aesthetic considerations seem to have influenced de Broglie's thinking towards the idea of matter waves. He evidently felt that nature should be symmetrical, so if *particles of light* (photons) were to be associated with electromagnetic radiation, then so should *waves of matter* be associated with electrons. Simply stated, his hypothesis is this: *There is associated with the motion of every material particle a "fictitious wave" that somehow guides the motion of its quantum of energy.*

In spite of its rather vague character, this idea was remarkably successful. For example, using the methods of classical optics (such as Fermat's principle) to describe the propagation of quanta, de Broglie was able to explain how photons (and, for that matter, electrons) diffract and interfere: It is not the particles themselves but rather their "guide waves" that diffract and interfere. In de Broglie's words, "the fundamental bond which unites the two great principles of geometrical optics and of dynamics is thus fully brought to light."

De Broglie proffered these ideas in his Ph.D. dissertation, which he wrote at age 31. His thesis did not fully convince his examiners, who were impressed but skeptical of the physical reality of de Broglie's matter waves. One examiner later wrote, "at the time of the defense of the thesis, I did not believe in the physical reality of the waves associated with the particles of matter. Rather, I regarded them as very interesting objects of imagination."[21] Nevertheless, de Broglie passed.

Beautiful Equations

De Broglie's equations for the wavelength and frequency of his matter waves are elegant and simple. Even their derivations are not complicated.[22] In his seminal paper of 1923,

[21]It is not clear that de Broglie himself knew what he meant by a matter wave. In his thesis he notes that "the definitions of the phase wave and of the periodic phenomenon were purposely left somewhat vague." In subsequent papers, de Broglie tried several different interpretations of his elusive waves. [For a thorough discussion of these interpretations, see Chap. V of *The Historical Development of Quantum Theory,* Volume I, Part 2 by J. Mehra and H. Rechenberg (New York: Springer-Verlag 1982).]

[22]See Cropper, op. cit., pp. 57–63.

de Broglie began with *light quanta*—photons—so I'll first recap the derivation of the equation relating the wavelength and momentum of a photon and then press on to material particles.

The photon is a relativistic particle of rest mass $m_0 = 0$. Hence the momentum p of a photon is related to its total energy E through the speed of light c as

$$p = \frac{E}{c}. \qquad [\text{for } m_0 = 0] \tag{2.10}$$

To introduce the frequency ν of the photon, we use Einstein's equation for the photon energy

$$E = h\nu \tag{2.11}$$

to write Eq. (2.10) as

$$p = \frac{h\nu}{c}. \tag{2.12}$$

For a wave in free space, the wavelength is $\lambda = c/\nu$, so Eq. (2.12) becomes

$$p = \frac{h}{\lambda}. \tag{2.13}$$

Now, in contrast to a photon, a material particle such as an electron has a non-zero rest mass m_0. Therefore the relationship between the energy and momentum of such a particle moving at relativistic velocities (in a region of zero potential energy) is not Eq. (2.10), but rather

$$E^2 = p^2c^2 + m_0^2c^4. \tag{2.14}$$

If the velocity of the particle is non-relativistic ($v \ll c$), then its kinetic energy is simply[23]

$$T = \frac{p^2}{2m_0}, \tag{2.15a}$$

where T is the kinetic energy,[24]

[23]The notational conventions of physics dictate a confusing ambiguity between Eqs. (2.14) and (2.15), because the rest mass energy m_0c^2 is included in E in (2.14) but not in (2.15). The relativistic kinetic energy, which is sometimes denoted by the symbol K, is defined as

$$K = \sqrt{p^2c^2 + m_0^2c^4} - m_0c^2.$$

It is this quantity which, in the non-relativistic limit, reduces to Eq. (2.15*a*). [See *Classical Mechanics*, 2nd ed. by Herbert Goldstein (Reading, Mass.: Addison-Wesley, 1983, pp. 307–309).] I'll adhere to standard practice by using E for the energy in the relativistic and non-relativistic cases.

[24]This equation is usually written with the symbol E standing for the kinetic energy, as

$$E = \frac{p^2}{2m_0}.$$

I'll follow this convention in this book whenever dealing with non-relativistic particles (*i.e.*, most of the time).

$$T = E - m_0 c^2. \tag{2.15b}$$

In either case, the derivation of Eq. (2.13) cannot be applied to a material particle.

Nonetheless, de Broglie proposed that Eqs. (2.11) and (2.13) be used for material particles as well as photons.[25] Thus, *for electrons, atoms, photons and all other quantum particles, the energy and momentum are related to the frequency and wavelength* by

$$\boxed{\begin{aligned} p &= h/\lambda \\ E &= h\nu \end{aligned}} \qquad \text{de Broglie-Einstein equations} \tag{2.16}$$

Notice that the de Broglie equation $\lambda = h/p$ implies an *inverse* relationship between the total energy E of a particle and its wavelength, *viz.*,

$$\lambda = \frac{hc/E}{\sqrt{1 - \left(\frac{m_0 c^2}{E}\right)^2}}. \tag{2.17}$$

If applied to a photon (by setting the rest mass to zero), this equation reduces to Eq. (2.10). Hence the larger the energy of a particle, the smaller is its wavelength, and *vice versa.*

Question 2–1

Derive a relationship between the wavelength of an electron and its *kinetic energy* T. **Prove** that in the non-relativistic limit, your result reduces, as it should, to

$$\lambda = \frac{h}{p} = \frac{h}{m_0 v}.$$

The Unanswered Question

De Broglie's notion that some sort of wave guides the motion of every quantum enabled him to explain a variety of hitherto inexplicable phenomena, such as quantization of the orbital angular momentum of an electron in a Bohr orbit of the hydrogen atom. It also led him to suggest that the existence of matter waves could be verified by looking for electron diffraction patterns. Sure enough, in a series of experiments carried out in the years after 1923, such patterns were found. Electrons were diffracting all over the place: at non-relativistic velocities, in the classic experiments of C. J. Davisson and L. H. Germer (which were later repeated more accurately by G. P. Thompson[26]), and at relativistic velocities, in experiments by J. V. Hughes and others.[27] These successes vindicated de Broglie's ideas. Matter waves were real.

But de Broglie's research left unanswered deep and important questions: *what is a matter wave?* Of what is λ the wavelength? How do these mysterious waves control the

[25] De Broglie based his argument on the special theory of relativity. First, he equated the rest energy $m_0 c^2$ of a material quantum to the energy $h\nu_0$ of its "periodic internal motion," where ν_0 is the "intrinsic frequency" of the particle. Next, he considered a quantum moving at velocity v with momentum $p = m_0 v/\sqrt{1-(v/c)^2}$ and used relativistic kinematics to show that the frequency and wavelength of such a particle are given by Eq. (2.16).

[26] *Amer. Jour. Phys.*, **29**, 821 (1961).

[27] A complete list of references can be found in the article "De Broglie's Relativistic Phase Waves and Wave Groups" by H. R. Brown and R. de A. Martins, *Amer. Jour. Phys.*, **52**, 1130 (1984)

propagation of a microscopic particle? The answers are not to be found in de Broglie's writings.[28]

Nonetheless, by raising such questions and focusing the attention of the physics community on the need to associate some sort of wave with microscopic material particles, de Broglie initiated the quest for a new physical theory. This quest led physicists on a torturous path during the early twenties, which the French physicist Oliver Costa de Beauregard has called the Stone Age of Quantum Mechanics (*Era Paléoquantique*). But rather than follow this path, we shall jump in Part II to the end of the story, the successful theory of Erwin Schrödinger, Werner Heisenberg, and the other fathers of modern quantum mechanics.

2.7 THE CONTROVERSIAL CONCEPT OF COMPLIMENTARITY

> As for knowledge, it's the scandal of our age
> that philosophy has been whittled away to
> a barebones epistomology, and thence to
> an even barer agnoiology.
> Have I found a word you don't know, Louis?
> Agnoiology is the philosophy of ignorance,
> a philosophy for philosophers.
>
> —from *Camp Concentration*
> by Thomas M. Disch

To explain the behavior of quantum particles, we evidently must use both the classical wave and particle models—in spite of the apparent contradictions between them. To say that this situation poses logical problems is to understate wildly. In the early twenties, this predicament lay heavy on the minds of physicists, until the Danish physicist Niels Henrik David Bohr (1885–1962) proposed a way out.

Bohr to the Rescue

Bohr was one of the intellectual giants of early quantum theory. His ideas and his personality were enormously influential. During the twenties and thirties, the Bohr Institute in Copenhagen (which was financially supported by the Carlsberg Brewery) became a haven for scientists who were developing the new physics.

Bohr was not always receptive to quantum ideas; like many of his colleagues, he initially rejected Einstein's photons. But by 1925 the overwhelming experimental evidence that light actually has a dual nature had convinced him. So for the next several years, Bohr concentrated on the logical problem implied by this duality, which he considered the central mystery of the interpretation of quantum theory.

Unlike many of his colleagues, Bohr emphasized the mathematical formalism of quantum mechanics. Like de Broglie, he considered it vital to reconcile the apparently contradictory aspects of quanta. Bohr's uneasy marriage of the wave and particle models was the **Principle of Complimentarity**. This principle entails two related ideas:

[28]If you are curious about de Broglie's views, you may want to read excerpts from his thesis. You'll find some selections in the anthology *Wave Mechanics*, edited by G. Ludwig (New York: Pergamon, 1968). More accessible is his Nobel Prize lecture, which you can find in *The World of the Atom*, edited by H. A. Boorse and L. Motz (New York: Basic Books, 1963). But beware: de Broglie's works do not make light reading.

1. *A complete description of the observed behavior of microscopic particles requires concepts and properties that are mutually exclusive.*
2. *The mutually exclusive aspects of quanta do not reveal themselves in the same observations.*

The second point was Bohr's answer to the apparent paradox of wave-particle duality: There is no paradox. In a given observation, *either* quanta behave like waves *or* like particles.

How, you may wonder, could Bohr get away with this—eliminating a paradox by claiming that it does not exist because it cannot be observed? Well, he has slipped through a logical loophole provided by the limitation of quantum physics that we mentioned in Chap. 1: *quantum mechanics describes only observed phenomena.* From this vantage point, the central question of wave-particle duality is not "can a thing be both a wave and a particle?" Rather, the question is "can a thing be *observed* behaving like a wave and a particle in the same measurement?" Bohr's answer is no: in a given observation, quantum particles exhibit *either* wave-like behavior (if we observe their propagation) *or* particle-like behavior (if we observe their interaction with matter). And, sure enough, no one has yet found an exception to this principle.

Notice that by restricting ourselves to observed phenomena, we are dodging the question, "what is the nature of the reality behind the phenomena?" Many quantum physicists answer, "there is no reality behind phenomena." But that is another story.[29]

Complimentary Observables

Before leaving complimentarity, I want to mention another aspect of this topic, one that relates it to the Heisenberg Uncertainty Principle.[30] This principle implies a special relationship between position and momentum. The uncertainties in these observables are related by $\Delta x \, \Delta p \geq \hbar/2$ so if we obtain (via measurement) *precise knowledge* of either observable, we do so at the expense of *any* knowledge of the other. Bohr considered this mutual uncertainty as a manifestation of complimentarity in the mathematical formalism of quantum mechanics.

The relationship between position and momentum, as expressed in the Heisenberg Uncertainty Principle, differs from the complimentary relationship of the wave and particle nature of quanta: position and momentum are both *particle-like* attributes of quanta. Rather than describe position and momentum as complimentary, perhaps it would be better to describe them as *incompatible*.

A Philosophical Cop-Out?

The Principle of Complimentarity may seem a little vague. Indeed, that is how it struck the audience of distinguished physicists at the international conference in 1927 where Bohr first presented his ideas. These ideas are undeniably subtle, and Bohr did not always express himself clearly. Nevertheless, in spite of some initial confusion, the physicists developing and using quantum theory adopted complimentarity, and today it

[29]Anyone even mildly interested in such questions should go out *now* and buy a copy of Nick Herbert's superb book *Quantum Reality: Behind the New Physics* (New York: Anchor Press/Doubleday, 1985).

[30]This dimension of complimentarity was first articulated by Bohr and more fully developed by Wolfgang Pauli (1900–1958). Pauli, whom we'll meet again, was in a sense the progenitor of this book; he wrote the first quantum mechanics text.

is one of the central tenets of the *Copenhagen Interpretation of Quantum Mechanics.* (In Part II, we shall discuss this interpretation in detail.) The Copenhagen interpretation has its detractors, but it *is* a way of making sense (more or less) of the physics of the microworld, and it remains the dominant interpretation of quantum mechanics.

In spite of the wide adoption of the Principle of Complimentarity, many thinkers have found it philosophically offensive, perhaps because rather than confront the problem of duality head on, Bohr's principle seems to sneak around it, taking recourse in the limitation of quantum mechanics to observed phenomena. James R. Newman has expressed the uneasiness some feel with complimentarity:

> In this century the professional philosophers have let the physicists get away with murder. It is a safe bet that no other group of scientists could have passed off and gained acceptance for such an extraordinary principle as complimentarity.

2.8 FINAL THOUGHTS: A FEW WORDS ABOUT WORDS

I suspect that some of the difficulties people have with complimentarity stem from classical thinking, from their understandable determination to cling to classical models. But after all, the wave and particle descriptions of nature *are merely models.* There is no reason to expect them to apply to the whole physical world. The results we have seen in this chapter tell us that in the microworld, each of these models is *part of the truth,* but each by itself is incomplete. In some cases, the wave model must be invoked to understand observations; in others, the particle model.

Electrons aren't particles. They aren't waves, either. They are something else, for which we neither have a name nor a classical model. Properly, we should follow de Broglie's lead and refer to them as "fragments of energy," but that would be awkward. Or we could follow Eddington and call them "wavicles," but that sounds silly. So we stick with "particles" (leaving off the quotation marks) and do the best we can.

But we mustn't overlook the influence of the language we use on how we think. In the microworld, nature transcends our familiar language and the technical language of the classical physicist. On this point, Werner Heisenberg wrote:

> The problems of language here are really serious. We wish to speak in some way about the structure of atoms ... But we cannot speak about atoms in ordinary language.

So *be careful.* Beginning in the next chapter we shall buckle down and take a serious look at the mathematical and physical structure of quantum mechanics, leaving behind for the moment purely qualitative concerns. But tread warily, lest you slip back into the black abyss of classical thinking and consequent confusion. With this *caveat* out of the way, we can proceed.

ANNOTATED SELECTED READINGS

The Beginnings

The saga of the development of modern physics during the "Quantum Stone Age" makes fascinating reading. You can get a glimpse of this history in a modern physics book, such as

1. Wehr, R., J. A. Richards, Jr., and T. W. Adair III, *Physics of the Atom*, 3rd ed. (Reading, Mass.: Addison-Wesley, 1978).
2. Tipler, P. A., *Modern Physics* (New York: Worth, 1978).
3. Eisberg, R., and R. Resnick, *Quantum Physics of Atoms, Molecules, Solids, Nuclei and Particles*, 2nd ed. (New York: Wiley, 1974).

But there is more to this topic than these introductory surveys let on. You can learn about the key experiments that led the way in this quest in the original papers collected in

4. Boorse, H. A., and L. Motz, eds., *The World of the Atom* (New York: Basic Books, 1963),

or in texts such as

5. Melissinos, A., *Experiments in Modern Physics* (New York: Academic Press, 1969).

Lively accounts of several of these experiments have been published in the five volumes of

6. Maglich, B., ed., *Adventures in Experimental Physics* (Princeton, N.J.: World Science Education, 1972).

Of the many delightful popular accounts of the development of modern physics, several of which are mentioned in the Selected Reading list for Chap. 1, I particularly like the treatment of history in

7. Gamow, George, *Thirty Years that Shook Physics* (New York: Dover, 1985).

One of the most imaginative treatments of this period is

8. McCormmach, Russell, *Night Thoughts of a Classical Physicist* (Cambridge, Mass.: Harvard University Press, 1982). This book (which is available in paperback) is a *fictional* account of the impact on an (imaginary) classical physicist of the coming of modern physics. It is superb.

You should certainly read some of the writings of the masters, the men who developed modern physics. A good selection of these can be found in the huge anthology (4). If you only read one author on this history, I would recommend Werner Heisenberg, who was a superb writer as well as a brilliant physicist. Heisenberg's essays are readable, entertaining, and informative. They are available in three (paperback) volumes

9. Heisenberg, W., *Physics and Philosophy: The Revolution in Modern Science* (New York: Harper & Brothers, 1958).
10. Heisenberg, W., *Physics and Beyond: Encounters and Conversations* (New York: Harper & Row, 1971).
11. Heisenberg, W., *Across the Frontiers* (New York: Harper & Row, 1974).

Want more? There are some excellent single-volume historical treatments available. One that you will find recommended throughout this book (and that is available in paperback) is

12. Cropper, William, H., *The Quantum Physicists and an Introduction to their Physics* (London: Oxford University Press, 1970).

Also recommended for its insight and anecdotes is

13. Segrè, E., *From X-Rays to Quarks: Modern Physicists and Their Discoveries* (San Francisco: W. H. Freeman & Co., 1980).

For its care and thoroughness, I highly recommend

14. Jammer, M., *The Conceptual Development of Quantum Mechanics* (New York: McGraw-Hill, 1966).

Finally, for sheer completeness you can't beat the massive set

15. Mehra, J., and H. Rechenberg, *The Historical Development of Quantum Theory* (New York: Springer-Verlag, 1982).

About Waves

Sections 2.3 and 2.4 contain rather cursory reviews of the irreducible minimum you need to know about wave phenomena. Several good texts on waves have been written, and you'll find reference to many of them in later chapters. For reviews of diffraction and interference (beyond those provided here), try

16. Giancoli, D., *Physics* (Englewood Cliffs, N.J.: Prentice Hall, 1984). A nice (and short) introductory treatment.

17. French, A. P., *Vibrations and Waves* (New York: W. W. Norton, 1971). (Introductory level, but far more detailed than you'd find in a freshman physics text like Giancoli's.)

18. Elmore, W. C., and Mark A. Heald, *The Physics of Waves* (New York: McGraw-Hill, 1979), Chap. 10. (Advanced).

Duality Explored

An excellent, detailed history of the development of wave-particle duality is given in

19. Wheaton, B. R., *The Tiger and the Shark: Empirical Roots of Wave-Particle Dualism* (New York: Cambridge University Press, 1983). This book is especially noteworthy for its treatment of the interface between theory and experiment during this period.

To find out what is going on today in duality research (yes, there is such research), dig into

20. Diner, S., S. Fargue, G. Lochak, and F. Selleri, eds., *The Wave-Particle Dualism: A Tribute to Louis de Broglie on his 90th Birthday* (Boston: D. Reidel, 1984).

EXERCISES AND PROBLEMS

> I hear and I forget.
> I see and I remember.
> I *do* and I understand.
>
> —Confucius

In these exercises and problems we confront the conceptual nature of quantum physics in a variety of situations, ranging from actual experiments to thought experiments to the real world. None of these problems requires much algebra or extensive calculations, but several will demand careful thought. (Some of these problems are review exercises. To solve them you may need to refresh your memory from one of the modern physics texts recommended in this chapter's Selected Readings list.)

Exercises

2.1 Wave-Particle Duality in Action.

Two kinds of microscopes are currently used in research: electron microscopes and light microscopes. Both exploit the wave nature of microscopic particles—electrons in the former apparatus, photons in the latter.

The limit of the resolving power of a microscope is the smallest distance separating two points in space that can be distinguished under ideal conditions. This quantity is roughly equal to the wavelength of the particles used in the microscope. Hence the resolving power is a measure of the uncertainty in the position of an observed object.

(a) Consider a light microscope in which 4 eV photons illuminate an object. **Calculate** the momentum of such a photon (in MKS units).

(b) **Evaluate** the minimum uncertainty in the position of a subatomic particle being observed with 4 eV photons.

(c) Repeat parts (a) and (b) for an electron microscope that uses 4 eV electrons.

(d) Suppose the position of a particle is determined with the light microscope of parts (a) and (b). What is the corresponding uncertainty in the *momentum* of the observed particle after the position measurement?

(e) Suppose the position of a particle is measured with the electron microscope of part (c). What is the uncertainty in the *momentum* of the observed particle after the measurement?

(f) What advantages do electron microscopes have over light microscopes?

2.2 Macroscopic Physics or Microscopic Physics?

(a) Consider a particle of mass 1 g that is moving at a speed $v = 10\,\text{m sec}^{-1}$. To study the physics of this particle, do we need to consider its wave nature? **Justify your answer with a simple calculation and a brief explanation of the significance of the result.**

(b) Here is a table of the diameters and ionization potentials of several atoms:

Atom	Diameter (Å)	Ionization Potential (eV)
He	1.16	24.6
Ne	1.36	21.6
Ar	2.64	15.7

An important experimental technique for studying some of the properties of an atom is to ionize the *valence electron*, causing it to be ejected from the atom so that its properties can be studied.[31] In such an experiment do we need to consider the wave nature of the electron? **As in part (a), be sure to justify your answer.**

Problems

2.1 Black-Body Radiation.

One of the early advances in the study of black-body radiation was the **Stefan-Boltzmann law**. This law is a simple formula for the power radiated from the surface of a black body. It expresses the radiated power P per unit surface area in terms of the temperature as

$$P = \sigma T^4,$$

where T is the temperature in degrees Kelvin of the black-body radiator and σ is the **Stefan-Boltzmann constant,**

$$\sigma = 5.67 \times 10^{-8}\,\text{W–m}^{-2}\text{–K}^{-4}.$$

In astrophysics, the upper atmospheres of stars are sometimes considered to be black-body radiators. Using such a model and the Stefan-Boltzmann law, **estimate** the atmospheric temperature of the sun.

Hint: For the incident radiant solar flux arriving at the earth's outer atmosphere use the value $1.4 \times 10^3\,\text{J–m}^{-2}\text{–sec}^{-1}$. The diameter of the sun is about 1.4×10^9 m, and the average radius of the earth's orbit is 1 astronomical unit—about 1.5×10^{11} m.

2.2 The Ultraviolet Catastrophe.

Another early advance in understanding the physics of black-body radiation was the derivation by Rayleigh and Jeans of a formula for the radiation density function ρ. These physicists used

[31] The *valence electron* is the least tightly bound electron in the atom. It is bound by an amount equal to the ionization potential of the atom. For example, it takes 24.6 eV to ionize a He atom.

classical arguments to obtain this quantity as a function of the wavelength λ of the radiation and the temperature T of the cavity.[32] They obtained

$$\rho(\lambda, T) = \frac{8\pi k_B T}{\lambda^4},$$

where k_B is the Boltzmann constant.

(a) From Planck's radiation formula (2.2) **derive** an expression for $\rho(\lambda, T)$.
Hint: Remember that the *total* energy density—*i.e.*, including all frequencies—must be the same whether ν or λ is used as the independent variable.

(b) Compare the Rayleigh-Jeans radiation formula given above with Planck's result as expressed in (a). Are the two in agreement for any ranges of λ and T? If so, determine *quantitatively* where the two expressions agree.

(c) Using Planck's result for $\rho(\lambda, T)$ show that the *total* energy density is proportional to T^4.

(d) Derive the *total* energy density from the Rayleigh-Jeans formula. Why was this result called *the ultraviolet catastrophe*?

2.3 An Actual Case of Electron Interference.

A double-slit experiment with electrons like the one discussed in § 2.5 was reported in 1961 by Claus Jönsson.[33] The apparatus he used is quite similar to that shown in Fig. 2.3. In this apparatus, two slits of width $w = 0.5\,\mu$ (where $1\,\mu = 10^{-6}\,\text{m}$) are cut in a diaphragm made of copper foil. These slits were separated by $s = 2\,\mu$. Jönsson accelerated electrons toward this diaphragm through a potential difference of $50\,\text{kV}$. The distance from the diaphragm to the detector was $D = 35\,\text{cm}$.

(a) Let p denote the momentum of the electrons. **Derive** an expression for the fringe spacing, i.e., the separation between *adjacent maxima* in the pattern of interference fringes that forms at the detector. (Express your answer in terms of p, D, s, and fundamental constants. Depending on how you approach this problem, you may need to read § 2.4).

(b) From the wavelength of the electrons in this experiment **evaluate** the fringe spacing.

(c) In the design of Jönsson's experiment, it was necessary to include some electrostatic lenses between the double-slit diaphragm and the detector. These lenses are shown schematically in Fig. 2.3.1. From your answer to part (b), **explain** why these lenses are necessary.

(d) Suppose we want to perform this experiment with visible light rather than with electrons. To obtain interference with *photons* in Jönsson's apparatus, we must scale its dimensions accordingly—*i.e.*, we must multiply each of the dimensions given above by the ratio of the wavelength of a visible photon to that of an electron. **Calculate** values of s, w, D, and the fringe spacing for such an experiment. Would it be practical to perform this experiment with photons? Why or why not?

2.4 Evaluating de Broglie Wavelengths.

(a) The de Broglie wavelength of a particle is an important parameter in the design of experiments based on phenomena such as particle diffraction. To see why, first **evaluate** the de Broglie wavelengths of
  (1) a 144 eV electron;
  (2) a 1 eV electron;
  (3) a 1 eV proton;
  (4) a 1 eV alpha particle.

[32] A recap of their derivation can be found in *Fundamentals of Modern Physics* by R. M. Eisberg (New York: Wiley, 1967), pp. 51–57.

[33] *Zeitschrift für Physik.* **161**, 454 (1961).

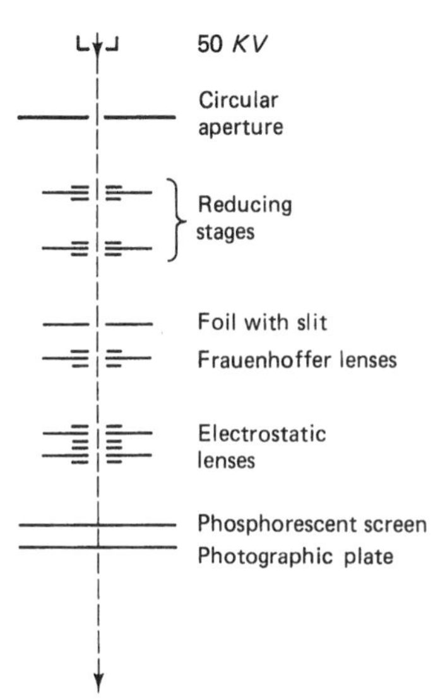

Figure 2.3.1 Arrangement of lenses used in the double-slit experiment with electrons.

Now, **decide** which of these particles would be most suitable for use in a crystal diffraction experiment in which the crystal spacing is typical—*i.e.*, on the order of 1.0 Å. **Explain your answer.**

(b) Neutron-crystal diffraction experiments are typically carried out using neutrons of energy $\approx 50\,\text{meV}$. Why?

(c) An important class of high-energy physics experiments entails accelerating protons to an energy large enough that their wavelength is 0.5 fermi. What is the energy of such a proton?
Hint: You cannot use the same equations to do parts (b) and (c) of this problem.

(d) In this problem, we have considered situations in which we could base our calculation of the de Broglie wavelength on the non-relativistic energy-momentum relation

$$E = \frac{p^2}{2m} \tag{2.4.1}$$

and situations in which we had to use the relativistic relation

$$E^2 = p^2c^2 + m_0^2c^4, \tag{2.4.2}$$

where m_0 is the rest mass of the particle in question. The justification for using (2.4.1) when appropriate is that this equation is the non-relativistic limit of (2.4.2). Of course, using (2.4.1) introduces some error into the resulting wavelength, but for non-relativistic velocities this error is negligible. Suppose, for example, we are willing to accept a 5% error in the de Broglie wavelength of a particle. **Determine** the *minimum energy* for which Eq. (2.4.1) will lead to a wavelength of this accuracy for

(1) an electron;
(2) a proton.

2.5 Diffraction in Everyday Life.

In each of the following tragic circumstances, **determine** whether or not particle diffraction will occur. *Justify your answer in each case with a quantitative result and a qualitative argument.*

(a) A 1986 Mercedez-Benz, of mass 4000 kg travelling at 200 mph, slams into a garage;
(b) A stone of mass 10 g hurtles at 10 m/sec into a 0.3 m window;
(c) A grain of pollen of diameter 10^{-5} cm enters a 1.0 cm nostril.

Hint: A reasonable density for a pollen grain is $2\,\text{gm/cm}^3$. Assume that the pollen is at room temperature ($27°$ C) and that its translational energy is equal to the thermal energy of the air molecules in the room.

2.6 Quantization and the Correspondence Principle.

In § 2.2 we saw Max Planck astound the world of physics by proclaiming that microscopic oscillators in a black body can have only discrete energies

$$E_n = nh\nu, \tag{2.6.1}$$

where ν is the frequency of the oscillator and $h = 6.626 \times 10^{-34}$ J–sec. Yet we know that *macroscopic* oscillators have continuous energies. In this problem we shall explore this apparent contradiction and show that it isn't a contradiction at all.

(a) Consider the following macroscopic oscillator: two masses $m = 1\,\text{g}$ are attached to a spring and oscillate with amplitude $A = 1.0\,\text{cm}$ at frequency $\nu = 100\,\text{Hz}$. Use classical physics to **calculate** the energy of the oscillator, expressing your result in joules, ergs, and in electron volts (eV). Now, using Planck's relation (2.6.1), **calculate** the corresponding value of n.

(b) From your answer to (a), **calculate** the *percentage change in energy* that results from *unit change* in n. Could you measure such a change?

(c) The H_2 molecule is a *microscopic* oscillator. The mass of each hydrogen atom is roughly 1.7×10^{-24} g. The frequency of oscillation of the two atoms in H_2 is $\nu \approx 1.0 \times 10^{14}\,\text{sec}^{-1}$. From this data, **calculate** the energy of this oscillator (in electron volts) for $n = 1$, 2, and 3.

(d) Using classical physics, **calculate** numerical values (in Å) for the amplitude of the three states considered in (c). We see that *the amplitude of a microscopic oscillator is quantized.*

2.7 An Equation for de Broglie's Matter Waves.

In § 2.6, we discussed the hypothesis of the irrepressible Prince Louis de Broglie, who in his Ph. D. thesis maintained that all matter has wave-like properties—so that, for example, material particles should display diffraction phenomena when they interact with objects whose size is of the order of the particle's wavelength. In this problem, we use the de Broglie and Einstein relations (2.16) to derive a "wave equation" for a particle moving through space in the absence of any forces—*i.e.*, a free particle.

> **NOTE**: It is convenient in this problem to write these relations in terms of the *angular frequency* $\omega = 2\pi\nu$ and the *wave vector* $k = 2\pi/\lambda$, to wit:
>
> $$E = h\nu = \hbar\omega$$
> $$p = h/\lambda = \hbar k,$$
>
> where $\hbar = h/(2\pi)$.

(a) **Write down** an expression in exponential form for a *traveling wave* $\Psi(x, t)$ propagating in the $+x$ direction. Use A for the amplitude of the wave, λ for its wavelength, and ν for its frequency. We shall suppose that this wave "represents" (in some sense) a free particle of mass m.

(b) **Rewrite** your wave function $\Psi(x, t)$ in terms of the particle's momentum p and its energy E.

At this point we ask the obvious question: what sort of equation does $\Psi(x, t)$ satisfy—*i.e.*, what is the "wave equation" for a free particle? From our experience with waves in classical physics, we expect this to be a differential equation in x and t. The general form of such an equation is

$$a_0 + a_1 \frac{\partial}{\partial x}\Psi + a_2 \frac{\partial^2}{\partial x^2}\Psi + \cdots = b_1 \frac{\partial}{\partial t}\Psi + b_2 \frac{\partial^2}{\partial t^2}\Psi + \cdots$$

We want to derive a general equation for the de Broglie wave, so we must impose a restriction on the coefficients $a_0, a_1, \ldots, b_1, b_2, \ldots$. These quantities can involve fundamental constants (such as m, e, and h) but not quantities that depend on the wave (such as λ and E).

(c) Figure out the *simplest* such differential equation by imposing the requirement that your equation be consistent with the classical relationship between energy E and momentum p of a free particle, $E = p^2/(2m)$. You have just derived the Schrödinger equation for a free particle!

(d) Argue from your answer to (c) that the wave function for a free particle must be a complex function of x and t.

2.8 An Equation for de Broglie's Matter Waves.

In § 2.1 we briefly discussed quantization of energy. But energy is not the only observable of microscopic systems that is quantized. To the contrary: *most* observables of quantum systems exhibit quantization. In this problem we shall illustrate *quantization of angular momentum* in a thought experiment (*Gedanken-experiment*). This experiment also illustrates the subtle, beautiful interplay of the fundamental ideas of quantum physics: in this case, of duality and quantization.

In Fig. 2.8.1 is shown a "quantum hamster cage." (Assume that the "quantum hamster" is out frolicking, so that the cage is empty.) This miracle of rare device consists of two end discs connected by N equally-spaced pins, where N is very large. The cage is mounted on frictionless bearings about the indicated rotation axis so that it can turn freely. (The rotation axis is perpendicular to the paper in this figure.) Let a denote the radius of the cage, as shown in the side view in this figure.

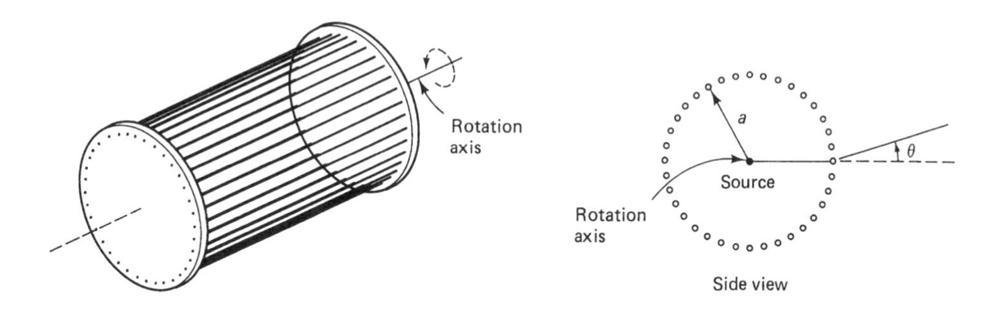

Figure 2.8.1 The quantum hamster cage. [Adapted from *Introduction to Quantum Mechanics* by R. H. Dicke and J. P. Wittke (Reading, Mass.: Addison-Wesley, 1960).]

Suppose a monochromatic light source projects a beam of photons radially outward from the rotation axis. The beam falls on a small portion of the circumference that, since N is large, includes many pins.

(a) To a good approximation, the pins in this set-up constitute a *plane-wave diffraction grating*, so light from the source will be diffracted through various discrete angles. Let θ denote one of the angles of diffraction. **Derive** an expression for $\sin\theta$ in terms of the photon wavelength λ, N, a, and an integer n. Your answer is an expression, in the language of mathematics, of the *wave-like* characteristics of the photons in the beam.

(b) Now let's explore the *particle-like* characteristics of the photons. After a photon leaves the source, it moves radially outward. Consider such a photon with linear momentum **p** *before it encounters the pins*. **Determine** the angular momentum about the drum axis of the moving photon. Justify your answer.

(c) Suppose the photon of part (b) is scattered by the pins through an angle θ. **Determine** the magnitude of the angular momentum of the photon *after it is scattered*.

(d) Using conservation of angular momentum, the de Broglie relations (2.16) and your results from parts (a)—(c), **show** that the angular momentum *of the drum* is an integral multiple of $N\hbar$, where $\hbar \equiv h/2\pi$. (Assume that the drum is initially at rest.)

2.9 Pair Annihilation.

Among Einstein's epochal contributions to physics are the ideas of (1) the quantization of energy associated with electromagnetic radiation, as expressed by $E = h\nu$, and (2) the mass-energy equivalence expressed by $E = mc^2$. In this problem, we shall explore the connection between these ideas in a particularly interesting physical situation.

The process of pair annihilation involves an electron and a positron. A **positron** is a particle with the same *mass* and *magnitude of charge* as the electron, but with the opposite *sign of the charge*—*i.e.*, a positron is a positively-charged "anti-electron".[34] In the annihilation process, an electron and a positron that are (essentially) at rest next to one another unite and are annihilated. The process entails complete conversion of mass energy to radiation energy.

(a) **Calculate** the amount of energy that results from such a pair annihilation, expressing your result in joules, ergs, and eV.

(b) In the most commonly observed annihilation event, two photons of equal energy are emitted. **Explain** physically why this result is expected.

(c) Photons, as you know, have zero rest mass. We can, however, assign a "pseudomass" $\bar{m}$ to the photon *provided we agree to use it only in equations for the momentum and relativistic self-energy*. Consider a photon of energy $h\nu$. **Derive** an expression for $\bar{m}$ and for the momentum p in terms of the photon wavelength λ and fundamental constants. **Calculate** values for $\bar{m}$ and p for the photons in part (b).

[34] The positron was the first anti-particle discovered. It was discovered by C. D. Anderson in 1932; four years later, Anderson received half a Nobel Prize for his efforts.

CHAPTER 3

First Postulates

The State Function and Its Interpretation

There was a moment's silence while everybody thought.
"I've got a sort of idea," said Pooh at last,
"but I don't suppose it's a very good one."
"I don't suppose it is either," said Eeyore.

—*The House at Pooh Corner*
by A. A. Milne

Like classical physics, quantum physics need not be able to explain *everything*. But it should enable us to qualitatively and quantitatively explain and predict the properties and behavior of a reasonable variety of systems. In the microscopic realm, the systems we are interested in range in complexity from a particle moving through space in the absence of external forces to complex bound aggregates of particles, such as nuclei, atoms, molecules, and solids. And like macroscopic (classical) systems, microscopic (quantum) systems can exist in a variety of *states*; each state is characterized by unique values for the physical properties of the system: energy, momentum, etc.

In Part I we reviewed the way in which a classical state of a single macroscopic particle is defined by the particle's *trajectory* $\{\mathbf{r}(t), \mathbf{p}(t)\}$.[1] We also argued that a state of a *microscopic* particle, such as an electron, cannot be represented by a trajectory, because according to the Heisenberg Uncertainty Principle we cannot specify simultaneous precise values of $\mathbf{r}(t)$ and $\mathbf{p}(t)$. Indeed, *trajectories must not appear in a physical theory for the microscopic universe*. In this chapter we'll discover and explore the way in which we do represent a state in this universe. We're going to meet *the wave function*.

Our guide in this quest for a "state-descriptor" for the microverse is the wave-particle duality that characterizes the behavior of microscopic particles. In particular, in § 3.1 we'll search for clues in the results of the *double-slit experiment* of Chap. 2. Doing so will lead us to the wave function and to the first postulate of quantum mechanics, the topic of § 3.2. In reexamining data from the double-slit experiment we'll also find a clue to the physical interpretation of the wave function. This interpretation is the essence of the second postulate of quantum mechanics, which we'll formulate in § 3.4. Finally, in § 3.6 we'll extend the probabilistic approach to two vital statistical quantities: the expectation value and the uncertainty. Discussing the uncertainty will bring us full circle to one of the key ideas we discussed in Chap. 2.

I want to keep our discussion of these new ideas simple at first, so in this chapter we'll consider *states of a single particle at a fixed time*. The time development of these states—according to the Schrödinger Equation—is the topic of Chap. 6. I also want to focus on a single physical property: *position*, because it's comparatively easy to extract information about this property from a wave function. We'll develop machinery for studying other physical properties in Chap. 5.

3.1 PATTERNS OF PROPAGATING POSSIBILITY: PROBABILITY WAVES

Any theory of the physics of microscopic particles must be able to explain the weird behavior of electrons in the double-slit experiment. In § 2.5 we saw that *en masse*

[1] More precisely, the state of a system in classical mechanics is defined by a set of generalized coordinates and their canonical momenta. See Jerry B. Marion, *Classical Dynamics of Particles and Systems*, 2nd ed. (New York: Academic Press, 1970).

electrons in this experiment create an interference pattern at the detector. Yet individual electrons arrive as localized "clumps" of energy. This dual nature must be built into any theory that purports to describe the microworld. But such a theory must also incorporate the limitation we discovered in Chap. 2: we cannot know which slit a particular electron passes through. In this section, we'll reconsider the double-slit experiment and formulate a couple of hypotheses to explain its results. This voyage of discovery will lead us to a fundamental element of quantum physics: the wave function.

Another Look at the Double-Slit Experiment with Light

Let's see if the Young interference experiment—the double-slit experiment with light (§ 2.4)—can give us a clue to the mystery of electron interference. According to the classical theory of electricity and magnetism, the interference pattern that forms when a (coherent) beam of *light* passes through the double-slit apparatus arises from the *superposition* of two electromagnetic waves, one diffracted from each slit. Let's look a bit more closely at this explanation.

The detector in the Young experiment measures *light intensity*. This quantity is proportional to the squared magnitude of the electromagnetic field at the detector.[2] Let's label the electric field vectors for each diffracted wave by a subscript that indicates which slit the wave passed through: $E_1(x)$ and $E_2(x)$.[3] In terms of these fields, the intensity $I_{12}(x)$ for the case of both slits open is

$$I_{12}(x) \propto |E_1(x) + E_2(x)|^2. \tag{3.1a}$$

Now, when we expand the right hand side of (3.1*a*), we find terms that involve *products* of the two electric fields:

$$I_{12}(x) \propto |E_1(x)|^2 + |E_2(x)|^2 + [E_1^*(x)E_2(x) + E_1(x)E_2^*(x)]. \tag{3.1b}$$

The two terms in square brackets in Eq. (3.1*b*)—the "cross terms"—are the mathematical manifestations of the physical superposition of diffracted waves that produces the interference pattern. Of course, this explanation rests on the fact that Maxwell's theory thoughtfully provides things to superpose, namely electromagnetic waves.[4]

[2]In fact, the intensity at a point x on the detector is

$$I = \epsilon_0 c\,|E(x)|^2,$$

where ϵ_0 is the permittivity of free space and c is the speed of light.

[3]For convenience, I will implement a very slight change of notation in this chapter from that in Chap. 2: to denote the slits I'll use numbers rather than the more primitive "upper" and "lower" labels of § 2.5. Of course, you can number the slits any way you like; just be consistent. Similarly, we use I_{12} rather than I_{double} for the double-slit intensity. Finally, we use x, the distance along the detector in Fig. 2.3, rather than θ as the independent variable.

[4]What allows us to add electromagnetic waves in this fashion and thereby obtain an intensity that explains the observed interference pattern in Young's experiment is the *linearity* of the fundamental equations for electromagnetic fields, Maxwell's equations. Similarly, the fundamental equation of quantum physics, the Schrödinger equation, turns out to be linear, and, as we'll soon see, quantum waves must obey superposition. The connection between superposition of quantum waves and the linearity of the equation that we solve to determine these waves will be an important topic in Chap. 6.

Back to Electrons

Now let's return to the double-slit experiment with electrons and try to apply the principle we used to explain the results of Young's experiment: superposition. Immediately, we run into a road-block. In classical mechanics we do not use waves to describe the states of particles. We use trajectories, and trajectories can't be superposed and don't interfere.

Aside: Superposition According to de Broglie. As we saw in § 2.6, de Broglie dealt with this difficulty by introducing a concept not included in conventional classical mechanics: the *matter wave*. His idea sounds good: we need something to superpose, and a matter wave is as good as any other. But it's hard to see how to make de Broglie's nebulous notion the basis of a workable physical theory. In fact, all straightforward efforts to associate a wave with an electron fail. For example, we might try to explain the behavior of the electrons by conjecturing that they actually *are* classical waves. But the results of the double-slit experiment rule out this hypothesis: the pattern of dots seen at the detector after a short time clearly shows the particle-like qualities of electrons. More subtly, these results also preclude the hypothesis that electrons are classical particles somehow guided through space by a wave. Such an explanation assumes that electrons follow trajectories, and we know from Chap. 2 that trajectories are forbidden. Evidently, a more subtle idea is needed.

Let's return to the results of the double-slit experiment and see if they suggest an approach. It is useful to think about this experiment in terms of events. We can define an **event** in this experiment to be *the arrival of an electron at the detector*. If you'll glance back at Fig. 2.8, you'll see that the first several events in the double-slit experiment seem to be random—*i.e.*, the locations of the spots formed by these arrivals appear to be random. Significantly, *it is only after a large number of events that we see a well-defined pattern*. Let's see if we can incorporate this feature of the results into a hypothesis.

One way to describe mathematically phenomena that are evident only after a large number of events have occurred is with *probability theory*. The results of the double-slit experiment suggest that we try thinking about the propagation of electrons using this approach. For example, we could ascribe to the electrons a *probability* of arriving at each location x on the detector. This conjecture, in turn, suggests that we associate with the electrons in the incident beam a "probability function." The idea is that this function somehow contains *information* about the probability that an electron in the beam will wind up at each x after it passes through the diaphragm. According to this interpretation, the detector implicitly measures this probability function, for the information carried by this function is manifested in the measurement by the number of electrons that arrive at each x.

The trick is to turn this vague idea into a hypothesis that is consistent with the experimental results. Here's our first attempt at formulating such a hypothesis: interpret the electron intensity measured by the detector as a *probability distribution*; *i.e.*,

> ***Trial Hypothesis 1***
>
> The motion of the electrons in the incident beam is governed by a probability distribution function that we'll denote by $P_{12}(x,t)$. That is, at each time t this function is large at values of x where an electron is likely to be found and small at values of x where it is unlikely to be found.

This formulation seems a step in the right direction; but does it work? Let's put it to the test of experimental verification by devising a variation on the double-slit apparatus of § 2.8. Suppose we close a shutter on slit 2, thereby forcing all incident electrons to pass through slit 1. We observe that at the detector these electrons form a *diffraction pattern* like the one in Fig. 2.2a; the corresponding intensity is shown in Fig. 2.6a. Let's denote the probability distribution function for detecting an electron at position x for this apparatus—with only slit 1 open—by $P_1(x,t)$. Similarly, we'll denote by $P_2(x,t)$ the probability function for electrons in an apparatus where slit 1 is closed and slit 2 is open. The intensity function for this case is also that of Fig. 2.6a.

Now, if our trial hypothesis is correct, we should be able to relate the probability function for the apparatus in which *both* slits are open, $P_{12}(x,t)$, to the single-slit functions $P_1(x,t)$ and $P_2(x,t)$. So let's return to the two-slit case. If we conjecture that in this case each electron that reaches the detector went through *either* slit 1 *or* slit 2, then these probability distribution functions should be related by[5]

$$P_{12}(x,t) = P_1(x,t) + P_2(x,t). \qquad \text{(An Incorrect Equation)}$$

But the experimental data we studied in Chap. 2 tells us that this equation is wrong. To convince yourself, go back to Fig. 2.6. Compare the sum of the intensities shown there to the intensity for the interference pattern. Evidently, trial hypothesis # 1 won't do the job. Can you figure out what's wrong with it?

Birth of a Notion

The trouble with our first hypothesis is that it ignores the lesson of the Young interference experiment: to explain an interference pattern we need something to superpose. But we can't superpose probability distributions; they don't interfere with one another. *But waves do.* This suggests that *at its most basic level, our theory must be formulated in terms of some sort of wave.*

So let's devise an alternative trial hypothesis, one that makes the fundamental element of the theory a "probability wave." This new wave will have all the attributes of a classical wave: a wavelength, frequency, amplitude, and so forth. I'll denote it by the Greek symbol $\Psi(x,t)$. We know little about this function at present (!), so we'll consider the most general case, and suppose that Ψ is *complex.*[6]

In our new hypothesis, it is the probability wave $\Psi(x,t)$ that contains information about the likelihood of an electron being detected at position x at time t. In this new approach, the detector measures the "intensity" of our probability wave, *i.e.*, the square of its magnitude. So the relationship between the wave and the probability distribution it describes is[7]

$$\boxed{\text{probability} \quad \propto \quad |\text{ wave }|^2} \qquad (3.2)$$

[5]I have here used a result from probability theory: Since probabilities of mutually exclusive events *add*, the probability function for the two-slit experiment must be the sum of those for each single-slit experiment. If this result isn't familiar, see the review of probability theory in § 3.3.

[6]For a review of complex numbers and functions, see Appendix K.

[7]Equation (3.2) and its successors in this section are written as proportionality relations rather than equalities because we are here discussing *relative* rather than *absolute* probabilities. This distinction will be explained in § 3.3–3.5 when we investigate normalization.

So the idea is that the function $\Psi(x,t)$ is to be interpreted as a *position probability amplitude*. Here's the new hypothesis:

> ***Trial Hypothesis 2***
>
> The propagation of the electrons is governed by a probability wave $\Psi(x,t)$. At any time t, this function is the *position probability amplitude* for detection of the electron at position x.

Does this hypothesis conform to observations? Let's see. Consider first the apparatus with slit 2 closed and slit 1 open, and let $\Psi_1(x,t)$ denote the probability wave for electrons that reach the detector in this case. According to Eq. (3.2), the probability of an electron in such a state showing up at location x on the detector at time t is determined from the squared modulus of this function, *viz.*[8]

$$P_1(x,t) \propto |\Psi_1(x,t)|^2. \tag{3.3}$$

In an apparatus with slit 1 closed and slit 2 open, the probability amplitude for electrons that reach the detector is $\Psi_2(x,t)$, and the corresponding intensity is

$$P_2(x,t) \propto |\Psi_2(x,t)|^2. \tag{3.4}$$

In the double-slit experiment, the functions $\Psi_1(x,t)$ and $\Psi_2(x,t)$ define two possible *states* of the electron, one corresponding to passage through slit 1, the other to slit 2. In this case, the appropriate probability wave is the superposition (*aha!*) of Ψ_1 and Ψ_2, *i.e.*,[9]

$$\Psi_{12}(x,t) = \Psi_1(x,t) + \Psi_2(x,t)\ . \tag{3.5}$$

Now this is important: *the probability amplitude $\Psi_{12}(x,t)$ represents a superposition state that includes the two possibilities represented by $\Psi_1(x,t)$ and $\Psi_2(x,t)$.* The measured intensity corresponds to the probability distribution, which according to (3.2) is the squared modulus of the intensity:

$$P_{12}(x,t) \propto |\Psi_{12}(x,t)|^2 \tag{3.6}$$

$$= |\Psi_1(x,t) + \Psi_2(x,t)|^2. \tag{3.7}$$

The last step in verification of trial hypothesis # 2 is demonstrating that Eq. (3.7) predicts the observed interference pattern. To simplify the algebra involved in this demonstration, I want to write the functions $\Psi_1(x,t)$ and $\Psi_2(x,t)$ in terms of their magnitudes and phases.[10] If $\alpha_1(x,t)$ and $\alpha_2(x,t)$ denote the phases, we can write the probability amplitudes as

$$\Psi_1(x,t) = |\Psi_1(x,t)|\, e^{i\alpha_1(x,t)} \tag{3.8a}$$

[8]Strictly speaking, $|\Psi_1(x,t)|^2$ is a *probability density function* and must be multiplied by an infinitesimal volume element dx in one dimension to determine a probability. We'll get to this subtle but important complication in due course (§ 3.4).

[9]Nothing in the apparatus singles out one slit over the other. That is, it's equally likely that an incident electron will go through slit 1 as through slit 2. That's why I assigned equal weights to each single-slit amplitude in this equation.

[10]Any complex number z can be written as the product of a magnitude $|z|$ times a phase factor $e^{i\alpha}$ (see Appendix K). Note carefully that in Eqs. (3.8) both magnitude and phase depend on x and t.

and

$$\Psi_2(x,t) = |\Psi_2(x,t)|\, e^{i\alpha_2(x,t)}. \tag{3.8b}$$

When we substitute (3.8) into (3.7), we obtain the probability function

$$P_{12}(x,t) \propto |\Psi_1|^2 + |\Psi_2|^2 + |\Psi_1|\,|\Psi_2|\left[e^{i(\alpha_1-\alpha_2)} + e^{-i(\alpha_1-\alpha_2)}\right] \tag{3.9a}$$
$$= |\Psi_1|^2 + |\Psi_2|^2 + 2|\Psi_1|\,|\Psi_2|\,\cos(\alpha_1 - \alpha_2). \tag{3.9b}$$

Look at the third term on the right-hand side of (3.9*b*): it's an interference term! [Recall the crucial role played by the interference term in Eq. (3.1*b*) in our explanation of the results of Young's experiment.] Depending on the phase difference $\alpha_1 - \alpha_2$ of the probability waves, this term produces constructive or destructive interference in the quantity measured at the detector.

To read the physical significance of this mathematical result, we turn to our proposed interpretation of the probability function $P_{12}(x,t)$. According to this interpretation, electrons congregate at values of x where constructive interference occurs (the high-intensity regions of the detector) and avoid places where destructive interference occurs (the low-intensity regions). These extremes (and the values inbetween) explain the observed interference patterns of the double-slit experiment. Evidently, trial hypothesis # 2 works!

Look at what we've accomplished. Reasoning by analogy with the theory underlying the Young interference experiment and clinging tenaciously to the idea of affiliating some sort of "detection probability" with quantum particles, we have discovered the fundamental element of quantum theory: the wave function. That the predictions of our new theory agree with the results of the double-slit experiment is reason to turn this hypothesis into a formal postulate, which we'll do in the next section.

3.2 SYSTEMS AND STATES THEREOF

Quantum physicists are interested in all sorts of physical systems—free protons, conduction electrons in metals, atoms ranging from the simple (hydrogen) to the dauntingly complicated (uranium), molecules from H_2 to DNA. Each of these **systems** consists of one or more microscopic particles that may be interacting among themselves via internal forces and may be exposed to external fields (*e.g.*, electromagnetic or gravitational fields). Remarkably, states of these diverse systems are represented by the same type of varmints: state functions.

Introducing the State Function

Position probability amplitudes like the waves we used in § 3.1 to explain electron interference are the basic elements of quantum theory. In this theory each state of a system is represented by a different wave function. Actually, most quantum physicists are interested only in states that can be studied in a laboratory—such states are said to be **physically realizable**. The wave function also goes by the aliases *state function* and *state vector*. I'll use capital Greek letters (see Appendix G) to symbolize wave functions and, in particular, will use Ψ for a "generic wave function."[11]

The stratagem of representing physical states by wave functions cannot be justified or proven. (But, as I suggested in § 3.1, its consequences can be verified experimentally.) This postulate is one of the foundations upon which quantum mechanical theory is constructed:

> ***The First Postulate of Quantum Mechanics***
>
> Every physically-realizable state of a system is described in quantum mechanics by a state function Ψ that contains all accessible physical information about the system in that state.

Note the word "accessible" in Postulate I. The information we can extract from a wave function is limited, and there is no getting around these limitations. For example, the uncertainties that lurk like little land mines in quantum theory prohibit us *in principle* from knowing the precise values of position and momentum of a microscopic particle. We cannot obtain precise values of these properties in any way—not by experimental measurement, not by theoretical calculation. But all we *can* know about such a particle is buried in its wave functions. And it is only in this sense that the theory of quantum mechanics can be described as "complete."

State functions come in many guises; they can be written as functions of momentum, of energy, of position ... of any observable you wish. But, particularly in this book, you'll usually see **wave functions** written as functions of the spatial coordinates of the particles that comprise the system and of the time, to wit:

$$\boxed{\Psi = \Psi\,(\text{position, time})} \qquad \text{wave function}$$

Aside: States in Classical Physics. Notice how different this prescription for specifying a state is from the trajectories of Newtonian mechanics. In classical physics a particle's spatial coordinates are functions of time. But, as we saw in Chap. 2, paths are not allowed in descriptions of quanta. Therefore, *in quantum physics position and time are independent quantities.*

Throughout Parts II and III of this volume, we'll consider several *one-dimensional single-particle systems*. For such a system, the position variable is x and the state function is simply

$$\Psi = \Psi(x, t). \tag{3.10}$$

But we can trivially generalize this form to a single particle in three dimensions with position variable $\mathbf{r}$, *viz.*,

$$\Psi = \Psi(\mathbf{r}, t), \tag{3.11}$$

or to a many-particle system, such as two particles in three dimensions, *viz.*

$$\Psi = \Psi(\mathbf{r}_1, \mathbf{r}_2, t). \tag{3.12}$$

Such systems will come front-and-center in Volume II.

[11]In more general and abstract treatments of quantum mechanics, a state is represented by a vector in an infinite-dimensional linear vector space called a *Hilbert space*. You need not worry now about this generalization, but if you are interested see the Selected Readings for this chapter.

Example 3.1. A Typical Wave Function

One of the most important single-particle one-dimensional systems is the simple harmonic oscillator: a particle of mass m moving in one dimension with potential[12] $V(x) = m\omega_0^2 x^2/2$. In Chap. 9 we'll find that one of the physically-realizable state functions for the simple harmonic oscillator has the functional form

$$\Psi(x,t) = Ae^{-\beta^2 x^2/2 - iEt/\hbar}. \tag{3.13}$$

Here A is the amplitude of the wave function, and E and β are constants related to various physical properties of the particle. (In particular, E is its total energy in the state represented by this wave function.) Notice that in general $\Psi(x,t)$ is *complex*—it is real only at particular times, such as $t = 0$. This illustrates a property to which we'll return shortly: all state functions are, in general, complex; they are real only at special times. The function in (3.13), along with a few other state functions we'll meet in subsequent chapters, are on display in Fig. 3.1.

Aside: Graphing State Functions. The complex nature of state functions poses some practical difficulties. For example, to graph a function such as (3.13) would require two three-dimensional graphs: one to show the real part of $\Psi(x,t)$ as a function of x and t and one for the imaginary part. For this reason we usually look at "snapshots" of state functions—graphs in which time is "frozen" and we show only the dependence of $\Psi(x,t)$ on its spatial coordinate.

In considering state functions such as those in Fig. 3.1, you must keep in mind that physically these functions are *position probability amplitudes*—that's the idea we proposed in § 3.1 and that we'll make formal in § 3.4. Hence the squared modulus of the state function, $|\Psi(x,t)|^2 = \Psi(x,t)^*\Psi(x,t)$, will play a crucial role in digging physical information out of this function.[13]

Look at Fig. 3.1 and try to envision the squared moduli of these functions. Notice that for each wave function you can find a (rather small) finite region of space where the probability of finding the particle is significantly larger than anywhere else. A state function that defines such a region of space is said to be **spatially localized**. Note this property well: *Spatial localization is an extremely important property of state functions, for only functions that are spatially localized can represent quantum states.* In § 3.5 I'll show you how to translate this physical requirement into a mathematical constraint: the normalizability condition.

The Principle of Superposition

In § 3.1 I emphasized the crucial role of superposition in explaining the behavior of electrons (and photons) in the double-slit experiment. Indeed, we introduced waves in the first place so we would have things to superpose. The identification of *wave functions* with *quantum states* lends a profound and very important significance to superposition in quantum mechanics:

[12] Strictly speaking, $V(x)$ is the potential *energy* of the particle. Quantum physicists, however, refer to such functions as *potentials*, and, to stay in step, so will I.

[13] We must take account of the complex nature of the wave function when we form the probability distribution function; that is why we cannot write $[\Psi(x,t)]^2$. For more on this matter, see § 3.4.

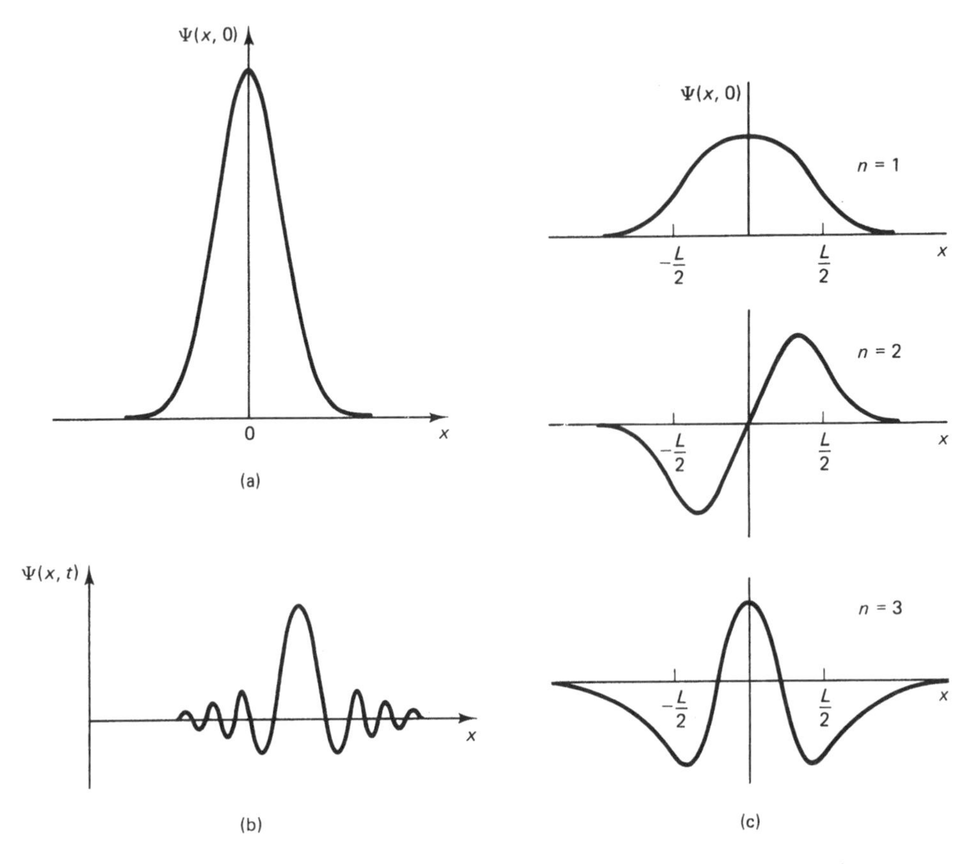

Figure 3.1 A complex of wave functions. "Snapshots" at $t = 0$ of the position probability amplitude for (a) a particle in a state represented by Eq. (3.13) with $A = (1/2\pi)^{1/4}$ and $\beta = \sqrt{2}$ (see also Examples 3.3–3.5); (b) a typical wave packet (Chap. 4); and (c) three states of a particle in a finite-square-well potential of width L (Chap. 9).

The Principle of Superposition

If Ψ_1 and Ψ_2 represent two physically-realizable states of a system, then the linear combination $\Psi = c_1\Psi_1 + c_2\Psi_2$, where c_1 and c_2 are arbitrary complex constants, represents a *third* physically-realizable state of the system.

Actually, the Principle of Superposition is not restricted to linear combinations of only two quantum waves. In its most general form, this principle says that an *arbitrary* linear combination of *any number*—even an infinite number—of quantum waves represents a new state of the system. More precisely: given N state functions of a system, $\{\Psi_1, \Psi_2, \Psi_3, \ldots, \Psi_N\}$, every linear combination

$$\Psi = \sum_{i=1}^{N} c_i \Psi_i \tag{3.14}$$

represents another state function. (The coefficients c_i are arbitrary and may be real or complex.) *The essential physical content of the Principle of Superposition is that for any choice of coefficients c_i, Eq. (3.14) represents a state that can be realized in the laboratory.* That is, we can construct from a given set of state functions an *infinite number* of other state functions, each new function corresponding to a different set of coefficients $\{c_1, c_2, \ldots, c_N\}$. And each linear combination we create represents a new physically-realizable state.

I hope it's clear that the Principle of Superposition follows directly from our decision to represent physical states of microscopic particles by *wave functions.* This principle towers over the microworld, appearing time and time again in quantum physics and its applications. [In Chap. 6, for example, it will reappear in a particularly important property of the Schrödinger equation (linearity).] Remember it.

*3.3 PROBABILITIES: A REVIEW

In § 3.1 we interpreted the quantum wave of a microscopic particle (an electron) in the double-slit experiment as a *position probability amplitude.* As we'll see in § 3.4, this interpretation forms the backbone of quantum physics. In fact, *the physical information contained in a state function is inherently probabilistic in nature.* Hence *probability theory* is one of the most important mathematical and conceptual tools of every quantum mechanic. In this section we'll briefly review this topic, as preparation for Postulate II.[14]

Ensembles

In the real world, a rock concert is an event. But in the world of quantum physics, the word **event** refers to obtaining *a particular result* in a measurement of an observable on a microscopic system. For example, if we're interested in measuring the position of a ball rolling in the x direction along a linear track, we might ask: what is the probability of finding the ball at $x = 10\,\text{m}$? The definition of probability we use in quantum mechanics is expressed in terms of such events.

A probability is an inherently *statistical quantity.* That is, you can't define a probability in terms of a single system; you must do so by referring to *a very large number of identical systems.* These systems must also be independent of one another (*i.e.*, not interacting with one another), and they must all be in the same physical state. A collection of identical systems that satisfies these criteria is called a **statistical ensemble**, or, more tersely, an **ensemble**. Take a moment to re-read, memorize, and ponder the concept of an ensemble; understanding ensembles is the key to understanding probabilities.

The individual systems that comprise an ensemble are called its **members**. Ensembles are of special importance in quantum physics because in this discipline the word *measurement* refers to an **ensemble measurement**. *To perform an ensemble measurement of an observable, we perform precisely the same experiment on the members of an ensemble of identical, non-interacting systems, all of which are in the same physical state.*

You may be wondering what, precisely, I mean by "identical systems". Well, this phrase means just what it sounds like: the members of the ensemble must be identical

[14]If you feel comfortable with probabilities, means, and standard deviations, skip or skim this section. If you want more information than you find here, try the excellent, detailed discussion of probability and statistics in Chap. 2 of *Statistical Physics* by F. Reif (New York: McGraw-Hill, 1965). In a lighter vein, see *Lady Luck* by W. Weaver (New York: Anchor Books, 1963).

in their composition and in all their intrinsic properties. For example, an ensemble for measuring the position of the ball on the linear track would consist of a huge number of identical balls on identical linear tracks. In setting up such an ensemble, we would have to ensure that not only the properties of particles in different members but also the external forces on each system (*e.g.*, gravity or external electromagnetic fields) are the same. *And we must ensure that all members of the ensemble are in the same state.* For example, all balls must follow the same trajectory on their respective tracks. For a microscopic system this condition requires that all members be represented by the same wave function.[15]

"Wait just a minute," you may be thinking, "why would anyone ever carry out the same measurement on a huge number of identical systems all in the same physical state? Were I to do so, I would just get the same answer over and over again."

Well, yes and no. To be sure, *in principle* all results of an ensemble measurement of an observable, such as position, on a *macroscopic* system would be the same.[16] But this is not true of a quantum system: *identical experiments performed on identical systems in identical quantum states need not — and, in general, do not — yield the same results.* Like the Heisenberg Uncertainty Principle, this extraordinary statement is an intrinsic fact of nature, not a consequence of experimental imperfection. It would be no less true if the apparatus and the experimentalist using it were perfect! We'll return to this disquieting notion in § 3.6.

Probability Defined

With all this chit-chat about ensembles out of the way, we are at last ready to define a probability. For the sake of clarity, we'll start with a *position probability*; we can trivially extend this definition to any observable.

Suppose we perform an ensemble measurement of the position of a particle that is constrained to move in the x direction. (For the moment, I won't specify whether the particle is microscopic or macroscopic; for our present purposes the size of the system doesn't matter.) Each member of the ensemble exhibits a particular value of x, and, since the ensemble consists of a huge number of members, we wind up with a lot of data.

To organize all this data, let's label *the values we observe* with an integral index j. Each different observed value of position gets a new integer, $j = 1, 2, 3, \ldots,$ — on and on and on until we have labeled all the distinct values we obtained. Let M denote the maximum value of j. For convenience, we'll arrange the resulting indexed values in ascending sequence, *viz.*,

$$x_1 < x_2 < x_3 < \ldots < x_M.$$

[15]In his insightful article "Probability Interpretation of Quantum Mechanics" [*Amer. Jour. Phys.*, **48**, 1029 (1980)] Roger Newton notes that since all members of an ensemble are physically identical, we can identify a *particular* ensemble by the physical properties of its members. Extending this point, Newton defines the state of a quantum system as "[the symbolic representation of] the ensemble to which it belongs."

[16]But in the laboratory, members of an ensemble of macroscopic systems will yield different results, because of imprecision in the design and construction of the apparatus and in the experimentalist's technique. Consequently, such experiments must be carried out on an ensemble in order to determine the experimental error inherent in the measurement. It is not, however, experimental error but rather "inherent imprecision" that infects measurements in the microverse.

According to this scheme x_M is the largest value of position exhibited by any member of the ensemble. Notice that if N is the *total number of members in the ensemble*, then $M \leq N$.

Now, if the ensemble is large enough ($N \gg 1$), then any *particular* value of x can (and probably will) occur several times. So a list of observed values doesn't fully describe the data. We must also note how many members exhibited each value x_j. Let's denote this number by n_j. We can now display the result of our ensemble measurement as follows:

$$x_1,\ x_2,\ x_3,\ x_4, \ldots \tag{3.15a}$$

$$n_1,\ n_2,\ n_3,\ n_4, \ldots \tag{3.15b}$$

But Eqs. (3.15) is a rather ungainly way to represent a large number of results—especially if our ensemble contained, say, 10,000 members. This information is more conveniently codified in a function: the *probability* of obtaining a particular value of x. Moreover, introducing such a function will make easier our subsequent generalization of probability theory to an infinite ensemble.

We determine the probability of obtaining the value x_j in an ensemble measurement of position from the *fraction of members that yielded this value.* To calculate this quantity we merely divide the number of systems that exhibit x_j—the number n_j in (3.15*b*)—by the total number of experiments N:

$$\frac{\text{number of experiments that yield } x_j}{\text{total number of experiments}} = \frac{n_j}{N}. \tag{3.16a}$$

Now, the fraction n_j/N depends explicitly on the number N of members in the ensemble. A more general quantity is the **absolute position probability** $P(x)$, the generalization of (3.16*a*) to an infinite ensemble:

$$P(x) \equiv \lim_{N \to \infty} \frac{n_j}{N} \qquad \text{Absolute Position Probability} \tag{3.16b}$$

Note carefully that I have written the absolute probability as a function of x, not of x_j; since the number of members in the ensemble described by (3.16*b*) is infinite, we must treat position as a *continuous* variable.

Mathematical Aside: Absolute versus Relative Normalization. Dividing each n_j by N in Eqs. (3.16) "normalizes" the values of n_j by putting them on a common scale. By itself, each integer n_j describes only the *relative* probability that we will obtain x_j. (For example, by comparing n_3 to n_7 we can ascertain only the likelihood that we will obtain x_3 as opposed to x_7.) But the fraction n_j/N tells us the "absolute" likelihood of obtaining the value x_j. That's why $P(x)$ is called the *absolute* probability.

How could we represent pictorially the results of an ensemble measurement? Since ensembles in the lab are finite, only a finite number of values would be obtained in such a measurement; we could represent these data on a histogram, as in Fig. 3.2a. But the absolute probability $P(x)$ is the extension of these data to an infinite ensemble. So this function is a continuous curve, as in Fig. 3.2b.

Combinations and Normalization

The algebra of probabilities in quantum mechanics is the same as in statistics. For example, suppose we must calculate the probability of obtaining one of two values

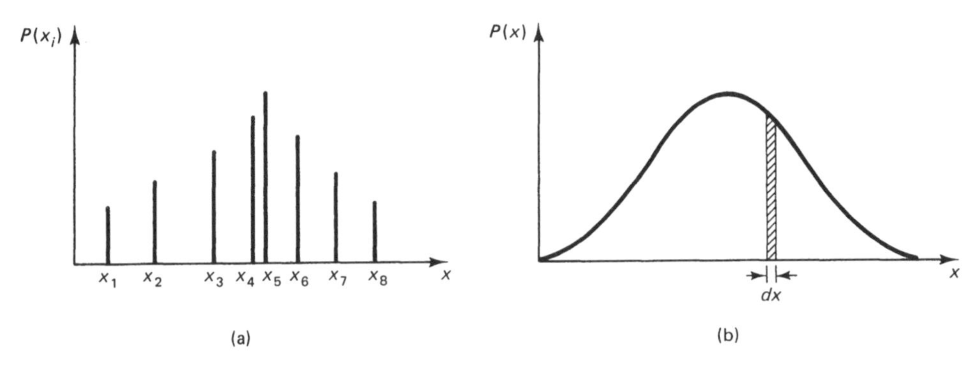

Figure 3.2 Probability portraits. (a) A histogram displaying the results of a position measurement in which only eight values—x_1 through x_8—were found. The vertical axis shows the frequency with which each value was obtained. (b) A plot of the absolute position probability $P(x)$, which is obtained in the limit $N \to \infty$. The shaded area is the probability of finding the particle in the region x to $x + dx$.

of position—say, x_3 or x_7. These events are *mutually exclusive—i.e.*, if you get x_3, you didn't get x_7. According to the rule of combination of probabilities, the desired probability is simply the sum of the probabilities for each event, *viz.*,

$$P(x_3 \text{ or } x_7) = P(x_3) + P(x_7). \tag{3.17}$$

To calculate the probability of finding one of several values, we just generalize (3.17): the probability of obtaining any of the values x_1 or x_2 or x_3 or ... or x_M is

$$\begin{aligned} P(x_1 \text{ or } \ldots \text{ or } x_M) &= P(x_1) + P(x_2) + \ldots + P(x_M) \\ &= \sum_{j=1}^{M} P(x_j), \end{aligned} \tag{3.18}$$

where *the summation includes all values of x_j in the list (3.15).*

This list includes every value obtained in the ensemble measurement. Since each member must exhibit one of these values, the "combined probability" (3.18) should equal 1. Let's see if it does. Using the definition of $P(x_j)$ in (3.16*a*) we can write (3.18) as

$$P(x_1 \text{ or } x_2 \text{ or } \ldots \text{ or } x_M) = \frac{1}{N} \sum_{j=1}^{M} n_j. \tag{3.19}$$

But the sum of n_j, the number of measurements that give x_j, over all possible values of x_j equals N. So, as advertised,

$$P(x_1 \text{ or } x_2 \text{ or } \ldots x_M) = 1 \qquad \text{[for a finite ensemble]}. \tag{3.20}$$

This result follows because we normalized the individual probabilities in (3.16*a*). For this reason, we call (3.20) a *normalization condition.*

The Position Probability Density

I noted above that in the limit $N \to \infty$ the probability must be written as a function of the continuous variable x. Properly, we should call the function $P(x)$ a **probability density,** because to calculate from this function the absolute probability, we must multiply it by an *infinitesimal* element of position dx. That is, the function $P(x)$ is actually the *position probability per unit length.* The **absolute probability** of finding the particle at a value of x between, say, x_0 and $x_0 + dx$ is $P(x_0)dx$.[17] When we take account of the continuous nature of x and the limit in the definition (3.16*b*) of $P(x)$, the **normalization condition** (3.20) becomes

$$\int_{-\infty}^{\infty} P(x)dx = 1. \qquad \text{Normalization Condition} \tag{3.21}$$

The Ensemble Average

The probability density $P(x)$ describes the results of an ensemble measurement of position. Appropriately, we characterize this kind of information as **probabilistic information** —a term which will become important later in this chapter. Another, more compact way to represent the same information is via **statistical quantities**, such as the *ensemble average*, which you may know under its colloquial name, the *mean value.*

You know what I mean by the mean: To calculate, say, the mean grade of N students on a calculus exam, we just add their individual scores and divide the result by N. In this chapter, we're concerned with the mean value of position—a quantity we denote by $\langle x \rangle$.[18]

It's not hard to calculate this quantity for a measurement on a finite ensemble: we just add each x_j in (3.15*a*) n_j times and divide the resulting sum by N. Alternately, we could calculate the ensemble average from the probability (3.16*a*); *viz.*,

$$\langle x \rangle = \sum_{j=1}^{M} x_j \, P(x_j) \qquad \text{[for a finite ensemble].} \tag{3.22}$$

The form of Eq. (3.22) makes sense. *For a finite ensemble $P(x_j)$ is the fraction of measurements that yield the value x_j—a quantitative measure of how important x_j is to $\langle x \rangle$.* So we can treat this probability as a "weighting factor" for the position that is its argument. That is, we construct the average by multiplying each x_j by its weighting factor $P(x_j)$ and summing over all (distinct) values.

But position is a continuous quantity, so the probability appropriate to a position measurement is the *position probability density* $P(x)$. To take this fact into account, we should alter slightly our definition (3.22) of the ensemble average $\langle x \rangle$, using an integral rather than a discrete sum. The probability of obtaining a value in the infinitesimal interval from x to $x + dx$ is $P(x)dx$, so our expression for the **ensemble average of position** is

$$\langle x \rangle \equiv \int_{-\infty}^{\infty} x \, P(x) \, dx. \qquad \text{ensemble average} \tag{3.23}$$

[17]You may be wondering what it means to talk about finding a macroscopic particle in an *infinitesimal* interval of position. In a classical position measurement, such as the one we're discussing, the particle is said to be "at x" if its center of mass (a point) is at x. With this understanding, we can meaningfully discuss finding such a particle between x_0 and $x_0 + dx$.

[18]The use of angle brackets is standard in quantum mechanics, where the mean—also known as the *ensemble average*—of an observable Q is symbolized by $\langle Q \rangle$. You'll see more means in Chap. 5, where we study the property of generic observables.

In quantum theory, the mean value of position is called its **expectation value**. You'll be formally introduced to the expectation value in § 3.6.

Aside: What Does "Mean" Mean? You probably already have a good feeling for the mean. But a word of caution: Many novice quantum mechanics confuse the ensemble average of an observable with that observable's *most probable value*. The most probable value of position is not the mean value; it is the value of x at which the particle is most likely to be found. You can figure out the most probable value of position from $P(x)$, but you probably won't get the value $\langle x \rangle$. The distinction between these two quantities is illustrated in Fig. 3.3.

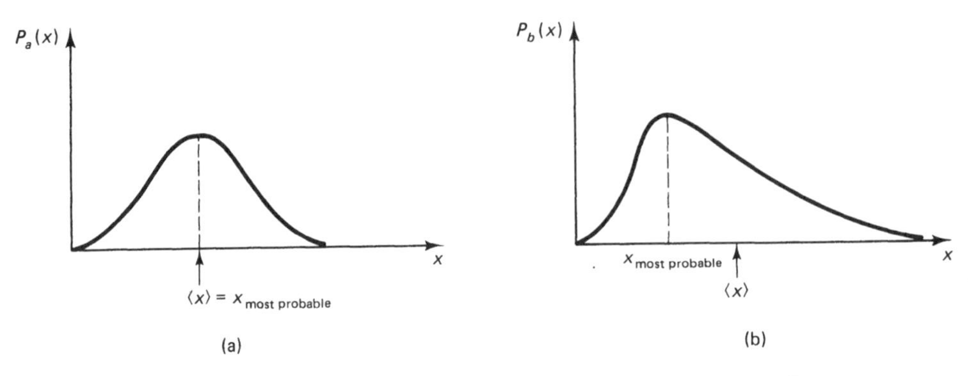

Figure 3.3 Two probability densities that illustrate the difference between the most probable value of x (the dashed line) and the average value of x (the arrow). In (a) the symmetry of the probability density about the dashed line causes $\langle x \rangle$ to occur at the same place as the most probable x. But in (b) the probability density is skewed, so $\langle x \rangle$ is displaced to the right of the most probable value.

The Dispersion

The ensemble average is a useful piece of information. But it doesn't tell the whole story. For example, the mean value of the temperature in a normal year in Oklahoma is pretty close to 70° F; but this number by itself doesn't tell you that over the course of a typical year the temperature ranges from a nippy 10° or so below zero in the winter to a toasty 110° in the summer—information you might well want if you're considering moving there. That is, by itself the mean temperature is deceptive.

Far more information is communicated by probability densities, such as those in Fig. 3.4. The ensemble averages of position for these densities are the same, yet the probability functions themselves are quite different. The results described by the curve in Fig. 3.4a are all roughly the same: $x \approx \langle x \rangle$. In contrast, the data represented in Fig. 3.4b embraces a wide range of values about $\langle x \rangle$. Clearly, to describe the results of an ensemble measurement, we need a second number, one that describes the *spread or fluctuation of individual results about their ensemble average*. Happily, the theory of statistics provides such a number: the dispersion.

The **dispersion** is a quantitative measure of the extent to which individual results fluctuate about the ensemble average.[19] Together with the mean value, the dispersion

[19]This quantity is also known as the *variance* or, more commonly, the *mean square deviation*. In many important applications it depends on time and so carries a t dependence; see Pblm 1.2.

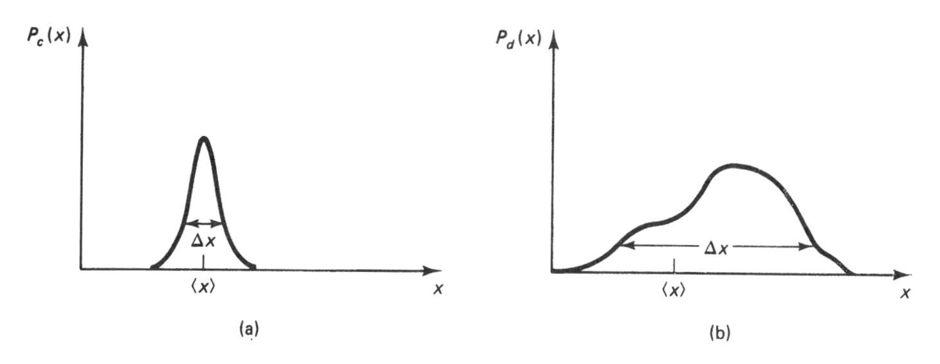

Figure 3.4 Two probability densities that illustrate the limitations of the ensemble average $\langle x \rangle$. Individual results that give the narrow probability density in (a) fluctuate very little about $\langle x \rangle$ and hence give a small standard deviation, in contrast to the broad distribution of (b).

accurately yet compactly characterizes a large amount of data. For instance, the dispersion of the temperature in Oklahoma is enormous; while that in Houston is small—it is the high mean value there that is the problem.

The definition of the dispersion of a quantity comes to us from the theory of probability. For a position measurement, the **dispersion** of the results about $\langle x \rangle$ is defined as

$$(\Delta x)^2 \equiv \left\langle \, (x - \langle x \rangle)^2 \, \right\rangle . \qquad \text{dispersion} \tag{3.24}$$

The right-hand side of (3.24) is not especially simple; it instructs us to *take the average of the square of the difference between x and $\langle x \rangle$*. But you might as well commit it to memory; you're going to need it.

This definition may seem somewhat arbitrary, but I think it makes sense. Since the dispersion is supposed to be a *single* number that characterizes the fluctuations of a large number of results about their mean, it is not surprising that this quantity be defined in terms of an average. But most other quantities we might consider, such as $x - \langle x \rangle$, are unsatisfactory. Can you deduce why?

Question 3–1

Show that the mean value of $x - \langle x \rangle$ is zero. Also show that the dispersion as defined in (3.24) is necessarily non-negative.

In statistics (and in quantum mechanics) fluctuations of data about their mean are usually described by *the square root of the dispersion.* This number is called the **standard deviation**:

$$\Delta x \equiv \sqrt{(\Delta x)^2} = \left[\, \langle (x - \langle x \rangle)^2 \rangle \right]^{1/2} . \qquad \text{standard deviation} \tag{3.25}$$

You need to get a feeling for the implications of a large (or small) standard deviation. A small Δx corresponds to the situation illustrated in Fig. 3.4a, where every member in the ensemble yields a value of x close to $\langle x \rangle$. A huge Δx, on the other hand, corresponds to the situation in Fig. 3.4b, where the data falls all over the place. Evidently, if the state of a system is characterized by a very large dispersion in position, we don't know what to expect in a measurement of this observable. We are *uncertain*.[20]

Question 3–2

Consider experiments to measure the *energy* of (a) a racquetball confined to a cubical court, and (b) a conduction electron in a metal. **Write down** prescriptions for calculating the ensemble average and standard deviation for these measurements.

3.4 THE GOSPEL ACCORDING TO BORN

Erwin with his psi can do
Calculations quite a few.
But one thing has not been seen
Just what does psi really mean.

—Walter Hückel
(translated by Felix Bloch)

According to Postulate I, the wave function is the basic element of quantum physics: Every physical state of a quantum system is represented by a wave function Ψ that contains all that we can know about the physical properties of the system in that state.[21] But this postulate does not address the physical meaning of the wave function whose existence it hypothesizes. The interpretation of the state function is crucial to its use: we must know how to extract physical information from this function. But at present we know only that Ψ is complex and depends on position and time; as such, it seems resistant to straightforward physical interpretation.

Since the early days of quantum physics this issue—just what does the state function mean?—has been a source of controversy. Since the 1930s physicists have known how to calculate state functions and use them to predict and explain the physics of a wide range of microscopic systems; but they have been unable to agree on an interpretation of these functions. In his excellent book *The Quantum Physicists and an Introduction to their Physics* (New York: Oxford University Press, 1970), William H. Cropper summarizes the situation thusly:

> The modern physicist is forced to admit, with some embarrassment, that although he can formulate a powerful and beautiful mathematical theory of atomic and molecular behavior, he cannot be sure he knows exactly what that theory means.

To be sure, the literature of quantum physics is replete with different interpretations of the state function—some of which are pretty wild (see the Selected Readings for

[20]You guessed it: In quantum mechanics Δx is the *position uncertainty*. As we'll learn in § 3.6, the uncertainty for a given state is calculated from the wave function of the state.

[21]Actually, there exist in nature states that cannot be represented by a wave function. These states are called *mixed states*; a quantum state that *can* be represented by a wave function is called a *pure state*. The distinction between mixed and pure states is beyond our scope at present, but I should note the condition for a pure state: A system is in a pure state if it is impossible, by any measurement, to distinguish one member of the ensemble from any other member. For more on mixed states, take a look at Compliment E of Chapter III in *Quantum Mechanics,* Volume I by Claude Cohen-Tannoudji, Bernard Diu, and Franck Laloë (New York: Wiley-Interscience, 1977) or at the more specialized and advanced treatise *Density Matrix Theory and Applications* by Karl Blum (New York: Plenum, 1981).

this chapter). But none can be *proven* correct. In this book, I'll show you the most widely held view of quantum mechanics: the **Copenhagen Interpretation**. This theory is founded on an interpretation of the state function due to Max Born (1882–1970).

We've already used this interpretation (although I didn't identify it as such) in § 3.1. There we explained the behavior of electrons in the double-slit experiment by associating with them a "quantum probability wave." This is Born's idea:[22]

$$\boxed{\text{wave function } \Psi = \text{position probability amplitude}}$$

Although it's easy to express this notion mathematically, to fully understand it we must delve more deeply into the meaning of *measurement* in quantum mechanics.

The Meaning of a Measurement on a Quantum System

Were we concerned with macroscopic (classical) physics, we could discuss experiments on individual systems. That is, because information about observables for a macroscopic system is not inherently probabilistic, we don't *have* to talk about such experiments in terms of ensembles—although it's often useful to do so. But, as already noted, the situation in the microscopic universe is quite different.

Quantum mechanics speaks the language of probabilities, and as discussed in the last section, probabilities are defined in terms of ensembles. This feature of quantum theory is of paramount importance:

> ***Rule***
>
> Quantum mechanics describes the outcomes of ensemble measurements, where an ensemble measurement consists of a very large number of identical experiments performed on identical, non-interacting systems all of which have been identically prepared so as to be in the same state.

In quantum physics, a state is specified by a wave function, so when we say "the members of the ensemble are in the same state," we mean that they are represented by the same wave function. This vital definition is summarized in Fig. 3.5.

Aside: Ensembles versus Individual Systems. It's not unrealistic for us to restrict quantum theory to statements about ensembles (rather than about individual systems). Most laboratory experiments are performed on ensembles of microscopic systems. For example, in a typical *collision experiment*, we fire a beam of particles—say, electrons—at a dilute gas of target particles—say, sodium (Na) atoms. Ideally, the particles in the beam would be mono-energetic (*i.e.*, they would all have the same energy) and would not interact with one another. (Actually, the particles' energy is uncertain, and the experimentalist tries to set up conditions that reduce interactions to a minimum.) Such a beam is the most famous example of a quantum ensemble, because prior to the collision all particles in it are in the same state.

The same must be true of the target atoms. We could enforce this condition by using a laser to ensure that all the Na atoms are in the same state before collision. We could then perform all sorts of interesting experiments: for example, we could

[22] In this chapter we'll explore the Born interpretation *only for position*. But as you work through Volume I, you'll learn how to construct probability amplitudes for other observables (*e.g.*, momentum in § 4.6). Remember that the only thing special about position is that it's the focus of this chapter.

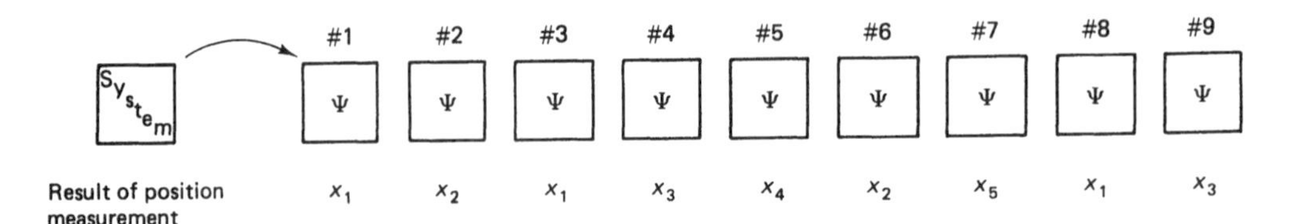

Figure 3.5 A schematic representation of a quantum ensemble. In this figure, the (microscopic) system is represented by a little box. The ensemble is constructed by replicating the system many times; the total number of members in the ensemble is N. (Note that N should be huge.) All member systems must be prepared in the same way so that prior to measurement they are in the same state. I have indicated this important condition by labelling each member (box) by the symbol Ψ. (This state function therefore labels the ensemble as a whole.)

measure the energy levels of Na atoms by observing the frequencies of photons emitted when excited Na atoms decay after excitation by electrons.

Each individual system consists of *an electron and a Na atom*, but we perform the experiment on an ensemble of such systems. So a physicist who says "we measure the energy of the Na atom" actually means "we simultaneously determine the energy of all members of the ensemble of Na atoms." Our mission is to calculate the state function for the system and use this function to predict and explain the results of such measurements.

The Born Interpretation

Let's now codify Born's solution to the problem of how to get information out of a quantum state function in another postulate. The plan, recall, is to interpret such a function as a *probability amplitude*—a quantity whose squared modulus is a probability density. Expressed in terms of position, this new postulate reads:

The Second Postulate of Quantum Mechanics

If a system is in a quantum state represented by a wave function Ψ, then $P\,dv = |\Psi|^2\,dv$ is the probability that in a position measurement at time t the particle will be detected in the infinitesimal volume element dv.

For a one-dimensional, single-particle system the wave function is simply a function of x and t, and the position probability density is $|\Psi(x,t)|^2$. So for such a system, the Born interpretation asserts that $P(x_0,t)\,dx$ is the probability of finding the particle at time t in an infinitesimal region dx about the particular point x_0.

It's easy to generalize this postulate to systems in (two or) three dimensions and even to systems that consist of many particles. For example, the probability density for a single particle in three dimensions in a state represented by the wave function $\Psi(\mathbf{r},t)$ is

$$P(\mathbf{r},t) = |\Psi(\mathbf{r},t)|^2 = \Psi^*(\mathbf{r},t)\Psi(\mathbf{r},t).$$

The probability of finding such a particle in an infinitesimal three-dimensional volume element d^3r about $\mathbf{r}_0$ is just what you would expect:

$$P(\mathbf{r}_0,t)\,dv = \Psi^*(\mathbf{r}_0,t)\Psi(\mathbf{r}_0,t)\,dv.$$

Note that the absolute value bars in Born's expression $|\Psi|^2$ for the probability density are essential: this density must be *real*, but in general the wave function Ψ is complex. The probability density must be real because it appears in quantum mechanical expressions for quantities such as the mean value of position (see § 3.6)—*quantities that could be measured in the laboratory and hence must be real numbers.*

Example 3.2. An Ensemble of Square Wells

In Part IV, we'll investigate several applications of quantum mechanics to single-particle, one-dimensional systems. One of the most important of these is the *symmetric infinite square well* shown in Fig. 3.6a.[23] The probability amplitudes (state functions) and corresponding position probability densities (at time $t = 0$) for several states of the square well are graphed in Fig. 3.6b.

The wave functions in Fig. 3.6b illustrate some important, peculiar features of quantum states. For example, in the state represented by Ψ_2, the probability of finding the particle at $t = 0$ in an infinitesimal "volume" element dx about the position $x_0 = L/4$ is quite large. This fact seems simple, but it is really rather subtle. Each *member* of the ensemble—i.e., each particle in its box—is either in the volume element centered on $x_0 = L/4$ or it isn't; what Ψ_2 tells us is that in an ensemble measurement of position, a large number of members will exhibit a value very near or equal to $x_0 = L/4$. In contrast, the probability of finding a particle in a volume element near $x_0 = 0$ is very small. [Notice that the wave function is zero at $x = 0$, so the probability $P_2(0,0)\,dx = |\Psi_2(0,0)|^2\,dx$ is also zero at that point.]

The Integrated Probability

In a real experiment we cannot locate a particle *at a point*; at best we may be able to determine if the particle is in a *very small but finite region* of position values—*e.g.*, in the interval from $x = a$ to $x = b$. So we need to know how to calculate the probability of finding the particle somewhere in a finite region. (Since we cannot measure position more accurately than this, we don't care *where* in the region the particle is.) Such a calculation poses no logical problem: we merely sum the individual probabilities $P(x,t)dx$ for all values of x in the interval $[a,b]$. Now, for a continuous variable such as x, "sum" means "integrate," so the desired probability is

$$\boxed{\begin{aligned}\text{probability of detection in [a,b]} &\equiv P(\,[a,b]\,,t)\\ &= \int_a^b P(x,t)\,dx = \int_a^b |\Psi(x,t)|^2\,dx\end{aligned}} \tag{3.26}$$

Interlude: What Quantum Mechanics Can't Do

Born's idea, as codified in Postulate II, provides the vital link between the mathematical function Ψ and the physical state it represents. Armed with this interpretation such physicists as Werner Heisenberg and Erwin Schrödinger rapidly developed the machinery for extracting from wave functions information about position, momentum, and a host of other observables. We'll explore their work in Chap. 5 and Part IV.

[23]The particle in a square well is no mere academic example; in current research it is used to model a variety of physical systems, such as conduction electrons in some solids and electrons in certain organic molecules. You'll see an example of what I mean in § 8.9.

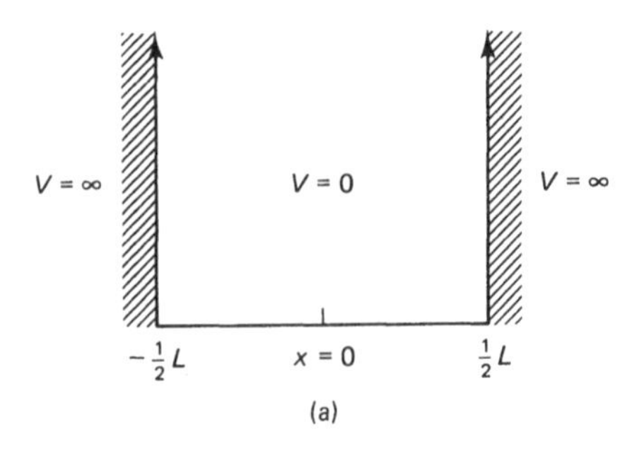

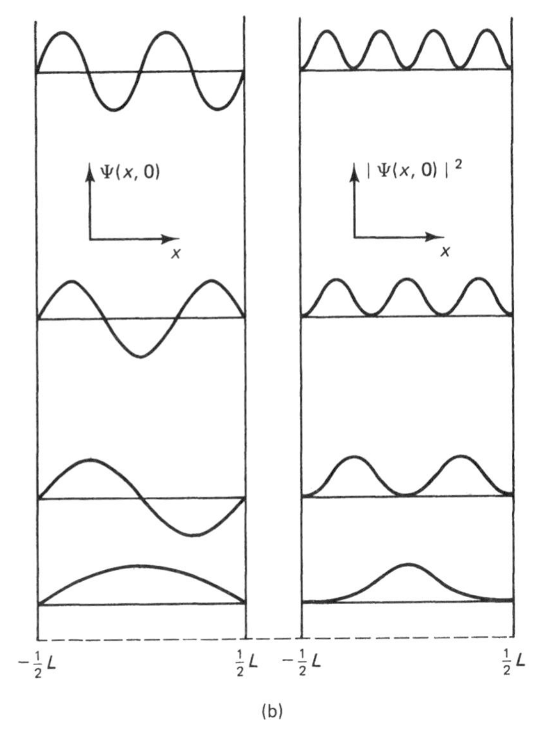

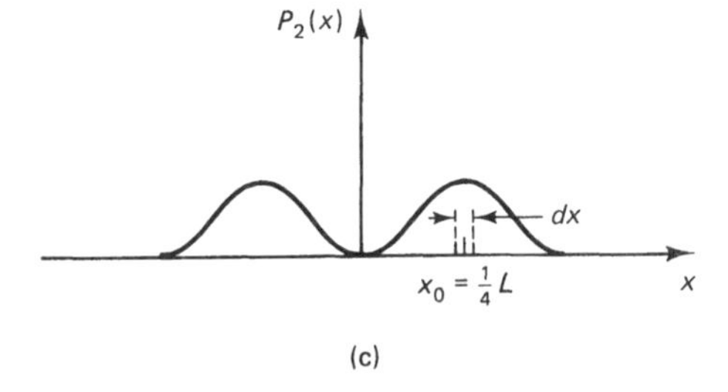

Figure 3.6 (a) The infinite square well potential energy. (b) Four "bound-state" wave functions of the square well and their probability densities (at $t = 0$). (c) At $t = 0$, a particle in state Ψ_2 is quite likely to be found in the "volume" element dx at $x_0 = L/4$.

Still, although quantum mechanics is quite powerful, there are some questions that it simply does not address. These limitations are best explained by example.

Consider an ensemble consisting of, say, Avogadro's number of radioactive nuclei. Each nucleus of each atom is in the same quantum state as every other nucleus. Nuclei decay. So, pondering the future of our ensemble, we might formulate two questions:

1. What is the likelihood that in a particular time interval *any* nucleus in the ensemble will decay?
2. Will a *particular* nucleus decay in this time interval?

The first of these queries addresses the behavior of the *collection* of identical systems that comprise the ensemble. Since quantum physics pertains to ensembles, it can, in principle, provide an answer to this question. But the second question concerns the behavior of a *specific* member of an ensemble. Such a question is beyond the scope of any theory that speaks only the language of ensemble probabilities. In fact, individual nuclei in such an ensemble decay at different times. But Copenhagen quantum physics cannot explain this fact, except to argue that the behavior of a particular member of an ensemble is governed by *random chance*. This viewpoint, you might guess, takes a little getting used to.

3.5 HOW TO NORMALIZE A STATE FUNCTION

In our review of probabilities (§ 3.3), I noted that the absolute probability density $P(x)$ obeys the normalization condition [Eq. (3.21)]

$$\int_{-\infty}^{\infty} P(x)\, dx = 1. \tag{3.27}$$

We must ensure that quantum-mechanical probability densities, as calculated from state functions, satisfy this condition. In this section, we'll see how to enforce the normalization condition—an ability we'll need throughout the rest of our journey through quantum physics.

The Normalization Condition

The importance of normalization follows from the Born interpretation of the state function as a position probability amplitude. According to § 3.4, we interpret the *integrated probability density* on the left-hand side of (3.27) as the probability that in a position measurement at time t we will find the particle *anywhere* in space. [This integral is just the generalization of the one in Eq. (3.26) to all space: $a = -\infty$ and $b = +\infty$.] Now, think physically for a moment: what should be the value of this integral?

The particle *does exist*, so this integral must equal 1. All this is to say that for the Born interpretation of a state function to make sense, the position probability density calculated from the state function must be an *absolute* probability density. As such, it must satisfy the **normalization condition** (3.27), which we write:

$$\boxed{\int_{-\infty}^{\infty} P(x,t)dx = \int_{-\infty}^{\infty} \Psi(x,t)^* \Psi(x,t)dx = 1} \quad \text{Normalization Condition} \tag{3.28a}$$

A bit of jargon: The integral over all space of the probability density is sometimes called the **normalization integral**; and a mathematical function Ψ whose normalization integral equals 1 is said to be **normalized**.[24]

[24]In mathematics, functions satisfying (3.28) are said to be "square-integrable." For more on the mathematical underpinnings of quantum theory, see J. M. Jauch, *Foundations of Quantum Mechanics* (Reading Mass.: Addison-Wesley, 1968).

You should have little trouble generalizing the normalization condition (3.28*a*) to simple three-dimensional systems. Consider a single particle in three dimensions in a state represented by the wave function $\Psi(\mathbf{r}, t)$. The normalization integral is the probability of finding the particle *anywhere in space*, so in this case we must integrate over all space—*i.e.*, we require a three-fold integral with respect to the infinitesimal volume element $dv = d^3r$:

$$\int \Psi(\mathbf{r}, t)^* \, \Psi(\mathbf{r}, t) \, dv = 1. \tag{3.28b}$$

Sometimes we find (or derive) a wave function whose normalization integral doesn't satisfy the normalization condition. If this integral is *finite*, then we can fix the wave function—we can normalize it, using a procedure I'll describe in a moment. A function that can be normalized is, sensibly, said to be **normalizable**.

In physics, normalizable functions are particularly important, for only normalizable functions can represent a quantum state. For this reason, we say that such functions are **physically admissible**.

But if the normalization integral is *infinite*, we're in trouble. Such a function cannot be normalized and hence cannot properly represent a quantum state.[25] The reason is that we can't meaningfully interpret the squared modulus of a function whose normalization integral is infinite as a probability density. The probability of finding the particle anywhere in space in a position measurement is not infinity; it's unity.

The normalization condition exemplifies the beautiful interplay of physics and mathematics that characterizes quantum mechanics. Here a *physical constraint*—that a quantum particle must exist—corresponds to a *mathematical limitation* on the class of functions that can represent states in nature. As we proceed, you should be on the alert for other strands in the fabric of mathematics and physics that is quantum mechanics. You may find, as I do, that these are among the most marvelous features of this remarkable physical theory.

Aside: The Constancy of Normalization. Another (equally sensible) consequence of the normalization conditions (3.28) is that the integral $\int_{-\infty}^{\infty} \Psi(x,t)^* \Psi(x,t) \, dx$ doesn't depend on the time. Obviously, the probability of detecting the particle anywhere is unity no matter when we perform the measurement. So *a function once normalized, stays normalized.* To prove this fact, we need the time-dependent Schrödinger Equation, which will appear in Chap. 6 (see § 6.5).

The Normalization Requirement and Boundary Conditions

The demand that the normalization integral equal 1 constrains the *mathematical form* of the wave function in a rather subtle way. Many novice quantum mechanics incorrectly infer from the normalization condition that the state function itself cannot extend throughout space—*i.e.*, that $\Psi(x, t)$ must be non-zero only in a *finite* region of space. Although reasonable, this deduction is wrong.

To avoid such logical pitfalls, we must always focus on precisely what the equations of quantum mechanics mean. *Equation (3.28a) doesn't prohibit a wave function* $\Psi(x, t)$

[25] Notice the qualifier *properly*. As we'll see in Chap. 8, there exist solutions to the time-dependent Schrödinger Equation that are mathematically valid but not physically admissible because their normalization integral is infinite. Yet, non-normalizable functions play an important role in quantum mechanics, especially in the study of scattering processes.

from having infinite extent, if as x approaches $\pm\infty$, the function decays to zero rapidly enough that the normalization integral is finite. So the condition that follows logically from this equation is

$$\Psi(x,t) \longrightarrow 0 \qquad \text{as} \qquad |\,x\,| \longrightarrow \infty \qquad \text{rapidly enough.} \tag{3.29}$$

We'll illustrate this condition in Example 3.3. Equation (3.29) is called a "boundary condition" on the function $\Psi(x,t)$, because it is a condition that function must satisfy at a "boundary."[26] I'll have more to say about boundary conditions in Chap. 6, and we'll explore their consequences throughout the remainder of this book.

How to Normalize

Enough abstract theory. Suppose we're confronted with a function $\Psi'(x,t)$ whose normalization integral is equal to a finite number other than 1:

$$\int_{-\infty}^{\infty} \Psi'(x,t)^* \Psi'(x,t)\, dx = M. \tag{3.30}$$

What to do?

I'm sure you're ahead of me. To obtain from $\Psi'(x,t)$ a function whose normalization integral is unity we just multiply the offending function by the constant $1/\sqrt{M}$. The resulting function

$$\Psi(x,t) = \frac{1}{\sqrt{M}}\,\Psi'(x,t) \qquad \text{[normalized state function]} \tag{3.31a}$$

satisfies the desired condition

$$\int_{-\infty}^{\infty} \left[\frac{1}{\sqrt{M}}\Psi'(x,t)\right]^* \left[\frac{1}{\sqrt{M}}\Psi'(x,t)\right] dx = 1 \tag{3.31b}$$

and so is physically admissible. [Of course, if the integral in Eq. (3.30) is infinite, this gambit fails spectacularly.]

Example 3.3. Subduing a Normalizable State Function

To see how the normalization procedure works, let's return to the wave function of a particle with a simple harmonic oscillator potential that we saw in Example 3.2 [Eq. (3.13)]:

$$\Psi'(x,t) = A e^{-\beta^2 x^2/2} e^{-iEt/\hbar}. \tag{3.32a}$$

Evaluating the normalization integral for the function (3.32a), we find

$$\int_{-\infty}^{\infty} \Psi'(x,t)^* \Psi'(x,t)\, dx = A^2 \int_{-\infty}^{\infty} e^{-\beta^2 x^2}\, dx = A^2 \sqrt{\frac{\pi}{\beta^2}}. \tag{3.32b}$$

[26] You may be wondering where the "boundary" in (3.29) is. The name "boundary condition" evokes a condition that a function must satisfy at some point. But Eq. (3.29) is a condition the wave function must satisfy *in the limit* $|x| \to \infty$. Nevertheless, in the parlance of the field, such a requirement is called a boundary condition.

(*Notice what happened to the time dependence in the normalization integral.*) Clearly, $\Psi'(x,t)$ is normalizable—and hence physically admissible (*i.e.*, this function can represent a quantum state). But, equally clearly, its normalization integral isn't equal to 1. This defect is easily remedied: we just choose the "normalization constant" A in (3.32*a*) so that $A^2 = \sqrt{\beta^2/\pi}$, *i.e.*,

$$A = \left(\frac{\beta^2}{\pi}\right)^{1/4} \tag{3.32c}$$

This choice gives as the normalized counterpart of $\Psi'(x,t)$, the function

$$\Psi(x,t) = \left(\frac{\beta^2}{\pi}\right)^{1/4} e^{-\beta^2 x^2/2} e^{-iEt/\hbar}. \tag{3.33}$$

[You can find a graph of this function (at $t = 0$) in Fig. 3.1a.] Since the Gaussian function (3.32*a*) can be normalized, it is a viable candidate for state-function-hood.[27] Before I leave this example, I want to reinforce its moral:

Rule

Unless you *know* that a state function is normalized, your first step—before you calculate anything physical from the function—should be to check that it is normalizable and, if so, to normalize it.

Quantum Arbitrariness

Example 3 illustrates one way to construct a normalized function. "But," you may be wondering, "is my newly normalized function

$$\Psi(x,t) = \frac{1}{\sqrt{M}} \Psi'(x,t) \tag{3.34}$$

unique: is it the *only* normalized function I can construct from $\Psi'(x,t)$?" The answer—which may surprise you—lurks in the Born interpretation.

Recall that the normalization condition is expressed in terms of the position probability density—the *product* $|\Psi(x,t)|^2 = \Psi(x,t)^*\Psi(x,t)$. Consequently we can multiply the state function by *any complex number of unit modulus* without changing the probability density. That is, we can multiply $\Psi(x,t)$ by $e^{i\delta}$, where δ is any real number, without changing the normalization integral.[28] In the probability density constructed from $e^{i\delta}\Psi(x,t)$, all dependence on δ disappears because

$$(e^{i\delta})^* e^{i\delta} = e^{-i\delta} e^{i\delta} = e^0 = 1. \tag{3.35}$$

[27] There are, however, restrictions other than normalizability that a prospective state function must satisfy if it is to represent a quantum state. I'll discuss some of these mathematical conditions at the end of this section. Beyond these, a state function must satisfy the equation of motion of the system—the time-dependent Schrödinger Equation (see Chap. 6).

[28] Any complex number z whose modulus is unity can be written as $z = e^{i\delta}$ for some real number δ. If this assertion isn't familiar, take a few minutes to review Appendix K.

Ergo all functions of the form

$$\Psi(x,t) = e^{i\delta} \frac{1}{\sqrt{M}} \Psi'(x,t) \tag{3.36}$$

lead to the same normalization integral:

$$\begin{aligned} |\Psi(x,t)|^2 &= \Psi(x,t)^* \, \Psi(x,t) \\ &= \left[\frac{e^{-i\delta}}{\sqrt{M}} \Psi'(x,t)^*\right] \left[\frac{e^{i\delta}}{\sqrt{M}} \Psi'(x,t)\right] \\ &= \frac{1}{M} \mid \Psi'(x,t) \mid^2. \end{aligned} \tag{3.37}$$

Regardless of the value of δ (which may, of course, be zero), the function (3.36) will obey the normalization condition (3.28).

We have discovered that the wave function—the state descriptor of quantum physics-is, to an extent, arbitrary! This result should strike you as a bit odd; it certainly doesn't pertain to the more familiar state descriptor of *classical physics*: the trajectory that represents a particular classical state of a particle is unique. The factor $e^{i\delta}$ is called a **global phase factor**, so what we've learned is: a state function is arbitrary to within a global phase factor.

I view this "quantum arbitrariness" as an artifact of the theory. (A *useful* artifact, because quantum arbitrariness often comes in handy in problem solving.) By this I mean that we cannot *detect* the arbitrariness of the state function via measurement. As we'll soon see, *the global phase factor vanishes from all quantum mechanical expressions for quantities we can measure*. This disappearing act works for the same reason the phase factor vanished from the probability density in (3.35): quantum mechanical expressions for measurable quantities inevitably involve the probability density, which is the *product* $\Psi^* \, \Psi$. Form this product, and Eq. (3.35) will eliminate the phase factor for you.

For instance, we cannot distinguish by any physical measurement the states represented by the functions

$$e^{i\delta} \, \Psi(x,t) = \Psi(x,t) \qquad (\delta = 0) \tag{3.38a}$$

and

$$e^{i\delta} \, \Psi(x,t) = -\Psi(x,t) \qquad (\delta = \pi) \tag{3.38b}$$

The point to remember is:

> ***Rule***
>
> Two state functions that differ only by a global phase factor, Ψ and $e^{i\delta}\Psi$, represent the same physical state of the system.

Additional Restrictions on the Wave Function

We have seen how the physical fact that (even in the weird world of quantum mechanics) a particle must exist *somewhere* restricts the class of physically admissible functions to those that are normalizable. But normalizability is not the only mathematical condition a function must satisfy if it is to represent a quantum state.

The first of these conditions follows from the Born interpretation of the state function as a position probability amplitude. At any time the value of the probability for finding the particle in an infinitesimal region of space must be unique. This implies that the state function must assume only one value at each time—*i.e.*, that it must be **single-valued**. Consider, for contrast, the *multi-valued* beast in Fig. 3.7a. Since the squared modulus of this function has two values, it can't be understood as a position probability density. Hence this double-valued function is useless as a state descriptor in quantum physics.

Additional restrictions follow from the equation of motion of quantum mechanics: the time-dependent Schrödinger Equation. We'll study this equation, which towers over the many and varied implementations and applications of quantum physics, in Chap. 6. But, by way of preview, here it is for a particle in one dimension with potential energy $V(x)$:

$$-\frac{\hbar^2}{2m}\frac{\partial^2}{\partial x^2}\Psi(x,t) + V(x)\Psi(x,t) = i\hbar\frac{\partial}{\partial t}\Psi(x,t). \tag{3.39}$$

What conditions follow from this equation? Look at its first term. See the second derivative of $\Psi(x,t)$ with respect to x? Speaking mathematically, we cannot even define this derivative unless the *first* derivative of the function in question, $\Psi(x,t)$, is continuous.[29] Mathematicians call a function whose first derivative is everywhere continuous a **smoothly varying function**. So we conclude that the wave function must be smoothly varying. It can't have any kinks (see Fig. 3.7c).

Another condition also follows from the presence of the second derivative $\partial^2\Psi(x,t)/\partial x^2$ in the Schrödinger equation (3.39). We define the *second* derivative in terms of the first derivative, and we can define the *first* derivative only if the function being differentiated is itself *continuous*. Incidentally, the Born interpretation provides another incentive to require that $\Psi(x,t)$ be continuous: if it weren't, then at a point of discontinuity, the value of its modulus squared would not be unique, in which case we could not meaningfully interpret this value as the position probability density.

To summarize:

> ***Rule***
>
> A function is physically admissible and hence may represent a state of a quantum system if it is single-valued, continuous, smoothly varying, and normalizable—and if it satisfies the system's Schrödinger Equation (3.39).

Aside: The Finiteness of Wave Functions. State functions are also finite at all points—usually. But don't misconstrue this observation as a *requirement*; a state function can be infinite at a point provided that the mathematical nature of this singularity is such that the normalization integral formed from the function is finite. We'll come across some examples of this curiosity when (in Part V) we focus on three-dimensional systems.

[29]For a proof of this assertion, see Chap. 2 of *Quantum Physics* by Rolf G. Winter (Belmont, Ca.: Wadsworth, 1979). If you want to read further on the mathematical requirements of wave functions, see the aforementioned book by Jauch.

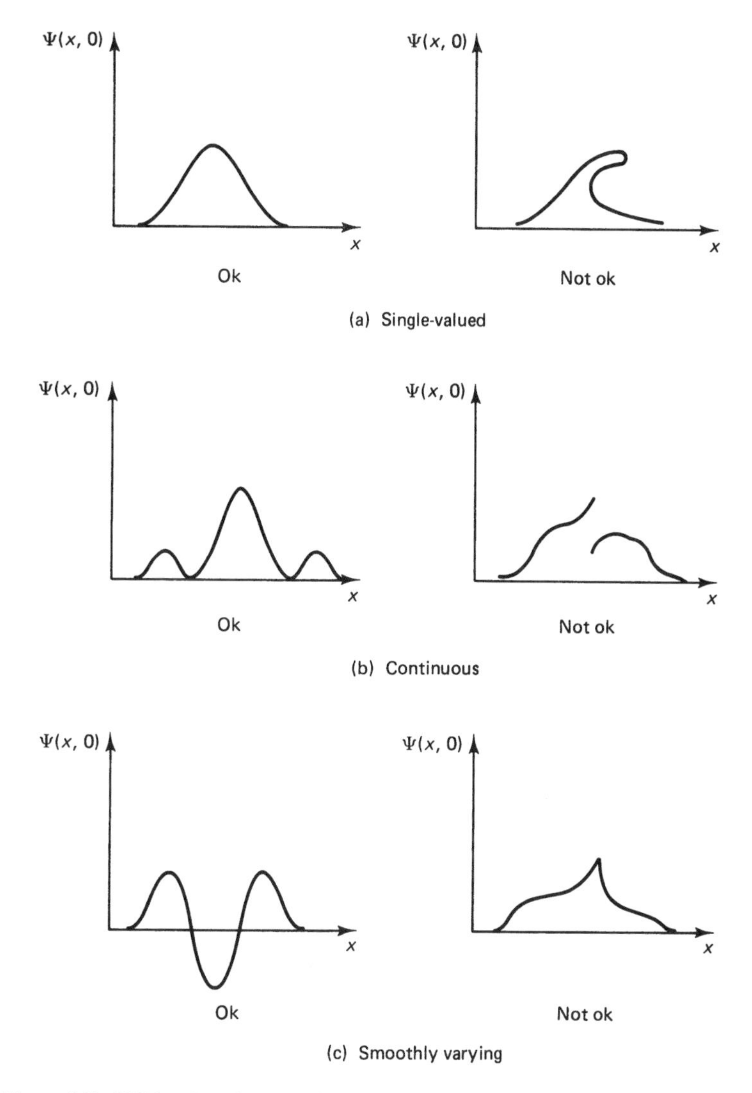

Figure 3.7 Valid and useless one-dimensional functions. In each case, the functions sketched on the right of this figure are not physically admissible. They fail because they are not (a) single-valued; (b) continuous; and (c) smoothly-varying. The functions on the left are Ok.

3.6 GREAT EXPECTATIONS (AND UNCERTAINTIES)

The Born interpretation enables us to determine from a wave function a particular kind of information: probabilistic information. We can use the probability density to answer the question: *in an ensemble measurement of position at time t, what is the probability that a member of the ensemble will exhibit a value in the infinitesimal range from x to $x + dx$?* But in § 3.3 I introduced another kind of information: statistical information. To characterize the results of an experiment, we use two statistical quantities—the *ensemble average* and the *standard deviation*. These are, respectively, the mean value of the results and a measure of the extent to which these results fluctuate around the mean. For the observable we're considering in this chapter, position, the ensemble average is denoted by $\langle x \rangle$ and the uncertainty by Δx. But how do we calculate these quantities from a state function? The answer follows easily from their definitions and the Born Interpretation.

The Expectation Value

In quantum theory, the ensemble average of an observable for a particular state of the system is called the **expectation value** of that observable. Thus, $\langle x \rangle$ answers the question: *what is the average of the results that would be obtained in an ensemble measurement of position performed at time t on a system in a state represented by* $\Psi(x,t)$?

To obtain an expression for $\langle x \rangle$ we need only adapt the definition (3.23) of the ensemble average (the mean) to quantum mechanics. This definition,

$$\langle x \rangle = \int_{-\infty}^{\infty} x\, P(x)\, dx,$$

instructs us to integrate over all values of x (*i.e.*, over the whole range of the spatial coordinate of a one-dimensional wave function), thereby accumulating the product of each possible value of the observable x and its "weighting function" $P(x,t)$. Using Postulate II to write the probability density in terms of $\Psi(x,t)$, we obtain[30]

$$\begin{aligned}\langle x \rangle &\equiv \int_{-\infty}^{\infty} x\, P(x,t)\, dx \\ &= \int_{-\infty}^{\infty} \Psi(x,t)^*\, x\, \Psi(x,t)\, dx.\end{aligned} \qquad \text{Expectation Value of Position} \qquad (3.40a)$$

Notice that *we must integrate over the whole "space" of the wave function.* For a one-dimensional system, this entails only a single integration from $-\infty$ to $+\infty$. But for a single particle in *three* dimensions, we must write the expectation value as an integration over *three* spatial coordinates, with the differential volume element dv; *viz.*,

$$\langle x \rangle = \int \Psi(\mathbf{r},t)^*\, x\, \Psi(\mathbf{r},t)\, dv. \qquad (3.40b)$$

Now, $\langle x \rangle$ is the average of data from a position measurement on an ensemble of particles in state Ψ. This value can be determined in the laboratory and hence must be a real number. And, sure enough, it is. Since the probability density $\Psi^*\Psi$ is a real function, and the spatial coordinate x assumes only real values, the integrals (3.40) are also real.[31]

Question 3–3

Write the state function $\Psi(x,t)$ as the sum of real and imaginary parts and **prove** that $P(x,t)$ is real.

[30] It may seem perverse to put x in between $\Psi(x,t)^*$ and $\Psi(x,t)$ in this definition. But this form for expectation values is standard—for reasons that will become clear in Chap. 5.

[31] Any number c with the property that $c^* = c$ is real. This property (see Appendix K) provides one way to prove that $\langle x \rangle$ is real.

Before turning to a sample calculation of an expectation value, I want to note some of the important properties of this quantity. First, in the definitions (3.40) we integrate *only* over spatial variables. So *in general, the expectation value $\langle x \rangle$ depends on time.* From time to time, I'll explicitly indicate this dependence, writing the expectation value with the slightly ungainly notation $\langle x \rangle (t)$.

A second important property of this quantity is its dependence on *the state of the system*: different quantum states, represented by different state functions, have different ensemble averages. This statement sounds obvious, but, perhaps because this dependence is not explicitly indicated by the symbol for the expectation value, it's easy to overlook. Figure 3.8 illustrates the state-dependence of $\langle x \rangle$ by comparing the probability densities and corresponding expectation values for two states of the *finite square well* (see Example 3.2).

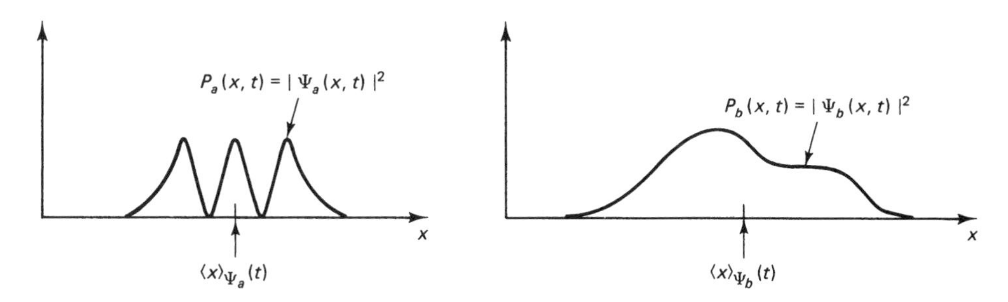

Figure 3.8 Snapshots at fixed times t of two probability densities for the finite square well and the corresponding expectation values. Clearly, $\langle x \rangle$ depends on the state of the system.

Example 3.4. Expectation Value of Position for a Particle

Consider a microscopic particle with a simple harmonic oscillator potential in a quantum state represented by the wave function we normalized in Example 3.3. To simplify the algebra in this, our first example, let's suppose that the properties of the oscillator are such that the constant $\beta = 1$, leaving the more general case for Chap. 9. In this case the wave function (3.33) is simply

$$\Psi(x,t) = \left(\frac{1}{\pi}\right)^{1/4} e^{-x^2/2} e^{-iEt/\hbar}. \tag{3.41}$$

Let's calculate the average value we would obtain in a measurement at $t = 0$ of the position of an ensemble of particles in this state.

According to the definition (3.40a), the expectation value of x at $t = 0$ for this state is

$$\langle x \rangle = \sqrt{\frac{1}{\pi}} \int_{-\infty}^{\infty} x\, e^{-x^2}\, dx. \tag{3.42}$$

To evaluate this integral, we could dive for our integral tables and start scribbling. Or we could think for a moment. I recommend the latter approach.

When faced with a manipulation of this sort, the first thing you ought to do is to check the *parity* of the integrand: *i.e.*, determine if it is an even or an odd function.[32] So let's

[32]Recall that the *parity* of a function $f(x)$ is even if $f(-x) = f(x)$ and is odd if $f(-x) = -f(x)$.

examine the integrand in Eq. (3.42), which I'll call $I(x)$. This function, $I(x) \equiv xe^{-x^2}$, is odd, because $I(-x) = -I(x)$. But the integral from $-\infty$ to $+\infty$ of any odd function is necessarily zero. Equation (3.42) contains just such an integral, so we conclude that

$$\langle x \rangle = 0 \qquad \text{for} \quad \Psi(x,0) = \left(\frac{1}{\pi}\right)^{1/4} e^{-x^2/2}. \tag{3.43}$$

Question 3–4

Does Eq. (3.43) hold for $t > 0$? **Prove** your answer.

By the way, be careful when you interpret a result such as Eq. (3.43). In particular, be wary of deducing from a *statistical quantity*, such as the mean value, conclusions about the nature of the ensemble on which the measurement was performed. For example, we can imagine two quite different ensembles that would yield $\langle x \rangle = 0$:

1. an ensemble all members of which exhibit $x = 0$ in the position measurement;
2. an ensemble such that each member that yields a non-zero value, say x_0, is "matched" by a member that yields $-x_0$.

In a measurement on the second ensemble, an equal fraction of members exhibit $+x$ and $-x$, so the mean is zero. We cannot distinguish between these two cases without additional information—which you will find later in this section.

The Expectation Value of a Generic Observable

Most observables can be written as functions of position and momentum. We'll tackle the quantum mechanical treatment of momentum in Chap. 5 and so will defer until then the question of how to calculate the expectation value of an arbitrary observable. But many observables are functions of position only, and we know enough to write down the form of the expectation value for such an observable.

Let's denote by $Q(x)$ a generic observable that depends only on position. Suppose the experimental physicists down the hall have built an apparatus that can carry out an ensemble measurement of $Q(x)$. Our job is to calculate from the state function, which in this scenario we've already obtained by solving the Schrödinger equation (Chap. 6), the average value that the folks down the hall ought to expect. To write down the expression for $\langle Q(x) \rangle$ we return, once again, to the probability density and the Born interpretation.

Consider a point x_0 in an infinitesimal element dx. The product of $P(x_0, t)\,dx$ is the probability of detecting a particle in the interval dx at time t. Now, the value of $Q(x)$ at this point is $Q(x_0)$, so the probability of obtaining this value in the measurement of $Q(x)$ is just $P(x_0, t)\,dx$. That is, *we can use the position probability density as the weighting factor in a calculation of the expectation value of any observable that depends only on position,* to wit:

$$\langle Q(x) \rangle = \int_{-\infty}^{\infty} \Psi^*(x,t)\, Q(x)\, \Psi(x,t)\, dx. \tag{3.44}$$

Note carefully that this simple generalization does not apply to an observable that depends on both position and momentum.

The Uncertainty

The other statistical quantity we use in quantum physics is the *standard deviation of an observable*—otherwise known as its **uncertainty**. For a position measurement, the uncertainty in x answers the question: *in an ensemble measurement at time t of the position of a particle in state $\Psi(x,t)$, what is the spread of individual results around the expectation value $\langle x \rangle$?* We reviewed the definition of the standard deviation in § 3.3. We can now adapt this definition to quantum states just as we did the definition of the mean.

The standard deviation is the square root of the dispersion. According to the definition (3.25), this quantity, written here for position, is

$$\Delta x \equiv \sqrt{(\Delta x)^2} = \left[\left\langle (x - \langle x \rangle)^2 \right\rangle \right]^{1/2}. \qquad \text{Position Uncertainty} \qquad (3.45a)$$

Unfortunately, this definition is easier to write down than to apply.

Let's ponder it in detail. Using our definitions for the expectation values, here are our instructions:

1. calculate the expectation value of position, $\langle x \rangle$;
2. form the function $x - \langle x \rangle$;
3. square the function determined in step (2);
4. calculate the expectation value of the result of step (3);
5. take the square root of the resulting number.

Mercifully, we don't have to do all this every time we need to calculate an uncertainty. There is an easier way: use the simpler, more convenient, equivalent expression

$$\boxed{(\Delta x)^2 = \langle x^2 \rangle - \langle x \rangle^2} \qquad (3.45b)$$

This result lets us calculate the position uncertainty for a quantum state from the expectation values of x and of x^2. The form of $\langle x^2 \rangle$ follows from the generalization (3.44), *viz.*,

$$\langle x^2 \rangle = \int_{-\infty}^{\infty} \Psi^*(x,t)\, x^2\, \Psi(x,t)\, dx. \qquad (3.46)$$

Much simpler, right? More good news: Eq. (3.45*b*) leads to a host of other time-saving results. One of the most useful is:

$$\Delta x = \sqrt{\langle x^2 \rangle} \qquad \text{if} \qquad \langle x \rangle = 0. \qquad (3.47)$$

Question 3–5

I hope it's clear how Equation (3.45*b*) simplifies the chore of figuring out uncertainties. This result is so important that you really ought to derive it. Do so.

Example 3.5. The Position Uncertainty for a Microscopic Particle

Let's return one last time to the state function (3.41) at $t = 0$. Recall that in Example 3.4 we evaluated the expectation value of position for this state and found that because the wave function is symmetric about $x = 0$, this quantity is zero. Now let's calculate the uncertainty Δx.

If we use Eq. (3.45*b*) rather than the definition (3.45*a*), we need evaluate only one additional integral, the expectation value of x^2:

$$\langle x^2 \rangle = \sqrt{\frac{1}{\pi}} \int_{-\infty}^{\infty} x^2 \, e^{-x^2} \, dx. \tag{3.48}$$

This integral isn't particularly challenging. First we exploit the even parity of its integrand to rewrite it as

$$\int_{-\infty}^{\infty} x^2 \, e^{-x^2} \, dx = 2 \int_{0}^{\infty} x^2 \, e^{-x^2} \, dx. \tag{3.49}$$

Then, ploughing through our trusty integral tables (or Appendix I), we find

$$\int_{0}^{\infty} x^2 \, e^{-x^2} \, dx = \frac{1}{4} \sqrt{\pi}. \tag{3.50}$$

This gives $\langle x^2 \rangle = 1/2$. We can now evaluate the position uncertainty for this state,

$$\Delta x = \sqrt{\frac{1}{2}} \qquad \text{for} \quad \Psi(x, 0) = \left(\frac{1}{\pi}\right)^{1/4} e^{-x^2/2}. \tag{3.51}$$

Equation (3.51) answers the question left dangling at the end of Example 3.4: what is the nature of the ensemble that gave $\langle x \rangle = 0$ for this state? Clearly it was *not* an ensemble all members of which exhibited the same value; had this been the case, all data points would have been the same and we'd have found $\Delta x = 0$. No, it was one in which the position of the particle was uncertain, and the quantity we have here calculated is a quantitative measure of that uncertainty. You can see this feature of the state in the snapshot of its wave function in Fig. 3.9.

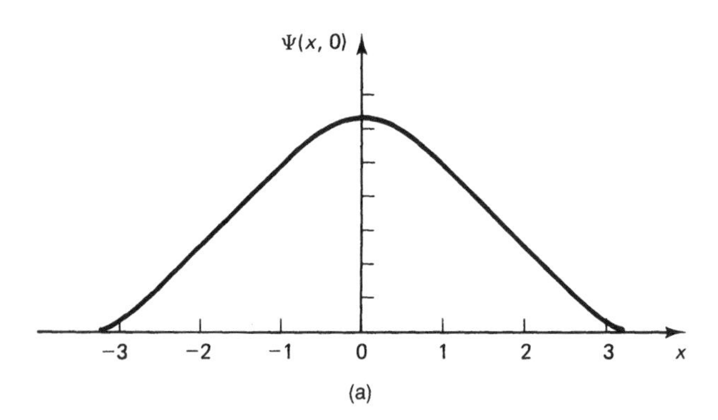

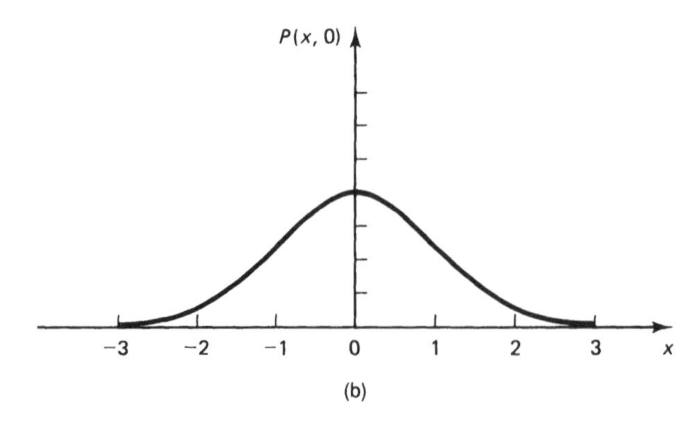

Figure 3.9 (a) The state function considered in Examples 3.4 and 3.5. (b) The corresponding probability density.

Just as in Eq. (3.44) we generalized the definition of $\langle x \rangle$ to an arbitrary observable that can be expressed as a function of x, so can we generalize the definition (3.45) of Δx to such an observable, *viz.*,

$$\Delta Q(x) = \left[\langle \left(Q(x) - \langle Q(x) \rangle \right)^2 \rangle \right]^{1/2}. \tag{3.52}$$

As you might guess, in actual calculations we usually implement the alternative expression

$$\left[\Delta Q(x) \right]^2 = \langle \left[Q(x) \right]^2 \rangle - \langle Q(x) \rangle^2. \tag{3.53}$$

Expectation values and uncertainties are of paramount importance to the user of quantum mechanics. Once we have learned how to calculate state functions and how to represent observables that depend on position and momentum, we'll be applying the definitions of this section over and over. So remember them.

3.7 FINAL THOUGHTS: WHAT THE STATE FUNCTION ISN'T

As if by magic, Postulates I and II have transformed de Broglie's vague notion of a matter wave into a beautiful mathematical construct: the state function. The Born Interpretation of this function, the bulwark of quantum theory, defines the very character of that theory as intrinsically probabilistic, statistical—and limited. From state functions, we can calculate probabilities for various outcomes of measurements and such statistical quantities as the average value and the uncertainty. But nothing more.

Even so simple a question as "Where is a particular electron" is beyond the purview of quantum mechanics. At most, a quantum physicist can determine the probability of finding *one of an ensemble* of electrons at a particular location, the mean value of the positions obtained in an ensemble measurement, the position where an electron is most likely to be found, and a rough measure of the size of the region where most of the electrons will appear. In this chapter we've begun to learn how to do such calculations.

Other, more philosophical questions also are beyond quantum mechanics. For example, in this chapter I have talked a lot about Ψ but have not addressed questions such as: what, precisely, is this function? Where does it exist? In ordinary, geometrical space, like an electromagnetic wave? And what is the nature of its connection to the material system whose state it represents? I've dodged such questions because they lie on the fringes of orthodox quantum theory, in the domain of speculations, rather than answers. But the human mind abhors a vacuum; we seem to need to relate important new ideas to familiar concepts, and nearly every newcomer to quantum mechanics must engage in the struggle to get rid of classically-based answers to these questions.

Our tendency to draw on classical physics as we grope for an understanding of quantum theory is certainly understandable. We're seduced by similarities. For example, the state function plays a role in quantum mechanics that is *analogous* to that played by the trajectory in classical mechanics: it is the basic mathematical element, the state descriptor, of the theory. But the identification must end with this analogue! A wave function is not a trajectory. My point, which you'll hear me rant about again, is that at both the conceptual and the mathematical level, quantum mechanics is not just a funny-looking reformulation of classical physics. The two physical theories are fundamentally, physically different.

Even words lead us astray. When thinking about a "wave function," we naturally tend to envision a wave in the classical sense: an actual, extant, three-dimensional waveform that propagates and carries energy. To be sure, we know from Chap. 2 that to picture a microscopic particle as a wave is wrong. But what of the "wave functions" that represent states of these particles? They seem rather like classical waves, don't they?

They do—but they aren't. Although the *properties* and *behavior* of quantum waves are familiar, these waves are totally unlike any you have ever studied. For example, unlike classical waves, *quantum waves don't carry energy*. So it's impossible to detect wave functions *per se*; these bizarre entities manifest themselves only indirectly, through their influence on the behavior of quantum particles as they propagate through space. (This feature led Einstein to refer to quantum waves as "ghost waves.") The foundation of your thinking about wave functions should be not the classical concept of a wave but rather the novel notion that *state functions carry information*—probabilistic and statistical information about how the particles whose states they represent would behave were their properties measured.

Moreover, wave functions do not exist in space, at least not in the sense that classical waves exist. Consider space. The space in which we walk around, frolic, study quantum mechanics, and perform experiments is called $\Re^3$; we specify locations in $\Re^3$ by the values of three real numbers in an orthogonal coordinate system.

Now consider a state function. Such a function is complex and so could not exist in $\Re^3$, for only real numbers and functions exist in this space. Moreover, the state function for a many-particle system depends on several coordinates. How could a function such as $\Psi(\mathbf{r}_1, \mathbf{r}_2, t)$ exist in geometrical space? It's nonsense to suggest that such a function exists, as does an electromagnetic wave, in a space of three spatial coordinates.

The moral of these ruminations is that whatever Ψ is, it's not a conventional undulation in space, like an electric field. So where does this leave us in our quest to understand the reality behind the state function? Do wave functions really exist? Here is Werner Heisenberg trying to answer this question:

> The probability wave ... [stands] in the middle between the idea of an event and the actual event, a strange kind of physical reality just in the middle between *possibility* and *reality*. [emphasis mine]

Heisenberg's answer is strange indeed, for it leaves us in a kind of metaphysical limbo without an experimental escape route. When we're not looking at microscopic particles, they behave in ways that correspond to the classical wave model; yet, when we measure their properties, they behave like classical particles. The postulates of this chapter provide a way out of this logical conundrum: we assume that complex functions exist that are somehow associated with the states of a microscopic system. These functions carry probabilistic information about the outcomes of measurements of the observables of the system. Finally, they can interfere according to the Principle of Superposition—and therein lies the explanation of the observed behavior of quantum particles.

As far as anyone knows, the behavior of all microscopic particles is governed by probability waves; every electron, proton, and neutron in every atom in every blob of matter in the universe (including your author) is guided by its state function. As Nick Herbert says in his *Quantum Reality* (Doubleday/Anchor Press, 1985):

> In each atom ... something seems smeared out to fill the atom, an indescribable something we call the "probability cloud," "realm of positional possibilities," "electron

wave function," or "quantumstuff" without really being very sure what we're talking about. Whatever it is, though, the whole world is made of it.

In our next chapter, we'll take a close look at the "quantumstuff" of the simplest imaginable system: one particle in one dimension doing nothing.

ANNOTATED SELECTED READINGS

Basic Tools

If you've read this far, surely you know that you're going to need a good mathematical handbook and a very good table of integrals. You need not, however, spend thousands of dollars on these basic tools. For essential mathematical data at a reasonable price, I recommend

1. Fischbeck, H. J., and K. H. Fischbeck, *Formulas, Facts, and Constants for Students and Professionals* (New York: Springer-Verlag, 1982) (available in paperback).
2. Dwight, Herbert B., *Tables of Integrals and Other Mathematical Data*, 4th ed. (New York: Macmillan, 1961). This book also contains a very good set of integral tables.

If you are willing to spend a little more—for a volume that you will use throughout your professional career—I can oblige. For my money, the ultimate mathematical handbook is

3. Korn, G. A., and T. M. Korn, *Mathematical Handbook for Scientists and Engineers*, 2nd ed. (New York: McGraw-Hill, 1968).

And the ultimate integral table is the truly awesome compendium

4. Gradshteyn, I. S., and I. M. Ryzhik, *Table of Integrals, Series, and Products,* Corrected and Enlarged Edition (New York: Academic Press, 1980).

Complex Variables

Complex numbers and functions are an essential part of the language of quantum physics. As we have seen, the basic element of this theory is a complex function. Actually, you won't need a sophisticated knowledge of these topics to cope with this book. (For example, neither residues nor complex integration appear herein.) And Appendix K contains a very brief survey for the initiated. For others, I recommend the sections on complex variables in the following texts (listed here in order of increasing complexity):

5. Hildebrand, F. B., *Advanced Calculus for Applications* (Englewood Cliffs, N.J. Prentice Hall, 1962), Chap. 10, especially §10.1–10.3.
6. Arfken, George, *Mathematical Methods for Physicists*, 2nd ed. (New York: Academic Press, 1970), Chap. 6, especially § 6.1.
7. Cushing, James T., *Applied Analytical Mathematics for Physical Scientists* (New York: Wiley, 1975), Chap. 7.

And for the enthusiast, I recommend the following books that are devoted exclusively to complex variables:

8. Kyrala, A., *Applied Functions of a Complex Variable* (New York: Wiley-Interscience, 1972). This book contains a particularly valuable set of worked examples.
9. Churchill, Ruel V., *Complex Variables and Applications*, 2nd ed. (New York: McGraw-Hill, 1960).

Interpretations of Quantum Mechanics

The endlessly provocative question of the meaning of the state function, which I lighted on briefly in § 3.7, has been argued in a rather intimidating number of works—some of them highly controversial. In addition to the popular-level books recommended in Chap. 2, I strongly advise the curious reader to dig up a copy of these items:

10. Jammer, Max, *The Philosophy of Quantum Mechanics: The Interpretations of Quantum Mechanics in Historical Perspective* (New York: Wiley, 1974). This fascinating and largely readable book is a clear survey of the controversy surrounding the meaning of the state function.
11. Belinfante, F. J., *Measurement and Time Reversal in Objective Quantum Theory* (New York: Pergamon, 1975). Nowhere near as intimidating as its title, this little volume—which is available in softcover at a reasonable price—contains a particularly useful analysis of the probability interpretation (§ 1) and a section provocatively entitled "Quantum Mechanics and God."
12. DeWitt, Bryce S., and R. Neill Graham, eds., *The Many-Worlds Interpretation of Quantum Mechanics* (Princeton, N.J.: Princeton University Press, 1973). Much of this (paperback) book consists of reprints of classic papers on this, the principle alternative to the Copenhagen interpretation. Many of these may be beyond most readers of this book. But at least Bryce DeWitt's "Quantum Mechanics and Reality" should be accessible to you, and the whole volume is a worthwhile and fairly inexpensive addition to your library.

Among the plethora of papers on the interpretation of quantum physics, I particularly want to encourage you to peruse the following:

13. Epstein, Paul S., "The Reality Problem in Quantum Mechanics" *Amer. Jour. Phys.* **13**, 127 (1945).
14. Landè, Alfred, "Quantum Fact and Fiction" *Amer. Jour. Phys.* **33**, 123 (1965).
15. Shimony, Abner, " Role of the Observer in Quantum Theory" *Amer. Jour. Phys.* **31**, 755 (1963).

Max Born has written several fascinating and lucid books on quantum theory and its interpretation. Especially recommended are

16. *Causality and Chance in Modern Physics* (Philadelphia: University of Pennsylvania, 1957),

and Born's autobiographical reflections

17. *My Life and Views* (New York: Scribners, 1958).

Born and Einstein argued at fascinating length over these matters, and you can find their correspondence in

18. *The Born-Einstein Letters* (New York: Walker and Company, 1971).

Finally, an invaluable guide to further reading on this subject is

19. DeWitt, Bryce S., and R. Neill Graham, "Resource Letter IQM-1 on the Interpretation of Quantum Mechanics" *Amer. Jour. Phys.* **39**, 724 (1971).

Alternative Postulations

One's choice of the postulates upon which one erects quantum mechanics is, to some extent, flexible. The graduate (and many of the undergraduate) texts in the Bibliography at the end of this book illustrate this point, but you may want to consider one of the following books. Both adopt a more formal approach to quantum theory than mine but write at the same level. And both are filled with valuable insights:

20. Gillespie, Daniel T., *A Quantum Mechanics Primer* (London: International Textbook, 1973). An exceptionally clear text, albeit limited in scope. What Gillespie covers, he covers very well.

21. Liboff, Richard L., *Introductory Quantum Mechanics* (San Francisco: Holden-Day, 1980); see especially Liboff's discussion in Chap. 3.

EXERCISES AND PROBLEMS

> If all else fails,
> immortality can always be assured
> by spectacular error.
>
> —John Kenneth Galbraith

Exercises

3.1 On the Interpretation of the State Function: Short-Answer Questions

Fig. 3.1.1 is a graph of the normalized state function and corresponding probability density for a quantum state of a single particle in one dimension at $t = 0$. The magnitude $|\Psi(x, 0)|$ of the wave function continues to decay smoothly to zero as $x \to \pm\infty$.

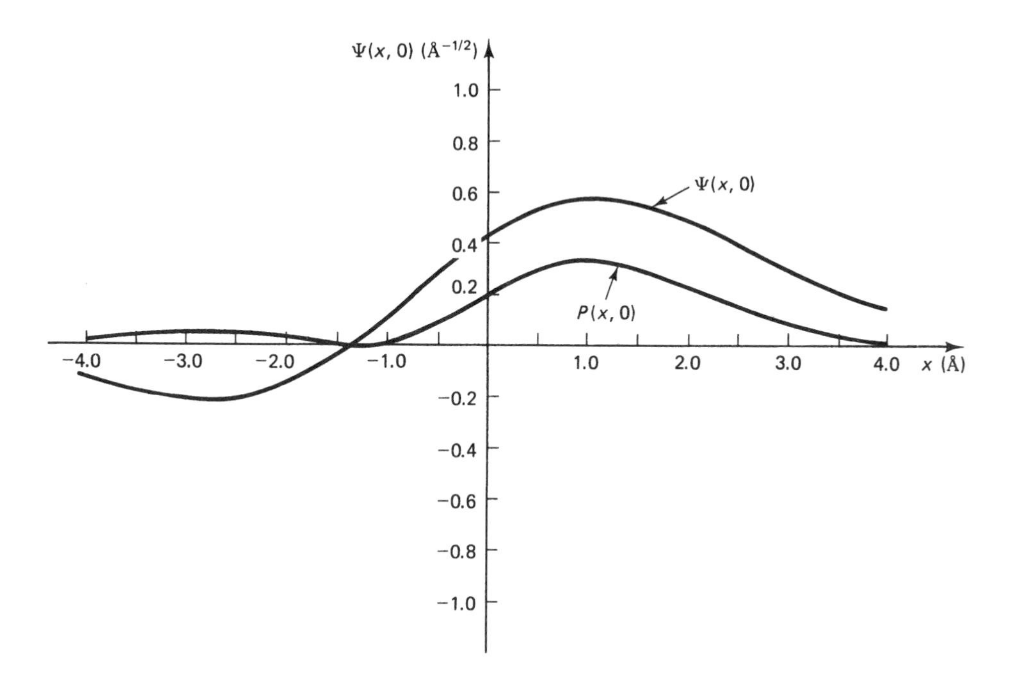

Figure 3.1.1 A sample state function and its probability density.

(a) Is this function physically admissible? Why or why not?
(b) In an ensemble measurement of the position of the particle in this state, where is the particle most likely to be found?
(c) Is the expectation value of position in this state equal to your answer to (b)? less than your answer? greater than your answer? Explain your reasoning carefully and fully.
(d) Is the uncertainty in momentum of this state equal to zero? Justify your answer.

3.2 A Probable-Position Problem

Consider a particle of mass m in a one-dimensional infinite square well, *i.e.*, whose potential energy is the one in Fig. 3.6a. Suppose that the system is in a quantum state represented by the normalized wave function of the *ground state* (the state of minimum allowed total energy):

$$\Psi_1(x,t) = \sqrt{\frac{2}{L}}\cos\left(\pi\frac{x}{L}\right)e^{-i\omega t} \qquad \left[-\frac{1}{2}L \le x \le +\frac{1}{2}L\right]$$

where ω is a constant.

(a) **Determine** an expression for the probability density of the ground state at time t. **Discuss** carefully and fully the physical interpretation of this quantity.

(b) At time $t = 0$, we perform an ensemble measurement of the position of the particle. **Evaluate** the probability that in this measurement we will find the particle in the right half of the well—*i.e.*, in the finite interval $[0, L/2]$.

(c) Compare your answer to part (b) with the predictions of classical physics for the probability of detecting a macroscopic particle of mass M in the right half of the well.

Hint: What is the classical probability density for a free particle?

3.3 Measurement in Quantum Mechanics: Short-Answer Questions

Consider a system whose state function $\Psi(x,t)$ at a fixed time t_0 is shown in Fig. 3.3.1.

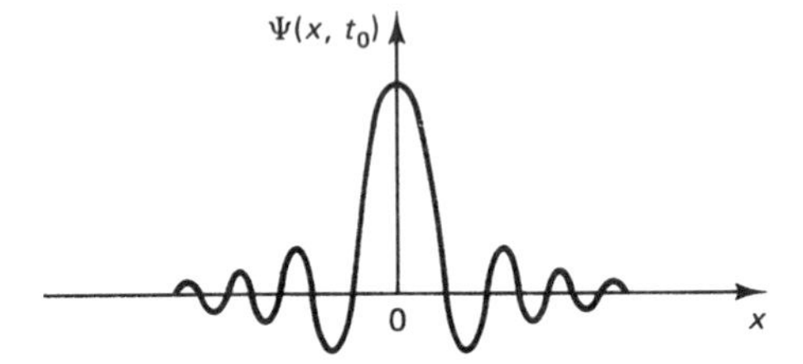

Figure 3.3.1

(a) **Describe** how you would calculate the expectation value of position for this state from the functional form of $\Psi(x,t)$.

(b) What is the value of ensemble average of position for this state at time t_0?

(c) **Describe** briefly how you would calculate the position uncertainty for this state from $\Psi(x,t)$.

(d) Is Δx zero, positive-but-finite, or infinite? Why?

(e) Is the momentum uncertainty Δp zero, positive but finite, or infinite? Why?

(f) Can this state function be normalized? If not, why not? If so, describe briefly how you would normalize $\Psi(x,t)$.

3.4 More on the Interpretation of the State Function

Consider the state function shown in Figure 3.4.1.

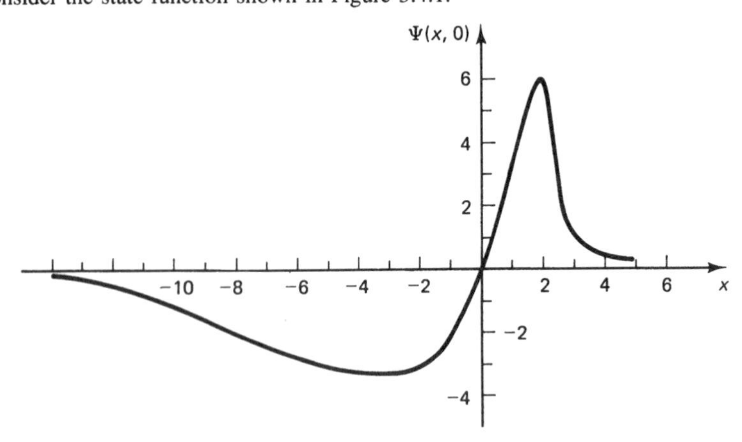

Figure 3.4.1

(a) Answer parts (a)–(d) of Exercise 3.1 for this function.
(b) **Draw** a rough sketch of the position probability density for this state.
(c) **Discuss** carefully the physical significance for a position measurement of the presence of the *node* in the state function at $x = 0$.

Problems

3.1 Fun with Probabilities

(a) Preparing for a Thanksgiving trip to Las Vegas, an enterprising student decides to study the likelihood of his becoming fantastically wealthy playing dice. Let s denote the number of spots shown by a (six-sided) die thrown at random. (Assume the unlikely idealization of an honest casino—no loaded dice.)
 (1) **Calculate** the average result of an ensemble measurement of s.
 (2) **Calculate** the standard deviation of the results of this study.
(b) A zoology graduate student has conned his thesis advisor into letting him do a dissertation on the distribution of hairs in a rare species of gnu. Prior to leaving for gnu country in the Kalahari, he finds in the library an earlier research study on this fascinating topic that found that the number of hairs on this species can only be the integer $N_\ell = 2^\ell$, where ℓ is any non-negative integer. Moreover, the probability of finding an animal with N_ℓ hairs has been found to be $e^{-1}/(\ell!)$. Worried that this might be bad gnus, he investigates further:
 (1) **Evaluate** and interpret the ensemble average $\langle N \rangle$.
 (2) **Evaluate** and interpret the standard deviation, ΔN.

3.2 Exploring the States of a Simple Quantum System

In this problem, we'll try to gain familiarity with several important new quantities introduced in Chap. 3 by considering three states of an extremely simple system: the particle in an infinite square well. This system consists of a particle of mass m in one dimension with potential

$$V(x) = \begin{cases} \infty & x \le -\frac{1}{2}L \\ 0 & -\frac{1}{2}L < x < \frac{1}{2}L; \\ \infty & x \ge \frac{1}{2}L. \end{cases}$$

Here are three physically admissible state functions for this system (which we'll derive in Chap. 7):

$$\begin{aligned} \Psi_1(x,t) &= N_1 \cos\left(\pi\frac{x}{L}\right) e^{-iE_1 t/\hbar} && \text{ground state} \\ \Psi_2(x,t) &= N_2 \sin\left(2\pi\frac{x}{L}\right) e^{-iE_2 t/\hbar} && \text{first excited state} \\ \Psi(x,t) &= N\left[\cos\left(\pi\frac{x}{L}\right) e^{-iE_1 t/\hbar} + \sin\left(2\pi\frac{x}{L}\right) e^{-iE_2 t/\hbar}\right]. && \text{another state} \end{aligned}$$

All state functions are zero outside the box. In these expressions, N_1, N_2, and N are normalization constants, and E_1 and E_2 are the energies of the ground and first excited states, respectively. As we'll discover in Chap. 7, these energies are given by

$$E_n = n^2 \frac{\pi^2\hbar^2}{2mL^2} \qquad \text{for } n = 1, 2, \ldots.$$

Notice that the third state, the one represented by $\Psi(x,t)$, is a superposition of those represented by the other two wave functions.

(a) **Normalize** each of these state functions.

(b) For **each** of these three wave functions, answer the following questions concerning the results of an ensemble measurement of position that is performed at some time $t > 0$:

  (1) What average value of x will be obtained? (If your answer is time-dependent, write the time dependence as a *real* function of time.)

  (2) Is your answer to (b1) equal to the **most probable** position for Ψ_1 and Ψ_2? Why or why not?

  (3) Does your answer to (b1) depend on time?

(c) **Evaluate** the uncertainty in position for the ground and first excited states.

(d) **Sketch carefully** the real part of $\Psi(x,t)$ for two times, $t = 0$ and $t = 2mL^2/(3\pi\hbar)$. Using your sketches, **explain** the behavior of $\langle x \rangle$ for this state as time passes (as found in part (b3) above). Plot these functions in terms of $\frac{x}{L}$. [Choose a convenient (but reasonable) value for L if you wish.]

3.3 Deducing Information from State Functions without Work: I

To actually solve for the state functions of a quantum mechanical system requires a fair bit of mathematical acuity—we must solve the time-dependent Schrödinger equation (3.39), a task we shall spend most of this book mastering. It is possible, however, to deduce lots of important physical information from such state functions without solving complicated equations provided you understand the meaning of the quantum mechanical quantities involved. In this problem (and the next) we shall do this for a couple of simple and important systems.

In Fig. 3.3.1 you'll find a graph of wave functions for four quantum states of a particle of mass m in one dimension with a simple harmonic oscillator potential energy

$$V(x) = \frac{1}{2}m\omega_0^2 x^2, \tag{3.3.1}$$

where ω_0 is the natural frequency of the oscillator. These special *stationary states* have several unusual properties; they are the subject of Chap. 7. Of immediate relevance is that *the energy of a stationary state has a well-defined value*. That is, *for a stationary state $\Delta E(t) = 0$ for all times t.*

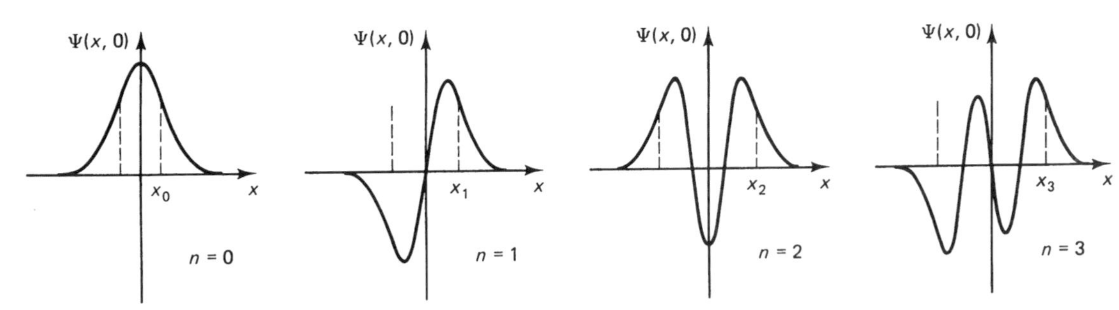

Figure 3.3.1 Snapshots of the first four stationary states of a one-dimensional simple harmonic oscillator at $t = 0$.

We'll denote the stationary-state wave functions of the one-dimensional simple harmonic oscillator by $\Psi_n(x,t)$, where n is an index that tells us the energy of the state: $n = 0$ is the *ground state* (the state of minimum energy), $n = 1$ the *first excited state*, and so forth. The energies of these states are

$$E_n = \left(n + \frac{1}{2}\right)\hbar\omega_0. \qquad \text{for } n = 0, 1, 2, \ldots \tag{3.3.2}$$

(a) **Sketch** the probability densities $P_n(x, 0)$ for the four stationary states whose wave functions are shown in Fig. 3.3.1. (Be careful to show correctly the relative magnitude of the peak heights of these functions.)

(b) By extrapolating from the shapes of the wave functions in Fig. 3.3.1, **sketch** the wave function for $\Psi_4(x, 0)$ and the corresponding probability density $P_4(x, 0)$.

(c) What is the mean value of position $\langle x \rangle_n$ (at time $t = 0$) for these four stationary states? **Think, don't calculate.** Support your answer with an argument using words and pictures, but no equations.

(d) The state functions for the first two of the states shown in Fig. 3.3.1 can be conveniently expressed in terms of the constant

$$\beta^2 \equiv \frac{m\omega_0}{\hbar}. \tag{3.3.3}$$

They are

$$\Psi_0(x, 0) = \left(\frac{\beta^2}{\pi}\right)^{1/4} e^{-\beta^2 x^2/2} \tag{3.3.4a}$$

$$\Psi_1(x, 0) = \left(\frac{\beta^2}{\pi}\right)^{1/4} \sqrt{2}(\beta x)e^{-\beta^2 x^2/2}. \tag{3.3.4b}$$

Using the definition of the ensemble average of position, support your answer to part (c) by **graphing** the *integrands* of $\langle x \rangle_0$ and $\langle x \rangle_1$.

(e) What *mathematical property* of the state functions $\Psi_n(x, 0)$, explains your answer to part (c)? Assume that this property holds for all stationary state wave functions of the simple harmonic oscillator potential (we'll verify this assumption in Chap. 8). Now, **guess** the mean value of position $\langle x \rangle_n$ for *any* stationary state of this system.

3.4 More Deductions from Simple Wave Functions

A particularly simple collection of wave functions are those for a particle in an infinite square well. Let's consider the first four of these functions, $\Psi_n(x, 0)$ for $n = 1, 2, 3$, and 4. Assume that the width of the potential well is L. Like the states of the simple harmonic oscillator in the previous problem, these states are *stationary*, with (well-defined) energies. For the infinite square well, these energies are

$$E_n = n^2 \frac{\pi^2\hbar^2}{2mL^2} \qquad \text{for } n = 1, 2, 3, \ldots \tag{3.4.1}$$

In Chap. 7 we'll derive the explicit mathematical forms for the stationary-state wave functions. For the first two, we'll find

$$\Psi_1(x, t) = \sqrt{\frac{2}{L}} \cos\left(\pi \frac{x}{L}\right) e^{-iE_1 t/\hbar} \tag{3.4.2}$$

$$\Psi_2(x, t) = \sqrt{\frac{2}{L}} \sin\left(2\pi \frac{x}{L}\right) e^{-iE_2 t/\hbar} \tag{3.4.3}$$

(a) By thinking about the symmetry of the infinite square well and the above expressions for $\Psi_1(x, 0)$ and $\Psi_2(x, 0)$, **guess** the *mathematical form* of $\Psi_3(x, 0)$. (You need not worry about the normalization constant N_3, although you probably can guess this too.)

(b) **Sketch** $\Psi_5(x, 0)$ and the corresponding probability density.

(c) What is $\langle x \rangle_n$ for $n = 1, 2, \ldots, 5$? (**Think, don't calculate.**) **Extrapolate** your answer to arbitrary n, providing a justification in words and pictures (but not equations).

Hint: Try graphing the integrands of $\langle x \rangle_1$ and $\langle x \rangle_2$.

(d) Notice that the stationary-state wave functions alternate *parity* with increasing n—*i.e.*, the ground state ($n = 1$) is an *even function of* x, the first excited state ($n = 2$) is an *odd function of* x, the second excited state ($n = 3$) is even, and so forth. Does this property of $\Psi_n(x, t)$ have any effect on the expectation value $\langle x \rangle_n$? Does it have any effect on the position uncertainty $(\Delta x)_n$ for these states? **Explain why or why not.**

(e) Do you expect $(\Delta x)_n$ to be independent of n? **Justify your answer, preferably with sketches.**

(f) For a stationary state $\Psi_n(x, t)$, do the expectation value $\langle x \rangle_n$ and the uncertainty $(\Delta x)_n$ depend on time? Justify your answer.

3.5 The Classical Limit is Not Necessarily Macroscopic

Many students of quantum mechanics fall into the seductive trap of thinking that the only limit in which classical mechanics pertains to a system is that in which the size of the particle(s) in the system is large—the macroscopic limit. This is not always the case. In this problem, we shall begin to explore this point using the infinite square well discussed in several prior problems and our understanding of the basic quantum principles. (Note: You need not have worked the earlier problems on the infinite square well to solve this one, but you may need to use some of the results included in the statements of those problems.)

(a) The wave function for the ground state of the infinite square well at $t = 0$ is

$$\Psi_1(x, 0) = \sqrt{\frac{2}{L}} \cos\left(\pi \frac{x}{L}\right). \tag{3.5.1}$$

This is the first of an infinite number of special states of this system, called *stationary states*, for which $\Delta E = 0$. **Evaluate** the position uncertainty $(\Delta x)_1$ for this state at $t = 0$.

(b) **Sketch** the probability density $|\Psi_n(x, 0)|^2$ for a stationary state with a *very large energy*; *i.e.*, for $n \gg 1$.

(c) **Evaluate** the position uncertainty $(\Delta x)_n$ for $n \gg 1$. (Express your answer in terms of L.)

Hint: Your sketch in (b) should suggest a nifty approximation that will greatly simplify the integral you need to evaluate in (c).

(d) Compare your answer to (c) with the *classical* standard deviation of position—*i.e.*, the value you would get for a macroscopic particle in an infinite square well of macroscopic width L. Do your answers to parts (b)–(d) depend on the mass m of the particle? What can you conclude from this problem about the classical limit of quantum mechanics?

3.6 Exposing the Non-Classical Behavior of a Quantum Particle

In Fig. 3.6.1 is shown the ground-state wave function $\Psi_1(x, 0)$ for a particle of mass m in a *finite* potential well of strength V_0. The particle in this state has a well-defined energy $E < 0$—*i.e.*, the energy in this state is certain and $(\Delta E)_1 = 0$.

If you look carefully at $\Psi_1(x, 0)$, you will note that the position probability amplitude for this state is non-zero in the *classically-forbidden regions* $x > L/2$ and $x < -L/2$. In such a region, $E < V$, so $T = E - V < 0$: a "classically-forbidden" situation. In Chap. 8 we'll solve for these wave functions and will see why they are non-zero in classically-forbidden regions. In this problem, we shall explore some ramifications of this fact from the vantage point of the key ideas of quantum physics.

Suppose we set out to *detect* experimentally particles in a classically-forbidden region. How might we do this? We could use a photon microscope, scattering photons from the particle. Of course, we must be sure that the particle from which the photon scatters is actually in a classically-forbidden region, *i.e.*, that is not at some $x < L/2$ or $x > -L/2$. For simplicity, let's concentrate in this problem on the *right-hand* classically forbidden region, $x > L/2$.

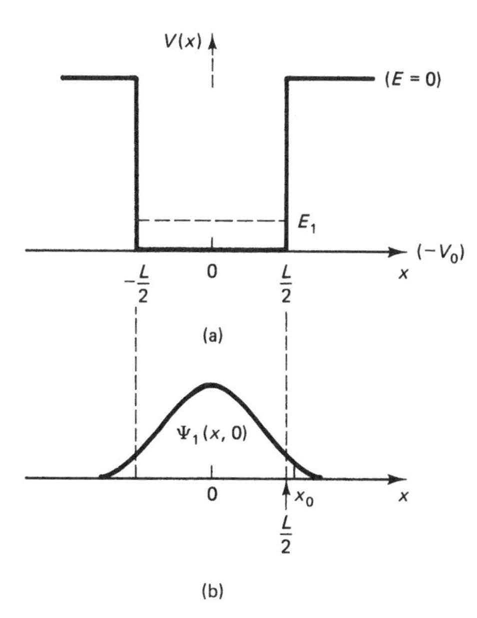

Figure 3.6.1 (a). A finite square well potential energy. In this problem we consider the ground-state of a particle of mass m with this potential energy. (b). The wave function at $t = 0$ for this state.

First we need to define this region more precisely. Looking at Fig. 3.6.1, we notice that the wave function decays *exponentially* in the classically-forbidden region. In fact, as we'll learn in Chap. 8, the form of the wave function in this region is

$$\Psi(x,0)\bigg|_{x > L/2} = Be^{-\kappa x}, \tag{3.6.1}$$

where the constant κ is defined as

$$\kappa \equiv \frac{1}{\hbar}\sqrt{2m(-E)}. \tag{3.6.2}$$

To avoid having to worry about the minus sign on the energy, which results from our choice of the top of the well in Fig. 3.6.1 as the zero of energy, we'll introduce the **binding energy** ϵ, which is defined as

$$\epsilon \equiv -E, \tag{3.6.3}$$

whence κ can be written

$$\kappa = \frac{1}{\hbar}\sqrt{2m\epsilon}. \tag{3.6.4}$$

Now, taking a clue from optics, let's define the classically forbidden region as $[L/2, x_0]$, where x_0 is defined as the position where the probability density is $1/e$ of its maximum value in this region,

$$P(x_0,0) = \tfrac{1}{e}\,P\left(\tfrac{L}{2},0\right). \tag{3.6.5}$$

(The total internal reflection of light at a glass-to-air interface behaves in a fashion analogous to the probability density in this problem.)

(a) To guarantee that we detect particles that are in the classical-forbidden region, we require that the position uncertainty be no greater than the size of this region, *i.e.*,

$$\Delta x \le x_0 - \tfrac{L}{2}. \tag{3.6.6}$$

What is the resulting uncertainty in the (x component of) momentum p of the particle?

(b) If p is uncertain, then we cannot know the precise value of the kinetic energy K of the particle. What is the uncertainty ΔK in this quantity?

(c) Now, suppose we find a particle in the classically-forbidden region. Can we know that its kinetic energy is negative? **Explain** how the uncertainty principle and wave-particle duality conspire in this measurement to resolve the apparent paradox of a negative kinetic energy particle.

(d) Two students, pondering the state function shown in Fig. 3.6.1, comment as follows:

> Student A: "A quantum particle in the classically-forbidden region has a negative kinetic energy."
>
> Student B: "There is a finite probability of detecting the particle in the classically-forbidden region."

Is either of these statements correct? If not, discuss the logical error in the thinking of the student(s) who is (are) wrong.

Hint: What condition must we place on the intensity of the photon beam used to illuminate the classically-forbidden region if we are to detect an electron inside this region? Thinking about this question should lead you to an upper limit on the wavelength of the photon we can use in this experiment and hence to a conclusion very much like that of part [c].

CHAPTER 4

An Illustrative Interlude

Wave Packets in One Dimension

I dare say that's an idea which
has already occurred to you,
but with the weight of my great mind
behind it, no doubt it strikes
the imagination more forcibly.

—-Lord Peter Wimsey
in *Strong Poison*
by Dorothy L. Sayers

Postulates I and II are the first bricks in the foundation of quantum mechanics. The first postulate identifies the basic element of the theory: the wave function. The second postulate interprets this function as a position probability amplitude. As befits the introduction, the discussion in Chap. 3 was general and rather abstract. So in this chapter, we'll take a breather from this development to explore at length a single, simple example: *a microscopic free particle in one dimension.* The physics of this system is especially simple because, as its name implies, such a particle is free of external forces or fields.[1]

Aside: Free Particles in the Real World. A free particle is, of course, an idealization; no system is truly isolated from external influences. Nevertheless, the example of this chapter finds wide application in quantum physics. The reason is that in many situations, the external forces that do act on a particle are so weak that we can neglect them. For example, in electron scattering experiments physicists use a beam of non-interacting electrons. Such a beam is, of course, subject to the force of gravity. But we need not take this force into account in a description of its physics, because the effect of this very weak force on the particles in the beam is negligible.

Our objective in this chapter is the wave function $\Psi(x,t)$ of a free particle in one dimension. Ordinarily, we would obtain this function by mathematical machinations—we'd solve the time-dependent Schrödinger Equation [Eq. (3.39)],

$$\left[-\frac{\hbar^2}{2m}\frac{\partial^2}{\partial x^2} + V(x)\right]\Psi(x,t) = i\hbar\frac{\partial}{\partial t}\Psi(x,t). \tag{4.1}$$

Since a free particle experiences no external forces, its potential energy is a constant the value of which I'll set equal to zero:[2]

$$V(x) = 0 \qquad \text{[for a free particle].} \tag{4.2}$$

So the Schrödinger Equation for a free particle is just (4.1) without the term in $V(x)$.

At this stage in our study, I want to avoid the algebraic chore of solving this second-order partial differential equation—even for so simple a system as the free particle. Instead, I want to lead you to a deeper understanding of the nature of wave functions. So our strategy (in § 4.1 and 4.2) will be to *deduce* rather than derive the state function for a free particle.

What we will find is that this function must be a *complex wave packet.* Wave packets play an important role in quantum mechanics, so we'll spend a little time (in

[1] A classical analogue of a free particle in one dimension is a macroscopic bead constrained to move along a (hypothetical) infinite, frictionless wire.

[2] Recall that the potential energy of a system is arbitrary to within an additive constant, since only *potential differences* are measurable. See Chap. 8 of *Fundamentals of Physics*, 2nd ed., by D. Halliday and R. Resnick (New York: Wiley, 1981).

§ 4.2–4.4) examining the mathematics of their construction via Fourier analysis. But, as always, physics remains paramount, and in § 4.5 we'll discover *uncertainty* in the representation of a state by a wave packet. Finally, in § 4.6–4.7, we'll see a tantalizing glimpse of how to extract information about observables other than position from a wave function. This analysis will hurtle us into Chap. 5, where we tackle momentum.

4.1 ON THE NATURE OF THE FREE-PARTICLE STATE FUNCTION

> Always trust the simplest explanation that fits all the facts unless there's a damn good reason not to do so.
>
> — Dr. Stuart Hay
> in *Incarnate*
> by Ramsey Campbell

What do we know about state functions? In particular, what properties should characterize the state functions of a particle in one dimension? Well, in § 3.5 I argued that such a function must be single-valued, continuous, and—most importantly—normalizable. That is, it must obey the boundary condition (3.29), according to which $\Psi(x,t)$ must go to zero (as $|x| \to \infty$) rapidly enough that its normalization integral $\int_{-\infty}^{\infty} \Psi^*(x,t)\Psi(x,t)\,dx$ is finite. Physically, this condition means that the wave function must be *spatially localized.* That's one property.

Another is evident from the Schrödinger Equation (4.1): $\Psi(x,t)$ must be complex. Here's why: the right-hand side of this equation has a multiplicative factor of $i = \sqrt{-1}$ that is absent from the left-hand side. Therefore *no purely real or purely imaginary function can satisfy this equation.* Were $\Psi(x,t)$ purely real, for example, then the left-hand side would be real and the right-hand side would be imaginary—which would violate the equal sign![3]

These two properties pertain to the state functions of *any* system. When we turn to the free particle (or any other *isolated system*) we find a third, quite different property: a fundamental symmetry of nature called the *homogeneity of free space*. I want to tell you about this property, which is the topic of the next subsection, to introduce you to the power and beauty of symmetry arguments. From this symmetry property of space itself, we'll derive a condition on the probability density $P = |\Psi|^2$ that *all free-particle state functions* must satisfy. Then we'll use this property—and a little imagination and logic—to deduce the wave function for the free particle in one dimension.

The Homogeneity of Free Space

What's the first step we must take if we want to investigate the physics of a system, be it microscopic or macroscopic? We must define a *reference frame*. The one you're probably most familiar with is the *rectangular (or Cartesian) coordinate system* shown in

[3]This argument does *not* imply that $\Psi(x,t)$ must be complex at all times; it can be real (or imaginary) at certain, discrete times. But it does imply that the wave function cannot be purely real or imaginary *at all times*.

Fig. 4.1. The directions of the three mutually orthogonal axes that define this reference frame are $\hat{e}_x$, $\hat{e}_y$, and $\hat{e}_z$.[4] These axes intersect at the origin O.

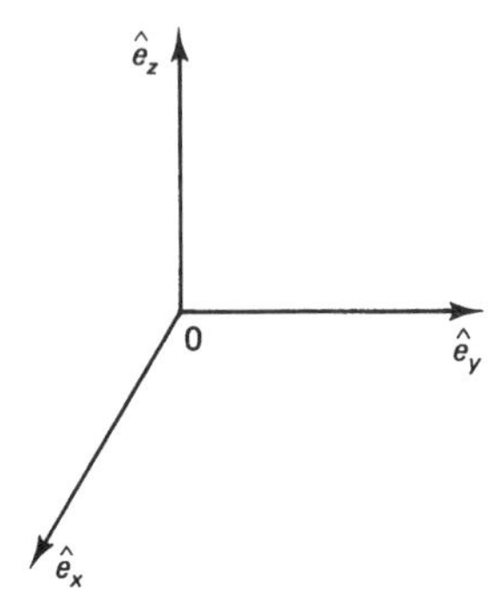

Figure 4.1 A Cartesian coordinate system. Within this system several sets of orthogonal coordinates can be used, among them *rectangular* and *spherical* coordinates.

The decision of where to put the origin of coordinates is very important, because all vectors are referred to this point. But the choice of origin is ours to make. After all, coordinates and reference frames are just mathematical artifices that physicists invent to solve problems.

A Historical Aside. This feature of theoretical physics was noted in 1885 in an essay by Charles Howard Hinton (1853–1907):

> To measure, we must begin somewhere, but in space there is no "somewhere" marked out for us to begin at. This measuring is something, after all, foreign to space, introduced by us for our convenience.
>
> And as to dimensions, in order to enumerate and realize the different dimensions, we must fix on a particular line to begin with, and then draw other lines at right angles to this one.
>
> But the first straight line we take can be drawn in an infinite number of directions. Why should we take any particular one?
>
> If we take any particular line, we do something arbitrary, of our own will and decision, not given to us naturally by space.

Hinton's comments lead us to expect physically measurable properties to be independent of such arbitrary choices. And, indeed, they are.[5]

In practice we have several ways to pick the origin of coordinates. We might choose an origin that will simplify the solution of whatever equations we confront. Or we might select as the origin the location of a measuring apparatus. But most often we look to nature for guidance. For example, we might identify some physical entity—say, a (point) particle in the system or the source of an electromagnetic field—as the origin.

But if the particle is free, then we cannot use the latter gambit, for nothing in free space distinguishes one point over another. That is, nature does not identify a preferred origin for a free particle. In this case we can put O anywhere we wish.

[4] Little hats do a lot of work in quantum mechanics. Throughout this book, a little hat on an e, such as $\hat{e}_i$, will denote a unit vector in the i direction. (For rectangular coordinates, $i = x, y, z$.) But from Chap. 5 onward, little hats on *other* symbols will denote operators. Be wary.

[5] Hinton was so interesting a character that I can't resist telling you a little about him. Many consider him one of the great philosophers of hyperspace; he was fascinated by the fourth dimension. No, fascinated is the wrong word; Hinton was obsessed. In an effort to construct a mental image of a four-dimensional universe, he memorized 46,656 one-inch cubes, each of which had been assigned a Latin name. Evidently, this daunting task was not challenging enough for Hinton, who next decided to learn his cubic yard of cubes in each of its twenty-four possible orientations. For more on this eccentric but brilliant mathematician, see Rudy Rucker's delightful book, *The Fourth Dimension: Toward a Geometry of Higher Reality* (Boston: Houghton Mifflin Co., 1984).

This discussion of the origin of coordinates leads us to the fundamental symmetry property I mentioned above: the **homogeneity of free space**. *When we say "free space is homogeneous" we mean that the physical properties of a system in free space are independent of the origin of coordinates.* The implications of homogeneity for a physical theory (such as quantum mechanics) are strict indeed:

> ***Rule***
> All physical quantities we calculate must be independent of the origin of coordinates.

You can see what this means if you imagine picking up the system and *translating* it through space to a new location. According to this rule, all measurable quantities that you calculate from a state descriptor of the system after translation must equal those calculated from the corresponding state descriptor before translation.

Since in quantum mechanics we calculate properties from wave functions, it should come as no surprise that the homogeneity requirement imposes a limitation on wave functions. But be careful—the nature of this limitation is rather subtle. *Homogeneity does not imply that the state function itself must be independent of the coordinate origin, because this function is not measurable.* But all measurable quantities are calculated from the probability density $\Psi^*\Psi$, so *the homogeneity of free space demands that the probability density be independent of the origin of coordinates.* That is, a *translation* of a microscopic free particle must not alter its probability density. Such a translation appears in Fig. 4.2.

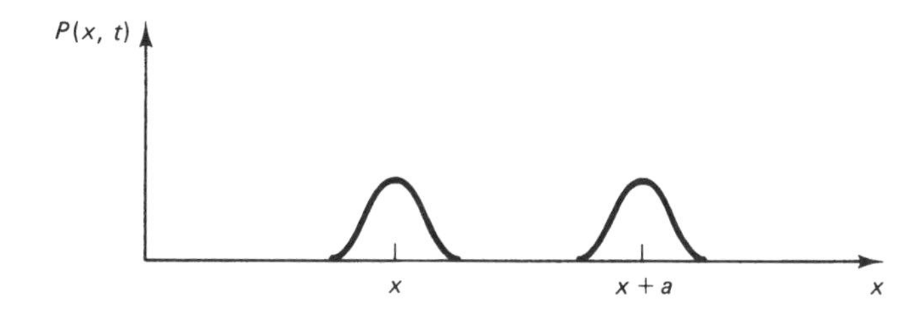

Figure 4.2 The effect on the probability density of the translation of a microscopic system by a distance a in the $+x$ direction. The state function of the system is changed by the translation and so depends on a. But the probability density must be unchanged; *i.e.*, it must be independent of a.

To use this property to deduce the free-particle state function, we'll need a mathematical expression of it. So let's consider translating a system that *before translation* is in a quantum state represented by Ψ. The state function *after translation* will, in general, be some other function of position and time. (We'll assume instantaneous translation, so t does not change.) This function will also depend on the translation distance a. So I'll represent the state after translation by Ψ'_a. Now, the relationship between the state functions of a one-dimensional system before and after translation in the x direction is simply

$$\Psi'_a(x,t) = \Psi(x-a,t) \qquad \text{[for translation by } a \text{ in the } +x \text{ direction].} \tag{4.3}$$

With this notation we can easily write down the homogeneity requirement that the probability densities before and after translation be equal, *viz.*,

$$\boxed{\text{homogeneity of free space} \quad \Longrightarrow \quad |\Psi'_a|^2 = |\Psi|^2}. \tag{4.4}$$

We are now ready to tackle the problem of the free particle. But before we begin, let's summarize the properties that the state function of such a particle must obey:

1. Ψ must be spatially localized (normalizable);
2. Ψ must be a complex function of position and time;
3. Ψ must not violate the homogeneity of free space.

Homogeneity and the Free-Particle State Function

To show you how to use the homogeneity requirement (4.4) to check out prospective free-particle state functions, I'll try one that we know is wrong. Consider a free particle in a state with momentum p and energy E.[6]

According to the de Broglie relations (2.16), the wave characteristics of particles in such a state are the associated wavelength $\lambda = h/p$ and frequency $\nu = E/h$, where h is Planck's constant. These relations suggest a *guess* at the wave function of the particle: a **real harmonic wave** with this wavelength and frequency:

$$f(x,t) = A\cos\left(\tfrac{2\pi}{\lambda}x - 2\pi\nu t + \alpha\right). \qquad \text{real harmonic wave} \tag{4.5}$$

In this generic harmonic wave, the amplitude A and phase constant α are real numbers that, in practice, are determined from the initial conditions on the problem.[7] In this section we are concerned only with the *functional form* of $f(x,t)$, so for convenience I'll set $\alpha = 0$.

Before proceeding to shoot down this guess, I want to get rid of the unsightly factors of 2π littering Eq. (4.5). Harmonic waves are usually written in terms of the **wave number** k and **angular frequency** ω, which are defined by

$$k \equiv \frac{2\pi}{\lambda} \qquad \text{wave number} \tag{4.6a}$$

$$\omega \equiv 2\pi\nu \qquad \text{angular frequency} \tag{4.6b}$$

The de Broglie relations specify the connection between these fellows and the momentum p and energy E of the state:

$$p = \hbar k \tag{4.7a}$$

$$E = \hbar\omega, \tag{4.7b}$$

where I've introduced the all-purpose 2π-absorber (otherwise known as the "rationalized Planck's constant")

[6]In nature, free-particle states in which energy and momentum can be known precisely don't exist. Statistically speaking, we say that the uncertainties in the energy and momentum are positive for all states: $\Delta E > 0$ and $\Delta p > 0$. But states do exist in which the energy and momentum are very nearly certain (*i.e.*, these uncertainties are very small), and we often approximate such a state by a function with a well-defined E and p.

[7]The initial conditions on $f(x,t)$ are just its values at a particular position and time, usually at $x = 0$ and $t = 0$. We can choose the phase constant α to force the general form (4.5) to agree with any initial conditions. If, for example, $f(0,0) = A$, we choose $\alpha = 0$. If, on the other hand, $f(0,0) = 0$, we choose $\alpha = 3\pi/2$. The latter often-used choice, by the way, transforms the cosine function into a sine function, so that the function becomes $f(x,t) = A\sin(kx - \omega t)$. And so forth.

$$\boxed{\hbar \equiv \frac{h}{2\pi}} \tag{4.8}$$

In its new clothes, our guess (4.5) for a free-particle state function looks like

$$\Psi(x,t) = A\cos(kx - \omega t). \tag{4.9}$$

This function is sketched in Fig. 4.3.

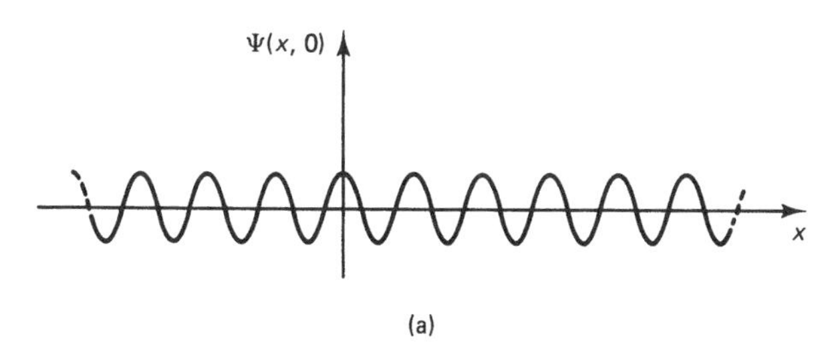

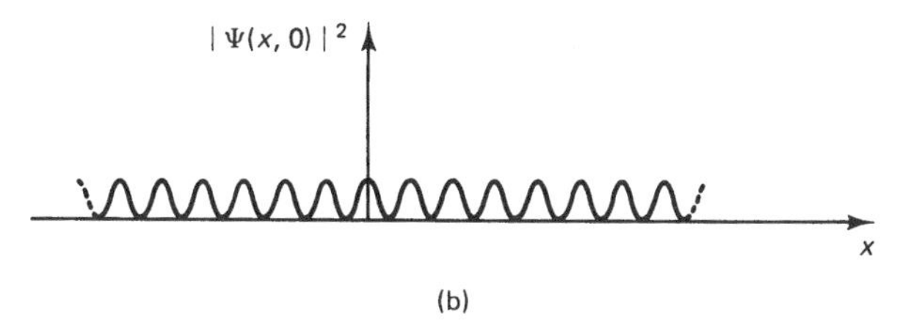

Figure 4.3 (a) The real harmonic wave of Eq. (4.9) at $t = 0$, and (b) its square. This function oscillates throughout space, as suggested by the dotted lines at the fringes of these curves. Consequently it cannot be normalized and is not physically admissible.

Aside: The Dispersion Relation. You probably remember from your study of waves that the angular frequency ω and the wave number k are related via a **dispersion relation**. In quantum mechanics, we can determine the dispersion relation from the relationship between the total energy and the momentum. For example, the energy of a free particle is purely kinetic, so provided its velocity is non-relativistic ($v \ll c$), the total energy E is related to the linear momentum p by the simple formula

$$E = \frac{p^2}{2m}. \qquad \text{[for a non-relativistic free particle]} \tag{4.10a}$$

Translating this into a dispersion relation via the definitions (4.7), we get[8]

$$\omega(k) = \frac{\hbar k^2}{2m}. \qquad \text{dispersion relation for a free particle} \tag{4.10b}$$

Now we all know why Eq. (4.9) won't do. First, we can see from Fig. 4.3 that for this guess the normalization integral—the area under the probability density curve—is infinite, so this function can't be normalized. Moreover, it's not complex and so couldn't

[8]If the particle is traveling at relativistic velocities, the total energy (including the rest-mass energy) and the linear momentum are related by Eq. (2.14), so (4.10*b*) is not applicable. For such a particle of rest mass m_0, the dispersion relation is

$$\omega(k) = \left[c^2k^2 + \left(\frac{m_0c^2}{\hbar}\right)^2\right]^{1/2}$$

satisfy the Schrödinger equation (4.1). But I want to show you yet another reason it won't work: it violates the homogeneity requirement.

Example 4.1. The Real Harmonic Wave and the Homogeneity Requirement

The demonstration is easy: we just determine the probability densities of the state (4.9) before and after translation and see if they're the same. Before translation, the probability is just the squared modulus of the function (4.9),

$$P(x,t) = A^2 \cos^2(kx - \omega t). \tag{4.11}$$

After translation, the wave function is given [a là Eq. (4.3)] by

$$\Psi'_a(x,t) = A \cos[k(x-a) - \omega t], \tag{4.12a}$$

which we can expand (using the handy trig identities in Appendix H) into

$$\Psi'_a(x,t) = A\left[\cos(kx - \omega t)\cos ka + \sin(kx - \omega t)\sin ka\right]. \tag{4.12b}$$

So the probability density after translation is

$$\begin{aligned}\Psi'^*_a(x,t)\,\Psi'_a(x,t) &= A^2 \cos^2(kx - \omega t)\cos^2 ka \\ &+ A^2\left[\sin^2(kx-\omega t)\sin^2 ka + 2\cos ka \sin ka \cos(kx-\omega t)\sin(kx-\omega t)\right].\end{aligned} \tag{4.13}$$

Now look carefully at the first term in (4.13). Compare it to (4.11). See? The first of the three terms in the probability density *after translation* is $P(x,t)\cos^2 ka$, the product of the probability density *before translation* and an additional factor! This result proves that a real harmonic wave, in addition to its other faults, violates the homogeneity requirement:

$$P'_a(x,t) \neq P(x,t). \qquad \text{[for a real harmonic wave]}. \tag{4.14}$$

This fact *alone* is enough to invalidate this function as a state descriptor for a free particle, for it is inconsistent with a fundamental symmetry property of nature itself. Back to the drawing board.

Question 4–1

As you probably know, any real function can be decomposed into a series of sine and cosine functions. Using this fact, present a brief argument that no real function of x and t can satisfy the homogeneity requirement.

Question 4–2

Show that the linear combination of plane harmonic waves

$$\Psi(x,t) = \cos(kx - \omega t) + i\sin(kx - \omega t) \tag{4.15}$$

does satisfy the homogeneity requirement.

A Second Guess: The Complex Harmonic Wave

Of the three defects of the real harmonic wave, the easiest to repair is its real character. A simple *complex* function that we might nominate as a candidate for the representative of a free-particle state with well-defined energy and momentum is the generalization suggested by Question 4–2: a **complex harmonic wave** of the form

$$\Psi(x,t) = A\,e^{i(kx-\omega t)}, \qquad \text{complex harmonic wave} \qquad (4.16)$$

where, as in the real harmonic wave, the frequency and wave number are related by the free-particle dispersion relation[9] $\omega(k) = \hbar k^2/2m$. Now let's see if this guess satisfies the demanding homogeneity requirement.

Example 4.2. The Complex Harmonic Wave and the Homogeneity Requirement

The state function (4.16) *after translation* is

$$\Psi'_a(x,t) = A\,e^{i(kx-ka-\omega t)} \qquad (4.17a)$$

$$= e^{-ika}\left[A\,e^{ik(x-\omega t)}\right] \qquad (4.17b)$$

$$= e^{-ika}\,\Psi(x,t). \qquad (4.17c)$$

I hope you're way ahead of me by now. According to (4.17*c*), the state function after translation is identical to that before translation except for the multiplicative factor e^{-ika}. So the state function depends on the origin of coordinates only through this *exponential* factor. [Notice that since ka is a constant, this is just a global phase factor (§ 3.5).] When we form the probability density after translation, this factor obligingly goes away:

$$\Psi'^*_a(x,t)\Psi'_a(x,t) = e^{ika}\Psi^*(x,t)\,e^{-ika}\Psi(x,t) \qquad (4.18a)$$

$$= \Psi^*(x,t)\,\Psi(x,t). \qquad (4.18b)$$

So the probability density formed from a complex harmonic wave is unchanged by a translation. We conclude that such a wave does satisfy the homogeneity requirement (4.4). It also satisfies another of our properties: it's complex. That's two out of three. Now, what's wrong with it?

Question 4–3

Verify that the complex harmonic wave (4.16) satisfies the time dependent Schrödinger Equation (4.1) for the free particle.

On the Horns of a Dilemma

Alas, the simple complex harmonic wave (4.16) is not spatially localized. This regrettable property is obvious from its probability density, *which is constant*:

$$P(x,t) = \Psi^*(x,t)\Psi(x,t) = |A|^2. \qquad \text{[for a complex harmonic wave]}. \qquad (4.19)$$

A constant probability density implies that we're equally likely to detect an electron anywhere in space. That is, the complex harmonic wave would represent a state in which we know *nothing* about the position of the particle; this observable is "infinitely

[9]Eq. (4.16) is the general form of a complex traveling wave that propagates in the $+x$ direction. The corresponding form for a wave propagating in the $-x$ direction is

$$\Psi(x,t) = A\,e^{-i(kx+\omega t)}.$$

You may be familiar with such waves from your study of electricity and magnetism, where they are used to represent electric and magnetic fields. It is important to remember, however, that complex electromagnetic waves are just problem-solving tools; we ultimately must take the real parts of such functions, because only real electromagnetic waves exist in the laboratory. For more on complex functions in electromagnetic theory, see Chap. 8 of *Introduction to Electrodynamics* by David J. Griffiths (Englewood Cliffs, N. J.: Prentice Hall, 1981).

indeterminate" ($\Delta x = \infty$). So in no sense is a such a state spatially localized. Indeed, it's hard even intuitively to see how we could associate such a wave with a particle.

Question 4–4

Is the infinite position uncertainty for the complex harmonic wave consistent with the Heisenberg Uncertainty Principle? If so, why? If not, why not?

The lack of spatial localization shows up in the normalization integral for the complex harmonic wave, *which is infinite*:

$$\int_{-\infty}^{\infty} \Psi^*(x,t)\Psi(x,t)dx$$

$$= |A|^2 \int_{-\infty}^{\infty} dx = \infty. \qquad \text{[for a complex harmonic wave]}. \qquad (4.20)$$

We know from § 3.5 that we cannot normalize such a function and hence must reject it as physically inadmissible to quantum physics.

Let's review. We argued that the free-particle wave function we seek must satisfy three innocent-looking properties: it must be spatially-localized, complex, and consistent with the homogeneity requirement. Our first guess, the real harmonic wave, violated all of these conditions. Our second, the complex harmonic wave, violated only the requirement of spatial localization. But that's not good enough.

Our explorations have not led to the answer—but they have led to valuable clues. The complex harmonic function is so nearly the wave function we seek that it makes sense at this point to try to *construct* a spatially-localized function by adding together a bunch of these harmonic waves with different wave numbers and frequencies. In the next section, we'll do just that: we're going to build a wave packet.

4.2 WAVE PACKET CONSTRUCTION FOR FUN AND PROFIT

Go back to the functions graphed in Fig. 3.1. These are "snapshots" of physically admissible wave functions (taken at times when they happen to be real). Each wave function is characterized by a *single dominant finite region of enhanced position probability*—that is the distinctive feature of spatial localization. The probability density for each of these functions defines a "high probability" region of space, a realm of possible positions at which the particle is likely to be found. Our goal is to construct such a function.

This point of view emphasizes why a single harmonic wave fails: such a wave is monochromatic (*i.e.*, characterized by a single frequency), but *regions of enhanced amplitude arise from the interference of harmonic waves with differing wave numbers and frequencies*. This fact will be our guiding principle as we set out to build a free-particle wave function. And our tool will be superposition.[10]

It's not hard to see how to use superposition to build a wave packet. In a superposition of monochromatic harmonic waves, two kinds of interference happen. *Constructive interference* occurs at positions where the interfering waves are *in phase*, leading to an

[10]In this section, I'll assume that you're familiar with the mathematical and physical properties of waves at roughly the sophomore undergraduate level. If you need a refresher, try one of the Selected Readings at the end of this chapter.

enhanced amplitude of the total wave. And *destructive interference* occurs where the interfering waves are *out of phase*, leading to a diminished amplitude. We can use these interference phenomena to construct a spatially localized function by forming a linear combination of complex harmonic waves with varying frequencies and amplitudes. Such a linear combination is called a **wave packet**.

Applying these facts about waves to the task of § 4.1, constructing a function that satisfies the three criteria discussed there, we conclude that what we need is a complex wave packet with a single region of enhanced amplitude:

$$\boxed{\text{free particle state function} = \text{a complex wave packet}} \tag{4.21}$$

A Digression: The Pure-Momentum Function

Before we look at such packets we must get out of the way a mundane but important matter of notation. In spite of their defects, plane harmonic waves such as (4.16) are widely used in quantum mechanics, where they are usually written in terms of momentum and energy, as

$$\Psi_p(x,t) = A\, e^{i(px-Et)/\hbar} \tag{4.22}$$

Such a function is called, sensibly, a **pure momentum function**. Notice that I've appended a subscript p to $\Psi(x,t)$ to remind us what value of momentum we're talking about. Similarly, when discussing the complex harmonic wave written in terms of the wave number k [Eq. (4.16)], I'll sometimes affix to Ψ a subscript k.

Aside: The Mathematics of Waves—Phase Velocity. The most important property of a harmonic wave is its propagation velocity, which is called **phase velocity**. The phase velocity is defined as the velocity of a point of constant phase.[11] For a one-dimensional wave the phase velocity is

$$v_{\text{ph}} \equiv \left.\frac{\partial x}{\partial t}\right|_{\Psi=\text{constant}} \qquad \text{phase velocity} \tag{4.23}$$

For example, the phase velocity of a complex harmonic wave is [see Exercise 4.1]

$$v_{\text{ph}} = \lambda\nu = \frac{\omega}{k} = \frac{E}{p}. \tag{4.24}$$

So far I have left one thing in the pure momentum function (4.22) unspecified: the amplitude A. We're at liberty to choose this constant to fit the physical conditions of our problem—and when we get to collision problems (Chap. 8) we'll take advantage of this flexibility. But there is a conventional choice for A that exploits a nifty property of complex plane waves.

If the problem we're solving doesn't suggest a value of A, a good choice for the complex harmonic wave $\Psi_k(x,t)$ is $A = 1/\sqrt{2\pi}$. The resulting function

[11] If you are a little bit rusty on the meaning of these quantities *please review*. You'll find useful information about phase velocities in such sources as Chap. 7 of *Vibrations and Waves* by A. P. French (New York: Norton,1966) and in *Waves* by Frank S. Crawford, Jr. (New York: McGraw-Hill, 1968).

$$\Psi_k(x,t) = \frac{1}{\sqrt{2\pi}} e^{i(kx-\omega t)} \tag{4.25}$$

is, for reasons I'll explain in a moment, called a **normalized plane wave**.

For the pure momentum function $\Psi_p(x,t)$, the optimum choice is $A = 1/\sqrt{2\pi\hbar}$. The so-called **normalized pure momentum function** is

$$\boxed{\Psi_p(x,t) = \frac{1}{\sqrt{2\pi\hbar}} e^{i(px-Et)/\hbar}} \tag{4.26}$$

From here on I'll use these choices of A unless there is a good reason to do otherwise.

You'll discover why these choices are desirable as you work with the pure momentum function (see especially § 4.7). Here's a preview: they enable us to take advantage of the fact that complex exponentials form a representation of the *Dirac Delta function*, *viz.*,[12]

$$\frac{1}{2\pi}\int_{-\infty}^{\infty} e^{-ik'x}\, e^{ikx}\, dx = \delta(k - k'). \tag{4.27}$$

You're probably wondering why I refer to these as "normalized" functions, when we know that a complex harmonic wave can't be normalized. This usage is, indeed, inconsistent. Historically, it entered colloquial parlance (at least among mathematicians) because two "normalized plane waves" satisfy a kind of "pseudo-normalization integral,"

$$\int_{-\infty}^{\infty} \Psi_k^*(x,0)\Psi_{k'}(x,0)\, dx = \delta(k - k'). \qquad \text{Dirac Delta function normalization} \tag{4.28}$$

The General Form of a Wave Packet

Equation (4.21) gives us our marching orders: to construct a free-particle wave function, we must add together (superpose) complex plane waves with different wave numbers k. Your first thought might be to try a linear combination of a *finite* number of plane waves with discrete wave numbers k_j and amplitudes A_j, such as

$$\Psi(x,t) = \frac{1}{\sqrt{2\pi}} \sum_{j=1}^{N} A_j\, e^{i(k_j x - \omega_j t)}.$$

This reasonable suggestion runs aground on a basic property of waves: *The superposition of a finite number of plane waves with discrete wave numbers has an infinite number of comparable regions of constructive interference. Each of these regions is separated from its neighbor by a region of destructive interference.* This property of wave packets is illustrated in Fig. 4.4 and Problem 4.1.

[12]You may or may not have run across the Dirac Delta function $\delta(x)$ in your studies. Actually, it's not a function at all; it's what mathematicians call a *functional*. You can get an idea what $\delta(x)$ is by envisioning a "spike" that is zero everywhere except the point where its argument is zero, $x = 0$. So in (4.28), $\delta(k - k')$ is zero everywhere except at at $k = k'$, where it's infinite. You can find a brief overview of the Dirac Delta function and its properties in Appendix L. For more information, I recommend the introductory treatments found on pp. 53–55 of *Quantum Physics* by R. G. Winter (Belmont, Ca.: Wadsworth, 1979), and in Appendix 2 of *Quantum Mechanics*, Volume I by Claude Cohen-Tannoudji, Bernard Diu, and Franck Laloë (New York: Wiley-Interscience, 1977).

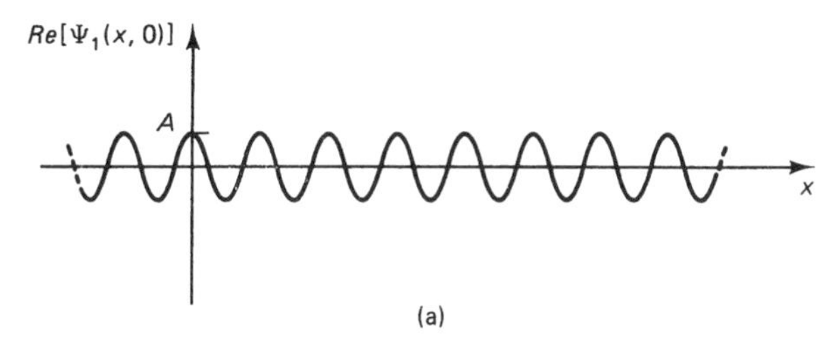

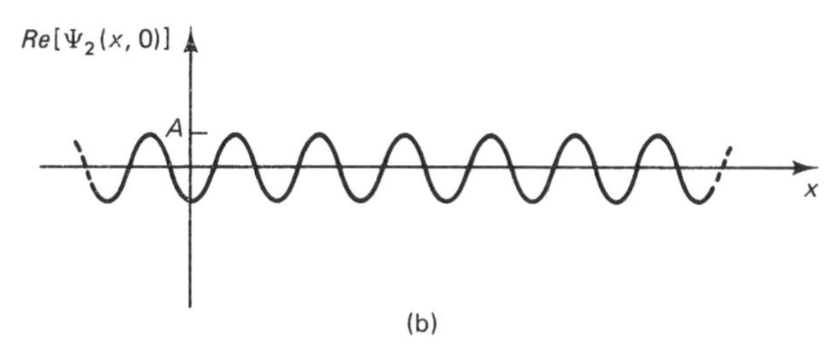

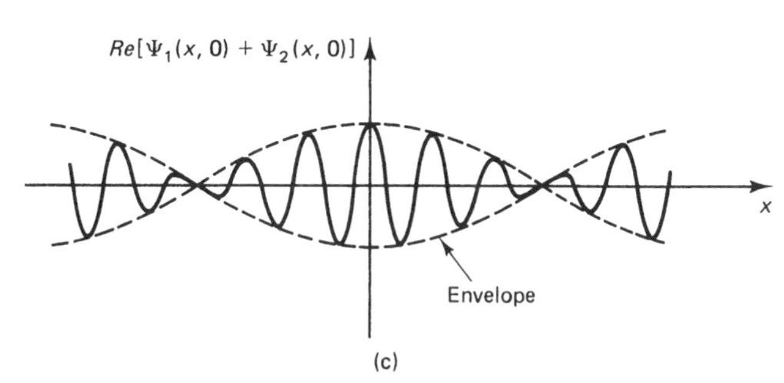

Figure 4.4 A simple (and unsuitable) wave packet constructed by superposing a mere *two* plane harmonic waves. Each wave has equal amplitude, $A(k_1) = A(k_2) = A$, but a different wave number: $k_1 = 12\pi,\ k_2 = 10\pi$. (a) The first wave, $\Psi_1(x,0)$. (b) The other wave, $\Psi_2(x,0)$. (c) Their superposition, $\Psi(x,0) = \Psi_1(x,0) + \Psi_2(x,0)$. The spread of wave numbers for the plane waves in $\Psi(x,0)$ is $\Delta k = 2\pi$. (In this figure, only the real parts of the three waves are shown.)

But we want a packet with one *dominant* region of enhanced amplitude—with a single biggest bump. [Additional regions of secondary amplitude (smaller bumps) are okay.] The solution is to use an infinite number of plane waves:

> ***Rule***
>
> To construct a wave packet with a single dominant region of enhanced amplitude we must superpose an infinite number of plane wave functions with infinitesimally differing wave numbers.

Since k is a continuous variable, our wave packet will have the form of an integral over functions of the form $e^{i(kx-\omega t)}$. To control the mixture of complex harmonic waves in the superposition, we'll assign each complex plane wave an "amplitude" $A(k)$. Using the normalized plane-wave functions (4.25), we write the **general form for a wave packet** as

$$\boxed{\begin{aligned}\Psi(x,t) &= \frac{1}{\sqrt{2\pi}}\int_{-\infty}^{\infty} A(k)\, e^{i(kx-\omega t)} dk \\ &= \int_{-\infty}^{\infty} A(k)\, \Psi_k(x,t)\, dk\end{aligned}} \qquad \text{wave packet} \qquad (4.29)$$

At present, the **amplitude function** $A(k)$ is arbitrary; in practice, we define this function to give $\Psi(x,t)$ physical properties that correspond to the quantum state it represents.[13]

Before going any further, I want to emphasize an extremely important point. *Equation (4.29) is applicable to many different (one-dimensional) quantum systems; it is not limited to the free particle.* In fact, most of the results of this chapter are not limited to the free particle. It is crucial to remember that in constructing wave packets we specify a particular system by our choice of the dispersion relation $\omega = \omega(k)$. For the free particle, the appropriate relation is (4.10*b*), $\omega(k) = \hbar k^2/2m$. To apply the general form to other systems, we just select the appropriate dispersion relation. Please don't forget this.

Aside: The Mathematician's View. A mathematician would probably interpret Eq. (4.29) as an expansion of $\Psi(x,t)$ in a "complete set" of plane wave functions $\Psi_k(x,t)$, with $A(k)$ the *expansion coefficient* corresponding to the plane wave with wave number k. [The amplitude function $A(k)$ is also sometimes called the **spectral distribution function**.] In Chap. 12 we'll discover the (considerable) advantages to adopting this more formal perspective on superpositions of quantum waves.

Do you understand how the superposition (4.29) forms a spatially localized function? The key is the *phases* $kx - \omega t$ of the constituent plane harmonic waves $\Psi_k(x,t)$. Consider a fixed time—say, $t = 0$. At this time all waves in the packet are *in phase* at one (and only one) point. This value of x is the *the point of total constructive interference*. At this point the probability density $P(x,0) = |\Psi(x,0)|^2$ attains its *global maximum*; for this reason, this point is called the **center of the wave packet**. We usually denote the center of the wave packet by x_0.[14]

Now, at any other $x \neq x_0$ the plane waves in the packet are *out of phase*, so some destructive interference occurs. As we move away from the center of the packet (*i.e.*, as $|x - x_0|$ increases), this interference becomes more and more destructive, until eventually the amplitude of the state function becomes negligibly small. This is why a wave packet is spatially localized [*c.f.*, Eq. (3.29)].[15]

When we construct a wave packet, we specify the *system* by the dispersion relation $\omega = \omega(k)$. To specify a particular *state* of this system, we choose the amplitude function $A(k)$. This function determines such properties of the packet as its "width"—the spread of values of x encompassed by $\Psi(x,0)$. Let's see how this works.

[13]There appears to be an inconsistency between the dimensions of the normalized plane wave (4.25), the normalized pure momentum state function (4.26), and the physically admissible state functions of Chap. 3. In one dimension, the latter must have units of inverse square root of length ($1/\sqrt{\text{L}}$), because the integral $\int_{-\infty}^{\infty} \Psi^*(x,t)\Psi(x,t)\,dx$ must be dimensionless [*c.f.*, Eq. (3.28*a*)]. Sure enough, neither $\Psi_k(x,t)$ nor $\Psi_p(x,t)$ has the correct dimensions. But this isn't a problem because these functions appear in wave packets where they are multiplied by amplitude functions that fix the dimensionality of the packet. For example, the units of $A(k)$ in (4.32) are (see § 4.3) $\sqrt{\text{L}}$. Similarly, in § 4.6 we'll construct a wave function out of pure momentum state functions with an amplitude function $\Phi(p)$ that, when multiplied by $\Psi_p(x,t)$, produces a packet with the correct dimensions.

[14]Be careful not to confuse the center of the wave packet with the average position, $\langle x \rangle (0)$; for some wave functions, these two points do coincide. But not for all. One other thing: the center of the packet at $t = 0$ need not be the origin of coordinates; we can construct a wave packet with its center at $x_0 \neq 0$ (see § 4.6).

[15]The argument I have sketched here is formally known as the *stationary phase condition*. This condition determines where the point of total constructive interference occurs for a superposition of plane waves. You can find a good introductory discussion of the stationary phase condition in § 2.5 of *Introduction to the Quantum Theory*, 2nd ed. by David Park (New York: McGraw-Hill, 1974) and more advanced discussions on p. 51 of *Quantum Mechanics*, Volume I by Albert Messiah (New York: Wiley, 1966) and on p. 19 of *Quantum Mechanics*, Volume I by Kurt Gottfried (Reading, Mass.: Benjamin/Cummings, 1966).

4.3 THE AMPLITUDE FUNCTION

The amplitude function $A(k)$ governs the mixture of plane harmonic waves in a particular wave packet; in this sense $A(k)$ distinguishes one wave packet from another. In practice, we usually determine the amplitude function from the value of the state function at $t = 0$ (§ 4.4). But in this section I'll take the amplitude function as given, because I want to show you—by example—how a superposition of an infinite number of complex harmonic waves with a particular amplitude function produces a wave packet with a single dominant region of enhanced position probability. We'll also examine the subtle relationship between the evolution of a wave packet and the motion of a classical particle and will peek into a laboratory where wave packets are used in collision experiments.

Figure 4.5 shows two typical amplitude functions. The function in Fig. 4.5a is peaked at $k = k_0$, so the dominant plane wave in the packet formed when we insert this function into the general form (4.29) is $\Psi_{k_0}(x,t)$. Plane wave functions $\Psi_k(x,t)$ with wave numbers far from k_0 make negligible contributions to the total packet $\Psi(x,t)$.

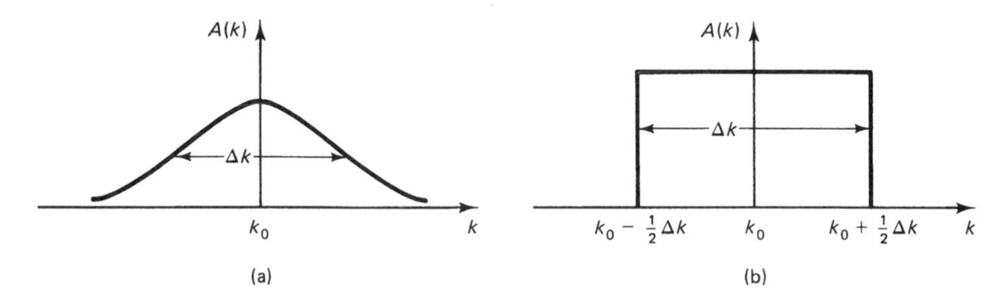

Figure 4.5 Two amplitude functions. Each function is characterized by its center k_0 and width Δk. The function in (a) is a Gaussian and will be discussed in § 4.5; the one in (b) is a square step and is used in Example 4.1. In (b), the value of $A(k_0) = 1/\sqrt{\Delta k}$ produces a properly normalized amplitude function (see § 4.4).

The function in Fig. 4.5b is more equitable: all harmonic waves with wave numbers in the range $k_0 - \frac{1}{2}\Delta k \le k \le k_0 + \frac{1}{2}\Delta k$ contribute equally to the resulting wave packet. In Example 4.1 we'll construct a wave packet from this amplitude function. Stay tuned.

Aside: On Complex Amplitude Functions. By the way, the sample amplitude functions in Fig. 4.5 are real. But $A(k)$ can be (and often is) complex. If so, we sometimes choose to write it in terms of a magnitude $|A(k)|$ and a phase $\alpha(k)$, *i.e.*,

$$A(k) = |A(k)|\, e^{i\alpha(k)}. \tag{4.30}$$

It's important to not lose sight of the physics amidst all the mathematics of wave packet construction. Try to think about these amplitude functions as controllers of the mixture of *pure momentum states* in the quantum state represented by the wave packet $\Psi(x,t)$. After all, each function $\Psi_k(x,t)$ in such a packet is associated (via the de Broglie relation) with a particular value of the momentum, $p = \hbar k$. This way of looking at the superposition $\int_{-\infty}^{\infty} A(k)\Psi_k(x,t)\,dx$ reveals that *the momentum of a quantum state represented by a wave packet is uncertain*, *i.e.*,

$$\Delta p > 0. \qquad \text{[for a wave packet]}. \tag{4.31}$$

This connection between the amplitude function $A(k)$ and the momentum uncertainty Δp is no accident, and we'll return to it in § 4.6 and 4.7.

Question 4–5

What amplitude function $A(k)$ would produce a packet with a well-defined momentum (*i.e.*, with $\Delta p = 0$)? What's wrong with the function $\Psi(x, t)$ you would get were you to insert this amplitude function into Eq. (4.29)?

Example 4.3. Construction of a Valid State Function

Let's see what sort of wave packet we get when we use the amplitude function of Fig. 4.5b,[16]

$$A(k) = \begin{cases} \dfrac{1}{\sqrt{\Delta k}} & k_0 - \frac{1}{2}\Delta k \le k \le k_0 + \frac{1}{2}\Delta k \\ 0 & \text{otherwise} \end{cases} \tag{4.32}$$

For simplicity, we'll consider only $t = 0$.

With this amplitude function, the wave packet (4.29) becomes a simple finite integral over the region of wave number where $A(k)$ is non-zero:

$$\Psi(x, 0) = \frac{1}{\sqrt{2\pi}} \int_{-\infty}^{\infty} A(k)\, e^{ikx}\, dk \tag{4.33a}$$

$$= \frac{1}{\sqrt{2\pi\, \Delta k}} \int_{k_0 - \frac{1}{2}\Delta k}^{k_0 + \frac{1}{2}\Delta k} e^{ikx}\, dk. \tag{4.33b}$$

Perusing our integral tables (Appendix I), we find the desired integral and easily derive

$$\Psi(x, 0) = \frac{1}{\sqrt{2\pi\, \Delta k}}\, e^{ik_0 x}\, \frac{2}{x} \sin\left(x \frac{\Delta k}{2}\right). \tag{4.34}$$

The corresponding probability density,

$$P(x, 0) = \frac{1}{2\pi\, \Delta k} \frac{4}{x^2} \sin^2\left(x \frac{\Delta k}{2}\right), \tag{4.35}$$

appears in Fig. 4.6.

Question 4–6

Evaluate the normalization integral for the state function (4.34).

The wave function (4.34) has an infinite number of local maxima. Still, this function doesn't violate the requirement of spatial localization, because the peak values of these humps decrease as x increases in either direction from $x = 0$. [This causes the *envelope* of the wave packet to die away to zero as $|x| \to \infty$.] Consequently the probability density defines only one *dominant* region of enhanced position probability, as desired.

Nevertheless, the position of a particle in an ensemble of particles represented by this state function is clearly uncertain: the function in Fig. 4.6 has a non-zero width ($\Delta x > 0$). Not surprisingly, the momentum is also uncertain ($\Delta p > 0$). This result follows from the Heisenberg Uncertainty Principle or, putting it mathematically, from the spread of wave numbers that characterizes the amplitude function (4.32).

Example 4.1 illustrates important features of the wave packets used in quantum physics. To summarize: *A wave packet is a normalizable function that is peaked about*

[16] I chose the value of the constant $A(k) = 1/\sqrt{\Delta k}$ so that the amplitude function would be normalized, *i.e.*, so that $\int_{-\infty}^{\infty} A^*(k) A(k)\, dk = 1$. That $A(k)$ satisfy this condition isn't mathematically essential, but it is convenient (see § 4.4), and it takes care of the dimensionality problem noted in footnote 13.

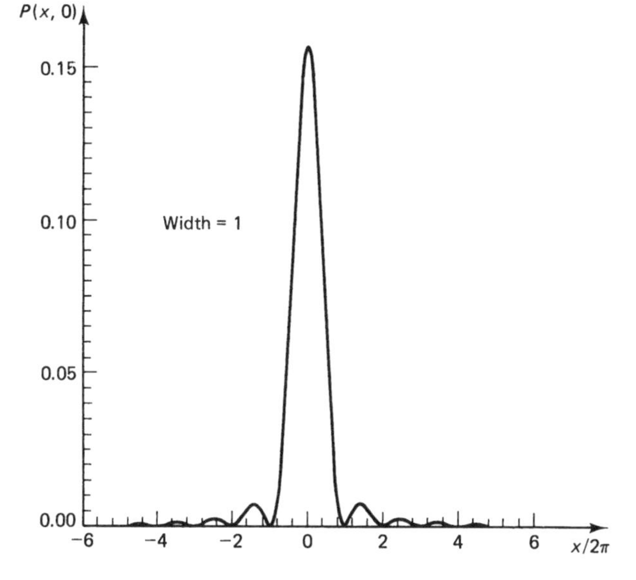

Figure 4.6 The position probability density $P(x, 0)$ for the wave packet erected in Example 4.3. This function is physically admissible because (among its other virtues) it is spatially localized.

a point of total constructive interference, its center. This function is characterized by a non-zero but finite spread in the variables x and k.

Wave Packet Motion and the Correspondence Principle

In Example 4.1 we considered a wave packet at a fixed time. The way such packets evolve as time goes by is the subject of Chap. 6 (especially § 6.7). But I want to show you one feature of wave packet motion now, because it illustrates the quaint and curious relationship between quantum mechanics and classical physics, a pervasive theme of this book.

Each of the monochromatic traveling waves that makes up a wave packet has a phase velocity $v_{\text{ph}} = \omega/k$. But the propagation velocity *of the packet* is, by definition, the velocity of its center. From your study of waves, you probably remember that this quantity is the **group velocity** and that (in one dimension) it is defined as

$$v_{\text{gr}} \equiv \left.\frac{d\omega}{dk}\right|_{k\,=\,k_0} . \qquad \text{group velocity} \tag{4.36}$$

In words: we can calculate v_{gr} by evaluating the first derivative of the dispersion relation $\omega = \omega(k)$ at the center of the amplitude function, $k = k_0$.[17]

We can easily figure out the group velocity of a free particle from the dispersion relation $\omega(k) = \hbar k^2/(2m)$. Applying the definition (4.36), we find that for the (general) wave packet (4.29),

$$v_{\text{gr}} = \frac{\hbar k_0}{m} = \frac{p_0}{m} \qquad \text{[for a free particle].} \tag{4.37a}$$

[17]With the passing of time, most wave packets disperse (spread). But this phenomenon is not described by v_{gr}, which provides information only about the propagation of the packet. Eventually, a packet may spread so much that it loses its definition; if this happens, the group velocity is no longer meaningful.

Does this result look familiar? It should: it's the *classical* velocity that a particle of mass m would have if its momentum was p_0:

$$v_{\text{classical}} = \left. \frac{dE}{dp} \right|_{p=p_0} = \frac{p_0}{m}. \qquad \text{[classical physics]} \tag{4.37b}$$

Isn't *that* interesting?

Be careful about how you interpret this little demonstration. It's very easy to be seduced by the similarity in Eqs. (4.37) into concluding that classical physics *does* pertain to microscopic particles. But this conclusion is not correct. What Eqs. (4.37) say is that a free particle wave packet with an amplitude function $A(k)$ peaked about $k = k_0$ evolves in such a way that *the position of its center*, the peak of the dominant region of enhanced probability, changes at a speed identical to that of a classical particle with the same mass and momentum $\hbar k_0$. It's important to note that (4.37*b*) pertains to a particle with a definite momentum, p_0, while the wave packet from which we calculated (4.37*a*) has an indeterminate momentum ($\Delta p > 0$).

This correspondence between the motion of a quantum-mechanical wave packet and the dynamics of a (related) classical particle is a manifestation of the *Correspondence Principle* (Chap. 2), according to which the predictions of quantum physics must reduce to those of classical physics "in the classical limit." "But where," I hear you wonder, "is 'the classical limit' in all this?"

Splendid question. Look, a classical particle has a well-defined position and momentum, so one way to take a wave packet to the classical limit is to reduce the position and momentum uncertainties to zero:

$$\Delta x \to 0 \quad \text{and} \quad \Delta p \to 0. \qquad \text{classical limit of a wave packet} \tag{4.38}$$

In this limit, the wave packet reduces to a probability amplitude that represents a state in which the position of the particle and its momentum are definite at each time [in fact, $\Psi(x,t)$ reduces to a Dirac Delta function]. What we showed in Eqs. (4.37) is that this point moves through space at the speed predicted by classical physics.[18]

This insight is but a glimmer of a far more profound result we'll discover in Chap. 11. There we'll explore the classical limit of the *laws of motion of quantum physics* and will find that in the classical limit the expectation value $\langle x \rangle (t)$ for a quantum state obeys Newton's Second Law, *i.e.*, the wave packet for a *macroscopic* particle follows that particle's classical trajectory.

Wave Packets in the Laboratory

Do you feel as if you're drowning, getting deeper and deeper into mathematics, leaving physics forever behind? Do you find it ominous that the next section is subtitled "A Mathematical Detour"? If so, don't fret. As we master all this mathematics we're really learning the language of quantum physics. But perhaps we should stop by the lab to see how an experimental physicist might produce a beam of free particles in a state that could be represented by a wave packet.

[18]The uncertainties in (4.38) are related by the Heisenberg Uncertainty Principle $\Delta x\, \Delta p \geq \hbar/2$. From this relation, we could argue that the classical limit is that in which $\hbar$ is negligible and can hence be approximated by zero. This limit is sometimes written $\hbar \to 0$—an expression that is probably perplexing, since $\hbar$ is not zero at all; it's a *constant* equal to 1.0546×10^{-34} J sec. If you want to read more about the correspondence principle, see the excellent discussion in Chap. 8 of *Introduction to Quantum Mechanics* by R. H. Dicke and J. P. Wittke (Reading, Mass.: Addison-Wesley, 1960).

Such a beam would most likely be used in a collision experiment like the one sketched in Fig. 4.7. The *source* emits a beam of nearly mono-energetic particles—here, electrons—with energy E_0 and momentum $p_0 = \sqrt{2m_0E_0}$, where m_0 is the rest mass of the projectile. (For an electron, $m_0 = m_e = 9.109 \times 10^{-31}$ kg.) The resulting beam travels a finite (but large) distance to the target, which might be a gas cell or a second beam directed at right angles to the electron beam. Scattered projectiles travel a finite (but large) distance to the detector, which we can move through space so as to sample the scattered particles at any angle.

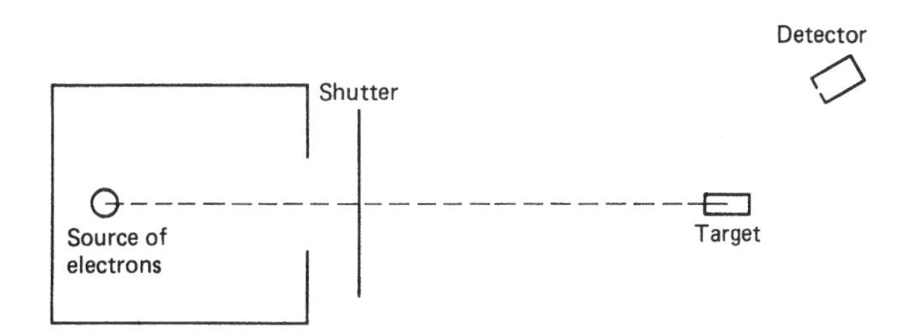

Figure 4.7 A (rather crude) schematic of a collision experiment. The target in this experiment is fixed. As described in the text, by judicious use of the shutter we can produce a *pulse* of electrons in a state represented by a wave packet with a known *width* Δx.

If the density of the beam is low enough that to a good approximation the particles don't interact and if the source "prepares" them so that all are in the same quantum state, then the beam is an ensemble (see Chap. 3). But what is the wave function that represents the particles in the beam?

The nature of some collision experiments permits us to *approximate* this wave function by a plane wave (as the limit of an extremely fat wave packet; see § 6.9 and 8.3). But in some experiments, the beam is "chopped"—rather like pepperoni.

We could generate a chopped beam of electrons by, for example, placing a movable shutter in front of the opening in the source, as in Fig. 4.7. By rapidly raising, then lowering the shutter, we generate a *pulse* of electrons of length L. (The value of L is determined by the energy of the electrons and by how long the shutter is open.) This pulse is an ensemble of free particles whose state is represented by a wave packet.

From the physical conditions in the laboratory, we can infer several features of this packet. For one thing, the position uncertainty of the particles in this beam is non-zero; in fact, Δx is roughly equal to the dimension of the beam: $\Delta x \approx L$. Consequently the momentum of the particles is not well-defined; rather, the state is characterized by a distribution of momenta that is peaked about p_0. According to the Heisenberg Uncertainty Principle, the momentum uncertainty is roughly $\Delta p \approx \hbar/(2L)$. All of these features conform to our understanding of the nature of a wave packet.

Aside: Wave Packets for Three-Dimensional Systems. In general, we must treat a wave packet in the lab as a three-dimensional function. The three-dimensional extension of the general form (4.29) is

$$\Psi(\mathbf{r}, t) = \frac{1}{\sqrt{(2\pi)^3}} \int A(\mathbf{k})\, e^{i(\mathbf{k}\cdot\mathbf{r}-\omega t)}\, d\mathbf{k}, \tag{4.39a}$$

where $d\mathbf{k}$ is a volume element in "momentum space" (*e.g.*, in rectangular coordinates, $d\mathbf{k} = dk_x\, dk_y\, dk_z$). The center of this packet, $\mathbf{r}_0(t)$, propagates at the group velocity, which is a vector determined by taking the gradient of the dispersion relation, *viz.*, $\omega = \omega(\mathbf{k})$ at $k = k_0$, *i.e.*

$$\mathbf{v}_{\rm gr} = \nabla_k \, \omega(\mathbf{k}) \bigg|_{k=k_0} . \tag{4.39b}$$

Fortunately, we usually need not consider explicitly two of the three dimensions of the packet's motion. This bit of legerdemain will be explained in Chap. 8.

The details of how to construct such a wave function are interesting, but discussing them would take us too far afield. But before we leave this section, I should note that in a real collision experiment, we can set the mean values and uncertainties of the position and momentum of the incident particles, to the precision allowed by nature and human fallibility. These characteristics *define* the wave packet that represents the state of the projectiles. And that is how we create and use wave packets in laboratory quantum physics.

*4.4 FOURIER ANALYSIS (A MATHEMATICAL DETOUR)

In § 4.2, we wrote down the general form (4.32) of a wave packet,

$$\Psi(x,t) = \frac{1}{\sqrt{2\pi}} \int_{-\infty}^{\infty} A(k) \, e^{i(kx-\omega t)} dk. \tag{4.40a}$$

At $t = 0$, this equation takes on a form that may be familiar:

$$\Psi(x,0) = \frac{1}{\sqrt{2\pi}} \int_{-\infty}^{\infty} A(k) \, e^{ikx} \, dk. \tag{4.40b}$$

Equation (4.40b) reveals that *the amplitude function $A(k)$ is the Fourier transform of the state function $\Psi(x,t)$ at $t = 0$.* That is, the amplitude function is related to the wave function at $t = 0$ by a Fourier integral.[19]

Fourier analysis—the generation and deconstruction of Fourier series and integrals—is the mathematical method that underlies the construction of wave packets by superposition. Fourier analysis is a vital element of our quantum mechanics' tool kit, for we can use it to zip back and forth between the amplitude function and the corresponding state function. Moreover, the study of Fourier analysis leads to a deeper understanding of waves *per se*. When this method is applied to *probability waves—i.e.*, to state functions—it leads to important, sometimes astonishing physical principles (see § 4.6). For all these reasons, we're going to take a short sojourn into Fourier analysis.

Mathematicians commonly use Fourier analysis to rip functions apart, representing them as sums or integrals of simple component functions, each of which is characterized by a single frequency. This method can be applied to any function $f(x)$ that is *piecewise continuous—i.e.*, that has at most a finite number of finite discontinuities. Wave functions, which must be continuous (see § 3.5), satisfy this condition and so are prime candidates for Fourier analysis. Moreover, Fourier analysis is a preview of an extremely powerful mathematical technique: the method of eigenfunction expansion (Chap. 12).

Whether we represent $f(x)$ via a Fourier series or a Fourier integral depends on whether or not this function is *periodic*. Any function that repeats itself is said to be

[19] If you already figured this out and don't feel the need of a review of Fourier series and integrals, skip to § 4.5. (You may, however, want to read the last subsection—on the definition of the width of a function.) If you've never studied Fourier analysis, you'll find a list of good references at the end of this chapter.

periodic. More precisely, if there exists a finite number L such that $f(x+L) = f(x)$, then $f(x)$ is **periodic** with **period** L. Perhaps the most familiar periodic functions are $\sin(2\pi x/L)$, $\cos(2\pi x/L)$, and $\exp(2\pi i x/L)$. We can write any function that is periodic (or that is defined on a *finite* interval) as a *Fourier series*. But if $f(x)$ is non-periodic or is defined on the *infinite* interval from $-\infty$ to $+\infty$, we must use a *Fourier integral*.

Fourier Series

Fourier series are not mere mathematical devices; they can be generated in the laboratory. For example, a *spectrometer* decomposes an electromagnetic wave into spectral lines, each of which has a different frequency and amplitude (intensity). Thus, a spectrometer decomposes a periodic function in a fashion analogous to the Fourier series. But our present concern is with the mathematics.

Suppose we want to write a periodic, piecewise continuous function $f(x)$ as a series of simple functions. Let L denote the period of $f(x)$, and choose as the origin of coordinates the midpoint of the interval defined by this period, $-L/2 \le x \le L/2$. (This choice allows us to exploit the symmetry (if any) of $f(x)$.)

The particular functions that appear in the Fourier series expansion of a periodic function are sines and cosines. Letting a_n and b_n denote (real) expansion coefficients, we write this series as[20]

$$f(x) = a_0 + \sum_{n=1}^{\infty} \left[a_n \cos\left(2\pi n \frac{x}{L}\right) + b_n \sin\left(2\pi n \frac{x}{L}\right) \right]. \qquad \text{real Fourier series} \qquad (4.41)$$

We calculate the coefficients in (4.41) from the function $f(x)$ as

$$a_0 = \frac{1}{L} \int_{-L/2}^{L/2} f(x)\, dx \qquad (4.42a)$$

$$a_n = \frac{2}{L} \int_{-L/2}^{L/2} f(x) \cos\left(2\pi n \frac{x}{L}\right) dx \qquad (n = 1, 2, \ldots) \qquad (4.42b)$$

$$b_n = \frac{2}{L} \int_{-L/2}^{L/2} f(x) \sin\left(2\pi n \frac{x}{L}\right) dx \qquad (n = 1, 2, \ldots) \qquad (4.42c)$$

Notice that the summation in (4.41) contains *an infinite number of terms*. In practice we retain only a finite number of these terms—this approximation is called **truncation**. Truncation is viable, of course, only if the sum converges to whatever accuracy we want *before* we chop it off.

A Mathematical Aside: On Truncation. Truncation is not as extreme an act as it may seem. If $f(x)$ is normalizable, then the expansion coefficients in (4.41) decrease in magnitude with increasing n, *i.e.*,

$$|a_n| \to 0 \text{ and } |b_n| \to 0 \text{ as } n \to \infty.$$

Under these conditions, which are satisfied by physically admissible wave functions, the sum (4.41) can be truncated with impunity at some finite maximum value $n_{\max}$ of the index n. Of course, a certain amount of trial and error is required to determine the value of $n_{\max}$ that is required for the desired accuracy.

[20] *Beware:* the precise form of the Fourier series (and integral) representations of a function depend on the interval of periodicity; many authors choose the limits of this interval to be $-L$ and $+L$.

Truncation is not the only simplification one stumbles across in working with Fourier series. If $f(x)$ is particularly simple, all but a small, finite number of terms in (4.41) may be zero. And you should always look for coefficients that are zero before you start grinding out integrals. This advice leads to the question: how can we spot such coefficients without evaluating them? That is, what (other than an accident) could make an expansion coefficient zero?

The Power of Parity

In Example 3.4, we saw that parity arguments can render integral evaluation trivial if the integrand is an odd function of the variable of integration. The same thing is true of the integrals in the expansion coefficients (4.42): they call for integration over a symmetric integral (from $-L/2$ to $+L/2$) of an integrand that involves trigonometric functions with well-known parity properties:[21]

$$\sin(-x) = -\sin x \qquad \text{(odd)} \tag{4.43a}$$

$$\cos(-x) = +\cos x \qquad \text{(even)} \tag{4.43b}$$

From Eqs. (4.43) it follows that if $f(x)$ has definite parity, then either the integrand in a_n or that in b_n will be odd:

Rule

If $f(x)$ is even or odd, then half of the expansion coefficients in its Fourier series are zero.

If, for example, $f(x)$ is *odd* [*i.e.*, $f(-x) = -f(x)$], then all the coefficients a_n are zero, and we're left with a series involving only the "sine coefficients" b_n. Such a series is called a *Fourier sine series.* The constant coefficient a_0 in (4.41) is zero for any function of definite parity. So if $f(x)$ is such a function, there are two possibilities:

$$f(x) \text{ is odd } \quad [f(-x) = -f(x)] \Longrightarrow \begin{cases} a_n = 0 \qquad (n = 1, 2, \ldots) \\ f(x) = \displaystyle\sum_{n=1}^{\infty} b_n \sin\left(2\pi n \frac{x}{L}\right) \end{cases} \qquad \text{Fourier sine series} \tag{4.44a}$$

$$f(x) \text{ is even } \quad [f(-x) = +f(x)] \Longrightarrow \begin{cases} b_n = 0 \qquad (n = 1, 2, \ldots) \\ f(x) = \displaystyle\sum_{n=1}^{\infty} a_n \cos\left(2\pi n \frac{x}{L}\right) \end{cases} \qquad \text{Fourier cosine series} \tag{4.44b}$$

[21] For more on parity in Fourier analysis, see § 5.10 of *Advanced Calculus for Applications* by F. B. Hildebrand (Englewood Cliffs, N.J.: Prentice Hall, 1962).

Aside: Functions That Don't Have Definite Parity. The Fourier sine and cosine expansions may come in handy even if $f(x)$ does not have definite parity, because we can easily write *any* function as the sum of an even function and an odd function. Suppose $f(x)$ is neither even nor odd [*i.e.*, $f(-x) \neq \pm f(x)$]. Let's define two new functions $g(x)$ and $h(x)$ as

$$\begin{aligned} g(x) &= \tfrac{1}{2}\left[f(x) + f(-x)\right] \\ h(x) &= \tfrac{1}{2}\left[f(x) - f(-x)\right]. \end{aligned}$$

By definition, $g(x)$ is even and $h(x)$ is odd. Using these definitions, we can write the function $f(x)$ as [see the next Question]

$$f(x) = \underbrace{g(x)}_{\text{even}} + \underbrace{h(x)}_{\text{odd}}.$$

So for the rest of this section, I'll assume that $f(x)$ is either even or odd.

Question 4–7

Write down the functions $g(x)$ and $h(x)$ for

$$f(x) = \frac{1}{2}e^{-x^2} + 7x$$

and prove that each has definite parity.

Fourier and His Series: A Historical-Mathematical Aside. As a brief respite from all this math, let's look back in time. The method of representing functions by series (or integrals) of sine and cosine functions was derived by a Frenchman named Jean Baptiste Joseph Fourier (1768–1830). When young, Fourier dreamed of serving in the French army. (Had these dreams not been dashed, Fourier might not have become much older.) But he was too adept at mathematics for the French government to let him become cannon fodder, so Fourier had to settle for a career as a world-renowned mathematician. In the early 1800's, Fourier set out to find a way to represent functions in terms of sines and cosines.

To develop a series representation of a periodic function $f(x)$, Fourier used a rather standard optimization technique. His first step was to write the function he wanted to represent as a *finite* series of the form

$$f_K(x) = a_0 + \sum_{n=1}^{K}\left[a_n \cos\left(2\pi n\frac{x}{L}\right) + b_n \sin\left(2\pi n\frac{x}{L}\right)\right].$$

The subscript K in this expansion indicates the number of terms included in it. Fourier then asked: what choices for the coefficients a_n and b_n lead to a function $f_K(x)$ that *most closely approximates* $f(x)$—*i.e.*, to the "best" such representation? His answer: the function with coefficients that *minimize the mean square deviation*,[22]

$$\frac{1}{2L}\int_{-L/2}^{+L/2}\left[f(x) - f_K(x)\right]^2\,dx.$$

Fourier easily implemented that minimization via the *orthogonality relations* for the sine and cosine:

$$\int_{-L/2}^{L/2} \cos\left(2\pi n\frac{x}{L}\right)\sin\left(2\pi m\frac{x}{L}\right)dx = 0 \qquad (4.45a)$$

[22]We have met this sort of beast before (in § 3.3); it is just a measure of the extent to which the series representation $f_K(x)$ differs from $f(x)$ over the interval $[-L/2, +L/2]$.

$$\int_{-L/2}^{L/2} \cos\left(2\pi n \frac{x}{L}\right) \cos\left(2\pi m \frac{x}{L}\right) dx = \int_{-L/2}^{L/2} \sin\left(2\pi n \frac{x}{L}\right) \sin\left(2\pi m \frac{x}{L}\right) dx$$

$$= \begin{cases} 0, & \text{if } m \neq n; \\ 1, & \text{if } m = n. \end{cases} \tag{4.45b}$$

What he found—the optimum expansion coefficients—are those of Eqs. (4.42).

Following this signal success, Fourier went on to write a classic study of heat (which he believed to be vital to man's health) and then died tumbling down a flight of stairs. But his theory lives on.

The Fourier series is by no means the only way to expand a function. (You probably know about the most common alternative, a Taylor series.) But its value derives in part from its wide applicability. Moreover, evaluation of the Fourier coefficients is usually comparatively easy. Because the integrals (4.42) involve sines and cosines, they can often be dug out of integral tables. Even if $f(x)$ is sufficiently nasty that the required integrals are too complicated to evaluate analytically, we can always dump them on the nearest computer. (Computers love sines and cosines.)

The Complex Fourier Series

If $f(x)$ does not have a definite parity, we can expand it in a **complex Fourier series.** To derive this variant on the Fourier series (4.41), we just combine the coefficients a_n and b_n so as to introduce the complex exponential function $e^{i2\pi nx/L}$; *viz.*,

$$f(x) = \sum_{n=-\infty}^{\infty} c_n \, e^{i2\pi nx/L}. \qquad \text{Complex Fourier Series} \tag{4.46}$$

Note carefully that in the complex Fourier series (4.46) the summation runs from $-\infty$ to ∞. The expansion coefficients c_n for the complex Fourier series are

$$c_n = \frac{1}{L} \int_{-L/2}^{L/2} f(x) \, e^{-i2\pi nx/L} \, dx. \tag{4.47}$$

As you might expect, in general these coefficients are complex.

Question 4–8

> **Derive** Eqs. (4.46) and (4.47) and thereby **determine** the relationship of the coefficients c_n of the complex Fourier series of a function to the coefficients a_n and b_n of the corresponding real series.

Fourier Integrals

I have emphasized that only a *periodic* function (or one defined on a finite interval) can be expanded in a Fourier series. To represent a *non-periodic* function in terms of sines and cosines, we resort to an integral: *any normalizable function can be expanded in an infinite number of sine and cosine functions that have infinitesimally differing arguments.* Such an expansion is called a **Fourier integral.**[23]

The Fourier integral representation of a non-periodic function looks a little like the complex Fourier series representation (4.46) of a periodic function. Its general form is

[23] A function $f(x)$ can be represented by a Fourier integral provided the integral $\int_{-\infty}^{\infty} |f(x)| \, dx$ exists. All wave functions satisfy this condition, for they are normalizable.

$$f(x) = \frac{1}{\sqrt{2\pi}} \int_{-\infty}^{\infty} g(k)\, e^{ikx}\, dk$$ inverse Fourier transform (Fourier integral) (4.48)

The function $g(k)$ in (4.48) plays the role analogous to that of the expansion coefficients c_n in the complex series (4.46). The relationship of $g(k)$ to $f(x)$ is more clearly exposed by the inverse of (4.48),

$$g(k) = \frac{1}{\sqrt{2\pi}} \int_{-\infty}^{\infty} f(x)\, e^{-ikx}\, dx$$ Fourier transform (4.49)

In mathematical parlance, $f(x)$ and $g(k)$ are said to be Fourier transforms of one another. More precisely, $g(k)$ is the **Fourier transform** of $f(x)$, and $f(x)$ is the **inverse Fourier transform** of $g(k)$. When convenient, I'll connote these relationships by the shorthand notation

$$g(k) = \mathcal{F}[f(x)] \qquad \text{and} \qquad f(x) = \mathcal{F}^{-1}[g(k)]. \tag{4.50a}$$

Our immediate application of Fourier transform theory is to the wave packet $\Psi(x,0)$ and its amplitude function $A(k)$. If we translate the relationship (4.40*b*) between these functions into our new notation, we have

$$A(k) = \mathcal{F}[\Psi(x,0)] \qquad \text{and} \qquad \Psi(x,0) = \mathcal{F}^{-1}[A(k)]. \tag{4.50b}$$

Many useful relationships follow from the intimate relationship between $f(x)$ and $g(k)$. For our purposes, the most important is the so-called **Bessel-Parseval relationship**:

$$\int_{-\infty}^{\infty} |f(x)|^2\, dx = \int_{-\infty}^{\infty} |g(k)|^2\, dk.$$ Bessel-Parseval relationship (4.51)

[Don't forget the absolute value bars in (4.51); they are essential if $f(x)$ is complex.] Among the many applications of this relationship in quantum mechanics is the *normalization of the amplitude function* (§ 4.7).

Question 4–9

Use the properties of the Dirac Delta function to prove the Bessel-Parseval relationship (4.51).

How Wide is My Wave Function? (A Short Discourse on Widths)

Mathematicians, scientists, and engineers often talk about the "width" of a function. Unfortunately, these folk have been unable to agree on precisely what they mean by "the width of a function," so in your reading you may come across diverse definitions of this quantity.

But the idea is simple and the name appropriate: The "width" w_x of a function $f(x)$ is just a quantitative measure of its spread, its spatial extent. We have already seen (in several examples) that the width of any normalizable function is finite and non-zero: *i.e.*, $0 < w_x < \infty$. But beyond this observation, confusion reigns. In Fig. 4.8 you'll find some widely used definitions of w_x.

To avoid further confusion, I want to get my definition up front right now. *In this book I shall define the width of a function $f(x)$ to be the standard deviation of its position,* *as defined in § 3.3 [see Eq. (3.25)].* Hence the width of a state function $\Psi(x,t)$ is its position uncertainty Δx, as illustrated in Fig. 4.8c.

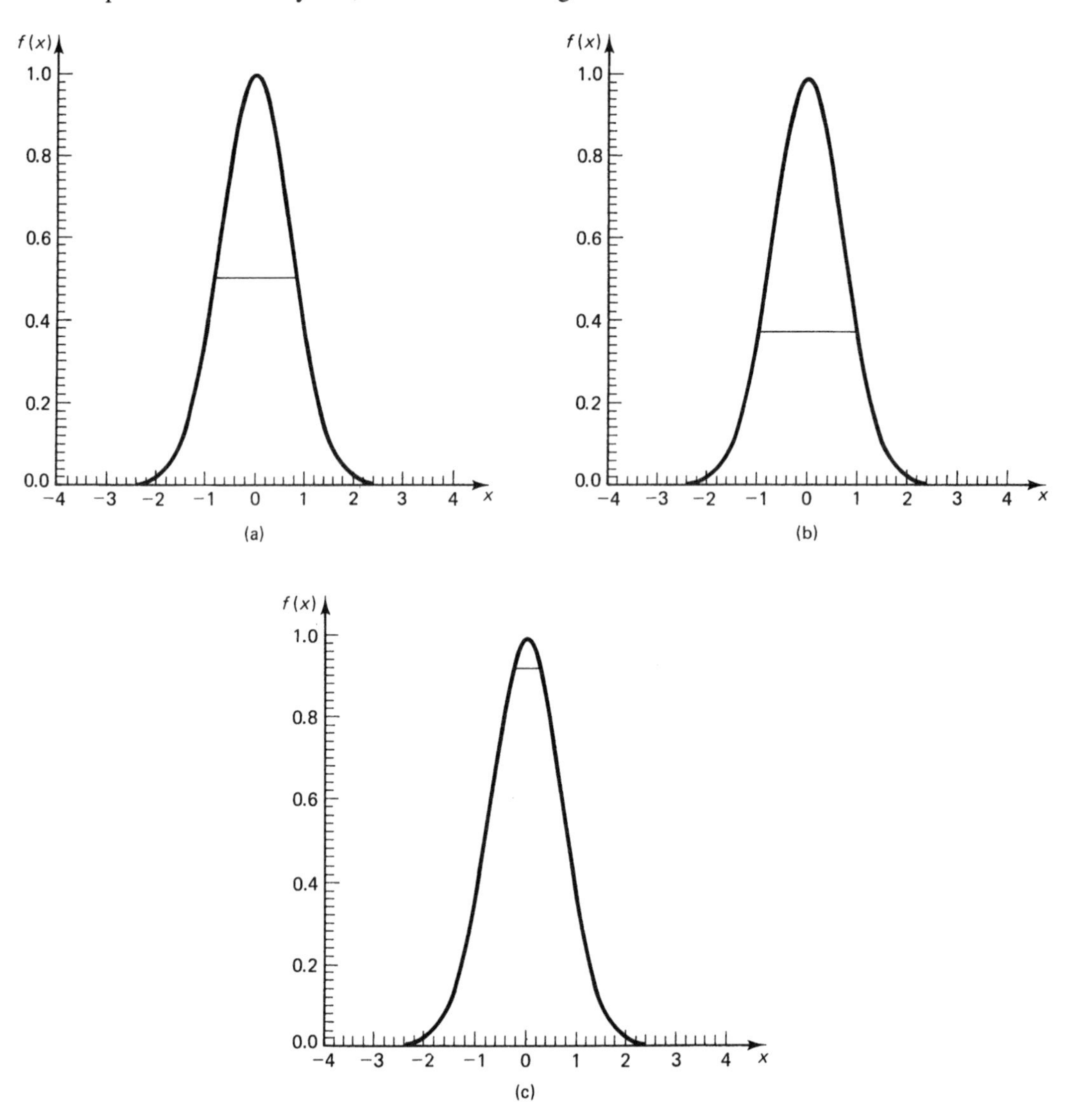

Figure 4.8 A cornucopia of definitions of the width w_x of a function $f(x)$. The function at hand is $f(x) = e^{-x^2}$. (a) The full-width-at-half-maximum, $w_x = 1.665$. (b) The extent of the function at the special point where $f(x) = [f(x)]_{\max}/e$. The function e^{-x^2} is equal to $1/e$ of its maximum value at $x = \pm 1$, so this definition gives $w_x = 1.736$. (c) My definition: the standard deviation of x; for this function, $w_x = \Delta x = 0.560$.

We can define a width for any function; for example, the width w_k of the amplitude function $A(k)$ is a measure of its extent in the variable k (see § 4.7). And, not surprisingly, if two functions are related, so may be their widths. For instance, there is an "inverse correlation" between the width w_x of a function $f(x)$ and the width w_k of its Fourier transform: the wider the function $f(x)$, the narrower its Fourier transform $g(k)$. From an analysis of the Fourier integral (4.48), mathematicians have shown that

$$f(x) = \mathcal{F}^{-1}[g(k)] \quad \Longrightarrow \quad w_x \, w_k \approx 1. \tag{4.52a}$$

Question 4–10

Explain briefly why Eq. (4.52*a*) is true. Base your argument on the amount of destructive interference in the wave packet as the distance from the point of *total constructive interference* increases.

Although Eq. (4.52*a*) is useful, it expresses this inverse relationship only qualitatively. But the *precise* value of the product $w_x \, w_k$ depends on the precise definition of w_x and w_k. With my definition of w_x and w_k as the standard deviations of position and wavenumber, this relation takes on the form

$$\boxed{w_x \, w_k = \Delta x \, \Delta k \geq \tfrac{1}{2}} \tag{4.52b}$$

In § 4.6, we'll translate this relationship into a principle (which you can probably guess now) that relates a critical physical property of a state function and its amplitude function.

4.5 AN EXTENDED EXAMPLE: THE GAUSSIAN WAVE PACKET UNMASKED

In Example 4.1 we used a known amplitude function $A(k)$ to construct a free-particle wave packet (at $t = 0$) as a superposition of complex harmonic waves. Mathematically speaking, we were *calculating the wave function as the inverse Fourier transform of the amplitude function*:

$$\Psi(x, 0) = \mathcal{F}^{-1}[A(k)] = \frac{1}{\sqrt{2\pi}} \int_{-\infty}^{\infty} A(k) \, e^{ikx} dk. \tag{4.53a}$$

In quantum physics, we usually want to go the other way, to *calculate the amplitude function as the Fourier transform of the wave function at the initial time* $t = 0$:

$$A(k) = \mathcal{F}[\Psi(x, 0)] = \frac{1}{\sqrt{2\pi}} \int_{-\infty}^{\infty} \Psi(x, 0) \, e^{-ikx} dx. \tag{4.53b}$$

In this section I want to illustrate this technique by examining an extremely important type of state function: the Gaussian function. In particular, we'll generate the amplitude function for a simple Gaussian function and then look at its width. What we discover when we do so will carry us back into the physics of the Heisenberg Uncertainty Principle (§ 4.6).

Example 4.4. The Amplitude Function for a Gaussian

The Gaussian function is a wave packet with a well-defined center and a single peak. Its amplitude function, which we're going to calculate, has the same properties. In § 6.7 we'll watch the time development of a Gaussian function; here we consider only $t = 0$. The Gaussian function contains one parameter, a real number L that governs its width.

The most general form of such a function has a center at $x_0 \neq 0$ and corresponds to an amplitude function that is centered at $k_0 \neq 0$:

$$\Psi(x, 0) = \left(\frac{1}{2\pi L^2}\right)^{1/4} e^{ik_0 x} \, e^{-[(x - x_0)/(2L)]^2}. \tag{4.54}$$

But for simplicity, I want to consider a Gaussian centered at $x_0 = 0$ with an amplitude function centered at $k_0 = 0$, *i.e.*,

$$\Psi(x, t=0) = \left(\frac{1}{2\pi L^2}\right)^{1/4} e^{-x^2/(4L^2)}. \tag{4.55}$$

In Fig. 4.9, you'll find two such wave packets (with different values of L). Each exhibits the characteristic shape of a Gaussian function: each has a single peak and decreases rather rapidly as x increases from zero. But because the Gaussian function decays *exponentially*, it never actually equals zero for finite x. (It is, nonetheless, normalizable.) You'll also find in Fig. 4.9 the probability density $|\Psi(x,0)|^2$ for a Gaussian. This figure illustrates one of the special properties of a Gaussian wave function: its probability is also a Gaussian function, one that has the same center but is narrower than the state function from which it is calculated.

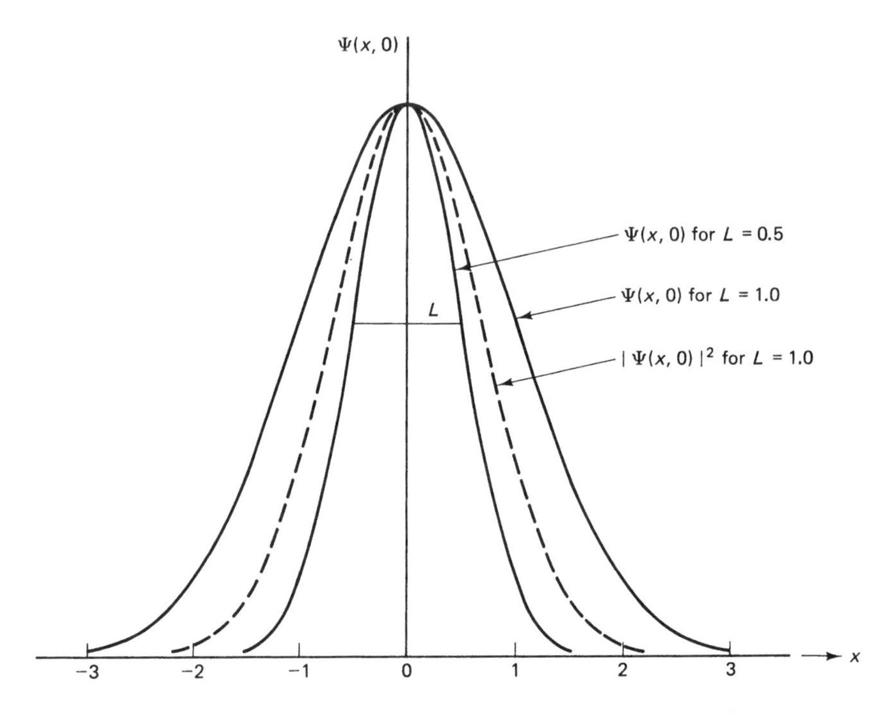

Figure 4.9 Two Gaussian wave packets of the form (4.55); the parameter L for these functions takes on the values $L = 0.5$ and $L = 1.0$ (solid curves). The corresponding probability density for $L = 1.0$ is shown as a dashed curve.

Question 4–11

Verify that the state function (4.55) is normalized.

To see precisely how individual plane harmonic waves $\Psi_k(x,t)$ combine to form a Gaussian wave packet, let's evaluate the amplitude function $A(k)$. We begin by substituting the state function (4.55) into the general expression (4.53*b*) for the amplitude function, *viz.*,

$$A(k) = \frac{1}{\sqrt{2\pi}}\left(\frac{1}{2\pi L^2}\right)^{1/4} \int_{-\infty}^{\infty} \exp\left[-\frac{1}{4L^2}x^2 - ikx\right] dx. \tag{4.56}$$

This rather nasty-looking integral is a standard (if perhaps unfamiliar) form:[24]

$$\int_{-\infty}^{\infty} e^{-\alpha x^2 - \beta x}\, dx = \sqrt{\frac{\pi}{\alpha}}\, e^{\beta^2/(4\alpha)}. \qquad (\alpha > 0) \tag{4.57}$$

This handy form is just what we need to conquer the integral in (4.56). Letting $\alpha = 1/(4L^2)$ and $\beta = ik$ in this expression, we obtain

$$A(k) = \left(\frac{2}{\pi}L^2\right)^{1/4} e^{-k^2 L^2}. \qquad \text{[for a Gaussian with } x_0 = 0\text{]}. \tag{4.58}$$

Compare carefully the *mathematical form* of the amplitude function (4.58) and the (initial) wave function (4.55). Lo and behold: *the Fourier transform of a Gaussian function of the variable x is a Gaussian of the variable k.* (Please don't generalize: this reciprocity is not a general property of functions and their Fourier transforms; it is a special feature of the Gaussian function.)

The Gaussian $\Psi(x, 0)$ and its amplitude function are graphed in Fig. 4.10. As advertised above, the center of $A(k)$—the value of the wave number at the peak of the amplitude function—is at $k_0 = 0$. Both the wave function and the amplitude function have a non-zero width, and from the (Fourier transform) relationship these functions bear to one another, it's not surprising that their widths are inversely related[25] [*c.f.*, Eqs. (4.52)]: $w_x\, w_k \approx 1$.

Normalization of the Amplitude Function

In Question 4–11 I asked you to verify that the Gaussian wave function is normalized. Happily, we need not explicitly normalize the amplitude function calculated from this wave function, because a property of Fourier transforms—the Bessel-Parseval relationship (4.51)—guarantees that *the Fourier transform of a normalized function is normalized*:

$$\int_{-\infty}^{\infty} |A(k)|^2\, dk = \int_{-\infty}^{\infty} |\Psi(x, 0)|^2\, dx = 1. \tag{4.59}$$

[24] Gaussian functions occur so often in quantum physics and chemistry that most of their mathematical properties have been derived and tabulated. You'll find enough information to work the problems in this book in Appendix J, which is based on the Appendix in *Elementary Quantum Mechanics* by David S. Saxon (San Francisco: Holden-Day, 1968). To find everything you could conceivably want to know about Gaussian functions (and more) get a copy of "The Gaussian Function in Calculations of Statistical Mechanics and Quantum Mechanics," which appears in *Methods of Computational Physics,* Volume 2, edited by B. Alder, S. Fernbach, and M. Rotenberg (New York: Academic Press, 1963).

[25] A word about widths of Gaussian functions: The conventional value of the width of a Gaussian function of the form (4.54) is $w_x = L/\sqrt{2}$. This differs from what my definition gives, $w_x = \Delta x = L$. The difference arises because many authors use L as the width not of the *wave function* but rather of the *probability density*. I further need to alert you that many authors define the Gaussian function in a slightly different fashion than I have, using $L/\sqrt{2}$ everywhere that I have used L. In a book that adopts this form, you'll see, instead of Eq. (4.54), the function

$$\Psi(x, t = 0) = \left(\frac{1}{\pi L^2}\right)^{1/4} e^{-x^2/(2L^2)}.$$

Both forms are OK—after all, it's the x-dependence of this function that identifies it as a Gaussian. I prefer (4.54) because it leads to simple expressions for the position and momentum uncertainties. But this multiplicity of conventional forms should cause no difficulty: you can regain our equations from those in a book that adopts this alternate form by replacing L everywhere by $L/\sqrt{2}$.

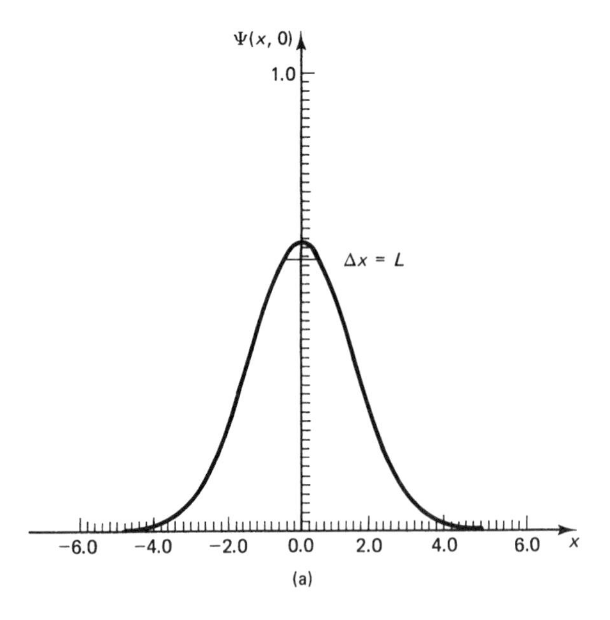

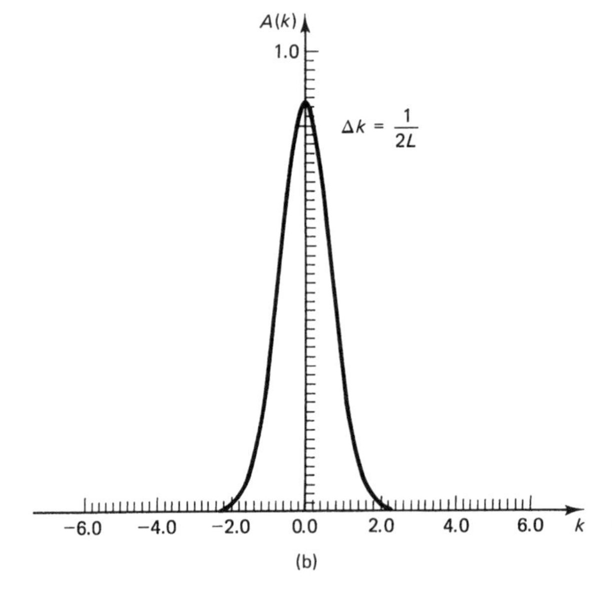

Figure 4.10 (a) A Gaussian wave packet and (b) its Fourier transform. Note that the two functions have the same *mathematical form* [see Eq. (4.58)]. Also note that the constant $1/(2L)$ plays the same role in $A(k)$ that L does in $\Psi(x, 0)$.

Of course, the Bessel-Parseval equality applies to any wave function and its amplitude function—not just to a Gaussian. And Eq. (4.59) comes in handy in problem solving, as a way to check the (sometimes considerable) algebra required to evaluate an amplitude function: you can be pretty sure that if you made a mistake in the evaluation of $A(k)$, then the resulting function will not be normalized. Consider this a hint.

The Width of a Gaussian and Its Amplitude Function

Before leaving the Gaussian wave function, I want to probe further the relationship between its width and that of its amplitude function. According to my definition, the widths of these Gaussian functions, w_x and w_k, respectively, are just the standard deviations Δx and Δk we calculate according to the expressions of § 3.3.

Example 4.5. The Product of the Widths

Consider a Gaussian of the form (4.55) with $L = 1$. Evaluating the standard deviation of such a function is quite easy (see Example 3.5), *viz.*,

$$\Delta x = 1 \qquad \text{for} \quad \Psi(x,0) = \left(\frac{1}{2\pi}\right)^{\frac{1}{4}} e^{-x^2/4}. \tag{4.60}$$

So all we need is Δk.

To evaluate this quantity from $A(k)$, we just generalize Eq. (3.45) for Δx to a function whose independent variable is k instead of x:

$$(\Delta k)^2 = \langle k^2 \rangle - \langle k \rangle^2. \tag{4.61a}$$

The mean values in this expression are

$$\langle k \rangle = \int_{-\infty}^{\infty} A^*(k)\, k\, A(k)\, dk \tag{4.61b}$$

$$\langle k^2 \rangle = \int_{-\infty}^{\infty} A^*(k)\, k^2\, A(k)\, dk. \tag{4.61c}$$

Now, substituting the amplitude function (4.58) with $L = 1$ into these equations and performing a little simple algebra, we find

$$\Delta k = \frac{1}{2} \qquad \text{for} \quad \Psi(x,0) = \left(\frac{1}{2\pi}\right)^{\frac{1}{4}} e^{-x^2/4}. \tag{4.62}$$

Combining this result with (4.59) for the width of the wave function, we obtain the product

$$\Delta x\, \Delta k = \frac{1}{2} \qquad \text{for} \qquad \Psi(x,0) = \left(\frac{1}{2\pi}\right)^{\frac{1}{4}} e^{-x^2/4}. \tag{4.63}$$

This result is a special case of Eq. (4.52*b*), according to which the product of the widths of two wave packets of any form is *greater than or equal to* 1/2. In fact, the Gaussian function is the *only* mathematical form for which this product *equals* 1/2. In qualitative terms, Eq. (4.63) just illustrates a familiar property of wave packets (§ 4.4): *the greater the spread of the amplitude function, the narrower the extent of the wave function and position probability density for the state.*

Question 4–12

The amplitude function for the *general form* of the Gaussian, Eq. (4.54), is

$$A(k) = \left(\frac{2}{\pi}L^2\right)^{1/4} e^{-ix_0 k}\, e^{-(k-k_0)^2 L^2}.$$

Prove that the standard deviations in x and k for the general form of a Gaussian packet—(4.54) with arbitrary L—are

$$\Delta x = L \qquad \text{and} \qquad \Delta k = \frac{1}{2L}.$$

Time Development: A Glimpse Ahead

Although I promised to wait until Chap. 6 to deal with the time development of wave functions, I can't resist showing you how easily we can write down a general form for $\Psi(x,t)$ once we have in hand the amplitude function. This function appears in the general form of a wave packet [Eq. (4.29)]:

$$\Psi(x,t) = \frac{1}{\sqrt{2\pi}} \int_{-\infty}^{\infty} A(k)\, e^{i(kx-\omega t)}\, dk. \tag{4.64}$$

Since the amplitude function doesn't depend on t, we just insert it into (4.64) and voilà: we've got an integral form of the wave function at any time. My point in showing you this is to suggest an important connection between the initial wave function and its subsequent form, a connection that *can* be made through the medium of the amplitude function:

$$\boxed{\Psi(x,0) \quad \Longrightarrow \quad A(k) \quad \Longrightarrow \quad \Psi(x,t)} \tag{4.65}$$

Notice, by the way, that to generate $\Psi(x,t)$ from $A(k)$ we must know the dispersion relation $\omega(k)$ for the system.

Question 4–13

Write down the state function $\Psi(x,t)$ that has a Gaussian with $L = 1$ as its initial value at $= t = 0$.

In Chap. 12, we'll discover lurking in the connection (4.65) a powerful problem solving strategy, one that can be applied to a wide variety of systems to bypass the task of solving the Schrödinger Equation.

4.6 UNCERTAINTY REGAINED

One of the defining characteristics of a wave packet

$$\Psi(x,t) = \int_{-\infty}^{\infty} A(k)\, \Psi_k(x,t)\, dk \tag{4.66}$$

is its non-zero (but finite) standard deviation: $\Delta x > 0$. If the packet represents a state of a particle at time t, then this mathematical fact has an important physical interpretation: the position of the particle is uncertain.

Similarly, the amplitude function $A(k)$, the Fourier transform of $\Psi(x,0)$, is characterized by a non-zero (finite) standard deviation Δk. And, as I emphasized in § 4.3, this mathematical fact has important consequences for the momentum: this observable is necessarily uncertain [Eq. (4.31)]: $\Delta p = \hbar \Delta k > 0$. Notice that this conclusion pertains to *any* state of *any* system, provided the state is represented by a wave packet. Were we to measure the momentum on an ensemble of particles in such a state, we'd obtain results that fluctuate around the mean value $\langle p \rangle$.

This summary illustrates how in quantum physics mathematical properties of waves translate into physical characteristics of particles in quantum states represented by wave functions. The lexicon of the simplest such translations is the two de Broglie relations $p = \hbar k$ and $E = \hbar\omega$.

Let's see what happens if we use these relations to translate the fundamental wave property that relates the standard deviations of a wave packet and its Fourier transform.

Return of the Heisenberg Uncertainty Principle

The standard deviations Δx and Δk are related by [Eq. (4.52*b*)] $\Delta x\,\Delta k \geq 1/2$. Translating these standard deviations into position and (via $p = \hbar k$) momentum uncertainties, we trivially regain the Heisenberg Uncertainty Principle (HUP):[26]

$$\Delta x\,\Delta p \geq \frac{1}{2}\hbar. \tag{4.67}$$

In quantum mechanics, we calculate the position uncertainty Δx from the state function $\Psi(x,t)$ and the momentum uncertainty Δp from the amplitude function $A(k)$ as

$$\Delta x = \sqrt{(\Delta x)^2} = \sqrt{\langle x^2\rangle - \langle x\rangle^2} \tag{4.68a}$$

$$\Delta p = \hbar\sqrt{(\Delta k)^2} = \hbar\sqrt{\langle k^2\rangle - \langle k\rangle^2} \tag{4.68b}$$

Note carefully the inequality in Eq.(4.67). Only one wave packet gives equality in Eq. (4.67), the Gaussian function of § 4.4:

$$\Delta x\,\Delta p = \frac{1}{2}\hbar \qquad \text{(Gaussian wave packet)} \tag{4.69}$$

For this reason a Gaussian is sometimes referred to as the *minimum uncertainty wave packet*.[27]

Our derivation of the HUP is so straightforward that you may be tempted to overlook its profound implications. Equation (4.67) appears here as an inevitable consequence of our decision (Postulate I) to represent quantum states by wave functions that (Postulate II) are interpreted as position probability amplitudes (and hence must be spatially localized). I hope this derivation convinces you once and for all that *uncertainty relations are intrinsic to the quantum-mechanical description of the microverse and as such are wholly unrelated to experimental error or defects in measuring apparatus.* Note also the generality of our argument: it shows that the HUP is not limited to a particular microscopic system or quantum state. This is heady stuff, and I suggest you pause and ponder the subtle and beautiful interplay of mathematics and physics in the remarkably simple derivation of Heisenberg's profound principle.

Aside: The Uncertainty Principle in Three Dimensions. For a particle in three dimensions with wave function $\Psi(\mathbf{r},t)$, the rectangular components of position and momentum—the components of $\mathbf{r}$ and $\mathbf{p}$ along the orthogonal axes $\hat{e}_x$, $\hat{e}_y$, and $\hat{e}_z$—obey uncertainty relations like (4.67):

[26] Because these uncertainties have explicit definitions, some authors call Eq. (4.67) "the precise uncertainty principle." Whatever you call it, this inequality is a special case of the *Generalized Uncertainty Principle* (the GUP). The GUP, which we'll meet in § 11.4, is an indeterminacy relationship between the uncertainties of any two observables that cannot simultaneously be measured to infinite precision. As noted in Chap. 2, two such observables are *complementary* in the sense that the more we know about one, the less we can know about the other.

[27] You can prove this assertion by minimizing the product $\Delta x\,\Delta p$. For the details, see § 10.9 of *Quantum Theory* by David Bohm (Englewood Cliffs, N. J.: Prentice Hall, 1953).

$$\begin{aligned}\Delta x\,\Delta p_x &\geq \tfrac{1}{2}\hbar \\ \Delta y\,\Delta p_y &\geq \tfrac{1}{2}\hbar \\ \Delta z\,\Delta p_z &\geq \tfrac{1}{2}\hbar.\end{aligned} \tag{4.70}$$

Fig. 4.11 summarizes graphically the correlation between the position and momentum uncertainties that is expressed mathematically in the HUP (4.67). For a narrow state function (Fig. 4.11a), the *position* of the particle is rather well-defined (*i.e.*, Δx is small). But, inevitably, the corresponding amplitude function $A(k) = \mathcal{F}[\Psi(x,0)]$ encompasses a wide range of pure momentum states (Fig. 4.11b). So we know very little about the *momentum* of the particle in such a state, and Δp is large. On the other hand, a broad state function (Fig. 4.11c) implies a more precise knowledge of momentum (Fig. 4.11d).

With the translation of wave properties of the Fourier transform $A(k)$ of $\Psi(x,0)$ into physical statements concerning momentum, we have broadened our insight into the physical content of the state function. Back in Chap. 3, we considered this function solely as a source of information about *position*. But in this chapter we've discovered how to extract statistical information concerning a *momentum measurement* (the uncertainty $\Delta p = \hbar\Delta k$) from the state function via its amplitude function. But this is not the only momentum information that lies hidden in $A(k)$; in the next section we'll see how to dig out detailed probabilistic information about a measurement of this observable.

4.7 (NOT QUITE) ALL ABOUT MOMENTUM

As you've doubtless deduced, there is buried in the amplitude function a wealth of physical information about momentum. In fact, we can determine from this function probabilistic and statistical information about momentum that is analogous to the information about position we can extract from $\Psi(x,t)$. To show you how, I first want to rewrite the general form (4.29) of the wave packet,

$$\Psi(x,t) = \frac{1}{\sqrt{2\pi}} \int_{-\infty}^{\infty} A(k)\, e^{i(kx-\omega t)}\, dk, \tag{4.71}$$

in terms of the *pure momentum state functions* I introduced in § 4.2 [Eq. (4.26)]:

$$\Psi_p(x,t) = \frac{1}{\sqrt{2\pi\hbar}}\, e^{i(px-Et)/\hbar}. \tag{4.72}$$

To do so requires only a simple change of variable—but even so simple a task can hide traps for the unwary.

Changing Variables in the Amplitude Function

At first, the necessary variable change seems trivial. The de Broglie relation $p = \hbar k$ shows us how to change from wave number to momentum, so to write $\Psi(x,t)$ as an integral over p we first define a new function—say, $\Phi(p)$—that is proportional to the amplitude function $A(k)$. We then just write the Fourier transform relation (4.71) as an integral over $\Phi(p)\Phi_p(x,t)$ expressing the differential volume element dk as $dk = dp/\hbar$. This is a good plan, but I'm going to vary it slightly so as to minimize later anguish.

I'll define the function $\Phi(p)$ in terms of the amplitude function $A(k)$ as

$$\Phi(p) \equiv \frac{1}{\sqrt{\hbar}}\, A\left(\frac{p}{\hbar}\right). \qquad \text{momentum probability amplitude} \tag{4.73}$$

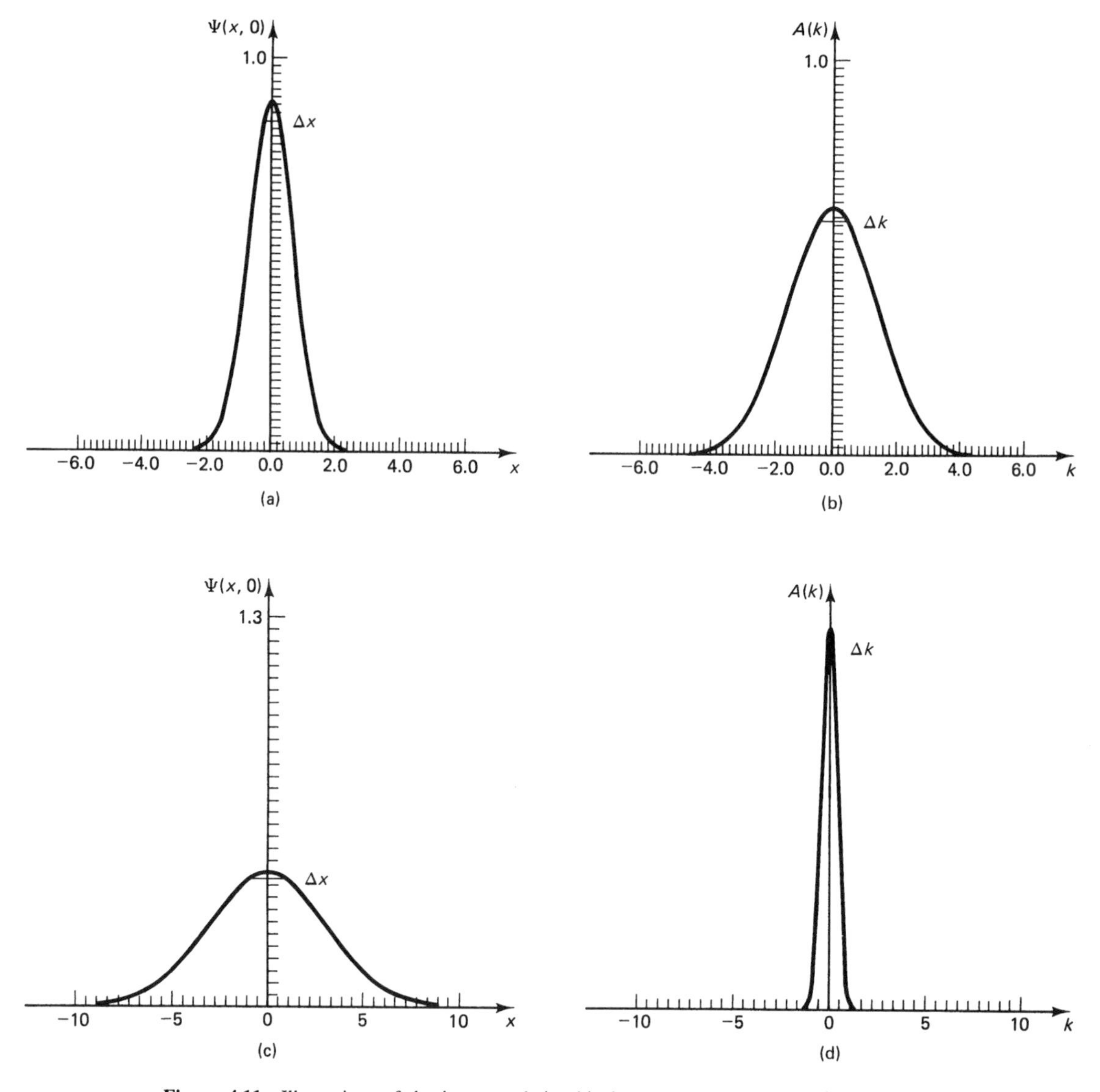

Figure 4.11 Illustrations of the inverse relationship between the widths of $\Psi(x, 0)$ and $A(k)$. Because of this relationship, the momentum uncertainty in the state represented by $\Psi(x, 0)$ is inversely correlated with the position uncertainty Δx via the HUP, Eq. (4.67).

Notice that, by analogy with the position probability amplitude $\Psi(x, t)$, I've dubbed $\Phi(p)$ the **momentum probability amplitude**.

In terms of this function, the wave packet (4.71) at $t = 0$ becomes the momentum integral

$$\Psi(x, 0) = \frac{1}{\sqrt{2\pi\hbar}} \int_{-\infty}^{\infty} \Phi(p)\, e^{ipx/\hbar}\, dp. \tag{4.74}$$

[Don't fret about the peculiar factors of $1/\sqrt{\hbar}$ in Eqs. (4.73) and (4.74); I'll explain what I am doing in the next aside.]

The inverse of (4.74), the expression for $\Phi(p)$ as an integral with respect to x of $\Psi(x,0)$, follows easily from the definition (4.73):

$$\Phi(p) = \frac{1}{\sqrt{2\pi\hbar}} \int_{-\infty}^{\infty} \Psi(x,0)\, e^{-ipx/\hbar}\, dx. \tag{4.75}$$

We find, not unexpectedly, that *the momentum probability amplitude is the Fourier transform of the position probability amplitude*, with x and p rather than x and k as the transformation variables:

$$\boxed{\begin{aligned} \Psi(x,0) &= \mathcal{F}^{-1}[\,\Phi(p)\,] \\ \Phi(p) &= \mathcal{F}[\,\Psi(x,0)\,] \end{aligned}} \tag{4.76}$$

I prefer Eqs. (4.76) to the Fourier relationships of § 4.5 for $\Psi(x,0)$ and $A(k)$, these equations emphasize the physically important variables position and momentum. So unless I tell you otherwise, please understand the shorthand relations (4.76) to refer to Eqs. (4.74) and (4.75) [see the summary in Fig. 4.12].

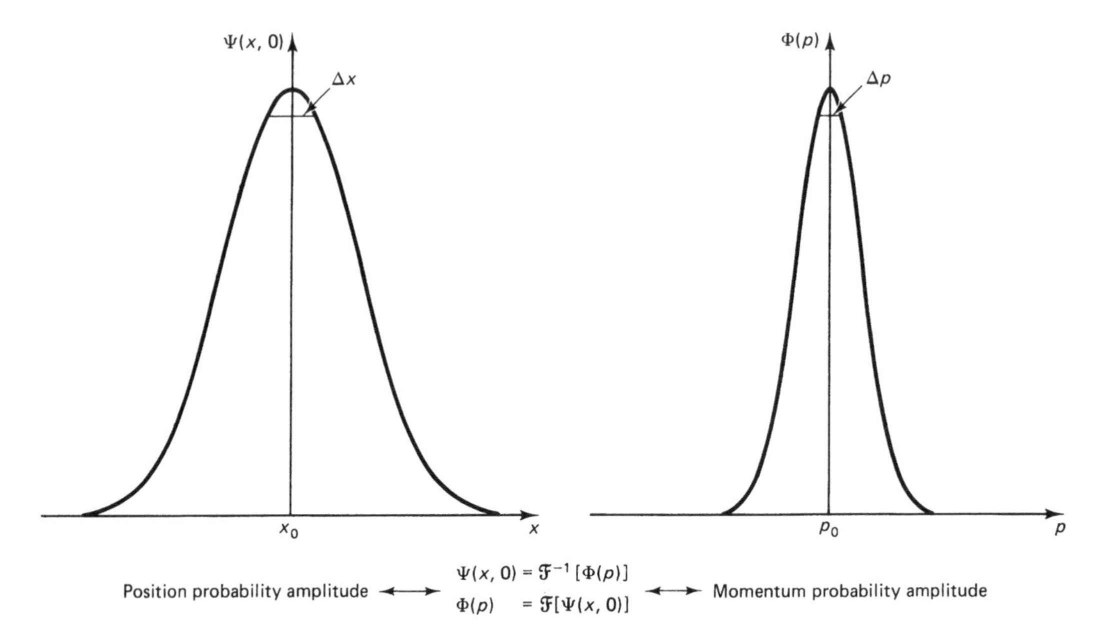

Figure 4.12 Summary of the Fourier relationships between a state function $\Psi(x,0)$ and the corresponding momentum probability amplitude $\Phi(p)$.

We can generalize the momentum-integration form of the wave packet, (4.74), to arbitrary time, just as we did to obtain the wave-number-integration form in Eq. (4.64) of § 4.5:

$$\Psi(x,t) = \frac{1}{\sqrt{2\pi\hbar}} \int_{-\infty}^{\infty} \Phi(p)\, e^{i(px-Et)/\hbar}\, dp \tag{4.77a}$$

$$= \int_{-\infty}^{\infty} \Phi(p)\, \Psi_p(x,t)\, dp. \tag{4.77b}$$

Of course, to apply Eqs. (4.77) to a particular system, we must relate ω and k via the appropriate dispersion relation. Once we have done so, we can express any state function as a superposition of pure momentum functions $\Psi_p(x,t)$. Notice, by the way, that the momentum probability amplitude $\Phi(p)$ does not depend on time.[28]

Aside: Defining the Momentum Probability Amplitude. You may be wondering why I defined the momentum probability amplitude in (4.73) with an extra factor of $1/\sqrt{\hbar}$, rather than as simply $A(p/\hbar)$. Well, when we change variables from k to p, we have to somehow deal with the factor of $1/\hbar$ that arises from the differential relationship $dk = dp/\hbar$; this $1/\hbar$ appears in the expression for the wave packet as an integral over momentum. By including a factor of $1/\sqrt{\hbar}$ in the definition (4.73) and absorbing the other $1/\sqrt{\hbar}$ in the prefactors in the Fourier transform relations (4.74) and (4.75), we preserve a useful property of the amplitude function: automatic normalization.

Recall (§ 4.4) that, courtesy of the Bessel-Parseval relation (4.51), an amplitude function $A(k)$ determined from a *normalized* state function $\Psi(x,0)$ is automatically normalized with respect to integration over k [*c.f.*, Eq. (4.59)]. This means that in problem solving, we need normalize only $\Psi(x,0)$. With the definition (4.73) of $\Phi(p)$, automatic normalization is assured for the momentum probability amplitude—applying our change of variables to (4.59), we find

$$\int_{-\infty}^{\infty} |\Phi(p)|^2\,dp = \int_{-\infty}^{\infty} |\Psi(x,0)|^2\,dx = 1. \tag{4.77c}$$

The Interpretation of $\Phi(p)$: Born Again

My appellation for $\Phi(p)$—the *momentum probability amplitude*—strongly suggests an analogy to the *position probability amplitude* $\Psi(x,0)$. In fact, these functions are different ways of representing the same quantum state (§ 4.7) that highlight information about different observables:

$$\left.\begin{array}{ll} \Psi(x,t) & \longrightarrow \text{position} \\ \Phi(p) & \longrightarrow \text{momentum} \end{array}\right\} \qquad \text{same quantum state}$$

This analogy suggests that we interpret $\Phi(p)$ for momentum like we interpreted $\Psi(x,t)$ for position. To jog your memory, here's the Born interpretation of $\Psi(x,t)$ as a position probability amplitude (Postulate II):

Interpretation of $\Psi(x,t)$: If the members of an ensemble are in a quantum state represented by the wave function $\Psi(x,t)$, then

$$P(x,t)dx = |\Psi(x,t)|^2\,dx$$

is the probability that in a position measurement at time t a particle will be detected in the infinitesimal region from x to $x+dx$.

[28]Some applications of quantum mechanics are based on a time-dependent momentum probability amplitude $\Phi(p,t)$. This function is defined, as you might have guessed, as the inverse Fourier transform of the wave function at time t:

$$\Phi(p,t) \equiv \frac{1}{\sqrt{2\pi\hbar}} \int_{-\infty}^{\infty} \Psi(x,t)\, e^{-ipx/\hbar}\,dx$$

In practice, a time-dependent momentum amplitude is useful only in a very small number of cases—because the integral in its definition is quite ferocious. You can find a good discussion of these time-dependent momentum amplitudes in *Elementary Quantum Mechanics* by David S. Saxon (San Francisco: Holden-Day, 1968).

Now, a reasonable generalization of this idea to the momentum amplitude $\Phi(p)$, which represents the same quantum state as $\Psi(x,t)$, is[29]

Interpretation of $\Phi(p)$: If the members of an ensemble are in a quantum state that is represented by a wave function $\Psi(x,t)$ with Fourier transform $\Phi(p) = \mathcal{F}[\,\Psi(x,0)\,]$, then

$$P(p)\,dp = |\Phi(p)|^2\,dp = \Phi^*(p)\Phi(p)\,dp$$

is the probability that in a momentum measurement at time t a particle's momentum will be found to have a value in the infinitesimal region from p to $p + dp$.

This generalization is the key to calculating *probabilistic information about momentum* from a wave function $\Psi(x,t)$. And it buttresses the extension of our definitions of the statistical properties $\langle x \rangle$ and Δx to momentum, *viz.*,

$$\boxed{\begin{aligned}\langle p \rangle &\equiv \int_{-\infty}^{\infty} p\,P(p)\,dp \\ &= \int_{-\infty}^{\infty} \Phi^*(p)\,p\,\Phi(p)\,dp\end{aligned}} \qquad \text{Expectation Value of Momentum} \qquad (4.78)$$

$$\boxed{\begin{aligned}\Delta p &\equiv \sqrt{(\Delta p)^2} \\ &= \left[\left\langle (p - \langle p \rangle)^2 \right\rangle\right]^{1/2} \\ &= \sqrt{\langle p^2 \rangle - \langle p \rangle^2}\end{aligned}} \qquad \text{Momentum Uncertainty} \qquad (4.79)$$

In (4.79), the **expectation value** of p^2, which we need to evaluate the **momentum uncertainty**, is just

$$\langle p^2 \rangle = \int_{-\infty}^{\infty} \Phi^*(p)\,p^2\,\Phi(p)\,dp. \qquad (4.80)$$

To illustrate the application of these extensions, I've gathered our results (from § 4.6) for a sample Gaussian wave function $\Psi(x,0)$ and its Fourier transform, suitably expressed them in terms of x and p, and tabulated the whole thing in Table 4.1.

Question 4–14

Write the expressions for $\Psi(x,0)$ and for $\Phi(p)$ in Table 4.1 in terms of Δx and Δp respectively (instead of L). Your results show the important dependence of the Gaussian function and its Fourier transform on their (initial) widths—a feature of the Gaussian to which we'll return in § 6.7.

[29] In more formal treatments of quantum mechanics, such as befits a graduate course, Postulate II is stated abstractly as a way to interpret the quantum state descriptor in terms of an arbitrary observable. To see how this goes, see *Quantum Mechanics*, Vol. I by C. Cohen-Tannoudji, B. Diu, and F. Laloë (New York: Wiley-Interscience, 1977).

TABLE 4.1 POSITION AND MOMENTUM INFORMATION FOR A GAUSSIAN STATE FUNCTION WITH $L = 1.0$, $x_0 = 0$, AND $p_0 = 0$

Position	Momentum
$\Psi(x,0) = \left(\frac{1}{2\pi L^2}\right)^{1/4} e^{-x^2/(4L^2)}$	$\Phi(p) = \left(\frac{2}{\pi}\frac{L^2}{\hbar^2}\right)^{1/4} e^{-p^2L^2/\hbar^2}$
$\langle x \rangle = 0$	$\langle p \rangle = 0$
$\Delta x = L$	$\Delta p = \frac{1}{2L}\hbar$

Question 4–15

We understand the existence of a center x_0 of a wave packet $\Psi(x,t)$ as due to constructive interference in the superposition of complex harmonic waves that is the packet. For a symmetric packet this point coincides with the expectation value at $t = 0$, *i.e.*, $x_0 = \langle x \rangle (0)$. Provide a similar explanation of the existence of the center $p_0 = \hbar k_0 = \langle p \rangle$ of the momentum probability amplitude.

With this generalization to $\Phi(p)$ of the expressions for the expectation value and the uncertainty, we've developed machinery for calculating all we are permitted to know about momentum from this probability amplitude. And an impressive menu it is: We can work out the probabilities of each possible outcome of an ensemble measurement of momentum as well as statistical quantities that characterize all the results of the measurement: the mean value ($\langle p \rangle$) and the extent to which individual results fluctuate about the mean (Δp). Underlying this scheme are a few basic principles: the de Broglie relation $p = \hbar k$, the Born interpretation of a state function as a probability amplitude, and basic properties of waves.

Still, the momentum probability amplitude $\Phi(p)$ is but one avenue to momentum information about a quantum state. There are others—which is good news, because in some applications, determining $\Phi(p)$ as the Fourier transform of the wave function $\Psi(x,0)$ is an algebraic nightmare. But fear not. Chap. 5 is in the offing.

4.8 FINAL THOUGHTS: A WEALTH OF REPRESENTATIONS

Our quest for a free-particle state function, which began in § 4.1 with the complex harmonic wave and culminated in § 4.2–4.3 with the wave packet, has taken us deep into the mysteries of momentum. But before I bring this observable center stage, in the next chapter, I want to reinforce the physics that underlies the mathematics of Fourier analysis—which led us into the world of momentum amplitudes in the first place.

Underlying the application of Fourier analysis to the construction and deconstruction of wave packets are physical principles familiar from Chaps. 2 and 3. Predominant among these is the omnipresent *Principle of Superposition*, which we saw reflected in our imagining a wave packet as a superposition of pure momentum states. Second, *wave-particle duality* is implicit in our use of the mathematics of waves to build and analyze functions that represent states of microscopic particles, as it is in our use of the de Broglie relations to interpret these states. And finally, *uncertainty* appeared naturally, almost as an afterthought, from our mathematical machinations.

This subtle interplay of the mathematics of complex functions and the physics of the microworld is one of the marvels of quantum physics. No one really knows *why* this mathematical artifice of man's mind happens to describe the real world so well; this is a question for philosophers and metaphysicists.[30]

The principal results of this chapter were the general form (4.29) of the wave packet, the Fourier transform relations (4.76) between the position and probability amplitudes, and our finding (§ 4.7) that a state function yields (probabilistic and statistical) information about position or about momentum—depending on how we treat it. The algebraic procedures we use to extract position information from $\Psi(x, 0)$ and momentum information from $\Phi(p)$ are identical; only the variables are changed. This powerful analogy is summarized in Table 4.2.

TABLE 4.2 POSITION AND MOMENTUM REPRESENTATIONS OF A QUANTUM STATE OF A PARTICLE IN ONE DIMENSION

	Position	Momentum
Observable	x	p
State function	$\Psi(x, 0)$	$\Phi(p)$
Probability density	$P(x, 0) = \lvert\Psi(x, 0)\rvert^2$	$P(p) = \lvert\Phi(p)\rvert^2$
Expectation value	$\langle x\rangle(0) = \int_{-\infty}^{\infty} \Psi^*(x, 0)\, x\, \Psi(x, 0) dx$	$\langle p\rangle = \int_{-\infty}^{\infty} \Phi^*(p)\, p\, \Phi(p) dp$
Uncertainty	$\Delta x = \sqrt{\langle(x - \langle x\rangle)^2\rangle}$	$\Delta p = \sqrt{\langle(p - \langle p\rangle)^2\rangle}$
	$\Delta x = \sqrt{\langle x^2\rangle - \langle x\rangle^2}$	$\Delta p = \sqrt{\langle p^2\rangle - \langle p\rangle^2}$

In more advanced formulations of quantum mechanics, the position and momentum amplitudes $\Psi(x, t)$ and $\Phi(p)$ are but two of an infinite number of "representations" of a more abstract state descriptor, the *state vector*. A state vector is not a function. Neither is it a vector in our geometrical space, $\Re^3$. Instead, it lives in an abstract vector space called a *Hilbert space*, where it is ruled by the mathematics of *linear algebra*. The state vector is usually denoted by the symbol $|\Psi(t)\rangle$.

Our wave function $\Psi(x, t)$ is the representation of this critter in (one-dimensional) "position space," where the coordinate is x. Our momentum probability amplitude $\Phi(p)$ is its representation in "momentum space," where the coordinate is p. In fact, one can devise an *infinite number of different representations of a quantum state*. Each representation corresponds to a different observable: there is an energy representation, an angular momentum representations, and so forth.

This more abstract formulation was developed by P. A. M. Dirac.[31] Although later in this volume I'll show you a wonderfully convenient shorthand notation Dirac introduced for integrals, I won't show you quantum mechanics from his powerful but austere vantage point. At this stage, I want to keep matters as concrete as possible.

[30]But if you're curious to read a physicist's thoughts on the matter, you should read Eugene Wigner's fascinating essay on the interconnections between mathematics and physics. It is reprinted as Chap. 17 of his collection *Symmetry and Reflections* (Cambridge, Mass.: MIT Press, 1970).

[31]Dirac's exposition of quantum mechanics, *The Principles of Quantum Theory*, 4th ed. (Oxford: Clarendon Press, 1958)—which is now available in paperback—is a genuine classic that eventually should be read and re-read by every serious student of quantum theory.

Before we leave the matter of incarnations of the state vector, I want to make one final important point:

> ***Rule***
>
> No representation of a state contains any more (or less) physical information than any other.

This being the case, our decision to represent a state by $\Psi(x, 0)$ or by $\Phi(p)$ or by a function of some other variable becomes a matter of personal proclivity. In practice, the choice of representation is guided by expediency: we can usually find a representation that is particularly appropriate to the problem at hand (*i.e.*, that saves us work).

ANNOTATED SELECTED READINGS

Waves

You probably studied waves in a freshman or sophomore general physics course or in a separate course devoted solely to this topic. If not, you should have a look at one of the following excellent texts (arranged in order of increasing difficulty):

1. French, A. P., *Vibrations and Waves* (New York: W. W. Norton, 1961) (available in paperback). This book contains an especially good treatment of *superposition* in Chap. 2 and of interference phenomena in Chap. 8.
2. Crawford, Frank S. Jr., *Waves: Berkeley Physics Course*, Volume 3 (New York: McGraw-Hill, 1968). An exhaustive treatment of waves that emphasizes their applications to nearly every area of physics. For interesting discussions of dispersion and the group velocity of de Broglie waves, see Supplementary Topics 2 and 4.
3. Elmore, W. C., and Mark A. Heald, *The Physics of Waves* (New York: McGraw-Hill, 1969).

Wave Packets and Uncertainty

One of the most intriguing things to emerge from our discussions of wave packets was the uncertainty principle. For a more detailed analysis of the relationship between this important physical principle and the mathematics of state functions, look in nearly any graduate-level quantum book, such as

4. Rabinowitz, A., *Quantum Mechanics* (New York: Elsevier, 1968), § 30 and 31.
5. Bohm, David, *Quantum Theory* (Englewood Cliffs, N. J.: Prentice Hall, 1953), § 10.9.

The same A. P. French who wrote the book on waves I recommended above has also co-authored an excellent quantum mechanics text (at a slightly lower level than this one):

6. French, A. P., and Edwin F. Taylor, *An Introduction to Quantum Physics* (New York: Norton, 1978) (available in paperback). This whole book is readable, insightful, and worth your time; but I particularly recommend their discussion of wave packets (§ 8.7).

Fourier Analysis

Fourier series and integrals are so important to the mathematics of science and engineering that any "math methods" book worth its salt will contain an exposition of this topic, usually with lots of illustrations. A treatment that is particularly oriented towards students of physics, albeit at a rather advanced level, is the one in

7. Mathews, Jon, and R. L. Walker, *Mathematical Methods of Physics* (New York: W. A. Benjamin, 1980), § 4.1 and 4.2.

You can find a good summary suitable for undergraduates in

8. Arfken, George, *Mathematical Methods for Physicists,* 2nd ed. (New York: Academic Press, 1970). Chap. 14 covers Fourier series, while § 15.2–15.6 discuss Fourier transforms; the latter section is devoted to the momentum representation.

EXERCISES AND PROBLEMS

> Little Jack Horner
> Sits in a corner
> Extracting cube roots to infinity.
> An assignment for boys
> That will minimize noise
> And produce a more peaceful vicinity.
>
> —-from *A Space-Child's Mother Goose*
> by Frederick Winsor

Exercises

4.1 More Defects of the Monochromatic Harmonic Wave

This exercise explores yet another reason why a plane harmonic wave cannot represent a state of a free particle.

(a) **Show** that the phase velocity of a complex harmonic wave is given by Eq. (4.24).

(b) Now, consider a very fast-moving particle of mass m—in fact, suppose the velocity v of the particle is near (but, of course, less than) the speed of light. **Evaluate** the phase velocity of the pure momentum state function for such a particle. Why is your result ludicrous and physically offensive?

4.2 Momentum Measurements on a Chopped Beam: I

Consider a pulse of free particles, the length of which is $2a$. At time $t = 0$ the (normalized) state function of the particles in this "chopped beam" is

$$\Psi(x,0) = \begin{cases} 0 & -\infty < x < -a; \\ \frac{1}{\sqrt{2a}} & -a \le x \le a; \\ 0 & a < x < \infty. \end{cases}$$

(a) **Determine** the amplitude function $A(k)$ for this state.

(b) **Write down** an expression for the state function $\Psi(x,t)$ for this state for $t > 0$. (You may write your result in terms of an integral ready to be looked-up; do not try to evaluate this integral—it's a mess!)

(c) In an ensemble measurement of the *momentum* of the particles in the pulse at $t = 0$, what average value would be obtained? **Justify your answer**.
Hint: Don't immediately start calculating; think.

(d) In the measurement of part (c), what value of p will *most* of the particles exhibit? Why?

(e) What is the probability of obtaining the value $p = \hbar\pi/a$ in the measurement of part (c)?
Hint: How important is the contribution of the pure momentum state with this value of p to the superposition that makes up the state function at $t = 0$?

4.3 Momentum Measurements on a Chopped Beam: II

Consider a free particle of mass m moving in one dimension with a speed $v \ll c$. We shall write the state function for this particle as a localized wave packet.

(a) **Write down** the general form of the wave packet $\Psi(x,t)$ as a superposition of pure momentum states. **Discuss** how you would control the mixture of pure momentum states so that the function with momentum $p_0 = \hbar k_0$ is the most important constituent of the wave packet and functions with $|\,p\,| \gg p_0$ contribute negligibly to the state function.

(b) **Evaluate** the group velocity v_{gr} of the wave packet and show that it is equal to the classical velocity of the particle, $v_{\text{cl}} = dE/dp$. **Discuss** the physical significance of this result.

Problems

4.1 Construction of a Simple Wave Packet

The general form of a real harmonic wave is given by Eq. (4.5). In this problem we shall actually construct a wave packet by superposing a finite number of such waves.

(a) **Draw** an *accurate* graph of the sum of six real harmonic waves with the following wavelengths:

$$\lambda = 1.5,\ 1.6,\ 1.7,\ 1.8,\ 1.9,\ \text{ and } 2.0 \text{ cm}.$$

Take the amplitude of each wave to be 1.0 cm. Adjust the phase constants so that *each harmonic wave* satisfies the condition

$$f(x_0, t_0) \; = \; 1.0 \text{ cm}.$$

Sketch the composite packet for $|x| \; \leq \; 10.0\,\text{cm}$. (Note: You will probably find it convenient to choose $x_0 = 0$ and $t_0 = 0$.)

(b) **Calculate** the product of the spread w_x of the wave packet you have formed and the range w_k of wave numbers included and **verify** that this product is a number of the order of unity.

Remark: Notice that we do not obtain a *single* region of enhancement in this case, since there are always several values of x where all the constituent plane harmonic waves are in phase. Several regions inevitably result when we superpose a *finite* number of plane waves with wave numbers that differ by *finite* amounts. To obtain a single region of enhancement, we must superpose an infinite number of harmonic waves with infinitesimally differing wave numbers, as in the following problem.

4.2 Wave Packet Construction Can Be Fun

Let's build a wave packet using a particularly simple amplitude function. The general form of the wave packet is given by Eqs. (4.29), where $\omega = \omega(k)$ is the dispersion relation. (*In this problem, we won't assume that particle is necessarily free.*)

Consider the amplitude function

$$A(k) \; = \; \begin{cases} 0 & k \; < \; k_0 \; - \; \frac{1}{2}\Delta k; \\ \frac{1}{\sqrt{\Delta k}} & k_0 - \frac{1}{2}\Delta k \leq k \leq k_0 + \frac{1}{2}\Delta k; \\ 0 & k \; > \; k_0 + \frac{1}{2}\Delta k. \end{cases}$$

PART A Consider the initial time, $t = 0$.

(a) **Evaluate** the state function $\Psi(x, 0)$.

(b) **Evaluate** the position probability density at $t = 0$.

(c) **Graph** the probability density vs. x, and show that this wave packet conforms to the Heisenberg Uncertainty Principle $\Delta x \; \Delta p \; \geq \; \hbar/2$. [For Δx use the location of the first node (zero) of the state function. This number is an adequate approximation to the uncertainty and is considerably easier to obtain than the accurate value—which you don't need anyway for this problem.]

(d) On your graph in part (c), **graph** a second curve showing the probability density at $t = 0$ for a value of Δk twice as large as the one you used in part (c).

Hint: In parts (c) and (d), you may find it easier to graph the probability density as a function of the scaled variable $x/(2\pi/\Delta k)$.

PART B In Chap. 6 we'll investigate wave packet propagation. As a preview, consider the behavior of the wave packet of part A at times $t > 0$, using Eq. (4.29) as a starting point.

(a) To simplify the integrand of $\Psi(x,t)$, expand $\omega(k)$ in a *Taylor series* about $k = k_0$, retaining only terms through first order. **Show** that your result can be written

$$\omega(k) = \omega_0 + v_{\rm gr}(k - k_0),$$

where $\omega_0 = \omega(k = k_0)$ and $v_{\rm gr}$ is the group velocity of the wave packet evaluated at $k = k_0$. Under what conditions on Δk is this approximation to $\omega(k)$ valid?
Hint: The Taylor series expansion of a function $f(x)$ about a point $x = x_0$ is

$$f(x) = f(x_0) + (x - x_0) f'(x_0) + \frac{(x - x_0)^2}{2!} f''(x_0) + \cdots + \frac{(x - x_0)^{(n-1)}}{(n-1)!} f^{(n-1)}(x_0) + \text{remainder},$$

where $f'(x_0)$ is the first derivative df/dx evaluated at x_0, etc. The usefulness of this expansion is somewhat limited, because it can be applied only to functions that satisfy a rather demanding criteria: the derivatives of $f(x)$ to all orders must exist and must be finite. Nonetheless, the Taylor series is invaluable, and if you haven't studied it I recommend that you read § 16.3 of *Calculus and Analytic Geometry*, 3rd ed. by George B. Thomas, Jr. (Reading, Mass.: Addison-Wesley, 1960).

(b) Now, **carry out** the integration over k and thereby **obtain** an explicit expression for the state function and thence for the probability density $P(x,t)$ in terms of x, t, $v_{\rm gr}$, and Δk. **Verify** that your result agrees with that of part (b) at $t = 0$. How is $P(x,t)$ related to $P(x - v_{\rm gr}t, 0)$? What does your answer to this question imply about the propagation of this wave packet? Does it retain its shape?

(c) Let $x_{\rm max}$ denote the position of the maximum of $P(x,t)$. **Write** an equation for the time dependence of $x_{\rm max}$.

(d) Now suppose that you are sitting on the x axis—say, at a point $x_0 > 0$—watching all the waves go by. **Describe** (in words or with sketches or both) what you see, *i.e.*, describe the probability density considered as a function of *time* at a fixed *position.*

(e) Let Δt denote the time interval that separates the arrival at x_0 of the central peak of $P(x,t)$ and the adjacent node. **Show** that this wave packet leads to the "energy-time uncertainty relation" $\Delta E\, \Delta t \approx h$.

Remark: This extremely important problem reinforces the point made in this chapter: uncertainty relations are inevitable consequences of the use of wave packets to represent states of subatomic particles. More precise definition of the uncertainties involved would have led to the "precise" uncertainty principles of this chapter and Chap. 11.

4.3 Coping with Complex Amplitude Functions

Using Eq. (4.30), use a stationary phase argument to **show** that the center of a wave packet constructed from a complex amplitude function occurs at

$$x_0 = -\left.\frac{d\alpha}{dk}\right|_{k=k_0}.$$

4.4 Unmasking the True Nature of a Wave Packet

Consider the state function

$$\Psi(x,0) = N \exp\left[-\tfrac{1}{L}|x|\right] e^{ik_0x}$$

(a) **Normalize** this function.

(b) **Derive** the corresponding amplitude function $A(k) = \mathcal{F}[\Psi(x,0)]$ and thereby demonstrate that this wave packet is actually a superposition of pure momentum states.

(c) **Verify** that your amplitude function is normalized.

(d) **Sketch** the *real part* of $\Psi(x,0)$ and the amplitude function $A(k)$. [Choose convenient (but reasonable) values for k_0 and L.]

4.5 Fooling Around with Gaussians

The extended example of § 4.5 concerned a Gaussian function (4.55), which is centered at $x_0 = 0$ with an amplitude function centered at $k_0 = 0$. In this problem, we'll consider the often-used more general form (4.54).

(a) **Evaluate** the expectation value of x for this state.

(b) **Evaluate** the position uncertainty for this state.

(c) Gaussian functions are widely used in the study of the physics of matter. Suppose that $\Psi(x,0)$ represents a particle of mass m moving in a gravitational field with potential energy

$$V(x) = mgx.$$

Evaluate $\langle V(x) \rangle$ for this state.

(d) Repeat part (c) for a *simple harmonic oscillator potential energy; i.e.*,

$$V(x) = \tfrac{1}{2}m\omega_0^2 x^2.$$

(e) **Evaluate** the expectation value of momentum for this state.
Hint: To simplify your algebra, let the center of the wave packet, x_0, be zero for part (e).

4.6 Momentum Information for the Particle in a Box

In Chap. 4 we used Fourier analysis to determine momentum information from free particle wave functions such as the Gaussian wave packet of § 4.5. But we can also apply the mathematical analysis and physical reasoning of this chapter to quantum mechanical systems with non-zero potential energies. To illustrate, let's consider the simplest single-particle system in which the particle is subject to external forces: the infinite square well.

In various problems in Chap. 3, we probed the *stationary states* of this system. The wave functions that represent these states have the form

$$\Psi_n(x,t) = \psi_n(x)e^{-iE_n t/\hbar}.$$

For a one-dimensional infinite square well of length L, the explicit form of these functions at $t = 0$ are

$$\Psi_n(x,0) = \begin{cases} \sqrt{\frac{2}{L}}\cos\left(n\pi\frac{x}{L}\right) & n = 1,3,5,\ldots \\ \sqrt{\frac{2}{L}}\sin\left(n\pi\frac{x}{L}\right) & n = 2,4,6,\ldots \end{cases} \tag{4.6.1}$$

The (well-defined) energies of these states are

$$E_n = n^2\frac{\pi^2\hbar^2}{2mL^2}, \tag{4.6.2}$$

which we can write in terms of the magnitude p_n of the linear momentum in the state n as

$$E_n = \frac{p_n^2}{2m} = \frac{\hbar^2 k_n^2}{2m}, \tag{4.6.3}$$

with

$$p_n = \hbar k_n \equiv n\frac{\hbar\pi}{L}. \tag{4.6.4}$$

Thus far, we have analyzed several of these functions as position probability amplitudes —*i.e.*, as sources of probabilistic and statistical information about *position*. For a change of pace, let's look at these wave functions from a different point of view: as sources of *momentum information.*

(a) **Derive** expressions for the momentum probability amplitudes $\Phi_n(p)$ for the stationary states of the infinite square well. Consider even and odd values of n separately. Write your amplitudes in terms of p, L, n, and fundamental constants.
Hint: To simplify the integrals you need to evaluate, first write the wave functions $\Psi_n(x, 0)$ in terms of *complex exponentials* such as $e^{ip_n x/\hbar}$. Evaluate the integral, then use the handy (but little known) trigonometric identity

$$\sin\left(\frac{n\pi}{2} \pm x\right) = \begin{cases} +i^{n-1}\cos x & \text{for n odd;} \\ \mp i^n \sin x & \text{for n even} \end{cases} \tag{4.6.5}$$

to put your momentum amplitude in a convenient form. Each amplitude should contain either the factor $\cos(pL/2\hbar)$ or $\sin(pL/2\hbar)$, accordingly as n is odd or even.

(b) **Derive** a *single* expression for the momentum probability density $P_n(p)$—*i.e.*, an expression that can be used for even or odd values of n. **Graph** the momentum probability density for $n = 1$, 2, 3, and 4.

(c) The following quotation appears on p. 245 of the textbook *Foundations of Modern Physics* by Paul A. Tipler (New York: Worth, 1969):

> ... if we measure the momentum of the particle in a box, we expect to find either $p = +\hbar k_n$ or $p = -\hbar k_n$, and the two possibilities are equally likely.

Discuss the validity of this statement. Under what circumstances, if any, is it correct?

(d) Using your momentum probability amplitude $\Phi_1(p)$, **evaluate** the *expectation value of momentum* and the *momentum uncertainty* for the $n = 1$ state of the infinite square well.

CHAPTER 5

Observables in Quantum Physics

A Pragmatist's Approach

When I could find voice, I shrieked aloud in agony,
"Either this is madness or it is Hell."
"It is neither," calmly replied the voice of the Sphere,
"it is Knowledge..."

—-A. Square,
in *Flatland: A Romance of Many Dimensions*
by Edwin A. Abbott

In Chap. 3 we learned a new way to represent one of the basic elements of physical theory: the *state* of a system. Because the state descriptor of classical physics, the trajectory, is inapplicable to microscopic particles, we had to postulate a state descriptor more suited to the microworld: the position probability amplitude $\Psi(x,t)$, the wave function.

The second class of basic elements of a physical theory consists of the mathematical representative of *observables*—the physical properties of a system. The idea of an observable is familiar: the brilliant yellow of a sunflower, the taste of a ripe apple, the rich odor of a compost heap—all are observables of everyday life. In physics, of course, we concentrate on more quantifiable observables, properties such as position, energy, linear and angular momentum. In physics, then, *observables are physical attributes of a system that can be measured in the laboratory.*

In classical physical theory, we represent observables by real functions of time; these are the "property descriptors" of that theory. Underlying this way of representing observables is a concept so intuitively obvious that you may never have pondered it: when we think about the physical properties of a classical system, we assume that the system "possesses" them. This thinking is mirrored in the way we talk about classical observables, in such statements as "a billiard ball of mass $0.1\,\text{kg}$ *has* a position specified by the coordinates $x = 4.0\,\text{m}$ and $y = 6.2\,\text{m}$ and an energy of $150\,\text{J}$."

Take a moment to think about this statement. It's based (implicitly or explicitly) on an experimental determination of the properties it addresses (mass, position, energy). It reveals that our sense that a particle can be said to "have" (*i.e.*, to possess) such properties presumes that the only incertitude infecting such a determination is experimental error. That is, we assume that well-defined, definite values of x, y, and E can be associated with the particle—regardless of whether we measure these properties, when we measure them, or even which ones we measure. The task of the experimental physicist is to measure them as accurately as possible.

But this notion is not appropriate to the microverse, for *in the absence of a measurement, a microscopic system does not necessarily have values of its physical properties*—at least not in the classical sense of possession of properties. This disquieting notion first appeared in our discussion of the double-slit experiment (in Chap. 2): from the moment electrons leave the source until the instant before they arrive at the detector (where their positions are measured), they cannot be said to "have a position." Rather, they are in a quantum state that allows *an infinity of possible positions*. One of these possibilities becomes actual when an electron arrives at the detector.

This strange way of thinking about observables was also implicit in the wave functions we examined in Chap. 4. Particles in an ensemble whose state is represented by a Gaussian wave packet, for example, don't "have" a position. Rather, their state encompasses an infinity of possibilities. We can think of these possibilities as *latent* in the state function until the observable, position, is actually measured, at which point one actual value is realized. The probability that one or the other possibility will be realized is calculated from the probability density $|\Psi(x,t)|^2$.

The consequences of the latency of observables in an (unobserved) microscopic ensemble are profound indeed. As we'll discover in § 5.2, it means that we cannot represent observables in quantum physics by the real functions of time familiar to us from classical theory. Instead, we must use mathematical instructions called operators—instructions that tell us how to extract information about observables from wave functions. The saga of operators in quantum mechanics begins in § 5.3, continues through this chapter, and is concluded in Chaps. 10 and 11.

My first job is to try to convince you that the representation of observables in Newtonian mechanics, which we'll review in § 5.1, isn't applicable to the microscopic universe. To this end, I'll focus (beginning in § 5.2) on a single, extremely important observable: *linear momentum.* Once I've shown you why the classical representation of momentum ($p = mdx/dt$) is incompatible with the physics of microscopic particles, I'll introduce the quantum representation of this property, the operator $\hat{p} = -i\hbar\partial/\partial x$, in § 5.3. The use of this strange beast to represent momentum is the thrust of our next fundamental postulate (number III, if you're counting), which I'll present in § 5.4. To begin the process of getting acquainted with operators, we'll examine some of their algebraic properties in § 5.5 and will test their compatibility with classical theory, in the appropriate limit, at the end of this chapter.

A final note: I appreciate that the idea of representing a physically measurable quantity by a mathematical instruction probably seems mighty peculiar. So throughout this chapter, I'll adopt a ruthlessly pragmatic stance towards the representation of observables. Our goal will be to come up with devices that we can use to extract information about physical properties from wave functions. So think about operators as a quantum physicist's tools—the quantum mechanical counterpart of a carpenter's screwdrivers, hammers, and lathes.

5.1 OBSERVABLES FOR MACROSCOPIC SYSTEMS

Observables are the essence of physics. Experimental physicists measure the values of observables in the laboratory. Theoretical physicists calculate them—either by hand or, more likely nowadays, on a computer. To many classical physicists, the goal of Newtonian physics is to understand the nature of macroscopic systems through the study of their observables; classical mechanics is a tool kit for predicting the values of these observables from the trajectories of the particles comprising the system.

For even the simplest physical system, it's easy to think of lots of observables. Some are dependent on other observables; some are not. If, for example, we know the linear momentum $\mathbf{p}(t)$ of a particle, we can determine its kinetic energy $T = p^2/2m$. But if we want to evaluate its potential energy $V(\mathbf{r})$, we must know the position $\mathbf{r}$. In any case, we can (in principle) measure the values of all the observables of a macroscopic system to infinite accuracy, secure in the knowledge that measuring one observable does not change the value of another (independent) observable. If, for example, we measure the position of a billiard ball, we do not thereby alter its kinetic energy (unless we are very maladroit).

In classical physics, observables are represented by functions of time. To understand why this strategy *doesn't* work for microscopic systems, let's see why it *does* work for macroscopic systems. Essential to the Newtonian description of a classical particle is the premise that at any time the particle *has* a position and a linear momentum—and a host of other properties that can be determined from these observables. This means that we can ascribe (real) numbers—the values of these properties—to the particle, whether or not we measure them. These values usually change with time, so it makes sense to establish a correspondence between *observables* and *functions*: *e.g.*, in one dimension we have

CLASSICAL PHYSICS

OBSERVABLE		FUNCTION
position	$\Longleftrightarrow$	$x(t)$
momentum	$\Longleftrightarrow$	$p(t)$
energy	$\Longleftrightarrow$	$E[x(t), p(t)]$

This correspondence is the bedrock of Newtonian mechanics, the physical theory of macroscopic objects.

Let's see how this correspondence works for a couple of observables for a single macroscopic particle of mass m. The position of the particle at time t is specified by the values of three real numbers, the spatial coordinates.[1] The most commonly used coordinates are

$$\mathbf{r} = (x, y, z) \qquad \text{rectangular coordinates}$$
$$\mathbf{r} = (r, \theta, \varphi) \qquad \text{spherical coordinates}$$

Classical physicists define the other fundamental property descriptor, the linear momentum $\mathbf{p}(t)$, in terms of the trajectory,

$$\mathbf{p}(t) = m\frac{d}{dt}\mathbf{r}(t). \tag{5.1}$$

One reason the momentum is so important is that we can write all other classical observables as functions of $\mathbf{r}$ and $\mathbf{p}$.

Some of these other observables are scalar quantities, like the total energy of a conservative system

$$E = T(\mathbf{p}) + V(\mathbf{r}) \tag{5.2a}$$
$$= \frac{p^2}{2m} + V(\mathbf{r}). \tag{5.2b}$$

For example, the total energy of a one-dimensional simple-harmonic oscillator with natural frequency ω_0 is

[1]Remember that a coordinate system is just an artifice. In this book, we'll use primarily rectangular or spherical coordinates. But practicing physicists work with lots of different coordinate systems, usually basing their choices on the symmetry properties of the physical system they're studying. For example, a central force problem has spherical symmetry and hence is more easily (and more physically) described in spherical than in rectangular coordinates. We can define an infinite number of independent coordinate systems, but physicists most often use one of eleven special systems: those in which the "Helmholtz equation"

$$\nabla^2\psi - k^2\psi = 0$$

can be separated. If you want to learn more about useful coordinate systems, see Chap. 2 of *Mathematical Methods for Physicists*, 2nd ed. by George Arfken (New York: Academic Press, 1970). If you want to have on hand absolutely everything you ever conceivably could need to know about coordinate systems, get a copy of *Field Theory Handbook*, 2nd ed. by P. Moon and D. E. Spencer (New York: Springer-Verlag, 1971). This awesome compendium of equations and facts is invaluable, although it isn't the sort of thing you would want to take as light reading on a vacation.

$$E = \frac{p^2}{2m} + \frac{1}{2}m\omega_0^2 x^2. \tag{5.3}$$

Others are vector quantities, like the angular momentum

$$\mathbf{L} = \mathbf{r} \times \mathbf{p}. \tag{5.4}$$

All of these functions play a role in the dynamics of classical physics; in fact, the *laws* of physics are relationships between these representations of observables. For example, the laws of motion of classical physics can be written as[2]

$$\frac{d}{dt}\mathbf{p} = -\nabla V(\mathbf{r}) \tag{5.5}$$

$$\frac{d}{dt}\mathbf{L} = \mathbf{r} \times \mathbf{F} = \mathbf{r} \times \frac{d}{dt}\mathbf{p}. \tag{5.6}$$

But if we try to use such functions in a physical theory for the microworld, we run into the conceptual obstacle I noted in the introduction to this chapter: A quantum system does not, in general, have *a* position, *a* momentum, *an* energy, etc., at a particular time. So it makes no sense to represent the physical properties of such a system by functions of time. Does this mean that we must abandon the very idea of physical properties?

Certainly not. The *concept* of position, for example, is certainly meaningful for a microscopic particle, for its position can be measured. The problem is that prior to such a measurement, the particle does not *have* a position in any meaningful sense. This is the essential difference between the position of a macroscopic particle and that of a microscopic particle: the position of a quantum particle is "latent." It becomes "actual" only when someone measures it. This is true in general of the observables of quantum physics: they manifest themselves only when measured, so we cannot meaningfully represent them by functions of time.[3]

To summarize our problem: A quantum physics worthy of the name must consist of physical laws. These laws must be expressed in terms of mathematical constructs that represent measurable quantities of the system they describe. But we cannot write these laws using the property descriptors of classical physics. To discover what sort of machinery we do use to represent observables in quantum mechanics, we must probe more deeply into the problems with the classical way of knowledge. So in the next section we shall concentrate on a single, very important observable: the linear momentum.

5.2 THE TROUBLE WITH MOMENTUM

What do we mean by *the momentum of a microscopic particle?* You may think I answered this question in Chap. 4, where I showed you how to use Fourier analysis to extract momentum information from a state function. But there is still a problem.

[2]To get through this book you won't need to know classical physics at a very sophisticated level. But if you didn't realize that Eq. (5.5) is an alternative form of Newton's Second Law, $\mathbf{F} = m\mathbf{a}$, you probably should spend a little time with your favorite classical physics book—or one of the Suggested Readings at the end of this chapter.

[3]A fine discussion of "latent" observables can be found in Chaps. 8 and 17 of Henry Margenau's superb book *The Nature of Physical Reality: A Philosophy of Modern Physics* (Woodbridge, Conn.: Ox Bow Press, 1977). Margenau refers to the observables of classical systems as "possessed" observables. (I am indebted to Margenau's writings on the meaning of physics for much of the philosophy in this book.)

Physics is a pragmatic science. Its fundamental quantities are introduced via *operational definitions*—definitions that specify, implicitly or explicitly, a series of steps that can be implemented in a laboratory. In her excellent survey of the physics of matter, Ginestra Amaldi discusses this important underlying principle of physics[4]

> ...one should not introduce into physics magnitudes which cannot be measured by means of an experiment, an experiment which can be executed—or if it cannot be executed *because of practical limitations of the apparatus available* is at least in principle feasible. ... This principle... goes to the heart of the problem of knowledge.

This principle is particularly germane to quantum theory, since the predictions of quantum mechanics are strictly limited to *measurable quantities*. In this sense, quantum mechanics is a narrowly pragmatic theory: it is mute about quantities we cannot study in the laboratory.

Momentum Measurements on a Macroscopic Particle

To illustrate the idea of an *operational definition*, which is crucial to understanding the trouble with the classical definition of momentum, let's contemplate a measurement of momentum. Consider first a macroscopic particle of mass m moving serenely along in one dimension. The classical definition of the momentum of the particle is

$$p(t) = m\frac{d}{dt}x(t). \tag{5.7a}$$

This is an operational definition, as we can see if we rewrite the derivative in (5.7*a*) as a limit. Letting x_1 and x_2 denote the positions of the particle at times t_1 and t_2, we have

$$p = m \lim_{t_2 \to t_1} \frac{x_2 - x_1}{t_2 - t_1}. \tag{5.7b}$$

Equation (5.7*b*) prescribes an operational procedure: we measure the position of the particle at two successive times (as close together as possible) and calculate p from Eq. (5.7*b*). Simple, right? Yes—if the particle is macroscopic.

Momentum Measurements in the Microworld

Now let's think about applying this definition to microscopic systems. The argument I want to make is a little subtle, so follow closely.

Consider an ensemble of microscopic particles whose state *the instant before the first position measurement* is represented by $\Psi(x, t_1)$. Like the states we saw in Chaps. 3 and 4, this one corresponds to an indeterminate position $[\Delta x(t_1) > 0]$. So before the first measurement, the situation of the particles in our ensemble is like that of the electrons in the double-slit experiment prior to their arrival at the detector: *they do not have a value of position.*

At time t_1, we perform an ensemble measurement of position. For each particle in the ensemble, one of the possibilities latent in $\Psi(x, t_1)$ becomes actual (*e.g.*, spots appear). Therefore our measurement has changed the state of the particles in the ensemble from one in which the position was uncertain to ones in which the positions are definite.

[4]Ginestra Amaldi, *The Nature of Matter: Physical Theory from Thales to Fermi* (Chicago: University of Chicago Press, 1966), pp. 102–103, emphasis mine.

Now, our definition (5.7*b*) calls for a second position measurement at time t_2. How might we perform such a measurement? Well, we could start by selecting all particles in the initial ensemble that in the first measurement exhibited the particular value x_1. (This collection of particles is a sub-ensemble all of whose members are, at time t_1, in the quantum state with definite position $x = x_1$.) As time passes, the state of these particles will evolve into one with an uncertain position, so the second position measurement after even a short interval will yield a range of values. (We'll explore such measurements in detail in Chap. 13.) We could pick one of these, call it $x_2(t_2)$, and calculate the momentum from (5.7*b*).

I hope this prescription makes you nervous; it's fraught with logical problems. For one thing, we don't know which of the positions exhibited at t_1—or, for that matter, at t_2—to use in (5.7*b*). You might suggest that we get around this difficulty by modifying this definition to use not *particular* values of position, but rather the expectation values $\langle x \rangle (t_1)$ and $\langle x \rangle (t_2)$. This suggestion is insightful, but it doesn't resolve a second, far more serious problem: the measurement at t_1 changed the state of the particles in the ensemble. So the subsequent behavior of the particle is not what it would have been had we not performed the first measurement. In fact, *the change of state induced by the first measurement alters the results of the second measurement.*

So if we use (5.7*b*), we don't really know what we're calculating! It's certainly not "the momentum of the particle." It's just a number: the difference between an unmeasured and a measured position divided by a time interval and multiplied by a mass. But this number does not describe a property *of the particle itself.*

In a nutshell: the formal problem with applying the classical definition of momentum to an ensemble of microscopic particles arises because *it is impossible to measure the position of such particles without altering their state.* This vicissitude is inevitable, because prior to measurement, the particles do not have a position or a momentum. (Once again we see that it makes no sense to ascribe trajectories to microscopic particles.)

That is, we cannot represent the observable momentum by $m\,dx/dt$, and we must resist the temptation to think about it in classical terms. Since the classical definition of momentum cannot be implemented for microscopic systems, this definition is, in quantum mechanics, meaningless.[5]

Limitations of the Fourier Road to Momentum

"Wait just a minute," you cry. "I remember that in Chap. 4 we calculated all sorts of information about the momentum of a free particle: the momentum probability distribution, the mean value of the momentum, the momentum uncertainty. Are you saying that this information is meaningless?!"

Certainly not. It's not momentum *information*, but rather the *classical definition* of this quantity that loses significance in the quantum realm.

"But," you continue, "it seems mighty strange that in Chap. 4 we could calculate information about an observable that we cannot even define."

It does indeed. But the reason is that in Chap. 4, I deftly dodged the question of *what we mean by momentum* by basing the development on the de Broglie relation $p = \hbar k$—in effect, I snuck up on momentum via functions of the wavenumber. In

[5]For this reason, velocity is not considered a fundamental attribute of microscopic particles. In classical physics, the velocity of a particle (in one dimension) is just $v = dx/dt$, but in quantum mechanics, velocity is *defined* as $v \equiv p/m$.

particular, we used this relationship to convert the Fourier transform of the wave function,

$$A(k) = \mathcal{F}[\Psi(x,0)] = \frac{1}{\sqrt{2\pi}} \int_{-\infty}^{\infty} \Psi(x,0)\, e^{-ikx}\, dx, \tag{5.8}$$

into the momentum probability amplitude

$$\Phi(p) = \frac{1}{\sqrt{\hbar}} A\left(\frac{p}{\hbar}\right). \tag{5.9}$$

Once we adopted the Born interpretation of this function, we could calculate various probabilistic and statistical quantities relating to momentum, such as the expectation value

$$\langle p \rangle = \int_{-\infty}^{\infty} \Phi^*(p)\, p\, \Phi(p)\, dp, \tag{5.10}$$

by simply generalizing our definitions for position.

The disadvantage to this route to momentum information is that it's circuitous and often awkward. Were Eq. (5.8) our only recourse, then in every application we would have to evaluate $A(k)$. The algebra involved in this little chore is often horrendous. So for practical purposes, if no other, we need a more direct way to extract momentum information from state functions.

But there's another reason for looking again at momentum: relying on the de Broglie relation to avoid the question of how to define momentum is really a cop-out. The lack of a definition of so fundamental an observable is a gaping hole in the edifice of our theory.

So in the next section we'll tackle this problem head on. Once we've devised a mathematical tool for extracting momentum information, our strategy will be to adopt this tool as the operational definition of momentum in quantum mechanics. Knowing how to represent the momentum (and position) of microscopic particles, we can write down tools for other observables. And thus shall we arrive at Postulate III.

5.3 AND NOW FOR SOMETHING COMPLETELY DIFFERENT: THE MOMENTUM OPERATOR

> In a contemplative fashion,
> In a tranquil state of mind,
> Free of every kind of passion
> Some solution we shall find.
>
> —from *The Gondoliers*
> by Gilbert and Sullivan

When I say "we extract momentum information from a state function," I mean, of course, *probabilistic or statistical information*—for that is all nature will allow. We know from Chap. 3 how to calculate such information for position—for example, the mean value of position is the expectation value

$$\langle x \rangle = \int_{-\infty}^{\infty} \Psi^*(x,t)\, x\, \Psi(x,t)\, dx. \tag{5.11}$$

But suppose we want the mean value of the *momentum* of a particle in the state $\Psi(x,t)$ and are forbidden (or, more likely, unable) to take the Fourier route of Chap. 4? We could reason by analogy, speculating that the mathematical form of $\langle p \rangle$ might be the same as that for $\langle x \rangle$; *i.e.*,

$$\langle p \rangle = \int_{-\infty}^{\infty} \Psi^*(x,t)\, p\, \Psi(x,t)\, dx. \tag{5.12}$$

But now what do we do? What do we insert in (5.12) for p? To proceed further, we need a "momentum extractor"—a mathematical device that represents p in such a way that we can evaluate expressions like (5.12). Evidently, our momentum extractor should be expressed in terms of x, the variable of integration in (5.12).

A Momentum Extractor for a Pure Momentum State Function

To simplify our quest for a momentum extractor, I want to consider for a moment a very simple function: one of the pure momentum functions

$$\Psi_p(x,t) = \frac{1}{\sqrt{2\pi\hbar}}\, e^{i(px-Et)/\hbar} \tag{5.13}$$

that make up the wave packet

$$\Psi(x,t) = \int_{-\infty}^{\infty} \Phi(p)\, \Psi_p(x,t)\, dp. \tag{5.14}$$

In particular, consider the pure momentum function

$$\begin{aligned}\Psi_p(x,t) \\ &= (5.375 \times 10^2 \text{m}^{-\frac{1}{2}}) \exp\{i[(4.9 \times 10^{12}\text{m}^{-1})\, x - (5.8 \times 10^{18}\text{sec}^{-1})\, t]\}.\end{aligned} \tag{5.15}$$

Our mission, should we decide to accept it, is to extract the value of the momentum p from this function. [No peeking; it's not fair to just read the value of p off Eq. (5.15).]

What sequence of operations will, when applied to the pure momentum function (5.13), produce the *value* of p. Well, since p appears in the argument of an exponential, we might get it out by differentiating with respect to x, *viz.*,

$$\frac{\partial}{\partial x}\, e^{i(px-Et)/\hbar} = \frac{ip}{\hbar}\, e^{i(px-Et)/\hbar}. \tag{5.16}$$

Notice that we must use a partial derivative in (5.16), because the function $\Psi_p(x,t)$ depends on two independent variables, x and t.

The first derivative almost does the job: this operation produces the momentum multiplied by $i/\hbar$. To get just the value of p, we should operate on $\Psi_p(x,t)$ with $(\hbar/i)\, \partial/\partial x$, to wit:

$$\frac{\hbar}{i}\frac{\partial}{\partial x}\Psi_p(x,t) = p\, \Psi_p(x,t). \tag{5.17}$$

Eureka! We have discovered an **operator**—a mathematical instruction—that extracts the value of momentum from a pure momentum function. I'll denote this operator by $\hat{p}$ and, because I prefer my factors of $\sqrt{-1}$ in the numerator, use $1/i = -i$ to write its definition as

$$\boxed{\hat{p} \equiv \frac{\hbar}{i}\frac{\partial}{\partial x} = -i\hbar\frac{\partial}{\partial x}} \qquad \text{momentum operator} \tag{5.18}$$

In words, this operator instructs us to do the following:

$$\hat{p} \quad \Longleftrightarrow \quad \text{differentiate with respect to } x\text{, then multiply by } -i\hbar. \tag{5.19}$$

Throughout this book, I'll put little hats on operators to distinguish them from the observables they represent.

Question 5–1

Use the tool defined in Eq. (5.18) to extract *the value* of the linear momentum from the pure momentum state (5.15).

A Leap of Faith: The Momentum Operator for an Arbitrary State

The little demonstration above—and Question 5–1—shows how to use the operator $-i\hbar\partial/\partial x$ to extract momentum information *from a pure momentum state function.* But such a function corresponds to a *definite* momentum and is not normalizable; at best it can represent a quantum state only approximately. So how do we represent momentum in calculations involving the arbitrary wave packet (5.14)?

The essential assumption of this chapter, which we'll make formal in the next section, is the obvious answer: use $-i\hbar\partial/\partial x$ to represent the observable momentum *for any quantum state* in one dimension. This assumption establishes the following correspondence:

QUANTUM PHYSICS

OBSERVABLE		OPERATOR
momentum	$\Longleftrightarrow$	$\hat{p}$

Aside: The Momentum Operator in Three Dimensions. For a particle in three dimensions, the momentum is represented by an operator with three components, $\hat{p}_x$, $\hat{p}_y$, and $\hat{p}_z$. These quantities are defined just like (5.18), *i.e.*,

$$\hat{p}_x = -i\hbar\frac{\partial}{\partial x} \qquad \hat{p}_y = -i\hbar\frac{\partial}{\partial y} \qquad \hat{p}_z = -i\hbar\frac{\partial}{\partial z}.$$

Such an operator is, sensibly, called a *vector operator.* The vector operator for linear momentum can be conveniently written in terms of the gradient operator ∇ as $-i\hbar\nabla$.

A Mathematical Aside: On Eigenvalue Equations. Our basic assumption is that $\hat{p}$ can be used to determine momentum information about any state. In general, the effect of $\hat{p}$ on a function $\Psi(x,t)$ is to produce some *other* function of x and t. But something special happens when $\hat{p}$ operates on a pure momentum state function:

$$\hat{p}\,\Psi_p(x,t) = p\Psi_p(x,t). \qquad \text{[for a pure momentum state only]} \tag{5.20}$$

Equation (5.20), *which is not a general relationship*, has a special mathematical form: an operator, in this case $\hat{p}$, acting on a function equals a number times the function. Equations of this form are called *eigenvalue equations.* We'll encounter such equations repeatedly in our journey through the wonderland of quantum theory and, in Chap. 11, will examine them in their own right.

Question 5–2

Show that when $\hat{p}$ acts on the function

$$\Psi(x,0) = \left(\frac{1}{2\pi L^2}\right)^{1/4} e^{-x^2/(4L^2)}$$

the result is not a constant times $\Psi(x,0)$. This means that particles in a state represented by this function do not have a well-defined value of the momentum (see Chap. 11).

Back to the Expectation Value of Momentum

Now we can answer the question I left dangling after Eq. (5.12): how can we evaluate the expectation value of momentum from the wave function $\Psi(x,t)$? Answer: merely insert the operator $\hat{p}$ in Eq. (5.12), *viz.*,

$$\langle p \rangle = \int_{-\infty}^{\infty} \Psi^*(x,t)\,\hat{p}\,\Psi(x,t)\,dx \qquad (5.21a)$$

$$= \int_{-\infty}^{\infty} \Psi^*(x,t)\left(-i\hbar\frac{\partial}{\partial x}\right)\Psi(x,t)\,dx. \qquad (5.21b)$$

Equation (5.21*b*) instructs us to carry out the following steps:

1. differentiate $\Psi(x,t)$ with respect to x;
2. multiply the result by $-i\hbar$ and then by $\Psi^*(x,t)$;
3. integrate the resulting integrand with respect to x.

Perform these three steps, and voilà, you have calculated the average value that would be expected in a measurement at time t of the linear momentum on an ensemble of particles in a state represented by $\Psi(x,t)$.[6] Let's carry out these instructions for a specific case.

Example 5.1. Expectation Value of Momentum for the Infinite Square Well

In Example 3.2 we looked at some of the wave functions for a single particle with a very simple potential: a symmetric finite square well. In various problems in Chaps. 3 and 4, we consider a simple variant on this potential, the infinite square well

$$V(x) = \begin{cases} -\infty & < x < -\frac{1}{2}L; \\ 0 & -\frac{1}{2}L \le x \le \frac{1}{2}L; \\ \infty & \frac{1}{2}L < x < \infty. \end{cases}$$

Notice that this potential is infinite everywhere except inside a region of length L, where it is zero. In Chap. 7 we'll study in detail the quantum states of this system, but for now I want to consider one particular state for purposes of applying the momentum operator to it.

Like all bound systems, the particle in an infinite square well has a state of minimum energy. This state is called the ground state, and in this state the energy is well-defined (*i.e.*, for this state $\Delta E = 0$). The wave function for the ground state at $t = 0$ is zero outside the region $-\frac{1}{2}L \le x \le \frac{1}{2}L$, and inside this region is proportional to a simple cosine function:

$$\Psi_1(x,t) = \begin{cases} 0 & -\infty < x < -\frac{1}{2}L; \\ \sqrt{\frac{2}{L}}\cos\left(\pi\frac{x}{L}\right) & -\frac{1}{2}L \le x \le \frac{1}{2}L; \\ 0 & \frac{1}{2}L < x < \infty. \end{cases} \qquad (5.22)$$

[6] In Chap. 7 we'll discover special quantum states called *stationary states* for which $\langle p \rangle$ is independent of time. But in general, the expectation value of any observable depends on time.

Question 5–3

What is the expectation value of *position* for this state? Does your answer imply that the position of a particle in this state is definite—*i.e.*, is the position uncertainty for this state zero? Why or why not? (Think, don't calculate!)

Now, suppose that at $t = 0$ we measure the linear momentum on an ensemble of these systems in the ground state. To figure out the average value we'll find in the measurement, we insert the wave function (5.22) into our operational expression (5.21) for the expectation value of momentum, *viz.*,

$$\langle p \rangle = \int_{-\infty}^{\infty} \Psi_1^*(x,0)\,\hat{p}\,\Psi_1(x,0)\,dx \tag{5.23a}$$

$$= \frac{2}{L}(-i\hbar)\int_{-L/2}^{L/2} \cos\left(\pi\frac{x}{L}\right)\left[\frac{d}{dx}\cos\left(\pi\frac{x}{L}\right)\right]dx. \tag{5.23b}$$

[Note very carefully the limits of integration in (5.23*b*). Be *sure* you understand why *for this system* these limits are $-L/2$ and $L/2$, not $-\infty$ and ∞.] We can evaluate the integral in (5.23*b*) either by brute force or by thought. By now, I suspect you can guess which strategy I prefer: we think.

Looking at the integrand in (5.23*b*), we notice immediately that one of the functions in it has *definite parity*: the cosine function is *even* with respect to the midpoint of the infinite square well, $x = 0$. Therefore the first derivative of this function,

$$\frac{d}{dx}\left[\cos\left(\pi\frac{x}{L}\right)\right] = -\frac{\pi}{L}\sin\left(\pi\frac{x}{L}\right), \tag{5.24}$$

is *odd* with respect to $x = 0$. And we all know what *the integral over a symmetric interval of the product of an even and an odd function* is, right? Right: the integral is zero. Ergo,

$$\langle p \rangle = 0 \qquad \text{(ground state of the infinite square well)}. \tag{5.25}$$

Question 5–4

Evaluate $\langle p \rangle$ by explicit integration (just this once). Isn't it better to *think before you integrate*?

Of course, we could have used the methods of Chap. 4 to calculate $\langle p \rangle$, first determining the momentum probability amplitude for the ground state of the infinite square well,

$$\Phi(p) = \frac{1}{\sqrt{2\pi\hbar}}\int_{-L/2}^{L/2} \Psi_1(x,0)\,e^{-ipx/\hbar}\,dx, \tag{5.26}$$

then substituting this function into

$$\langle p \rangle = \int_{-\infty}^{\infty} \Phi^*(p)\,p\,\Phi(p)\,dp. \tag{5.27}$$

But this procedure involves a lot more work (as you discovered if you worked Problem 4.6).

Before leaving this example, I want you to take a moment to ponder the physical consequences of the fact that in the ground state of the infinite square well, $\langle p \rangle = 0$. This result does not imply that the momentum in this state is well-defined, *i.e.*, it does not mean that in the ensemble measurement, all members yield the same value of p. *Be sure you understand this point.*

Question 5–5

The energy of the ground state of a particle in an infinite square well is well-defined—its value (see Chap. 7) is

$$E_1 = \frac{\pi^2 \hbar^2}{2mL^2}. \tag{5.28}$$

Now, $V = 0$ inside the box, so in the region where the state function is non-zero, the energy and momentum are related by $E = T = p^2/2m$. Does the fact that the energy of the state is well-defined contradict the fact that the momentum is not sharp? If not, explain why not. (If so, we are in big trouble.)

For future reference, here's the handy mathematical property we used to subdue the integral in Example 5.1. Mark it well:

$$\boxed{\begin{aligned} \frac{\partial}{\partial x}\,[\text{even function of } x] &= \text{odd function of } x \\ \frac{\partial}{\partial x}\,[\text{odd function of } x] &= \text{even function of } x \end{aligned}} \tag{5.29}$$

Momentum: Final Ruminations

I don't want to leave you with the wrong impression: in this section, I have not actually *derived* the momentum operator.[7] But I *have* argued that in quantum theory we must treat momentum differently than we do in classical physics and that it is reasonable to use the operator $\hat{p} = -i\hbar\partial/\partial x$ to determine momentum information from a state function $\Psi(x,t)$.

In fact, we are in the happy situation of having *two* ways to calculate the expectation value of momentum. We can calculate $\langle p \rangle$ from $\Psi(x,t)$ directly, a la Eq. (5.21), *or* we can evaluate the momentum probability amplitude $\Phi(p) = \mathcal{F}_p\left[\Psi(x,0)\right]$ and use Eq. (5.27). If we execute the required algebra correctly, both procedures will give the same answer, for they are entirely equivalent. In a few cases, the Fourier road is the easier of the two. But for most problems, evaluating the momentum amplitude is such a pain that we prefer to use the momentum operator.

But enough about momentum. Lots of other physical properties are of interest in physics—energy, angular momentum, etc. Practitioners of Newtonian mechanics can represent these observables by functions of position and momentum and (perhaps) time. But we need to know how to represent observables in quantum physics. In the next section, we'll unearth a simple procedure for devising a quantum mechanical operator for any observable.

[7]Different authors sometimes choose a different set of postulates in their formulation of quantum mechanics. Some sets of postulates permit the *derivation* of the operator identification $\hat{p} = -i\hbar\partial/\partial x$ from Fourier analysis of $\Psi(x,t)$, rather than postulation of this correspondence. Still other authors set up their postulates so they can obtain this correspondence from the Correspondence Principle (see § 5.7). Time permitting, you should sample several different ways of laying out quantum theory; *e.g.*, Chap. 3 of *Introductory Quantum Mechanics* by Richard L. Liboff (San Francisco: Holden-Day, 1980); Chap. 4 of *The Quantum Physicists and An Introduction to Their Physics* by W. H. Cropper (London: Oxford University Press, 1970); and *Elementary Quantum Mechanics* by David Saxon (San Francisco: Holden-Day, 1968).

5.4 THE OPERATOR POSTULATE

In § 5.3 we established the fundamental correspondence between the observable *momentum* and the mathematical operator $\hat{p}$: in the quantum theory of a single particle in one dimension, $\hat{p} = -i\hbar\partial/\partial x$ *represents* momentum. The third postulate of quantum mechanics, the topic of this section, is a generalization of this idea to an arbitrary observable.

The Position Operator

The importance of the observable position in physical theory is comparable to that of momentum. In classical physics, we can write the functions that represent observables in terms of these observables, and in quantum mechanics we can express (most) operators in terms of the operators that represent position and momentum. In § 5.3, we developed the operator $\hat{p} = -i\hbar\partial/\partial x$ for momentum, and in Chap. 3 used the operator for position (without so identifying it). Let's review.

Look again at the expression for the expectation value of position,

$$\langle x \rangle = \int_{-\infty}^{\infty} \Psi^*(x,t) x \Psi(x,t)\, dx. \tag{5.30}$$

What does this expression tell us to do? Well, the first step in calculating $\langle x \rangle$ is evaluating $x\Psi(x,t)$. This is an *instruction* (*aha!*): it says "multiply $\Psi(x,t)$ by x."

This suggests that we define the **position operator** $\hat{x}$ as

$$\hat{x} \iff \text{multiply by } x. \qquad \text{position operator} \tag{5.31}$$

This definition established a second correspondence,

OBSERVABLE		OPERATOR
position	$\iff$	$\hat{x}$

Formally, we indicate the use of the position operator in Eq. (5.30) for $\langle x \rangle$ by just replacing the x between $\Psi^*(x,t)$ and $\Psi(x,t)$ by $\hat{x}$, as

$$\langle x \rangle = \int_{-\infty}^{\infty} \Psi^*(x,t)\, \hat{x}\, \Psi(x,t)\, dx. \tag{5.32}$$

You may be wondering why I didn't define $\hat{x}$ in Chap. 3? It seemed to me that introducing so formal a definition so early in our introduction to quantum mechanics might appear to be a sadistic complication. But now we are far enough along to see the usefulness of introducing the position operator: since any classical observable can be written as a function of x and p, we can use the operators $\hat{x}$ and $\hat{p}$ to translate classical observables into quantum mechanical operators. That is precisely what we are about to do.

Generating Operators

Consider an arbitrary observable. I'll denote a generic observable by the symbol Q and will write the function that represents this observable in classical physics $Q(t)$ if I want to emphasize the time dependence and $Q(x,p)$ otherwise. To obtain the operator that represents Q in quantum mechanics, we just replace the variable x in $Q(x,p)$ by the position operator $\hat{x}$ and the variable p by the momentum operator $\hat{p}$, *i.e.*,[8]

CLASSICAL OBSERVABLE		QUANTUM OPERATOR
$Q(x,p)$	$\Longleftrightarrow$	$\hat{Q}(\hat{x},\hat{p})$

In practice we merely insert $\hat{p} = -i\hbar\partial/\partial x$ everywhere in $Q(x,p)$ that p appears. (What should we insert everywhere that x appears?) To summarize:

$$\boxed{\begin{array}{ccc} x & \Longleftrightarrow & \hat{x} \\ p & \Longleftrightarrow & \hat{p} = -i\hbar\dfrac{\partial}{\partial x} \end{array}} \tag{5.33}$$

That's all there is to it![9]

Example 5.2. The Energy Operator for the Simple Harmonic Oscillator

To apply our new tool, let's work out the energy operator for a simple harmonic oscillator (SHO): a microscopic particle of mass m with natural frequency ω_0 and potential energy

$$V(x) = \tfrac{1}{2}m\omega_0^2x^2. \tag{5.34}$$

The first step in developing a quantum mechanical operator for any observable is to write down the classical function that represents that observable.

In classical mechanics, the function that represents the total energy is the *Hamiltonian* $H(\mathbf{r},\mathbf{p})$. For a single particle with a velocity-independent potential energy, the classical Hamiltonian is just the familiar function[10]

$$H[\mathbf{r}(t),\mathbf{p}(t)] = T(\mathbf{p}) + V(\mathbf{r},t).$$

The operator that represents the total energy of a physical system in quantum mechanics is also called the **Hamiltonian**. *It is represented by the symbol* $\hat{\mathcal{H}}$.

To determine the Hamiltonian of a microscopic simple harmonic oscillator, we start with the classical function

$$H(x,p) = T(p) + V(x) \tag{5.35}$$

$$= \frac{p^2}{2m} + \frac{1}{2}m\omega_0^2x^2. \tag{5.36}$$

[8]This procedure will handle nearly any operator. But in a few cases, the operator it produces isn't unique. In others, the operator it produces fails to satisfy a vital property called Hermiticity (see Chap. 11). Quantum mechanics has (advanced) machinery to deal with either instance, and such cases are so rare that you needn't worry about them. But you should be aware of them.

[9]Well, almost all. If the state functions we're working with are written as functions of position (and time), Eqs. (5.33) will generate appropriate operators. But in some problems we prefer to represent quantum states by functions of momentum—or of some other observable. In such cases, we simply translate the operators $\hat{x}$, $\hat{p}$, and $\hat{Q}$ into the appropriate "representation" (see § 5.8).

[10]Formally, the Hamiltonian is defined in terms of the Lagrangian of the system and the conjugate coordinates and momenta of the particles that comprise the system. This function is used in *Hamilton's formulation* of classical physics, an alternative to Newton's theory. For an introduction to classical Hamiltonian mechanics, see § 7.12 of *Classical Dynamics of Particles and Systems*, 2nd ed. by Jerry B. Marion (New York: Academic Press, 1970), or § 6–14 of *Introduction to the Principles of Mechanics* by Walter Hauser (Reading, Mass: Addison-Wesley, 1965). At a more advanced level, you should peruse the beautiful development in Chap. VII of *Mechanics*, 2nd ed. by L. D. Landau and E. M. Lifshitz (Reading, Mass.: Addison-Wesley, 1969). There are close, subtle, and provocative connections between the Hamiltonian formulation of classical mechanics and Schrödinger wave mechanics, as you'll see if you read Chap. 8 of the fine text *Introduction to Quantum Mechanics* by R. H. Dicke and J. P. Wittke (Reading, Mass.: Addison-Wesley, 1960).

To generate *the Hamiltonian operator* for the simple harmonic oscillator we just apply the prescription (5.34) to each occurrence of x and p in this function, *viz.*,

$$\hat{\mathcal{H}} = \hat{T} + \hat{V} = \frac{\hat{p}^2}{2m} + \frac{1}{2}m\omega_0^2\hat{x}^2. \qquad \text{(SHO)} \tag{5.37}$$

Notice that the **kinetic energy operator**

$$\hat{T} = \frac{\hat{p}^2}{2m} \qquad \text{kinetic energy operator} \tag{5.38}$$

is independent of the potential and so applies to any single-particle one-dimensional system.

Now, suppose we want to use the Hamiltonian operator (5.37) to evaluate, say, the mean value of the energy of an ensemble of simple harmonic oscillators in a state represented by a known wave function $\Psi(x,t)$. By analogy with our expressions for $\langle x \rangle$ and $\langle p \rangle$, we write

$$\langle E \rangle = \int_{-\infty}^{\infty} \Psi^*(x,t)\, \hat{\mathcal{H}}\, \Psi(x,t)\, dx. \tag{5.39}$$

The first step in evaluating this integral is working out the effect of the operator $\hat{\mathcal{H}}$ on $\Psi(x,t)$. So we write $\hat{\mathcal{H}}$ in terms of x by inserting the explicit forms of $\hat{x}$ and $\hat{p}$ into (5.37):

$$\hat{\mathcal{H}} = \frac{1}{2m}\left(-i\hbar\frac{\partial}{\partial x}\right)^2 + \frac{1}{2}m\omega_0^2\hat{x}^2. \tag{5.40}$$

To proceed to the next step, we must evaluate $(-i\hbar\partial/\partial x)^2\, \Psi(x,t)$.

Depending on your mathematics background, you may or may not know how to do this, how to act on a function with the square of an operator. In case you don't know, I'll defer the rest of this example until I've shown you a little about operator algebra (§ 5.5). To be continued.

The Operator Postulate on Stage

To close this section, I want to make formal the scheme we have deduced for representing observables. This postulate is on a par with Postulate I, which specified how to represent the other class of elements of quantum theory: states of a system. Postulate III specifies how to represent the observables of a system:

> ***The Third Postulate of Quantum Mechanics***
>
> In quantum mechanics, every observable is represented by an operator that is used to obtain physical information about the observable from state functions. For an observable that is represented in classical physics by the function $Q(x,p)$, the corresponding operator is $\hat{Q}(\hat{x},\hat{p})$.

A final note: In Part IV, where we probe deeper into the formalism of quantum mechanics, we'll learn that not just any mathematical operator can represent an observable. Like state functions, the class of observables of use in quantum mechanics is restricted. Just as the functions that represent quantum states must be complex and normalizable, so must the operators that represent observables satisfy a condition called *Hermiticity* (see Problem 5.1).

Moreover, some operators in quantum mechanics do not correspond to classical observables. Indeed, in examining microscopic systems, we'll meet some observables, such as spin, that have no classical counterpart. And we'll make the acquaintance of operators, such as the parity operator, that can't be written as functions of $\hat{x}$ and $\hat{p}$. All this is yet to come.

Operators: A Historical Aside. The idea that the laws of quantum physics could be formulated in terms of operators was first propounded by Max Born, whom we met in Chap. 3, and Norbert Weiner, a mathematician and child prodigy. In late 1925, Born visited MIT, where he gave a series of guest lectures on atomic structure and solid state physics. There he renewed his acquaintance with Weiner, who was an instructor in mathematics. At this time Schrödinger had not published his wave mechanics, and Born and the rest of the physics community were using a theory, due to the ubiquitous Werner Heisenberg, in which observables were represented by *matrices.*[11]

Weiner, who received his Ph.D. at the ripe old age of 18, was already an expert on "the operator calculus," which he learned while using Fourier analysis to unravel the complicated electrical oscillations that make up information in telephone lines.[12] Listening to Born lecture about Heisenberg's ideas, Weiner realized that these ideas could be usefully expressed in the language of operators. This realization sparked a collaboration that ultimately led to a formulation of quantum theory that, although incomplete, was more powerful and more widely applicable than matrix mechanics—especially for Born's beloved atomic collision problems.

Unfortunately, Born and Weiner did not fully develop their theory, and hence their contribution did not have the impact it might have. Although they used an operator $\hat{p}$ for linear momentum, they did not spot the vital connection $\hat{p} = -i\hbar\, \partial/\partial x$. And their timing was rotten, for within a few months (in 1926) Erwin Schrödinger stunned the physics community with his wave mechanics, which immediately became the formulation of choice. Born has commented on his oversight: "I shall never forgive myself, because if we had done this, we would have had the whole wave mechanics at once, a few months before Schrödinger."[13]

Born and Weiner applied their theory successfully to the microscopic simple harmonic oscillator (Example 5.2) and to uniform motion in one dimension. And, although their work was almost immediately overshadowed by Schrödinger's, their idea of using operators in quantum mechanics was here to stay. A host of physicists further developed operator mechanics—notably, Paul A. M. Dirac—and their work is now an essential part of the repertoire of every practicing quantum mechanic.

5.5 THE MATHEMATICS OF OPERATORS: AN INTRODUCTION

We need education in the obvious
more than investigation into the obscure.

—Oliver Wendell Holmes

Norbert Weiner, who with Max Born devised the operator postulate, learned *operator mechanics* as part of his training as a mathematician. And this subject is properly considered a branch of mathematics. In the 18th century, differential calculus was all the

[11]Heisenberg's theory is still in use today, under the appropriate name *matrix mechanics*. For an introduction to this alternate approach to quantum mechanics, see *Matrix Mechanics* by H. S. Green (The Netherlands: P. Noordhoff, Ltd., 1965).

[12]Weiner was a fascinating man, and if you are at all interested in the lives of the shapers of contemporary physics, look up his autobiography, which he modestly (and appropriately) entitled *I Am A Mathematician: The Later Life of a Prodigy* (Garden City, N. Y.: Doubleday, 1956).

[13]Quoted in *The Historical Development of Quantum Theory,* Volume 3 by J. Mehra and H. Rechenberg (New York: Springer-Verlag, 1982).

rage. But many mathematicians were uncomfortable with the "metaphysical implications" of a mathematics formulated in terms of infinitesimal quantities (like dx). This disquiet was the stimulus for the development of the operator calculus.[14]

Two 18th-century mathematicians, Gottfried Wilhelm Leibniz (1646–1716) and Joseph-Louis Lagrange (1736–1813) founded the operator method. But the formal theory that Born and Weiner adapted to quantum mechanics was developed later by three men: Francois Joseph Servois, Robert Murphy, and George Boole. You may have heard of Boole for his "Boolean algebra"—a theory based on operator methods in which Boole, who was nothing if not ambitious, tried to explain all of mathematics.

To anyone who has studied calculus, operators are familiar critters. Indeed, the essential elements of calculus are operators: the derivative $\partial/\partial x$ and the integral $\int^x \cdots dx'$. (The centered dots in this integral represent the function being integrated; *e.g.*, $\sin x'$ in $\int^x \sin x'\, dx'$.)

An **operator**, then, is just an *instruction*, a symbol that tells us to perform one or more mathematical acts on a function—say, $f(x)$. The result is another function $g(x)$, which may be a constant. For this reason, the effect of operators is sometimes denoted by the symbol $\mapsto$. Here are two equivalent ways of writing the effect of an operator $\hat{Q}$ on a function $f(x)$:

$$g(x) = \hat{Q}f(x) \qquad \Longleftrightarrow \qquad \hat{Q} : f(x) \mapsto g(x).$$

The essential point about operators is that they act on functions; we might think of operators as analogous to active verbs in English, with functions as their objects.

The instruction we are told to perform may be as trivial as "multiply $f(x)$ by the constant 4," *e.g.*,

$$\hat{4}f(x) = 4f(x), \tag{5.41}$$

or as complicated as "integrate $f(x)$ along a specified contour in the complex plane." A sampler of operators is presented in Table 5.1.

TABLE 5.1 A SAMPLER OF OPERATORS

Symbol	Instructions	Example
$\hat{3}$	multiply by 3	$\hat{3}\, f(x) = 3f(x)$
$\hat{A}$	differentiate once	$\hat{A}\, f(x) = \dfrac{d}{dx}\, f(x)$
$\hat{B}$	form complex conjugate	$\hat{B}\, f(x) = f^*(x)$
$\hat{C}$	destroy	$\hat{C}\, f(x) = 0$
$\hat{D}$	expand in Taylor series	$\hat{D}\, f(x) = \sum_{n=0}^{\infty} \dfrac{1}{n!}\, x^n\, \dfrac{d^n f}{dx^n}$

Working with Operators

Every quantum mechanic must be able to manipulate, combine, and simplify expressions containing operators—expressions such as $\hat{x}\hat{p} - \hat{p}\hat{x}$. To flawlessly manipulate such

[14]For a history of the development of the operator calculus, see the article by Salvatore Pincherle in *Encykl. d. math. Wiss. II/1*, Part 2 (issue no. 6, March 1906, pp. 761–817).

expressions, you must remember that like surgeons, operators must operate on something. An operator without a function is like a verb without an object. You're much more likely to do your operator algebra correctly if you will follow this rule:

> ***Rule***
>
> Always write operator expressions and equations in terms of a general function $f(x)$.

Suppose, for example, we want to simplify the operator expression $\hat{x}\hat{p} - \hat{p}\hat{x}$ (as we shall in Example 5.4). Our first step should be to write this expression in terms of a generic function $f(x)$, as $\hat{x}\hat{p}f(x) - \hat{p}\hat{x}f(x)$. *Then* we try to simplify.

In such manipulations, it's also important to keep in mind that[15]

> ***Rule***
>
> Operators act on everything to their right unless their action is constrained by brackets.

Thus, when we write $\hat{p}f_1(x)f_2(x)$, we *mean* $\hat{p}[f_1(x)f_2(x)]$, not $[\hat{p}f_1(x)]$ times $f_2(x)$. (Do you see the difference?)

Operator Algebra

Mastering the algebra of operators isn't particularly difficult. But you must be wary lest you inadvertently slip into patterns of thought familiar from your experience with the mathematics of numbers and functions. Operators are not just funny-looking numbers or functions: the algebra of operators is, in some important respects, quite different from that of numbers or functions.

To appreciate the distinction between operators and other mathematical species, let's consider a simple example. Let the symbol A represent a *real number*, say $A \equiv 3$. Suppose that for some reason we need to evaluate the combination $Af(x)g(x)$. Does the *position* of A in this expression make any difference?

No—real numbers *commute* with functions; *e.g.*, $3f(x) = f(x)3$. So we can write the expression $Af(x)g(x)$ in several equivalent ways:

$$Af(x)g(x) = f(x)Ag(x) = g(x)Af(x) \tag{5.42a}$$

$$3f(x)g(x) = f(x)3g(x) = g(x)3f(x) \tag{5.42b}$$

This trivial example illustrates the principle of *commutativity of real numbers*.

Does this principle apply to operators? To find out, let's try one—say, the simple operator $\hat{A} \equiv d/dx$. The first expression in (5.42*a*) is

[15]There is an important exception to this rule. The infinitesimal in an integral operator is part of the instruction. For example, dx' in $\hat{A} \equiv \int^x \cdots dx'$ is part of the integration operator, so

$$\hat{A}\, f(x) = \int^x f(x')\, dx'.$$

This is one reason mathematicians sometimes write the integrand after the infinitesimal, as $\int^x dx' \cdots$.

$$\hat{A}\, f(x)g(x) = \frac{d}{dx} f(x)g(x) \tag{5.43a}$$

$$= f(x)\frac{d}{dx}g(x) + g(x)\frac{d}{dx}f(x). \tag{5.43b}$$

And the second is

$$f(x)\hat{A}\, g(x) = f(x)\frac{d}{dx}g(x). \tag{5.44}$$

Except in certain special cases (5.44) is not the same as (5.43*b*). (And for $A = d/dx$, the third expression, $f(x)g(x)\hat{A}$ is meaningless.) The moral of this trifling exercise is:

> ***WARNING***
>
> **Relations that are "obvious" from our experience with real numbers and functions may be incorrect when extended (without thought) to operators.**

How to Cope with Sums, Differences, and Products

In this book, you'll only need mastery of the simplest features of the algebra of operators. The simplest operator manipulations—addition and subtraction—rarely pose problems. As you can readily verify, two operators $\hat{Q}_1$ and $\hat{Q}_2$ can be added or subtracted according to the familiar rules

$$\begin{aligned}\left(\hat{Q}_1 \pm \hat{Q}_2\right) f(x) &= \hat{Q}_1\, f(x) \pm \hat{Q}_2\, f(x) \\ &= \pm\hat{Q}_2 f(x) + \hat{Q}_1 f(x)\end{aligned} \qquad \begin{bmatrix}\text{addition}\\ \text{subtraction}\end{bmatrix} \tag{5.45}$$

Mathematicians refer to this property as *distributivity*.

More interesting is what happens when we form the *product* of two (or more) operators. The product $\hat{Q}_1\, \hat{Q}_2$ cannot be treated like the products of two numbers or of two functions. Instead, the rules of operator algebra assign to the product combination the following meaning:

> ***Rule***
>
> The product of operators implies successive operation.

For instance, to evaluate $\hat{Q}_1\, \hat{Q}_2 f(x)$, we first act on $f(x)$ with $\hat{Q}_2$ to form a new function of x—call it $g(x)$. We then act on $g(x)$ with $\hat{Q}_1$ to form the desired function. That is, we can break this product down into *two operations*:

$$h(x) = \hat{Q}_1\, \hat{Q}_2 f(x) \qquad \Longleftrightarrow \qquad \begin{cases}\hat{Q}_2\colon f(x) \mapsto g(x)\\ \hat{Q}_1\colon g(x) \mapsto h(x)\end{cases}$$

Notice that I've used the mathematician's symbol $\mapsto$ to vivify the *active* role of the operators $\hat{Q}_1$ and $\hat{Q}_2$.

This little analysis demonstrates the following rule:

> ***Rule***
>
> The product of two operators is a third operator.

So we could write the product considered above by defining yet a third operator, $\hat{Q}_3$, that is *equivalent* to the product, *viz.*,[16]

$$\hat{Q}_3 \equiv \hat{Q}_1 \hat{Q}_2 \qquad \text{and} \qquad h(x) = \hat{Q}_3 f(x). \tag{5.46}$$

Please pay special attention to the meaning of operator products. This is the facet of operator algebra that most often trips up newcomers to quantum mechanics. So etch the above rule in your memory.

Example 5.3. The Kinetic Energy Operator Revisited

Using the simple rules of this section in hand, we can finish Example 5.2. There we set out to develop an expression for the Hamiltonian of a simple harmonic oscillator in terms of operators and functions of x. We developed the following expression for this Hamiltonian [Eq. (5.40)]:

$$\hat{\mathcal{H}} = \frac{1}{2m}\left(-i\hbar\frac{\partial}{\partial x}\right)^2 + \frac{1}{2}m\omega_0^2\hat{x}^2. \tag{5.47}$$

To complete our labors, we must simplify the first term in (5.47), the kinetic energy operator; *i.e.*, we must work out the effect on a function $f(x)$ of the *square* of the momentum operator $\hat{p}^2$.

According to the above guidelines, the *wrong* way to evaluate $\hat{p}^2$ is to treat it as though $\hat{p}$ were a function, *i.e.*,

$$\hat{p}^2 f(x) \neq \left[-i\hbar\frac{d}{dx}f(x)\right]^2.$$

The *right* way is to follow instructions: operate on $f(x)$ twice in succession with $\hat{p}$:

$$\hat{p}^2 f(x) = \hat{p}\hat{p}f(x) \tag{5.48a}$$

$$= -\hbar^2\frac{d^2}{dx^2}f(x). \tag{5.48b}$$

Ergo the form of the **kinetic energy operator** we use to act on functions of x is

$$\boxed{\hat{T} = \frac{\hat{p}^2}{2m} = -\frac{\hbar^2}{2m}\frac{\partial^2}{\partial x^2}.} \qquad \text{kinetic energy operator} \tag{5.49}$$

With this operator, the desired Hamiltonian (5.47) is simply

$$\hat{\mathcal{H}} = -\frac{\hbar^2}{2m}\frac{\partial^2}{\partial x^2} + \frac{1}{2}m\omega_0^2\hat{x}^2. \qquad \text{[one-dimension simple harmonic oscillator]} \tag{5.50}$$

In Chap. 9, we'll study the state functions of the simple harmonic oscillator that we obtain from this Hamiltonian.

[16]Two operators $\hat{Q}$ and $\hat{R}$ are said to be *equivalent* if for all functions $f(x)$, they satisfy

$$\hat{Q}f(x) = \hat{R}f(x). \qquad \text{equivalent operators}$$

The Commutator: A Sneak Preview

What are **commuting operators**? Ones that drive to work? No, no. Two operators $\hat{Q}_1$ and $\hat{Q}_2$ commute if they obey the simple operator expression

$$\boxed{\hat{Q}_1 \hat{Q}_2 = \hat{Q}_2 \hat{Q}_1.} \qquad \text{commuting operators} \qquad (5.51)$$

Ponder this equation carefully. If it seems to you that *any* two operators should obey it, then you're thinking about operators as though they were real numbers, and are in trouble.

The critical feature of Eq. (5.51) is that it's an *operator equation.* Since operators must operate on functions, it follows that

> ***Rule***
>
> To be valid an operator equation must hold for an arbitrary (well-behaved) function $f(x)$.

This rule is the key to verifying valid operator equations—and exposing invalid ones. I can't overemphasize this point: if an operator equation is true, it must hold for *any* function. But if it fails for *a single* function, then it's false.

If, for example, you are investigating two operators $\hat{Q}_1$ and $\hat{Q}_2$ and can find *one function* for which $\hat{Q}_1\hat{Q}_2 f(x) \neq \hat{Q}_2\hat{Q}_1 f(x)$, then you have proven that $\hat{Q}_1$ and $\hat{Q}_2$ do not commute.

Let's get specific. Consider the operators

$$\hat{Q}_1 = \frac{d}{dx} \qquad (5.52)$$

$$\hat{Q}_2 = \hat{x}. \qquad (5.53)$$

Let these guys act on the function $f(x) = e^{2x}$. What happens? To find out, first evaluate $\hat{Q}_1\hat{Q}_2 f(x)$, then evaluate $\hat{Q}_2\hat{Q}_1 f(x)$.

You should have found that the functions that result from these operator products differ depending on the *order* in which $\hat{Q}_1$ and $\hat{Q}_2$ act, *i.e.*,

$$\hat{Q}_1\hat{Q}_2 e^{2x} \neq \hat{Q}_2\hat{Q}_1 e^{2x}. \qquad (5.54)$$

The fact that Eq. (5.51) does not hold *for the single function* $f(x) = e^{2x}$ is enough to prove that the operators d/dx and x do not commute, *i.e.*,

$$\frac{d}{dx}\hat{x} \neq \hat{x}\frac{d}{dx}. \qquad (5.55)$$

As always, my examples have a moral. This one is:

> ***WARNING***
>
> **The order in which the operators operate is vital; never assume that two operators commute. Always check their commutativity.**

Mathematical Aside: Commutativity of Functions. Non-commutativity isn't a problem for functions such as $f(x)$ and $g(x)$ or real numbers (such as a and b), since the algebras of real numbers and of functions are commutative, *i.e.*,

$$a \cdot b = b \cdot a \tag{5.56a}$$

$$f(x)\, g(x) = g(x)\, f(x) \tag{5.56b}$$

Question 5–6

Dream up some pairs of operators that *do* commute. Now come up with some pairs that *do not* commute.

Of all the mathematical peculiarities of operators we've examined, their commutativity (or non-commutativity) is most important physically. In Chap. 11, we'll unearth a remarkable connection between this property of operators and uncertainty relations involving observables. As a sneak preview, I'll close this section with an important example.

Example 5.4. Do Position and Momentum Commute?

Consider the operator $\hat{Q} \equiv \hat{x}\hat{p} - \hat{p}\hat{x}$. Were we to encounter this operator in a quantum mechanics problem, our first thought would be to try to simplify it—to reduce it, if possible, to a simpler operator.[17] But before tackling this chore, we'd write its definition in terms of a generic function, as

$$\hat{Q} f(x) = \hat{x}\hat{p} f(x) - \hat{p}\hat{x} f(x). \tag{5.57}$$

We begin the task of simplifying $\hat{Q}$ by writing Eq. (5.57) in terms of x by inserting into this equation explicit forms for $\hat{p}$ and $\hat{x}$, *viz.*,

$$\hat{Q} f(x) = x\left[-i\hbar\frac{\partial}{\partial x} f(x)\right] - \left(-i\hbar\frac{\partial}{\partial x}\right)\left[x f(x)\right] \tag{5.58a}$$

$$= -i\hbar x\frac{df}{dx} + i\hbar\frac{d}{dx}\left[x f(x)\right]. \tag{5.58b}$$

Now we expand the first derivative in the second term in Eq. (5.58*b*) to give two terms, one of which cancels the *first* term in (5.58*b*). This manipulation leads to a delightfully simple form for $\hat{Q}$:

$$\hat{Q} f(x) = i\hbar f(x). \tag{5.59}$$

To write (5.59) as an *operator* equation, I must introduce the identity operator $\hat{1}$, which does very little:

$$\hat{1} f(x) = f(x). \qquad \text{identity operator} \tag{5.60}$$

Voilà:

$$\boxed{\hat{x}\hat{p} - \hat{p}\hat{x} = i\hbar\hat{1}} \tag{5.61}$$

We have discovered the curious fact that operating on a function with this particular combination of $\hat{x}$ and $\hat{p}$ is equivalent to multiplying the function by a constant. Amazingly, this fact leads to the Heisenberg Uncertainty Principle! (Isn't quantum mechanics fantastic!) But to see how, you must wait for Chap. 11.

[17] $\hat{Q}$ is a critter called a *commutator* (Chap 11). In particular, $\hat{x}\hat{p} - \hat{p}\hat{x}$ is the commutator of position and momentum, which is denoted $[\hat{x}, \hat{p}]$.

5.6 GENERALIZING OUR EXPECTATIONS (AND UNCERTAINTIES)

What do we do with operators once we have found them? Well, one common application is the calculation of **expectation values** and **uncertainties** for various quantum states. You now know how to use Postulate III to design the operator $\hat{Q}$ that represents whatever observable Q we're interested in. So if I give you the wave function $\Psi(x,t)$ that represents such a state, you can evaluate these two pieces of statistical information about the state from the following straightforward generalizations of familiar equations:[18]

$$\boxed{\langle Q \rangle = \int_{-\infty}^{\infty} \Psi^*(x,t)\hat{Q}\Psi(x,t)\,dx} \quad \text{expectation value of an arbitrary observable} \tag{5.62}$$

$$\boxed{\begin{aligned}\Delta Q &\equiv \sqrt{\left\langle (Q - \langle Q \rangle)^2 \right\rangle} \\ &= \sqrt{\langle Q^2 \rangle - \langle Q \rangle^2}.\end{aligned}} \quad \text{uncertainty of an arbitrary observable} \tag{5.63}$$

To illustrate the use of these expressions, let's return to the ground state of the infinite square well (Example 5.1).

Example 5.5. The Momentum Uncertainty for the Ground State of the Infinite Square Well

The wave function of this state, $\Psi_1(x,0)$, is given in Eq. (5.22). In Example 5.1, we discovered that the expectation value of momentum for this state is $\langle p \rangle = 0$. So in a measurement of the momentum performed at $t = 0$ on an ensemble consisting of a huge number of identical particle-in-a-box systems, each in the ground state, the average of the values we'd get is zero. To predict how these values are distributed about this mean, we must calculate the momentum uncertainty

$$\Delta p = \sqrt{\langle p^2 \rangle - \langle p \rangle^2}, \tag{5.64}$$

which for this state simplifies to

$$\Delta p = \sqrt{\langle p^2 \rangle} \qquad \text{for a state in which } \langle p \rangle = 0. \tag{5.65}$$

So to determine the momentum uncertainty, we need only evaluate

$$\left\langle p^2 \right\rangle = \int_{-\infty}^{\infty} \Psi^*(x,t)\,\hat{p}^2\,\Psi(x,t)\,dx \tag{5.66a}$$

$$= -\hbar^2 \int_{-\infty}^{\infty} \Psi^*(x,t)\frac{\partial^2}{\partial x^2}\Psi(x,t)\,dx. \tag{5.66b}$$

This isn't hard. For the wave function $\Psi_1(x,0)$ in Eq. (5.22) we have

$$\left\langle p^2 \right\rangle = -\hbar^2 \frac{2}{L}\int_{-L/2}^{L/2} \cos\left(\pi\frac{x}{L}\right)\left[\frac{d^2}{dx^2}\cos\left(\pi\frac{x}{L}\right)\right]dx \tag{5.67a}$$

$$= \frac{2\pi^2\hbar^2}{L^3}\int_{-L/2}^{L/2} \cos^2\left(\pi\frac{x}{L}\right)dx. \tag{5.67b}$$

[18] We could justify these expressions formally via extensions of the arguments used in Chap. 3 [*c.f.*, the discussion surrounding Eq. (3.40)]. I should note, however, that some authors treat Eq. (5.62) as a part of Postulate III.

Note carefully that I changed the partial derivative with respect to x in (5.66) to an ordinary derivative, because x is the only variable in Eqs. (5.67)—and watch those limits of integration!

Now, can we use parity to simplify the integral in (5.67*b*)? Well, the integrand has definite parity, but it's even—not odd. So the integral isn't equal to zero. We can, however, write this integral as

$$\int_{-L/2}^{L/2} \cos^2\left(\pi \frac{x}{L}\right) dx = 2 \int_{0}^{L/2} \cos^2\left(\pi \frac{x}{L}\right) dx. \tag{5.68}$$

We can now integrate by parts. If we define

$$u = \pi \frac{x}{L}, \tag{5.69}$$

we can write Eq. (5.67b) in terms of u as

$$\langle p^2 \rangle = \frac{2\pi^2\hbar^2}{L^3} 2 \frac{L}{\pi} \int_0^{\pi/2} \cos^2 u \, du. \tag{5.70}$$

A trip to the nearest integral table yields

$$\int_0^{\pi/2} \cos^2 u \, du = \frac{\pi}{4}. \tag{5.71}$$

And, using this integral in (5.70) and taking the square root of the result, as prescribed by (5.65), we find the momentum uncertainty

$$\Delta p = \frac{\pi\hbar}{L} \qquad \text{(ground state of infinite square well).} \tag{5.72}$$

Since $\Delta p > 0$, the momentum of particles in the ground state is not sharp. So in an ensemble measurement, various members yield different values of p. These values are distributed about the mean momentum for this state, $\langle p \rangle = 0$.

Example 5.6. The Heisenberg Uncertainty Principle—Verified.

With very little additional work, we can for the first time test the Heisenberg Uncertainty Principle. According to this fundamental tenet of quantum physics, the product of the position and momentum uncertainties *for any quantum state of any system* must be greater than or equal to $\hbar/2$. We just evaluated Δp for a particular quantum state of a particular system: the ground state of the infinite square well. So let's evaluate Δx for this state and thence the uncertainty product.

The position uncertainty is no more difficult to evaluate than was the momentum uncertainty. From the form of the wave function (5.22) we see that in the region where it's non-zero (inside the "box") this function is a simple cosine. This function is symmetric about $x = 0$, so the expectation value of position in this state is zero:

$$\langle x \rangle = \int_{-\infty}^{\infty} \Psi_1^*(x,0)\, \hat{x}\, \Psi_1(x,0)\, dx = 0. \tag{5.73}$$

Question 5–7

Argue from the fact that the probability density $|\Psi_1(x,0)|^2$ is symmetric and from the *physical* meaning of $\langle x \rangle$ that this quantity must be zero. Now carry out the integral in (5.73) explicitly and prove that your argument is correct.

This being the case, the position uncertainty simplifies to

$$\Delta x = \sqrt{\langle x^2 \rangle}. \tag{5.74}$$

So all we need do now is evaluate the mean value of x^2,

$$\langle x^2 \rangle = \frac{2}{L} \int_{-L/2}^{+L/2} x^2 \cos^2\left(\pi \frac{x}{L}\right) dx. \tag{5.75}$$

A second implementation of integration by parts [a là Eq. (5.69)] yields $\langle x^2 \rangle = 0.0327\, L^2$, whence the position uncertainty is

$$\Delta x = 0.1808\, L, \tag{5.76}$$

Question 5–8

Fill in the steps leading to Eq. (5.76).

Now we can evaluate the uncertainty product:

$$\Delta x\, \Delta p = 0.57\hbar. \tag{5.77}$$

Sure enough, for the ground state of the infinite square well, this product is greater than $0.5\hbar$, which verifies the prediction of the Heisenberg Uncertainty Principle. In practice, we don't often bother to substantiate this principle (which we'll prove in Chap. 11). But a verification like Eq. (5.77) is a good way to check your algebra in the evaluation of uncertainties. Consider this a hint.

Aside: A Discouraging Word about Uncertainties. Evaluating uncertainties for observables other than position and momentum usually involves pretty formidable algebra. Suppose, for example, that you want to evaluate the mean value of the energy of a particle in the infinite square well,

$$\Delta E = \sqrt{\langle E^2 \rangle - \langle E \rangle^2}. \tag{5.78}$$

To be sure, *formally* this task poses no problems. The two expectation values are just

$$\langle E \rangle = \int_{-\infty}^{\infty} \Psi^*(x,t)\, \hat{\mathcal{H}}\, \Psi(x,t)\, dx \tag{5.79a}$$

$$\langle E^2 \rangle = \int_{-\infty}^{\infty} \Psi^*(x,t)\, \hat{\mathcal{H}}^2\, \Psi(x,t)\, dx. \tag{5.79b}$$

Inside the box $V = 0$, so the Hamiltonian simplifies to the kinetic energy operator $\hat{T}$. Since the wave function is zero outside the box, these expectation values are just the finite integrals

$$\langle E \rangle (t) = \int_{-L/2}^{L/2} \Psi^*(x,t) \hat{T} \Psi(x,t)\, dx$$

$$\langle E^2 \rangle (t) = \int_{-L/2}^{L/2} \Psi^*(x,t) \hat{T}^2 \Psi(x,t)\, dx.$$

But look at these integrals. The expectation value doesn't look too bad:

$$\langle E \rangle (t) = \int_{-L/2}^{L/2} \Psi^*(x,t) \frac{\hat{p}^2}{2m} \Psi(x,t)\, dx$$
$$= \int_{-L/2}^{L/2} \Psi^*(x,t) \left(-\frac{\hbar^2}{2m} \frac{\partial^2}{\partial x^2} \right) \Psi(x,t)\, dx.$$

But to calculate the uncertainty ΔE we must evaluate

$$\langle E^2 \rangle = \int_{-L/2}^{L/2} \Psi^*(x,t) \left(\frac{1}{4m^2} \hat{p}^4 \right) \Psi(x,t)\, dx.$$

For most wave functions, evaluating $\hat{p}^4\, \Psi(x,t)$, if possible at all, is quite arduous — it's certainly not the sort of thing you would want to do unless you absolutely had to.

Yet, all is not lost. There are many ways to skin the proverbial cat.[19] In Part IV we'll develop diverse strategies for evaluating expectation values and uncertainties, strategies that are more clever (and *much* easier) than brute force application of the equations in this section.

5.7 OPERATORS AND THE CORRESPONDENCE PRINCIPLE

One of the most curious things about the *operators* that represent observables in quantum mechanics is that they seem to have no clear physical relationship to the *functions of time* that represent observables in classical physics. To be sure, there is a formal relationship between the two: the operator $\hat{Q}$ that represents an observable Q is related to the classical function $Q[x(t), p(t)] = Q(x, p)$ via the replacements

$$x \iff \hat{x} = x \qquad \text{and} \qquad p \iff \hat{p} = -i\hbar \frac{\partial}{\partial x}.$$

Still, we intuitively suspect that if, as we required in Chap. 1, quantum physics is to be consistent with classical physics, then when applied to a state of a *macroscopic* system, the operator $\hat{Q}$ should behave like the function $Q(t)$. This consistency requirement is the essence of the Correspondence Principle (see § 1.2):

> ***Rule***
>
> **The Correspondence Principle**: In the classical limit, the laws of quantum mechanics must reduce to those of Newtonian Mechanics.

But precisely *how* does this principle work, in practice? That is, what is the nature of the physical correspondence between the quantum mechanical operator $\hat{Q}$ and the classical function $Q(t)$? Well, here's an example: if we compute the value at time t of the property Q for a macroscopic system from the operator $\hat{Q}$ and from the function $Q(t)$, we should get the same number to within any measurable accuracy. (Remember that the statements of quantum physics are limited to what can be measured in the lab.) In order to see if this consistency requirement is satisfied, we must look closely at the meaning of such evaluations—and that is the thrust of this section.

[19]Cat lovers: please don't write letters. This is a metaphor. I love cats.

The Classical Limit—And How to Get There

The key phrase in the statement of the Correspondence Principle is "the classical limit." You probably have a feeling for what this phrase means: an electron is *not* a classical system; a racquetball is. But to understand the relationship between classical and quantum physics, you need to understand the precise meaning of this limit.

Many quantum mechanics texts define the classical limit in a way that, to newcomers, often sounds absurd. Consider, for example, the following statement of the Correspondence Principle:[20] "In the limit where $\hbar \to 0$, the laws of Quantum Mechanics must reduce to those of Classical Mechanics."

Think about this statement. What could it possibly mean? We know that $\hbar$ is an immutable fundamental constant of nature; its value isn't zero, it's 1.0546×10^{-34} J-sec. (If $\hbar$ were zero, this book, its author, and its readers would not exist.)

The statement of the classical limit as "the limit in which $\hbar \to 0$" is not wrong; it's just a little terse. To understand it, we must recall the limitation of quantum physics to *measurable phenomena.* That is, quantum effects—which follow from the fact that $\hbar \neq 0$—are of no significance *unless they can be measured.* So what the quoted statement really means is *the classical limit is the limit in which all quantum effects are immeasurably small.*

Example 5.7. The Classical Limit and Uncertainties

To illustrate this essential idea, let's ponder an aspect of quantum theory that is blatantly non-classical: the limitations on knowledge we see reflected in the Heisenberg Uncertainty Principle $\Delta x\,\Delta p \geq \hbar/2$. I want to apply this principle to an ensemble of macroscopic systems, to show you that for such systems, quantum mechanical uncertainties are not measurable and hence are irrelevant—*i.e.*, in this sense, a macroscopic system behaves according to our classical expectations.

Suppose we want to simultaneously measure the position and velocity of an ensemble of identical particles of mass m. *Whether the particles are microscopic or macroscopic, these measurements are subject to the Heisenberg Uncertainty Principles.* To conveniently express this restriction, I'll define a *velocity uncertainty* as

$$\Delta v \equiv \frac{1}{m}\Delta p \tag{5.80}$$

and write the uncertainty principle as

$$\Delta x\,\Delta v \geq \frac{1}{2}\frac{\hbar}{m}. \tag{5.81}$$

Now, suppose the particles are *macroscopic*, say, racquetballs with $m = 0.1\,\text{kg}$. This large mass makes the product (5.81) very small,

$$\Delta x\,\Delta v \approx 10^{-33}. \tag{5.82}$$

True, it's non-zero—but we can't verify this fact experimentally. So for all practical (*i.e.*, experimental) purposes, the uncertainty product is zero:

$$\Delta x\,\Delta v \approx 0 \qquad \text{[for macroscopic particles]}. \tag{5.83}$$

[20] From *Quantum Mechanics* by Albert Messiah (New York: Wiley, 1966), p. 214.

The point of this example is that *although the uncertainty principles of quantum mechanics do apply to macroscopic particles, they in no way limit the accuracy of a measurement of position and velocity.*

The demonstration in Example 5.7 can be put on a more rigorous footing by writing down a wave packet that represents a state of the particles in the ensemble and calculating from this function the uncertainties Δx and Δp. One finds that the individual uncertainties Δx and Δp are negligibly small, *i.e.*,[21]

$$\Delta x \approx 0 \qquad \text{and} \qquad \Delta p \approx 0 \qquad \text{[for macroscopic particles]}. \tag{5.84}$$

Equation (5.84) makes perfect sense. At any time, a macroscopic particle *has* a well-defined value of x and of p, so we should be able to measure these observables to any desired precision (allowing for experimental inaccuracy). In the language of quantum physics, this implies that quantum indeterminacies Δx and Δp should be immeasurably small. As they are.

Statistical Quantities in the Classical Limit

We've seen that in the classical limit, quantum mechanical uncertainties become negligible. I want you to think for a moment about the following question: what should the expectation value $\langle Q \rangle$ of an observable Q reduce to in this limit?

***** Pause while reader ponders *****

A clue to the answer is to be found in the definition of $\langle Q \rangle$. Remember that, like the uncertainty, the expectation value is a *statistical quantity*; *i.e.*, it describes the results of an ensemble measurement—a series of identical measurements performed, at the same time t, on identical systems in the same state. If the state can be described by classical physics, all members of the ensemble exhibit the same result (ignoring, for the moment, experimental error): the value of the function $Q(t)$ at the measurement time. (That's why the standard deviation of these results, ΔQ, is zero.) So the average of these results, the expectation value $\langle Q \rangle$, is $Q(t)$. We conclude that in the classical limit, $\langle Q \rangle (t)$ should become equal, to any measurable accuracy, to $Q(t)$.

Aside: The Classical Limit of a Microscopic System. I don't want to leave you with the impression that "the classical limit" is limited to macroscopic particles. For some *microscopic* systems, all quantum mechanical uncertainties are immeasurably small *for certain states*. These particular states, therefore, behave according to the laws of classical physics. For example, highly energetic states of a microscopic simple harmonic oscillator behave in this fashion, because (Chap. 9) for such states all quantum effects are too small to be measured. The point is that *the classical limit may pertain either to all states of a macroscopic system or to certain states of a microscopic system.*

We have discovered an extremely important key to how the Correspondence Principle works:

[21]For details, see § 12-7 of *Elementary Quantum Mechanics* by P. Fong (Reading, Mass: Addison-Wesley, 1960) and Chap. 8 of *Introduction to Quantum Mechanics* by R. H. Dicke and J. P. Wittke (Reading, Mass: Addison-Wesley, 1960).

Rule

The element of quantum theory that in the classical limit corresponds to the classical function $Q(t)$ is the expectation value of the operator $\hat{Q}$ with respect to the wave function $\Psi(x,t)$ that represents the state.

That is, we get from the state functions of quantum mechanics to the classical limit via the statistical quantities $\langle Q \rangle$ and ΔQ:

$$\begin{aligned} \langle Q \rangle (t) &\xrightarrow[\text{the classical limit}]{} Q(t) \\ \Delta Q(t) &\xrightarrow[\text{the classical limit}]{} 0 \end{aligned} \tag{5.85}$$

The Rules of the Game

In this chapter, we've explored the elements of quantum theory that represent physical properties, operators, and how they are related to the corresponding elements of classical theory, functions of time. What we have not investigated are the *laws* we construct from these elements. But the Correspondence Principle demands that *the laws of quantum physics reduce to those of classical physics in the classical limit.* How, you may be wondering, does this work?

Here's a preview of a result we'll prove in Chap. 11. Once we have studied the time dependence of state functions and expectation values, we'll easily verify that $\langle p \rangle$ and $\langle x \rangle$ obey the provocatively familiar equation

$$\langle p \rangle = m \frac{d}{dt} \langle x \rangle . \tag{5.86a}$$

You can probably see already what this equation translates into when we take the classical limit. In this limit, the expectation values in (5.86*a*) behave according to Eq. (5.85), *i.e.*,

$$\langle x \rangle (t) \xrightarrow[\text{the classical limit}]{} x(t) \tag{5.86b}$$

$$\langle p \rangle (t) \xrightarrow[\text{the classical limit}]{} p(t). \tag{5.86c}$$

So the quantum mechanical *law* (5.86*a*) reduces (to within the accuracy of any measurement) to the familiar equation

$$\langle p \rangle = m \frac{d}{dt} \langle x \rangle \xrightarrow[\text{the classical limit}]{} p(t) = m \frac{d}{dt} x(t). \tag{5.87}$$

Isn't that satisfying?

5.8 FINAL THOUGHTS: OPERATORS IN MOMENTUM SPACE

> Let's review the situation,
> Solve the complicated plot.
> Quiet calm deliberation
> Disentangles every knot.
>
> —-from *The Gondoliers*
> by Gilbert and Sullivan

With this chapter, we've devised ways to represent the two essential elements of physical theory: states and observables. And we've begun assembling a dictionary of the "property descriptors" of quantum physics: observables (see Table 5.2).

TABLE 5.2 A PARTIAL DICTIONARY FOR OBSERVABLES AND CORRESPONDING OPERATORS

Observable	Operator	Instructions
position	$\hat{x}$	multiply by x
momentum	$\hat{p}$	$-i\hbar\dfrac{\partial}{\partial x}$
Hamiltonian	$\hat{\mathcal{H}}$	$\hat{T}+\hat{V}$
kinetic energy	$\hat{T}$	$-\dfrac{\hbar^2}{2m}\dfrac{\partial^2}{\partial x^2}$
potential energy	$\hat{V}$	multiply by $V(x,t)$

In the examples of this chapter, we used operators to evaluate expectation values and uncertainties. But operators tell us far more than just the values of statistical quantities. In Chap. 11, we'll examine the *eigenvalues* of an operator. In quantum mechanics, these numbers have a profound, startling physical significance: *the eigenvalues of an operator are the only values that can be obtained in a measurement of the observable that operator represents.* Nothing remotely like this occurs in classical physics.

The systems we've investigated have consisted of a single particle in one dimension, so I wrote the operators in Table 5.2 in terms of the position variable x. But we can generalize all of them to more complicated systems. For instance, the kinetic-energy operator for a single particle in three dimensions is

$$\hat{T} = \frac{1}{2m}\left(\hat{p}_x^2 + \hat{p}_y^2 + \hat{p}_z^2\right) \tag{5.88a}$$

$$= -\frac{\hbar^2}{2m}\nabla^2 \tag{5.88b}$$

$$= -\frac{\hbar^2}{2m}\left(\frac{\partial^2}{\partial x^2} + \frac{\partial^2}{\partial y^2} + \frac{\partial^2}{\partial z^2}\right). \tag{5.88c}$$

The position variables x, y, and z appear explicitly in Eq. (5.88c), which we therefore call "the kinetic energy operator in the position representation."

But the position representation is but one of a multitude of "representations" used in quantum mechanics. The idea of alternative representations of the elements of quantum theory is not new: in § 4.8, I remarked that the position probability amplitude $\Psi(x, 0)$ is but one of many ways to represent a state at $t = 0$. Another, entirely equivalent descriptor for this state is the momentum probability amplitude $\Phi(p)$. We call this function the wave function in the *momentum representation.*

As you'll discover in your *next* course in quantum mechanics, the various representations of a state we use are actually explicit functional forms of a more abstract entity called a *state vector*. This beast, which I've mentioned before, lives in an abstract vector space called a *Hilbert space*. Similarly, you'll meet operators in Hilbert space, where they too have an abstract form.

Operators, like state descriptors, can be written in a variety of representations. In this chapter, we've worked exclusively in the position representation, where x is the independent variable. In this representation the operators for position and momentum have the form

$$\hat{x} = x \qquad \hat{p} = -i\hbar\frac{\partial}{\partial x} \qquad \text{[position representation]}. \tag{5.89}$$

In the *momentum representation*, where p is the independent variable, these operators look like (see Pblm. 5.6)

$$\hat{x} = i\hbar\frac{\partial}{\partial p} \qquad \hat{p} = p \qquad \text{[momentum representation]}. \tag{5.90}$$

In the momentum representation, these operators act on functions of momentum—the momentum probability amplitudes $\Phi(p)$—not on functions of x. For example, we evaluate the expectation value of position at $t = 0$ from $\Phi(p)$ as

$$\langle x \rangle (0) = \int_{-\infty}^{\infty} \Phi^*(p)\, \hat{x}\, \Phi(p)\, dp \tag{5.91a}$$

$$= \int_{-\infty}^{\infty} \Phi^*(p) \left(i\hbar\frac{\partial}{\partial p}\right) \Phi(p)\, dp. \tag{5.91b}$$

All representations are equivalent, in the sense that they all lead to the same physical information about states and properties. The choice of representation is up to you. But you needn't be concerned about this choice now; except for these occasional end-of-the-chapter rambles, I'll work in the position representation.

There is, after all, lots left to do in this representation. For example, we've got to figure out how to describe the evolution of a quantum state. On to Chap. 6.

ANNOTATED SELECTED READINGS

Operator Mathematics

Several useful texts on the mathematical properties of operators are available. Most of the topics covered in these books are somewhat advanced, and we won't get to them until Chap. 10. Nonetheless, here they are. My personal favorite—and an excellent investment at the price—is

1. Jackson, J. D., *Mathematics for Quantum Mechanics* (New York: W. A. Benjamin, Inc., 1962) (available in paperback). See especially § 4.2.

A useful book at a more advanced level is

2. Jordan, Thomas F., *Linear Operators for Quantum Mechanics* (New York: Wiley, 1969). Chap. II contains an introduction to operators, and Chap. IV discusses advanced features of operator algebra.

In addition to these books, most texts on mathematical methods for physicists include discussions of basic operator mathematics, and nearly any undergraduate quantum mechanics text includes bushels of exercises involving operators. Of the latter, I particularly recommend

3. Gillespie, Daniel T., *A Quantum Mechanics Primer* (New York: Intext, 1970), Chap. 2.

And, for general insight into operators and their role in quantum mechanics, you can't do better than

4. Dirac, P. A. M., *The Principles of Quantum Mechanics*, 4th ed. (Oxford: Clarendon Press, 1958), Chap. 2.

EXERCISES AND PROBLEMS

> The tasks of filling in the blanks
> I'd rather leave to you.
>
> —The Mikado,
> in *The Mikado*
> by Gilbert and Sullivan

Most of the problems for this chapter exercise your ability to manipulate operators. But Pblm. 5.2 shows you a path to the all-important momentum operator that is different from the one I used in § 5.2. And Pblm. 5.3, which you should at least read, reveals a subtle difficulty that can infect the process of generating operators using the prescription in § 5.3. Finally, the last several explore the momentum representation of operators.

Exercises

5.1 Operators—and All That

The operators $\hat{Q}_1$, $\hat{Q}_2$, and $\hat{Q}_3$ are defined to act as follows upon a (well-behaved) function $f(x)$:

- $\hat{Q}_1$ squares the first derivative of $f(x)$;
- $\hat{Q}_2$ differentiates $f(x)$ twice;
- $\hat{Q}_3$ multiplies $f(x)$ by x^4.

[a] For each of these operators, **write down** an explicit expression for $\hat{Q}_i\, f(x)$.

[b] **Simplify** the operators expression $\hat{Q}_2\,\hat{Q}_3 - \hat{Q}_3\,\hat{Q}_2$, writing your result in terms of the variable x. Do $\hat{Q}_2$ and $\hat{Q}_3$ commute?

[c] **Derive** an expression for

$$\hat{Q}_2\,\hat{Q}_3 - \hat{Q}_3\,\hat{Q}_2$$

in terms of the operators $\hat{x}$ and $\hat{p}$.

[d] An operator $\hat{Q}$ is **linear** if, for arbitrary well-behaved functions $f_1(x)$ and $f_2(x)$, the following property holds:

$$\hat{Q}\,\left[c_1\, f_1(x) + c_2\, f_2(x)\right] = c_1\,\hat{Q}\, f_1(x) + c_2\,\hat{Q}\, f_2(x).$$

Prove whether or not each of the three operators defined above is linear.

Problems

5.1 A "Derivation" of the Momentum Operator

One way to deduce the form of the momentum operator $\hat{p}$ is to work from the necessary correspondence between expectation values and classical observables. Suppose, as in § 5.2, we seek an expression for the mean value of the momentum, $\langle p \rangle$. Reasoning by analogy with our expression for $\langle x \rangle$, we could use $P(x, t)$ as a weighting factor and write

$$\langle p \rangle = \int_{-\infty}^{\infty} pP(x,t)\, dx = \int_{-\infty}^{\infty} p\Psi^*(x,t)\Psi(x,t)\, dx. \tag{5.1.1}$$

To use this result, we need to express $p\Psi^*(x,t)\Psi(x,t)$ as a function of x. Let's begin with Eq. (5.86), which I'll ask you to take on faith:

$$\langle p \rangle = m\frac{d}{dt}\langle x \rangle. \tag{5.1.2}$$

[a] Expand the first time derivative of $\langle x \rangle$ and use the time-dependent Schrödinger Equation (3.39) to show that

$$\frac{d}{dt}\langle x \rangle = \frac{i\hbar}{2m}\int_{-\infty}^{\infty}\left[x\Psi^*(x,t)\frac{\partial^2}{\partial x^2}\Psi(x,t) - x\Psi(x,t)\frac{\partial^2}{\partial x^2}\Psi^*(x,t)\right] dx. \tag{5.1.3}$$

[b] Next, show that

$$\langle p \rangle = \int_{-\infty}^{\infty}\left[-i\hbar\frac{\partial}{\partial x}\Psi(x,t)\right]\Psi^*(x,t)\, dx \tag{5.1.4a}$$

$$= \int_{-\infty}^{\infty}\Psi^*(x,t)\hat{p}\Psi(x,t)\, dx \tag{5.1.4b}$$

where $\hat{p} = -i\hbar\partial/\partial x$.
Hint: Perform *two* integrations by parts on the second term on the right-hand side of Eq. (5.1.3).

[c] Repeat the derivation of the momentum operator starting from the expression (5.1.4*a*) for the expectation value. Use integration by parts and the Dirac delta function.

5.2 A Correspondence Conundrum (Symmetrization)

The expectation value $\langle Q \rangle$ of an observable Q can be measured in the laboratory—it is just the mean value obtained in an ensemble measurement of Q. This quantity, therefore, must be real. This fact of nature has interesting implications for the generation of quantum mechanical observables, as we shall now see.

Suppose we need to evaluate the mean value of the *product of position and momentum*. The *observable* in question is xp, and, in classical physics, it makes no difference whether we represent this observable by $x(t)p(t)$ or by $p(t)x(t)$. In quantum physics we have a choice of two operators: $\hat{x}\,\hat{p}$ or $\hat{p}\,\hat{x}$, and it does make a difference.

You can see why by writing down expressions for the mean value of xp for a state represented by a wave function $\Psi(x,t)$ using these two operators. In the first instance, the mean value of xp is

$$\langle xp \rangle = \int_{-\infty}^{\infty}\Psi^*(x,t)x\left(-i\hbar\frac{\partial}{\partial x}\right)\Psi(x,t)\, dx, \tag{5.2.1}$$

but in the second, it's

$$\langle px \rangle = \int_{-\infty}^{\infty}\Psi^*(x,t)\left(-i\hbar\frac{\partial}{\partial x}\right)x\Psi(x,t)\, dx. \tag{5.2.2}$$

[a] **Show** that *neither* $\hat{x}\hat{p}$ nor $\hat{p}\hat{x}$ is an acceptable operator representation of the observable xp, because both violate the requirement that the expectation value of a physically measurable quantity must be real.

[b] Now try the operator

$$\hat{C} \equiv \frac{1}{2}\left[\hat{x}\,\hat{p} + \hat{p}\,\hat{x}\right]. \tag{5.2.3}$$

Is this an acceptable operator representation of the observable xp? **Prove** your answer. (*Hint:* Feel free to integrate by parts.) Why do we need the factor of 1/2 in this expression? What, if anything, does your answer have to do with the Correspondence Principle?

Remark: This problem illustrates a difficulty that occasionally crops up in determining operators from classical functions of time. If two operators $\hat{Q}_1$ and $\hat{Q}_2$ do not commute, *i.e.*, if $\hat{Q}_1\hat{Q}_2 \neq \hat{Q}_2\hat{Q}_1$, then there is an ambiguity in the operator that represents $Q_1 Q_2$: should we use $\hat{Q}_1\hat{Q}_2$ or $\hat{Q}_2\hat{Q}_1$? The answer, as you saw in this problem, is *neither*. In Chap. 10 we'll learn why both of these forms fail: all quantum mechanical operators must satisfy a property called *Hermiticity*. Neither $\hat{Q}_1\hat{Q}_2$ nor $\hat{Q}_2\hat{Q}_1$ satisfies this property, so neither can be used in quantum mechanics. However, the combination (5.2.3) is Hermitian and is the correct operator representation of the observable xp. In quantum mechanics, the process of constructing an observable for a product of two non-commuting observables is called *symmetrization.*

5.3 Probing the General Properties of Wave Packets

In Chap. 4, we considered specific wave packets that could represent states of the free particle at $t = 0$. All of these functions have the same mathematical structure:

$$\Psi(x,0) \;=\; f(x)\, e^{ip_0 x/\hbar},$$

where $f(x)$ is a *real*, relatively smooth function of x. For convenience let's suppose that $\Psi(x,0)$ has been properly normalized and that $f(x)$ is centered at $x = 0$ with width Δx. Examining the properties of this general form is useful, since a huge number of commonly--used wave packets can be described in this way—with appropriate choice of $f(x)$, of course.

[a] **Evaluate** the expectation value of p for this wave packet.
Hint: The normalization requirement on the state function means that $|\Psi(x,0)| \to 0$ as $x \to \pm\infty$. Consider integration by parts.

[b] Now, consider a free-particle state function that is *real* at $t = 0$. What is $\langle p \rangle$ for a state represented by such a function? Is the mean value of the kinetic energy, $\langle T \rangle$, *necessarily* equal to zero for such a state? Is $\langle x \rangle$ necessarily equal to zero for such a state? If not, under what conditions on the real wave function $\Psi(x,0)$ is $\langle x \rangle \;=\; 0$?

[c] Consider the amplitude function $A(k)$ for a general wave function that is *real* at $t = 0$. Under what condition on this wave function is $A(k)$ a *real function of* k? Suppose this condition is met; what is $\langle x \rangle$ for a state that is represented by such a function?

5.4 Practicing Operator Algebra

In this chapter, we've formulated a prescription for determining the operator that represents an observable in quantum mechanics starting with the classical function of x and p that represents it in classical theory. We usually derive these operators in the position representation; but, as I noted in § 5.8, we can do so in the momentum representation. Although we won't use the momentum representation extensively in this book, we shall work with it in this problem in order to obtain some practice with operator algebra.

[a] **Write down** the operators in the position and momentum representations for each of the following observables of the specified *one-dimensional* systems:
(1) the kinetic energy of a free particle;
(2) the potential energy of a particle in a gravitational field $V(x) = mgx$;
(3) the potential energy of a particle with a simple harmonic oscillator potential energy.

[b] In classical physics, the total energy E of a conservative system (*i.e.*, one whose potential energy does not depend on time) is related to the Hamiltonian by

$$H(x,p) - E = 0 \qquad \text{[for a conservative system]}$$

This is just the classical expression of the law of *conservation of energy*. Consider a one-dimensional simple harmonic oscillator. **Write down** the operator equivalents of this equation in the position and momentum representation. Let your equations operate on $\Psi(x,t)$ or $\Phi(p)$ as appropriate.

5.5 The Position Operator in the Momentum Representation

Let $\Psi(x,0)$ denote an arbitrary free-particle state function at $t = 0$. Starting from the expression (5.11) for the position expectation value, and using the Fourier transform relations of Chap. 4 to zip back and forth between the position and momentum representations, **derive** Eq. (5.90) for the position operator in the momentum representation.

5.6 How to Choose a Representation

NOTE: This problem uses results from Pblm. 4.6.

In § 5.8, I noted that quantum mechanical operators can be written in a variety of different *representations*. For example, Eqs. (5.89) give the operators $\hat{x}$ and $\hat{p}$ in the *position representation*; Eqs. (5.90) give these operators in the *momentum representation*. Position and momentum operators are the fundamental observables of physics, so once we know $\hat{x}$ and $\hat{p}$ in a given representation, we can work out the form of any observable in that representation.

For example, the form of the *kinetic energy operator* $\hat{T}$ for a single particle in one dimension in the position representation is developed in Example 5.3 [see Eq. (5.49)]. In the momentum representation, the operator $\hat{p}$ instructs us to simply multiply its operand by p. Hence the kinetic energy operator in the momentum representation is simpler than in the position representation:

$$\hat{T} = \frac{\hat{p}^2}{2m} = \begin{cases} -\dfrac{\hbar^2}{2m}\dfrac{\partial^2}{\partial x^2} & \text{position representation;} \\[2ex] \dfrac{p^2}{2m} & \text{momentum representation} \end{cases} \tag{5.6.1}$$

Comparing these forms, you might guess that it's easier to evaluate expressions that include $\hat{T}$ in the momentum representation, where this operator calls for a simple multiplication, than in the position representation, where it requires evaluating a second derivative. Let's see if this is true.

Suppose we want to evaluate the expectation value $\langle T \rangle_n (0)$ of the kinetic energy at $t = 0$ for the nth stationary state of the infinite square well of Example 5.1. We can use the *momentum probability amplitude* $\Phi_n(p)$ from Problem (4.6) to determine this quantity in the momentum representation.

[a] **Write down** expressions for $\langle T \rangle_n (0)$ in the position and momentum representations. In each case, write down an *explicit integral*, inserting the appropriate amplitude functions and carefully indicating the limits of integration, but **do not evaluate the integrals.**

[b] Choose one of your expressions in part [a] and **evaluate** $\langle T \rangle_n (0)$. (One of your expressions is much easier to evaluate than the other. Obviously, I want you to choose the easiest of the two expressions.)

[c] **Repeat** parts [a] and [b] for the *uncertainty in the kinetic energy*, $(\Delta T)_n(0)$.

5.7 Efficient Evaluation of Expectation Values

We can evaluate the expectation value of position $\langle x \rangle$ in either the position representation [a là Eq. (3.40*a*)] or in the momentum representation [a là Eq. (5.91*b*)]. Almost invariably, one representation is to be preferred over the other—if for no other reason than that the required algebra is easier in one representation. Therefore, it is useful to consider the conditions under which the position or momentum representation of this quantity is preferable.

[a] Consider a Gaussian wave packet Eq. (4.54) centered on $x = 0$ (see § 4.5). **Evaluate** $\langle x \rangle (t)$ and $\langle p \rangle (t)$ for this wave packet in the position representation.

[b] Evaluate $\langle x \rangle (t)$ and $\langle p \rangle (t)$ for a Gaussian wave packet in the momentum representation (see Question 4–12).

[c] The wave function of a *square wave packet* is

$$\Psi(x, 0) = \begin{cases} 0 & -\infty < x < L; \\ \dfrac{1}{\sqrt{2L}} e^{ip_0 x/\hbar} & -L \leq x \leq L; \\ 0 & L < x < \infty. \end{cases}$$

Write down explicit expressions for $\langle x \rangle$ and $\langle p \rangle$ in terms of this wave function (but don't evaluate the integrals yet).

[d] **Derive** an expression for the momentum probability amplitude $\Phi(p)$ for the square wave packet. Using your function, **write down** expressions for $\langle x \rangle$ and $\langle p \rangle$ in terms of the momentum probability amplitude.

[e] Examine your expressions for $\langle x \rangle$ and $\langle p \rangle$ from parts [c] and [d]. Choose the easiest expression for each expectation value and evaluate the necessary integrals to determine the mean values of position and momentum for the square wave packet.

CHAPTER 6

A Quantum Equation Of Motion

The Schrödinger Equation At Last

My natural instinct teaches me
(And instinct is important, O!)
You're everything you ought to be,
and nothing that you oughtn't O!

—Melissa,
in *Princess Ida*
by Gilbert and Sullivan

We all know intuitively what we mean by *motion*. Motion is change, evolution, development. A dove soars across a turquoise noon-day sky; a meteor plunges earthward from the night-black heavens; a train derails and falls off a bridge, plummeting to the jagged rocks below. The world is alive with motion.

It is only a slight exaggeration to say that physics is the study of motion. Physicists spend much of their lives studying the *evolution of physical systems*: how the properties of these systems change with time.

In classical physics, too, we know what we mean by motion. To study the motion of, say, a single macroscopic particle, we solve Newton's Second Law for the particle's **orbit** $\mathbf{r}(t)$ for all times after some initial time t_0. Once we know the orbit, we can calculate other physical properties of the particle. So, in principle at least, we can fully describe the evolution of a state of the particle provided we know the potential $V(\mathbf{r}, t)$ that describes the forces acting on it and the initial conditions $\mathbf{r}(t_0)$ and $\mathbf{v}(t_0)$ that specify the state. Of course, we must also be able to solve the equations of motion. This mathematical and numerical task is no picnic—and therein lies the principle difficulty of classical physics.

An additional problem rears its head in the study of *microscopic* systems, for we must answer the subtle conceptual question, *what does it mean to speak of the motion of a microscopic particle?* This conundrum arises because of a feature of microscopic particles that we discussed in Chap. 5: such a particle does not have a position until we measure this observable. That is, the conceptual difficulty with motion in quantum mechanics derives from the *latent character* of the particle-like properties of microscopic particles.

In quantum physics we describe observables and their evolution in the language of *probabilities*. We have already discovered (Chap. 3) that knowledge about the position of microscopic particles (in an ensemble) is represented by the probability density for their quantum state, $P(x, t) = \Psi^*(x, t)\Psi(x, t)$, and by such statistical quantities as the expectation value $\langle x \rangle$. So it should come as no surprise that we talk about the *motion* of such particles in probabilistic terms.

So the very nature of knowledge in quantum physics gives rise to an unusual, counter-intuitive notion of motion that is difficult to picture. So physicists call on *pictorial analogies* based on their (macroscopic) experience. For example, one way to *envision* motion in the microworld is as the propagation—or *flow*—of a region of enhanced position probability. This peculiar idea is illustrated in Fig. 6.1; we'll explore it in detail later in this chapter. But *beware*: these analogies are seductive. They can lead you down a garden path to impure classical thoughts. So, as you contemplate Fig. 6.1, keep in mind the *caveats* of Chap. 3 concerning the meaning of the probability density. For example, each of the density functions shown in this figure has a finite spread (or width); but this does not imply that the corresponding particles are of finite size. The particles whose state these functions represent are, in our model, *indivisible mass-points*, without dimension or internal structure. Rather, the finite spread in Fig. 6.1 means that the positions of these particles cannot be known precisely—*i.e.*, they're not meaningfully defined in the absence of measurement.

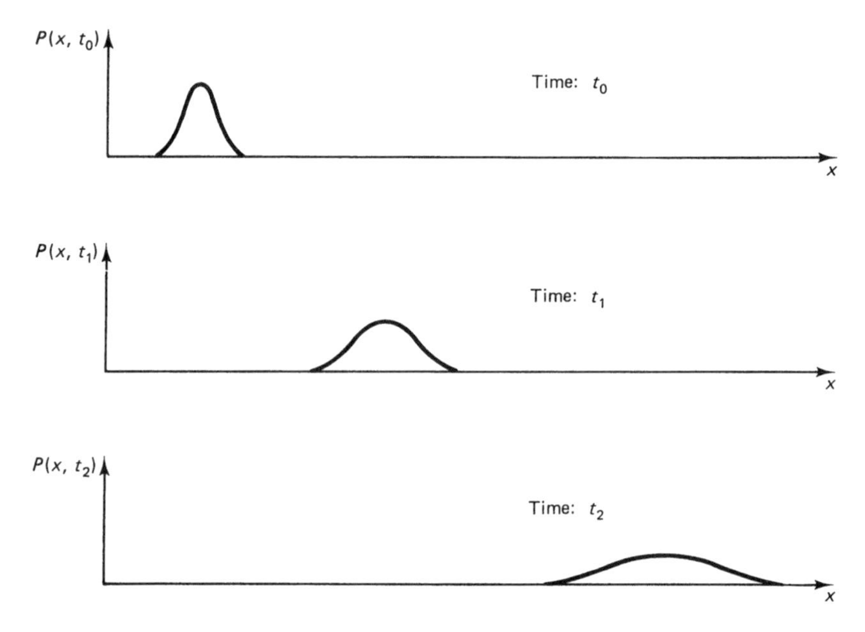

Figure 6.1 Snapshots of a one-dimensional wave packet at three times. The propagation and spreading illustrated here will be explored in § 6.6 and 6.7.

You also need to remember that neither the wave function nor the probability density exists in ordinary, geometrical space. A probability density $P(x,t)$ *is information incarnate*: it speaks to the likelihood of obtaining one of a variety of possible results in a position measurement. As the particles in an ensemble move through space, their wave function $\Psi(x,t)$ evolves with time; consequently $P(x,t)$ changes, and so does the (probabilistic) position information it contains. Thus *in quantum physics, the "motion of a particle" is described by the evolution of the state function, which is interpreted at each time from the position probability density.*

We begin our exploration of quantum motion by familiarizing ourselves with the equation that describes the time development of state functions: the time-dependent Schrödinger equation (3.39) (§ 6.1 and 6.2). We're going to spend most of the rest of this book tracking down solutions of this important equation, so we'll take the time here to get to know it well. The "personality" of the Schrödinger equation is defined by its mathematical characteristics, which we'll scrutinize in § 6.3–6.4. Then, in § 6.7, we'll meet our paradigm of quantum motion: the propagation of a free-particle wave packet.

A final note: Left to its own devices, a quantum state changes with time according to the Schrödinger equation. But a measurement also alters a quantum state, in an indeterminate way that is not described by this equation. We'll learn how to treat measurement-induced changes of state in Chap. 13. But here we'll consider only unobserved particles. No measurements allowed.

In Newtonian mechanics, we study the evolution of the state descriptor of a particle—its trajectory—by solving the *classical equation of motion*, Newton's Second Law (5.5). Similarly, to study the evolution of the quantum state descriptor—the state function—we need a "quantum equation of motion." In 1926 the Austrian physicist Erwin Schrödinger (1887–1961) published such an equation. Schrödinger was an extraordinary scientist and gifted thinker who contributed to fields as diverse as psychology and biology. He also wrote lucidly on quantum physics for both technical and lay readers.[1] But Schrödinger's greatest contribution was undoubtedly the equation that bears his name: the *time-dependent Schrödinger Equation* (TDSE), [Eq. (3.39)],

$$-\frac{\hbar^2}{2m}\frac{\partial^2}{\partial x^2}\Psi(x,t) + V(x)\Psi(x,t) = i\hbar\frac{\partial}{\partial t}\Psi(x,t). \tag{6.1}$$

This equation fully describes the evolution of the quantum states of a system—if, that is, no one sneaks into the lab and performs a measurement on it.

Although the TDSE was not—indeed, cannot be—*derived*, neither did it spring from empty space into Schrödinger's mind.[2] In the 1920's, Schrödinger came to the study of the then nascent quantum theory via research in statistical mechanics and the physics of continuous media, work that inculcated in him a thorough understanding of the mathematics and physics of vibrations and waves. Stimulated by a 1925 paper by Einstein on the quantum theory of an ideal gas, Schrödinger erected his theory on the foundation of de Broglie's matter waves (Chap. 2). Guided by geometrical and wave optics, Schrödinger generalized a fundamental equation of classical mechanics, the Hamilton equation, to formulate his wave theory of microscopic particles.[3]

The road we'll follow to the TDSE is more direct than the one Schrödinger travelled. Our first step on this road is to find an operator to represent energy, which we'll do using a strategy we devised in Chap. 5.

The Energy Operator

In § 5.3 we sought an operator $\hat{p}$ that represents the observable momentum. Our game plan was to find an operator that, when applied to a pure momentum function with momentum p and energy E,

$$\Psi_p(x,t) = \frac{1}{\sqrt{2\pi\hbar}}e^{i(px-Et)/\hbar}, \tag{6.2}$$

[1]Readers interested in pursuing some of Schrödinger's best technical writings should seek out his *Four Lectures on Wave Mechanics* (London: Blackie and Sons, 1928). Noteworthy among his non-technical works are *My View of the World* (London: Cambridge University Press, 1964) and *Science and Humanism* (London: Cambridge University Press, 1951). A particularly good biography is William T. Scott's *Erwin Schroedinger* (Amherst: University of Mass. Press, 1967).

[2]You can find a fascinating account of the genesis of the TDSE—and of Schrödinger's other trail-blazing research in quantum theory—in Chap. 5 of *The Conceptual Development of Quantum Mechanics* by Max Jammer (New York: McGraw-Hill, 1966). But you'll have to look for this book in a library rather than a bookstore; it has, inexcusably, been allowed to go out of print.

[3]Hamilton's equation looks a bit like the TDSE. It is written in terms of a quantity called the *action function* that is usually denoted by W. The action function is the integral of $T - V$ along a path of the system, and Hamilton's equation has the form $-\partial W/\partial t = T + V$. For more on the Hamiltonian formulation of classical mechanics, see Chap. 7 of *Classical Dynamics of Particles and Systems* by Jerry B. Marion (New York: Academic Press, 1970).

would yield the *value* of the momentum, *i.e.*,

$$\hat{p}\Psi_p(x,t) = p\Psi_p(x,t). \tag{6.3}$$

We then *postulated* (Postulate III) that this operator could be used to extract momentum information from all quantum states:

$$\hat{p} = -i\hbar\frac{\partial}{\partial x}. \tag{6.4}$$

We now need an *energy operator* $\hat{E}$ that extracts information concerning the energy. So let's find an operator that, when applied to a pure momentum function, yields the *value* of the energy, *i.e.*,

$$\hat{E}\Psi_p(x,t) = E\Psi_p(x,t). \tag{6.5a}$$

The appearance of this value in the exponential in (6.2), where it's on a par with the momentum, suggests that, as in (6.4), we should *differentiate*. But in this exponential, E multiplies t. So we differentiate *with respect to time*, *viz.*,

$$\frac{\partial}{\partial t}\Psi_p(x,t) = \frac{1}{i\hbar}E\Psi_p(x,t). \tag{6.5b}$$

From (6.5*a*) we can easily isolate the operator that satisfies (6.5*a*):

$$\boxed{\hat{E} \equiv i\hbar\frac{\partial}{\partial t}} \qquad \text{energy operator} \tag{6.6}$$

Question 6–1

Use this operator to extract the *value* of the energy from the pure momentum function (5.15).

Still following the approach of Chap. 5, we now *hypothesize* that we can identify this operator with the energy of an arbitrary quantum state. This supposition establishes the correspondence

QUANTUM PHYSICS

OBSERVABLE		OPERATOR
energy	$\Longleftrightarrow$	$\hat{E}$

(6.7)

The Energy Equation

Using the energy operator, we can translate classical equations containing the observable energy into their quantum-mechanical counterparts. The most fundamental of these equations is (§ 5.4) the equality of E and the classical Hamiltonian,

$$E = H. \tag{6.8}$$

By definition H (for a single particle in one dimension) is[4]

$$H(x,p) = T(p) + V(x,t) \tag{6.9}$$

$$= \frac{p^2}{2m} + V(x,t). \tag{6.10}$$

The quantum-mechanical counterpart of Eq. (6.8) is the operator equation

$$\hat{E} = \hat{\mathcal{H}} = \hat{T} + \hat{V}. \tag{6.11}$$

To obtain an explicit form of this equation for a particular system, we just replace the abstract operators in it with the appropriate instructions. For example, using Table 5.2 and Eq. (6.6), we find that in the position representation the Hamiltonian operator (6.10) is

$$\hat{\mathcal{H}} = -\frac{\hbar^2}{2m}\frac{\partial^2}{\partial x^2} + V(x,t). \tag{6.12}$$

So the quantum-mechanical *energy equation* (6.11) is

$$i\hbar\frac{\partial}{\partial t} = -\frac{\hbar^2}{2m}\frac{\partial^2}{\partial x^2} + V(x,t). \tag{6.13}$$

This is the operator form of the **time-dependent Schrödinger Equation**.

To get the form (6.1) of this equation, we just let its operators act on an *arbitrary* wave function $\Psi(x,t)$, *viz.*,

$$\boxed{-\frac{\hbar^2}{2m}\frac{\partial^2}{\partial x^2}\Psi(x,t) + V(x,t)\Psi(x,t) = i\hbar\frac{\partial}{\partial t}\Psi(x,t)} \quad \text{TDSE} \tag{6.14}$$

This omnipresent equation is often written in a shorthand form, such as

$$\hat{\mathcal{H}}\Psi(x,t) = i\hbar\frac{\partial}{\partial t}\Psi(x,t) \tag{6.15a}$$

or, by people who really want to conserve pencils, as

$$\hat{\mathcal{H}}\Psi(x,t) = \hat{E}\Psi(x,t). \tag{6.15b}$$

I don't care how you write the Schrödinger equation, but *commit it to memory.*

The Fundamental Postulate of Quantum Dynamics

Note carefully that I did not *derive* the TDSE. Like the operator identifications in it, this equation is one of our fundamental postulates. More precisely, the fourth postulate of quantum physics avers that the TDSE describes the evolution of *any* state of any quantum system:

[4]The potential energy V may depend on time—as it does, for example, for a particle in a time-dependent electromagnetic field. For most of this volume, however, we shall concentrate on systems with time-independent potentials—so-called *conservative systems*. (See Chap. 7.)

The Fourth Postulate of Quantum Mechanics

The time development of the state functions of an *isolated* quantum system is governed by the time-dependent Schrödinger Equation $\hat{\mathcal{H}}\Psi = i\hbar\frac{\partial}{\partial t}\Psi$, where $\hat{\mathcal{H}} = \hat{T} + \hat{V}$ is the Hamiltonian of the system.

Notice carefully the word *isolated* in this postulate—which serves to remind us that the TDSE describes the evolution of a state *provided no observations are made.* An observation (usually) alters the state of the observed system; and, versatile as it is, the TDSE cannot describe such changes.[5]

The role of the TDSE in problem solving is paramount: once we know the Hamiltonian $\hat{\mathcal{H}}$ of a system, we can use this equation to discover the space- and time-dependence of its state functions. In this chapter we'll probe the *time dependence* of these functions; in Chap. 7, we'll investigate their spatial dependence.

Aside: On Wave Equations. Schrödinger initially referred to his theory by the slightly repellent name *undulatory mechanics*, which he later changed to *wave mechanics.* So the TDSE is often called the *wave equation* of quantum mechanics. But I won't use this colorful language, lest it mislead you into imagining that the state function is a wave in ordinary, geometrical space, like an electromagnetic wave. Strictly speaking, the TDSE does fall into the mathematical class of equations called *wave equations.* But it differs in form from the classical wave equations that we use to describe oscillations on a string, electromagnetic fields, and the like. For a three-dimensional wave $\Gamma(\mathbf{r}, t)$ with phase velocity v_{ph}, the *classical* equation of propagation is

$$\nabla^2\,\Gamma(\mathbf{r}, t) = \frac{1}{v_{\text{ph}}}\frac{\partial^2}{\partial t^2}\Gamma(\mathbf{r}, t).$$

Unlike the TDSE, this equation is *second order in time.* There are other classical wave equations, however, that are first-order in time, such as the classical heat equation; but unlike the TDSE, these equations do not require a complex solution. [For more on classical wave equations, see *The Physics of Waves* by W. C. Elmore and M. A. Heald (New York: McGraw-Hill, 1969).]

The TDSE in 3D

Our immediate applications of Schrödinger's theory will be to simple systems in one dimension, so we'll use Eq. (6.14). But it's easy to generalize this equation to a single particle in three dimensions or to a many-particle system. For the former system, the state function is $\Psi(\mathbf{r}, t)$ and the Hamiltonian operator is

$$\hat{\mathcal{H}} = -\frac{\hbar^2}{2m}\nabla^2 + V(\mathbf{r}, t). \qquad \text{[single particle in three dimensions]} \tag{6.16}$$

In this operator, the Laplacian ∇^2 in rectangular coordinates is

$$\nabla^2 = \frac{\partial^2}{\partial x^2} + \frac{\partial^2}{\partial y^2} + \frac{\partial^2}{\partial z^2}. \tag{6.17}$$

[5] But not *all* measurements induce a change of state. For example, an energy measurement on an ensemble in a state with *a well-defined energy* won't change the state of the members. This is, however, a very special case. The point to retain is that *in general*, a measurement induces an instantaneous, uncontrollable change of state.

So the TDSE for a single particle in three dimensions is

$$-\frac{\hbar^2}{2m}\nabla^2\Psi(\mathbf{r},t) + V(\mathbf{r},t)\Psi(\mathbf{r},t) = i\hbar\frac{\partial}{\partial t}\Psi(\mathbf{r},t). \tag{6.18}$$

The Hamiltonian of a many-particle system—say, an atom—contains kinetic and potential energy operators for each constituent particle *and* interaction potentials for all pairs of particles; we shall meet and subdue this monster in Volume II.

Even in one dimension, the TDSE in all its glory may appear rather fearsome: all those ominous partial derivatives, and an $i = \sqrt{-1}$ lurking, menacingly, on the right-hand side. But it's just another partial differential equation. And mathematicians, numerical analysts, and theoretical physicists have spent decades devising schemes to solve such equations. The remaining chapters of this book are a guided tour through some of their best work.

Getting Started: Initial Conditions

For any Hamiltonian $\hat{\mathcal{H}}$ there exists an infinite number of *mathematically valid* solutions to the TDSE. This wealth of wave functions raises a question: how do we specify a *particular* solution—that special function which represents the state we're interested in? Evidently, the TDSE alone is not sufficient to identify a particular state; additional information is called for. For a clue to the missing information, let's look to the theory of differential equations.[6]

Peering at Eq. (6.14), we espy a *second-order position derivative* and a *first-order time derivative*. These derivatives prescribe the additional information needed to specify a particular solution: The second-order partial with respect to x requires the values of $\Psi(x,t)$ at two positions x; these (complex) numbers are called **boundary conditions**. The first-order partial derivative with respect to t requires the wave function at a single time t_0, which we usually choose as $t_0 = 0$. This (complex) *function* is called the **initial condition**:

$$\{\Psi(x,t_0)\colon -\infty < x < \infty\} \qquad \text{initial condition} \tag{6.19}$$

In this section we'll discuss initial conditions, leaving boundary conditions for § 6.3.

The idea of defining a state by feeding initial conditions to a differential equation is hardly new. In Newtonian mechanics, we solve a differential equation: Newton's Second Law, which for a single particle in one dimension with potential energy $V(x,t)$ is

$$m\frac{d^2}{dt^2}x(t) = -\frac{\partial}{\partial x}V(x,t). \tag{6.20}$$

This differential equation is second order in time, so specifying a particular solution requires two pieces of information: the position and velocity (or, equivalently, the momentum) at time t_0.[7]

[6]To become a reasonably proficient quantum mechanic, you'll need a modicum of information about differential equations. Several worthwhile texts on this subject are recommended in this chapter's Selected Readings. You do not, however, need to understand the mathematical theory of *partial* differential equations; in the few instances where we must consider such equations, I'll provide the necessary mathematics.

[7]For convenience, physicists often prefer to define the initial state of a classical system by the position at two successive times, $x(t_0)$ and $x(t_1)$. Since the velocity is just the first derivative of position, knowing these two positions is equivalent to knowing $x(t_0)$ and $v(t_0)$, provided the difference $t_1 - t_0$ is small.

So classical and quantum physics are alike in that both require the state descriptor at a single time. The difference between these theories is the *kind of information* they require: in classical physics, the *real numbers* of the trajectory $\{x(t_0), p(t_0)\}$ at t_0; in quantum physics, the *complex state function* $\Psi(x, t_0)$ at t_0. In practice, we usually figure out the initial conditions from the physical circumstances of the system at t_0—*e.g.*, from the way an experiment is initially set up. We'll see a very simple example of this in Example 6.6.

I've illustrated these differences schematically in Fig. 6.2 and summarized them below:

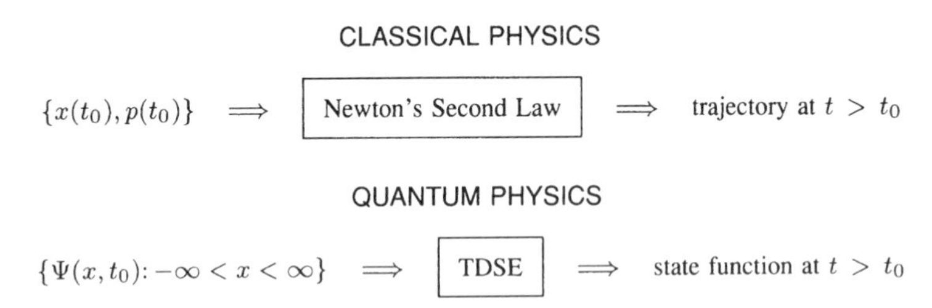

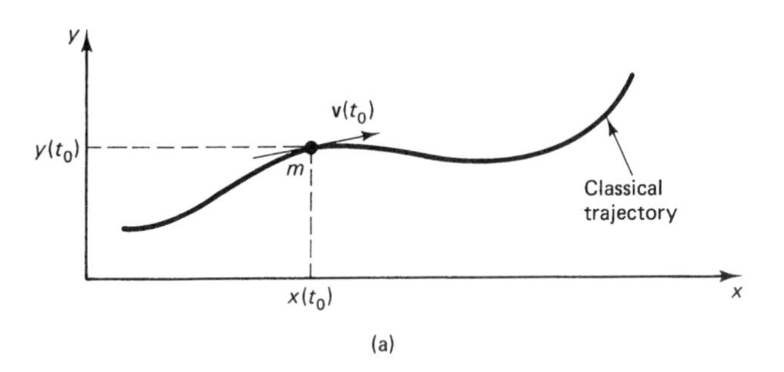

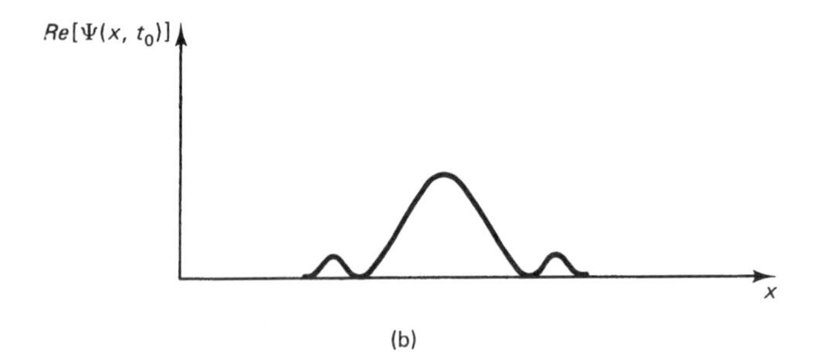

Figure 6.2 Initial conditions in (a) classical and (b) quantum physics. In (a) a macroscopic particle moves through a *two-dimensional* space along the trajectory shown. In (b) a microscopic particle in *one dimension* is represented by a (complex) state function $\Psi(x, t_0)$, the real part of which is sketched.

The physical implications of these differences are profound indeed. For example, *uncertainty* is implicit in the evolution of a quantum state, for its initial condition is a *function* characterized by a non-zero position uncertainty, $\Delta x(t_0) > 0$. And this uncertainty $\Delta x(t)$ increases as the state evolves (§ 6.7). No such indeterminacy plagues the classical description of motion.

In § 6.3, we'll examine other consequences of the mathematical form of the TDSE. But first: some examples.

And we are right, I think you'll say,
To argue in this kind of way
And I am right,
And you are right,
And all is right—too-looral-lay!.

—from *The Mikado*
by Gilbert and Sullivan

In Chaps. 3–5, I showed you several alleged state functions. But each of these functions is a valid state descriptor only if it satisfies the TDSE of its system. In this section, I want to tie up a couple of loose ends by demonstrating that two of the wave functions we've used most often, the free particle wave packet and the ground-state wave function of the infinite square well, are in fact legitimate.

Example 6.1. The Ground State Wave Function of the Particle in a Box

In two examples in Chap. 5 we calculated properties of the ground state of a particle of mass m whose potential was infinite everywhere except in a region of length L centered on $x = 0$. This system is the **particle in a box**. In Example 5.1, we evaluated $\langle p \rangle$ and, in Example 5.5, Δp for this state at $t = 0$.

In these examples I used (without proof) the wave function

$$\Psi_1(x, 0) = \sqrt{\frac{2}{L}} \cos\left(\pi \frac{x}{L}\right) \qquad \left[-\frac{1}{2}L \le x \le \frac{1}{2}L\right]. \tag{6.21}$$

[The label to the right of this function reminds us that it's zero everywhere except in the box—a condition that arises from the infinite potential at the walls (§ 7.4).] This function is the $t = 0$ counterpart of the ground-state wave function, which we'll derive in Chap. 7. The time-dependence we'll discover there is a simple exponential involving t and the constant

$$E_1 = \frac{\pi^2 \hbar^2}{2mL^2}. \tag{6.22}$$

This time-dependent wave function is

$$\Psi_1(x, t) = \sqrt{\frac{2}{L}} \cos\left(\pi \frac{x}{L}\right) e^{-iE_1 t/\hbar} \qquad \left[-\frac{1}{2}L \le x \le \frac{1}{2}L\right]. \tag{6.23}$$

The constant E_1 is the (well-defined) energy of this state.

To check that the function (6.23) satisfies the TDSE for this system, we need worry only about the region inside the box, for the state function is zero elsewhere. In this region $V = 0$, so $\hat{\mathcal{H}} = \hat{T}$, and the TDSE (6.14) is

$$-\frac{\hbar^2}{2m} \frac{\partial^2}{\partial x^2} \Psi(x, t) = i\hbar \frac{\partial}{\partial t} \Psi(x, t) \qquad \left[-\frac{1}{2}L \le x \le \frac{1}{2}L\right]. \tag{6.24}$$

Now we just insert $\Psi_1(x, t)$ into each side of (6.24). The left-hand side of this equation contains the second derivative of $\Psi_1(x, t)$, which we calculate from the first derivative

$$\frac{\partial}{\partial x} \Psi_1(x, t) = \sqrt{\frac{2}{L}} \left(-\frac{\pi}{L}\right) \sin\left(\pi \frac{x}{L}\right) e^{-iE_1 t/\hbar} \tag{6.25}$$

as

$$\frac{\partial^2}{\partial x^2}\Psi_1(x,t) = \frac{\partial}{\partial x}\left[\frac{\partial}{\partial x}\Psi_1(x,t)\right] \tag{6.26a}$$

$$= \sqrt{\frac{2}{L}}\left(-\frac{\pi}{L}\right)\left(\frac{\pi}{L}\right)\cos\left(\pi\frac{x}{L}\right)e^{-iE_1t/\hbar} \tag{6.26b}$$

$$= -\frac{\pi^2}{L^2}\Psi_1(x,t). \tag{6.26c}$$

So the left-hand side of (6.24) is

$$-\frac{\hbar^2}{2m}\frac{\partial^2}{\partial x^2}\Psi_1(x,t) = \frac{\pi^2\hbar^2}{2mL^2}\Psi_1(x,t). \tag{6.27}$$

[Do you see a familiar factor on the right-hand side of Eq. (6.27)? If not, glance at Eq. (6.22), then look again.]

Evaluating the right-hand side is even easier: we just differentiate $\Psi_1(x,t)$ once with respect to time:[8]

$$\frac{\partial}{\partial t}\Psi_1(x,t) = -\frac{iE_1}{\hbar}\Psi_1(x,t). \tag{6.28}$$

So the right-hand side of Eq. (6.24) is

$$i\hbar\frac{\partial}{\partial t}\Psi_1(x,t) = E_1\Psi_1(x,t) = \frac{\pi^2\hbar^2}{2mL^2}\Psi_1(x,t). \tag{6.29}$$

This expression is equal to $\hat{\mathcal{H}}\Psi_1(x,t)$ in Eq. (6.27). The ground state function (6.23) does indeed solve the TDSE for the system. Postulate IV is satisfied.

Return of the Free Particle

In addition to the infinite square well, we've exhaustively examined the free particle of Chap. 4. The TDSE of this system is simply

$$-\frac{\hbar^2}{2m}\frac{\partial^2}{\partial x^2}\Psi(x,t) = i\hbar\frac{\partial}{\partial t}\Psi(x,t). \qquad \text{[for a free particle]} \tag{6.30}$$

In § 4.1, I argued that a reasonable state function for a free particle in one dimension is a *wave packet*—a superposition of an infinite number of pure momentum functions that looks like

$$\Psi(x,t) = \int_{-\infty}^{\infty}\Phi(p)\Psi_p(x,t)\,dp.$$

Although this packet is spatially localized (and hence normalizable), it cannot represent a quantum state of a free particle unless it satisfies the TDSE (6.30). I want to take you through this verification in two stages: first I'll show you that the pure momentum state functions that make up the packet satisfy this equation (Example 6.2). Then we'll turn to the packet itself (Example 6.3).

[8] You can see now why I used the full x- and t-dependent form of the proposed function in the TDSE. Had we tried to validate this function *at a particular time*, say, at $t = 0$, we'd have been unable to work out the time derivative in this equation.

Example 6.2. The Pure Momentum Function

As presented in Chap. 4, the pure momentum function has the form

$$\Psi_p(x,t) = \frac{1}{\sqrt{2\pi\hbar}}\, e^{i(px-Et)/\hbar}.$$

To make clear the derivation to follow, I want to write this function in terms of k. To this end, I'll use the free-particle dispersion relation to relate E to k:

$$E = \frac{p^2}{2m} = \frac{\hbar^2 k^2}{2m}. \qquad \text{[for a free particle]} \tag{6.31}$$

Inserting this relation into the general form of the pure momentum function gives a function that pertains *only to the free particle*,

$$\Psi_p(x,t) = \frac{1}{\sqrt{2\pi\hbar}} e^{i[kx-\hbar k^2 t/(2m)]}. \qquad \text{[for a free particle]} \tag{6.32}$$

The derivatives we need when we insert this function into the TDSE are[9]

$$\frac{\partial^2}{\partial x^2} e^{i[kx-\hbar k^2 t/(2m)]} = (ik)(ik) e^{i[kx-\hbar k^2 t/(2m)]} \tag{6.33a}$$

$$= -k^2 e^{i[kx-\hbar k^2 t/(2m)]} \tag{6.33b}$$

and

$$\frac{\partial}{\partial t} e^{i[kx-\hbar k^2 t/(2m)]} = -\frac{i\hbar k^2}{2m} e^{i[kx-\hbar k^2 t/(2m)]}. \tag{6.33c}$$

Now, the left-hand side of the TDSE is $-\hbar^2/(2m)$ times (6.33*b*),

$$\hat{\mathcal{H}}\Psi_p(x,t) = \frac{\hbar^2 k^2}{2m}\Psi_p(x,t). \tag{6.34}$$

The right-hand side is $+i\hbar$ times (6.33*c*), which is equal to (6.34). This equality verifies the TDSE for the pure momentum function! Again, Postulate IV is happy.

In Example 6.2, we've proven only that the pure momentum function for a free particle satisfies the appropriate TDSE. Yet, as we know, this function is not a valid state function, for it fails to satisfy the conditions of § 3.5. This example reminds us that

> ***Rule***
>
> To represent a quantum state, a function must satisfy the TDSE of the system and be normalizable, single-valued, continuous, and smoothly-varying.

[9]The general rules for differentiating exponents I'm using are handy problem-solving tools. Here are their general forms:

$$\frac{\partial^n}{\partial x^n} e^{i(kx-\omega t)} = (ik)^n e^{i(kx-\omega t)}$$

$$\frac{\partial^n}{\partial t^n} e^{i(kx-\omega t)} = (-i\omega)^n e^{i(kx-\omega t)}$$

Example 6.3. The Gaussian Free-Particle State Function

Now let's tackle the wave packet

$$\Psi(x,t) = \frac{1}{\sqrt{2\pi}} \int_{-\infty}^{\infty} A(k) e^{i(kx-\omega t)}\, dk. \tag{6.35}$$

To simplify the proof that this function satisfies the TDSE (6.15*b*), let's write this equation more compactly, as $(\hat{\mathcal{H}} - \hat{E})\Psi(x,t) = 0$. For the wave packet (6.35), the operator $(\hat{\mathcal{H}} - \hat{E})\Psi(x,t)$ is

$$\begin{aligned}(\hat{\mathcal{H}} - \hat{E})\Psi(x,t) = &-\frac{\hbar^2}{2m}\frac{\partial^2}{\partial x^2}\left[\frac{1}{\sqrt{2\pi}}\int_{-\infty}^{\infty} A(k) e^{i[kx-\hbar k^2/(2m)t]}\, dk\right] \\ &- i\hbar\frac{\partial}{\partial t}\left[\frac{1}{\sqrt{2\pi}}\int_{-\infty}^{\infty} A(k) e^{i[kx-\hbar k^2/(2m)t]}\, dk\right].\end{aligned} \tag{6.36}$$

To prove that the right-hand side is zero, we must first get rid of the unsightly partial derivatives parading through it. The amplitude function $A(k)$ is independent of x and t, so we can treat it as a constant when we evaluate $\partial^2/\partial x^2$ and $\partial/\partial t$. Let's interchange the order of integration and differentiation in both terms in (6.36) and slide these derivatives through the integrals:[10]

$$\begin{aligned}(\hat{\mathcal{H}} - \hat{E})\Psi(x,t) = &-\frac{\hbar^2}{2m}\int_{-\infty}^{\infty} A(k)\left[\frac{\partial^2}{\partial x^2} e^{i[kx-\hbar k^2/(2m)t]}\right] dk \\ &- i\hbar\int_{-\infty}^{\infty} A(k)\left[\frac{\partial}{\partial t} e^{i[kx-\hbar k^2/(2m)t]}\right] dk.\end{aligned} \tag{6.37}$$

[I have also done a little housecleaning in (6.37), eliminating factors of $1/\sqrt{2\pi}$ from each term.]

But we've already evaluated the derivatives in Eq. (6.37) in Example 6.2. Using Eqs. (6.33) we find

$$(\hat{\mathcal{H}} - \hat{E})\Psi(x,t) = \int_{-\infty}^{\infty} A(k)\left[\frac{\hbar^2 k^2}{2m} - i\hbar\left(-\frac{i\hbar k^2}{2m}\right)\right] e^{i[kx-\hbar k^2/(2m)t]}\, dk. \tag{6.38}$$

Now, the expression in square brackets in (6.38) is zero—which proves that the wave packet (6.35) satisfies the TDSE of a free particle. This wave packet also is normalizable, continuous, and smoothly-varying, so it can represent a state of this system. Notice that this verification applies to *any* free-particle wave packet, because nowhere did we specify a *particular* amplitude function.

In practice, we never need explicitly verify that a superposition such as the free-particle wave packet satisfies the TDSE of a system. Once we know that the *constituents* of the superposition satisfy this equation, we can conclude that their linear combination does. This vital property—the Principle of Superposition we first saw in Chap. 3—follows from the mathematical characteristics of the Schrödinger equation, which we're going to investigate right now.

[10]What lets us do this is a theorem (from freshman calculus) called *Leibniz's Rule*. According to this rule, we can legitimately interchange the order of integration and differentiation in terms like those appearing in Eq. (6.35) provided two conditions are met: (1) the integrand must be *sectionally continuous*, *i.e.*, it must have at most a finite number of point discontinuities; and (2) the integral must converge uniformly before and after the differentiation.

6.3 A PERSONALITY PROFILE OF THE TDSE

Just as our friends have identifiable personality traits—grumpy, chummy, psychotic, etc.—so do equations have distinctive characteristics. One of the best ways to become familiar with a new equation is to identify its "mathematical personality." In this section, we'll "matho-analyze" the time-dependent Schrödinger Equation (TDSE).

A physicist is interested in the TDSE as the equation that determines the space and time dependence of state functions. But a mathematician, looking at this equation, sees *a linear, homogeneous partial differential equation of second order.*[11]

Each of these *mathematical* attributes of the TDSE has important *physical* consequences. For instance, in § 6.1, I noted that the (mathematical) fact that the TDSE contains a first-order time derivative requires us to feed it (physical) information concerning the quantum state at the initial time:

MATHEMATICS	$\Longleftrightarrow$	PHYSICS
first-order time derivative	$\Longleftrightarrow$	initial conditions

In this section we'll discover several more examples of the beautiful interrelationship between mathematics and physics that permeates quantum mechanics.

Question 6–2

We've already noted another attribute of the TDSE: the presence of an $i = \sqrt{-1}$ implies that its solutions must, in general, be complex. Prove that *no real function of* x and t can solve the TDSE. [*Hint*: What differential equation does $\Psi^*(x,t)$ satisfy?]

On Boundary Conditions

In § 6.1, I remarked that in order to specify a particular solution of the TDSE, we must provide the values of the state function at two locations; these values are **boundary conditions**. [Alternatively, we could provide the value of $\Psi(x,t)$ and its first derivative $\partial\Psi(x,t)/\partial x$ at one location.] For example, the following conditions must apply *at all times* to most wave functions for a single particle in one dimension that is confined (by the potential of the system) to a finite region of space:

$$\begin{aligned} \Psi(x,t) &\to 0 \qquad \text{as } x \to -\infty \\ \Psi(x,t) &\to 0 \qquad \text{as } x \to +\infty. \end{aligned} \qquad \text{[at all times } t\text{]} \tag{6.39}$$

These conditions are familiar from Chap. 3 as statements that the wave function must be *spatially localized* [see Eq. (3.29)]. But Eqs. (6.39) may not correspond to your notion of boundary conditions, for they prescribe the values of $\Psi(x,t)$ not at two (finite) points in space, but rather in two *limits*: $x \to -\infty$ and $x \to +\infty$.

Only in special model problems are boundary conditions *for a one-dimensional system* specified at finite points. A familiar and important example is the particle-in-a-box of Example 6.1. For the states of this system, the boundary conditions are the values of $\Psi(x,t)$ at the "walls":

$$\Psi(x = -\tfrac{1}{2}L, t) = 0 \tag{6.40a}$$

$$\Psi(x = +\tfrac{1}{2}L, t) = 0. \tag{6.40b}$$

[11]Recall that a differential equation is said to be of *n-th order* if the highest derivative it contains is of order n. Thus, although the TDSE contains a *first-order* time derivative, it is a *second-order* equation, because its highest order derivative is $\partial^2/\partial x^2$.

These boundary conditions are a consequence of the *physical environment* of the particle. The "walls" of the box are impenetrable, because the potential energy outside is infinite. [This model is the limit of the potential of a physical system in which a particle experiences extremely strong impulsive forces at these two points (see § 8.1 and Pblm. 8.11).] So the particle cannot "escape," and the probability that we will detect it at $x \leq -L/2$ or at $x \geq +L/2$ in a position measurement is zero. Mathematically, this means the probability density $P(x,t)$ is zero outside the box, and so is the wave function. The requirement that this function be continuous everywhere (§ 3.5) leads us to the boundary conditions (6.40).[12]

Aside: Boundary Conditions in Three Dimensions. How we specify boundary conditions for three-dimensional quantum systems depends on the coordinate system we choose to work in. For instance, a widely-used three-dimensional model potential is a "spherical box" of radius a:

$$V(\mathbf{r}) = \begin{cases} -V_0 & r \leq a; \\ 0 & r > a \end{cases}.$$

This potential is *spherically symmetric*; that is, it's independent of the angles θ and φ. So the chore of determining the state functions of a particle with this potential will be simplified if we choose to work in a coordinate system that conforms to this symmetry property: the *spherical coordinates* (r, θ, φ).

In spherical coordinates (see Volume II), the kinetic energy operator $\hat{T}$ contains second-order partial derivatives with respect to r, θ, and φ. To specify a particular state function, therefore, we must provide *six* boundary conditions—two for each variable. The boundary conditions on r, for example, are specified as follows:

$$\Psi(r = 0, \theta, \varphi) = \text{constant}; \tag{6.41a}$$

$$\Psi(r, \theta, \varphi) \xrightarrow[r \to \infty]{} 0 \tag{6.41b}$$

(In Part V, we'll delve into the boundary conditions for the variables θ and φ.) Notice that in (6.41) we state one boundary condition *at a point*, $r = 0$, and the other *in a limit*, $r \to \infty$.

Linearity and Homogeneity: A Mathematical Side-Trip

Our quantum equation of motion, the TDSE, is a special type of second-order partial differential equation: it is *linear* and *homogeneous*. I want to show you the physical consequences of these properties, but first we should review their definitions. For simplicity, I'll begin with an *ordinary* differential equation.

The general form of an **ordinary differential equation of order** n in the variable x is

$$a_0(x)\frac{d^n}{dx^n}f(x) + a_1(x)\frac{d^{n-1}}{dx^{n-1}}f(x) + a_2(x)\frac{d^{n-2}}{dx^{n-2}}f(x) + \cdots + a_n(x)f(x) = g(x). \tag{6.42}$$

[12] You may be concerned by the fact that the first derivative of $\Psi(x,t)$ is not continuous at the walls—*i.e.*, the state functions of the infinite square well are not *smoothly varying* at $x = \pm L/2$. Not to worry: this property is an artifice of the *infinite* potential in this model. The function $\partial\Psi(x,t)/\partial x$ is discontinuous at the walls because there is a discontinuous change in the potential energy at these points. Actual potentials may change sharply and drastically, but they invariably do so continuously.

The prefactors $a_i(x)$ may be constants or functions and can be real or complex. If the function $g(x)$ on the right-hand side is zero, then this equation is said to be **homogeneous**. But if $g(x) \neq 0$, the equation is said to be **inhomogeneous**.[13]

Notice that a linear equation cannot contain terms involving *powers of* $f(x)$ *or of its derivatives*; terms such as

$$f^3(x) \qquad \text{or} \qquad \left[\frac{d}{dx}f(x)\right]^3$$

destroy the linearity of a differential equation. Similarly, a linear equation must not contain terms involving *products of* $f(x)$ *and one or more of its derivatives*; terms such as

$$f(x)\frac{d}{dx}f(x) \qquad \text{or} \qquad \left[\frac{d}{dx}f(x)\right]\frac{d^3}{dx^3}f(x)$$

are also forbidden. If any such terms befoul the purity of a differential equation, then that equation is **non-linear**—and is to be avoided if possible. Here are a couple of examples:

$$\frac{d}{dx}f(x) + \sin x = 0 \tag{6.43}$$

is a linear equation. But

$$\left[\frac{d}{dx}h(x)\right]^2 + h(x)\sin x = 0 \tag{6.44}$$

is not.

"I don't care about *ordinary* differential equations," you mutter. "The TDSE is a *partial* differential equation. Tell me about *partial differential equations*!"

Okay. The general form of a partial differential equation of arbitrary order in several variables is a mess—lots and lots of terms. So I'll show you the general form for a **second-order, linear, partial differential equation in two independent variables**. Since we're interested in the TDSE, I'll use x and t as the independent variables:

$$\begin{aligned} a_0(x,t)\frac{\partial^2}{\partial x^2}f(x,t) + a_1(x,t)\frac{\partial}{\partial x}f(x,t) + a_2(x,t)\frac{\partial}{\partial t}f(x,t) + a_3(x,t)\frac{\partial^2}{\partial t^2}f(x,t) \\ + a_4(x,t)\frac{\partial^2}{\partial x \partial t}f(x,t) + a_5(x,t)f(x,t) = g(x,t). \end{aligned} \tag{6.45}$$

The multiplicative factors $a_i(x,t)$—one or more of which may be constant or zero—are, in general, complex functions of the independent variables x and t. If $g(x,t) = 0$, Eq. (6.45) is **homogeneous**; otherwise it is **inhomogeneous**. The kind of terms that can populate a *linear* partial differential equation are limited. Absolutely forbidden are terms containing *powers of derivatives of* $f(x,t)$ or *products of factors involving* $f(x,t)$ *and its derivatives*: *e.g.*,

$$\frac{\partial}{\partial x}f(x,t)\frac{\partial}{\partial t}f(x,t), \qquad \left[\frac{\partial}{\partial x}f(x,t)\right]^2, \qquad \text{or} \qquad f(x,t)\frac{\partial}{\partial x}f(x,t).$$

[13]Mathematicians have a slightly different—but equivalent—definition of a homogeneous differential equation: Suppose a function $f(x)$ satisfies a linear differential equation such as (6.42). If $cf(x)$ satisfies the same equation, where c is an arbitrary complex constant, then the equation is said to be **homogeneous**. Take a moment to convince yourself that this definition is equivalent to the one given in the text.

It is easy to see that the TDSE (6.14) for a single particle in one dimension conforms to the form (6.45). If we gather all its terms on the left-hand side, we get

$$-\frac{\hbar^2}{2m}\frac{\partial^2}{\partial x^2}\Psi(x,t) + V(x,t)\Psi(x,t) - i\hbar\frac{\partial}{\partial t}\Psi(x,t) = 0, \tag{6.46}$$

which form makes it easy to read off the prefactors $a_i(x,t)$ for the TDSE:

$$\begin{aligned} a_0(x,t) &= -\frac{\hbar^2}{2m} \\ a_1(x,t) &= 0 \\ a_2(x,t) &= -i\hbar \\ a_3(x,t) &= 0 \\ a_4(x,t) &= 0 \\ a_5(x,t) &= V(x,t) \end{aligned} \tag{6.47}$$

A Little About Linear Differential Operators

Because of the rather ungainly complexity of the general form (6.45), mathematicians often use a shorthand notation for its operators. I'll use $\hat{D}(x,t)$ to denote a general linear, second-order differential operator in x and t.[14] The general form of a linear operator is

$$\begin{aligned} \hat{D}(x,t) \equiv a_0(x,t)\frac{\partial^2}{\partial x^2} &+ a_1(x,t)\frac{\partial}{\partial x} + a_2(x,t)\frac{\partial}{\partial t} \\ &+ a_3(x,t)\frac{\partial^2}{\partial t^2} + a_4(x,t)\frac{\partial^2}{\partial x \partial t} + a_5(x,t). \end{aligned} \tag{6.48}$$

This notation enables us to write the general second-order linear differential equation (6.45) as

$$\hat{D}(x,t)f(x,t) = g(x,t) \tag{6.49}$$

and the TDSE as

$$\hat{D}(x,t)\Psi(x,t) = 0. \tag{6.50}$$

For the latter, the differential operator $\hat{D}(x,t)$ is

$$\hat{D}(x,t) = \hat{\mathcal{H}} - \hat{E} \tag{6.51a}$$

$$= \hat{T} + \hat{V} - \hat{E} \tag{6.51b}$$

$$= -\frac{\hbar^2}{2m}\frac{\partial^2}{\partial x^2} + V(x,t) - i\hbar\frac{\partial}{\partial t}. \tag{6.51c}$$

Note carefully that (6.50) is a homogeneous differential equation.

[14]The operator $\hat{D}(x,t)$ belongs to a general class of mathematical critters called *linear operators.* Such operators play a very important role in quantum theory (see Chap. 10). By definition, an operator $\hat{A}$ is **linear** if for any two functions f_1 and f_2,

$$\hat{A}\left[c_1 f_1(x) + c_2 f_2(x)\right] = c_1\,\hat{A}\,f_1(x) + c_2\,\hat{A}\,f_2(x).$$

By the way, please don't misconstrue my notational whimsy. The argument (x,t) I've appended to the general linear operator is there just to remind us what the independent variables are. *The notation $\hat{D}(x,t)$ does not imply that $\hat{D}$ acts on x and/or on t.*

Aside: The TDSE in Three Dimensions. The TDSE (6.18) for a single particle in three dimensions is a linear, homogeneous differential equation in *four variables*. For this case, the linear operator corresponding to (6.51) is

$$\hat{D}(\mathbf{r},t) = -\frac{\hbar^2}{2m}\nabla^2 + V(\mathbf{r},t) - i\hbar\frac{\partial}{\partial t}. \qquad \text{[three-dimensions]} \tag{6.52}$$

Similarly, we can generalize the operator (6.52) to several particles in three dimensions (Volume II).

Justifying the Normalization of State Functions

To start our investigation of the properties of a linear homogeneous differential equation such as the TDSE (6.51), we'll consider a feature that is related to the normalization procedure we developed in § 3.5. Suppose a function $f(x,t)$ solves $\hat{D}(x,t)f(x,t) = 0$. Then, for any complex constant c, the function $cf(x,t)$ also solves this equation; *i.e.*,

$$\boxed{\hat{D}(x,t)f(x) = 0 \qquad \Longrightarrow \qquad \hat{D}(x,t)[cf(x,t)] = 0} \tag{6.53}$$

(Do you see why? Here's a hint: What happens to the constant c when it is acted upon by a linear differential operator?)

This simple fact justifies our procedure for normalizing state functions. Remember how we do this: If the function $\Psi'(x,t)$ satisfies the TDSE but has a normalization integral equal to a finite number $M \neq 1$, then we construct the corresponding *normalized* function by multiplying $\Psi'(x,t)$ by a complex constant:

$$\Psi(x,t) = \frac{e^{i\delta}}{\sqrt{M}}\Psi'(x,t). \tag{6.54}$$

The constant δ in the global phase factor $e^{i\delta}$ is an arbitrary real number. The new function satisfies the normalization condition

$$\int_{-\infty}^{\infty} \Psi^*(x,t)\Psi(x,t)\,dx = 1, \tag{6.55}$$

and therefore can be meaningfully interpreted as a position probability amplitude. The property (6.53) assures us that if $\Psi'(x,t)$ satisfies the TDSE, so does its normalized counterpart $\Psi(x,t)$.

The second mathematical property we're going to explore leads to a physical principle so important that it gets its own section.

6.4 SUPERPOSITION DERIVED

Like (6.53), which we used to justify normalization, our second property follows from the linearity and homogeneity of the TDSE. Suppose $f_1(x,t)$ and $f_2(x,t)$ solve $\hat{D}(x,t)f(x,t) = 0$. Because this equation is homogeneous, the *sum* of these functions is also a solution; *i.e.*,

$$\boxed{\left.\begin{aligned}\hat{D}(x,t)f_1(x,t) &= 0\\ \hat{D}(x,t)f_2(x,t) &= 0\end{aligned}\right\} \quad \Longrightarrow \quad \hat{D}(x,t)\left[f_1(x,t) + f_2(x,t)\right] = 0.} \tag{6.56}$$

Taken together, properties (6.53) and (6.56) reveal that

> ***Rule***
>
> Any linear combination of solutions of a linear, homogeneous differential equation is another solution of that equation.

So if $f_1(x,t)$ and $f_2(x,t)$ solve $\hat{D}(x,t)f(x,t) = 0$ and if c_1 and c_2 are arbitrary complex constants, then $c_1 f_1(x,t) + c_2 f_2(x,t)$ also solves this equation:

$$\left.\begin{aligned} \hat{D}(x,t)f_1(x,t) &= 0 \\ \hat{D}(x,t)f_2(x,t) &= 0 \end{aligned}\right\} \quad \Longrightarrow \quad \hat{D}(x,t)\left[c_1 f_1(x,t) + c_2 f_2(x,t)\right] = 0. \tag{6.57}$$

This property can be extended to an arbitary linear combination of solutions; when applied to the TDSE, it leads to the Principle of Superposition.

Let Ψ_1 and Ψ_2 represent quantum states of a system with Hamiltonian $\hat{\mathcal{H}}$,[15]

$$\begin{aligned} (\hat{\mathcal{H}} - \hat{E})\Psi_1 &= 0 \\ (\hat{\mathcal{H}} - \hat{E})\Psi_2 &= 0 \end{aligned}. \tag{6.58}$$

From the linearity and homogeneity of the TDSE we conclude [see Eq. (6.57)] that for arbitrary constants c_1 and c_2,

$$(\hat{\mathcal{H}} - \hat{E})[c_1\Psi_1 + c_2\Psi_2] = 0. \tag{6.59}$$

Therefore—and most importantly—the linear combination

$$\Psi = c_1\Psi_1 + c_2\Psi_2 \qquad \text{[a superposition state]} \tag{6.60}$$

represents a quantum state that is different from the ones represented by Ψ_1 and Ψ_2. This is a mathematical way of stating the **Principle of Superposition**. Our little demonstration shows that this cornerstone of quantum physics is a manifestation of the mathematical character of the TDSE.[16]

Example 6.4. A Superposition State of the Particle in a Box

To see a superposition state of a particular system, let's return to our simple paradigm, the particle in a box. At $t = 0$, the wave function of the ground state of this system is [Eq. (6.21)]

$$\Psi_1(x,0) = \sqrt{\frac{2}{L}}\cos\left(\pi\frac{x}{L}\right) \qquad \left[-\frac{1}{2}L \leq x \leq \frac{1}{2}L\right]. \tag{6.61a}$$

[15] In this section, I have omitted the position and time dependence from the wave function to emphasize that *these results pertain to any quantum system.*

[16] The superposition state (6.60) is physically realizable (*i.e.*, it represents a state that could exist in the laboratory) only if $\Psi(x,t)$ is continuous, smoothly-varying, single-valued, and normalizable. Its normalizability follows automatically if the composite functions Ψ_1 and Ψ_2 are themselves normalizable. But they need not be; we saw in § 4.2 (on wave packets) that we can form a normalizable superposition of an infinite number of *non-normalizable* functions.

The well-defined energy of this state ($\Delta E = 0$) is

$$E_1 = \frac{\pi^2 \hbar^2}{2mL^2}. \tag{6.61b}$$

The stationary state nearest in energy to the ground state is called the **first excited state**. At $t = 0$ the wave function of this state is

$$\Psi_2(x, 0) = \sqrt{\frac{2}{L}} \sin\left(2\pi \frac{x}{L}\right), \qquad \left[-\frac{1}{2}L \leq x \leq \frac{1}{2}L\right], \tag{6.62a}$$

and its well-defined energy is

$$E_2 = 4\frac{\pi^2 \hbar^2}{2mL^2}. \tag{6.62b}$$

Question 6–3

Verify that $\Psi_2(x,t)$ satisfies the TDSE for the particle in a box. [You will have to generalize the function in (6.62a) to times $t > 0$, just as we did for the ground state in Example 6.1.]

Now, according to the Principle of Superposition, *any* linear combination of $\Psi_1(x,t)$ and $\Psi_2(x,t)$ represents a third physically realizable state of this system. Suppose we mix 2 parts $\Psi_1(x,t)$ to 7 parts $\Psi_2(x,t)$ to brew the normalized state function

$$\Psi(x,t) = \sqrt{\frac{1}{53}}\left[2\Psi_1(x,t) - 7\Psi_2(x,t)\right]. \tag{6.63}$$

This function, which is graphed at $t = 0$ along with its constituents in Fig. 6.3, represents a physically realizable state of the particle in an infinite square well. You can see how it differs from the ground and first-excited states by examining the position information in the probability amplitudes.

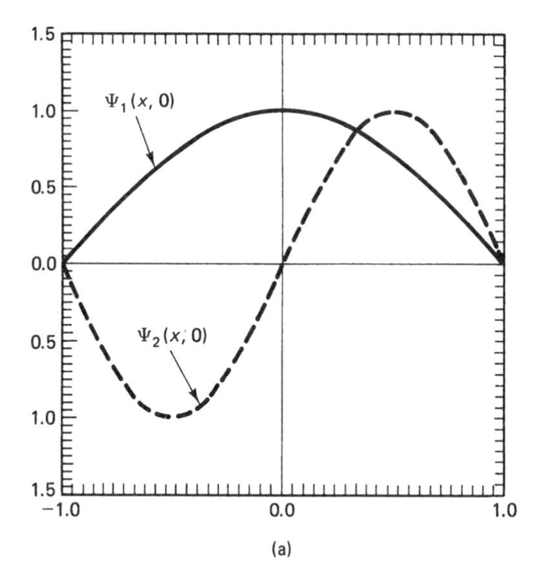

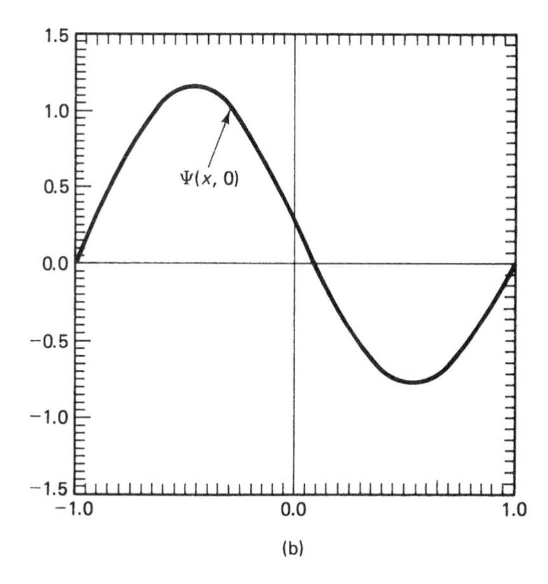

Figure 6.3 A superposition state of a particle in a one-dimensional infinite square well. (a) The ground and first-excited states [Eqs. (6.61a) and (6.62a)]. (b) The superposition state of Eq. (6.63).

Question 6–4

Verify my claim that Eq. (6.63) is normalized. [You can use the fact that $\Psi_1(x,0)$ and $\Psi_2(x,0)$ are individually normalized. Oh yes ... Watch those limits of integration!]

The superposition state (6.63) differs from its constituents in another important way: its energy is not sharp. Were we to calculate the energy uncertainty $\Delta E(t)$ from (6.63), we would obtain $\Delta E > 0$. I'll spare you the algebraic contortions of this evaluation; in Chap. 12, I'll show you an easy way to calculate such uncertainties.

In Example 6.4 we constructed a non-stationary state from two stationary states. But, as I noted in § 3.2, we can extend the Principle of Superposition to more than two stationary states. Or we can combine two or more *non-stationary states* to form yet another non-stationary state. But *all of the functions we combine to form such a superposition must solve the same Schrödinger equation.*

6.5 IS PROBABILITY CONSERVED?

The TDSE has many uses. Foremost is its role in the determination of state functions. But we can also use this fundamental equation to address physical questions about the microworld. To illustrate (and to develop a result we'll need in § 6.8), I want to consider the question of particle conservation: is the total number of particles conserved, or can particles be created and destroyed?[17]

In the language of probabilities, this question becomes a query about the position probability density: is the probability of finding a particle anywhere in space independent of time? We express this question mathematically in terms of the *integrated probability* $P([-\infty,\infty],t)$ of § 3.4. This quantity is the sum of the infinitesimal position probabilities $P(x,t)\,dx$ for all values of x, *i.e.*, [Eq. (3.26)]

$$P([-\infty,\infty],t) = \int_{-\infty}^{\infty} P(x,t)\,dx = \int_{-\infty}^{\infty} \Psi^*(x,t)\Psi(x,t)\,dx. \tag{6.64}$$

If microscopic particles are neither created nor destroyed, then this quantity must be independent of time—which is precisely the result I want to prove:

Proclamation 6.1

The probability of finding a microscopic particle anywhere in space does not change with time—*i.e.*, (non-relativistic) particles are neither created nor destroyed.

[17] Since the TDSE applies only to *non-relativistic* systems, our answer will pertain only to particles whose velocities are much less than the speed of light. Particle creation and destruction does occur at relativistic energies. If, for example, an electron and a positron (a particle with the same mass as an electron but with the opposite charge) collide, they can mutually annihilate, creating photons. The reverse process, in which photons create an electron-positron pair, also occurs. For a readable, non-technical introduction to these seemingly magical events, see pp. 210–267 of Gary Zukav's *The Dancing Wu Li Masters: An Overview of the New Physics* (New York: William Morrow, 1979). For a technical introduction, see § 4.6 of David Bohm's textbook *Quantum Theory* (Englewood Cliffs, N. J.: Prentice Hall, 1951) or, at a more advanced level, *Relativistic Quantum Mechanics* by J. D. Bjorken and S. D. Drell (New York: McGraw-Hill, 1964).

Here is the mathematical expression of this proclamation:

$$\boxed{\frac{d}{dt}P([-\infty,\infty],t) = \frac{d}{dt}\int_{-\infty}^{\infty} P(x,t)\,dx = 0} \qquad \text{conservation of position probability} \tag{6.65}$$

Novice quantum mechanics sometimes fall into the subtle trap of arguing that this result follows from the fact (*c.f.*, § 3.5) that the probability of finding a microscopic particle somewhere in space at a fixed time, say, t_0, is constant [Eq. (3.29)]

$$P\left([-\infty,\infty],t_0\right) = 1. \tag{6.66}$$

But Eq. (6.66) does not imply Eq. (6.65). To prove conservation of probability, we must look to the Schrödinger equation.

Question 6–5

Explain why we cannot conclude from (6.66) that (6.65) is true. [*Hint*: If $\Psi(x,t_0)$ is normalized at time t_0, can we conclude that it is normalized at any other time $t_1 \neq t_0$? If not, why not?]

Argument: We're going to prove Proclamation 6.1 for an *arbitrary* state of an *arbitrary* system. The proof uses only two facts:

[1] the wave function $\Psi(x,t)$ satisfies the TDSE

$$\hat{\mathcal{H}}\Psi(x,t) = i\hbar\frac{\partial}{\partial t}\Psi(x,t); \tag{6.67a}$$

[2] the wave function is *spatially localized—i.e.*, it satisfies the boundary conditions [Eq. (6.39)]

$$\Psi(x,t) \xrightarrow[x\to\pm\infty]{} 0. \qquad \text{[at any time } t\text{]} \tag{6.67b}$$

To prove that the time derivative of the integrated probability (6.64) is zero, we first move the derivative inside the integral, where it must become a partial derivative, *viz.*,[18]

$$\begin{aligned}\frac{d}{dt}P([-\infty,\infty],t) &= \frac{d}{dt}\int_{-\infty}^{\infty} \Psi^*(x,t)\Psi(x,t)\,dx \\ &= \int_{-\infty}^{\infty} \frac{\partial}{\partial t}\left[\Psi^*(x,t)\Psi(x,t)\right]\,dx.\end{aligned} \tag{6.68}$$

To further simplify this integral, we use the chain rule, obtaining

$$\frac{d}{dt}P([-\infty,\infty],t) = \int_{-\infty}^{\infty}\left[\Psi^*(x,t)\frac{\partial}{\partial t}\Psi(x,t) + \Psi(x,t)\frac{\partial}{\partial t}\Psi^*(x,t)\right]dx. \tag{6.69}$$

Next we must eliminate the partial derivatives in (6.69). At this point we avail ourselves of the TDSE (6.67*a*) to express these derivatives in terms of the system Hamiltonian, *e.g.*,

$$\frac{\partial}{\partial t}\Psi(x,t) = \frac{1}{i\hbar}\hat{\mathcal{H}}\Psi(x,t) = -\frac{i}{\hbar}\hat{\mathcal{H}}\Psi(x,t). \tag{6.70}$$

[18] Interchanging the order of differentiation and integration in (6.68) can be rigorously justified, because the limits of the integration (with respect to x) do not depend on the variable of the partial derivative, t. (Remember that in quantum mechanics x and t are independent variables.)

The partial derivative of $\Psi^*(x,t)$ is easily obtained from (6.70), because for any operator $\hat{Q}$,

$$\left[\hat{Q}f(x)\right]^* = \hat{Q}^* f^*(x). \tag{6.71}$$

Applying (6.71) to (6.70), we get

$$\frac{\partial}{\partial t}\Psi^*(x,t) = \frac{i}{\hbar}\left[\hat{\mathcal{H}}\Psi(x,t)\right]^*. \tag{6.72}$$

And substituting Eqs. (6.70) and (6.72) into (6.69), we find

$$\frac{d}{dt}P([-\infty,\infty],t) = \int_{-\infty}^{\infty}\left\{-\frac{i}{\hbar}\Psi^*(x,t)\hat{\mathcal{H}}\Psi(x,t) + \frac{i}{\hbar}\Psi(x,t)\left[\hat{\mathcal{H}}\Psi(x,t)\right]^*\right\} dx \tag{6.73a}$$

$$= -\frac{i}{\hbar}\int_{-\infty}^{\infty}\left\{\Psi^*(x,t)\hat{\mathcal{H}}\Psi(x,t) - \Psi(x,t)\left[\hat{\mathcal{H}}\Psi(x,t)\right]^*\right\} dx. \tag{6.73b}$$

Now, the Hamiltonian is $\hat{\mathcal{H}} = \hat{T} + \hat{V}$, so we can write Eq. (6.73*b*) as the sum of two integrals, one each for $\hat{T}$ and for $\hat{V}$:

$$\begin{aligned}\frac{d}{dt}P([-\infty,\infty],t) = &-\frac{i}{\hbar}\int_{-\infty}^{\infty}\left\{\Psi^*(x,t)\hat{T}\Psi(x,t) - \Psi(x,t)\left[\hat{T}\Psi(x,t)\right]^*\right\} dx \\ &-\frac{i}{\hbar}\int_{-\infty}^{\infty}\left\{\Psi^*(x,t)\hat{V}\Psi(x,t) - \Psi(x,t)\left[\hat{V}\Psi(x,t)\right]^*\right\} dx.\end{aligned} \tag{6.74}$$

But the potential energy $V(x,t)$ is real, so $\Psi(x,t)[\hat{V}\Psi(x,t)]^* = \Psi^*(x,t)\hat{V}\Psi(x,t)$, and the second integral in (6.74) is zero. We're left with

$$\frac{d}{dt}P([-\infty,\infty],t) = -\frac{i}{\hbar}\int_{-\infty}^{\infty}\left\{\Psi^*(x,t)\hat{T}\Psi(x,t) - \Psi(x,t)\left[\hat{T}\Psi(x,t)\right]^*\right\} dx. \tag{6.75}$$

Look carefully at the two terms in (6.75): do they obviously cancel? That is, can we use the argument that eliminated the potential-energy integral from (6.73) to conclude that $\Psi^*(x,t)\hat{T}\Psi(x,t)$ is equal to $\Psi(x,t)\left[\hat{T}\Psi(x,t)\right]^*$?

Absolutely not! The kinetic energy operator $\hat{T}$ is a *differential* operator,

$$\hat{T} = -\frac{\hbar^2}{2m}\frac{\partial^2}{\partial x^2}, \tag{6.76}$$

so it can alter the mathematical form of the function it operates on. The terms in (6.75) do cancel, but to prove this point we must take account of boundary conditions. And we must be crafty.

We start this last step in our demonstration by inserting the explicit form of $\hat{T}$ from (6.76) into (6.75):

$$\frac{d}{dt}P([-\infty,\infty],t) = \frac{i\hbar}{2m}\left[\int_{-\infty}^{\infty}\Psi^*(x,t)\frac{\partial^2}{\partial x^2}\Psi(x,t) - \Psi(x,t)\frac{\partial^2}{\partial x^2}\Psi^*(x,t)\,dx\right]. \tag{6.77}$$

To subdue the rather nasty integral in (6.77), I'm going to introduce a clever (but not necessarily obvious) mathematical trick.[19]

We're going to invoke a theorem familiar from first-year calculus that you probably know as **The Fundamental Theorem of Integral Calculus**:

$$\int_a^b \left(\frac{df}{dx}\right) dx = f(b) - f(a). \tag{6.78}$$

To apply this theorem to (6.77), we must first rewrite its integrand as a differential,

$$\begin{aligned} \Psi^*(x,t)\frac{\partial^2}{\partial x^2}\Psi(x,t) - \Psi(x,t)\frac{\partial^2}{\partial x^2}\Psi^*(x,t) = \\ \frac{\partial}{\partial x}\left[\Psi^*(x,t)\frac{\partial}{\partial x}\Psi(x,t) - \Psi(x,t)\frac{\partial}{\partial x}\Psi^*(x,t)\right], \end{aligned} \tag{6.79}$$

which transforms the integral into the desirable form

$$\begin{aligned} \frac{d}{dt}\int_{-\infty}^{\infty} P(x,t)\,dx \\ = -\frac{i\hbar}{2m}\int_{+\infty}^{-\infty}\frac{\partial}{\partial x}\left[\Psi^*(x,t)\frac{\partial}{\partial x}\Psi(x,t) - \Psi(x,t)\frac{\partial}{\partial x}\Psi^*(x,t)\right] dx. \end{aligned} \tag{6.80}$$

[Notice carefully that I switched the limits of integration in this equation. This is not mere caprice; the reason for it will become clear in § 6.7.]

We can now use Eq. (6.78) to eliminate the integral in our expression (6.77) for the rate of change, which becomes

$$\frac{d}{dt}\int_{-\infty}^{\infty} P(x,t)dx = -\frac{i\hbar}{2m}\left[\Psi^*(x,t)\frac{\partial}{\partial x}\Psi(x,t) - \Psi(x,t)\frac{\partial}{\partial x}\Psi^*(x,t)\right]_{\infty}^{-\infty}. \tag{6.81}$$

But according to the boundary condition (6.67*b*), each term on the right-hand side of (6.81) is zero at the upper and lower limits. So the right-hand side of (6.81) is zero, which proves conservation of probability (6.65).[20]

Q.E.D.

Aside: Introducing Hermiticity. From Eq. (6.81) we can derive an important property of the Hamiltonian operator. Take a look at Eq. (6.73*b*). We have proven that the left-hand side of this equation is zero. Therefore, *for any state function* $\Psi(x,t)$, the Hamiltonian obeys

$$\int_{-\infty}^{\infty} \Psi^*(x,t)\hat{\mathcal{H}}\Psi(x,t)\,dx = \int_{-\infty}^{\infty} \Psi(x,t)\left[\hat{\mathcal{H}}\Psi(x,t)\right]^* dx. \tag{6.82}$$

[19] Don't worry about how you would divine this trick without being shown it. Most of us stumble onto such tricks after lots of false starts and errors. As you work through hundreds of physics problems, you will discover tricks of your own. But a word to the wise: when you find one, commit it to memory, so it will emerge from your subconscious mind when you need it.

[20] This proof can be extended to three-dimensional one-particle systems and to many-particle systems. The essential tool is a mathematical result called *Green's Theorem*, which can transform an integral over a finite volume into an integral over the surface that bounds the volume. You can find the proof for a single-particle in three dimensions in § 3-5 of Robert H. Dicke and James P. Wittke, *Introduction to Quantum Mechanics* (Reading, Mass: Addison-Wesley, 1960). For the many-particle case, see § 54 (ii) of *Quantum Mechanics*, Volume II by Sir-Itiro Tomonaga (Amsterdam: North-Holland, 1966).

This property is called **Hermiticity**, and an operator that satisfies such a condition is said to be **Hermitian**. As we shall learn in Chap. 11, Hermiticity is an important property of all quantum mechanical operators.

Conservation of probability is another brick in the wall of self-consistency that supports quantum theory. Without it, the interpretation of $\Psi(x,t)$ as a position probability amplitude would be on shaky ground, because we would have no guarantee that an increase in the probability inside a finite region corresponds to a decrease outside the region.

Question 6–6

Explain why the normalization integral $\int_{-\infty}^{\infty} \Psi^*(x,t)\Psi(x,t)\,dx$ is independent of time. This result guarantees that we need normalize a state function only once.

Question 6–7

What would happen to conservation of probability if the TDSE were *inhomogeneous*? That is, suppose $(\hat{\mathcal{H}} - \hat{E})\Psi$ were equal to a non-zero (but well-behaved) function of x, so that the partial time derivative of the state function were given not by Eq. (6.70), but by

$$\frac{\partial}{\partial t}\Psi(x,t) = -\frac{i}{\hbar}\hat{\mathcal{H}}\Psi(x,t) + f(x). \qquad \text{[an incorrect equation]} \tag{6.83}$$

By repeating the proof of Proclamation 6.1 with this expression for $\partial\Psi(x,t)/\partial t$, show that if the TDSE were inhomogeneous, probability would not be conserved.

INTERMISSION

This chapter is a long one; it has 10 sections and 139 equations! You have arrived at a way station: a good place to interrupt your study. I recommend taking a break—go to a movie, have dinner, go to bed, whatever; then come back, *review* the first five sections, and press on.

*6.6 THE FREE-PARTICLE WAVE PACKET REVISITED

To illustrate some of the more abstract features of motion in quantum mechanics, I'm going to lead you through a detailed study of the evolution of a single state of the free particle. The wave function that represents the state we'll consider is a *Gaussian free-particle wave packet.* We proved in Example 6.3 that such a packet satisfies the free-particle TDSE. We're now going to explore its propagation and change of shape. I've split this extended example into two sections: here we'll derive the time dependence of $\Psi(x,t)$, and in § 6.7 we'll consider its physical interpretation.[21]

Wave Packets: A Quick Recap

Let's briefly remind ourselves what a wave packet is and how it works. Mathematically speaking, a wave packet of the form

$$\Psi(x,t) = \frac{1}{\sqrt{2\pi}} \int_{-\infty}^{\infty} A(k) e^{i(kx-\omega t)}\, dk \tag{6.84}$$

is a linear combination of an infinite number of monochromatic plane waves with infinitesimally differing wave numbers. The amplitude function $A(k)$ determines the mixture of plane waves in the packet. This function is characterized by its center $k_0 = \langle k \rangle$ and its width Δk. We calculate $A(k)$ as the Fourier transform of the initial wave packet, $A(k) = \mathcal{F}[\Psi(x,0)]$.

The constituent plane waves of the packet have different wave lengths, and they interfere. All of these waves are in phase at the point where total constructive interference occurs: the center of the packet $\langle x \rangle(t)$. At all other values of x, the plane waves interfere destructively. The extent of this interference increases as $|x - x_0|$ increases, so the amplitude of the packet *decreases* with increasing $|x - x_0|$. This interference gives the packet its essential spatially-localized character.

The Evolution of a Wave Packet

As the packet (6.84) evolves, its center $\langle x \rangle(t)$ propagates at the group velocity (Chap. 4). Let's place the coordinate origin, $x = 0$, at the center of the packet at the initial time $t = 0$. Then the propagation of the packet as a whole is characterized by the simple equation

$$\langle x \rangle (t) = v_{\text{gr}} t. \tag{6.85}$$

But something more interesting than propagation happens as time passes: the wave packet spreads.

A simple qualitative argument shows why. Think about the propagating wave packet as a bundle of monochromatic plane harmonic waves, each moving at its phase velocity [Eq. (4.24)]

$$v_{\text{ph}} = \frac{\omega(k)}{k} = \frac{\hbar k}{2m}. \qquad \text{[for a free particle]} \tag{6.86}$$

[21] Neither of these sections contains *new* physical results, so if you are pressed for time, you can jump to § 6.8.

Since the waves in the packet move at *different* velocities, they *disperse* as the packet propagates. This dispersion inevitably causes the packet to spread as time increases. Therefore the position uncertainty $\Delta x(t)$ increases with t. In this section and the next, we'll back up this argument with a (rather elaborate) calculation.

In Example 6.5, we're going to solve the TDSE for a state of the free particle defined by the (Gaussian) initial condition (see § 4.4)

$$\Psi(x,0) = \left(\frac{1}{2\pi L^2}\right)^{1/4} e^{ik_0 x} e^{-x^2/4L^2}. \tag{6.87}$$

The corresponding amplitude function is[22]

$$A(k) = \left(\frac{2}{\pi}L^2\right)^{1/4} e^{-(k-k_0)^2 L^2}. \tag{6.88}$$

This state is characterized by the statistical quantities (Table 4.1)

$$\langle x \rangle (0) = 0 \tag{6.89a}$$

$$\langle p \rangle (0) = \hbar k_0 \tag{6.89b}$$

$$\Delta x(0) = L \tag{6.89c}$$

$$\Delta p(0) = \frac{\hbar}{2L}. \tag{6.89d}$$

Question 6–8

Prove mathematically (or justify via a careful, rigorous argument) that (6.88) is the correct amplitude function for the packet (6.87).

Example 6.5. The Gaussian Free-Particle State Function

We must first decide on strategy: Do we attack the TDSE for a free particle (6.30) directly, trying to solve it using methods from the theory of partial differential equations? Or is there an easier way?

We know from Chap. 4 that there is an *alternate* way. (Whether it's easier remains to be seen.) Since we know the amplitude function (6.88), we can write the wave function for $t > 0$ as [Eq. (4.64)]

$$\Psi(x,t) = \frac{1}{\sqrt{2\pi}} \int_{-\infty}^{\infty} A(k) e^{i[kx - \hbar k^2 t/(2m)]}\, dk \tag{6.90a}$$

$$= \left(\frac{L^2}{2\pi^3}\right)^{1/4} \int_{-\infty}^{\infty} \exp\left[-L^2(k-k_0)^2 + ixk - i\frac{\hbar t}{2m}k^2\right] dk. \tag{6.90b}$$

[In writing (6.90*a*), I used the dispersion relation for a free particle, (6.31).] If we can evaluate this integral analytically, we needn't bother directly solving the TDSE. For a Gaussian initial state, this strategy works like a champ, because we can convert the integral in (6.90*b*) into a standard form that we can look up in tables (or work out analytically).

To this end, we first transform this integral into one that is independent of the center k_0 of the amplitude function. We just shift the origin of the integral from $k = 0$ to $k = k_0$ by changing the dummy variable of integration from k to a new variable $u \equiv k - k_0$. The

[22]The initial wave function (6.87) differs in one small but important respect from Eq. (4.54): the center of its amplitude function (6.88) occurs at a non-zero wave number, k_0, rather than at $k = 0$. Compare this amplitude function to Eq. (4.58).

differentials for these variables are simply related by $dk = du$, so in terms of u, the integral in (6.90b) is

$$\Psi(x,t) = \left(\frac{L^2}{2\pi^3}\right)^{1/4} \int_{-\infty}^{\infty} \exp\left[-L^2u^2 + ixu + ixk_0 - i\frac{\hbar t}{2m}(u^2 + 2k_0u + k_0^2)\right] du. \tag{6.91}$$

Equation (6.91) may not look like an improvement over (6.90b), but it is. The change of variable lets us remove from the integral several factors that do not contain u, *viz.*,

$$\Psi(x,t) = \left(\frac{L^2}{2\pi^3}\right)^{1/4} e^{i[k_0x - \hbar k_0^2 t/(2m)]} \int_{-\infty}^{\infty} \exp\left[-\left(L^2 + i\frac{\hbar}{2m}t\right)u^2 - \left(i\frac{\hbar k_0}{m}t - ix\right)u\right] du. \tag{6.92}$$

At first glance, this integral looks absolutely ghastly. But with a little work we can force it into an integral of the (standard) form (see Appendix J)

$$\int_{-\infty}^{\infty} e^{-\alpha u^2 - \beta u}\, du = \sqrt{\frac{\pi}{\alpha}}\, e^{\beta^2/4\alpha}. \qquad [\text{Re}\ \alpha > 0] \tag{6.93a}$$

All we need do is define α and β in this form as

$$\alpha = L^2 + i\frac{\hbar}{2m}t \tag{6.93b}$$

$$\beta = i\frac{\hbar k_0}{m}t - ix = -i\left(x - \frac{\hbar k_0}{m}t\right). \tag{6.93c}$$

Then we can apply (6.93a) to the wave function for a free-particle wave packet whose initial state is the Gaussian (6.87) to get

$$\Psi(x,t) = \left(\frac{L^2}{2\pi^3}\right)^{1/4} e^{i[k_0x - \hbar k_0^2 t/(2m)]} \sqrt{\frac{\pi}{L^2 + i\frac{\hbar}{2m}t}} \exp\left[\frac{-\left(x - \frac{\hbar k_0}{m}t\right)^2}{4\left(L^2 + i\frac{\hbar}{2m}t\right)}\right]. \tag{6.94}$$

This is the function we seek, the solution of the TDSE (6.30) subject to the initial condition (6.87).

Question 6–9

Show that you can evaluate the integral in Eq. (6.92) without recourse to integral tables by completing the square in the exponential.

We can write our hard-earned wave function in a more useful form by cleaning up the pre-factor; introducing the group velocity $v_{\text{gr}} = \hbar k_0/m$ and the frequency at the peak of the amplitude function, $\omega_0 = \hbar k_0^2/2m$; and rearranging slightly the exponential, to wit:

$$\boxed{\Psi(x,t) = \left[\frac{1}{2\pi L^2\left(1 + i\frac{\hbar}{2mL^2}t\right)^2}\right]^{1/4} e^{i(k_0x - \omega_0 t)} \exp\left[-\frac{1}{4L^2}\frac{(x - v_{\text{gr}}t)^2}{\left(1 + i\frac{\hbar}{2mL^2}t\right)}\right]} \tag{6.95}$$

This is the form we're going to investigate in the next section.

But first, we'd better check our work. In so complicated a derivation, it's easy to make mistakes. So once you've got your answer, ask it lots of questions. For example, try evaluating it at a point, a time, or in a limit where you know the correct answer and see if it agrees.

Question 6–10

Show that the wave function (6.95) at $t = 0$ reduces to the correct initial condition, Eq. (6.87).

You should also think about whether your result looks sensible. The wave function is not merely a mathematical function—it's supposed to represent a state of a free particle with the properties in Eq. (6.89): an (uncertain) momentum, peaked at $p_0 = \hbar k_0$, and an (uncertain) energy, peaked at $E_0 = \hbar\omega_0$. Are these properties reflected in the form of Eq. (6.95)?

Finally, we should see what we can deduce from the mathematical result we've labored over. Staring at the clutter of symbols in Eq. (6.95), we espy the plane harmonic wave $e^{i(k_0 x - \omega_0 t)}$, a pure momentum function with momentum p_0 and energy E_0. We also see that this plane wave is modulated by a x- and t-dependent exponential and by a t-dependent prefactor. Although these factors are complicated, if we stare long enough at (6.95) we discover that its form is that of a Gaussian. This is a very elegant result:

Rule

A function that is a Gaussian at $t = 0$ remains a Gaussian forever after.

This rule is illustrated in Fig. 6.4, which shows *the real part of the Gaussian state function*, $\text{Re}\{\Psi(x,t)\}$, at three times.

Question 6–11

Try to see the Gaussian forest through all the algebraic trees in Eq. (6.95): Can you pick off the width $\Delta x(t)$ from this form? (If not, don't fret. We'll discuss the width soon.)

*6.7 THE LIFE AND TIMES OF A GAUSSIAN WAVE PACKET

To study the "motion" of a microscopic particle, we must investigate the change with time of its probability density $P(x,t)$, expectation value $\langle x \rangle (t)$, and uncertainty $\Delta x(t)$. We'll first evaluate the probability density from the wave function (6.95)—an assignment that will require some algebraic adeptness.

The Probability Density for a Gaussian State

The hard way to obtain an expression for $P(x,t)$ is to substitute $\Psi(x,t)$ from (6.95) directly into $P(x,t) = |\Psi(x,t)|^2$. But we can easily derive a handy form for this density by first simplifying the exponential factor in square brackets in the wave function, which for convenience I'll call F:

$$F = -\frac{1}{4L^2} \frac{(x - v_{\text{gr}} t)^2}{\left(1 + i\frac{\hbar}{2mL^2}t\right)}. \tag{6.96}$$

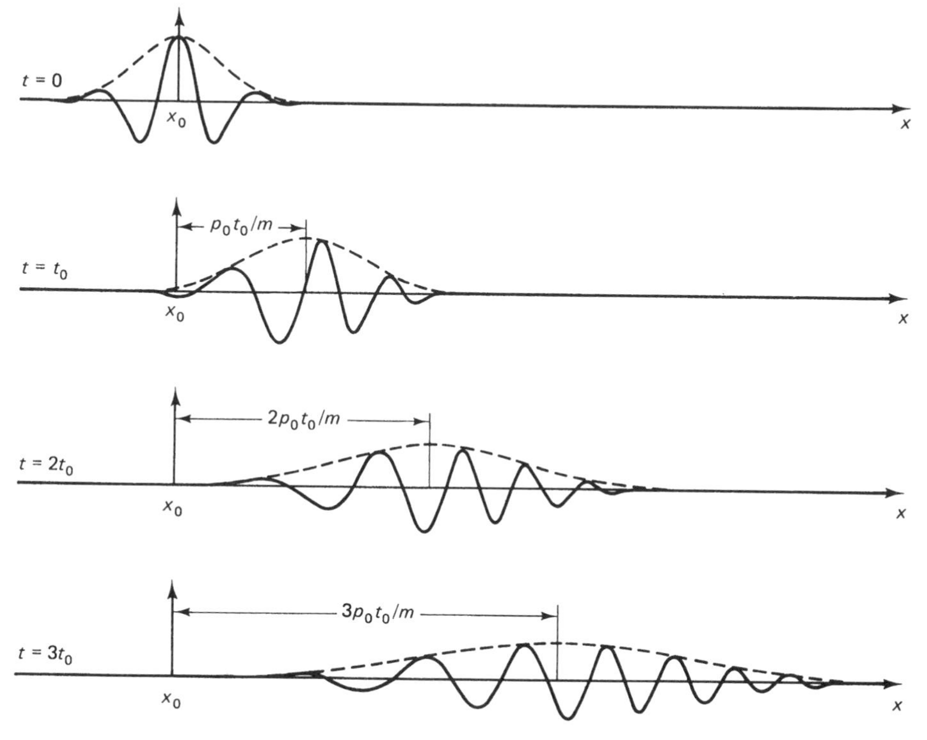

Figure 6.4 A schematic plot of the real part of the Gaussian wave packet (6.95) at the initial time $t = 0$ and at three subsequent times, t_0, $2t_0$, and $3t_0$. The center of this packet moves at the group velocity $v_{\text{gr}} = p_0/m$, and the width $\Delta x(t)$ increases with time. [From R. B. Leighton, *Principles of Modern Physics* (New York: McGraw-Hill, 1959).]

The troublemaker is the innocent-looking $i = \sqrt{-1}$ in the *denominator* of this factor.

Let's move this i into the numerator by multiplying both numerator and denominator by the complex conjugate of the offending denominator, *viz.*,

$$F = -\frac{1}{4L^2}\frac{(x - v_{\text{gr}}t)^2}{\left(1 + i\frac{\hbar}{2mL^2}\,t\right)}\frac{\left(1 - i\frac{\hbar}{2mL^2}\,t\right)}{\left(1 - i\frac{\hbar}{2mL^2}\,t\right)}. \tag{6.97}$$

Multiplying the factors in the denominator of (6.97) and putting our new form for F into (6.95), we obtain the following form for the free-particle, Gaussian state function:[23]

$$\begin{aligned}\Psi(x,t) = &\left(\frac{1}{2\pi L^2}\right)^{1/4}\left(\frac{1}{1 + i\frac{\hbar}{2mL^2}\,t}\right)^{1/2} \\ &\times e^{i(k_0x - \omega_0 t)}\exp\left[-\frac{1}{4L^2}\frac{(x - v_{\text{gr}}t)^2\left(1 - i\frac{\hbar}{2mL^2}\,t\right)}{1 + \frac{\hbar^2}{4m^2L^4}\,t^2}\right]\end{aligned} \tag{6.98}$$

Now we can easily take the complex conjugate of (6.98) and construct the probability density

[23]The part of the prefactor in this wave function that normalizes it is $(2\pi L^2)^{1/4}$. The rest of the prefactor is part of the time dependence of the wave function.

$$P(x,t) = \left[\frac{1}{2\pi L^2\left(1+\frac{\hbar^2}{4m^2L^4}t^2\right)}\right]^{1/2} \exp\left[-\frac{1}{2L^2}\frac{(x-v_{\text{gr}}t)^2}{1+\frac{\hbar^2}{4m^2L^4}t^2}\right]. \tag{6.99}$$

Question 6–12

Before pressing on, check my algebra. First fill in the gap between Eqs. (6.98) and (6.99). Then confirm from the initial wave function (6.87) that

$$P(x,0) = \sqrt{\frac{1}{2\pi L^2}}\, e^{-x^2/2L^2} \tag{6.100}$$

Then show that (6.99) reduces to this at $t = 0$.

In Fig. 6.5, I've graphed this probability density at three equally-spaced times. These graphs show how the region in which the particle is most likely to be found changes with time. The position of the *center* of this region, $\langle x\rangle(t)$, changes linearly with t, as in Eq. (6.85). Such figures are as close as we can come to a pictorial representation of the motion of a member of an ensemble of microscopic particles.

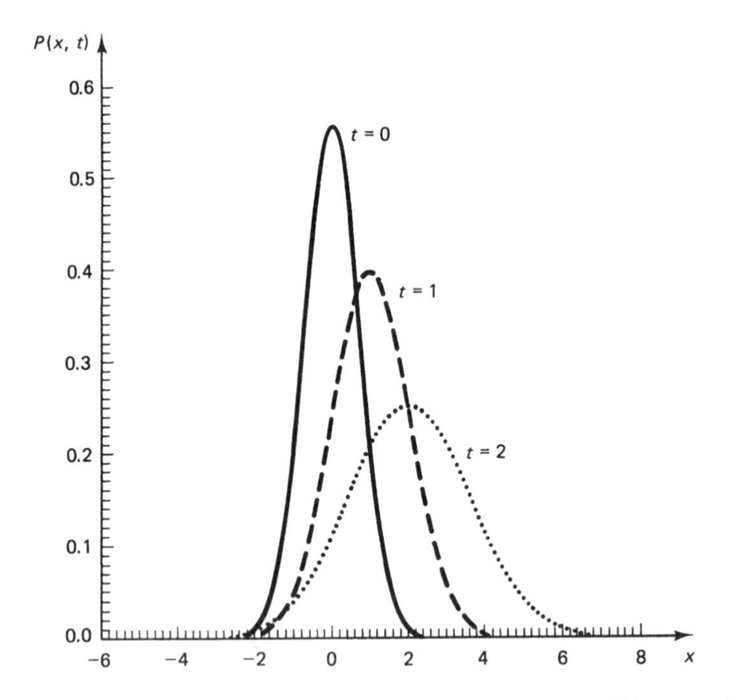

Figure 6.5 The probability densities for the Gaussian wave packet (6.98) at $t = 0$, $t = 1$, and $t = 2$. The wave number and initial ($t = 0$) width are $k_0 = 1$ and $L = 1/\sqrt{2}$, respectively.

Figure 6.5 also shows that as $\Psi(x,t)$ evolves, $P(x,t)$ changes shape. This is the **wave packet spreading** I alluded to in § 6.6. This odd—and wholly quantum mechanical—phenomenon can be seen in the laboratory: the later the time at which we perform such a measurement, the more likely we are to detect the particle far from the center of the packet, $\langle x\rangle(t)$.

Extracting the Width

Figure 6.5 graphically illustrates the increase with time of the *position uncertainty* of a free particle with a Gaussian wave function. This phenomenon is the physical consequence of wave packet spreading. To scrutinize it quantitatively, we need an expression for this uncertainty, $\Delta x\,(t) = \sqrt{\langle x^2\rangle\,(t) - \langle x\rangle^2\,(t)}$. We could, of course, just substitute the state function (6.98) into each of the expectation values in this expression and start integrating. But the resulting algebra is a nightmare. [It's $\langle x^2\rangle\,(t)$, by the way, that generates the awkward algebra; we already know $\langle x\rangle\,(t) = \hbar k_0 t/m$.] So I'd like to describe an alternative approach: extracting the width.

This stratagem doesn't always work, but it may save so much work that it's worth a try. The idea is to read the width function directly from the wave function or, in this case, from the probability density. First we'll write the probability density at $t = 0$ in a form that contains the *initial* width [Eq. (6.89c)], $\Delta x(0) = L$. Then we'll compare $P(x, 0)$ to $P(x, t)$ and *deduce* the width $\Delta x(t)$.

You derived the probability density at $t = 0$ for a Gaussian state function, Eq. (6.100) when you worked Question 6-12. Replacing L in this equation by $\Delta x(0)$, we have

$$P(x,0) = \sqrt{\frac{1}{2\pi[\Delta x(0)]^2}}\; e^{-x^2/2[\Delta x(0)]^2}. \tag{6.101}$$

Now, look at the density for $t > 0$, Eq. (6.99). Comparing this function to its $t = 0$ form (6.101), we find that we can write it as

$$P(x,t) = \sqrt{\frac{1}{2\pi[\Delta x(t)]^2}}\;\exp\left[-\frac{(x - v_{\text{gr}}t)^2}{2[\Delta x(t)]^2}\right] \qquad \text{Probability Density for a Gaussian Wave Packet} \tag{6.102}$$

provided the square of the width function is

$$[\Delta x(t)]^2 = L^2\left(1 + \frac{\hbar^2}{4m^2L^4}\,t^2\right) = [\Delta x(0)]^2\left(1 + \frac{\hbar^2}{4m^2[\Delta x(0)]^4}\,t^2\right). \tag{6.103}$$

This comparison gives for *the position uncertainty for a particle in a state represented by a Gaussian wave packet with initial uncertainty* $\Delta x(0)$ the simple result

$$\Delta x(t) = \Delta x(0)\sqrt{1 + \frac{\hbar^2}{4m^2[\Delta x(0)]^4}\,t^2}. \qquad \text{[for a Gaussian wave packet]} \tag{6.104}$$

Notice that at $t = 0$, Eq. (6.104) reduces to $\Delta x(0) = L$, as it should.

Using the identification (6.104), we have written $P(x, t)$ in a form that corresponds to a Gaussian state function of width $\Delta x(t)$ and that correctly reduces to Eq. (6.100) at $t = 0$. [The form (6.102) also shows explicitly how the probability density depends on the statistical quantities that characterize the wave packet, $\langle x\rangle\,(t)$ and $\Delta x(t)$.] This argument does not, of course, constitute a rigorous proof that (6.104) is the correct width, but it is highly suggestive.

Question 6–13

Evaluate the position uncertainty from the expectation values of $\langle x^2\rangle$ and $\langle x\rangle$. (You may not want to do this exercise; you will have to grind through considerable algebra to obtain Eq. (6.104). It should, however, convince you of the enormous amount of work you can save by *thinking carefully before you calculate*, asking yourself the question: can I get the result I need by reasoning rather than manual labor?)

Middle-Age Spread

In Fig. 6.6, I've graphed the position uncertainty (6.104). This graph vivifies our speculation based on the probability density graphs in Fig. 6.5: *the position uncertainty for a free particle in a Gaussian state increases with time.* And it reveals something new: this increase is at first quite gradual, and then becomes linear. The increasing imprecision in the position of the particle is a physical consequence of the dispersion of the constituent waves in the wave packet. It is one of the marvels of quantum mechanics.

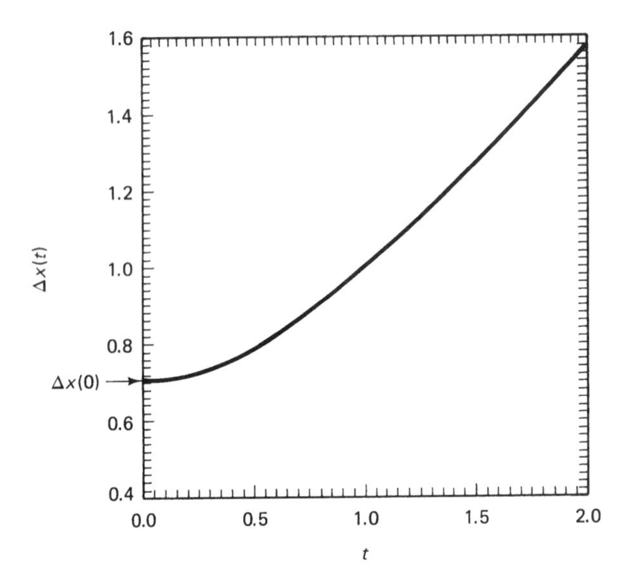

Figure 6.6 The width $\Delta x(t)$ of the Gaussian wave packet as a function of time. Initially the packet retains its shape, but once the width gets on the linear part of this curve, its spreading becomes noticeable.

Let's look more closely at the life of a wave packet. Suppose that initially it has a well-defined shape, corresponding to a small (but non-zero) width $\Delta x(0)$. We let the packet go, monitoring its career by ensemble measurements of position.

We can use Eq. (6.104) to predict the results of these measurements. Immediately after $t = 0$, the second term under the square root, $\hbar^2 t^2/(4m^2 L^4)$, will remain small compared to 1. So at these times the width is, to a good approximation, unchanged: $\Delta x(t) \approx \Delta x(0) = L$. At such times, the spreading of our Gaussian packet is barely noticeable: a youthful packet retains its shape.

But eventually, at times $t \approx 2mL^2/\hbar$, the width begins to increase more sharply (on the linear part of the curve in Fig. 6.6), and the packet, now entering middle-age, begins to spread noticeably.

The rest of the life of our wave packet is not a happy one. For $t \gg 2mL^2/\hbar$, it is bloated and nearly shapeless: $\Delta x(t)$ is enormous. At such times, we know almost nothing about where the particle is likely to be found, because the position uncertainty is huge. In the limit $t \to \infty$, the packet loses all definition. In this limit, we can no longer even meaningfully define its center, and we know *nothing* about the particle's location. This spreading and ultimate loss of definition of a wave packet is inevitable—a consequence not of external forces or even poor dietary habits, but of natural law.

So far I've focused on the position uncertainty, partly because this quantity is directly related to the spreading of the wave packet (and partly because I think it's more interesting than the momentum uncertainty). Still, to complete the story of the Gaussian function, we ought to look at Δp and the uncertainty product $\Delta x(t)\,\Delta p$.

The momentum uncertainty is a little less interesting than Δx because it doesn't change. Since the amplitude function $A(k)$ [and hence the momentum probability amplitude $\Phi(p)$] for a free particle is independent of t (Chap. 4), so is the momentum uncertainty,

$$\Delta p(t) = \Delta p(0) = \frac{\hbar}{2L}. \tag{6.105}$$

[From now on I'll drop the superfluous argument from Δp.]

The position and momentum uncertainties must obey the omnipresent Heisenberg Uncertainty Principle (HUP), $\Delta x(t)\,\Delta p \geq \hbar/2$, which requirement provides a dandy way to verify our work in this section. Using Eqs. (6.98), we find that the uncertainty product at $t = 0$ is $\Delta x(0)\,\Delta p = \hbar/2$, the smallest value the HUP allows! This is a very unusual occurrence—equality in the HUP—and it occurs *only* for the Gaussian wave packet at $t = 0$. It is why physicists refer to this packet as the **minimum uncertainty wave packet**.

We know that the HUP will be satisfied for $t > 0$, because $\Delta x(t)$ increases with time and Δp doesn't change. More quantitatively, their product increases as

$$\Delta x(t)\,\Delta p = L\,\frac{\hbar}{2L}\sqrt{1 + \frac{\hbar^2}{4m^2L^4}\,t^2} \tag{6.106a}$$

$$= \frac{\hbar}{2}\sqrt{1 + \frac{\hbar^2}{4m^2L^4}\,t^2} \quad \text{[for a Guassian Wave Function]} \tag{6.106b}$$

[Notice that *only at* $t = 0$ is the uncertainty product independent of the *initial* width of the packet, $\Delta x(0) = L$.]

Question 6–14

Gazing intently at the state functions in Fig. 6.4, we notice that the *local wavelength* of the oscillations on either side of the center of the packet at the times shown,

$$\langle x\rangle_n(t_0) = \frac{np_0t_0}{m} \tag{6.107}$$

are smaller on the right than on the left. **Explain** this observation. (*Hint*: The wavelength is *inversely* proportional to the momentum.)

In this section we have plumbed the depths of a Gaussian wave packet. I hope this extended example has helped you develop an *intuition* for how a typical state function evolves. The phenomena we've seen—propagation of the packet at the group velocity and spreading—are characteristic of nearly all (non-stationary) wave functions. But in some systems, spreading takes on bizarre characteristics: if a microscopic particle is bound in the potential of a simple harmonic oscillator potential, the packet *sloshes* back and forth in the well. But to see this, you'll have to wait 'til Chap. 9.

The pictures in Figs. 6.5 of the Gaussian probability density at three times suggests an *analogy* to the evolution of a quantum state. It is as though there is a current of "probability fluid" that flows through space. At any fixed time, the *density* of this fluid is largest where the particle is most likely to be found. This analogy is the basis of an extremely useful concept—the probability current density—which affords us new insight into quantum motion.[24]

A Notion of Motion

To introduce the idea of a probability current, let's think about an ensemble measurement of position. Consider a detector fixed somewhere on the x axis, as in the region $[a, b]$ in Fig. 6.7a. Suppose an ensemble of particles is incident on the detector from the left, *i.e.*, *in the direction of increasing* x, in a quantum state represented by a wave packet $\Psi(x, t)$. The detector measures the position probability by counting the number of particles that it registers per unit time.

An actual detector measures the probability in a *finite interval*, which I'll denote $[a, b]$. The quantum mechanical quantity this device measures is the *integrated probability* of § 3.4 [Eq. (3.26)],

$$\text{probability of detection in } [a,b] \tag{6.108}$$
$$= P\left([a,b],t\right) = \int_a^b P(x,t)\,dx = \int_a^b |\Psi(x,t)|^2\,dx.$$

We interpret this quantity (in the gospel according to Born) as the probability that in an ensemble measurement of position at time t, we'll find the particle in the *detector region* $[a, b]$. In practice, then, $P([a, b], t)$ is what we measure when we measure position.

Let's see how the integrated probability changes with time. Initially (*i.e.*, at $t = 0$), the state function is localized far from the detector at $\langle x \rangle (0)$ (Fig. 6.7a). At this time, then, the detector registers $P([a, b], 0) \approx 0$.[25] As time passes, the particle moves in the direction of increasing x, first into and then out of the detector region—a process shown in the "snapshots" of Figs. 6.7b–6.7d. What happens to $P([a, b], t)$ while this is going on?

Well, the first change in this quantity occurs at the time $t_1 > 0$ when the particle is close enough to the detector that this device measures a non-zero probability $P([a, b], t_1)$. Thereafter, $P([a, b], t)$ increases until the time t_2 when the maximum of $P(x, t)$ is very

[24]Probability flow is included here because the associated derivations and concepts follow closely from the material in the previous sections. But if you want to get on to Chap. 7, you can defer § 6.8 and 6.9 until you begin studying one-dimensional scattering in Chap. 8. If you choose this option, insert these two sections just before § 8.2.

[25]Strictly speaking, there might be a minute position probability within $[a, b]$—if, for example, the wave packet is a Gaussian, which never *equals* zero for finite x. But since the accuracy of real detectors is limited, we can always arrange the apparatus so that $\langle x \rangle(0)$ is large enough to give a negligible reading. Note, by the way, that throughout this discussion we're ignoring the details of the interaction of the particle with the detector. The difficult question of how a measuring apparatus interacts with the particle whose properties it is measuring, is treated in advanced books, such as *Measurement and Time Reversal in Objective Quantum Theory* by F. J. Belinfante (New York: Pergamon, 1975).

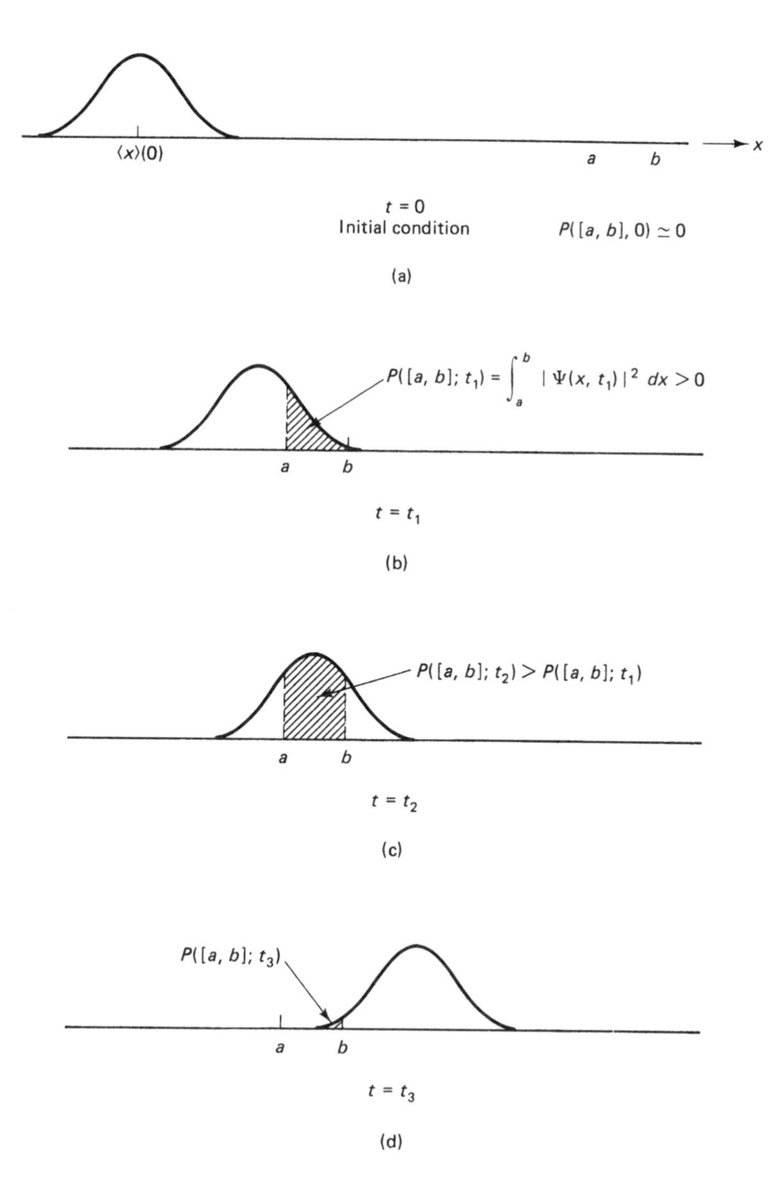

Figure 6.7 One-dimensional probability flow in a beam. (a) Initially, the particles are so far from the detector region $[a, b]$ that the detector registers zero. Sketches (b)–(d) show the gradual encroachment by the packet on the space defined by the detector: (b) As ensemble nears the detector, more and more of the probability density is within the detector region $[a, b]$. Consequently the integrated probability $P([a, b]; t)$ increases until (c) time t_2, when it attains its maximum. (d) As the particle moves on, the measured probability dies to zero.

near the detector (Fig. 6.7c). After this, $P([a, b], t)$ decreases, ultimately returning to an immeasurably small value when the packet no longer significantly overlaps the region $[a, b]$ (*i.e.*, $t > t_3$ in Fig. 6.7d). These qualitative changes in the integrated probability are sketched in Fig. 6.8. Notice that the probability density for $0 < t < t_3$ is definitely not in steady state: $P([a, b], t)$ changes with time. So the *rate of change* of this quantity, $dP([a, b], t)/dt$, is not zero.

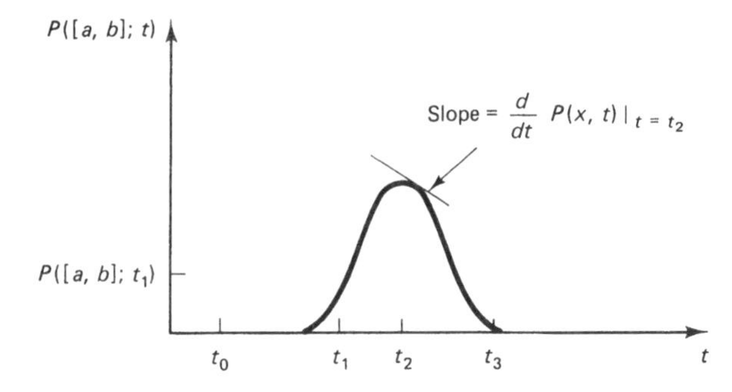

Figure 6.8 The integrated probability measured by the detector in Fig. 6.7. Note that the rate of change of the probability in the region $[a, b]$ is not constant. This change is what we mean by probability flow.

Go with the Flow

The notion of *probability flow* is implicit in our qualitative discussion of the change in the integrated probability. We can imagine a "probability fluid" that flows into and then out of the detector region. As the amount of probability fluid in $[a, b]$ increases (or decreases), the detector is more (or less) likely to register the presence of a particle. In this quaint picture, probability flows as the state function $\Psi(x, t)$ evolves according to the time-dependent Schrödinger Equation.

Be careful not to get carried away with this appealing analogy. We use the notion of probability flow to *envision* rather subtle physical behavior and to cast the equations of quantum mechanics into a form similar to those of other fields: *e.g.*, electricity and magnetism (flow of charge in a wire) and hydrodynamics (flow of gunk in a sewer pipe). But remember that *neither the state function nor the probability density actually exists in the laboratory.* If anything can rigorously be said to "flow" in quantum theory, it is *information*, for that is the physically significant content of $\Psi(x, t)$ and $P(x, t)$. So take heed:

> *WARNING*
>
> **Thou must not take probability fluid too seriously. Probability fluid does not exist in real space; "probability flow" is colorful language designed to create in your mind a picture, an analog of the evolution of a quantum system.**

With this caution out of the way, we turn to the quantitative incarnation of the concept of probability flow, the probability current density.

Introducing the Probability Current Density

We've found that a key quantity for describing probability flow is *the rate of change of the integrated probability within a finite region* $[a, b]$,

$$\frac{d}{dt} P\left([a, b], t\right) = \frac{d}{dt} \int_a^b P(x, t)\, dx = \frac{d}{dt} \int_a^b \Psi^*(x, t)\, \Psi(x, t)\, dx. \tag{6.109}$$

To introduce the probability current density, I want to rewrite Eq. (6.109) in a form very similar to the equation we derived in § 6.5 for the rate of change of the integrated probability *for infinite interval*, Eq. (6.81). Because the little derivation that follows draws heavily on the proof of Proclamation 6.1 (conservation of probability), I want you to stop now, go back, and review the algebra leading to Eq. (6.81). The vital point I want you to appreciate is that *at no point in the derivation of this equation did we use the fact that the interval was infinite.*

This observation makes it easy to adapt Eq. (6.81) to the finite interval $[a, b]$, as required by Eq. (6.109). We just replace the limits $-\infty$ and $+\infty$ in that equation by the finite values a and b, leaping over mounds of algebra to land on

$$\frac{d}{dt}\int_a^b P(x,t)\,dx = -\frac{i\hbar}{2m}\left[\Psi^*(x,t)\frac{\partial}{\partial x}\Psi(x,t) - \Psi(x,t)\frac{\partial}{\partial x}\Psi^*(x,t)\right]_b^a. \tag{6.110}$$

Look at the structure of Eq. (6.110). Writing it out in full, we find the right-hand side to be the difference of two terms, the first of which is evaluated at the *left* boundary of the detector region, $x = a$, and the second at the *right* boundary, $x = b$

$$\begin{aligned}\frac{d}{dt}\int_a^b P(x,t)\,dx = &\left\{-\frac{i\hbar}{2m}\left[\Psi^*(a,t)\frac{\partial}{\partial x}\Psi(x,t)\bigg|_{x=a} - \Psi(a,t)\frac{\partial}{\partial x}\Psi^*(x,t)\bigg|_{x=a}\right]\right\}\\ &-\left\{-\frac{i\hbar}{2m}\left[\Psi^*(b,t)\frac{\partial}{\partial x}\Psi(x,t)\bigg|_{x=b} - \Psi(b,t)\frac{\partial}{\partial x}\Psi^*(x,t)\bigg|_{x=b}\right]\right\}.\end{aligned} \tag{6.111}$$

This structure suggests that we define a function of x and t to represent these terms. This function is the **probability current density**:

$$j(x,t) \equiv -\frac{i\hbar}{2m}\left[\Psi^*(x,t)\frac{\partial}{\partial x}\Psi(x,t) - \Psi(x,t)\frac{\partial}{\partial x}\Psi^*(x,t)\right] \quad \text{probability current density} \tag{6.112}$$

Armed with this definition, we can write Eq. (6.111) in a clearer form, which is called the **rate equation for probability flow**:

$$\frac{d}{dt}P([a,b],t) = j(a,t) - j(b,t) \quad \text{rate equation for probability flow} \tag{6.113}$$

Note that

1. the probability current density $j(x,t)$ is evaluated from the state function $\Psi(x,t)$ and, in general, depends on x and on t;
2. $j(x,t)$ is used as in Eq. (6.113) to calculate the rate at time t of probability flow through a finite region.

After a brief example, we'll discuss the physical meaning of the probability current density and the rate equation.

Question 6–15

To gain familiarity with the definition of the probability current density, show that the following alternate forms of $j(x,t)$ are equivalent to Eq. (6.112):

$$j(x,t) = \frac{\hbar}{m}\text{Im}\left[\Psi^*(x,t)\frac{\partial}{\partial x}\Psi(x,t)\right], \tag{6.114a}$$

where Im[...] means "take the imaginary part of the function in square brackets," and

$$j(x,t) = \frac{1}{m}\text{Re}\left[\Psi^*(x,t)\hat{p}\Psi(x,t)\right], \tag{6.114b}$$

where $\hat{p}$ is the momentum operator and Re[...] means "take the real part of the complex function in square brackets."

Question 6–16

True or False: The function $-[i\hbar/(2m)]\partial P(x,t)/\partial t$ is equivalent to the right-hand side of Eq. (6.112).

Example 6.6. Probability Current for a Pure Momentum Function

The definition (6.112) of the probability current density is rather complicated and abstract, so let's apply it to a very simple function of x and t: a *pure momentum function.*[26] Specifically, consider a pure momentum function corresponding to momentum $p_0 = \hbar k_0$ and energy $E_0 = \hbar\omega_0$,

$$\Psi_{p_0}(x,t) = Ae^{i(p_0 x - E_0 t)/\hbar}. \tag{6.115}$$

[In this example, I won't choose $A = 1/\sqrt{2\pi\hbar}$, as I did in Chap. 4. Letting A remain arbitrary frees us to renormalize the pure momentum function in a way that is useful in the study of collisions (see § 6.9).] To evaluate the probability current density for $\Psi_{p_0}(x,t)$, we just substitute the form (6.115) into the definition (6.112) of $j(x,t)$, obtaining

$$j(x,t) = |A|^2\,\frac{\hbar k_0}{m}. \qquad \text{[for a pure momentum state]} \tag{6.116}$$

The probability current density (6.116) for a pure momentum function thus corresponds to *steady state* probability flow, as you can see from the rate equation (6.113). Since the current density is independent of position, *i.e.*, $j(a,t) = j(b,t)$ for any finite region $[a,b]$, this equation becomes

$$\frac{d}{dt}P\left([a,b],t\right) = 0. \qquad \text{[for a pure momentum function]} \tag{6.117}$$

So the integrated probability *for a pure momentum state* does not change with time *and* is independent of position. That is, the probability fluid associated with a pure momentum function is homogeneous and uniform, and its flow rate is constant. Please note that *the pure momentum function is a very special case*; in general the probability current density is a complicated function of x and t—see Pblm. 6.6.

Aside: An Artificial Analogy. I remarked above that we can use the probability flow (and its current) to write equations of quantum mechanics in a form similar to those of other fields. To show you how, I'll transform Eq. (6.116). For a pure momentum function, we have

$$\hbar k_0 = p_0, \qquad p_0/m = v_{\text{gr}} \qquad \text{and} \qquad |A|^2 = P(x,t).$$

So we can write our expression for the probability current density for this function in the (slightly artificial) way

$$j(x,t) = P(x,t)\,v_{\text{gr}}.$$

This form shows that we can consider the probability current density as the product of a probability *fluid density* and the *flow velocity* v_{gr} of this fluid. Similar results are obtained in hydrodynamic and electromagnetic theory.

[26]This is no mere academic exercise. To see how we might use this function in an application, consider again Fig. 6.6. Suppose that when the particles reach the detector, their wave function is *very* fat [*e.g.*, $\Delta\,x(t)$ is *huge* for $t_1 \le t \le t_2$ in Fig. 6.7.] Then the *momentum* distribution of this state will be sharply peaked about the value $p_0 = \hbar k_0$ (*i.e.*, Δp is teeny). To the detector, which can "see" only the finite region $[a,b]$, such a packet looks like a pure momentum function with wave number k_0 and frequency $\omega_0 = \hbar k_0^2/2m$. For more on modeling a wave packet with a pure momentum function, see Chap. 8.

To make this example concrete, let's evaluate the probability current density for a beam of electrons in a state that can be approximated by the pure momentum function we investigated in § 5.3, Eq. (5.15):

$$\Psi(x,t) = (5.375 \times 10^2\,\mathrm{m}^{-\frac{1}{2}})\,\exp\{i[(4.9 \times 10^{12}\,\mathrm{m}^{-1})\,x \;-\; (5.8 \times 10^{18}\,\mathrm{sec}^{-1})\,t]\}. \tag{6.118}$$

To evaluate $j(x,t)$ for this state, we read off (6.118) the values $|A| = 5.375 \times 10^2\,\mathrm{m}^{-1/2}$ and $k_0 = 4.9 \times 10^{12}\,\mathrm{m}^{-1}$. Inserting these values and the rest mass of the electron, $m_e = 9.109 \times 10^{-31}\,\mathrm{kg}$, into Eq. (6.116), we get $j = 1.639 \times 10^{-10}\,\mathrm{sec}^{-1}$. (*Notice the units.*)

But What Does the Probability Current Density Mean?

At this point, our understanding of the probability current density is based on qualitative, intuitive arguments. Such arguments are fine as far as they go—but they don't go far enough. We need a precise physical interpretation.

To find one, we'll go back to the rate equation (6.113), integrate it over a finite time interval, and then think about the resulting equation in terms of the Born interpretation and probability flow. This rate equation,

$$\frac{d}{dt}P\left([a,b],t\right) \;=\; j(a,t) - j(b,t), \tag{6.119}$$

describes the change in $P\left([a,b],t\right)$ per *infinitesimal* time element dt. If we integrate this equation over the finite time interval $\Delta t \equiv t_2 - t_1$, we obtain

$$\int_{t_1}^{t_2} \frac{d}{dt}P\left([a,b],t\right)\,dt \;=\; \int_{t_1}^{t_2} j(a,t)\,dt - \int_{t_1}^{t_2} j(b,t)\,dt. \tag{6.120}$$

Now, the left-hand side of (6.120) is the integral of a total derivative, so it's just [see Eq. (6.78)]

$$\int_{t_1}^{t_2} \frac{d}{dt}P\left([a,b],t\right)dt \;=\; P\left([a,b],t_2\right) - P\left([a,b],t_1\right). \tag{6.121}$$

Physically, this quantity is *the change in the integrated probability in the interval* $[a,b]$ *during the interval* Δt. I'll denote this change by the rather baroque symbol $\Delta P([a,b], \Delta t)$. The resulting time-integrated rate equation (6.120) is the key that unlocks the interpretation of the probability current density:

$$\begin{aligned}\Delta P\left([a,b],\Delta t\right) \;&=\; P\left([a,b],t_2\right) - P\left([a,b],t_1\right)\\ &=\; \int_{t_1}^{t_2} j(a,t)\,dt - \int_{t_1}^{t_2} j(b,t)\,dt.\end{aligned} \tag{6.122}$$

This equation says that the change in the (integrated) probability in the detector region during the finite interval Δt is equal to the difference of two quantities. To see what these quantities mean, we must consider what could cause in this change. There are two possibilities:

1. probability could flow *into* $[a,b]$ at $x = a$;
2. probability could flow *out of* $[a,b]$ at $x = b$.

We find on the right-hand side of the integrated rate equation (6.122) a term that represents each of these causes: the first term, $\int_{t_1}^{t_2} j(a,t)\,dt$, is the amount of probability that *enters* the region $[a,b]$ at $x = a$ during the interval Δt. And $\int_{t_1}^{t_2} j(b,t)\,dt$ is the amount that *leaves* the region at b during this interval.[27]

This way of thinking about (6.122) suggests that we interpret the probability current density $j(x_0,t)$ at the point x_0 and time t as the *instantaneous rate of probability flow in the direction of increasing x*. Be sure you understand the *directional* quality of the probability current density. This feature is implicit in the integrated rate equation (6.122). Suppose, for example, that the change in the integrated probability, $\Delta P\left([a,b],\Delta t\right)$, is positive. Then the net probability in the detector region *increases* during Δt. In this case $\int_{t_1}^{t_2} j(a,t)\,dt$ must be greater than $\int_{t_1}^{t_2} j(b,t)\,dt$. So in order to make our definition of $j(x,t)$ consistent with Eq. (6.122), we must ensure that a *positive* current density corresponds to probability flow in the $+x$ direction. Similarly, a negative probability current density must correspond to probability flow in the $-x$ direction.

Summary

I began this section by deriving the rate equation for probability flow through a finite region, Eq. (6.113), the mathematical structure of which led us to introduce the probability current density of Eq. (6.112). Then, reasoning from the integrated rate equation (6.122), I argued for an interpretation of $j(x,t)$ that is consistent with this equation and Born's idea of basing the interpretation of quantum mechanical quantities on probability. To summarize our findings, here's a recap of the key quantities of this section:

$$j(x,t) = \text{the instantaneous rate at which position probability crosses the point } x \text{ at time } t\text{, moving in the } +x \text{ direction} \tag{6.123a}$$

$$\int_{t_1}^{t_2} j(x,t)\,dt = \text{the amount of position probability that passes } x \text{ during the interval } \Delta t = t_2 - t_1 \tag{6.123b}$$

You'll find an accompanying summary of the key equations of this section (and of § 6.9) in Table 6.1.

TABLE 6.1 SUMMARY OF KEY RESULTS FOR THE PROBABILITY CURRENT DENSITY IN ONE AND THREE DIMENSIONS

One dimension	Three dimensions
for a finite interval $[a,b]$	for a volume V bounded by surface S
$j(x,t)$ is a signed scalar equal to	$\mathbf{j}(\mathbf{r},t)$ is a vector equal to
$-\frac{i\hbar}{2m}\left[\Psi^*(x,t)\frac{\partial}{\partial x}\Psi(x,t) - \Psi(x,t)\frac{\partial}{\partial x}\Psi^*(x,t)\right]$	$-\frac{i\hbar}{2m}\left[\Psi^*(\mathbf{r},t)\nabla\Psi(\mathbf{r},t) - \Psi(\mathbf{r},t)\nabla\Psi^*(\mathbf{r},t)\right]$
$\frac{\partial}{\partial t}P(x,t) + \frac{\partial}{\partial x}j(x,t) = 0$	$\frac{\partial}{\partial t}P(\mathbf{r},t) + \nabla\cdot\mathbf{j}(\mathbf{r},t) = 0$
$\frac{d}{dt}P\left([a,b],t\right) = j(a,t) - j(b,t)$	$\frac{d}{dt}\int_V P(\mathbf{r},t)\,dv + \int_S \mathbf{j}(\mathbf{r},t)\cdot d\mathbf{a} = 0$

[27] Remember that we're considering particles moving in the direction of *increasing* x. So we needn't worry about probability fluid *leaving* the detector region at $x = a$ or *entering* this region at $x = b$. These cases would be appropriate, however, to particles moving to the left.

Aside: Pointwise Conservation of Probability. The rate equation (6.119) shows that in general, the probability *in a finite region* is not conserved (the exception is the pure momentum function of Example 6.6), *i.e.*,

$$\frac{d}{dt} P\left([a, b], t\right) \neq 0. \tag{6.124}$$

But by recasting the rate equation in *differential form*, we discover a mathematical result that proves that probability is conserved *at a point*:

$$\frac{\partial}{\partial t} P(x,t) + \frac{\partial}{\partial x} j(x,t) = 0. \quad \text{pointwise conservation of probability} \tag{6.125}$$

This law is the most elemental form of the relationship between the probability density and the probability current density.

Since we can regain the rate equation (6.119) from this "pointwise" form by integrating (6.125) over the detector region $[a, b]$ (see Pblm. 6.1), it makes sense to consider Eq. (6.125) as the bottom rung of a "hierarchical ladder" of rate equations. These equations are summarized in Table 6.1, which also includes their three-dimensional counterparts from § 6.9.

Aside: More on Artificial Analogies. The mathematical structure of the differential equation (6.125) may look familiar from your study of electricity and magnetism. In the physics of the flow of electric charge, the fundamental equation relating the charge at point x and time t, $Q(x, t)$, to the electric current density $J(x, t)$ is

$$\frac{\partial}{\partial t} Q(x,t) + \frac{\partial}{\partial x} J(x,t) = 0, \tag{6.126}$$

which bears a striking structural similarity to Eq. (6.125). The current through a region $[a, b]$ is just

$$I = J(b,t) - J(a,t). \tag{6.127}$$

This is the flux associated with this flux of charge. Equations (6.125) and (6.126) belong to an important class of physical relations called **equations of continuity.**[28]

6.9 PROBABILITY FLOW IN THE LABORATORY

In this section, we'll use the probability current density to describe a beam of electrons in a collision experiment such as the one shown schematically in Fig. 6.9. Laboratories are three-dimensional, so we expect that we'll need the generalization to three dimensions of the one-dimensional equations of § 6.8. But once we have these generalizations in hand, we'll discover that in many experiments we can use their one-dimensional counterparts after all.

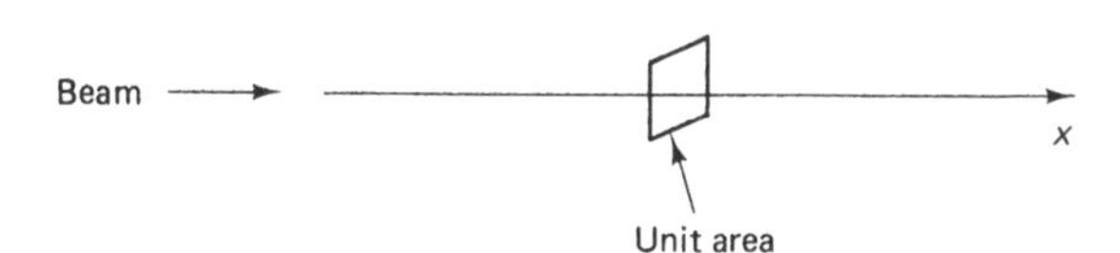

Figure 6.9 A beam experiment in the laboratory. The beam travels along the x direction perpendicular to the unit area shown; this unit area plays a key role in the interpretation of the three-dimensional probability current density, Eq. (6.133).

[28]For a more detailed discussion of this analogy see § 9.3 of *An Introduction to Quantum Physics* by A. P. French and E. F. Taylor (New York: W. W. Norton, 1978).

The Probability Current Density in 3D

There are three essential differences between the situation in one dimension and that in the laboratory. The first concerns the wave function that appears in the definition of the probability current density. In § 6.8, this wave function depended on only one spatial coordinate, x. But a quantum state of a system in three dimensions is represented by a wave function that depends on the three spatial coordinates x, y, and z, so we must calculate the current density from such a function.

Second, we must take account of the three independent orthogonal directions in the lab, as specified by the unit vectors $\hat{e}_x$, $\hat{e}_y$, and $\hat{e}_z$. To describe probability flow in the lab we need *three signed quantities*: $j_x(x,t)$ for flow in the x direction, $j_y(x,t)$ for the y direction, and $j_z(x,t)$ for the z direction. Hence the probability current density for a quantum state in three dimensions is a vector:

$$\mathbf{j}(\mathbf{r},t) = \hat{e}_x\, j_x(\mathbf{r},t) + \hat{e}_y\, j_y(\mathbf{r},t) + \hat{e}_z\, j_z(\mathbf{r},t). \tag{6.128}$$

Finally, in three dimensions we specify a finite region, such as the one that appears in the rate equations for probability flow, by a finite volume V delimited by a boundary surface S, as shown in Fig. 6.10.

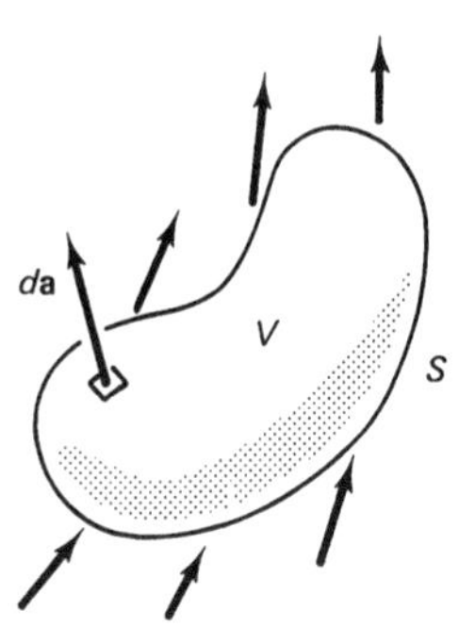

Figure 6.10 In three dimensions, probability flow is defined in terms of a finite volume V. The surface S bounds this volume. The unit area $d\mathbf{a}$ of this surface is assigned a direction: normal to and outward from the surface. The arrows outside the volume in this figure *symbolically* suggest the flow of "probability fluid." [Adapted from *Introduction to Quantum Mechanics* by R. H. Dicke and J. P. Whittke (Reading, Mass.: Addison-Wesley, 1960).]

To generalize the equations of § 6.8, we merely translate them from one to three dimensions, taking into account these three differences. Central to the description of probability flow is the *integrated probability inside a finite region*, Eq. (6.108). In three dimensions, this quantity is a three-fold integral over the finite volume V, *i.e.*,

$$\begin{aligned}\text{probability of detection in volume } V &= P([V],t) \\ &= \int_V P(\mathbf{r},t)\,dv = \int_V |\Psi(\mathbf{r},t)|^2\,dv.\end{aligned} \tag{6.129}$$

In this volume integral, dv is an infinitesimal volume element. (For example, in rectangular coordinates $dv = dx\,dy\,dz$; and in spherical coordinates, $dv = r^2\,dr\,\sin\theta d\theta\,d\phi$.)

The definition of the probability current density for the state $\Psi(\mathbf{r},t)$ is the generalization of Eq. (6.112). The only tricky thing about this generalization is accommodating the dependence of the wave function on $\mathbf{r}$ rather than only on x. We accomplish this by replacing the partial derivative $\partial/\partial x$ in (6.112) with the gradient operator

$$\nabla = \hat{e}_x\,\frac{\partial}{\partial x} + \hat{e}_y\,\frac{\partial}{\partial y} + \hat{e}_z\,\frac{\partial}{\partial z}. \tag{6.130a}$$

With this change, the probability current density for a single particle in three dimensions is

$$\boxed{\mathbf{j}(\mathbf{r},t) \equiv -\frac{i\hbar}{2m}\left[\Psi^*(\mathbf{r},t)\nabla\Psi(\mathbf{r},t) - \Psi(\mathbf{r},t)\nabla\Psi^*(\mathbf{r},t)\right]} \tag{6.130b}$$

Finally, we want to generalize the rate equation (6.113). Here's the one-dimensional equation, with all terms collected on the left-hand side:

$$\frac{d}{dt}P\left([a,b],t\right) + \left[j(b,t) - j(a,t)\right] = 0. \qquad \text{[the one-dimensional rate equation]} \tag{6.131}$$

Look at the quantity in square brackets. It's the difference of two terms. One represents probability flowing *out* of the one-dimensional "volume" $[a,b]$ at $x = b$; the other, probability flowing *into* this volume at $x = a$. If you think of the boundaries of this interval as a one-dimensional surface and examine Fig. 6.10 carefully, you can probably guess the counterpart of this quantity for a three-dimensional system. It's just the *surface integral* of the *component* of the three-dimensional probability current density on the infinitesimal element of area $d\mathbf{a}$ in this figure. With this modification, the rate equation for a three-dimensional quantum state becomes

$$\frac{d}{dt}\int_V P(\mathbf{r},t)dv + \int_S \mathbf{j}(\mathbf{r},t)\cdot d\mathbf{a} = 0. \tag{6.132}$$

Finally, we must generalize our interpretation of the probability current density, Eq. (6.123*a*). To this end, I want to remind you of how we describe a beam of *macroscopic* particles in classical physics. Imagine that the beam in Fig. 6.9 consists of such particles. The *particle density* for such a beam is the average number of particles per unit length of the beam. But in the classical theory of scattering, the most useful quantity is the *classical flux density*, the *number of particles* that crosses a unit area perpendicular to the beam per unit time. (The units of the classical flux density are particles/sec$-$m^2.) By calculating the classical flux through a surface, we can quantitatively determine the *rate of flow* of the particles in the beam.[29]

Now, the important aspect of this description for our generalization of the interpretation (6.131) is the recognition that in three dimensions we must define quantities related to a current (rate of flow) in terms of a unit area normal to the direction of the current. So for a three-dimensional system, we have

$$\mathbf{j}(\mathbf{r},t) = \text{the instantaneous rate (at position } \mathbf{r} \text{ and time } t\text{) at which position probability crosses a unit area perpendicular to the direction of flow.} \tag{6.133}$$

To clarify this interpretation and to show how $\mathbf{j}(\mathbf{r},t)$ is used in practice, we'll look at a real-world example in a moment.

[29]For an introduction to fluid flow in classical physics, see Chap. 8 of *Mechanics* by Keith R. Symon (Reading, Mass: Addison-Wesley, 1971). The flux density, which is also known as the *intensity*, is the basis of the definition of the classical cross section. For a clear description of classical scattering, see § 3.10 of *Classical Mechanics*, 2nd ed. by Herbert Goldstein (Reading, Mass: Addison-Wesley, 1980), or, if you have the first edition of this classic text, § 3.7.

Happily, we usually don't always have to cope with vectors, volume integrals, and gradients when we deal with probability flow in the analysis of an experiment. In many cases, we can use only the x component of $\mathbf{j}(\mathbf{r}, t)$ and, in fact, can use the equations in § 6.8. The justification for this wondrous simplification rests on the form of the three-dimensional equations we just wrote down and on the physical nature of the beams used in many experiments.

Suppose that as particles flow from the source to the target, their beam undergoes very little dispersion along directions perpendicular to its axis of flow. Then we can treat the beam (to a very good approximation) as flowing in one direction only—which, as in Fig. 6.9, we define to be the x axis. This assumption is well within the realm of modern experimental techniques; *e.g.*, in state-of-the-art low energy electron scattering experiments, the dispersion of the incident electron beam is less than three degrees about the beam axis. In this case, the y and z components of $\mathbf{j}(\mathbf{r}, t)$ are negligible, and we need contend with only $j_x(\mathbf{r}, t)$.

But this situation does not automatically reduce the equations of this section to their one-dimensional counterparts. The x-component of the probability current density still depends on x, y, and z because, according to Eq. (6.131), it's determined from the wave function $\Psi(\mathbf{r}, t)$, as

$$j_x(\mathbf{r}, t) = -\frac{i\hbar}{2m}\left[\Psi^*(\mathbf{r}, t)\frac{\partial}{\partial x}\Psi(\mathbf{r}, t) - \Psi(\mathbf{r}, t)\frac{\partial}{\partial x}\Psi^*(\mathbf{r}, t)\right]. \tag{6.134}$$

But *this expression contains only partial derivatives with respect to* x. If we could neglect the dependence of $\Psi(\mathbf{r}, t)$ on y and z, then Eq. (6.134) would reduce to the one-dimensional expression (6.112), which would justify our using the equations of § 6.8 without modification. Under certain conditions, neglect of the y and z dependence of the wave function is a very good approximation. Here's an example.

Suppose the conditions of the experiment justify our modeling the wave function $\Psi(\mathbf{r}, t)$ by the plane wave

$$\Psi(\mathbf{r}, t) \approx A(\mathbf{k}_0)\, e^{i(\mathbf{k}_0 \cdot \mathbf{r} - \omega_0 t)}. \tag{6.135a}$$

If the beam is well collimated along the x axis, as assumed above, then we can approximate $\mathbf{k}_0$ in (6.135*a*) by $k_0 \hat{e}_x$. With this approximation, the dot product in the exponent simplifies as $\mathbf{k}_0 \cdot \mathbf{r} = k_0 x$, and the plane wave function depends only on x and t, *i.e.*,

$$\Psi(\mathbf{r}, t) \approx A(k_0)\, e^{i(k_0 x - \omega_0 t)}. \tag{6.135b}$$

So for such a plane wave, the probability current density (6.134) depends only on x and t, and we can fully describe probability flow using the one-dimensional equations of § 6.8. In Example 6.7, I'll apply this model to an actual laboratory situation.

Aside: More on the Plane Wave Model. The plane wave function (6.135*a*) is a good approximation to a very fat wave packet—*i.e.*, one whose amplitude function $A(\mathbf{k})$ is very sharply peaked about the average wave vector $\mathbf{k}_0$ of the particles in the beam. This requirement is met by a *nearly monoenergetic beam*, because the energy uncertainty of the state of the particles in such a beam is very small. This implies a small momentum uncertainty. [Since the particles are free (*i.e.*, non-interacting), these uncertainties are related by $\Delta p = \sqrt{m/(2E_0)}\,\Delta E$.] And a small momentum uncertainty implies a large position uncertainty: a fat packet. Moreover, if this uncertainty, Δx, is much larger than the dimensions of the target, then the further approximation of (6.135b) seems reasonable—and it can be justified by rigorous analysis.

Example 6.7. A Model Wave Function for a Collision Experiment

The most we usually know about a beam in the laboratory is its electrical current and the mean energy E_0 of the particles in it. For example, in the 1980's atomic physicists can produce nearly monoenergetic beams of electrons of (mean) kinetic energy 0.01 eV in which a current of 3.0×10^{-5} amp flows through a 1 m^2 area perpendicular to the beam.[30] To represent the quantum state of the beam in such an experiment by a pure momentum function (6.135b), we must determine from this data the physical quantities that appear in this function: ω_0, k_0, and A_0.

We can easily calculate the values of ω_0 and k_0 from the energy E_0:

$$\omega_0 = \frac{E_0}{\hbar} = 1.520 \times 10^{14}\,\text{sec}^{-1}. \tag{6.136a}$$

And from the momentum,

$$p_0 = \sqrt{2mE_0} = 1.708 \times 10^{-25}\,\text{kg-m-sec}^{-1}, \tag{6.136b}$$

we get the wave number

$$k_0 = \frac{p_0}{\hbar} = 1.621 \times 10^{9}\,\text{m}^{-1}. \tag{6.136c}$$

All that remains is the amplitude A of the pure momentum function. It's convenient to choose this constant so that $P(x,t) = |A|^2$ is equal to the *relative probability density* in the beam. To this end, let's pick $|A|^2$ to be the *average particle density* in the beam—i.e., the average number of particles per unit length of the beam that passes through a unit area perpendicular to the beam. With this normalization, we can easily determine $|A|$ from the beam current.[31]

To do so, we must relate the beam current, which in this example is $3.0{\times}10^{-5}$ C/sec^{-1} to the probability current density j. Then we can easily calculate $|A|$, because Eq. (6.116) gives us this number in terms of j as

$$A = \sqrt{\frac{j}{p_0/m}} = \sqrt{\frac{mj}{p_0}}. \tag{6.137}$$

Now the electric current I is a measure of the rate of flow of *charge* in the beam, while j is a measure of the rate of flow of *position probability density*. In our interpretation for the pure momentum function, j is the number of particles per second incident on a unit area perpendicular to the beam (*i.e.*, along the x axis and in the yz plane). In this experiment an electrical charge of 3.0×10^{-5} C-sec^{-1} through a 1 m^2 area in the yz plane, so the number of electrons (of charge $q = 1.602 \times 10^{-19}$C) corresponding to this current is

$$j = \frac{3 \times 10^{-5}}{1.602 \times 10^{-19}} = 1.873 \times 10^{14}\text{electrons/sec}. \tag{6.138}$$

If we substitute this value of j, Eq. (6.136b) for p_0, and the value of the electron rest mass $m_e = 9.109 \times 10^{-31}$ kg into Eq. (6.137), we obtain $A = 3.161 \times 10^4\,\text{m}^{-1/2}$.

We can now write down the form of a plane wave that approximates this beam:

$$\Psi(\mathbf{r},t) = (3.161 \times 10^4\,\text{m}^{-3/2}) \exp\{i(1.621 \times 10^9\,\text{m}^{-1})x \\ -(1.520 \times 10^{14}\,\text{sec}^{-1})t\}. \tag{6.139}$$

[30]See H. S. W. Massey, *Atomic and Molecular Collisions* (London: Taylor and Francis LTD, 1979).

[31]The disadvantage to this choice is that the resulting pure momentum function no longer satisfies the Dirac Delta function normalization condition (4.28). But this is a small price to pay for the convenience of modeling the beam with the plane wave.

As promised, this plane wave depends only on x and t.

A final note: I have taken pains in this example to include *units* on all the physical quantities and in the plane wave (6.139). Newcomers to quantum physics often focus so intently on the equations and functions of the theory that they lose sight of these important dimensional indicators. To help you avoid such waywardness, I've summarized the units of key quantities in Table 6.2.

TABLE 6.2 UNITS

	One dimension	Three dimensions
State function	$\text{m}^{-1/2}$	$\text{m}^{-3/2}$
Probability density	m^{-1}	m^{-3}
Probability current density	sec^{-1}	$\text{sec}^{-1}\text{m}^{-2}$

Question 6–17

Figure out the units of the state function, probability density, and probability current density of a single particle in two dimensions (*e.g.*, confined to move in a two-dimensional plane). Try to generalize the equations in Table 6.1 to this case.

6.10 FINAL THOUGHTS: QUANTUM CAUSALITY

The time-dependent Schrödinger Equation is undeniably one of the great intellectual achievements of science. It laid the foundation on which physicists built our current theories of the physics of matter. Other physicists have performed a vast number of experiments the results of which have verified these theories. Armed with this hindsight, it's interesting to see how the TDSE was initially received.

The distinguished physicist Wolfgang Pauli (1900–1958) has vividly described the attitude of the physics community during the years prior to the publication of Schrödinger's paper:[32] "Physics is very muddled again at the moment; it is much too hard for me anyway, and I wish I were a movie comedian or something like that and had never heard anything about physics!" From this comment you might guess that Schrödinger's achievement was greeted with cheers. Wrong.

Not all members of the physics community expressed delight when confronted with Schrödinger's work. Although Max Planck and the renowned Dutch physicist H. A. Lorentz (1853–1938) waxed enthusiastic over his theory, no less a luminary than Werner Heisenberg, in a letter to Pauli, wrote,[33] "The more I ponder about the physical part of Schrödinger's theory, the more disgusting it appears to me." Still, Heisenberg's revulsion notwithstanding, the TDSE quickly dominated quantum theory, eclipsing for a time Heisenberg's alternative, matrix mechanics.

[32]Quoted in *Theoretical Physics in the Twentieth Century*, edited by M. Fierz and V. F. Weisskopf (New York: Interscience, 1960), p. 22.

[33]Quoted in Max Jammer, *The Conceptual Development of Quantum Mechanics* (New York: McGraw-Hill, 1966), p. 272. Writing about Heisenberg's work in a paper in the *Annalen der Physik* [**79**, 734 (1926)] Schrödinger comments that he was "discouraged if not repelled by what appeared to me a rather difficult method of transcendental algebra, defying any visualization."

Now that we have completed our study of the four basic postulates of quantum theory, let's step back momentarily from the *mechanics* of quantum physics to consider its philosophical implications.

In Chap. 1, I noted that the new physics of Heisenberg, Schrödinger, and Born dealt a severe blow to traditional philosophical tenets such as determinism and causality. But the TDSE restores to physics a strange kind of causality. Before explaining this curious comment, I want to summarize the philosophical situation before the advent of this equation.

The notion of cause and effect in the evolution of individual systems was a mainstay of the *mechanistic philosophy* that dominated the nineteenth century. This philosophy was based on Newton's Laws, but it extrapolated the purview of those laws to the whole of existence. In the 19th century, as today, Newtonian *physics* described not the whole universe, but rather any closed (*i.e.*, finite) mechanical system. The crucial leap from physics to philosophy was taken by the French astronomer and mathematician (and Napoleon's minister of the interior) Pierre Simon, marquis de Laplace (1749–1827). In his treatise "Théorie analytique des probabilitiés" Laplace ascribed *universal* applicability to Newton's Laws.

Laplace conjured up a super mathematician, armed with total mastery of all mathematical techniques. He argued that such a mathematician could *solve the universe*: Knowing the state of the universe at a given instant—the masses and trajectories of all the particles comprising all mechanical systems—and knowing the forces acting on and within these systems, this mythical creature could, according to Laplace, determine the physical properties of all the systems in the universe at any subsequent time. To infinite precision. With certainty.

Laplace acknowledged the superhuman—nay, *supernatural*—power of a creature who could *actually* solve the universe by calling his super-mathematician a **demon**. (In contemporary demonology, we might call it an *infinite generational computer*—the CRAY 9000, perhaps.) To his demon, wrote Laplace, "nothing would be uncertain, both future and past would be present before its eyes." Laplace's point was not that the task of solving the universe was *viable*, but rather that it was possible *in principle*. In this philosophical conviction, Laplace was wrong.

The Heisenberg Uncertainty Principle tells us so. This relationship between the inherent indeterminacy in our knowledge of the position and momentum of microscopic particles implies that we cannot specify the precise values of these observables at any instant. So it's impossible to predict the precise subsequent behavior of the system. The Heisenberg Principle casts an impenetrable fog over the line between *cause* (the initial conditions) and *effect* (the subsequent evolution) for an individual system and demolishes forever the doctrine of Laplacian mechanism.

Henry Margenau has vividly illustrated the consequences of the demise of mechanistic causality in his book *The Nature of Physical Reality* (Woodbridge, Conn.: Oxbow Press, 1977, p. 420):

> Single events of great magnitude cannot be said to have the picturesque single-event causes which classical mechanics envisaged: even though I know the exact state of a neutron (having a sharp energy) that is capable of setting off an atomic bomb, and know precisely the location of the block of plutonium that is ready to be exploded, I cannot predict whether the disaster will occur. The fate of the globe, as a single event, may hide itself within atomic uncertainty.

Yet, all is not lost.[34] Even in the quantum domain, a *limited causality* applies. After all, the TDSE does establish a causal relationship between the state of a system at the initial time—as described mathematically by the initial conditions $\{\Psi(x, t_0): -\infty < x < \infty\}$—and its state at a later time, $\Psi(x,t)$ for $t > t_0$ (assuming that the system is allowed to evolve free of measurement). But this is a different kind of causality from that of Laplace's demon.

The mechanistic causality of Laplace does not follow from quantum physics because the structure of quantum theory is fundamentally different from that of Newtonian mechanics: the latter theory describes *individual systems*; the former, *ensembles*. From the initial probability density $P(x, t_0)$, we can calculate various probabilistic and statistical data about position at a later time, such as $\langle x \rangle (t)$ and $\Delta x(t)$. *But these data describe only the results of a position measurement performed on as ensemble of identical systems in the state* $\Psi(x,t)$. Quantum physics does not provide a mechanism for predicting the specific outcome of a *single* measurement on a *single* system. Indeed, such a measurement cannot be described by a theory whose essential elements are interpreted in terms of probabilities.

Of course, Laplacian causality is still a viable philosophical stance from which to contemplate the macroscopic world. And perhaps some day, someone will describe the physics of a "sub-quantum domain"—a world of sub-microscopic entities over which the HUP does not hold dominion.[35] But in the quantum domain—the world of electrons, atoms, molecules, and the like—the only cause and effect relationships are those allowed by the TDSE.

ANNOTATED SELECTED READINGS

Alternate Approaches to the Schrödinger Equation

One of the best ways to familiarize yourself with new concepts is to read about them from several viewpoints. The Schrödinger equation is so important that you should try to find the time to see how other authors introduce it. I particularly recommend the following:

1. Cropper, William H., *The Quantum Physicists and an Introduction to Their Physics* (London: Oxford University Press, 1970), Chap. 4. (Available in paperback.) Cropper's approach is historical and his style is delightful.
2. Eisberg, Robert, and Robert Resnick, *Quantum Physics of Atoms, Molecules, Solids, Nuclei, and Particles*, 2nd ed., (New York: Wiley, 1985), § 5-2. The level of Eisberg and Resnick's eclectic book is lower than this one, and I strongly recommend it as a source of alternative introductions to several of the topics we'll study.
3. Fong, P., *Elementary Quantum Mechanics* (Reading, Mass.: Addison-Wesley, 1949), Chap. 2.

[34] Margenau goes on to reassure us that the odds of encountering a neutron with a sharp energy are quite remote and, moreover, that if the energy of the neutron "is only approximately known, ... the prospect of its encounter with the plutonium block can usually be predicted with a certainty sufficient for practical men."

[35] David Bohm has elegantly made the case for the existence of a sub-quantum domain in his *Causality and Chance in Modern Physics* (Philadelphia: University of Pennsylvania Press, 1957). This book contains a marvelous, largely non-technical discussion of the problem of causality. I highly recommend it.

Differential Equations

Many of the "math methods" books listed in the Reading Lists for Chaps. 4 and 5 contain fine introductions to the theory of differential equations. Here are some good sources for more complete and specialized treatments:

4. Hildebrand, Francis B., *Advanced Calculus for Applications* (Englewood Cliffs, N. J.: Prentice Hall, 1962), Chaps. 1, 3, and 4 of this wonderful book discuss ordinary differential equations; Chaps. 8–10 contain most of what you should know about partial differential equations.
5. Coddington, Earl A., *An Introduction to Ordinary Differential Equations* (Englewood Cliffs, N. J.: Prentice Hall, 1961). This book treats only ordinary equations, but it does that well.
6. Rainville, Earl D., and Phillip E. Bedient, *Elementary Differential Equations*, 4th ed. (New York: Macmillan, 1970).

A fine introductory book on *partial* differential equations is

7. Berg, Paul W., and James L. McGregor, *Elementary Partial Differential Equations* (San Francisco: Holden-Day, 1966).

Finally, I should mention an invaluable source for the intrepid differential equations solver:

8. Murphy, George M., *Ordinary Differential Equations and Their Solutions* (New York: Van Nostrand Reinhold Co., 1960). This book, which has unconscionably been allowed to go out of print, not only contains a splendid summary of methods for solving ordinary differential equations, it also includes tables of solutions to *hundreds* of such equations. If you find it in a used book store, grab it!

On the Philosophy of Quantum Physics

9. Bohm, David, *Causality and Chance in Modern Physics* (Philadelphia: University of Pennsylvania Press, 1957).
10. Jammer, Max, *The Conceptual Development of Quantum Mechanics* (New York: McGraw-Hill, 1966).
11. Margenau, Henry, *The Nature of Physical Reality* (Woodbridge, Conn.: Oxbow Press, 1977).
12. Przibram, K., ed., *Letters on Wave Mechanics*, (New York: Philosophical Library, 1961).
13. Winter, R. G., *Quantum Physics* (Belmont, Ca: Wadsworth, 1979).

EXERCISES AND PROBLEMS

> " . . . the main thing, however, is the algorithm!"
> " . . . Any child knows that!
> What's a beast without an algorithm?"
>
> —Trurl
> in "The Second Sally"
> from *The Cosmic Carnival of Stanislaw Lem*
> by Stanislaw Lem

This chapter's collection includes the usual practice exercises and straightforward applications of the material in the chapter you just studied. In addition, I've thrown in a couple of problems that ask you to work with the stationary-state wave functions for the infinite square well, which we're going to derive in the next chapter, an application of the probability current density to a practical (laboratory) situation, and a couple of fairly challenging puzzles at the end. The last problem in this set is probably the hardest yet—but it's worth your time.

Exercises

6.1 A Possible State Function for an Oscillator

The potential energy of a one-dimensional simple harmonic oscillator of mass m and natural frequency ω is $V(x) = \frac{1}{2}m\omega_0^2x^2$. Here (adapted from Chap. 9) are the normalized wave functions that represent the *ground state* and *first excited state* of this system:

$$\Psi_1(x,t) = \left(\frac{\alpha}{\pi}\right)^{1/4} e^{-\alpha x^2/2}e^{-iE_1t/\hbar} \tag{6.1.1a}$$

$$\Psi_2(x,t) = \left(\frac{\alpha}{\pi}\right)^{1/4} \sqrt{2}(\sqrt{\alpha}\,x)e^{-\alpha x^2/2}e^{-iE_2t/\hbar} \tag{6.1.1b}$$

In these expressions, α is a real constant and E_n is the real energy of the nth state. Each of these functions satisfies the time-dependent Schrödinger equation of the system.

Now, consider the function

$$\Psi(x,t) = \frac{1}{\sqrt{61}}\left[5\Psi_1(x,t) + 6i\Psi_2(x,t)\right]. \tag{6.1.2}$$

[a] Is Eq. (6.1.2) a valid state function for the system—*i.e.*, does it satisfy the time-dependent Schrödinger Equation? **Justify your answer either with a derivation or a complete, rigorous argument.** Is it normalized? **Prove your answer.**

[b] Suppose we need to evaluate the *energy uncertainty* $\Delta E(t) = \sqrt{\langle E^2\rangle - \langle E\rangle^2}$. Consider the task of evaluating $\langle E^2\rangle$. In Chaps. 5 and 6 we developed two operators that can represent the observable *energy*: the Hamiltonian $\hat{\mathcal{H}}$ and the *energy operator* $\hat{E} = i\hbar\partial/\partial t$. So we should be able to evaluate $\langle E^2\rangle$ using either $\hat{\mathcal{H}}$ or $\hat{E}$, *i.e.*,

$$\langle E^2\rangle = \int_{-\infty}^{\infty} \Psi^*(x,t)\hat{\mathcal{H}}^2\Psi(x,t)\,dx \tag{6.1.3a}$$

$$\langle E^2\rangle = \int_{-\infty}^{\infty} \Psi^*(x,t)\hat{E}^2\Psi(x,t)\,dx \tag{6.1.3b}$$

Use the time-dependent Schrödinger equation to prove rigorously that these two forms are equivalent, *i.e.*, **prove** that

$$\int_{-\infty}^{\infty} \Psi^*(x,t)\hat{\mathcal{H}}^2\Psi(x,t)\,dx = \int_{-\infty}^{\infty} \Psi^*(x,t)\hat{E}^2\Psi(x,t)\,dx. \tag{6.1.4}$$

6.2 A Lorentzian State Function

Consider a *free particle* of mass m in one dimension. Suppose the *momentum probability amplitude* of the initial state (*i.e.*, the state at $t = 0$) is

$$\Phi(p) = \sqrt{\frac{a}{\hbar}}\,e^{\frac{-a}{\hbar}|p|}, \tag{6.2.1}$$

where a is a real, positive constant and, of course, $|p|$ is the absolute value of the momentum variable p.

[a] **Show** that the initial wave function $\Psi(x,0)$ is

$$\Psi(x,0) = \sqrt{\frac{a}{2\pi}}\,\frac{2a}{a^2 + x^2} \tag{6.2.2}$$

Hint: $\Psi(x,0)$ is real. Subsequent parts of this problem will be easier if you express it as a real function. Here's a bonus hint:

$$|p| = \begin{cases} -p & p \le 0 \\ +p & p \ge 0 \end{cases} \tag{6.2.3}$$

By the way, you can work most of the rest of this exercise even if you can't do this part.

[b] What is the *parity* of $\Psi(x,0)$? What is the ensemble average of *position* in this state at $t = 0$? **Justify your answer.**

[c] **Evaluate** the mean value of the *momentum* (at $t = 0$) for this state. To do so, first **write down** *two* expressions for this quantity—one in terms of the amplitude function $\Phi(p)$, the other in terms of the initial wave function $\Psi(x,0)$ and the momentum operator $\hat{p}$ in the position representation. Now, choose one of your expressions and evaluate it.
Hint: Draw a sketch of $\Phi(p)$.

[d] In this part we consider the state at $t > 0$ that evolves from this initial state. **Describe** in words and equations *two ways* to determine $\Psi(x,t)$ for $t > 0$. In your discussion, include (but do not try to solve) all relevant equations and set up (but do not evaluate) all necessary integrals. (You'll have lots of opportunities to solve equations and evaluate integrals in the problems below.)

[e] **Show** that the *ensemble average of the energy* at $t = 0$ is $\hbar^2/4ma^2$.
Hint: You can evaluate this quantity either from $\Phi(p)$ or $\Psi(x,0)$, so your first step on this part should be to **write down** expressions for the mean value of the energy in terms of each of these functions. Then think carefully about which one will be easier to evaluate and use the easier expression.

Problems

6.1 Normalization—and Other Like Matters

In Question 6.6 in § 6.5 you showed that the normalization integral for a state function $\Psi(x,t)$ is independent of time, *i.e.*, that once a state function is normalized (*e.g.*, by the procedure in § 3.5) it stays normalized:

$$\int_{-\infty}^{\infty} \Psi^*(x,t_0)\Psi(x,t_0)\,dx = 1 \quad \Longrightarrow \quad \int_{-\infty}^{\infty} \Psi^*(x,t)\Psi(x,t)\,dx = 1 \qquad \text{for all } t > t_0. \tag{6.1.1}$$

(If you didn't answer Question 6.6, please do so now. Then press on.)

[a] The equation that describes the propagation of a *classical* wave $\Gamma(x,t)$ in one dimension has the form

$$\frac{\partial^2}{\partial x^2}\Gamma(x,t) = \frac{1}{v_{\text{ph}}^2}\frac{\partial^2}{\partial t^2}\Gamma(x,t) \tag{6.1.2}$$

where v_{ph} is the phase velocity of the wave. This equation is second-order in time. By contrast, the time-dependent Schrödinger equation for a free particle, Eq. (6.30), includes a *first-order* time derivative. Now, suppose things had been different: suppose the TDSE had the form of the wave equation (6.1.2)—*i.e.*, with a second-order time derivative in it. Would the normalization of the solutions of such an equation necessarily be independent of time—*i.e.*, would Eq. (6.1.1) hold if the TDSE had the form of Eq. (6.1.2)? **Prove your answer.**

[b] In applications of quantum mechanics to problems in atomic and molecular physics, one frequently must evaluate integrals of the form

$$S \equiv \int_{-\infty}^{\infty} \Psi_1^*(x,t)\Psi_2(x,t)\,dx, \tag{6.1.3}$$

where $\Psi_1(x,t)$ and $\Psi_2(x,t)$ represent two physically-realizable quantum states. Such integrals are called **overlap integrals**. (They arise primarily in the calculation of wave functions for atoms and molecules.) Evaluating overlap integrals would be an intolerable chore if these integrals had to be evaluated at each time. **Show** that the integral S is independent of time—*i.e.*, show that

$$\frac{d}{dt}\int_{-\infty}^{\infty}\Psi_1^*(x,t)\Psi_2(x,t)\,dx = 0. \tag{6.1.4}$$

Hint: Use the Principle of Superposition and the answer to Question 6.6.

6.2 The Principle of Superposition: A Non-Stationary State

The normalized wave functions and corresponding (well-defined) energies for the ground and first-excited states of a particle of mass m whose potential is an infinite square well of length L are

$$\Psi_1(x,t) = \sqrt{\frac{2}{L}}\cos\left(\pi\frac{x}{L}\right)e^{-iE_1t/\hbar} \qquad E_1 = \frac{\pi^2\hbar^2}{2mL^2} \tag{6.2.1a}$$

$$\Psi_2(x,t) = \sqrt{\frac{2}{L}}\sin\left(2\pi\frac{x}{L}\right)e^{-iE_2t/\hbar} \qquad E_2 = \frac{4\pi^2\hbar^2}{2mL^2} = 4E_1 \tag{6.2.1b}$$

According to the Principle of Superposition, the linear combination

$$\Psi(x,t) = c_1\Psi_1(x,t) + c_2\Psi_2(x,t), \tag{6.2.2}$$

where c_1 and c_2 are arbitrary complex constants, represents a state of this system. But, unlike the energies of the states represented by $\Psi_1(x,t)$ and $\Psi_2(x,t)$, the energy of the state represented by $\Psi(x,t)$ is not well-defined. Moreover, the probability density $|\Psi(x,t)|^2$ depends on time: this state is *non-stationary.*

[a] Actually, the constants c_1 and c_2 are not truly arbitrary, because the constraint that $\Psi(x,t)$ must be normalized imposes a limitation on them. **Derive** the relationship between c_1 and c_2 in Eq. (6.2.2) that must be satisfied if the wave function of the superposition state is to be normalized.

[b] At time $t > 0$ we perform an ensemble measurement of the total energy of the particle whose state is represented by $\Psi(x,t)$. **Derive** an expression for the average value we would obtain in such a measurement. Express your answer in terms of E_1, E_2, c_1, and c_2.

[c] **Prove** that $E_1 < \langle E\rangle(t) < E_2$. Does $\langle E\rangle$ depend on time?

[d] In some studies of nuclear physics, a *proton* in a nucleus can be modeled by a particle (with the mass of a proton) in an infinite square well of width L, where L is chosen to be a sensible approximation to the size of a nucleus. The "motion" of such a proton is often described in terms of a "charge distribution" $eP(x,t)$, where e is the proton charge and $P(x,t)$ the quantum-mechanical position probability density, that oscillates at the same frequency as $P(x,t)$. Now, suppose the state of the proton is represented by $\Psi(x,t)$ in Eq. (6.2.2) with $c_1 = c_2$. **Calculate** the frequency (in Hz) of a proton in a nucleus of dimension $L = 10\,\text{f}$.

6.3 More Explorations of States of the Infinite Square Well

Here is yet another state of the particle in a box (the infinite square well). The state represented by

$$\Psi(x,t) = \frac{1}{\sqrt{L}}\left[\cos\left(\pi\frac{x}{L}\right)e^{-iE_1t/\hbar} + \sin\left(2\pi\frac{x}{L}\right)e^{-iE_2t/\hbar}\right] \tag{6.3.1}$$

is a superposition state constructed from the wave functions that represent two of the states in the previous problem. In this problem, we'll use our understanding of operators and the Schrödinger equation to further investigate these three states.

[a] In Example 6.1, I proved that $\Psi_1(x,t)$ satisfies the TDSE. Do the same for $\Psi_2(x,t)$ and $\Psi(x,t)$.

[b] **Evaluate** the expectation value of momentum $\langle p \rangle (t)$ for the state represented by $\Psi(x,t)$. Write your answer as a real function involving E_1, E_2, L, and fundamental constants. Briefly **discuss** the time dependence of your result and contrast it to the time dependence of $\langle x \rangle$ from Pblm. 3.2.

[c] **Evaluate** the position and momentum uncertainties for $\Psi_2(x,t)$. (If you worked Pblm. 3.2 you can get one of these from your answer to it.) **Calculate** the uncertainty product $(\Delta x)_2(t)(\Delta p)_2(t)$ and **verify** that the Heisenberg Uncertainty Principle is satisfied for this state.

[d] Suppose that at some time $t > 0$ we perform an ensemble measurement of the energy of the system in the superposition state represented by $\Psi(x,t)$. **Evaluate** the average value of the energy that would be obtained in such a measurement. Does your answer depend on time? Now **evaluate** the uncertainty in the energy for this state. **Comment** on the physical implications of the fact that $(\Delta E)(t) > 0$.

Hint: You can evaluate the $\langle E \rangle (t)$ using either $\hat{\mathcal{H}}$ or $\hat{E}$. In this case, using the latter operator will greatly simplify the algebra.

6.4 Probability Flux in a "Real World" Example

Protons are widely used in high-energy collision experiments. For example, consider a beam of (noninteracting) protons travelling in the $+x$ direction. How might we represent the state of the protons in this ensemble? Well, if we are willing to surrender all information about the *position* of the particles in the beam ($\Delta x = \infty$), we can approximate their state function by a pure momentum state of the form (6.135*b*).

In experiments to study the physics of subatomic particles, counters are used that actually measure the *probability flux* $j(x,t)$ as *counts per second*. Typically, the cross-sectional area such a counter presents to a particle beam is $0.2\,\text{m}^2$. An important step in understanding such an experiment is writing down a state function to represent the particles in question. For obvious reasons, we prefer to use the pure momentum function model, rather than a complicated wave packet, when such a model is justified physically.

Consider a beam of protons with 2.7 MeV of kinetic energy. Suppose our counter measures 4300 counts per second. **Write down** an explicit expression for a pure momentum function that models the wave function of a particle in this proton beam. (Explicitly include *the values* of A, k_0, and ω_0.)

6.5 More Properties of the Gaussian Packet

In § 6.7 we investigated the phenomenon of wave packet spreading and showed that it led to considerable changes in the Gaussian wave function as time passes. But, in spite of spreading, the wave packet retains one characteristic until its ultimate total dispersion in the $t \to \infty$ limit. The state function is a Gaussian at all times: *once a Gaussian, always a Gaussian.*

[a] Use this peculiarity of the Gaussian wave packet to predict, without resort to graphical or algebraic methods, one final property of this free-particle wave function: its peak amplitude, the value of $\Psi(x,t)$ at the center $\langle x \rangle (t)$, decreases with time. Use these two facts:

[1] $\Psi(x,t)$ remains normalized at all times;

[2] $\Psi(x,t)$ spreads, so the region of x that is encompassed by $\Delta x(t)$ grows.

Hint: If the area under the probability curve $P(x,t)$ remains equal to 1 while the width of this curve grows, what must happen to the peak value of $\Psi(x,t)$?

[b] Now show that this behavior is a consequence of the multiplicative prefactor in the wave function,

$$\left[\frac{1}{2\pi L^2\left(1+\frac{\hbar^2}{4m^2L^4}t^2\right)}\right]^{1/2},$$

by graphing this factor as a function of time.

6.6 The Probability Current Density for a Wave Packet

In Example 6.6 I evaluated the one-dimensional probability current density for the simple case of a plane wave function. Our result, the constant in Eq. (6.116), was delightfully simple and, alas, quite atypical of quantum states. For a more realistic example, consider the Gaussian wave packet (6.98) for $k_0 = 0$,

$$\Psi(x,t) = \left[\frac{1}{2\pi L^2\left(1+i\frac{\hbar}{2mL^2}t\right)^2}\right]^{1/4} \exp\left[-\frac{1}{4L^2}\frac{x^2\left(1-i\frac{\hbar}{2mL^2}t\right)}{1+\frac{\hbar^2}{4m^2L^4}t^2}\right]. \tag{6.6.1}$$

Derive the following expression for the probability current density for this wave packet:

$$j(x,t) = \frac{1}{\sqrt{2\pi L^2}}\frac{\hbar^2}{4m^2L^4}xt^2\left(1+\frac{\hbar^2}{4m^2L^4}t^2\right)^{-3/2} \exp\left[-\frac{1}{2L^2}\frac{x^2}{1+\frac{\hbar^2}{4m^2L^4}t^2}\right]. \tag{6.6.2}$$

Hint: You'll find it useful to define two intermediate quantities that are independent of x:

$$a \equiv \left[\frac{1}{2\pi L^2\left(1+i\frac{\hbar}{2mL^2}t\right)^2}\right]^{1/4} \tag{6.6.3a}$$

$$b \equiv \frac{1}{4L^2}\frac{\left(1-i\frac{\hbar}{2mL^2}t\right)}{1+\frac{\hbar^2}{4m^2L^4}t^2} \tag{6.6.3b}$$

Write the wave packet and its complex conjugate in terms of a and b and evaluate the partial derivatives you need to determine j. One more thing: remember that the complex number b can be written as $b = \operatorname{Re} b + i \operatorname{Im} b$.

6.7 A Martian Wave Packet

In this chapter, we considered the evolution of a free particle wave packet: *i.e.*, its propagation and spreading with time. Let's see what happens when we try to apply these ideas to a (rather large) macroscopic particle.

Consider the planet Mars as a particle, the motion of which is to be described by a quantum mechanical wave packet. Suppose that at $t = 0$ (the birth of the universe) the width of the packet was 0.01 m—a ridiculously small number on an astrophysical scale. Now, neglecting the gravitational field of the sun, **calculate** the increase in the width of this packet from its birth (about 5 billion years ago) to the present.

6.8 A Probability Amplitude for Particle Propagation

The TDSE is the partial differential equation that must be satisfied by all wave functions of whatever system we're studying. To obtain a *particular* wave function, we solve this equation subject to the initial conditions $\{\Psi(x,t_0)\colon -\infty \le x \le \infty\}$ appropriate to whatever state we're interested in. We can make clearer this vital dependence of the wave function at any time $t > t_0$ on the function $\Psi(x,t_0)$ by looking at the TDSE from a different point of view. In this problem, we'll show that $\Psi(x,t)$ is related to its initial condition by

$$\Psi(x,t) = \int_{-\infty}^{\infty} K(x,t;x',t_0)\Psi(x',t_0)\,dx'. \tag{6.8.1}$$

The critter $K(x,t;x',t_0)$ in this expression is called the **propagator**.[36]

[a] We've already developed the machinery to derive Eq. (6.8.1). In our study of wave packets (in Chap. 4) we discovered that such a wave function can be written as an integral over pure momentum functions

[36] Actually, $K(x,t;x',t_0)$ is called **the retarded propagator** because it is zero for $t \le t_0$.

$$\Psi_p(x,t) = \frac{1}{\sqrt{2\pi\hbar}} e^{i[px - E(p)t]/\hbar} \tag{6.8.2a}$$

i.e., as a wave packet

$$\Psi(x,t) = \int_{-\infty}^{\infty} \Phi(p)\Psi_p(x,t)\, dp. \tag{6.8.2b}$$

The function $\Phi(p)$ is the momentum probability amplitude. We calculate this amplitude, in turn, from the initial wave function $\Psi(x, t_0)$ using the Fourier transform relationship (here written for $t_0 = 0$)

$$\Phi(p) = \frac{1}{\sqrt{2\pi\hbar}} \int_{-\infty}^{\infty} \Psi(x,0)\, e^{-ipx/\hbar}\, dx. \tag{6.8.3}$$

Now, by substituting this expression for $\Phi(p)$ into Eq. (6.8.2*b*), **demonstrate** that $\Psi(x,t)$ can be written as an integral over the initial function, as in Eq. (6.8.1). Also, **develop** the following *integral* expression for the propagator $K(x,t;x',t_0)$ in Eq. (6.8.1). [Write your propagator as an integral over p including $E(p)$ in the time-dependent factor. For convenience, let $t_0 = 0$.]

$$K(x,t;x',0) = \frac{1}{2\pi\hbar} \int_{-\infty}^{\infty} \exp\left[-i(x'-x)\frac{p}{\hbar} - i\frac{E(p)}{\hbar}t\right] dp. \tag{6.8.4}$$

[b] Now consider *the free particle*, for which you know the dispersion relation $E(p)$. **Evaluate** the integral in your propagator and thereby obtain the following analytic expression for the free-particle propagator:

$$K(x,t;x',0) = \left(-\frac{im}{2\pi\hbar t}\right)^{1/2} \exp\left[i\frac{(x-x')^2 m}{2\hbar t}\right]. \quad \text{[for a free particle]} \tag{6.8.5}$$

[c] Following Max Born's ideas, we interpret the wave function $\Psi(x', t_0)$ as a probability amplitude for finding the particle in an infinitesimal region dx' about x' in a position measurement performed at t_0. Similarly, we interpret $\Psi(x,t)$ as the position probability amplitude for the time $t > t_0$. These statements, in turn, suggest an interpretation for $K(x,t;x',t_0)$. **Discuss** the interpretation of the propagator as a probability amplitude and the physical significance of Eq. (6.8.1).

[d] One of the (many) important properties of the propagator is its simple "initial value"—*i.e.*, at $t = t_0$. **Prove** that

$$K(x,0;x',0) = \delta(x'-x), \tag{6.8.6}$$

where $\delta(x'-x)$ is the Dirac Delta function (see § 4.2). How would you interpret this result physically?

[e] The propagator also satisfies a differential equation that looks suspiciously like the TDSE:

$$\left[i\hbar\frac{\partial}{\partial t} - \hat{\mathcal{H}}(\hat{x},\hat{p})\right] K(x,t;x',t_0) = i\hbar\delta(x-x')\delta(t-t_0), \tag{6.8.7}$$

where $\hat{\mathcal{H}}$ is the Hamiltonian of the system. For the free particle, of course, $\hat{\mathcal{H}} = \hat{T}$. **Verify** that your free particle propagator (from part [b]) satisfies Eq. (6.8.7). Any function that satisfies a differential equation of this form—*i.e.*, with a Dirac Delta function on the right-hand side—is called a **Green's function**; so you have just proven that the free-particle propagator is just a particular kind of Green's function.

Remark: To keep the algebra in this problem as simple as possible, I've asked you to consider only the propagator for a free particle. But Eqs. (6.8.1), (6.8.6), and (6.8.7) also hold for a particle with a non-zero potential. If the particle is not free, it is sometimes easier to obtain the propagator by solving Eq. (6.8.7), subject to the (reasonable) boundary conditions:

$$K(x,t;x',t_0) = 0 \qquad \text{for } t \le t_0. \tag{6.8.8}$$

CHAPTER 7

Simplifying Matters

The Time-Independent Schrödinger Equation

No, there are still difficulties,
but the outline of the plot is clear
and will develop itself more clearly
as we proceed.

—M. Rozer to Lord Peter Wimsey
in *The Nine Tailors*
by Dorothy L. Sayers

In 1926 Erwin Schrödinger published four papers that laid the foundation for the wave theory of quantum mechanics. The *time-dependent Schrödinger Equation* (Chap. 6) was the subject of the *second* of these papers. In the first, Schrödinger introduced a simpler *time-independent* equation, which is usually derived from the TDSE.

The reason for this apparent inversion of natural order lies in Schrödinger's background. From research early in his career on the physics of continuous media, Schrödinger knew well the mathematical properties of the special class of equations called *eigenvalue equations*. The time-independent Schrödinger Equation (TISE) belongs to this class, so Schrödinger was primed to discover it first.

In this book, we are following the logic of mathematics, not history, so we began with Schrödinger's time-dependent equation, as codified in Postulate IV. From the TDSE, it's but a short hop to the *time-independent* Schrödinger equation. The solutions of the latter are functions of position but not of time; they describe the spatial dependence of an important class of quantum states called *stationary states*. These states play a vital role in quantum physics, even though they are rarely found in nature.

For any microscopic system, however complicated, the essential problem of quantum theory is the same: *Given the system Hamiltonian $\hat{\mathcal{H}}$—the operator that mathematically represents the system—solve the TDSE for state functions that fit prescribed initial conditions.* These conditions are usually established by experimental circumstances at the initial time. In many cases, the initial state is non-stationary. But non-stationary states are too complicated mathematically to be obtained directly from the TDSE. Stationary states provide a way out of this predicament, for they are the building blocks of non-stationary wave functions (Chap. 12).

In prior examples and problems, I've shown you wave functions for a few **stationary states**. All stationary-state wave functions have a common, simple mathematical form:

$$\Psi(x,t) = (\text{function of } x)\,(\text{function of } t). \qquad \text{stationary state} \tag{7.1}$$

The time-independent Schrödinger Equation determines the *function of* x in a stationary-state wave function.

Not all microscopic systems have stationary states—but most do. In § 7.1, we'll deduce the conditions under which these states exist and will derive equations for the functions of x and t that comprise their wave functions. Then in § 7.2, we'll solve the time equation and, in the process, discover one of the characteristics of a stationary state: its energy is sharp. We can dig this and other properties out of the mathematical character of the TISE, which we'll probe in § 7.3 and 7.5.

In Part III, we'll solve the TISE for several important one-dimensional systems and in Volume II will tackle systems in three dimensions and many-particle systems. As the number of dimensions and particles increases, the mathematical contortions we must perform to solve this equation become more complicated. But, happily, I can show you the basic steps in its solution in the context of a very simple example: the particle in a box. We shall solve this problem (at last) in § 7.4.

As you plow through this example, and others in chapters to come, be careful to not lose sight of the special nature of stationary states. *In general, quantum states are non-stationary*. The beautiful and powerful connection between stationary and non-stationary states is so important that it gets its own chapter (Chap. 12)—but I'll provide a sneak preview in § 7.6.

7.1 IN SEARCH OF SEPARABLE WAVE FUNCTIONS

It is easy to see why solving the time-dependent Schrödinger Equation for an arbitrary system is so hard. Even for a single particle in one dimension, this equation is rather formidable:

$$-\frac{\hbar^2}{2m}\frac{\partial^2}{\partial x^2}\Psi(x,t) + V(x,t)\,\Psi(x,t) = i\hbar\frac{\partial}{\partial t}\,\Psi(x,t). \tag{7.2}$$

The underlying problem is not just that it's a partial differential equation of second-order in x and first-order in t, but that we must contend with both of these variables at once. We cannot solve for the x dependence of the state function without taking account of its t dependence, because in general x and t are inextricably linked in the potential energy function $V(x,t)$.

As a consequence, the space and time dependence of a wave function may be very complicated. We have seen one example of this complexity in § 6.6: the Gaussian free-particle state function (6.98) is an involved mixture of x- and t-dependent factors:

$$\Psi(x,t) = \left[2\pi L^2\left(1+i\frac{\hbar}{2mL^2}\,t\right)^2\right]^{-1/4} \times e^{i(k_0 x-\omega_0 t)}\exp\left[-\frac{1}{4L^2}\,\frac{\left(x-\frac{p_0}{m}\,t\right)^2}{\left(1+i\frac{\hbar}{2mL^2}\,t\right)}\right]. \tag{7.3}$$

Only the comparative simplicity of the *free-particle* TDSE lets us solve this equation analytically.

Strategy: Separation of Variables

One way to subdue recalcitrant multi-variable differential equations is to seek solutions that have a particularly simple form. We know that wave functions must be complex functions of x and t; and we might guess that the simplest such solutions are of the product form (7.1):

$$\Psi(x,t) = \psi(x)\zeta(t). \qquad \text{[stationary state]} \tag{7.4}$$

I have let the *lower-case* Greek letter psi (ψ) represent the x dependence of $\Psi(x,t)$, and zeta (ζ) its t dependence. Mathematicians call such product functions **separable solutions** of the partial differential equation (7.2); we quantum mechanics call them **stationary-state wave functions**.

The idea of basing quantum theory on stationary states sounds good, but it raises several questions. Do separable wave functions exist for *all* physical systems? If not, for what kinds of systems *do* they exist? For such systems, how can we get our hands on the "pieces" $\psi(x)$ and $\zeta(t)$? And for these systems, how are the stationary-state wave functions related to the more general non-stationary wave functions? These are the focal questions of this chapter.

The separable form (7.4) implies that we can solve for $\psi(x)$ and $\zeta(t)$ separately. To do so, we must derive from the TDSE an equation for $\psi(x)$ that does not depend on t and a *separate* equation for $\zeta(t)$ that does not depend on x. This is the goal of the mathematical method called **separation of variables**, to reduce a partial differential equation in several variables to separate, simpler equations. We have no guarantee that this method will succeed for the TDSE, but it's worth a try.

To apply this strategy, let's substitute the separated form (7.4) into Eq. (7.2) and try to isolate on one side of the resulting equation *all terms that depend on* x and on the other, *all terms that depend on* t. That is, we are going to try to reduce the TDSE to an equality between two *single-variable algebraic expressions*:

$$\text{terms in } x, \frac{\partial}{\partial x}, \frac{\partial^2}{\partial x^2}, \ldots = \text{terms in } t, \frac{\partial}{\partial t}, \frac{\partial^2}{\partial t^2}, \ldots . \tag{7.5}$$

If such an arrangement is possible, then the expressions on the left and right sides of (7.5) must each equal a constant and this equation will *separate* (*aha!*) into two ordinary differential equations, one in x, the other in t.

Separation of Space and Time in the TDSE

Optimistically assuming that a solution of the form (7.4) exists, we write the TDSE (7.2) as

$$-\frac{\hbar^2}{2m}\frac{\partial^2}{\partial x^2}\psi(x)\zeta(t) + V(x,t)\,\psi(x)\zeta(t) = i\hbar\frac{\partial}{\partial t}\psi(x)\zeta(t)\,. \tag{7.6}$$

We can simplify (7.6) by exploiting the t-independence of $\psi(x)$ and the x-independence of $\zeta(t)$ to transform the partial derivatives into ordinary derivatives, *viz.*,

$$\zeta(t)\left[-\frac{\hbar^2}{2m}\frac{d^2}{dx^2}\psi(x)\right] + V(x,t)\psi(x)\zeta(t) = \psi(x)\left[i\hbar\frac{d}{dt}\zeta(t)\right]. \tag{7.7}$$

To begin arranging (7.7) in the form (7.5), let's divide both sides by the product $\Psi(x,t) = \psi(x)\zeta(t)$:[1]

$$\frac{1}{\psi(x)}\left[-\frac{\hbar^2}{2m}\frac{d^2}{dx^2}\psi(x)\right] + V(x,t) = \frac{1}{\zeta(t)}\left[i\hbar\frac{d}{dt}\zeta(t)\right]. \tag{7.8}$$

Equation (7.8) is *almost* separable; but in quantum mechanics, almost isn't good enough. In general, the potential energy can depend on x and t, and, if it does, our grand strategy fails, for we cannot further separate Eq. (7.8).

[1]You may be worried that $\psi(x)$ might be zero for some value of x; at such an x, the factor $1/\psi(x)$ is infinite and Eq. (7.8) ceases to make sense. Don't fret. Division by $\psi(x)$ is an intermediate step in this derivation, designed to clarify the logic of separation of variables. We shall soon restore $\psi(x)$ to its rightful position—see Eq. (7.10*a*).

Okay; let's scale down our ambitions: Rather than try to find separable state functions for *any* quantum system, let's *consider a system whose potential energy depends only on position variables.* If V in the TDSE depends only on x, then Eq. (7.8) has the magic form (7.5), *i.e.*,

$$\frac{1}{\psi(x)}\left[-\frac{\hbar^2}{2m}\frac{d^2}{dx^2}\psi(x)\right]+V(x) = \frac{1}{\zeta(t)}\left[i\hbar\frac{d}{dt}\zeta(t)\right]. \tag{7.9}$$

Comes now the crucial step. *Think* about Eq. (7.9): how could the left hand side—an expression in x that is *independent of* t—equal the right-hand side—an expression in t that is *independent of* x? Answer: only if *both sides are independent of x and t.* We can therefore equate each side of (7.9) to a constant—say, α. Doing so, and rearranging terms slightly, we obtain two separate equations:

$$-\frac{\hbar^2}{2m}\frac{d^2}{dx^2}\psi(x)+V(x)\psi(x) = \alpha\psi(x) \tag{7.10a}$$

and

$$i\hbar\frac{d}{dt}\zeta(t) = \alpha\zeta(t). \tag{7.10b}$$

For obvious reasons, mathematicians call α the **separation constant**.

Great. Equations (7.10) are *ordinary* differential equations—(7.10*a*) is second-order in x, (7.10*b*) is first-order in t—and as such are much easier to solve than the TDSE. *These two single-variable equations are equivalent to the TDSE for a product wave function* $\Psi(x,t) = \psi(x)\zeta(t)$, *i.e., for a stationary state.* The fact that we can obtain such equations at all, provided V does not depend on t, *proves* that stationary-state wave functions exist for systems with a time-independent potential energy.

We'll spend the rest of this chapter poking and prodding at Eqs. (7.10). But a couple of important features leap out at us immediately. First, although we can solve these x and t equations separately, they are not completely independent of one another: they are linked by the separation constant α. Second, the x equation (7.10*a*) contains the system Hamiltonian; indeed, we can write this equation as $\hat{\mathcal{H}}\psi(x) = \alpha\psi(x)$. [Since $\hat{\mathcal{H}}$ is an operator that represents the total energy (Chap. 5), you may suspect that α is the value of this observable in the stationary state. You would be right—see § 7.2].

The spatial function $\psi(x)$ therefore depends explicitly on the system we are studying *i.e.*, on details of its potential energy. But the time equation (7.10*b*) does not contain $\hat{\mathcal{H}}$: by solving this equation, we find, once and for all, the time dependence of *any* stationary-state wave function for *any* system! Equation (7.10*b*) is refreshingly simple, and you can probably guess the solution (see § 7.2). The x equation (7.10*a*) is a bit more formidable; it is the subject of much of the rest of this book.

Conservative Systems

The derivation of Eqs. (7.10) shows that the nature of the potential energy determines whether or not stationary states exist:

V independent of t	$\Longrightarrow$	stationary states exist

A system whose potential energy is time-independent is said to be **conservative.**[2] So the vital point to remember is

> ***Rule***
>
> Stationary states exist only for conservative systems.

To study a non-conservative system (which is not, incidentally, called a "liberal system") we must deal with the TDSE directly, a challenge we'll take up in Volume II.

The microscopic universe is full of conservative systems. Isolated atoms, molecules, and solids are conservative, as are many collision complexes formed by scattering a microscopic particle—an electron, say—from one of these targets. The class of conservative systems does, however, exclude some important problems, *e.g.*, those in which a time-dependent external field is present. (Lasers fall into this category.) For example, in Volume II we'll probe the quantum states of a microscopic particle of charge q exposed to a time-varying external electric field of magnitude $\mathcal{E}$ and frequency ω. We'll find that the Hamiltonian of the particle in such a field is

$$\hat{\mathcal{H}} = -\frac{\hbar^2}{2m}\frac{\partial^2}{\partial x^2} - q\mathcal{E}x\cos\omega t. \tag{7.11}$$

The time dependence of the potential energy $-q\mathcal{E}x\cos\omega t$ renders the TDSE for this problem inseparable, so stationary states for a particle in a time-varying electromagnetic field do not exist.[3]

You may recall from classical physics an important conservation law for a conservative *macroscopic* system: if the forces on the system are conservative, then its energy is conserved, *i.e.*, energy is a constant of the motion. In the next section we shall discover the form this law takes for microscopic systems.

7.2 TIME AND ENERGY

In § 7.1, I noted that *the time equation* (7.10*b*) is independent of the system's potential energy. This simple, first-order ordinary differential equation is easy to solve, and its solution has some surprising physical consequences.

The Time Factor for a Stationary-State Wave Function

To solve Eq. (7.10*b*), I'll invoke a standard trick from the theory of ordinary differential equations—one you'll find in the books in § II of the Suggested Readings for

[2]This nomenclature is taken from classical physics, where a system is *conservative* if the forces on it are such that the work done around a closed orbit is zero. Any conservative force is derivable from a potential energy that is independent of time—hence the connection between the classical and quantum definitions of a conservative system. If the forces acting on a classical particle are conservative, then the energy of the particle is conserved. For more on conservative systems, see § 1–1 of *Classical Mechanics*, 2nd ed. by Herbert Goldstein, (Reading, Mass: Addison-Wesley, 1980).

[3]Do not conclude from this example that the mathematical technique separation of variables *necessarily* fails if the potential energy depends on time. If, for example, $V(x, t)$ happens to be the sum of two terms, one of which depends only on x, the other of which depends only on t, then we can determine separable solutions to the Schrödinger Equation.

Chap. 6. This trick is based on the *logarithmic derivative* of a function. By definition, the **logarithmic derivative** of a function $\zeta(t)$ is

$$\text{logarithmic derivative of } \zeta(t) \equiv \frac{d}{dt}\ln\zeta(t) = \frac{1}{\zeta(t)}\frac{d}{dt}\zeta(t). \tag{7.12}$$

We can write Eq. (7.10*b*) in terms of the logarithmic derivative of $\zeta(t)$ by simply moving $\zeta(t)$ to the left-hand side, *viz.*,

$$\frac{1}{\zeta(t)}\frac{d}{dt}\zeta(t) = -\frac{i\alpha}{\hbar}. \tag{7.13}$$

If we use the definition (7.12) and then integrate the result from 0 to t, we have

$$\int_0^t \frac{d}{dt}\ln\zeta(t')\,dt' = -\frac{i\alpha}{\hbar}\int_0^t dt'. \tag{7.14}$$

'Tis now an easy matter to evaluate both sides of (7.14); to wit

$$\ln\zeta(t) - \ln\zeta(0) = \ln\frac{\zeta(t)}{\zeta(0)} = -\frac{i\alpha}{\hbar}t. \tag{7.15}$$

To obtain the time function, we just exponentiate both sides of (7.15) *viz.*,

$$\zeta(t) = \zeta(0)e^{-i\alpha t/\hbar}. \tag{7.16}$$

Question 7–1

Verify by direct substitution that the function in (7.16) solves the time equation (7.10*b*).

The factor $\zeta(0)$ in (7.16) is just the *initial condition* for the differential equation (7.10*b*). It is convenient to pick $\zeta(0) = 1$, so that the general form of a stationary-state wave function is[4]

$$\Psi(x,t) = \psi(x)e^{-i\alpha t/\hbar}. \qquad \text{[stationary state]} \tag{7.17}$$

We've accomplished a great deal with very little work: we have found the functional form of the time factor *for any stationary state of any system.* Isn't that terrific?! Yes, except for one little thing: what is the meaning of α?

Question 7–2

You can anticipate the answer to this question by applying a problem-solving gambit called *dimensional analysis*: The exponent in $\zeta(t)$ must be dimensionless, so what *must* be the dimension of α? Now, what observable do you think α stands for?

Identifying the Separation Constant

The Born interpretation of $\Psi(x,t)$ provides a clue to the physical significance of the separation constant α. According to Postulate II, this function is the amplitude of a probability wave. And now, in Eq. (7.17), we've discovered that for a stationary state, this wave *oscillates* at frequency $\alpha/\hbar$.

[4]This choice entails no loss of generality. The physically relevant initial condition is the one on $\Psi(x,t)$, *i.e.*, the function $\Psi(x,0)$. The choice $\zeta(0) = 1$ allows us to specify this initial condition in terms of the spatial function, *i.e.*, $\Psi(x,0) = \psi(x)$.

But wait a minute. From the de Broglie–Einstein relation $E = \hbar\omega$, we already know the "oscillation frequency of a particle": if E is the particle's energy, then this frequency is $E/\hbar$. To make our time function (7.17) consistent with this fundamental relation, *we interpret α as the total energy of the particle in the state represented by this function.* With this identification, the general form of a stationary state wave function becomes

$$\boxed{\Psi(x,t) = \psi(x)e^{-iEt/\hbar}}. \qquad \text{[stationary state]} \tag{7.18}$$

At this point, we know only the *time* dependence of $\Psi(x,t)$. We determine the (allowed) values of the energy and the spatial dependence of $\Psi(x,t)$ by solving Eq. (7.10*a*). Replacing α by E, we can write this equation as[5]

$$\boxed{-\frac{\hbar^2}{2m}\frac{d^2}{dx^2}\psi_E(x) + V(x)\psi_E(x) = E\psi_E(x)}. \qquad \text{TISE} \tag{7.19}$$

This is the **time-independent Schrödinger Equation.** Since the energy E appears explicitly in this equation, its solutions will depend on the value of this parameter. That's why I labeled the spatial function with E.

We have discovered that *the energy of a stationary state is well-defined (sharp).* Indeed, we can simply read the value of E off a stationary-state function of the form (7.18), such as Eq. (6.23) for the ground state of the infinite square well. In § 7.5, we'll build a rigorous foundation under this observation by *proving* that $\Delta E = 0$ for a stationary state.

The essential result of this chapter (so far) is: *if a microscopic system is conservative, then there exist special quantum states of the system, called stationary states, in which the energy is sharp.* This important result is the quantum mechanical counterpart of the classical law mentioned at the end of § 7.1: the energy of a conservative *macroscopic* system is a constant of the motion.[6]

Buoyed by our success with the time equation, we're ready to tackle the spatial equation, the time-independent Schrödinger Equation.

7.3 A PROFILE OF THE TISE

> "So did I," said Pooh, wondering what a Heffalump was like.
>
> —*Winnie the Pooh*
> by A. A. Milne

The TISE for $\psi_E(x)$ in a stationary state of energy E, Eq. (7.19), can be written in terms of the Hamiltonian $\hat{\mathcal{H}}$ as

[5]Only for bound states does the TISE determine the allowed energies. As we'll see in Chap. 8, for convenience, we admit to quantum mechanical problem-solving *continuum stationary states.* These functions are not rigorously admissible, but they are valuable models for the study of collisions of microscopic particles. The energy of a continuum state need not be determined by solution of the TISE.

[6]In quantum mechanics an observable Q is said to be a *constant of the motion* if its expectation value does not change with time *for any state* (see Chap. 11).

$$\boxed{\hat{\mathcal{H}}\psi_E(x) = E\psi_E(x)} \qquad \text{TISE} \tag{7.20}$$

[Be careful *not* to use $\hat{\mathcal{E}}$ instead of E in (7.20). What a difference a little hat makes!] Because $\hat{\mathcal{H}} = \hat{T} + \hat{V}$, the explicit form—and hence the solutions—of this differential equation depend on the potential energy $V(x)$.

For example, the potential of a particle of mass m in a uniform, one-dimensional gravitational field with gravitational constant g is mgx, so the TISE for a particle in such a gravitational well is

$$\left[-\frac{\hbar^2}{2m}\frac{d^2}{dx^2} + mgx\right]\psi_E(x) = E\psi_E(x). \tag{7.21}$$

Another important example is the simple harmonic oscillator: if ω_0 denotes the natural frequency of the oscillator, then the potential energy of the oscillator is $m\omega_0^2 x^2/2$. The TISE for this system is therefore

$$\left[-\frac{\hbar^2}{2m}\frac{d^2}{dx^2} + \frac{1}{2}m\omega_0^2 x^2\right]\psi_E(x) = E\psi_E(x). \tag{7.22}$$

To lay a foundation for our study in Part III of these and other systems, we'll here scrutinize the general TISE, much as we did the TDSE in § 6.3. We'll put on our mathematician's hat for awhile and try to deduce general mathematical properties of Eq. (7.20). In particular, we're going to investigate the *mathematical structure* of this equation, the normalization condition for its solutions, limitations on normalizable solutions and the implications of these limitations for the energy of the stationary state, and, finally, a new mathematical property called *degeneracy*.

The TISE Considered as an Eigenvalue Equation

For a single particle in one dimension, the TISE is a *second-order, ordinary differential equation* in the variable x. It's *linear* and *homogeneous* and has two *distinct* (*i.e.*, linearly independent) solutions.[7] Because it is of second order, the TISE requires two *boundary conditions*. These are just the boundary conditions on the state function $\Psi(x,t)$ that we discussed in § 6.3; for a stationary state, they show up as conditions on the spatial function $\psi_E(x)$. For example, the spatial function *for a state in which the particle is bound by a potential well* must satisfy the conditions

$$\psi_E(x) \xrightarrow[|x|\to\infty]{} 0. \tag{7.23}$$

The TISE is a differential equation of a very special type; it contains a *parameter*—the energy E—that appears in a specific way. Look at the structure of (7.20): a differential operator ($\hat{\mathcal{H}}$) operates on the unknown function [$\psi_E(x)$] to produce, not a *different* function of x, but rather the same function, $\psi_E(x)$, times the parameter E. Mathematicians call an equation of this form an **eigenvalue equation** and say "$\psi_E(x)$ is the

[7]A reminder: Two functions $f_1(x)$ and $f_2(x)$ are said to be *linearly independent* if the equation $c_1 f_1(x) + c_2 f_2(x) = 0$ is satisfied *only by* $c_1 = c_2 = 0$. That is, two functions that are proportional to one another are linearly dependent. A common instance of linearly dependent functions in the quantum mechanics of conservative systems is two spatial functions $\psi_1(x)$ and $\psi_2(x)$ that differ only by a a global phase factor $e^{i\delta}$. These functions are not distinct, and physically they represent the same quantum state. (Remember from Chap. 3 that state functions are indeterminate to within such a factor.)

eigenfunction of the differential operator $\mathcal{H}$, and E is the corresponding **eigenvalue.**" Thus, when feeling mathematically inclined, I may refer to the spatial function $\psi_E(x)$ as the **Hamiltonian eigenfunction**.

The TISE is by no means the only important eigenvalue equation in quantum mechanics, and in Chap. 11 we shall consider such equations in their own right. For now, though, let's stay with the TISE.

Aside: The TISE in 3D. The form of the TISE for a single particle in three dimensions—or, for that matter, for a many-particle three-dimensional system—is identical to (7.20). Such equations differ, of course, in the number of independent variables. For example, the TISE for a single particle in a three-dimensional isotropic harmonic oscillator potential $V(\mathbf{r}) = m\omega_0^2 r^2/2$ is

$$\hat{\mathcal{H}}\psi_E(\mathbf{r}) = \left[-\frac{\hbar^2}{2m}\nabla^2 + \frac{1}{2}m\omega_0^2 r^2\right]\psi_E(\mathbf{r}) = E\psi_E(\mathbf{r}), \tag{7.24}$$

where the Laplacian ∇^2 is given in Eq. (6.17). Notice that, unlike the one-dimensional TISE (7.20), Eq. (7.24) is a *partial* differential equation.

Normalization of the Spatial Function

For a function $\Psi(x, t)$—stationary or otherwise—to be valid, it must, of course, satisfy the TDSE of the system. But this function must also satisfy the additional demands of § 3.5: it must be single-valued, continuous, and smoothly varying. And, finally, it must be normalizable. That is, it must be possible to scale the function so that its normalization integral equals one:

$$\int_{-\infty}^{\infty} \Psi^*(x,t)\Psi(x,t)\,dx = 1. \tag{7.25}$$

If the state is stationary, then each of these conditions on $\Psi(x,t)$ translates into a condition on the Hamiltonian eigenfunction $\psi_E(x)$. The resulting normalization condition is particularly simple, because the time factor in a stationary state wave function is the *exponential* function $e^{-iEt/\hbar}$. The integrand in the normalization integral (7.25) is $\Psi^*(x,t)\Psi(x,t)$. The required complex conjugate is

$$\Psi^*(x,t) = \psi_E^*(x)e^{+iEt/\hbar}, \tag{7.26}$$

so the product $\Psi^*(x,t)\Psi(x,t)$ becomes $\psi_E^*(x)\psi_E(x)$. Hence the normalization condition on the wave function reduces to a similar condition on the Hamiltonian eigenfunction,[8]

$$\int_{-\infty}^{\infty} \psi_E^*(x)\psi_E(x)\,dx = 1. \tag{7.27}$$

We can therefore normalize *spatial functions* using the same procedure we applied to the time-dependent wave functions in § 3.5. Notice that the requirement (7.27) rules out "exponentially exploding" spatial functions, such as those in Fig. 7.1a.

[8]We need the complex conjugate of $\psi_E(x)$ in (7.27) only if this function is complex. A real, single-variable ordinary differential equation has real solutions. Moreover, a solution that satisfies boundary conditions such as (7.25) is unique to within the usual global phase factor. Thus, when studying *bound states of a one-dimensional system*, we're always at liberty to use real spatial functions. Such solutions are desirable, because they are easier to work with than complex solutions. Note that even if $\psi_E(x)$ is real, the wave function is complex, because the spatial function must be multiplied by the complex time factor.

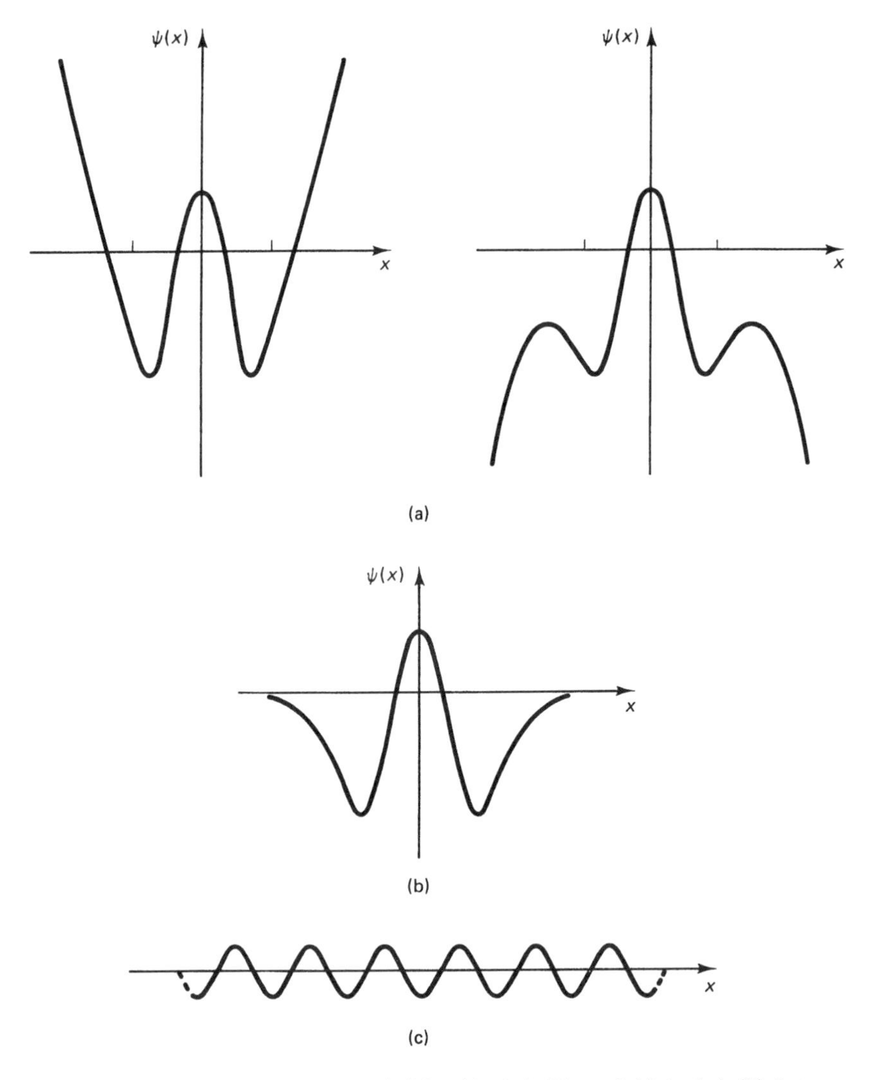

Figure 7.1 Examples of (a) inadmissible, (b) admissible, and (c) inadmissible-but-useful spatial functions. Neither of the functions in (a) can be normalized, because they blow up as $|x| \to \infty$. The oscillatory function in (c) does not properly go to zero in this limit, but we admit it to quantum theory as a useful model (see Chap. 8).

The disappearance of the time factor in the product $\Psi^*(x,t)\Psi(x,t)$ is a common incident in quantum mechanics. Most important quantities involve products containing the wave function and its complex conjugate. If such a product is present—and if the wave function represents a stationary state—then this cancellation will eliminate the time dependence (see §7.5)

Existence of Normalizable Spatial Functions

You may be wondering, "If I am given a value of E, can I be assured that a normalizable solution of the TISE for that energy exists?"

In general, no. To be sure, Eq. (7.20) can be solved for any value of E—using a host of mathematical and numerical methods (see Part III)—but *normalizable* solutions of this equation can be found only for certain values of E. Even if the number of normalizable solutions is infinite, their energies form a *discrete* list. That is, between any two valid energies E_{i-1} and E_i, there are an infinite number of "invalid energies," invalid because for values of E in the interval $E_{i-1} < E < E_i$ the solution $\psi_E(x)$ blows up as $|x| \to \infty$. Several such offenders appear in Fig. 7.1a.

Thus the requirement that the spatial functions be normalizable leads inexorably to **quantization of energy**. Only for special "allowed" energies do physically admissible stationary state functions, with spatial functions that satisfy the TISE *and* boundary conditions such as (7.23), exist. One of the most powerful corroborations of quantum theory is the fact that these energies and only these energies are ever found in energy measurements on microscopic systems. In § 7.4, we'll work an example that illustrates how boundary conditions lead to energy quantization.

Uniqueness and Degeneracy

Suppose we have found an energy E for which a normalizable solution $\psi_E(x)$ to the TISE exists. Might there be other, distinct solutions for this energy, or is this solution unique? The answer depends on the nature of the system we are studying. *A solution of any second-order, single-variable differential equation that satisfies two boundary conditions is unique.*[9] There follows from this mathematical fact a uniqueness condition:

> ***Rule***
>
> Bound-state solutions to the TISE of a one-dimensional single-particle system are unique.

This rule establishes a *one-to-one correspondence* between the quantized energies of a one-dimensional single-particle quantum system and its *bound* stationary state wave functions:

$$\text{valid quantized energy } E \Longleftrightarrow \text{unique } \Psi(x,t) = \psi_E(x)e^{-iEt/\hbar} \qquad \text{[bound state in one dimension]} \tag{7.28}$$

We describe this property by saying that a *bound state energy* E is **non-degenerate**.

The simple correspondence (7.28) does not necessarily apply to more complicated quantum systems. For example, in studying a microscopic particle in *three dimensions*, we'll find stationary-state energies to which there correspond *more than one* distinct spatial function. Such energies are said to be **degenerate**. In the jargon of quantum physics, a stationary state energy that has n distinct spatial functions is said to be **n-fold degenerate**.

Example 7.1. Stationary States of a Free Particle

In Chaps. 4–6 we exhaustively investigated a function of separable form, the pure momentum function[10]

[9] See, for example, *An Introduction to Ordinary Differential Equations* by Earl A. Coddington (Englewood Cliffs, N. J.: Prentice Hall, 1961.)

[10] Here, as in § 6.8, I'll leave A arbitrary. If we want to ensure Dirac Delta function normalization, as in Chap. 4, we can always choose $A = 1/\sqrt{2\pi\hbar}$.

$$\Psi_p(x,t) = Ae^{ipx/\hbar}e^{-iEt/\hbar}. \tag{7.29}$$

The corresponding energies are given by the free particle dispersion relation $\omega(k) = \hbar k^2/(2m)$ as

$$E = \frac{\hbar^2 k^2}{2m}. \qquad \text{[for a free particle]} \tag{7.30}$$

The spatial dependence of the pure momentum function (7.29) is

$$\psi_E(x) = Ae^{ipx/\hbar}. \tag{7.31}$$

This function satisfies the time-independent Schrödinger equation for a free particle,

$$-\frac{\hbar^2}{2m}\frac{d^2}{dx^2}\psi_E(x) = E\psi_E(x), \qquad \text{[for a free particle]} \tag{7.32}$$

if E is given by (7.30). But the spatial function (7.31) has several peculiar features, the most obvious of which is that it does *not* obey the boundary conditions (7.23).

Question 7–3

By substituting the function $\psi_E(x)$ in Eq. (7.31) into Eq. (7.32), derive Eq. (7.30) for the corresponding energies.

This failure comes about because, rather than decay to zero, the function $e^{ipx/\hbar}$ oscillates between $+1$ and -1 as $|x| \to \infty$. Consequently the spatial function (7.31) is not normalizable, *i.e.*, we cannot scale it to obtain a function that satisfies the condition (7.27). This property of the pure momentum function means that it's not physically admissible—although it's nonetheless useful as a model of a fat wave packet.

Two properties of the energies (7.30) follow from the impossibility of normalizing these functions. First, *the energies are not quantized*; all values $E > 0$ are allowed. [Remember: quantization follows from boundary conditions like (7.23), which the free-particle spatial functions do not obey.] Second, *these energies are degenerate.* In fact, for any value of $E > 0$, we can easily find *two* distinct solutions of (7.32). One of these solutions is (7.31); its x dependence is $e^{ipx/\hbar}$. And there is a second solution that is proportional to $e^{-ipx/\hbar}$.

It's easy to see from the free-particle energies (7.30) why *there are two distinct spatial functions for each energy*. These functions depend on the energy through their dependence on p. But *two* values of p correspond to each value of E, *i.e.*, $p = +\sqrt{2mE}$ and $p = -\sqrt{2mE}$. So each energy $E > 0$ is *two-fold degenerate.*

Because of this degeneracy, the notation in (7.31) is inadequate: it does not tell us *which* linearly independent solution for energy E we are talking about. We can, however, fully specify a particular spatial function by appending a superscript $(+)$ or $(-)$ to $\psi_E(x)$:

$$\psi_E^{(+)}(x) = Ae^{ipx/\hbar} \tag{7.33a}$$

$$\psi_E^{(-)}(x) = Be^{-ipx/\hbar}. \tag{7.33b}$$

The state functions $\psi_E^{(+)}(x)e^{-iEt/\hbar}$ and $\psi_E^{(-)}(x)e^{-iEt/\hbar}$ represent an ensemble of free particles of energy E moving in the $+x$ ($p = \sqrt{2mE}$) or $-x$ ($p = -\sqrt{2mE}$) direction, respectively.

An interesting consequence of the existence of two linearly-independent spatial functions for a single value of E is that we can easily construct *other* solutions to the TISE for this energy, such as

$$\psi_E(x) = A\psi_E^{(+)}(x) + B\psi_E^{(-)}(x), \tag{7.34}$$

where A and B are complex constants. In fact, since A and B are arbitrary, we can construct in this way *an infinite number of additional solutions.*

This flexibility is often useful in problem solving. For example, we can pick A and B in (7.34) so that this linear combination is real, *e.g.*,

$$A = B = \frac{C}{2} \qquad \Longrightarrow \qquad \psi_E^{(e)}(x) = C\cos\left(\frac{p}{\hbar}x\right) \tag{7.35a}$$

$$A = -B = \frac{D}{2i} \qquad \Longrightarrow \qquad \psi_E^{(o)}(x) = D\sin\left(\frac{p}{\hbar}x\right). \tag{7.35b}$$

[The $(+)$ and $(-)$ superscripts of (7.33) are not appropriate to the new spatial functions of (7.35), so I have labelled these functions by (e) and (o) to indicate their parity (even or odd). I'll explain this apparently quixotic choice of notation in § 7.4.] These choices provide *real*, rather than complex, spatial functions.

Note carefully that *the flexibility afforded us by degeneracy is not a general property of the stationary states of a single particle in one dimension.* For example, to any valid *bound state* energy, there is but one spatial function. In the next section, we'll study an example of just such a situation.

7.4 A PARTICLE IN AN ESCAPE-PROOF BOX

Most of the important properties of stationary states and the steps we carry out to obtain their wave functions appear in the solution of a very simple problem: the particle in a box. The potential energy of such a particle, which is shown in Fig. 7.2, is

$$V(x) = \begin{cases} \infty & -\infty < x < -\frac{1}{2}L \\ 0 & -\frac{1}{2}L \le x \le +\frac{1}{2}L \\ \infty & +\frac{1}{2}L < x < \infty \end{cases}. \tag{7.36}$$

Because this potential is *infinite* at the walls of the box, all state functions must obey the boundary conditions (§ 6.3)

$$\Psi(x = -\tfrac{1}{2}L,\ t) = 0 \tag{7.37a}$$

$$\Psi(x = +\tfrac{1}{2}L,\ t) = 0. \tag{7.37b}$$

And, since $V = 0$ inside the box—the only region in which the wave functions are non-zero—the Hamiltonian there is just the kinetic energy operator,

$$\hat{\mathcal{H}} = \hat{T} = -\frac{\hbar^2}{2m}\frac{d^2}{dx^2} \qquad \left[-\frac{1}{2}L \le x \le \frac{1}{2}L\right]. \tag{7.38}$$

So in this region, the TISE is

$$-\frac{\hbar^2}{2m}\frac{d^2}{dx^2}\psi_E(x) = E\psi_E(x) \qquad \left[-\frac{1}{2}L \le x \le \frac{1}{2}L\right] \tag{7.39}$$

In this section, we'll solve this equation and explore key properties of the resulting stationary states.

In earlier examples, we investigated the wave function for the *ground state* of this system: the stationary state with the *minimum total energy*. The value of this energy (which we'll derive in this section) is

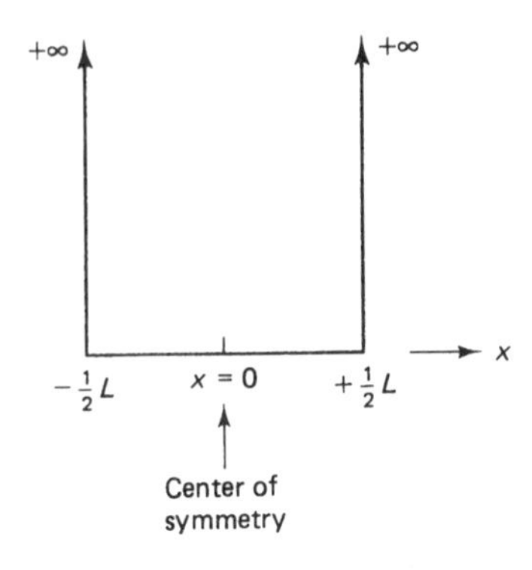

Figure 7.2 The Infinite Square Well. Note the center of symmetry at the midpoint of the box, the point $x = 0$.

$$E_1 = \frac{\pi^2\hbar^2}{2mL^2}. \tag{7.40}$$

And, in Example 6.1 (§ 6.2), we proved that the corresponding time-dependent function,

$$\Psi_1(x,t) = \sqrt{\frac{2}{L}}\cos\left(\pi\frac{x}{L}\right)e^{-iE_1t/\hbar} \qquad \left[-\frac{1}{2}L \le x \le \frac{1}{2}L\right], \tag{7.41}$$

satisfies the TDSE for this system.

The function $\Psi_1(x,t)$ clearly has the characteristic form $\psi_E(x)\,e^{-iEt/\hbar}$ of a stationary state, with spatial function

$$\psi_{E_1}(x) = \sqrt{\frac{2}{L}}\cos\left(\pi\frac{x}{L}\right). \tag{7.42}$$

And it's easy to verify that this function satisfies the time-independent Schrödinger equation of this system, Eq. (7.39), with $E = E_1$ given by (7.40). But in practice we must be able to obtain this spatial function—and those for all the other stationary states—by solving the TISE. We need a game plan.

I'm going to lay out and then illustrate a systematic strategy for solving the TISE for the bound-state spatial functions and energies of *any one-dimensional system.* The particular algebraic machinations required to solve this equation depend, of course, on the particular potential energy—but this overall strategy does not. There are only three steps:

> **Step 1.** Write down the general solution to the time-independent Schrödinger Equation.
>
> **Step 2.** Apply the boundary conditions imposed by the potential energy to the general solution of Step 1.
>
> **Step 3.** Normalize the bound-state spatial functions from Step 2.

It is the second step—imposition of the boundary conditions—that leads to quantization of the energies of the stationary states, as you'll see in the following example.

Example 7.2. The Stationary States of a Particle in a Box

Since the particle is free inside the box (where $V = 0$), we could simply write down a general solution to the TISE in this region as a linear combination of the complex exponential functions for a free particle in Eqs. (7.33). I prefer, however, to use the alternate, equivalent *real* trigonometric functions $\cos(px/\hbar)$ and $\sin(px/\hbar)$. Because they are real, these functions are easier to work with than their exponential cousins. Here, then, is a general solution to the TISE in the region where the spatial functions are non-zero:

$$\psi_E(x) = C\cos\left(\frac{p}{\hbar}x\right) + D\sin\left(\frac{p}{\hbar}x\right). \tag{7.43}$$

Note that the momentum and energy are related by $p = \sqrt{2mE}$. At this stage, the constants C and D in this linear combination are arbitrary, and p and E are unconstrained—*e.g.*, the energy can take on any value greater than zero.

That's step 1. Following our three-step procedure, we now implement the boundary conditions (7.37). Let's start at the right wall ($x = +\frac{1}{2}L$). Applying condition (7.37*b*) to the general solution (7.43), we get

$$C\cos\left(\frac{p}{\hbar}\frac{L}{2}\right) + D\sin\left(\frac{p}{\hbar}\frac{L}{2}\right) = 0. \tag{7.44}$$

How can we pick C and D to enforce this requirement? We certainly can't choose $C = 0$ *and* $D = 0$, for this choice would imply that $\psi_E(x) = 0$, the position probability density is zero everywhere, and there is no particle. And there is no value of p such that *both* the sine and cosine functions in (7.44) are zero, because sine and cosine are linearly independent functions. Evidently we must set one and only one of the arbitrary constants equal to zero—*either* $C = 0$ *or* $D = 0$. To ensure that the remaining sine or cosine function in (7.44) is non-zero, we must constrain the *argument* of this function. Here are the two possibilities:

$$C = 0 \quad (D \neq 0) \qquad \text{and} \qquad \sin\left(\frac{p}{\hbar}\frac{L}{2}\right) = 0 \tag{7.45a}$$

or

$$D = 0 \quad (C \neq 0) \qquad \text{and} \qquad \cos\left(\frac{p}{\hbar}\frac{L}{2}\right) = 0. \tag{7.45b}$$

In either case, the resulting spatial functions will obey the boundary condition at the right wall. Surprisingly, they will also obey the boundary condition at the left wall.

Question 7–4

> Argue from the symmetry properties of the spatial function (7.43) that a solution of the TISE that satisfies (7.37*b*) must also satisfy (7.37*a*). Now apply the boundary condition at the left wall (7.37*a*), to the general solution (7.43). Show that you obtain the same restrictions on C, D, and p, Eqs. (7.45).

The requirements (7.45) imply that *valid spatial functions for the particle in a box exist only for values of p such that the sine or the cosine of $pL/(2\hbar)$ is zero.* The zeroes (or *nodes*) of these trigonometric functions occur at discrete, equally spaced values of their argument:

$$\cos\theta = 0 \qquad \text{for} \qquad \theta = \frac{\pi}{2}, \frac{3\pi}{2}, \frac{5\pi}{2}, \ldots \tag{7.46a}$$

$$\sin\theta = 0 \qquad \text{for} \qquad \theta = 0, \pi, 2\pi, \ldots. \tag{7.46b}$$

Applying Eqs. (7.46) to the functions in (7.45), we see that[11]

$$\sin\left(\frac{p}{\hbar}\frac{L}{2}\right) = 0 \qquad \text{if} \qquad p = \frac{2\pi\hbar}{L}, \frac{4\pi\hbar}{L}, \frac{6\pi\hbar}{L}, \ldots \tag{7.47a}$$

$$\cos\left(\frac{p}{\hbar}\frac{L}{2}\right) = 0 \qquad \text{if} \qquad p = \frac{\pi\hbar}{L}, \frac{3\pi\hbar}{L}, \frac{5\pi\hbar}{L}, \ldots \tag{7.47b}$$

[11] Although $\sin\theta = 0$ for $\theta = 0$, we exclude $p = 0$ from (7.47*b*), because the only zero-energy solution of (7.39) corresponds to the uninteresting case $\psi(x) = 0$ (no particle). The TISE for $E = 0$ becomes $d^2\psi(x)/dx^2 = 0$. The solution of this equation that is continuous at the walls is $\psi(x) = ax + b$, and this solution can satisfy the boundary conditions (7.37) only if $a = b = 0$, *i.e.*, if $\psi(x) = 0$. For more on this point see the articles by M. Bowen and J. Coster, *Amer. Jour. of Phys.*, **49**, 80 (1981) and by R. Sapp, *Amer. Jour. of Phys.* **50**, 1159 (1982).

Equations (7.47) imply that *the momentum of the particle in a box is quantized*; *i.e.*, spatial functions that satisfy the boundary conditions imposed by the potential exist only for *discrete* values of p in the list

$$p = \left\{ \frac{\pi\hbar}{L}, \frac{2\pi\hbar}{L}, \frac{3\pi\hbar}{L}, \frac{4\pi\hbar}{L}, \ldots \right\}. \qquad (7.47c)$$

We can conveniently index these values by introducing a positive integer $n = 1, 2, 3, \ldots$ and writing (7.47c) as

$$p_n = n\frac{\pi\hbar}{L} \qquad \text{for} \quad n = 1, 2, 3, \ldots. \qquad \text{[for the infinite square well]} \qquad (7.48)$$

Notice that each allowed value of the momentum corresponds to one and only one type of spatial function: *either* to Eq. (7.47a) [for n = 2, 4, 6, ...] *or* to Eq. (7.47b) [for n = 1, 3, 5, ...].

Since $E = p^2/(2m)$, these restrictions on p imply restrictions on E; to wit: the allowed stationary-state energies of a particle of mass m in a box of width L are

$$\boxed{E_n = n^2 \frac{\pi^2\hbar^2}{2mL^2} \qquad \text{for} \quad n = 1, 2, 3, \ldots} \qquad \text{[for the infinite square well]} \qquad (7.49)$$

Our derivation of Eq. (7.49) exemplifies an extremely important general principle of bound states:

> ***Rule***
>
> The two boundary conditions imposed by the potential energy that confines a particle to a finite region of space (in a bound quantum state) force quantization of energy.

But what about the spatial functions? We discovered [Eqs. (7.45)] that these functions come in two types. To write these down explicitly, let's use (7.48) for p_n in Eq. (7.45):

TYPE I:

$$\left.\begin{aligned} \psi_E^{(e)}(x) &= C\cos\left(\frac{p_n}{\hbar}x\right) \\ &\text{or} \\ \psi_n^{(e)}(x) &= C\cos\left(n\pi\frac{x}{L}\right) \end{aligned}\right\} \qquad n = 1, 3, 5, \ldots \qquad (7.50a)$$

TYPE II:

$$\left.\begin{aligned} \psi_E^{(o)}(x) &= D\sin\left(\frac{p_n}{\hbar}x\right) \\ &\text{or} \\ \psi_n^{(o)}(x) &= D\sin\left(n\pi\frac{x}{L}\right) \end{aligned}\right\} \qquad n = 2, 4, 6, \ldots \qquad (7.50b)$$

In Eqs. (7.50) I've introduced two labels on the spatial functions: a subscript n and a superscript, (e) or (o). The index n is called a **quantum number**. Notice that once we know n, we know the values of the energy [Eq. (7.49)] and the momentum [Eq. (7.48)] for the corresponding stationary state. The superscript indicates the *parity* of the function. All **TYPE I** functions are even,

$$\psi_n^{(e)}(-x) = \psi_n^{(e)}(x) \qquad [n = 1, 3, 5, \ldots] \qquad (7.51a)$$

and all **TYPE II** functions are odd,

$$\psi_n^{(o)}(-x) = -\psi_n^{(o)}(x) \qquad [n = 2, 4, 6, \ldots]. \qquad (7.51b)$$

Notice that we can accommodate both types of spatial functions with a single index, since each $n = 1, 2, 3, \ldots$ corresponds to one and only one type: odd values to cosine functions [Eq. (7.50a)], even values to sine functions [Eq. (7.50b)].

Step 3 of the solution of the TISE is to normalize the spatial functions. Let's do the odd-parity functions first. From Eq. (7.27), the normalization condition is[12]

$$1 = \int_{-L/2}^{+L/2} \psi_n^{(o)*}(x)\psi_n^{(o)}(x)\,dx \qquad n = 2, 4, 6, \ldots \tag{7.52}$$

$$= |D|^2 \int_{-L/2}^{+L/2} \sin^2\left(n\pi\frac{x}{L}\right)\,dx \tag{7.53}$$

$$= |D|^2\,\frac{L}{2}. \tag{7.54}$$

We might as well choose D to be real so the spatial functions will be real. We conclude that all the odd-parity eigenfunctions are normalized by the same factor, $D = \sqrt{2/L}$. Remarkably, this factor also normalizes the even-parity spatial functions.

Question 7–5

Consider the normalization of the even-parity functions and verify the claim in the last sentence. That is, show that to normalize $\psi_n^{(e)}(x)$ of Eq. (7.50b), we use $C = \sqrt{2/L}$.

The fact that the same normalization constant can be used for all spatial functions is a peculiarity of the infinite square well. In general, normalization constants are different for different stationary states.

To summarize our findings, here are the normalized spatial functions for the particle in a box:

$$\boxed{\psi_E(x) = \begin{cases} \psi_n^{(e)}(x) = \sqrt{\frac{2}{L}}\cos\left(n\pi\frac{x}{L}\right) & n = 1, 3, 5, \ldots \\ \psi_n^{(o)}(x) = \sqrt{\frac{2}{L}}\sin\left(n\pi\frac{x}{L}\right) & n = 2, 4, 6, \ldots \end{cases}} \tag{7.55}$$

For each spatial function, the corresponding energy is given by Eq. (7.49). If we want to highlight the variation in energy with n, we can write E_n in terms of the ground state energy E_1, as

$$\boxed{E_n = n^2 E_1 \text{ with } E_1 = \frac{\pi^2\hbar^2}{2mL^2}} \qquad \text{[for the infinite square well]} \tag{7.56}$$

These energies are sketched in Fig. 7.3. Diagrams such as this one are called, sensibly, **energy-level diagrams**. The list of allowed stationary-state energies is called—by spectroscopists, among others—the **energy spectrum**.

The full time-dependent *wave function* for each stationary state is easily constructed from the spatial functions (7.55), as

$$\Psi_n(x,t) = \psi_n(x)e^{-iE_nt/\hbar}. \tag{7.57}$$

[12]Watch the limits of integration; the spatial function is zero outside the box, so the integrals run not from $-\infty$ to $+\infty$ but rather from $-L/2$ to $+L/2$. To evaluate the integral in (7.53), just change the dummy variable of integration from x to $u = n\pi x/L$.

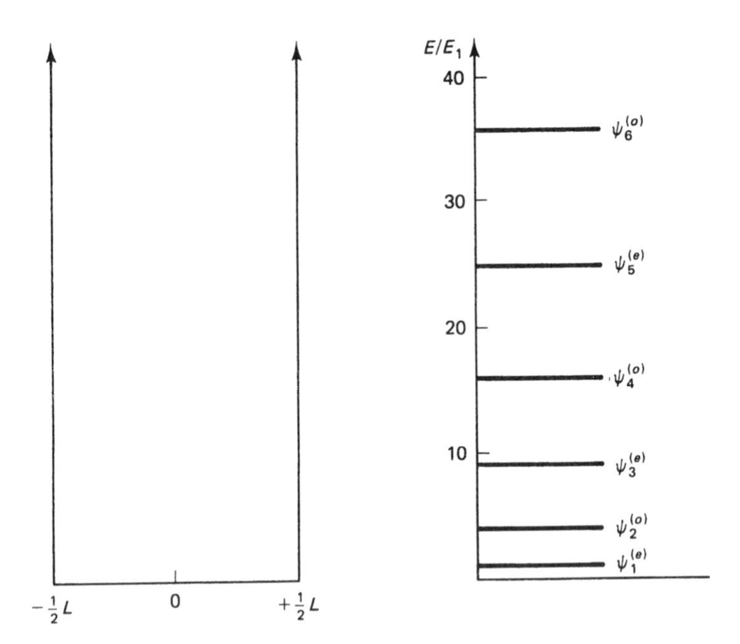

Figure 7.3 An energy-level diagram for the particle in a box. For convenience, I've plotted the (first six) stationary-state energies in units of E_1 [see Eq. (7.56)]. Beside each energy is the symbol of the corresponding spatial function.

For example, the first three such functions are

$$\Psi_1(x,t) = \sqrt{\frac{2}{L}}\cos\left(\pi\frac{x}{L}\right)e^{-iE_1t/\hbar} \qquad E_1 = \frac{\pi^2\hbar^2}{2mL^2}, \tag{7.58a}$$

$$\Psi_2(x,t) = \sqrt{\frac{2}{L}}\sin\left(2\pi\frac{x}{L}\right)e^{-iE_2t/\hbar} \qquad E_2 = \frac{4\pi^2\hbar^2}{2mL^2}, \tag{7.58b}$$

$$\Psi_3(x,t) = \sqrt{\frac{2}{L}}\cos\left(3\pi\frac{x}{L}\right)e^{-iE_3t/\hbar} \qquad E_3 = \frac{9\pi^2\hbar^2}{2mL^2}, \tag{7.58c}$$

These fellows are displayed along with their probability densities in Fig. 7.4.

Remarkable Symmetries

As we relax, feet up, drink in hand, reviewing the results of our first quantum triumph, we notice an interesting pattern in the stationary-state spatial functions of Fig. 7.4: *as the energy E_n increases, the functions $\psi_n(x)$ alternate parity, beginning with the even ground state $\psi_1^{(e)}(x)$.* This pattern is typical of stationary-state spatial functions for one-dimensional potentials that are symmetric with respect to a center of symmetry. To see what I'm talking about, have another look at this potential in Fig. 7.2.

The **center of symmetry** of a potential is the point on the x axis with respect to which $V(x)$ is *even*. In Fig. 7.2, the midpoint of the infinite square well is its center of symmetry. That is, if we define this point as the origin of coordinates, $x = 0$, then the function $V(x)$ is symmetric with respect to *inversion* through this point, the *symmetry operation* $x \mapsto -x$. This is the definition of a **symmetric potential**:

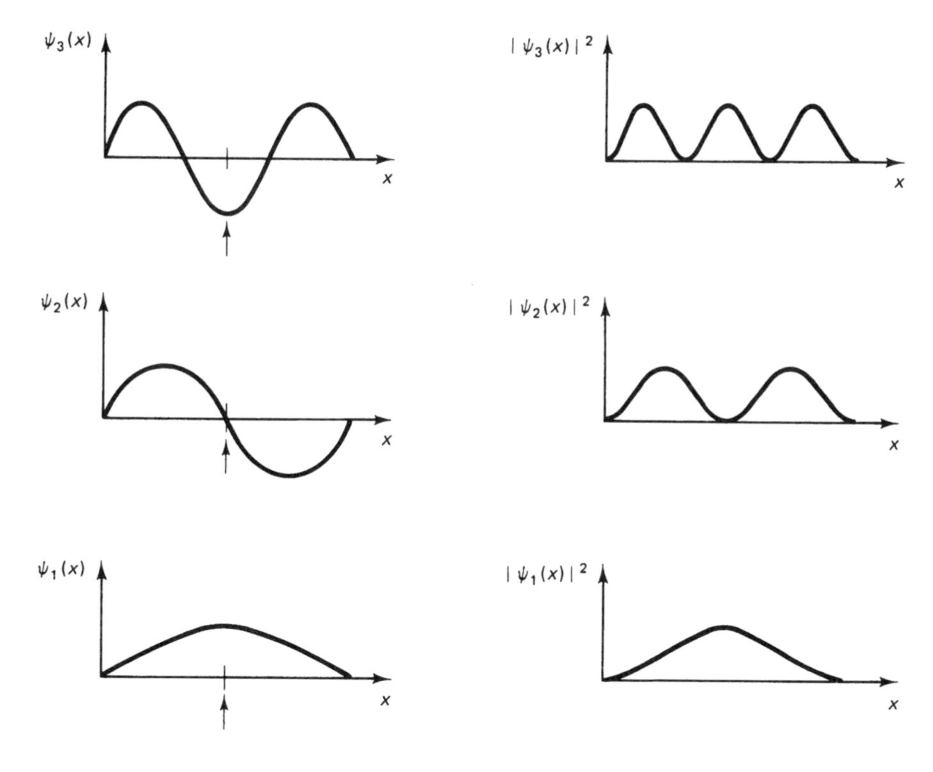

Figure 7.4 Spatial functions and corresponding probability densities for the $n = 1, 2,$ and 3 stationary states of the infinite square well. See Eqs. (7.58) for the analytic form of these spatial functions. Note the symmetry of these functions about the center of symmetry (arrow).

$$V(-x) = V(x). \qquad \text{symmetric potential} \tag{7.59}$$

In Fig. 7.5a you'll find a few other symmetric potentials. These examples differ in a visually obvious way from those in Fig. 7.5b, which do not have a center of symmetry. As you might guess, these unfortunates are called **asymmetric potentials.**

Why am I making a big deal out of the symmetry of the potential? Simply because the symmetry properties of a system (*i.e.*, of the potential energy of the system) has great power to simplify problem solving in quantum mechanics and to aid us in gaining profound insight into the physics of physical systems. For example, shortly (§ 9.4) we'll prove that the pattern in Fig. 7.4 pertains to all symmetric potentials:

Rule

All bound-state spatial functions of a one-dimensional system with a symmetric potential have definite parity. For such a system, the ground state is even, and the higher-energy functions alternate parity with increasing energy.

As we apply quantum mechanics to various systems, in Part III, you'll see how greatly this simple, seemingly magical rule eases the algebraic (and, in complicated problems, numerical) difficulties that confront us.

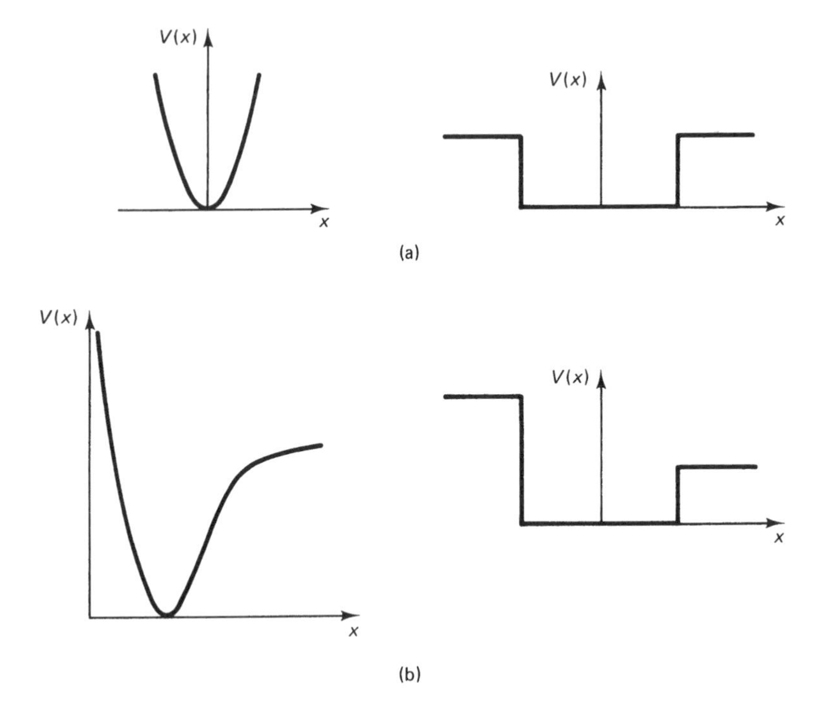

Figure 7.5 Examples of (a) symmetric and (b) asymmetric one-dimensional potentials. Shown in (a) is the harmonic oscillator potential $V(x) = m\omega_0^2 x^2/2$ and a symmetric finite square well, and in (b) an anharmonic oscillator (the Morse potential) and an asymmetric square well. The symmetry of the potentials in (a) markedly simplifies the chore of calculating their spatial functions.

Question 7–6

Repeat my solution of the TISE for the particle in a box, starting with a linear combination of complex exponential functions, (7.34). Show that you wind up with the same spatial functions (7.55), albeit after a bit more work.

Question 7–7

Suppose that in setting up the TISE for the particle in a box, we had chosen the origin of coordinates ($x = 0$) at the left wall, rather than at the midpoint of the well. What are the spatial functions in these coordinates? (*Hint*: Solve this problem two ways; first by directly solving the TISE, and second by implementing a change of variable $x \mapsto x - L/2$.) Why is this choice of origin a bad idea?

Aside: An Introduction to Curvature Arguments. It isn't hard to understand the alternating parity of the spatial functions of the infinite square well. The key to such insight is a mathematical property of each function called its *curvature*. The curvature of $\psi_E(x)$ is simply its second derivative, $d^2\psi_E(x)/dx^2$. (Loosely speaking, this quantity describes how rapidly the function "turns over" as x changes.) The TISE shows that the curvature of $\psi_E(x)$ is related to the value of E, since

$$\frac{d^2}{dx^2}\psi_E(x) = -\frac{2m}{\hbar^2}E\psi_E(x).$$

So the greater the curvature of $\psi_E(x)$, the greater the magnitude of E. Now, of the spatial functions in (7.55), the one with the *smallest* curvature is $\sqrt{2/L}\cos(\pi x/L)$.

(This function fits half of a cosine wave into the box.) So it is the spatial function of minimum energy. The function with the next smallest curvature fits a full sine wave into the box. And so forth.

The Stationary State of Minimum Energy

Of the many non-classical features exhibited by the particle in a box, perhaps none is more striking than that it is never at rest. A particle at rest in a region of zero potential energy would have zero total energy. But we've just proven [see Eq. (7.56)] that the minimum allowed energy for a particle in a such a region—inside the infinite square well—is greater than zero. What causes this incessant motion? The potential is zero, so no forces act on the particle (except the impulsive forces at the wall). Therefore *the non-zero kinetic energy of the particle in the infinite square well must be due to the simple fact that it is confined to a finite region*. This astonishing conclusion is quite contrary to any classical intuition we may still be carrying around.

Although this result fell out of our solution of the TISE for the infinite square well, it hinges on the Heisenberg Uncertainty Principle. To appreciate the essential connection between this principle and the non-zero minimum energy of a confined particle, let's consider an *ensemble* of identical particle-in-a-box systems, each in the stationary state with the minimum allowed total energy. We'll pretend that we don't know the value of this energy, which I'll call E_{min}. The question is: could E_{min} be zero?

What do we know about the energy? Since the state of the members is stationary, its energy is sharp. So *all members have the same value of the energy, E_{min}, which they exhibit in an energy measurement.* The ensemble average and uncertainty in the measured energy are

$$\langle E \rangle = E_{\text{min}} \qquad \text{and} \qquad \Delta E = 0. \tag{7.60}$$

To argue that the minimum energy must be positive, we need to look briefly to a different observable, position. From position, we'll return to momentum via the Heisenberg Uncertainty Principle and thence to the energy.

The position probability densities graphed in Fig. 7.4 show that in this stationary state, position is not sharp—*i.e.*, the position uncertainty Δx is positive. But neither is Δx infinite, because the particles can't escape from their boxes. So, without performing any calculations, we can put upper and lower bounds on the position uncertainty:

$$0 < \Delta x \leq L. \qquad \text{[for the particle in a box]} \tag{7.61}$$

These limits imply that the *momentum* uncertainty in this state must be positive, because, according to Heisenberg, $\Delta x\, \Delta p \geq \hbar/2$. (Recall from Example 5.6 that the momentum uncertainty for the ground state is $\Delta p = \pi\hbar/L$.)

Now, think carefully about what $\Delta p > 0$ implies about the individual results of the ensemble measurement: they must fluctuate about the mean momentum, $\langle p \rangle$. [For this stationary state, the magnitude of the mean momentum is related to E_{min} by $\langle p \rangle = p_{\text{min}} = \sqrt{2mE_{\text{min}}}$.] But if various individual results differ from one another, then at least one member must exhibit a non-zero momentum, $|p|$. The energy of this member is $|p|^2/(2m) > 0$.

But since $\Delta E = 0$, all members of the ensemble have the same (sharp) energy, E_{min}. We conclude, therefore, that E_{min}, the smallest possible sharp energy of a particle in an infinite square well, must be positive.

Question 7–8

Thought Question: Consider an ensemble of particle-in-a-box systems in an *arbitrary* state—*i.e.*, one that is not necessarily stationary. In an energy measurement, can any member of the ensemble exhibit the value $E = 0$? Why or why not?

Aside: Uncertainty and Quantization. Does the non-zero momentum uncertainty for stationary states of the infinite square well seem to contradict Eq. (7.48), which gives the quantized values of p_n? It doesn't, really, because *momentum is a directed quantity*. Equation (7.48) specifies the *magnitude* of p_n, but not its direction. To see the significance of the directed nature of momentum, we need only note that each function (7.58) is a linear combination of two travelling waves moving in opposite directions. For example, we can write the ground-state spatial function as

$$\psi_1(x) = \sqrt{\frac{2}{L}}\frac{1}{2}\left(e^{i\pi x/L} + e^{-i\pi x/L}\right).$$

So the momentum in such a state is not sharp, even though its magnitude is well defined.

Question 7–9

Deduce the value of $\langle p \rangle_n$ for *any* stationary state of the particle in a box. (*Hint*: Think, don't calculate, and re-read the above aside if you have trouble.)

The infinite square well is beloved of quantum pedagogues (such as your author), because it provides an algebraically simple problem that illustrates general problem-solving strategies and a host of quantum phenomena. And, as you'll discover as you solve the problems at the end of this chapter, it's a surprisingly good model of many "real-world" microscopic systems.

Still, the particle in a box is artificial in two noteworthy respects. First, *potential differences in the real world may be very large, but they are never infinite*. Second, *real potentials are never discontinuous, as is the infinite square well at the walls*. Because of these features, the spatial functions of the infinite square well are discontinuous at the walls—a difficulty that in the real world never rears its ugly head.[13]

7.5 A CLOSER LOOK AT STATIONARITY

In § 7.3 we explored the mathematical properties of stationary states and the equation from whence they come, the time-independent Schrödinger equation. In this section we'll focus on the *physical properties* of these states and, in particular, on the meaning of the adjective "stationary." Colloquially, this word means unmoving, static, time-independent. So let's investigate the *time-dependence* of the quantities that describe a bound stationary quantum state: the expectation value, uncertainty, and probability density.

The probability density for an arbitrary state $P(x,t) = \Psi(x,t)\Psi^*(x,t)$ does depend on time. But for a stationary state, this quantity is trivially time-independent, because the exponential factors in the wave function $\Psi(x,t) = \psi_n(x)e^{-iE_nt/\hbar}$ and its complex conjugate cancel:

[13]These discontinuities have generated considerable discussion in the pedagogical literature of quantum physics, including one paper with the marvelously ominous title "The Particle in a Box Is Not Simple," by F. E. Cummings, *Amer. Jour. of Phys.* **45**, 158 (1977). See also D. Home and S. Sengupta, *Amer. Jour. of Phys.* **50**, 522 (1982).

$$P_n(x,t) = P_n(x) = \psi_n^*(x)\psi_n(x). \tag{7.62}$$

Be not misled: although the probability density doesn't change with time, the wave function does. The point is that the time-dependent factor in the wave function is just a *global phase factor*, and (Chap. 3) all state functions are arbitrary to within such a factor. So $\Psi(x,t)$ and $\Psi(x,0)$ *represent the same physical state*. In this light, it's not surprising that all measurable properties of such a state are time independent.

The time-independence of $\Psi^*\Psi$ plays a vital role in evaluating statistical quantities for a stationary state. For an *arbitrary* quantum state (one that is not stationary), these quantities for an observable Q do depend on time (see § 5.6):

$$\langle Q\rangle(t) = \int_{-\infty}^{\infty} \Psi^*(x,t)\hat{Q}\Psi(x,t)\,dx. \tag{7.63a}$$

$$\Delta Q(t) = \sqrt{\langle Q^2\rangle - \langle Q\rangle^2}. \tag{7.63b}$$

But does this time dependence persist if the state represented by $\Psi(x,t)$ is stationary? To find out, let's return to the observable we've been investigating in this chapter.

The Energy

From our calculations on the stationary states of the infinite square well (§ 7.4), we conclude that for these states, the quantities (7.63) do *not* depend on time. Their values are

$$\begin{aligned} \langle E\rangle_n &= E_n \\ (\Delta E)_n &= 0 \end{aligned} \qquad [n^{\text{th}} \text{ stationary state}] \tag{7.64}$$

I want to show you now that this time-independence is a *general property of stationary states*.

Example 7.3. The Energy of a Stationary State

For $\hat{Q} = \hat{\mathcal{H}}$ and $\Psi(x,t) = \psi_n(x)e^{-iE_nt/\hbar}$, the ensemble average (7.63*a*) of the energy is

$$\langle E\rangle_n = \int_{-\infty}^{\infty} \psi_n^*(x)e^{+iE_nt/\hbar}\hat{\mathcal{H}}\psi_n(x)e^{-iE_nt/\hbar}\,dx. \tag{7.65}$$

The Hamiltonian for a conservative system,

$$\hat{\mathcal{H}} = -\frac{\hbar^2}{2m}\frac{d^2}{dx^2} + V(x), \tag{7.66}$$

does not alter a function of time such as $e^{-iEt/\hbar}$, so the time factors in this expectation value cancel, leaving the time-independent value

$$\langle E\rangle_n = \int_{-\infty}^{\infty} \psi_n^*(x)\hat{\mathcal{H}}\psi_n(x)\,dx. \tag{7.67a}$$

This proves that the time-independence seen in (7.64) for the infinite square well holds for the stationary states of any conservative system.

It also proves the generality of the relationship between the $\langle E\rangle$ and the allowed energies in (7.64). To see this, we just apply the TISE, $\hat{\mathcal{H}}\psi_n(x) = E_n\psi_n(x)$, to (7.67a) to get

$$\langle E \rangle_n = E_n \int_{-\infty}^{\infty} \psi_n^*(x)\psi_n(x)\,dx, \tag{7.67b}$$

and use the normalization integral (7.27) to conclude

$$\boxed{\langle E \rangle_n = E_n} \qquad [n^{\text{th}} \text{ stationary state}] \tag{7.68}$$

The same reasoning applies to the uncertainty

$$(\Delta E)(t) = \sqrt{\langle E^2 \rangle - \langle E \rangle^2}, \tag{7.69}$$

which, according to (7.68), is simply

$$(\Delta E)_n(t) = \sqrt{\langle E^2 \rangle_n - E_n^2}. \tag{7.70}$$

In the expectation value $\left\langle E^2 \right\rangle_n$, the time factors cancel, just as they did in (7.67*a*), leaving

$$\left\langle E^2 \right\rangle_n = \int_{-\infty}^{\infty} \psi_n^*(x)\hat{\mathcal{H}}^2\psi_n(x)\,dx. \tag{7.71}$$

Applying the TISE twice to $\psi_n(x)$ (and remembering that the product of two operators means successive operation), we can simplify the integrand in (7.71), to wit:

$$\begin{aligned} \hat{\mathcal{H}}^2\psi_n(x) &= \hat{\mathcal{H}}\hat{\mathcal{H}}\psi_n(x) \\ &= E_n\,\hat{\mathcal{H}}\psi_n(x) \\ &= E_n^2\psi_n(x). \end{aligned} \tag{7.72a}$$

So the mean value (7.71) is just

$$\left\langle E^2 \right\rangle = E_n^2, \tag{7.72b}$$

and from (7.70), the uncertainty *in any bound stationary state of any system* is

$$\boxed{(\Delta E)_n = 0} \qquad [n^{\text{th}} \text{ stationary state}] \tag{7.73}$$

Question 7–10

In Chap. 6, I noted that the Hamiltonian is equivalent to the operator $\hat{\mathcal{E}} = i\hbar\partial/\partial t$ —indeed, we can write the time-*dependent* Schrödinger equation in operator form as $\hat{\mathcal{H}} = \hat{\mathcal{E}}$. Consequently, we should be able to evaluate the expectation value and uncertainty of the energy *for any quantum state* either from $\hat{\mathcal{H}}$ or from $\hat{\mathcal{E}}$, *e.g.*,

$$\langle E \rangle_n = \langle \mathcal{E} \rangle_n = \int_{-\infty}^{\infty} \Psi^*(x,t)\hat{\mathcal{E}}\Psi(x,t)\,dx.$$

Derive Eqs. (7.68) and (7.73) for a stationary state using $\hat{\mathcal{E}}$ rather than $\hat{\mathcal{H}}$; *e.g.*, show that

$$\langle \mathcal{E} \rangle_n = E_n \qquad \text{and} \qquad (\Delta\mathcal{E})_n = 0.$$

Further Generalizations, Other Operators

What made the evaluation of $\langle E \rangle$ and of (ΔE) for a stationary state so trivial was the special relationship between the Hamiltonian and the wave function for such a state, $\hat{\mathcal{H}}\Psi_n = E_n \Psi_n$. This simple relationship doesn't apply to an arbitrary operator $\hat{Q}$, and *evaluating* these statistical quantities for an arbitrary observable does require some algebra (see the multitude of examples in Part III). But we can easily generalize the proof in Example 7.3 that $\langle E \rangle$ and ΔE for a stationary state are time-independent to an *arbitrary observable*.

Suppose the observable Q doesn't *explicitly* depend on time. The expectation value (7.63*a*) for the nth stationary state is

$$\langle Q \rangle (t) = \int_{-\infty}^{\infty} \psi_n^*(x) e^{+iE_n t/\hbar} \, \hat{Q} \, \psi_n(x) e^{-iE_n t/\hbar} \, dx. \tag{7.74}$$

Look at the time factors in this equation. Regardless of the form of $\hat{Q}$, these factors cancel, leaving a *time-independent expectation value*,

$$\langle Q \rangle_n = \int_{-\infty}^{\infty} \psi_n^*(x) \hat{Q} \psi_n(x) \, dx. \tag{7.75a}$$

The same argument applies to the uncertainty,

$$(\Delta Q)_n = \sqrt{\langle Q^2 \rangle_n - \langle Q \rangle_n^2}. \tag{7.75b}$$

We now understand precisely what is meant by **stationarity**:

> ***Rule***
>
> In a stationary state all physically measurable properties of a (time-independent) observable are static; *i.e.*, probabilities, expectation values, and uncertainties do not change with time.

Aside: Rates of Change. You will sometimes see the time-independence of quantum-mechanical quantities expressed in terms of the rate of change of the probability density, as

$$\frac{d}{dt} P(x,t) = 0. \qquad \text{[in a stationary state]} \tag{7.76}$$

The counterpart of this form for an observable Q is

$$\frac{d}{dt} \langle Q \rangle_n = 0. \tag{7.77}$$

For example, in a stationary state, the mean value of the energy is constant,

$$\frac{d}{dt} \langle E \rangle_n = 0. \tag{7.78}$$

This is the quantum-mechanical counterpart of the classical law of conservation of energy in a conservative (macroscopic) system (Chap. 11).

To Flow or Not to Flow?

Before leaving this discussion, let's look at the time dependence of the *probability current density* [Eq. (6.112)]

$$j(x,t) = -\frac{i\hbar}{2m}\left[\Psi^*(x,t)\frac{\partial}{\partial x}\Psi(x,t) - \Psi(x,t)\frac{\partial}{\partial x}\Psi^*(x,t)\right]. \tag{7.79}$$

In general, $j(x,t)$ is a function of time—indeed, it is this time dependence that gives rise to the colorful (and useful) notion of a probability fluid that flows through space. But if the state represented by $\Psi(x,t)$ is stationary, the time factors in $\Psi(x,t)$, its complex conjugate, and their derivatives all cancel, leaving

$$j_n(x) = -\frac{i\hbar}{2m}\left[\psi_n^*(x)\frac{d}{dx}\psi_n(x) - \psi_n(x)\frac{d}{dx}\psi_n^*(x)\right]. \tag{7.80}$$

So *the probability current density in a stationary state is independent of time.* Such a current density describes *steady state* flow—like that of the pure momentum function (Example 6.6).

Question 7–11

We can arrive at this conclusion via a different path. The *rate equation* for the integrated position probability in a finite region $[a,b]$ is [Eq. (6.119)]

$$\frac{d}{dt}P_n\big([a,b];t\big) = j_n(a,t) - j_n(b,t).$$

Starting from this equation, prove that the probability current density for a stationary state is independent of time.

For a bound state in one dimension, a further simplification to the probability current density (7.80) ensues. Because the TISE and the boundary conditions on a bound-state wave function are *real*, the spatial function that solves this equation, $\psi_n(x)$, is real. So $j_n(x) = 0$. This means that *there is no probability flow in a one-dimensional bound state.*

Aside: Probability Flow in Three Dimensions. To glimpse the nature of probability flow for a three-dimensional system, consider the equation for pointwise conservation of probability [see Eq. (6.125) and Table 6.1] for a stationary-state wave function. The probability density for such a state doesn't depend on time,

$$\frac{\partial}{\partial t}P(\mathbf{r},t) = 0,$$

so the probability current density obeys

$$\nabla \cdot \mathbf{j}(\mathbf{r},t) = 0.$$

This result shows that the probability flow in a stationary state of a three-dimensional system is *solenoidal*. The probability current density is zero only if the spatial function $\psi(\mathbf{r})$ is real—which is not necessarily the case for a bound state in three dimensions.

A final cautionary note: Some novice quantum mechanics slip into the trap of thinking that the name "stationary state" implies that the *particles* in such a state are at rest. We know (§ 7.4) that for a bound state, this is false: the minimum kinetic energy of a (confined) particle is always positive. It is, of course, not the particle, but rather its *physical properties* that are stationary, in the sense discussed in this section.

The essential point of this section is that (for observables that don't explicitly depend on time) we can just forget about the time dependence of a stationary-state wave function and calculate probabilities and statistical quantities directly from the spatial function $\psi_n(x)$. But for a state that is not stationary, the probability density and $\langle Q \rangle (t)$ and $(\Delta Q)(t)$ may well depend on time. In the next section of this chapter, we'll put stationarity in its place by looking briefly at such a state.

*7.6 SUPERPOSTIONS OF STATIONARY STATES

I want to be sure you don't overlook the very special, limited nature of stationary states.[14] First, not all systems *have* stationary states—separable solutions to the TDSE exist only for conservative systems. Second, *all* systems—conservative or not—exist in states that aren't stationary.

In general, some states of an arbitrary system are bound and some are not. But in either case, the non-stationary wave functions of such a system are not simple products like $\psi(x)\zeta(t)$; rather, like the Gaussian (7.3), they are complicated beasties in which position and time variables are inextricably intertwined.

Personally, I'm more comfortable with non-stationary states, for their physical properties are more akin to what I expect of a particle in motion. For example, the expectation value and uncertainty of position for a microscopic particle in a non-stationary state change with time, as does the position of a macroscopic particle in motion.

So I want to close this chapter on stationary states with a preview of their non-stationary siblings. For future reference, I've collected the properties of these two kinds of states in Table 7.1. In this section, we'll look briefly at how non-stationary states are constructed and at the properties of a single observable—the *energy*—in such a state.

TABLE 7.1 STATIONARY VS. NON-STATIONARY STATES.

	Stationary	Non-Stationary
State Function	$\Psi(x,t) = \psi_E(x)e^{-iEt/\hbar}$	$\Psi(x,t)$ not separable
Energy	sharp: $\Delta E = 0$	not sharp: $\Delta E(t) > 0$
Average Energy	$\langle E \rangle = E$, independent of t	$\langle E \rangle$ independent of t
Position Probability Density	$P(x) = \lvert\psi_E(x)\rvert^2$	$P(x,t) = \lvert\Psi(x,t)\rvert^2$
Observable Q	$\langle Q \rangle$ independent of t	$\langle Q \rangle (t)$ may depend on t

The Energy in a Non-Stationary State

One way to construct a wave function for a non-stationary state is to superpose some stationary-state wave functions, controlling the mixture of each contributing stationary state in our non-stationary brew by assigning each a coefficient. We did just this in Chap. 4, where we constructed a continuum wave function (a wave packet) for a free

[14]This section is an advance look at material we'll cover in depth in Chap. 12, so you can skip it if you want to.

particle by superposing pure momentum functions, using the amplitude function to control the mixture in the packet. Because the pure momentum functions are functions of infinite extent, we had to superpose an infinite number of them to produce a spatially localized wave packet.

Similarly, we can use superposition to construct non-stationary wave functions for bound states. But unlike their continuum counterparts, bound-state wave packets are spatially localized even if they include only a finite number of contributors, because each of these contributors is itself spatially localized. I wrote down one such function for the infinite square well in Example 6.4 [Eq. (6.63)]:

$$\Psi(x,t) = \sqrt{\frac{1}{53}}\left[2\,\psi_1(x)e^{-iE_1t/\hbar} + 7\,\psi_2(x)e^{-iE_2t/\hbar}\right]. \tag{7.81}$$

Notice that two values of the energy appear in this function, E_1 and E_2. This duplicity of energies suggests that the energy of the state represented by the superposition function $\Psi(x,t)$ is not sharp. Indeed, it isn't. The energy uncertainty for this non-stationary state is positive, as we'll now demonstrate.

Example 7.4. The Strange Case of the Uncertain Energy

The energy uncertainty $(\Delta E)(t) = \sqrt{\langle E^2\rangle - \langle E\rangle^2}$ depends on two expectation values, $\langle E\rangle(t)$ and $\langle E^2\rangle$. The first of these is the mean value of the energy; for the state represented by (7.81), this value is

$$\begin{aligned}\langle E\rangle(t) = \frac{1}{53}\int_{-L/2}^{+L/2} &\left[2\psi_1(x)e^{+iE_1t/\hbar} + 7\psi_2(x)e^{+iE_2t/\hbar}\right] \\ &\times \hat{\mathcal{H}}\left[2\psi_1(x)e^{-iE_1t/\hbar} + 7\psi_2(x)e^{-iE_2t/\hbar}\right] dx.\end{aligned} \tag{7.82}$$

The (real) spatial functions in (7.82) are given by Eqs. (7.55), and their energies by (7.56). The *worst* path to take at this point is to substitute these functions, their energies, and the explicit form of the Hamiltonian into this equation; this path leads immediately into a thorny patch of messy algebra.

We can avoid the briar patch by leaving the spatial functions and energies unspecified as long as possible. For example, from the TISE we know what the Hamiltonian does to the spatial functions in the integrand:

$$\begin{aligned}&\hat{\mathcal{H}}\left[2\psi_1(x)e^{-iE_1t/\hbar} + 7\psi_2(x)e^{-iE_2t/\hbar}\right] \\ &\qquad = 2E_1\psi_1(x)e^{-iE_1t/\hbar} + 7E_2\psi_2(x)e^{-iE_2t/\hbar}.\end{aligned} \tag{7.83}$$

So we can write (7.82) as four similar integrals,

$$\begin{aligned}\langle E\rangle(t) = \frac{1}{53}\Bigg\{&4E_1\int_{-L/2}^{+L/2}\psi_1(x)\psi_1(x)\,dx + 14E_1\int_{-L/2}^{+L/2}\psi_2(x)\psi_1(x)\,dx\,e^{i(E_2-E_1)t/\hbar} \\ &+14E_2\int_{-L/2}^{+L/2}\psi_1(x)\psi_2(x)\,dx\,e^{-i(E_2-E_1)t/\hbar} + 49E_2\int_{-L/2}^{+L/2}\psi_2(x)\psi_2(x)\,dx\Bigg\}.\end{aligned} \tag{7.84}$$

These four integrals fall into two classes. Two are *normalization integrals* of the form (7.27). These are easy to evaluate:

$$\int_{-L/2}^{+L/2} \psi_1(x)\psi_1(x)dx = 1$$
$$\int_{-L/2}^{+L/2} \psi_2(x)\psi_2(x)dx = 1. \tag{7.85}$$

The other two are integrals of two *different* spatial functions. Since the spatial functions (7.55) are real, these integrals are equal:

$$\int_{-L/2}^{+L/2} \psi_2(x)\psi_1(x)\,dx = \int_{-L/2}^{+L/2} \psi_1(x)\psi_2(x)\,dx = \frac{2}{L}\int_{-L/2}^{+L/2} \cos\left(\pi\frac{x}{L}\right)\sin\left(2\pi\frac{x}{L}\right)\,dx. \tag{7.86}$$

I hope you recognize this as the *orthogonality integral* for the sine and cosine functions—and remember that this integral equals zero. If not, please go refresh your memory with a glance at Eq. (4.45).[15] Inserting Eqs. (7.85) and (7.86) into the expectation value (7.84), we get (with refreshingly little work) the ensemble average of the energy in the non-stationary state (7.81),

$$\langle E\rangle = \frac{1}{53}(4E_1 + 49E_2) = \frac{200}{53}E_1 = 3.77E_1. \tag{7.87}$$

Notice that this is not equal to any of the eigenvalues E_n of the Hamiltonian. This exemplifies a general result for all non-stationary states: *the average value of the energy of a particle in a non-stationary state is not one of the energy eigenvalues.* Nevertheless, in an energy measurement each member of the ensemble exhibits one and only one of these eigenvalues (Chap. 13). That is, *only the eigenvalues E_n are observed in an energy measurement, whatever the state of the system.* The mean of these values is not equal to one of the eigenvalues because, as we're about to discover, individual measurement results fluctuate around $\langle E\rangle$.

Notice also that the time dependence has vanished from the expectation value of the energy of a non-stationary state, (7.87), as it did (§ 7.5) from this quantity for a stationary state. This time-independence is a consequence of the structure of Eq. (7.84): each time-dependent factor in this equation is multiplied by an orthogonality integral that is equal to zero. No such simplification occurs for an arbitrary observable Q, because the integrals that multiply time-dependent factors in $\langle Q\rangle(t)$ contain factors other than the spatial functions. Thus, for this state, $\langle x\rangle(t)$ and $\langle p\rangle(t)$ do depend on time.

Now back to the energy uncertainty. The other expectation value we need is

$$\left\langle E^2\right\rangle = \frac{1}{53}\int_{-L/2}^{+L/2} \left[2\psi_1(x)e^{+iE_1t/\hbar} + 7\psi_2(x)e^{+iE_2t/\hbar}\right] \times \hat{\mathcal{H}}^2\left[2\psi_1(x)e^{-iE_1t/\hbar} + 7\psi_2(x)e^{-iE_2t/\hbar}\right]dx. \tag{7.88}$$

The evaluation of $\left\langle E^2\right\rangle$ proceeds just like that of $\langle E\rangle$. We work out the effect of $\hat{\mathcal{H}}^2$ on the wave function (7.81) using $\hat{\mathcal{H}}^2 = \hat{\mathcal{H}}\hat{\mathcal{H}}$, as in Eq. (7.72). There result four integrals very much like those of (7.84). As before, two are normalization integrals and two

[15] Such orthogonality integrals are omnipresent in quantum mechanics (Part IV). The general form of such an integral is $\int_{-\infty}^{\infty}\psi_n^*(x)\psi_m(x)dx$. It is a general property of the eigenfunctions of the Hamiltonian that for two different spatial functions (*i.e.*, for $n \neq m$), this integral is zero.

are orthogonality integrals. And the latter are zero. Working through a simple arithmetic calculation, we find

$$\begin{aligned}\langle E^2\rangle &= \frac{1}{53}(4E_1^2 + 49E_2^2) \\ &= \frac{1}{53}\left[4E_1^2 + 49(4E_1)^2\right] \\ &= \frac{788}{53}E_1^2.\end{aligned} \tag{7.89}$$

Using this result and (7.87) for $\langle E\rangle$, we obtain the uncertainty for the state (7.81),

$$\Delta E = \sqrt{\frac{788}{53}E_1^2 - \left(\frac{200}{53}\right)^2 E_1^2} = 0.79\,E_1. \tag{7.90}$$

Question 7–12

Fill in the steps completely and carefully to obtain Eq. (7.90). Understanding this simple analysis will make solving later problems involving non-stationary states much easier.

So the energy for this non-stationary state is not sharp. Individual results of an ensemble energy measurement fluctuate about $\langle E\rangle = 3.77\,E_1$ with a standard deviation of $\Delta E = 0.79\,E_1$. This behavior differs strikingly from that of a *stationary* state like the one in Example 7.3. For the latter, all members give the same value, $\langle E\rangle_n = E_n$ with no fluctuations, and $(\Delta E)_n = 0$.

General Non-Stationary States

I included in the non-stationary wave function (7.81) only two constituents, the wave functions for the ground and first-excited states. The obvious generalization of this example is an *infinite* superposition, the most **general non-stationary state**:

$$\boxed{\Psi(x,t) = \sum_{n=1}^{\infty} c_n \psi_n(x) e^{-iE_nt/\hbar}} \qquad \text{general non-stationary state} \tag{7.91}$$

For the infinite square well, this wave function looks like

$$\begin{aligned}\Psi(x,t) &= \sum_{\substack{n=1\\ \text{odd}}}^{\infty} c_n \sqrt{\frac{2}{L}}\cos\left(n\pi\frac{x}{L}\right)e^{-iE_nt/\hbar} \\ &\quad + \sum_{\substack{n=2\\ \text{even}}}^{\infty} c_n \sqrt{\frac{2}{L}}\sin\left(n\pi\frac{x}{L}x\right)e^{-iE_nt/\hbar}.\end{aligned} \tag{7.92}$$

The coefficients c_n, which may be complex, control the mixture of stationary states. For the state (7.81) of the infinite square well, these guys are

$$c_1 = \frac{2}{\sqrt{53}}, \quad c_2 = \frac{7}{\sqrt{53}}, \quad c_3 = c_4 = \ldots = 0.$$

These numbers are called, quite reasonably, **mixture coefficients** or **expansion coefficients**.

Let's look for a moment at the form of the wave function (7.92) at the initial time, $t = 0$:

$$\Psi(x,0) = \sum_{\substack{n=1\\ \text{odd}}}^{\infty} c_n \sqrt{\frac{2}{L}} \cos\left(n\pi\frac{x}{L}\right) + \sum_{\substack{n=2\\ \text{even}}}^{\infty} c_n \sqrt{\frac{2}{L}} \sin\left(n\pi\frac{x}{L}\right). \tag{7.93}$$

Structurally, Eq. (7.93) is very like the Fourier series expansion (4.41) for the initial wave function $\Psi(x,0)$. We calculate the Fourier coefficients in a series expansion of a function $f(x)$ (§ 4.3) as integrals of sine or cosine functions times the function. Similarly, in quantum mechanics, we calculate the expansion coefficients in a non-stationary state wave function of the form (7.91) as integrals over the initial state function, to wit:

$$c_n(0) = \int_{-\infty}^{\infty} \psi_n^*(x)\Psi(x,0)dx. \tag{7.94}$$

Because these coefficients are intimately related to the wave function *at the initial time*, I've appended to them the argument (0). The parallel between (7.94) and the techniques of Fourier analysis is striking and no accident (Chap. 12).

Deducing the State Function

Equation (7.94) shows that from the initial wave function we can determine the expansion coefficients $\{c_n(0)\}$ and thence the state function $\Psi(x,t)$. But in the real world of laboratory physics, we often cannot proceed in this fashion. Instead we determine these coefficients experimentally—*e.g.*, by measuring the *energy* of the system in its initial state. To show you how this goes, I want to consider a final example.[16]

Example 7.5. Creation of a State Function

You are in a laboratory. The reigning experimentalists are showing you the results of an energy measurement on an ensemble of infinite-square-well systems. These measurements unveil the state at $t = 0$ to be a mixture of the two stationary states with quantum numbers $n = 2$ and $n = 6$. Further, the data reveals that the *relative mixture* of these states is $1:5$ for $n = 2 : n = 6$. Your mission: determine the wave function for subsequent times, $\Psi(x,t)$. You know the spatial functions $\psi_n(x)$ and energies E_n in the general expansion (7.92). All you need is the coefficients $c_n(0)$.

Translating the experimentalists' information into quantum mechanics, we first write down the form of the initial wave function,

$$\Psi(x,0) = c_2(0)\psi_2(x) + c_6(0)\psi_6(x), \tag{7.95}$$

and the ratio of the expansion coefficients,

$$\frac{c_2(0)}{c_6(0)} = \frac{1}{5}. \tag{7.96}$$

Since $c_6(0) = 5c_2(0)$, we can write the initial function (7.95) as

$$\Psi(x,0) = c_2(0)\psi_2(x) + 5c_2(0)\psi_6(x) = c_2(0)\left[\psi_2(x) + 5\psi_6(x)\right]. \tag{7.97}$$

[16]From such a measurement we can determine only the *ratio* of the magnitudes of these coefficients—not their real and imaginary parts. But we can discover their relative phases via other experiments, thereby fully determining the initial state function (to within the usual global phase factor). For more on such matters, see Chap. 13.

All that remains is to evaluate $c_2(0)$ in (7.97) and to write the resulting function in the general form (7.92) for $t > 0$. But what quantum-mechanical relationship can we use to calculate this coefficient? Well, there's only one property of the wave function $\Psi(x,0)$ that we haven't already used. Can you think what it is?

Right: we haven't *normalized* the initial function. The coefficient $c_2(0)$ provides the flexibility we need to enforce the condition

$$\int_{-\infty}^{\infty} \Psi^*(x,0)\Psi(x,0)dx = \int_{-L/2}^{+L/2} \Psi^*(x,0)\Psi(x,0)dx = 1. \tag{7.98}$$

A little algebra later, we find that this condition is satisfied by $c_2(0) = 1/\sqrt{26}$. So the initial wave function is

$$\Psi(x,0) = \frac{1}{\sqrt{26}}\left[\psi_2(x) + 5\psi_6(x)\right]. \tag{7.99}$$

In trying to generalize an initial wave function such as (7.99) to times $t > 0$, many newcomers to quantum mechanics come to grief. One of their most common mistakes is to write the $t > 0$ wave function as $\frac{1}{\sqrt{26}}\left[\psi_2(x) + 5\psi_6(x)\right]e^{-iEt/\hbar}$. I hope it's clear why this form is grossly in error. If not, please ponder and *commit to memory* the following:

WARNING

You cannot extend the initial state function for a non-stationary state to times $t > 0$ by multiplying by a single factor $e^{-iEt/\hbar}$. There is no single value of E that characterizes the state, because the energy of the state is not sharp.

The *right* way to proceed is indicated by the general form (7.91): we multiply *each term* in the expansion by the corresponding time factor. For the state represented by (7.99) at $t = 0$, we get

$$\Psi(x,t) = \frac{1}{\sqrt{26}}\left[\psi_2(x)e^{-iE_2t/\hbar} + 5\psi_6(x)e^{-iE_6t/\hbar}\right], \tag{7.100}$$

where the energies are $E_n = n^2\pi^2\hbar^2/(2mL^2)$ for $n = 2$ and $n = 6$.

I want to leave you thinking about the similarity between the procedure we used in Example 7.5 to determine the wave function for a bound non-stationary state of a particle in a box and the one we used in § 4.4 to determine the wave packet for an unbound non-stationary state of a free particle. Schematically, these procedures look like

$$\Psi(x,0) \Longrightarrow A(k) \Longrightarrow \Psi(x,t) = \frac{1}{\sqrt{2\pi}}\int_{-\infty}^{\infty} A(k)e^{i(kx-\omega t)}\,dk \qquad \text{[continuum state]} \tag{7.101}$$

$$\Psi(x,0) \Longrightarrow \{c_n(0)\} \Longrightarrow \Psi(x,t) = \sum_n^{\infty} c_n(0)\psi_n(x)e^{-iE_nt/\hbar} \qquad \text{[bound state]} \tag{7.102}$$

The similarity between these procedures suggests that $A(k)$ and $\{c_n(0)\}$ play analogous roles in the quantum theory of continuum and bound states. Germinating in this comparison are the seeds of a powerful generalization of these procedures, as we'll discover in Chap. 12.

7.7 FINAL THOUGHTS: DO STATIONARY STATES REALLY EXIST?

In this chapter, I've introduced the "other" Schrödinger Equation, the TISE

$$\hat{\mathcal{H}}\psi_E = E\psi_E. \tag{7.103}$$

The first Schrödinger Equation you met (in Chap. 6) was the TDSE,

$$\hat{\mathcal{H}}\Psi = \hat{\mathcal{E}}\Psi. \tag{7.104}$$

It's vital that you never lose sight of the difference between these equations and their roles in quantum theory.

The time-*dependent* Schrödinger Equation (7.104) is a second-order partial differential equation in space and time variables. We do not see in it *a value* of the energy because in general, we cannot associate a single, well-defined value of this observable with a quantum state. Instead, we find the energy operator $\hat{\mathcal{E}}$. We use this equation to study the evolution of *any* state, whether or not it is stationary.

The time-*independent* Schrödinger Equation (7.103) is an eigenvalue equation in the position variable x (for a one-dimensional, single-particle system). The eigenvalue is E, *the value* of the (sharp) energy. We use this equation to solve for the spatial dependence of a stationary-state wave function $\Psi = \psi_E\, e^{-iEt/\hbar}$.

Like many elements of physical theory, stationary states are an idealization. In the real world, one can prepare a system, via ensemble measurements, to be in a stationary state at an initial time. But the system won't remain in such a state indefinitely. Only a truly isolated system would, in principle, remain in a stationary state for all time–and in the real world a system cannot be isolated from all external influences. Even if we take great pains to remove from the vicinity of a particle all other objects and sources of fields, the particle interacts with the *vacuum electromagnetic field* that permeates all space. Understanding the origin of this field requires knowledge of relativistic quantum theory, which is beyond our current expertise. But the important point is that this field causes mixing of other stationary states with the initial state. This mixing may induce the system to eventually undergo a *transition* to another state.

The key word here is "eventually"—in many cases, a microscopic system remains in a stationary state long enough for us to perform an experiment. For example, excited stationary states of atoms are stable for periods ranging from 10^{-9} to 10^{-2} sec, depending on the type of transition the atom eventually undergoes. On the time scale of atomic processes, such durations are fairly long. So, provided we're interested in an event that occurs on an atomic time scale, we can safely treat the system as being in a stationary state.

In spite of their slightly idealized nature, stationary states have become the heart and soul of quantum mechanical calculations. Their usefulness—and the ease of solution of the TISE compared to the TDSE—fully justifies the time we shall spend on them, beginning in the next chapter.

EXERCISES AND PROBLEMS

"Is not my logic absolute?
Would not a three-year-old child
of most deficient intellect
be convinced by it?"

—Jules de Grandin
in *Satan's Stepson*
by Seabury Quinn

The problems in this chapter explore several aspects of stationarity and the properties of stationary-state wave functions. In addition, you'll further your understanding of the probability current density, examine a non-stationary "sloshing" state, and discover how to use the simple infinite square well to approximate a very complicated organic molecule.

Exercises

7.1 Momentum Uncertainties — The Easy Way

Derive a general expression for the momentum uncertainty $(\Delta p)_n$ for an arbitrary *stationary state* of a particle in an infinite square well of width L.

Hint: Consider carefully the way in which $\hat{p}$ appears in the expression for the Hamiltonian operator of this system. You should have to do very little work to get the right answer.

Problems

7.1 Interesting Properties of Stationary-State Functions

Like all wave functions, a stationary-state wave function is arbitrary to within a global phase factor $e^{i\delta}$, where δ is a real constant (see § 3.5).

[a] **Prove** that with no loss of generality, we can always choose δ so that the spatial functions $\psi_E(x)$ are real.

[b] **Prove** that the probability current density for any *bound* stationary state is zero.

7.2 The Probability Current Density for a Sloshing State

In Pblm. 3.2 we studied a non-stationary state of the infinite square well. The wave function for this state is a linear combination of $\Psi_1(x,t)$ and $\Psi_2(x,t)$ of Eqs. (7.58):

$$\Psi(x,t) \equiv \frac{1}{\sqrt{2}}\Big[\Psi_1(x,t) + \Psi_2(x,t)\Big]. \tag{7.2.1}$$

[a] **Derive** an expression for the probability current density $j(x,t)$ *at the midpoint of the well* ($x = 0$).

[b] Find the (time-dependent) probability that a particle in an ensemble of systems in this state will be found in a position measurement at time t to be in the *left half* of the box.

[c] **Verify** that the *rate* at which probability leaves the left half of the box is equal to $j(0,t)$. Provide a physical explanation of this result, basing your discussion on the interpretation of the probability current density.

7.3 The Probability Current Density for the Free Particle

In Pblm. 5.3, we wrote a general form for a free-particle wave packet at time $t = 0$, *i.e.*,

$$\Psi(x,0) = f(x)\,e^{ip_0x/\hbar}. \tag{7.3.1}$$

Recall that the function $f(x)$ was specified to be *real*. Suppose this function is centered on the origin ($x = 0$) and is zero outside the region $[-L/2, +L/2]$.

[a] **Derive** an expression for $j(x, 0)$ for a state that is represented by a wave packet of this form.

[b] **Prove** that for such a state

$$\int_{-\infty}^{\infty} j(x, 0)\, dx = \frac{\hbar k_0}{m}. \tag{7.3.2}$$

Discuss carefully the physical significance of this result.

[c] **Briefly discuss** how you would **qualitatively** expect $j(x, t)$ to change with time at two positions: $x_1 \ll -L/2$, and $x_2 \gg +L/2$.

7.4 Properties of the Infinite Square Well

In § 7.4 we solved the TISE for a particle that is confined to a one-dimensional infinite potential well of width L. In exercises and problems earlier in this book, we have investigated several properties of this system. In this problem, we'll gather together and generalize some of these properties.

[a] **Derive** a general expression for the *position uncertainty* $(\Delta x)_n$ for the nth stationary state of this system. **Calculate** $(\Delta x)_n$ for $n = 1$, 2, 3, and 4. **Discuss** briefly the physical significance of the variation of the position uncertainty as n increases.

[b] **Repeat** part (a) for the *momentum uncertainty* $(\Delta p)_n$.

[c] **Evaluate** $\langle T \rangle_n$ for the nth stationary state.

7.5 An Application of the Infinite Square Well

A surprising number of physical systems can be modelled by the infinite square well of § 7.4. For instance, the **hexatriene molecule** is a fairly complicated organic system. Some of the electrons in this molecule (the six "π electrons") are delocalized (*i.e.*, are free to move about inside the molecule). We can model this system by treating the electrons as if they were in an infinite potential well with walls located a distance of one-half the bond length from each of the terminal carbon nuclei, as shown in the Fig. 7.5.1. Using such a model, the width of the box turns out to be $L = 7.3$ Å.

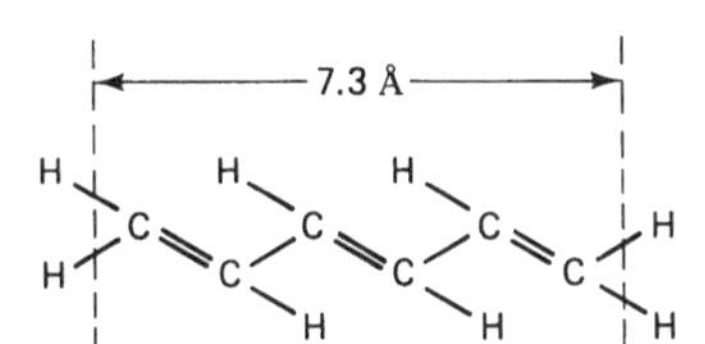

Figure 7.5.1 A ludicrously simple model of a very complicated organic molecule.

[a] **Evaluate** the first *four* energy levels of an electron in the (model) hexatriene molecule. Express your answers in joules and in eV.

As we'll learn in Volume II, electrons in *many-electron systems* are subject to some unusual constraints. For example, if a *one-dimensional* system contains more than two electrons, then only two electrons can have the same energy. We describe this situation rather colorfully, by saying that two electrons with the same energy "occupy the same energy level." So in a multi-electron one-dimensional "atom," no more than *two* electrons can occupy the same energy level. (This is the essence of **The Pauli Exclusion Principle**.) Applying this principle to the hexatriene molecule, we find that the *ground state* of this system is the state in which the lowest three energy levels are completely "filled," each with two electrons per level.

[b] Now, suppose we irradiate a hexatriene molecule in its ground state with radiation of just the right wavelength to induce a transition of *one electron* from the $n = 3$ to the $n = 4$ energy level. **Calculate** the wavelength (in angstroms) of the photon that is absorbed when this transition occurs. The experimentally observed absorption band occurs at $\lambda = 2580$ Å. Compare your answer to the observed wavelength and decide whether the infinite square well is a good model of this system.

7.6 Probability Flux and Standing Waves

In Example 7.1 we saw that the Hamiltonian eigenfunction for a free particle can be written either in terms of complex travelling waves [Eq. (7.34)] or real standing waves [such as Eqs. (7.35)]. In solving bound-state problems, the standing-wave solutions are by far the more convenient choice. The general form of the spatial function for a free particle with energy E in terms of standing waves is

$$\psi_E(x) = A \sin kx + B \cos kx, \tag{7.6.1}$$

where

$$k = \frac{\sqrt{2mE}}{\hbar} \tag{7.6.2}$$

and A and B are arbitrary constants.

[a] Suppose (for simplicity) that $B = 0$. **Evaluate** and sketch the *probability density* for the resulting state of the free particle. How does your result differ from the position probability of a *classical* free particle?

[b] **Evaluate** the *probability current density* for a state $\Psi(x, t)$ with spatial function given by Eq. (7.6.1) with $B = 0$. Provide a physical explanation for your result.

7.7 The Infinite Square Well in the "Classical Limit"

The stationary-state energies of the particle in a box are given by Eq. (7.56). Consider two adjacent energy levels, with quantum numbers n and $n + 1$.

[a] **Derive** an expression for the *fractional difference* in energy between these levels:

$$\frac{(\delta E)_n}{E_n} \qquad \text{where} \qquad (\delta E)_n \equiv E_{n+1} - E_n.$$

[b] **Discuss** the behavior of your result in the limit $n \to \infty$. Why is this limit called the "classical limit"?

CHAPTER 8

States of a Particle in One Dimension I

Mainly Piecewise-Constant Potentials

'Do you not think that there are things
which you cannot understand,
and yet which are;
that some people see things
that others cannot?'

—Dr. Van Helsing
in *Dracula*
by Bram Stoker

Enough abstract theory; it's time to solve problems. In Part II, while learning the rudiments of quantum physics, we examined two extremely simple systems: the free particle (Chap. 4 and § 6.6–6.7) and the particle in a box (Chap. 7). Each of these systems is an idealization that we can use to model real systems. For example, we can approximate a state of particles in a low-intensity beam by a state of the free particle, and a state of conduction electrons in a metallic solid by a state of an electron in an infinite square well.

The free particle was the focus of Chaps. 4 and 6. In Chap. 4, we deduced the form of the free-particle wave function: a *wave packet*. Then, in Chap. 6, we explored the *motion* of a free particle—the evolution of this wave function—and discovered that its probability density moves and spreads. Since such a particle is unconfined, its states are unbound.

The particle in a box was the centerpiece of Chap. 7. In this system, the "walls" of the infinite square well at $x = \pm L/2$ constrain the particle's motion. Whether in a stationary or a non-stationary state, it cannot escape from the well, from the interval $[-L/2 \leq x \leq L/2]$. Its states are bound.

But these examples are extreme idealizations. Potentials in the real world are neither zero everywhere nor infinite anywhere. And typically, realistic potentials have both *bound states* (like those of the particle in a box) and *unbound states* (like those of the free particle). (In the parlance of quantum physics, unbound states are called *continuum states* or *scattering states*.) In this chapter and the next, we shall begin the study of the general properties of these two classes of quantum states.

You will notice almost immediately that in these chapters we shift gears, emphasizing problem-solving strategies and real-world applications rather than historical sidelights and theoretical formulations. (Enthusiasts of theory have only to wait for Part IV, where we return to the formalism of quantum mechanics.) Nevertheless, our focus will remain on such physical questions as what kinds of quantum states are found in nature, how can we understand their physical properties, and what do these properties tell us about the microworld?

Our quest for answers to these questions will lead us through a general discussion of the nature of bound and continuum states (§ 8.2), a close look at continuum states (§ 8.3), and a step-by-step analysis of how to solve the Schrödinger Equation for a special, restricted class of potentials: those that are *piecewise constant* (§ 8.4–8.5). (We'll master the more challenging case of a continuous potential in Chap. 9.) We then launch into two famous examples: tunnelling through a potential barrier (§ 8.6) and bound states of a *finite* square well (§ 8.7 and 8.8). We close (§ 8.9) with a glimpse at a research problem in which piecewise-constant models find application.

8.1 WHAT DO ONE-DIMENSIONAL PIECEWISE-CONSTANT POTENTIALS HAVE TO DO WITH THE REAL WORLD?

In the next few chapters we'll study in detail several one-dimensional single-particle systems. But *real* physical systems are, of course, three-dimensional: their state functions depend on three spatial variables (*e.g.*, the Cartesian coordinates x, y, and z) and must satisfy the three-dimensional Schrödinger Equation. So you may be wondering: are the one-dimensional wells and barriers of these chapters merely hyped-up examples or are they relevant? To find out, let's look briefly at three-dimensional quantum mechanics.

The problem posed by a *three-dimensional single-particle system* is by now familiar. If the system is conservative, its potential is independent of time. So its Hamiltonian (in Cartesian coordinates) is

$$\hat{\mathcal{H}} = -\frac{\hbar^2}{2m}\left(\frac{\partial^2}{\partial x^2} + \frac{\partial^2}{\partial y^2} + \frac{\partial^2}{\partial z^2}\right) + V(x,y,z), \tag{8.1}$$

and its wave functions satisfy the TDSE

$$\left[-\frac{\hbar^2}{2m}\left(\frac{\partial^2}{\partial x^2} + \frac{\partial^2}{\partial y^2} + \frac{\partial^2}{\partial z^2}\right) + V(x,y,z)\right]\Psi(x,y,z,t) = i\hbar\frac{\partial}{\partial t}\Psi(x,y,z,t). \tag{8.2}$$

We can simplify Eq. (8.2) via the technique of *separation of variables*, just as we did the one-dimensional TDSE in § 7.1, because the separation of space and time does not depend on the *dimensionality* of the potential. The resulting separable solutions of (8.2) are products of a time factor $e^{-iEt/\hbar}$ and a spatial factor. The latter is the Hamiltonian eigenfunction, the solution of the three-dimensional TISE

$$\begin{aligned}&\hat{\mathcal{H}}\psi_E(x,y,z)\\ &= \left[-\frac{\hbar^2}{2m}\left(\frac{\partial^2}{\partial x^2} + \frac{\partial^2}{\partial y^2} + \frac{\partial^2}{\partial z^2}\right) + V(x,y,z)\right]\psi_E(x,y,z) = E\psi_E(x,y,z).\end{aligned} \tag{8.3}$$

For many three-dimensional systems we can reduce (8.3) to one (or more) *one-dimensional* equations. Under what physical conditions might such a simplification be possible? Well, suppose the forces on the particle act in only one dimension. For example, imagine a *classical* system that consists of two billiard balls connected by a spring whose force constant is so large that gravity is negligible. The restoring forces on these balls acts along their line of centers, which implicitly defines a single dimension. [The quantum-mechanical counterpart of this example is the vibration of two nuclei in a diatomic molecule (§ 9.6).] This system illustrates the point that *if the forces on a particle are uni-directional, then the potential energy that derives from these forces depends on only one spatial variable*. Thus, if $\mathbf{F} = F\,\hat{e}_x$, then $V = V(x)$. This happy circumstance enables us to dramatically simplify the TISE (8.3).

We effect this simplification by again invoking separation of variables. Just as in Chap. 7 we separated the t dependence of a (stationary-state) wave function $\Psi(x,t)$ from its x dependence, now we can separate the x dependence of the three-dimensional Hamiltonian eigenfunction $\psi_E(x,y,z)$ from its y and z dependence. The details of this marvelous simplification will be left for Part V; but here's a glimpse ahead.

Separation of Variables in the Three-Dimensional TISE

If the potential in Eq. (8.3) depends only on x, then there exist three-dimensional eigenfunctions of $\hat{\mathcal{H}}$ that are products of three *one-dimensional functions*, each of which is associated with a constant energy: *e.g.*, $\psi^{(x)}_{E_x}(x)$ with E_x. These **separated Hamiltonian eigenfunctions** look like

$$\psi_E(x, y, z) = \psi^{(x)}_{E_x}(x)\,\psi^{(y)}_{E_y}(y)\,\psi^{(z)}_{E_z}(z). \tag{8.4}$$

Each factor in this product eigenfunction depends on a single variable and satisfies a TISE in that variable. For instance, the x-dependent function in Eq. (8.4) satisfies the *one-dimensional TISE*

$$\left[-\frac{\hbar^2}{2m}\frac{d^2}{dx^2} + V(x)\right]\psi^{(x)}_{E_x}(x) = E_x\psi^{(x)}_{E_x}(x). \tag{8.5}$$

But what about the y- and z-dependent factors in (8.4)? Well, by hypothesis the forces act only in the x direction, so *the particle's motion in the y and z directions is that of a free particle.* Therefore, the corresponding factors in the Hamiltonian eigenfunction, $\psi^{(y)}_{E_y}(y)$ and $\psi^{(z)}_{E_z}(z)$, are *one-dimensional free-particle wave packets.*

So we can reduce the determination of the stationary states of a three-dimensional system with uni-directional forces to the problem of solving a one-dimensional Schrödinger Equation.[1]

$$\boxed{\begin{array}{c}\text{3-D TISE}\\ \hat{\mathcal{H}}\psi_E(x,y,z) = E\psi_E(x,y,z)\end{array}} \longrightarrow \boxed{\begin{array}{c}\text{free-particle functions}\\ \text{for } y \text{ and } z\end{array}} + \boxed{\begin{array}{c}\text{1-D TISE}\\ \hat{\mathcal{H}}\psi_E(x) = E\psi_E(x)\end{array}} \tag{8.6}$$

The simple separation (8.4) is not the only (or the most common) way to reduce the number of variables in the Schrödinger Equation. But other methods must wait for Part V.

Question 8–1

Fill in the steps leading from Eq. (8.4) to (8.5) and thereby discover the relationship between the total energy of the particle, E, and the three constants E_x, E_y, and E_z introduced in Eq. (8.4).

On Piecewise-Constant Potentials

In this chapter we'll consider conservative, one-dimensional single-particle systems whose potentials are piecewise constant. A **piecewise-constant potential** is one that is constant for all values of x except at a finite number of discontinuities—points where it changes from one constant value to another. You can see a simple piecewise-constant potential in Fig. 8.1c, where the finite square well (which we'll come to know and love in § 8.7) is on display.

[1] Novice quantum mechanics are sometimes seduced by this wonderful simplification into thinking that the *state functions* for such a system depend only on a single spatial variable. Not true. Equation (8.4) shows us that even if the potential depends only on x, the *wave functions* depend on x and y and z. The simplification is that these dependences are those of a free particle.

Of course, real potentials are continuous, finite functions. But many widely-used *model potentials* are discontinuous at one or more points in space (see § 8.9). Provided the potential we're modeling changes rapidly only in one or more *very small* regions of space, such a model may be an accurate approximation.

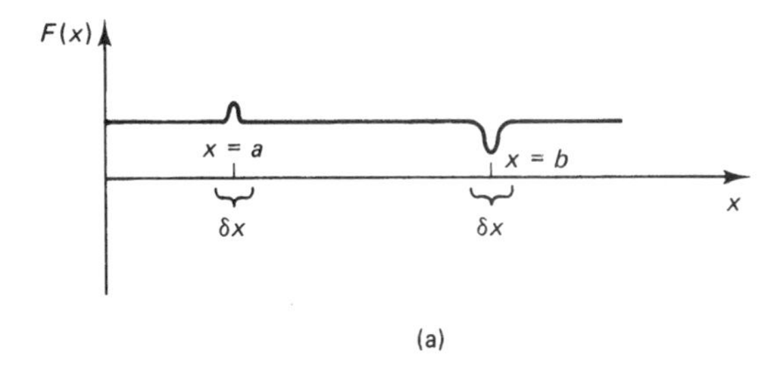

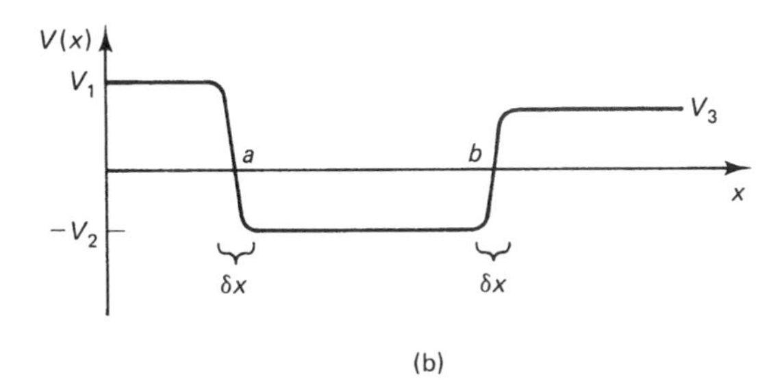

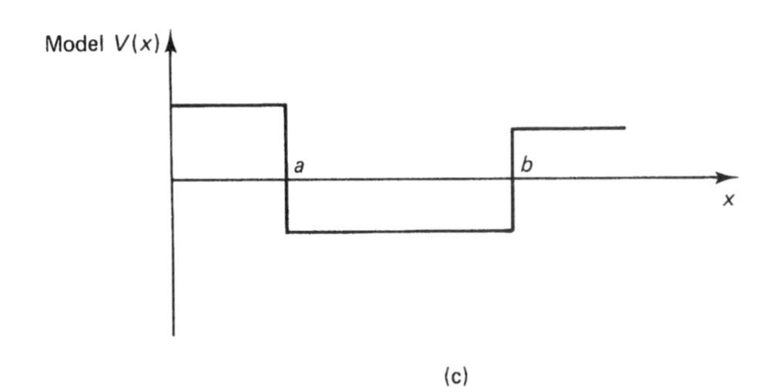

Figure 8.1 From impulsive forces to piecewise-constant model potential. (a) Uni-directional forces that change only in two very narrow regions, where they are impulsive. (b) The continuous potential energy corresponding to these forces. (c) A reliable model of the potential in (b), the TISE of which we'll solve in § 8.7.

Aside: More on the Validity of Piecewise-Constant Models. To keep piecewise-constant potentials in perspective, it's important to understand the particular physical conditions under which a real potential can be modeled accurately by a piecewise-constant function. The key to such insight is the *forces* that act in the system. The force in Fig. 8.1a is uni-directional and constant everywhere except in two small regions of size δx centered at $x = a$ and at $x = b$. In these regions, the force is impulsive. The resulting potential in Fig. 8.1b is constant everywhere except in these regions, where it changes abruptly. This potential is a prime candidate for modeling by a piecewise-constant approximation.

But the mere fact that the changes in $V(x)$ are abrupt does *not* mean that the model in Fig. 8.1c is a good approximation for all states. For example, the stationary states of the finite square well in this figure accurately approximate states of the potential in Fig. 8.1b only if the de Broglie wavelength of the particle is large compared to δx. This wavelength, λ, is the characteristic dimension of the state. If

δx is a small fraction of this dimension, then the details of *how* the potential changes in this region are irrelevant. To summarize this important guideline:

> ***Rule***
>
> An abrupt change in a potential can be replaced by a discontinuous change provided the de Broglie wavelength of the particle is large compared to the size δx of the region where the change occurs: $\lambda \gg \delta x$.

Question 8–2

Drawing on your understanding of classical wave phenomena (*e.g.*, a water wave passing through the entrance to a harbor), explain why a piecewise-constant model is an accurate approximation if δx is much smaller than λ.

Question 8–3

The particle-in-a-box is a common model of a system whose (continuous) potential corresponds to impulsive forces. As I remarked in Chap. 7, a good example of such a system is a conduction electron in a (solid) metal. [The forces on such an electron near the surface of the solid are (to a good approximation) impulsive.] Is this model likely to be a better approximation for a stationary state with a very high or a very low energy E?

Actually, one can use a piecewise-constant model even if the potential being modeled is continuous. For example, Fig. 8.2 shows a famous potential from molecular physics called **the Morse Potential**.[2] This potential pertains to *diatomic molecules*, such as H_2 or N_2 (*i.e.*, molecules with two nuclei): it is an approximate representation of the potential energy (due to the molecular electrons) in which these nuclei vibrate about their equilibrium separation. One way to solve the TISE for the Morse potential is to approximate it by a succession of constant "steps," as shown in the figure. Such a piecewise-constant model is called a **quadrature approximation** to $V(x)$. (In practice, we can more simply approximate the Morse potential by another continuous function, the simple-harmonic-oscillator potential of Chap. 9.)

In any case, if we can use a piecewise-constant potential, we are in luck, for this model renders the solution of the TISE comparatively trivial. As we'll see in § 8.4, we can simply write down the form of the Hamiltonian eigenfunction in each of the regions where $V(x)$ is constant. Then we "hook" these pieces together (at the points of discontinuity) using a straightforward, well-defined procedure.

Alas, life is rarely so simple. Many important and interesting potentials cannot be modeled accurately by a piecewise-constant approximation. And even if such a model is a good approximation for some states, it may be lousy for others. So we must augment our bag of problem-solving tricks with methods for subduing continuous potentials—which will come center stage in Chap. 9.

[2]The Morse potential was named after the man who dreamed it up, Philip Morse. You can find an account of how Morse arrived at the potential that bears his name by reading his delightful autobiography, *In at the Beginnings: A Physicist's Life* (Cambridge, Mass., M.I.T Press, 1977).

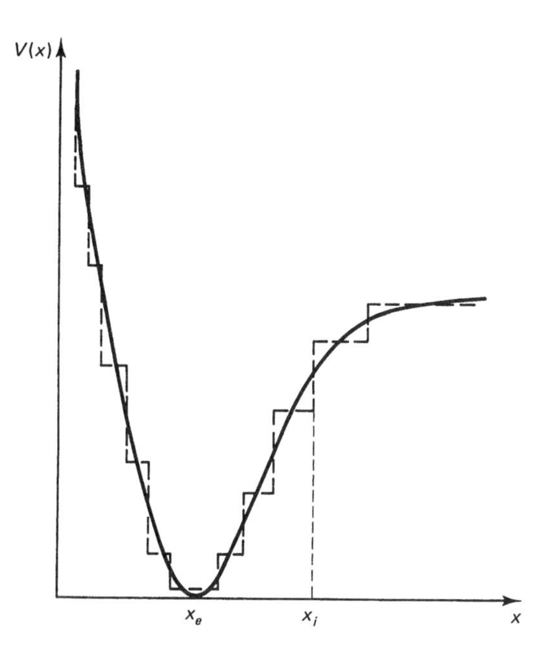

Figure 8.2 The Morse potential, an accurate model of the electronic potential of a diatomic molecule. With x denoting the separation of the nuclei, the form of this potential is $V(x) = D_e \left[1 - e^{-a(x-x_e)}\right]^2$, where D_e and a are parameters and x_e is the position of the minimum of the Morse well.

8.2 A DUPLICITY OF STATES: BOUND VERSUS UNBOUND STATES

Because no forces confine a free particle, it can roam throughout space. In quantum mechanics, this behavior is mirrored in the particle's wave functions. We learned in Chap. 6 that no matter where we look for a free particle, we might find it. That is, since such a particle is unconfined, the probability of detecting it at *any* location x could, at some time, be non-zero. That's why we call the states of a free particle **unbound states**.

A quite different situation confronts a particle in an infinite square well: because it is confined by the forces represented by the potential, the particle is trapped. At any time, the probability of detecting the particle at a position outside the well is zero. The states of a confined particle are, sensibly, called **bound states**.

In essence, the differences between bound and unbound states follow from their probability densities at very large values of $|x|$. To figure out whether a state is bound, we just ask: is the probability of finding the particle at a very large value of x ever non-zero? If not, then the particle is in a bound state.[3]

We can express these differences mathematically via the following conditions:

$$\boxed{|\Psi(x,t)|^2 \xrightarrow[x\to\pm\infty]{} 0 \qquad \text{at all times } t} \qquad \text{[bound state]} \tag{8.7}$$

$$\boxed{|\Psi(x,t)|^2 \xrightarrow[x\to\pm\infty]{} 0 \qquad \text{at any particular time } t} \qquad \text{[unbound state]} \tag{8.8}$$

[3]In one-dimensional scattering theory, the region $x \longrightarrow \pm\infty$ is called *the asymptotic region.* In three-dimensional scattering theory, the asymptotic region is defined in terms of the radial coordinate (of spherical coordinates r, θ, and φ): $r \longrightarrow \infty$.

Equation (8.8) is just a mathematical statement of an old friend: the spatial localization of an unbound wave packet (*i.e.*, the normalization condition). Note, by the way, that this condition is not limited to free particles; unbound states of a system may exist even if $V(x) \neq 0$.

Be sure you appreciate the difference between the conditions (8.7) and (8.8). The clue is in the qualifiers, "at all times" for bound states versus "at any particular time" for unbound states. The wave function of an unbound state is spatially localized *at any particular time*; but the wave function of a bound state is localized *in the same region of space (the vicinity of the potential) at all times.*

Question 8–4

Write Eqs. (8.7) and (8.8) as boundary conditions on the Hamiltonian eigenfunction $\psi_E(x)$.

Another important difference between a bound state and an unbound state is that the energy of a bound state is quantized, but the energy of an unbound state is not. [The boundary condition (8.7) leads to energy quantization for any bound stationary state (see, for example, the infinite square well in Chap. 7); the condition (8.8), however, has no such consequences.] Thus an energy measurement on an ensemble of particles in a bound state will yield only energies drawn from a discrete list, the eigenvalues: $E_1, E_2, E_3, \ldots$ of the Hamiltonian. In contrast, an energy measurement on an ensemble in an unbound state may yield any of a "continuum" of values. That is why unbound states are called **continuum states.**

What bound and continuum states have in common is, of course, the time-dependent Schrödinger equation. Both types of wave functions satisfy

$$\left[-\frac{\hbar^2}{2m}\frac{\partial^2}{\partial x^2} + V(x)\right]\Psi(x,t) = i\hbar\frac{\partial}{\partial t}\Psi(x,t). \tag{8.9}$$

Aside: Review. Before looking more closely at the properties of bound and continuum states, let's summarize their basic physics:

1. A particle in a **bound state** is confined by its potential. Although there may be a finite probability of detecting the particle far from the region of the potential, that probability dies to zero as $x \to \pm\infty$. As a consequence of these boundary conditions, the energies of *stationary* bound states are quantized.
2. A particle in a **continuum state** is unconfined. Although its state function is affected by the potential, the particle can escape the confines of that potential. Therefore it is possible to detect the particle in the asymptotic region.

To complement this summary, I've gathered together in Table 8.1 the properties of bound and continuum states we deduced in earlier chapters.

TABLE 8.1 BOUND VS. CONTINUUM STATES

	Bound State	Continuum State
energies	E_n are quantized	$E > 0$ are not quantized
degeneracy	E_n are non-degenerate	E are 2-fold degenerate
stationary-state	$\Psi(x,t) = \psi_n(x)e^{-iE_nt/\hbar}$	$\Psi(x,t) = \psi_E(x)e^{-iEt/\hbar}$
spatial function	$\psi_n(x)$ is normalizable	$\psi_E(x)$ is not normalizable
boundary conditions	$\psi_n(x) \xrightarrow[x\to\pm\infty]{} 0$	$\lvert\psi_E(x)\rvert \leq \infty$ (all x)
general state function	$\sum_n^\infty c_n\psi_n(x)e^{-iE_nt/\hbar}$	$\frac{1}{\sqrt{2\pi}}\int_{-\infty}^{\infty} A(k)e^{i(kx-\omega t)}\,dk$

Bound and Continuum States Together

In one important respect, the zero potential of a free particle, which supports only continuum states, and the infinite square well, which supports only bound states, are anomalous. *Most microscopic systems support both bound and continuum states.* In Fig. 8.3a, I've sketched a typical (continuous) potential and its continuum and bound-state energies. The collection of all energies (bound and continuum) of a system is called its **spectrum**.

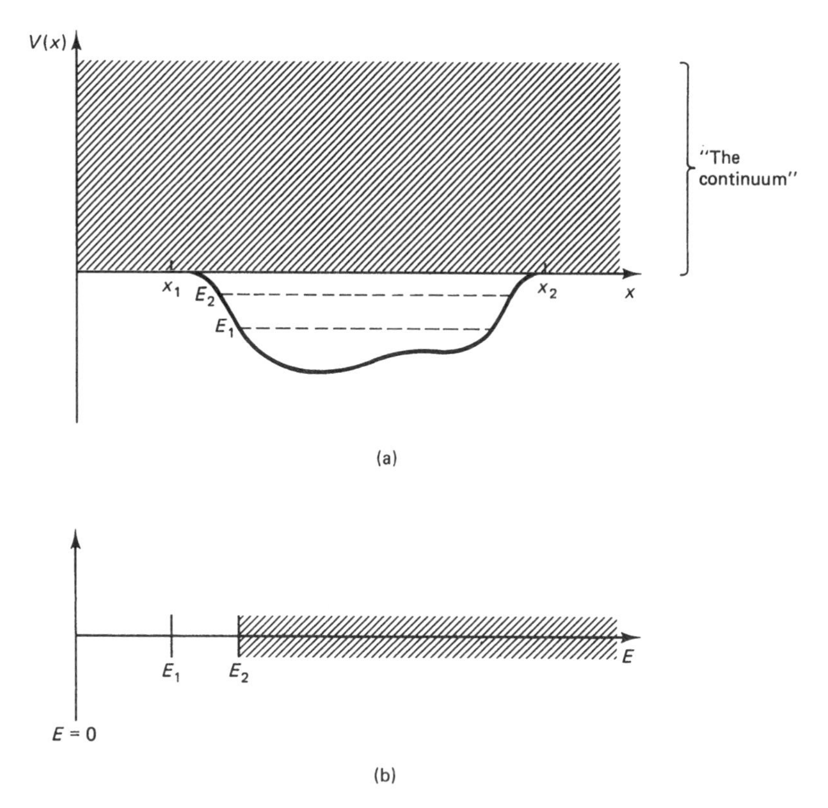

Figure 8.3 (a) A prototype of a one-dimensional finite-range potential. The dotted lines show the bound stationary-state energies, which in (b) are shown on an energy-level diagram. Above the x axis is the continuum of energies available to unbound states.

The potential in Fig. 8.3a is one of an important class of *finite-range potentials*. This name is a bit misleading; it doesn't really mean that the range of the potential is finite—*i.e.*, it doesn't imply that there exists a value of x, say $x = a$, such that $V = 0$ for $x > a$. In quantum physics, a **finite-range potential** is one that dies to zero as the spatial coordinate approaches infinity:

$$V(x) \xrightarrow[x \to \pm\infty]{} 0 \qquad \text{finite-range potential} \tag{8.10}$$

The potential in Fig. 8.3 satisfies this condition, because $V(x)$ is vanishingly small for $x \ll x_1$ and $x \gg x_2$. We've discussed the general properties of the bound and continuum states supported by such a potential, but it's worth taking a minute to review the *kinds of bound states* such a potential might support.

Aside: The Classical Connection. If the Hamiltonian of the microscopic system we're studying has a classical counterpart, then we can determine if the system supports bound states, continuum states, or both. (In many cases, we can figure out the classical counterpart of the quantum-mechanical Hamiltonian by simply making the masses macroscopic.) If there exist *bound orbits* of the classical Hamiltonian, then *bound states* of the corresponding quantum-mechanical Hamiltonian exist.[4]

Question 8–5

(a) Deduce whether each of the following systems supports bound states, continuum states, or both, justifying your answer with physical arguments: an oscillator with potential $V(r) = kr^2/2$; a microscopic planet moving according to the Kepler potential $V(r) = -\kappa/r$ (where κ is a constant). (b) Describe the spectrum (discrete? continuous? both?) of these systems.

As we know from Part II, bound states come in two flavors: stationary and non-stationary. A *stationary* state is characterized by a sharp (quantized) energy, so the uncertainty ΔE is zero (see the dotted lines in Fig. 8.3.) A stationary-state wave function is separable in space and time, and the corresponding probability density is independent of time.

But a *non-stationary* state must be represented by a wave packet, which is not separable. The probability density of such a state changes with time. (Recall, for example, the non-stationary state of the infinite square well of § 7.6.) And the energy of a non-stationary state is not sharp, so $\Delta E > 0$ at any time. But, like a stationary bound state, a non-stationary bound state is forever localized in the vicinity of the potential.

To illustrate the nature of the states of some more-or-less realistic one-dimensional finite range potentials, consider Fig. 8.4. The potential in Fig. 8.4a arises from the Coulomb forces between two oppositely-charged particles, such as the electron and proton in a hydrogen atom. In Volume II, we'll find that this potential supports an *infinite* number of bound states, as well as a continuum of unbound states. An electron in a continuum state of this potential is one that has been ionized—*i.e.*, it is localized an "infinite distance" from the proton, having left behind a bare, positively-charged proton.) By contrast, the purely repulsive potential in Fig. 8.4b, which resembles the interaction potential of atoms in a molecule whose electrons are in an excited state, supports no bound states. Indeed, there are no stationary states of this system. Finally, the potential in Fig. 8.4c represents an alpha particle (He^{++}) in the nucleus of a helium atom. This potential supports a finite number of stationary bound states and an infinite number of continuum states.

Most practicing physicists study both bound and continuum states. But some specialists emphasize one or the other type: spectroscopists study principally bound states, and scattering theorists (*e.g.*, your author) concentrate on continuum states. The kinds of questions these scientists ask and the way they seek answers are fundamentally different. These differences mirror the contrasts between the physics of bound and unbound states. For example, a typical calculation by a molecular *scattering* theorist might entail determining the probability that an atom colliding with a molecule will be scattered in a particular direction. A typical problem for a molecular *structure* theorist, on the other

[4] You remember *orbits* of classical particles from freshman physics, don't you? If not, you might want to review, say, Chap. 8 of *Classical Dynamics of Particles and Systems*, 2nd ed. by Jerry B. Marion (New York: Academic Press, 1970).

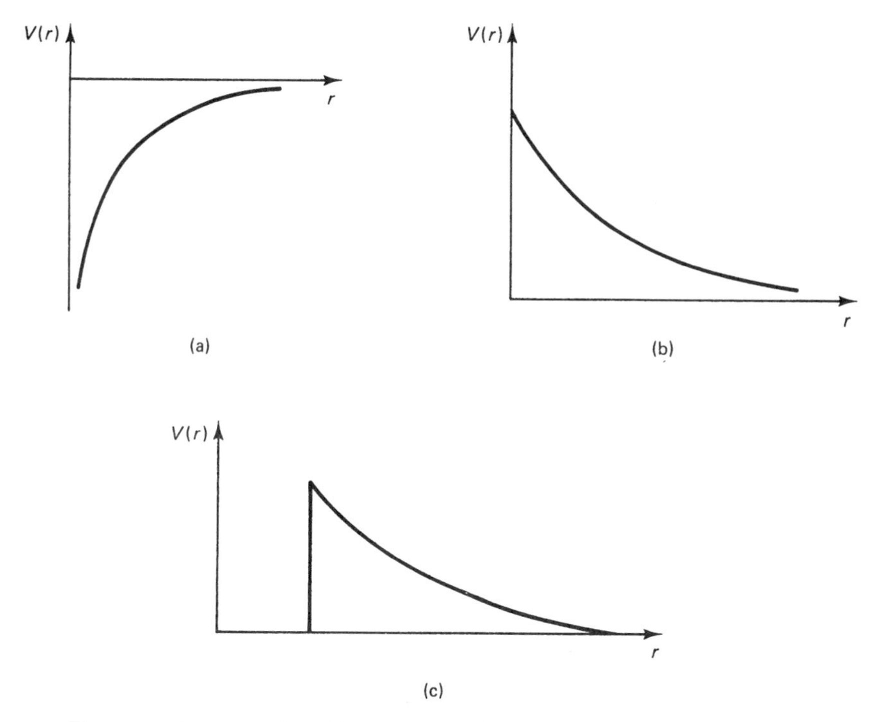

Figure 8.4 A sampler of one-dimensional model potentials. (a) The Coulomb potential between an electron and a proton. (b) A purely repulsive potential, patterned after a repulsive state of the CH molecule. (c) A potential well formed by nuclear and Coulomb forces that traps an alpha particle inside a Helium nucleus. [See Chap. 4 of *Elements of Nuclear Physics* by W. E. Meyerhoff (New York, McGraw-Hill, 1967).] The process of alpha decay proceeds via *tunnelling* (§ 8.6).

hand, might entail calculating the stationary-state energies of an electron in a molecule. The experimentalists across the hall might be determining scattering data by colliding a beam of atoms into a beam of molecules or by measuring structural information, such as bound-state energies, using the techniques of spectroscopy: *e.g.*, measuring the energy of photons emitted when a hydrogen molecule undergoes a transition between stationary states.

In this chapter we shall embrace both worlds, examining bound and continuum states (for rather simple systems). For future reference, then, let's state our goals:

Goals

1. **For the Study of Bound States:** Solve the TISE for the stationary-state energies and wave functions—the eigenvalues and eigenfunctions of the Hamiltonian—by imposing the boundary conditions appropriate to a bound state.
2. **For the Study of Continuum States:** Use the Schrödinger Equation to calculate the probability that a particle of known (but uncertain) energy will (at large times t) be found at a large positive value of x *or* at a large negative value of x.

*8.3 PARTICLES IN COLLISION: AN INTRODUCTION TO QUANTUM SCATTERING

> There is reason that all things
> are as they are,
> and did you see with my eyes
> and know with my knowledge,
> you would perhaps better understand.
>
> —Count Dracula
> in *Dracula*
> by Bram Stoker

When a microscopic particle is scattered by a microscopic target, the particle's linear momentum (and other properties, such as energy) may be changed. In quantum (as in classical) physics, the complicated interaction of a projectile and a target is represented by a continuous potential energy. In this section, we'll model this interaction (for a collision in one dimension) by a piecewise-constant potential $V(x)$.[5]

By contrast to a quantum collision, let's ponder a *classical* collision. In particular, let's model the projectile-target interaction by the barrier in Fig. 8.5a. Suppose a *macroscopic* particle travels toward the target in the direction of positive x with a well-defined energy E. The outcome of the ensuing encounter depends on whether this energy is less than or greater than the barrier height V_0. If $E < V_0$ (*e.g.*, E_1 in Fig. 8.5a), the particle has insufficient kinetic energy to "scale" the barrier and is repelled—we say it is **reflected**. So after such a collision, the particle is travelling backward (in the $-x$ direction) with the same energy, E. In this *elastic* collision, only the *direction* of the projectile's linear momentum is changed. If, on the other hand, $E > V_0$ (*e.g.*, E_2 in Fig. 8.5a), the particle passes through the region where $V \neq 0$—we say it is **transmitted**. Its speed is reduced only while it is in the interval $[-L/2 \leq x \leq L/2]$, so after the collision, the particle is travelling forward (in the $+x$ direction) with energy E. Thus, in a classical collision there are two *mutually-exclusive* outcomes: either the projectile is reflected (if $E < V_0$) or it is transmitted (if $E > V_0$).

But in a quantum collision, possibilities abound.[6] In fact (§ 8.6), *regardless of the proximity of E to V_0, both reflection and transmission are possible*. Even if the mean energy of the projectile is significantly less than the barrier height, the particle may "penetrate (tunnel through) the barrier," appearing after the collision at a large *positive* value of x. And even if its mean energy is much greater than V_0, the particle may be reflected, appearing at a large *negative* value of x. Of these two phenomena, perhaps the most remarkable is tunnelling; we'll explore this quintessential quantal marvel in § 8.6.

[5]This section contains a qualitative justification for using continuum stationary states (*e.g.*, pure momentum functions) in the study of collisions. This justification is preceded by a time-dependent description of a quantum mechanical collision that you may want to skip. But if you do so, please review § 6.9 (especially Example 6.6) before proceeding to the next section.

[6]The energy of a microscopic projectile in an unbound state cannot be sharp. In Fig. 8.5b I've acknowledged this fact by a fuzzy blur around the energies $E_1 < V_0$ and $E_2 > V_0$. So E_1 and E_2 represent the *mean* energies (the expectation values) of the projectile in quantum states $\Psi_1(x,t)$ and $\Psi_2(x,t)$, for both of which $\Delta E > 0$.

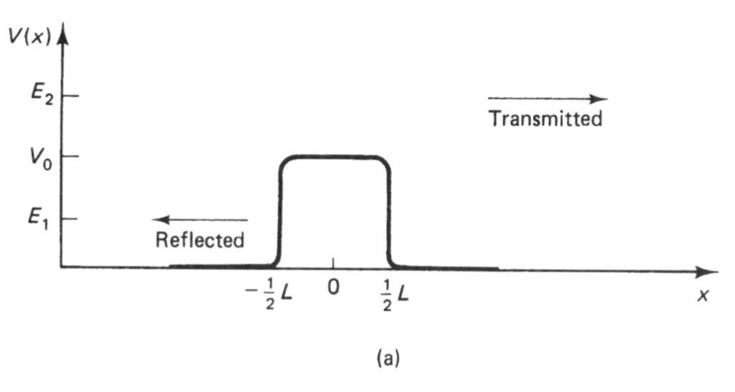

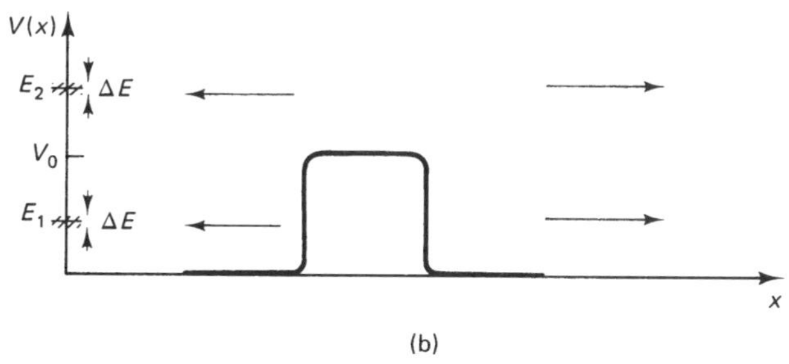

Figure 8.5 (a). A barrier potential of width L that represents the interaction of a macroscopic particle and a target in a *classical* collision. (b). The same for a microscopic particle of uncertain energy incident from the left on a microscopic target. Note that $E_1 = \langle E \rangle = \left\langle \Psi_1(x,t) \mid \hat{\mathcal{H}} \mid \Psi_1(x,t) \right\rangle$ and similarly for E_2.

A Microscopic Collision in a Macroscopic Laboratory

In Fig. 8.6a you'll find a (very) simple schematic of an experiment to study quantum collisions. This apparatus consists of a source, a target, and a detector. To be specific, suppose we want to study the scattering of electrons from a species of atom. As a **source** we would use an electron gun, like the one in the double-slit experiment (Chap. 2). This device produces a beam of *nearly monoenergetic* electrons of sufficiently low density that (to a good approximation) the particles in the beam do not interact with one another. The target in our experiment might be a bottle containing a gas of atoms, all of which (we'll assume) are in the same quantum state.[7] Finally, the detector, which is located far from the source and the target, is any device that can measure the number of electrons scattered at a particular angle and their energies. (One such device is an electron multiplier.)

In the experiment, we measure (a) properties of the projectiles in the beam that is emitted by the source; and (b) properties of the scattered particles that arrive at the detector. By so doing, we can investigate the structure and physical properties of the target and the nature of the interaction of microscopic particles. From such experiments has come much of our knowledge of the nature of matter—not to mention reassuring evidence that quantum mechanics is valid.

The Wave-Packet Description of a Microscopic Collision

To predict, confirm, or understand the results of a quantum collision experiment, we must calculate the probability that after the collision we would detect a (scattered) particle in the asymptotic region, where the detector is. Initially, all we know about the particle is its state function before the collision, $\Psi(\mathbf{r}, t)$.

[7]In practice, conditions like the temperature of the apparatus determine the distribution of states of the atoms in the gas. Another complication is the possibility of *multiple collisions*; *e.g.*, a projectile bounces off an atom only to collide with one or more additional atoms before it leaves the target region.

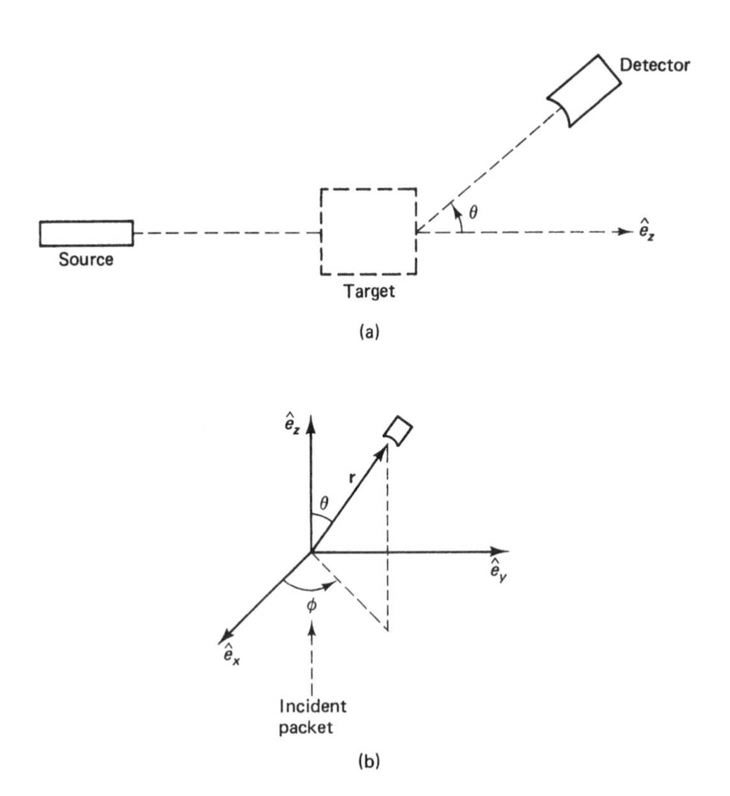

Figure 8.6 (a) An incredibly simplified idealization of a scattering experiment. Such experiments are described using the spherical coordinates shown in (b)

Let's denote the duration of the collision by the time interval Δt and suppose that this interval is centered around the **collision time** t_{coll}. So the particle leaves the source at $t_0 \ll t_{\text{coll}}$ and is detected at $t_d \gg t_{\text{coll}}$.[8]

Now, collisions are intrinsically *time-dependent phenomena.* So the most direct way to calculate the probabilities we need would be to solve the time-*dependent* Schrödinger equation for the wave packet that represents the state of the projectiles. We would initialize this equation with the function that describes the (known) state of the projectile at the initial time t_0, when the particle is localized in the vicinity of the source. We would then solve the TDSE for the state function at some time $t \gg t_{\text{coll}}$ and from the solution, evaluate the probability density for various locations of the detector.

[8]Ideally, we'd define the *beginning* of the interval Δt as "the time when the collision begins"—*i.e.*, the time at which the particle is sufficiently near the region of non-zero potential that its wave function begins to deviate from that of a free particle. And we'd define the *end* of the interval as the time at which the particle is sufficiently far from the potential that its wave function is altered no further. Unfortunately, this unambiguous definition becomes muddled if, as is often the case, the range of the potential $V(\mathbf{r})$ is not finite. A common instance of this irritating behavior is a potential that, as $r \to \infty$, dies to zero as a power of r. The preeminent example is the Coulomb potential, which dies to zero as $1/r$. We shall leave such troublesome ambiguities to more advanced treatments of quantum collisions. (See, for example, the excellent presentation in Chap. 2 of *Introduction to the Quantum Theory of Scattering* by L. S. Rodberg and R. M. Thaler (New York: Academic Press, 1967). For a very nice introduction at a lower level, see the qualitative discussion in § 9-11 of *An Introduction to Quantum Physics* by A. P. French and E. F. Taylor (New York: W. W. Norton, 1978). Come to think of it, all of Chap. 9 of this fine text would serve as useful supplementary reading for this chapter. You should also peruse the following papers, which supplement the discussions in this section: H. Eberly, *Amer. Jour. Phys.* **33**, 771 (1965); A. N. Kamal, ibid., **46** (1984); and I. Galbraith, Y. S. Ching, and E. Abraham, ibid., **60** (1984).)

This plan of attack sounds fine. But it leads to severe mathematical complications. In fact, these difficulties are so severe that we can't solve the TDSE except for very simple problems. Even for such a problem, the algebraic details of the solution would take us far astray. So here I want to show you the *qualitative* solution of one such problem: scattering by the potential barrier in Fig. 8.5b.

Suppose that before the collision, the particles in the incident beam are in a state that can be represented by a *Gaussian wave packet* (see Chaps. 4 and 6). For convenience, let's denote the initial time (which is some $t \ll t_{\text{coll}}$) by $t_0 = 0$. To conform to the experimental situation, the center of the initial wave packet, the expectation value $\langle x \rangle(0)$, is at a *large negative* value of x near the source (see Fig. 8.7a). To ensure that the particle is travelling to the right (toward the target), the mean momentum of the initial packet must be positive. (In fact, let's choose $\langle p \rangle (0) > \sqrt{2mV_0}$.) Finally, to ensure that the beam is nearly monochromatic, the initial packet must have a very small non-zero momentum uncertainty $(\Delta p)(0) \gtrsim 0$. (The mathematical form of a wave packet with these properties appears in Pblm. 4.6.) These and other properties of the incident wave packet are summarized in the box in Fig. 8.7a.

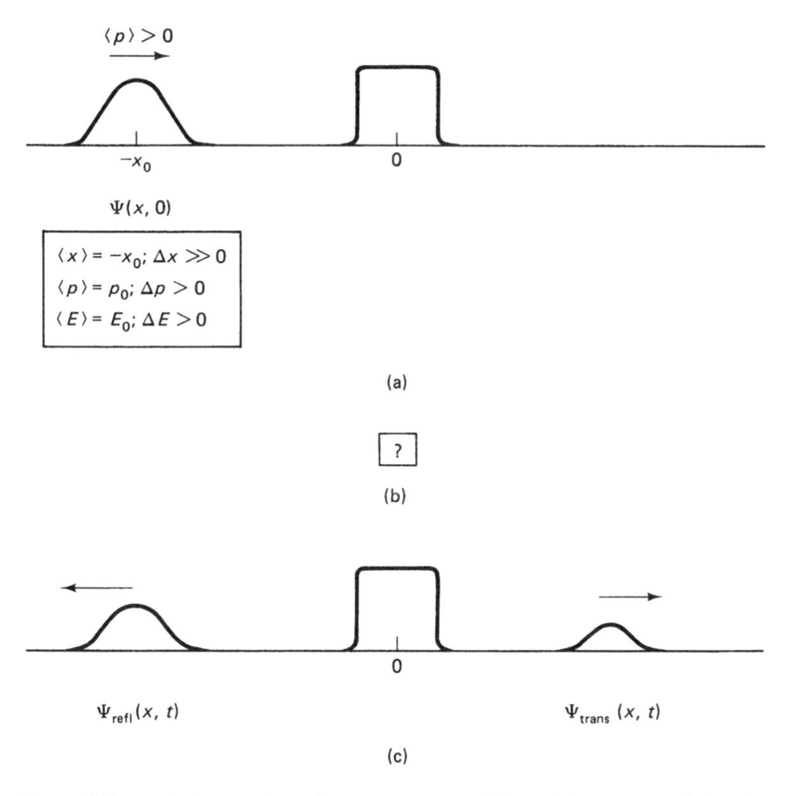

Figure 8.7 (a) Before and (c) after a quantum collision. What happens during the collision [(b)] is a mystery, but a clue appears in Fig. 8.8.

Now consider times $t > 0$. Until the incident particles near the target, the probability density of their state is localized in regions where $V(x) \approx 0$. So from t_0 to the "beginning of the collision" (a time $t \sim t_{\text{coll}}$), the projectiles propagate to the right *as free particles*—*i.e.*, according to § 6.7. The center of the probability density, for example, moves to the right at the group velocity, whose value we know to be $v_{\text{gr}} = \langle p \rangle (0)/m$.[9]

Eventually, the particles move into range of the potential, and their state functions begin to distort. The evolution of the wave packet during the collision (*i.e.*, during the interval Δt) is described by the TDSE (8.9) with the barrier potential of Fig. 8.7. What happens then is nothing short of remarkable. Figure 8.8 shows the results of a computer's solution of this equation for two barrier potentials of different widths. As a consequence of the incident particles' encounter with the barrier, the incident packet seems to "break up," forming two packets, one reflected, the other transmitted. But we in the laboratory cannot observe what happens *during* the collision, for the interaction takes place on a length scale of atomic dimensions, typically on the order of 10^{-10} m. So we wait.

Once the collision is "over" (*i.e.*, at some $t \gg t_{\text{coll}}$), we look to the detector to discover what happened.[10] After a large number of collisions have occurred, we tote up the fraction of particles that were reflected and the fraction that were transmitted. From these fractions, we then calculate the measured probability for reflection or transmission of a particle in the incident beam.

These "measured probabilities" agree with quantum-mechanical predictions based on the reflected and transmitted wave packets obtained from solution of the TDSE at large $t \gg t_{\text{coll}}$. So both theory and experiment predict that the probability density after the collision consists of two pieces: one due to the **reflected wave packet**, which I'll denote by $\Psi_{\text{refl}}(x,t)$, the other due to the **transmitted packet**, $\Psi_{\text{trans}}(x,t)$. The mean momentum of the *reflected* packet is negative (*i.e.*, it corresponds to particle motion in the $-x$ direction), and that of the *transmitted* packet is positive (motion in the $+x$ direction). *But neither of these packets individually represents the state of the scattered particles*—the wave function is the (unnormalized) superposition

$$\Psi(x,t) = \Psi_{\text{refl}}(x,t) + \Psi_{\text{trans}}(x,t). \qquad t \gg t_{\text{coll}} \tag{8.11}$$

After the collision, the reflected and transmitted packets propagate according to the free-particle Schrödinger equation, because, like the incident packet, they are localized in regions of zero potential.[11]

Your understanding of this collision as a quantum-mechanical phenomenon hinges on your grasp of the Born interpretation of the wave packets in (8.11). Each packet is a *probability amplitude*. The reflected packet, for instance, is the amplitude for the *possibility* that in a position measurement after the collision, we would detect a particle at some $x \ll -L/2$ moving in the $-x$ direction (towards the source)—a possibility we've described as reflection. The *probability* of this possibility actually occurring in an experiment can be calculated as $|\Psi_{\text{refl}}(x,t)|^2$ [although there is a much easier way to determine this information (*c.f.*, § 8.4)]. Note, by the way, that each particle in the

[9]The probability density also spreads a bit, but since wave packet spreading is not germane to our concerns here, I'll neglect it.

[10]Actually, we need two detectors. In our one-dimensional paradigm, both reflection and transmission are possible, so we put one detector at large negative x and one at large positive x. In a three-dimensional collision, reflection and transmission are but two of an infinity of possible outcomes, corresponding to all angles θ and φ in Fig. 8.6b. Reflection corresponds to $\theta = \pi$ and transmission to $\theta = 0$.

[11]This analysis holds until a scattered particle is detected. An encounter with a detector constitutes a *measurement* and actualizes one of the possibilities implicit in the wave function. Such a measurement therefore alters the state function (at the instant of detection), changing it into a function that is localized at the detector. For more on the fascinating subject of measurement in quantum theory see Chap. 13.

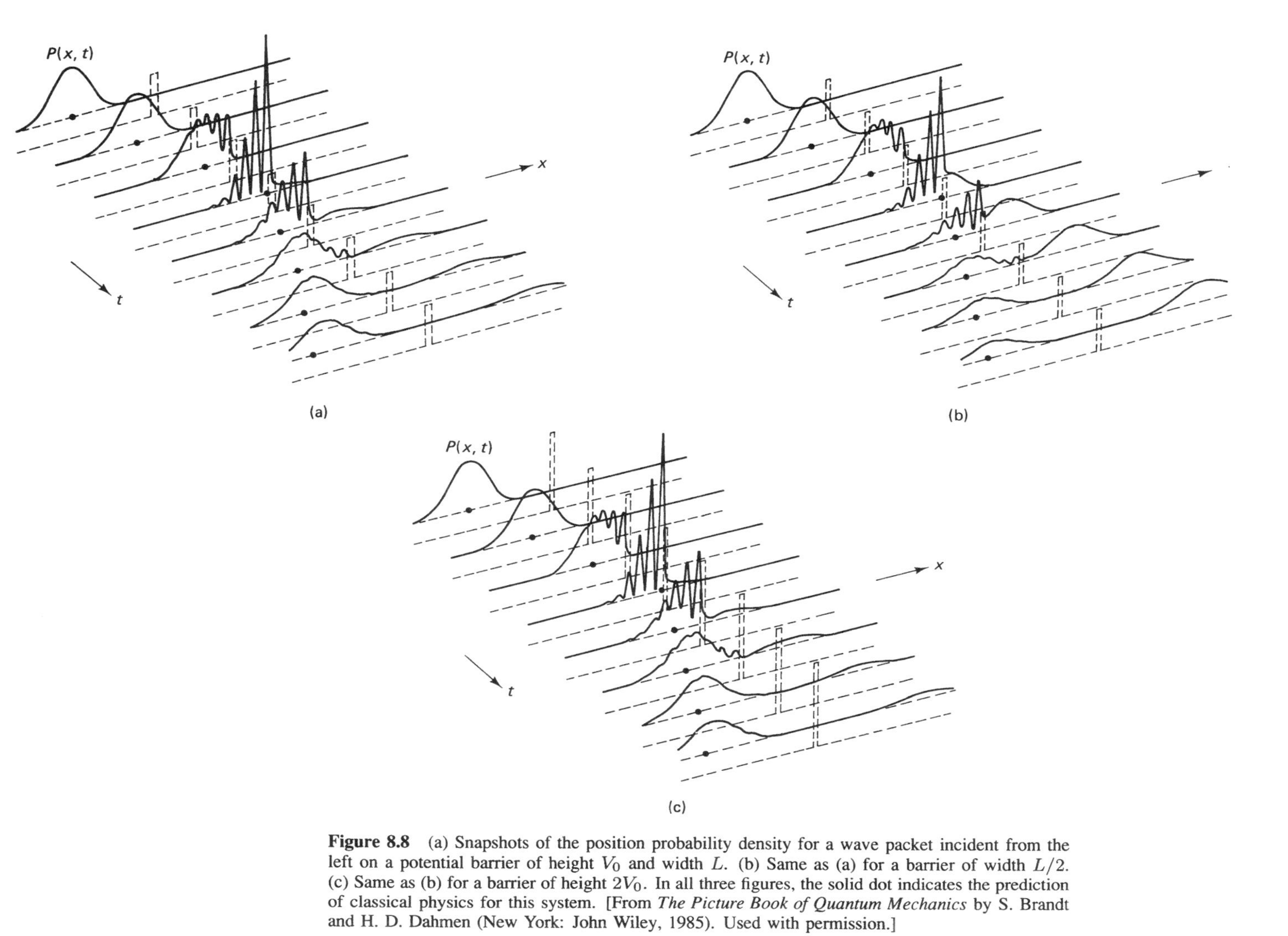

Figure 8.8 (a) Snapshots of the position probability density for a wave packet incident from the left on a potential barrier of height V_0 and width L. (b) Same as (a) for a barrier of width $L/2$. (c) Same as (b) for a barrier of height $2V_0$. In all three figures, the solid dot indicates the prediction of classical physics for this system. [From *The Picture Book of Quantum Mechanics* by S. Brandt and H. D. Dahmen (New York: John Wiley, 1985). Used with permission.]

incident beam is either reflected or transmitted. In non-relativistic quantum mechanics, particles do not divide.

Question 8–6

By my choice of the mean momentum of the initial packet, I arranged the above example so that the mean momentum *of the particles* after the collision is negative, *i.e.*,

$$\langle p \rangle(t) = \langle \Psi(x,t) \mid \hat{p} \mid \Psi(x,t) \rangle < 0 \qquad (t \gg t_{\text{coll}}). \tag{8.12}$$

What does this fact imply about the relative magnitude of the integrated probabilities for the reflected and transmitted packets?

Solving Collision Problems with Continuum Stationary States

Few practicing quantum physicists carry out the procedure I've just sketched. The numerical solution of the TDSE for a real collision remains impossible even on present-day supercomputers—which sad fact would seem to make the study of real scattering processes impossible. Happily, there is a way out of this conundrum.

This gambit requires relaxing *very slightly* the normalizability requirement on state functions. To facilitate calculation of scattering probabilities, physicists have admitted into quantum mechanics a class of functions called *continuum stationary-state functions.* These functions are finite everywhere, but they are not normalizable. Hence a continuum stationary-state function is not, strictly speaking, physically admissible. Nevertheless, we can use such functions to solve scattering problems, *provided we are very careful about their use and interpretation.*[12]

You've already seen a continuum stationary-state function: the *pure momentum function*

$$\Psi_p(x,t) \;=\; A\, e^{i(kx-\omega t)}, \tag{8.13a}$$

which I used in § 6.9 to represent the state of particles in a beam. For instance, in Example 6.7 we learned how to choose the constants in this form to represent a beam of 0.01 eV particles with a current of 3.0 nanoamp–cm^2, *viz.*,

$$\begin{aligned} A &= 3.161 \times 10^{-4}\,\text{m}^{-3/2}, \\ k &= 1.621 \times 10^{9}\text{m}^{-1}, \\ \omega &= 1.520 \times 10^{14}\,\text{sec}^{-1}. \end{aligned} \tag{8.13b}$$

And we found that the *probability current density* for this beam is $j = 1.873 \times 10^{14}\,\text{sec}^{-1}$.

I mention the probability current density because, beginning in the next section, this quantity will play a crucial role in our study of quantum collisions. There we'll discover how to calculate reflection and transmission probabilities from current densities for the various wave packets shown in Fig. 8.5. But before turning to problem solving, let's look further at the use in quantum collision theory of functions that are not physically admissible.

Consider again the incident packet in the collision experiment of Fig. 8.7. I defined the initial parameters of this packet (which are shown in the figure) so as to minimize

[12]One can come to grief, for example, calculating expectation values of some observables with a continuum stationary-state function, for such expectation values may be infinite. And the integrated probability density (over all space) for a continuum stationary state is infinite; so this quantity cannot be interpreted as in § 3.4.

the energy uncertainty at $t_0 = 0$, $(\Delta E)(0)$. For a free particle, this quantity is related to the position uncertainty by

$$(\Delta E)(0) = \frac{p}{m}\Delta p(0) \geq \frac{\hbar}{2}\frac{p}{m}\frac{1}{\Delta x(0)}. \tag{8.14}$$

Nature won't let us produce a beam with $(\Delta E)(0) = 0$, so we must settle for $(\Delta E)(0) \ll 1$. But according to Eq. (8.14), this condition implies that the packet is very fat: $(\Delta x)(0) \gg 1$, *i.e.*, we know little about the incident particle's position. [We don't need to know the initial position to analyze our experiment, provided the particle is initially far from the target, which we guarantee by choosing $\langle x \rangle (0)$ to be of macroscopic size, say $\langle x \rangle (0) = 10\,\text{m}$.]

So the incident wave packet is a superposition of pure momentum states (8.13a), with momenta chosen so the momentum probability amplitude is narrowly peaked about p_0. An excellent *approximation* to this packet is the limit of an *infinitely large packet*: a pure momentum state with momentum $p_0 = \langle p \rangle (0)$ and energy $E_0 = p_0^2/2m$. To summarize this argument:

$$\boxed{\text{wave-packet description}} \quad \xrightarrow[(\Delta x)(0)\to\infty]{(\Delta E)(0)\to 0} \quad \boxed{\text{continuum stationary state description}}$$

Thus, *a continuum stationary-state function is an idealized model of a huge wave packet, with a very small spread of energies about* E_0. In the $(\Delta E)(0) \to 0$ limit, we can approximate the state of the particles in the beam by this monoenergetic continuum stationary-state function.

Question 8–7

Thought question: The energy uncertainty is related to the duration of the collision by the *energy-time uncertainty relation* (§ 11.8), $(\Delta E)(\Delta t) \geq \hbar/2$. Based on this relation, discuss the consequences for our experiment of the requirement that $(\Delta E)(0) \gtrsim 0$.

Question 8–8

With a little imagination, one can dream up a classical analogue of the time-independent picture of a microscopic collision. Consider the Rock of Gibraltar in a hurricane. Water waves (of considerable amplitude) scatter from the Rock. To detect them, we put the Queen Mary nearby and measure her vertical displacement. Consider first a short incident group of waves that encounter the Rock at t_{coll}. Sketch the wave pattern in the ocean for $t \gg t_{\text{coll}}$. Now suppose a *huge* wave group is (continuously) incident on the Rock; sketch the steady-state condition that obtains at $t \gg t_{\text{coll}}$. How is this classical example analogous to the quantum collisions of this section?

8.4 SOLVING THE TISE FOR A PIECEWISE-CONSTANT POTENTIAL

Although the details of how to solve bound- and continuum-state problems differ, certain characteristics are shared by both kinds of states. What makes solving the TISE for a piecewise-constant potential so easy is the simple *analytic* form of the "pieces" of the Hamiltonian eigenfunction in each region of space where the potential is constant. The particular functional form of these pieces depends on whether the energy of the state is less than or greater than the value of the potential in the region, but in either case, the pieces of $\psi_E(x)$ are easy to write down. Indeed, the only non-trivial part of solving the problem is hooking the pieces together.

The piecewise-constant nature of the potentials we'll consider in the rest of this chapter allow us to develop a simple five-step procedure for solving the TISE for such a system. Before we explore this procedure, here's an overview, to which you can refer as you work through this section:

Procedure for Piecewise-Constant Potentials.

Step 1. Divide space ($-\infty < x < \infty$) into *regions* according to the values of $V(x)$. In each region, $V(x) = $ constant. Then classify each region as *classically allowed* or *classically forbidden* according to the energy of the state.

Step 2. Write down the general solution to the TISE in each region.

Step 3. Impose physical conditions (as appropriate) on the pieces of $\psi_E(x)$ in the *first* and *last* regions. These conditions will depend on whether the state is bound or unbound: for bound states, they ensure that the eigenfunction goes to zero as $x \to \pm\infty$; for unbound states, they correspond to circumstances in the laboratory.

Step 4. Hook the pieces of the eigenfunction together, using *continuity conditions*.

Step 5. Finish the problem (methodology differs for bound and unbound states):

Bound States: Impose boundary conditions (usually in the limits $x \to \pm\infty$) to obtain *quantized energies*. Then normalize the Hamiltonian eigenfunctions.

Unbound States: Evaluate the *probability current densities* in the first and last regions and from them the probabilities for transmission and reflection.

Regions and How to Classify Them

To *define* the regions of a piecewise-constant potential we need only look at it. But to *classify* these regions, we must consider the energy of the state we are interested in. Thus the definition of a region is independent of the state; its classification is not.

A **region** is simply a finite (or semi-infinite) range of x in which the potential is constant. Thus, in the qth region

$$V(x)\Big|_{\text{region } q} = V_q \qquad [x_{q-1} \le x \le x_q]. \tag{8.15}$$

(We allow for semi-infinite ranges, because the first region includes the limit $x \to -\infty$ and the last includes $x \to +\infty$.) This definition is illustrated in Fig. 8.9.[13]

The equation we must solve in region q is the TISE

$$\left(-\frac{\hbar^2}{2m}\frac{d^2}{dx^2} + V_q\right)\psi_E^{(q)}(x) = E\psi_E^{(q)}(x) \qquad [x_{q-1} \le x \le x_q]. \tag{8.16}$$

Clearly, the TISE in a particular region may differ from that in any other region. So, like Frankenstein's monster, the *Hamiltonian eigenfunction* for energy E is assembled out of

[13]For consistency and convenience, I have chosen the zero of energy *in this section* so that all energies and potentials are positive: $E > 0$ and $V_i > 0$. But for some problems (*e.g.*, the finite square well of § 8.9) it's more convenient to choose the zero of energy at the *top* of a potential well. The latter choice is analogous to the convention used in atomic physics.

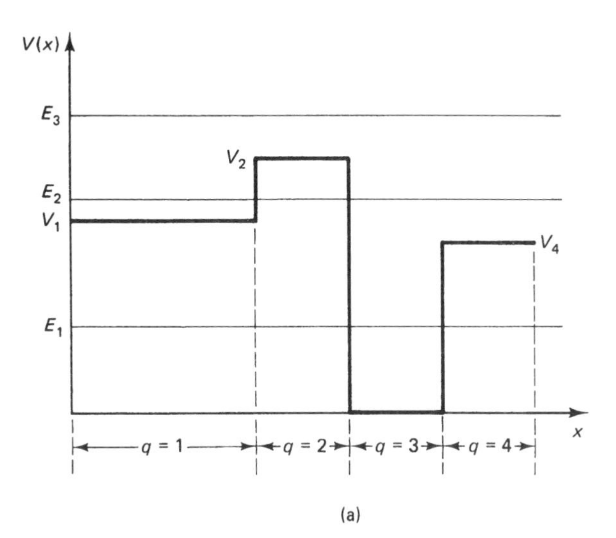

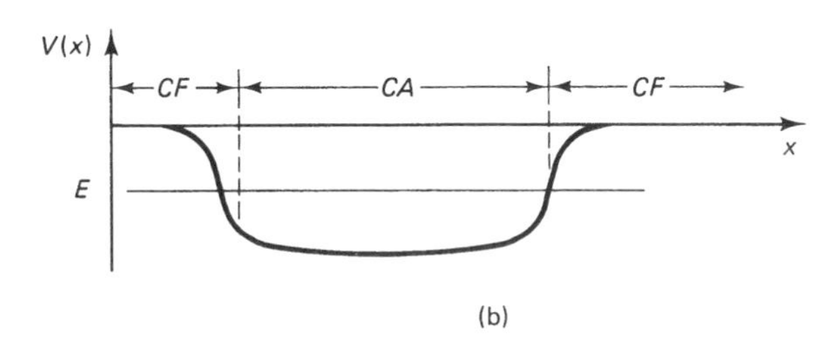

Figure 8.9 How to define and classify regions for (a) a piecewise-constant potential with 4 regions, and (b) a continuous potential. The classifications of regions for the potential in (a) can be found in Table 8.2.

pieces, one per region. The relationship of the *piece* $\psi_E^{(q)}(x)$ to the full eigenfunction $\psi_E(x)$ is simple but crucial:

$$\psi_E(x) = \psi_E^{(q)}(x) \qquad \text{for } [x_{q-1} \le x \le x_q]. \tag{8.17}$$

For example, the system in Fig. 8.9a has four regions, so we construct its Hamiltonian eigenfunction as

$$\psi_E(x) = \begin{cases} \psi_E^{(1)}(x) & -\infty < x \le x_1 \\ \psi_E^{(2)}(x) & x_1 \le x \le x_2 \\ \psi_E^{(3)}(x) & x_2 \le x \le x_3 \\ \psi_E^{(4)}(x) & x_3 \le x < \infty. \end{cases} \tag{8.18}$$

For a more general potential with N regions, we assemble $\psi_E(x)$ as[14]

[14]Some students find "segmenting" the eigenfunction in this way confusing, probably because they try to ascribe overmuch physical significance to the pieces $\psi_E^{(q)}(x)$. If you prefer, just think of this segmentation as a problem-solving strategy. The point to remember is: when all is said and done, physical information about the state is contained in the Hamiltonian eigenfunction $\psi_E(x)$. This procedure is similar to one in the study of transmission and reflection of electromagnetic waves through media with different indexes of refraction. (See, for example, Chap. 8 of *Introduction to Electrodynamics* by D. J. Griffiths (Englewood Cliffs, N. J.: Prentice Hall, 1981) or § 4.3 of *Optics*, 4th ed. by E. Hecht and A. Zajac (Reading, Mass: Addison-Wesley, 1979).

$$\psi_E(x) = \begin{cases} \psi_E^{(1)}(x) & -\infty < x \le x_1 \\ \psi_E^{(2)}(x) & x_1 \le x \le x_2 \\ \cdot & \\ \cdot & \\ \cdot & \\ \psi_E^{(N)}(x) & x_{N-1} \le x < \infty. \end{cases} \tag{8.19}$$

Now that we've identified the regions, we must classify them. We classify each region according to the *kinetic energy* that a particle with energy E (the energy of the state we are interested in) would have were it in the region. Since $E = T + V$, the kinetic energy $T = E - V_q$ in region q would be positive or negative depending on the magnitudes of E and V_q. The two possible cases, $E \ge V$ and $E < V$, are dubbed **classically allowed** (CA) and **classically forbidden** (CF), because a (classical) particle would be "forbidden" from a region where its kinetic energy would be zero. We also use this classification scheme in studying continuous potentials $V(x)$ (Chap. 9), so I'll summarize it here for this general case:

$$T = E - V(x) \quad \begin{cases} < 0 & \text{if } E < V(x) & \text{classically forbidden region} \\ \ge 0 & \text{if } E \ge V(x) & \text{classically allowed region} \end{cases}$$

This scheme is illustrated for the potential of Fig. 8.9a in Table 8.2. A preview of its application to continuous potentials appears in Fig. 8.9b.

TABLE 8.2 CLASSIFICATIONS OF REGIONS FOR THE POTENTIAL IN FIG. 8.9a.

Energy	Region 1	Region 2	Region 3	Region 4
E_1	$E_1 < V_1$ CF	$E_1 < V_2$ CF	$E_1 > V_3$ CA	$E_1 < V_4$ CF
E_2	$E_2 > V_1$ CA	$E_2 < V_2$ CF	$E_2 > V_3$ CA	$E_2 > V_4$ CA
E_3	$E_3 > V_1$ CA	$E_3 > V_2$ CA	$E_3 > V_3$ CA	$E_3 > V_4$ CA

The Hamiltonian Eigenfunction in a Classically-Allowed Region

Our next step is to solve the TISE in each region. The solutions in all CA regions have the same functional form, as do those in all CF regions. So we'll consider each case separately, starting with the CA regions.

But first, let's rearrange terms in the TISE (8.16) as

$$\frac{d^2}{dx^2}\psi_E^{(q)}(x) = -\frac{2m}{\hbar^2}(E - V_q)\psi_E^{(q)}(x) \qquad [x_{q-1} \le x \le x_q]. \tag{8.20}$$

Equation (8.20) is just the TISE *for a free particle* with E replaced by $E - V_q$. Whether region q is CA or CF, this equation has the simple form

$$\frac{d^2}{dx^2} \text{ function of } x = \text{constant} \times \text{function of } x. \tag{8.21}$$

This form makes it easy to write down the solutions in each case.

Suppose the region is classically allowed, so $E > V_q$. Then Eq. (8.21) is

$$\frac{d^2}{dx^2} \text{ function of } x = - \text{ positive constant} \times \text{function of } x. \tag{8.22}$$

The solution of such an equation, the "function of x" in (8.22), is an *oscillatory plane wave*. To write down such a function, we need to define the **wave number** k_q for the region q as

$$k_q \equiv \sqrt{\frac{2m}{\hbar^2}(E - V_q)} \qquad \text{wave number for CA region.} \tag{8.23}$$

In terms of this wave number, the TISE (8.20) looks like

$$\frac{d^2}{dx^2}\psi_E^{(q)}(x) = -k_q^2 \psi_E^{(q)}(x) \qquad \text{CA region } [x_{q-1} \leq x \leq x_q]. \tag{8.24}$$

Now, the *second-order ordinary differential equation* (8.24) has two linearly independent solutions. In fact, it has several *sets* of two linearly independent solutions. For example, it's satisfied by the real functions

$$\left\{ \sin k_q x, \ \cos k_q x \right\} \qquad \text{real, oscillatory solutions.} \tag{8.25}$$

We can write the *general solution* of Eq. (8.24) as an arbitrary linear combination of the functions (8.25), with coefficients that I'll denote by $C^{(q)}$ and $D^{(q)}$:

$$\psi_E^{(q)}(x) = C^{(q)} \cos k_q x + D^{(q)} \sin k_q x \qquad \text{CA region } [x_{q-1} \leq x \leq x_q]. \tag{8.26}$$

But we need not use the particular linearly independent solutions (8.25). Indeed, we often prefer to use *complex* exponential functions, because doing so makes it easier to hook the pieces together (Step 4 of our procedure). In a CA region, these alternate solutions are

$$\left\{ e^{ik_q x}, \ e^{-ik_q x} \right\} \qquad \text{complex, oscillatory solutions.} \tag{8.27}$$

In terms of complex functions, we write the general solution—with, of course, different multiplicative coefficients than those in Eq. (8.26)—as[15]

$$\psi_E^{(q)}(x) = A^{(q)} e^{ik_q x} + B^{(q)} e^{-ik_q x} \qquad \text{CA region } [x_{q-1} \leq x \leq x_q]. \tag{8.28}$$

[15]Be sure you understand why these alternatives are physically and mathematically equivalent (see Table 8.3). If you don't, have a look at any undergraduate differential equations text, such as Chaps. 1–3 of *An Introduction to Ordinary Differential Equations* by E. A. Coddington (Englewood Cliffs, N. J.: Prentice Hall, 1961) or Chap. 8 of *Mathematical Methods in the Physical Sciences*, 2nd ed. by M. L. Boas (New York: Wiley, 1983).

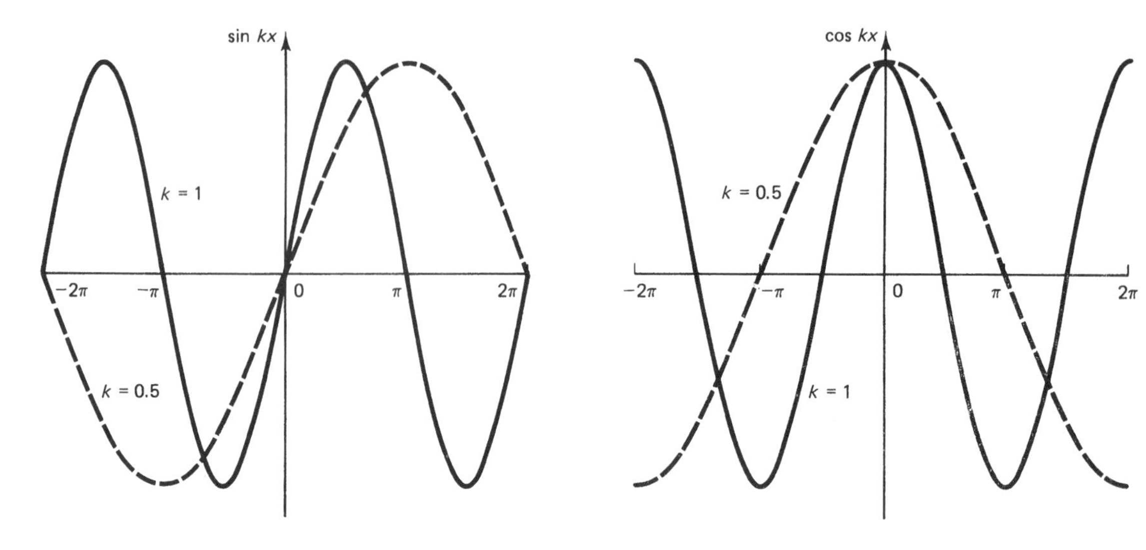

Figure 8.10 Real oscillatory solutions of the TISE (used in a CA region) for $k = 0.5$ and $k = 1.0$.

For convenience, I've collected several handy properties of the functions in (8.25) and (8.27) in Table 8.3. Useful graphs *which you should examine carefully* appear in Fig. 8.10.

TABLE 8.3 FUNCTIONS USED IN PIECING TOGETHER THE HAMILTONIAN EIGENFUNCTION FOR PIECEWISE-CONSTANT POTENTIALS

$f(x)$	relations	$f(-x)$	$\dfrac{df}{dx}$
Oscillatory Functions			
$\sin kx$	$\frac{1}{2i}\left(e^{ikx} - e^{-ikx}\right)$	$-\sin kx$	$k\cos kx$
$\cos kx$	$\frac{1}{2}\left(e^{ikx} + e^{-ikx}\right)$	$\cos kx$	$-k\sin kx$
e^{ikx}	$\cos kx + i\sin kx$	e^{-ikx}	ike^{ikx}
e^{-ikx}	$\cos kx - i\sin kx$	e^{ikx}	$-ike^{-ikx}$
Exponential Functions			
$\sinh \kappa x$	$\frac{1}{2}\left(e^{\kappa x} - e^{-\kappa x}\right)$	$-\sinh \kappa x$	$\kappa\cosh \kappa x$
$\cosh \kappa x$	$\frac{1}{2}\left(e^{\kappa x} + e^{-\kappa x}\right)$	$\cosh \kappa x$	$\kappa\sinh \kappa x$
$e^{\kappa x}$	$\sinh \kappa x + \cosh \kappa x$	$e^{-\kappa x}$	$\kappa e^{\kappa x}$
$e^{-\kappa x}$	$\cosh \kappa x - \sinh \kappa x$	$e^{\kappa x}$	$-\kappa e^{-\kappa x}$

The Hamiltonian Eigenfunction in a Classically-Forbidden Region

It's no more difficult to determine the pieces of the eigenfunction in a CF region than in a CA region. In a CF region, $E - V_q < 0$, so the TISE (8.20) has the form

$$\frac{d^2}{dx^2} \text{ function of } x = + \text{ positive constant} \times \text{function of } x. \tag{8.29}$$

The solutions of this equation are *real exponential functions.* [Compare Eqs. (8.29) and (8.22) and marvel at what a difference a little sign makes!] To write these solutions, we need to define the **decay constant** in the qth region,

$$\kappa_q \equiv \sqrt{\frac{2m}{\hbar^2}(V_q - E)} \qquad \text{decay constant for a CF region.} \tag{8.30}$$

In terms of this decay constant, the TISE becomes

$$\frac{d^2}{dx^2}\psi_E^{(q)}(x) = \kappa_q^2 \psi_E^{(q)}(x) \qquad \text{CF region } [x_{q-1} \le x \le x_q]. \tag{8.31}$$

By the way, the decay constant (8.30) is also called the **attenuation constant**, for reasons that are evident from Fig. 8.11. Whatever you call it, κ_q is a quantitative measure of the *rate of decay* of $\psi_E^{(q)}(x)$ with increasing distance into the CF region $[x_{q-1} \le x \le x_q]$. For this reason, the quantity $1/\kappa_q$ is called the **characteristic length** for decay in region q. The piece $\psi_E^{(q)}(x)$ attains $1/e$ of its maximum value $\psi_E^{(q)}(x_q)$ a distance $1/\kappa_q$ from the (left) boundary of the region.

Now, one set of linearly independent solutions to Eq. (8.31) is[16]

$$\left\{e^{-\kappa_q x}, e^{\kappa_q x}\right\} \qquad \text{real, exponential solutions.} \tag{8.32}$$

In terms of these functions, we write the *general solution in a CF region* as

$$\psi_E^{(q)}(x) = A^{(q)} e^{-\kappa_q x} + B^{(q)} e^{\kappa_q x} \qquad \text{CF region } [x_{q-1} \le x \le x_q]. \tag{8.33}$$

An alternative form of the general solution can be written in terms of *hyperbolic sine and cosine functions.* The properties of these functions, along with those of the real exponentials (8.32), are summarized in Table 8.3, and handy graphs of both types of functions appear in Fig. 8.11.

Question 8–9

What is the counterpart of Eq. (8.26) for a CF region? (Hint: $\cos ix = \cosh x$ and $\sin ix = i \sinh x$.)

Using the handy compendia in Table 8.2, we can easily analyze each region of a piecewise-constant potential and write down the piece of the Hamiltonian eigenfunction for each region. Of the steps in the procedure at the beginning of this section, there remains only to determine the coefficients in these pieces (and, for a bound state, the quantized energies)—matters we shall get to soon. But first, a couple of examples.

Example 8.1. The Potential Step for $E < V_0$: Act I (The Set-Up)

One of the simplest piecewise-constant potentials is the **potential step** in Fig. 8.12. In this system, a beam of particles incident from the left encounters an *impulsive* force at $x = 0$.

For this potential, we immediately espy two regions: one in which $V(x)$ is zero and one in which $V(x)$ equals a positive constant V_0, *i.e.*,

$$V(x) = \begin{cases} 0 & x < 0 \\ V_0 & x \ge 0. \end{cases} \tag{8.34}$$

[16]Notice that were we to define a *wave number* k_q, à la Eq. (8.23), for a CF region, this quantity would be imaginary. It would, in fact, be related to the decay constant κ_q by $k_q = i\kappa_q$. With this substitution the oscillatory functions in (8.27) become the real functions in (8.32).

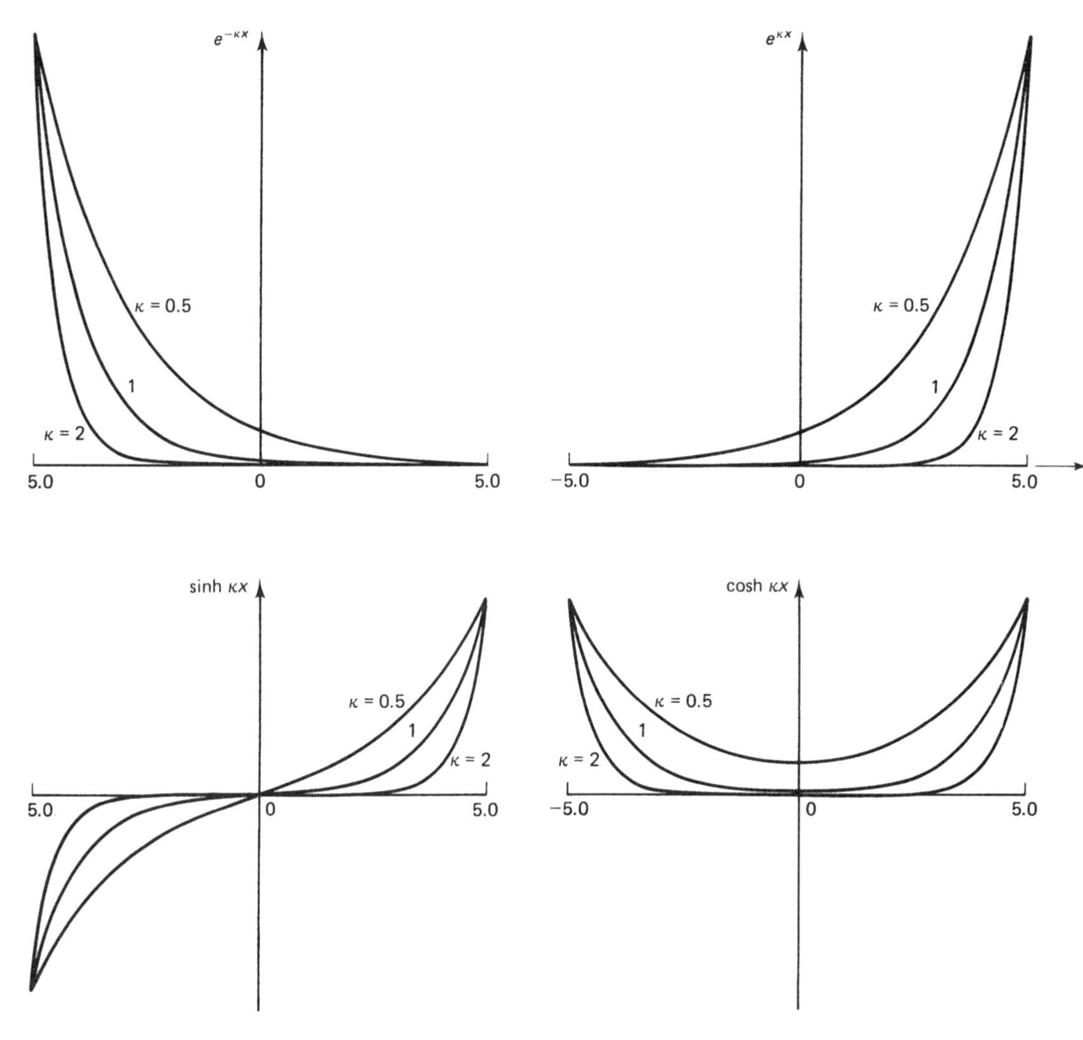

Figure 8.11 Solutions of the TISE for a CF region: exploding and decaying exponentials $e^{\kappa x}$ and $e^{-\kappa x}$ and hyperbolic functions $\sinh \kappa x$ and $\cosh \kappa x$ for $\kappa = 0.5$, 1.0, and 2.0. (See also Table 8.3.)

In this example, I'll consider an energy $E < V_0$, so I can show you how to handle both CA and CF regions. (For $E > V_0$ both regions are CA, right?) The wave numbers for each region are shown in Fig. 8.12, as is the general solution written in terms of $\psi_E^{(1)}(x)$ and $\psi_E^{(2)}(x)$. I've used the complex exponential form (8.28) in Region 1 to facilitate a later step, hooking the pieces together at the boundary $x = 0$ (see Example 8.3).

Now that we have the general solution (Step 2), we must impose physical conditions in the first and last regions (Step 3). To see what physical conditions we need, consider the solution in Region 2,

$$\psi_E^{(2)}(x) = A^{(2)} e^{-\kappa_2 x} + B^{(2)} e^{\kappa_2 x}. \tag{8.35}$$

Immediately, we spot a problem: one of the linearly independent real exponentials, $e^{\kappa_2 x}$, blows up in the limit $x \to \infty$, which limit is included in Region 2. Such unsightly explosions are common: pieces of $\psi_E(x)$ in semi-infinite CF regions are often plagued by exploding

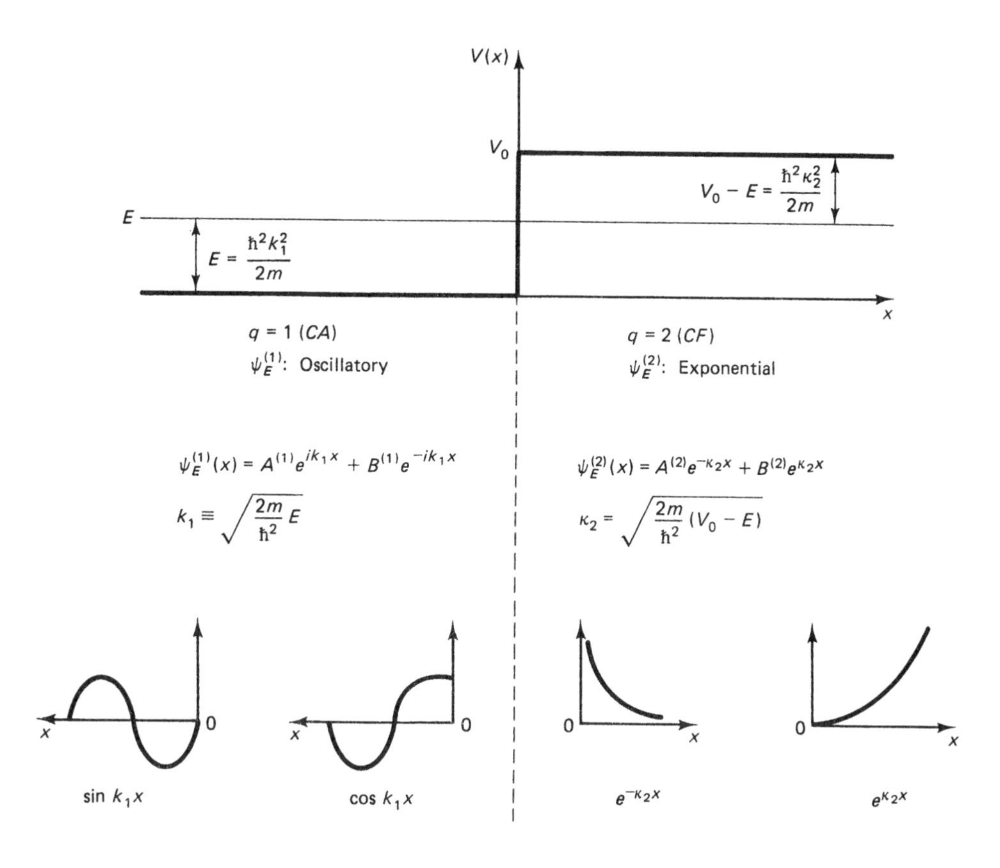

Figure 8.12 Analysis of a continuum stationary state of the potential step with $E < V_0$.

exponentials, because the real functions from which they are assembled have the unfortunate properties that

$$e^{\kappa x} \xrightarrow[x \to \infty]{} \infty,$$
$$e^{-\kappa x} \xrightarrow[x \to -\infty]{} \infty.$$

Exploding Exponentials

This behavior is just not acceptable: if we are to interpret the wave function $\Psi(x,t) = \psi_E(x)e^{-iEt/\hbar}$ as a position probability amplitude, we must *never* allow $\psi_E(x)$ to become infinite—not at a finite value of x, nor in either asymptotic region.

To eliminate this exploding exponential, we exploit the arbitrariness of the coefficients in (8.35). In particular, we set the coefficient that multiplies the offending term in $\psi_E^{(2)}(x)$ equal to zero:

$$B^{(2)} = 0 \qquad \Longrightarrow \qquad \psi_E^{(2)}(x) = A^{(2)}e^{-\kappa_2 x}. \tag{8.36}$$

Equation (8.36) is called an **asymptotic condition**. *Asymptotic conditions are constraints we must impose on the spatial function to make it satisfy a physical requirement in an asymptotic region*. In this instance, the asymptotic condition requires that $\psi_E(x)$ not blow up as $x \to \infty$ (in Region 2). (We'll discover another kind of asymptotic condition in Example 8.4, where we consider states with $E > V_0$.) Such conditions impose *physical* constraints on the allowable form of a *mathematical* solution; they are of such importance that we dignify them with a rule:

> ***Rule***
>
> Once you have written down the pieces of a Hamiltonian eigenfunction, immediately set to zero all coefficients that multiply exploding exponentials in either asymptotic region (the first and the last regions).

We can now assemble the solution to the TISE for the potential step for any energy $E < V_0$,

$$\psi_E(x) = \begin{cases} A^{(1)}e^{ik_1x} + B^{(1)}e^{-ik_1x} & x \le 0 \\ A^{(2)}e^{-\kappa_2 x} & x \ge 0. \end{cases} \tag{8.37}$$

[Incidentally, some authors refer to a decaying exponential such as $\psi_E^{(2)}(x)$ by the more appealing name **evanescent wave**.]

This completes the set-up of the problem of the potential step. Shortly, in Example 8.3, we'll finish this problem by connecting the pieces of the eigenfunction. And we'll return to the potential step for $E > V_0$ in several examples in § 8.5.

Question 8–10

Prepare a figure like Fig. 8.12 for the case $E > V_0$. For such an energy do you need to worry about exploding exponentials?

Example 8.2. The (Symmetric) Finite Square Well (The Set-Up)

The symmetric finite square well in Fig. 8.13 is as famous in the study of bound states as the potential step is in the study of continuum states. This potential is the simplest (realistic) example that illustrates the essentials of bound-state quantum mechanics. In Fig. 8.13 is the set-up for solving the TISE for a bound state of this potential, *i.e.*, for an energy $E < V_0$.

Notice several points:

1. We must enforce *two* asymptotic conditions, setting coefficients equal to zero in the *two* CF regions so as to stamp out exploding exponentials.
2. Because the potentials in these CF regions, V_1 and V_3, are equal, so are the decay constants: $\kappa_1 = \kappa_3$.
3. We could write the solution in Region 2, $\psi_E^{(2)}(x)$, either in terms of real trigonometric functions ($\sin k_2 x$ and $\cos k_2 x$) or in terms of complex exponentials (e^{ik_2x} and e^{-ik_2x}). The real functions simplify the ensuing algebra (Example 8.7), so let's choose them.

We're now ready to put the pieces together to form the Hamiltonian eigenfunction *for a bound state* ($E < V_0$),

$$\psi_E(x) = \begin{cases} B^{(1)}e^{\kappa_1 x} & -\infty < x \le -L/2 \\ C^{(2)}\cos k_2 x + D^{(2)}\sin k_2 x & -L/2 \le x \le L/2 \\ A^{(3)}e^{-\kappa_3 x} & L/2 \le x < \infty \end{cases} \tag{8.38}$$

To complete the solution of the TISE, we must hook these pieces together. This step determines all but one of the coefficients in Eq. (8.38). But since the state we're investigating is *bound*, its energies are quantized, and we must perform an additional chore: we must

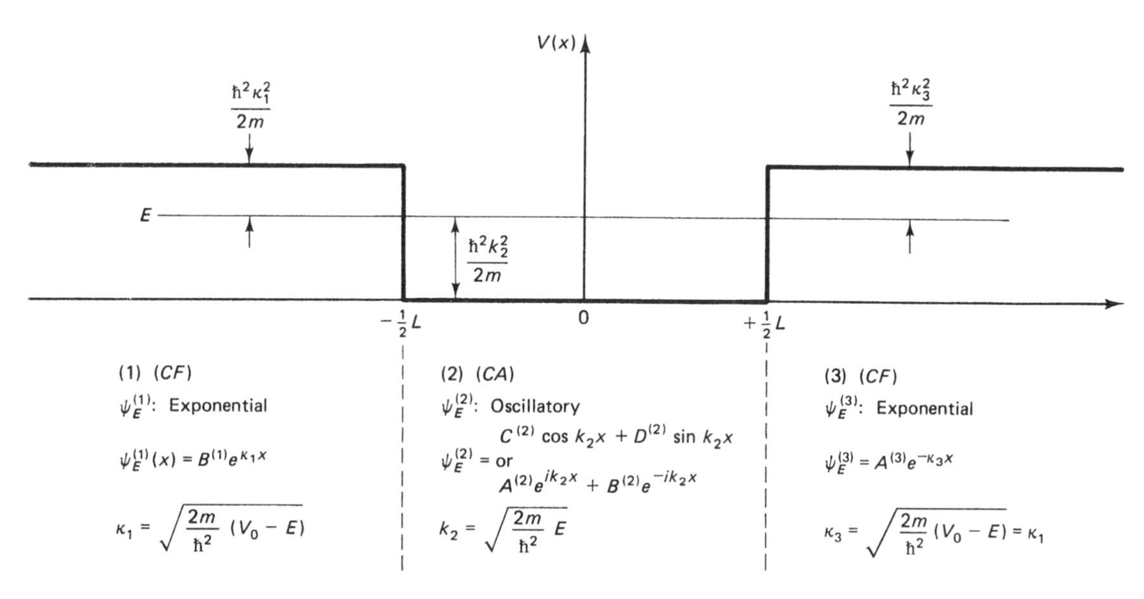

Figure 8.13 Analysis of a bound stationary state of a symmetric finite square well of height V_0 and width L.

find the values of these *quantized energies*, the particular eigenvalues of the Hamiltonian for which bound stationary-state eigenfunctions exist. I'll defer this chore until § 8.7.

Question 8–11

Replicate Fig. 8.13 for the case $E > V_0$. Write down the pieces of the Hamiltonian eigenfunction for such an energy. Is such a state bound or unbound? Why?

Aside: Regions and the Nature of Stationary States. The nature of a stationary state (bound or continuum) is intimately related to whether the first and last regions are CA or CF. This connection follows from the conditions (8.7) or (8.8), only *one* of which is in force for a particular state:

> ***Rule***
>
> A stationary state is *bound* if and only if the first and last regions are CF. If either region is CA, then there could be a non-zero probability of finding the particle in the state represented by $\Psi(x,t) = \psi_E(x)e^{-iEt/\hbar}$ in the corresponding asymptotic region, so the state is not bound.

This rule explains why the state in Example 8.1 is a *continuum* state even though its probability density decays to zero in Region 2 (*i.e.*, as $x \to +\infty$).

A Tale of the Hook

Now we come to Step 4 in our procedure for solving the TISE for a piecewise-constant potential. A piecemeal creature like (8.38) is obviously not a complete solution, for we don't know the coefficients multiplying all the x-dependent factors in it. And there may be a lot of these as-yet-unknown coefficients, for in general

$$N \text{ regions} \quad \Longrightarrow \quad 2N \text{ constants: } \left\{A^{(q)}, B^{(q)}\right\} \qquad \text{for } q = 1, 2, \ldots, N \tag{8.39}$$

The TISE *within* each region tells us nothing about the values of these coefficients. So we must consider the *region boundaries*.

We determine the coefficients (8.39) by "hooking together" the pieces of the Hamiltonian eigenfunction at these boundaries. The key to this procedure is found in the familiar mathematical requirements that *any* state function must satisfy (Chap. 3):

$\Psi(x,t)$ must be continuous, finite, single-valued, and smoothly-varying at all x.

The consequences of these requirements are the following conditions:

$$\begin{array}{lcl} \text{continuity} & \Longrightarrow & \psi_E(x) \text{ continuous at } x_q \\ \text{smoothly-varying} & \Longrightarrow & \dfrac{d}{dx}\psi_E(x) \text{ continuous at } x_q \end{array} \qquad \text{Continuity Conditions} \tag{8.40}$$

Within each region, these conditions are automatically satisfied by $\psi_E^{(q)}(x)$ (see Table 8.3). But, as shown in Fig. 8.14, at the region boundaries, we must *enforce* them. We do so by again exploiting the arbitrariness of the multiplicative constants in the Hamiltonian eigenfunction (see Example 8.3).[17]

Enforcing the continuity conditions (8.40) determines all but one of the constants in the eigenfunction. The role of this remaining constant depends on the nature of the state: if the state is unbound, this constant reflects conditions in the scattering experiment; if it's bound, this constant enforces normalization of the eigenfunction. We'll see continuity conditions in action for bound states in § 8.7. For now, let's apply them to a continuum state of the potential step.

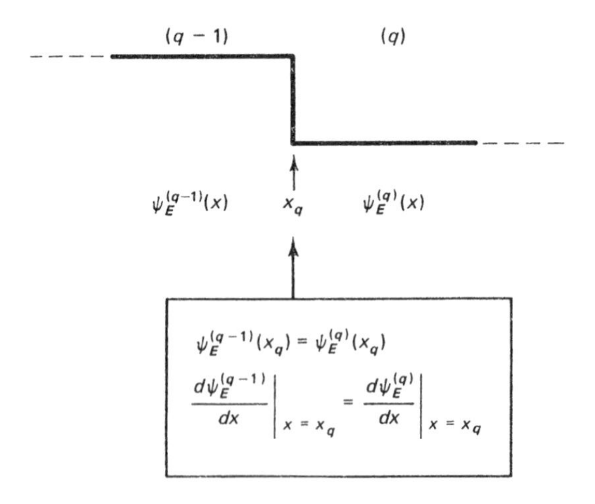

Figure 8.14 Continuity conditions at a region boundary for a piecewise-constant potential.

[17]Mathematical difficulties arise when we try to justify these continuity equations at the (unphysical) points of discontinuity of a piecewise-constant potential. These are not problems to be worried about, but if you'd like to read about them, see D. Branson, *Amer. Jour. Phys.* **47**, 100 (1979) and M. Andrews, ibid. **49**, 281 (1981).

Example 8.3. The Potential Step for $E < V_0$: Act II (The Hook)

In Example 8.1 we considered a continuum state of the potential step with energy $E < V_0$ and deduced the spatial function (8.37). Applying the continuity conditions (8.40) at the (finite) region boundary $x = 0$, we obtain two relations between the coefficients $A^{(1)}$, $B^{(1)}$, and $B^{(2)}$, *viz.*,

$$\psi_E^{(1)}(x=0) = \psi_E^{(2)}(x=0) \quad \Longrightarrow \quad A^{(1)} + B^{(1)} = A^{(2)}, \tag{8.41a}$$

$$\left.\frac{d\psi_E^{(1)}(x)}{dx}\right|_{x=0} = \left.\frac{d\psi_E^{(2)}(x)}{dx}\right|_{x=0} \quad \Longrightarrow \quad ik_1 A^{(1)} - ik_1 B^{(1)} = -\kappa_2 A^{(2)}. \tag{8.41b}$$

Equations (8.41) are *two* simultaneous algebraic equations in *three* unknowns. Solving them, we can evaluate *two unknowns in terms of the third*. For example, we can determine $B^{(1)}$ and $A^{(2)}$ in terms of $A^{(1)}$. Here's how.

Multiplying (8.41*a*) by κ_2 and adding the result to (8.41*b*), we find

$$\frac{B^{(1)}}{A^{(1)}} = \frac{ik_1 + \kappa_2}{ik_1 - \kappa_2} = \frac{k_1 - i\kappa_2}{k_1 + i\kappa_2}. \tag{8.42a}$$

(I've rearranged the terms in this result to obtain a cleaner form). Similarly, we find

$$\frac{A^{(2)}}{A^{(1)}} = \frac{2k_1}{k_1 + i\kappa_2}. \tag{8.42b}$$

Now, using Eqs. (8.42), we can write the eigenfunction (8.37) in terms of one (still-unknown) coefficient, to wit:

$$\psi_E(x) = \begin{cases} A^{(1)}\left[e^{ik_1 x} + \left(\dfrac{k_1 - i\kappa_2}{k_1 + i\kappa_2}\right)e^{-ik_1 x}\right] & x \le 0 \\[2ex] A^{(1)}\left(\dfrac{2k_1}{k_1 + i\kappa_2}\right)e^{-\kappa_2 x} & x \ge 0. \end{cases} \qquad (E < V_0) \tag{8.43}$$

Note that the Hamiltonian eigenfunction for $E < V_0$ is complex and that the (evanescent) wave in Region 2 is a decaying exponential, not a travelling wave.

As desired, the eigenfunction (8.42) is everywhere continuous and smoothly varying. In the next section, I'll finish this example and will unveil the physical significance of the remaining coefficient, $A^{(1)}$, and explain how we choose it to match the experimental (scattering) situation. We'll also learn how to determine from $\psi_E(x)$ the probabilities for reflection and transmission in continuum-state problems.

I want to emphasize that nowhere (thus far) have I distinguished the *problem-solving strategy* we use for bound states from that for continuum states. To be sure, the differences between bound and continuum states (Table 8.1) have influenced several of our actions, *e.g.*, specification of the piece of the eigenfunction in the first and/or last regions, and the asymptotic conditions that may have to be imposed on these pieces. But the basic procedure outlined at the start of this section holds for either kind of state: identify regions, classify them, write down the piece for each region, and hook the pieces together. For either kind of state, the solution in each region is a linear combination of simple, analytic functions (Table 8.3), although the particular form of each of these functions depends on whether the region is CA or CF. So you don't really have to learn *two* different problem-solving methods to handle the two different kinds of states. Isn't that nice?

8.5 THE INTERPRETATION OF CONTINUUM-STATE EIGENFUNCTIONS

"There is work to be done."

—Count Dracula
in *Dracula*
by Bram Stoker

To solve scattering problems, we use *continuum stationary-state functions* such as the pure momentum function of Chap. 4 (*c.f.*, § 8.3). Two quantities are essential to the *interpretation* of continuum stationary states: the reflection and transmission coefficients $R(E)$ and $T(E)$. Although these quantities are called "coefficients," they're actually *probabilities*. They answer the question, what is the probability that an electron incident from the left with energy E will be reflected (or transmitted) by its encounter with a potential $V(x)$? In this section, I'll introduce these coefficients via the corresponding *probability amplitudes*.

The reflection and transmission coefficients are defined in terms of the *incident, reflected, and transmitted fluxes*. These quantities are the probability current densities for the possible outcomes of the collision. We can calculate these current densities from the pieces of the continuum stationary-state function $\psi_E(x)$, provided each piece corresponds to a *travelling wave* or to a *standing wave*, which is just a superposition of travelling waves.

For example, the piece of the Hamiltonian eigenfunction (8.43) (for the potential step) that corresponds to Region 1 is a sum of two terms. One corresponds to a travelling wave propagating in the $+x$ direction (the e^{ik_1x} term), the other to a wave propagating in the $-x$ direction (the e^{-ik_1x} term). These are the *incident* and *reflected* waves, respectively.

Aside: A quick review. The pure momentum functions $Ae^{i(px-Et)/\hbar}$ and $Ae^{-i(px+Et)/\hbar}$ represent travelling waves that propagate in the $+x$ and $-x$ directions, respectively. This interpretation is consistent with the fact that the momentum of the first wave is $+p$ and that of the second is $-p$. For a pure momentum state that represents a state of a *beam of non-interacting particles*, the probability density is (*c.f.*, § 6.8–6.9)

$$P(x,t) = |A|^2 = \text{number of particles per unit length} \tag{8.44a}$$

$$j(x,t) = \pm\frac{p}{m}|A|^2 = \begin{array}{l}\text{number of particles incident on } 1\,\text{m}^2\\ \text{area perpendicular to the beam}\end{array} \tag{8.44b}$$

Example 8.4. The Potential Step for $E > V_0$: Act III (Reflection and Transmission Coefficients)

Let's work out the reflection and transmission coefficients for the eigenfunction (8.37) of Example 8.3. The piece of this function in Region 1 of the potential step consists of two travelling waves:

$$\psi_E^{(1)}(x) = \underbrace{A^{(1)}e^{ik_1x}}_{\text{incident}} + \underbrace{B^{(1)}e^{-ik_1x}}_{\text{reflected}}. \tag{8.45}$$

This function is a *superposition* of an *incident wave* of amplitude $|A^{(1)}|$ and a *reflected wave* of amplitude $|B^{(1)}|$.

We can use the ratio $B^{(1)}/A^{(1)}$ from Eq. (8.42a) to calculate the amplitude of the reflected wave. We find, interestingly, that $|B^{(1)}| = |A^{(1)}|$. That is, in Region 1 two waves *of equal amplitude* travel in opposite directions *at the same phase velocity*. [That's why $\psi_E^{(1)}(x)$, the superposition of these waves, is an (oscillatory) *standing wave*.] Because the amplitudes of the incident and reflected waves are equal, the total probability flow into or out of this region is zero.[18]

This means that every particle in the incident beam (with $E < V_0$) will be *reflected* by its encounter with the step at $x = 0$. The probability of reflection is therefore unity, and that of transmission is zero:

$$\boxed{\begin{aligned} R(E) &= 1 \\ T(E) &= 0 \end{aligned}} \qquad \text{[potential step for } E < V_0\text{]}. \tag{8.46}$$

This result is just what we'd expect were we applying classical physics to this problem. But don't be misled by this coincidence. As we'll see in Examples 8.5 and 8.6, the reflection and transmission coefficients for $E > V_0$ violate our classical expectations.

The interpretation we've applied to $\psi_E^{(1)}(x)$ doesn't fit the piece of $\psi_E(x)$ in Region 2. This function isn't a travelling wave; it is a decaying—or evanescent—wave. Its amplitude

$$|A^{(2)}| = |A^{(1)}| \frac{2k_1}{\sqrt{k_1^2 + \kappa_2^2}} e^{-\kappa_2 x} \tag{8.47}$$

decreases with increasing x in Region 2 at a "decay rate" given by the decay constant κ_2, as you can see from the probability density in Fig. 8.15. This finding is consistent with our discovery [Eq. (8.46)] that all incident particles with $E < V_0$ are reflected. No probability flow is associated with an evanescent wave.

Question 8–12

Prove that the probability current density for the function $e^{-\kappa x} e^{-iEt/\hbar}$ is zero.

Reflection and Transmission of Particles by the Potential Step

The predictions of quantum mechanics for the behavior of a beam of particles with energy $E < V_0$ that impinges on the potential step conform to the predictions of classical physics that such particles will be reflected but not transmitted. But how about particles with $E > V_0$? Well, classical physics predicts that such particles will be transmitted but not reflected. Let's see what quantum physics predicts.

Example 8.5. A Second Look at the Potential Step: ($E > V_0$)

In Question 8–9, you determined the general form of the Hamiltonian eigenfunction for a state of the potential step with $E > V_0$, obtaining

$$\psi_E(x) = \begin{cases} \psi_E^{(1)}(x) = A^{(1)} e^{ik_1 x} + B^{(1)} e^{-ik_1 x} & -\infty < x \le 0 \\ \psi_E^{(2)}(x) = A^{(2)} e^{ik_2 x} + B^{(2)} e^{-ik_2 x} & 0 \le x < \infty. \end{cases} \tag{8.48}$$

[18]Another way to see this feature of $\psi_E^{(1)}(x)$ is to write this function in terms of real oscillatory solutions (8.25), *viz.*,

$$\psi_E^{(1)}(x) = \frac{2A^{(1)}}{k_1 + \kappa_2} \left[k_1 \cos k_1 x - \kappa_2 \sin k_1 x \right].$$

The x dependence in this result is that of a standing wave.

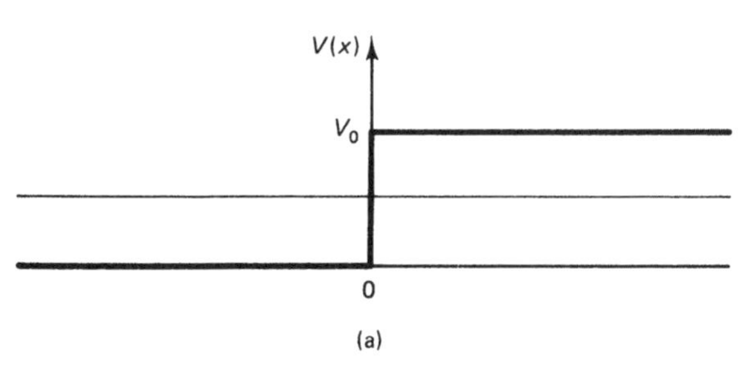

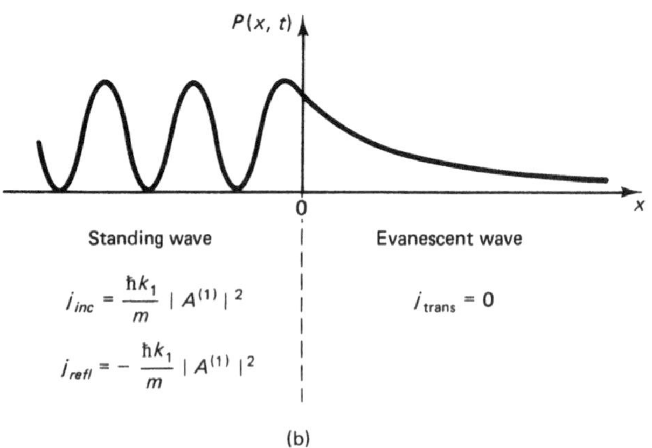

Figure 8.15 The probability density for a continuum stationary state of the potential step with $E < V_0$.

For such a state, both regions are classically allowed, and the wave numbers in the pieces of the eigenfunction (8.48) are

$$k_q = \sqrt{\frac{2m}{\hbar^2}\left(E - V_q\right)}. \qquad (q = 1, 2). \tag{8.49}$$

In this example, we'll hook these pieces together. Then, in Example 8.6, we'll calculate the reflection and transmission coefficients for this state.

Before applying the continuity conditions, we must consider the *asymptotic conditions*. Each piece of (8.48) is oscillatory, so we needn't cope with exploding exponentials. But there is nonetheless an asymptotic condition we must enforce if the general form (8.48) is to conform to the experiment.

To see the origin of this condition, note first that we can interpret each *term* of each piece as a travelling wave. To clarify this point, let's write the time-*dependent* state function corresponding to Eq. (8.49),

$$\Psi_E(x,t) = \begin{cases} A^{(1)}e^{i(k_1x-\omega_1t)} + B^{(1)}e^{-i(k_1x+\omega_1t)} & -\infty < x \le 0 \\ A^{(2)}e^{i(k_2x-\omega_2t)} + B^{(2)}e^{-i(k_2x+\omega_2t)} & 0 \le x < \infty. \end{cases} \tag{8.50a}$$

The angular frequencies in this function correspond to the wave numbers (8.49); they are

$$\omega_q = \frac{\hbar k_q^2}{2m} = \frac{E - V_q}{\hbar} \quad (q = 1, 2). \tag{8.50b}$$

Now, consider Eq. (8.50*a*) in light of Table 8.4. Do you see the problem?

TABLE 8.4 PIECES OF THE EIGENFUNCTION FOR THE POTENTIAL STEP WITH $E > V_0$

Term in $\psi_E(x)$	Interpretation
$A^{(1)}e^{ik_1x}$	incident wave (momentum $+\hbar k_1$)
$B^{(1)}e^{-ik_1x}$	reflected wave (momentum $-\hbar k_1$)
$A^{(2)}e^{ik_2x}$	transmitted wave (momentum $+\hbar k_2$)
$B^{(2)}e^{-ik_2x}$	incident wave (momentum $-\hbar k_2$)

The problem is that *one term in Eq. (8.50a) does not conform to the experimental arrangement* in this example: a beam of particles *incident from the left* on the potential step $V(x)$. This arrangement does not include a *source* of particles travelling to the left (with momentum $-\hbar k_2$). So to make our *mathematical* solution conform to the *physical* experiment, we must set $B^{(2)} = 0$ in Eqs. (8.48) and (8.50*a*).

Like the asymptotic condition that prohibits exploding exponentials, the requirement that the eigenfunction we construct must conform to the experimental set-up is *a consequence of the physics that we impose on the mathematics*:

> ***Rule***
>
> Once you have written down the pieces of the Hamiltonian eigenfunction, set to zero any coefficients that multiply incident travelling-wave terms for which there is no source in your experiment.

Putting together the remaining pieces and using the continuity conditions (8.40) to hook them together, we obtain for the continuum function for a state with energy $E > V_0$ the astounding result

$$\psi_E(x) = \begin{cases} A^{(1)}\left[e^{ik_1x} + \left(\dfrac{k_1 - k_2}{k_1 + k_2}\right)e^{-ik_1x}\right] & x \le 0 \\ A^{(1)}\left(\dfrac{2k_1}{k_1 + k_2}\right)e^{ik_2x} & x \ge 0 \end{cases} \qquad (E > V_0) \qquad (8.51)$$

What is astounding in this result is the *reflected travelling wave in Region 1*. The presence of this term means that for $E > V_0$, we'll find $T(E) < 1$ and $R(E) \neq 0$. These predictions are quite different from those of classical physics. Indeed, it's impossible to understand from classical arguments how a particle with $E > V_0$ could be *reflected* by the step. We'll *evaluate* these probabilities in the grand finale, Example 8.6.

Question 8–13

Fill in the steps leading to Eqs. (8.51). You can do this using our straightforward hooking procedure or, more easily, by thinking about how this state differs from the one in Eq. (8.43).

Question 8–14

Graph the probability density for a state with $E > V_0$ (*c.f.*, Fig. 8.15).

The Reflection and Transmission Coefficients

The form of the Hamiltonian eigenfunction in Example 8.5 is typical of continuum stationary states (for a beam incident from the left):

$$\psi_E(x) = \begin{cases} \text{incident-wave term} + \text{reflected-wave term} & \text{"source region"} \\ \text{transmitted-wave term} & \text{"detector region"} \end{cases} \tag{8.52}$$

The terms **source region** and **detector region** in (8.52) explicitly remind us where the source and the detector are. We note two "scattered waves," the reflected wave and the transmitted wave, each of which represents one possible outcome of the encounter of the particles in the incident beam with the potential step.

To clarify the role of these scattered waves, let's express them in terms of the *incident amplitude*. To this end, we'll define a quantity $\rho(E)$ that relates the amplitude of *the reflected wave* to that of the incident wave:

$$B^{(1)} = \rho(E)A^{(1)}. \tag{8.53a}$$

The counterpart of this quantity for *the transmitted wave* is

$$A^{(2)} = \tau(E)A^{(1)}. \tag{8.53b}$$

These quantities (which may be complex) are *probability amplitudes*: $\rho(E)$ is the **reflection probability amplitude** and $\tau(E)$ the **transmission probability amplitude**. In terms of these amplitudes, we can write the general eigenfunction (8.52) as

$$\psi_E(x) = \begin{cases} A^{(1)}e^{ik_1x} + A^{(1)}\rho(E)e^{-ik_1x} & x \le 0 \\ A^{(1)}\tau(E)e^{ik_2x} & x \ge 0 \end{cases} \quad (E > V_0). \tag{8.54}$$

Now, from these probability amplitudes we can calculate the reflection and transmission coefficients. To do so, we return to the probability current densities—the fluxes—for the incident, reflected, and transmitted waves. For each wave, we'll obtain the probability current density by substituting the appropriate piece of $\psi_E(x)$ into the definition of this quantity. Then we'll call upon Eq. (8.52) to help us interpret these densities. Let's do it.

By definition, the probability current density for $\psi_E^{(q)}(x)$ is

$$j^{(q)} = -\frac{i\hbar}{2m}\left[\psi_E^{(q)\,*}(x)\frac{d}{dx}\psi_E^{(q)}(x) - \psi_E^{(q)}(x)\frac{d}{dx}\psi_E^{(q)\,*}(x)\right]. \tag{8.55}$$

[Don't forget that each of the pieces $\psi_E^{(q)}(x)$ is only part of the Hamiltonian eigenfunction; the probability current density for the full stationary-state eigenfunction, $j(x,t)$, is zero.] Inserting the pieces from (8.54) into this definition, we obtain

$$j = \begin{cases} j^{(1)} = \dfrac{\hbar k_1}{m}\left[|A^{(1)}|^2 - |A^{(1)}|^2\,|\rho(E)|^2\right] & \text{source region} \\ j^{(2)} = \dfrac{\hbar k_2}{m}|A^{(1)}|^2\,|\tau(E)|^2 & \text{detector region.} \end{cases} \tag{8.56}$$

What does this result mean physically? Well, look first at Region 1. In this region the probability current density j is the sum of an *incident flux* j_{inc} and a *reflected flux* j_{refl}. In Region 2, the current density j is the *transmitted flux* j_{trans}. If we identify these fluxes as

$$j_{\text{inc}} \equiv \frac{\hbar k_1}{m}|A^{(1)}|^2 \tag{8.57a}$$

$$j_{\text{refl}} \equiv -\frac{\hbar k_1}{m}|A^{(1)}|^2|\rho(E)|^2 \tag{8.57b}$$

$$j_{\text{trans}} \equiv \frac{\hbar k_2}{m}|A^{(1)}|^2|\tau(E)|^2, \tag{8.57c}$$

we can write the current density as

$$j = \begin{cases} j_{\text{inc}} + j_{\text{refl}} & \text{source region} \\ j_{\text{trans}} & \text{detector region.} \end{cases} \tag{8.58}$$

Notice that the reflected flux j_{refl} must carry a minus sign, because the momentum of the reflected wave is $-\hbar k_1$ (*i.e.*, this wave "travels" in the $-x$ direction). Also note the presence in Eqs. (8.57) of the *squared moduli* of the reflection and transmitted *probability amplitudes* $\rho(E)$ and $\tau(E)$. Such factors are just what we'd expect to find in probabilities.

And, indeed, the step to the reflection and transmission probabilities is a small one. The reflection probability $R(E)$, for example, is equal to the *fraction* of particles in the incident beam that are reflected. This fraction is simply the ratio of the reflected flux to the incident flux [see Eq. (8.44*b*)]. Making a similar argument for the transmission coefficient, we arrive at the definitions

$$\boxed{\begin{aligned} R(E) &= \frac{|j_{\text{refl}}|}{|j_{\text{inc}}|} = |\rho(E)|^2 \\ T(E) &= \frac{|j_{\text{trans}}|}{|j_{\text{inc}}|} = \frac{k_N}{k_1}|\tau(E)|^2. \end{aligned}} \tag{8.59}$$

Note carefully the ratio of wave numbers k_N/k_1 in the transmission coefficient. Many novice problem solvers forget this factor and wind up with grossly erroneous transmission coefficients (unless the potential is such that $V_1 = V_N$).

Coefficients like those in (8.59) are the cornerstones of quantum collision theory (in one dimension). For more complicated problems, we can develop expressions for these coefficients using the procedure of this section: we work from the probability current densities and from our interpretation of the terms in the Hamiltonian eigenfunction as incident, reflected, and travelling waves. In the next section, we'll work through a more strenuous example of this procedure. But first, we've got to finish off the potential step.

Example 8.6. The Last of the Potential Step for $E > V_0$

Little remains to be done. To determine the reflection and transmission coefficients for this case, we just read the corresponding amplitudes off the eigenfunction (8.51) and substitute them into the definitions (8.59), obtaining

$$\begin{aligned} R(E) &= |\rho(E)|^2 = \left(\frac{k_1 - k_2}{k_1 + k_2}\right)^2 \\ T(E) &= \frac{k_2}{k_1}|\tau(E)|^2 = \frac{4k_1k_2}{(k_1 + k_2)^2}. \end{aligned} \qquad (E > V_0) \tag{8.60}$$

Since each incident particle must be either reflected or transmitted, the sum of these probabilities must equal one,

$$R(E) + T(E) = 1. \tag{8.61}$$

This requirement is a useful check on our algebra.

In contrast to Eq. (8.46) for $E < V_0$, our results (8.60) for $E > V_0$ do *not* conform to the predictions of classical physics. They do, however, agree with laboratory findings.

Question 8–15

In Eq. (8.46) we *deduced* the reflection and transmission coefficients for states of the potential step with $E < V_0$. Verify these deductions with a calculation based on the eigenfunction (8.43) and the definitions (8.59).

Question 8–16

Solve the TISE and calculate the reflection and transmission coefficients for scattering by the potential step of a particle with $E = V_0$.

Generalization and Summary

We have seen how to solve the TISE for continuum states of a piecewise-constant potential and how to interpret the resulting eigenfunctions to obtain scattering information: the probabilities of reflection and transmission. Although we have worked only with the potential step, which contains but a single point of discontinuity, the method of this section easily generalizes to more complicated piecewise-constant potentials, such as those in Fig. 8.16.

In fact, the only difference between the solution of the present example and the procedure for finding the reflection and transmission coefficients for a piecewise-constant potential with an arbitrary number of discontinuities is the quantity and complexity of the algebra involved. If we label the final region (where the detector is located) by N, we can write the solution of the TISE in the source and detector regions as a simple generalization of the form (8.54) for the potential step.[19]

For example, if, as in Fig. 8.16, the energy of the state satisfies $E > V_1$ and $E > V_N$, then the eigenfunction is

$$\psi_E(x) = \begin{cases} A^{(1)}\left[e^{ik_1x} + \rho_1(E)e^{-ik_1x}\right] & \text{source region} \\ \cdot & \\ \cdot & \\ \cdot & \\ A^{(1)}\tau_N(E)e^{ik_Nx} & \text{detector region.} \end{cases} \tag{8.62}$$

If, as in (8.62), the pieces of the eigenfunction in the source and detector regions correspond to travelling waves, then both reflection *and* transmission are possible, and both $R(E)$ and $T(E)$ are non-zero and less than one. But if $E < V_N$, then the wave in the detector region is evanescent, and the probability of transmission is zero.

[19] In problems with many regions, we may not need to solve for all the pieces of the eigenfunction. We may instead be able to use continuity conditions to determine the reflection and transmission probabilities, bypassing evaluation of the forms of $\psi_E^{(q)}(x)$ for intermediate regions $q = 2, \ldots, N-1$. Herein lies the *art* of solving problems with piecewise-constant potentials.

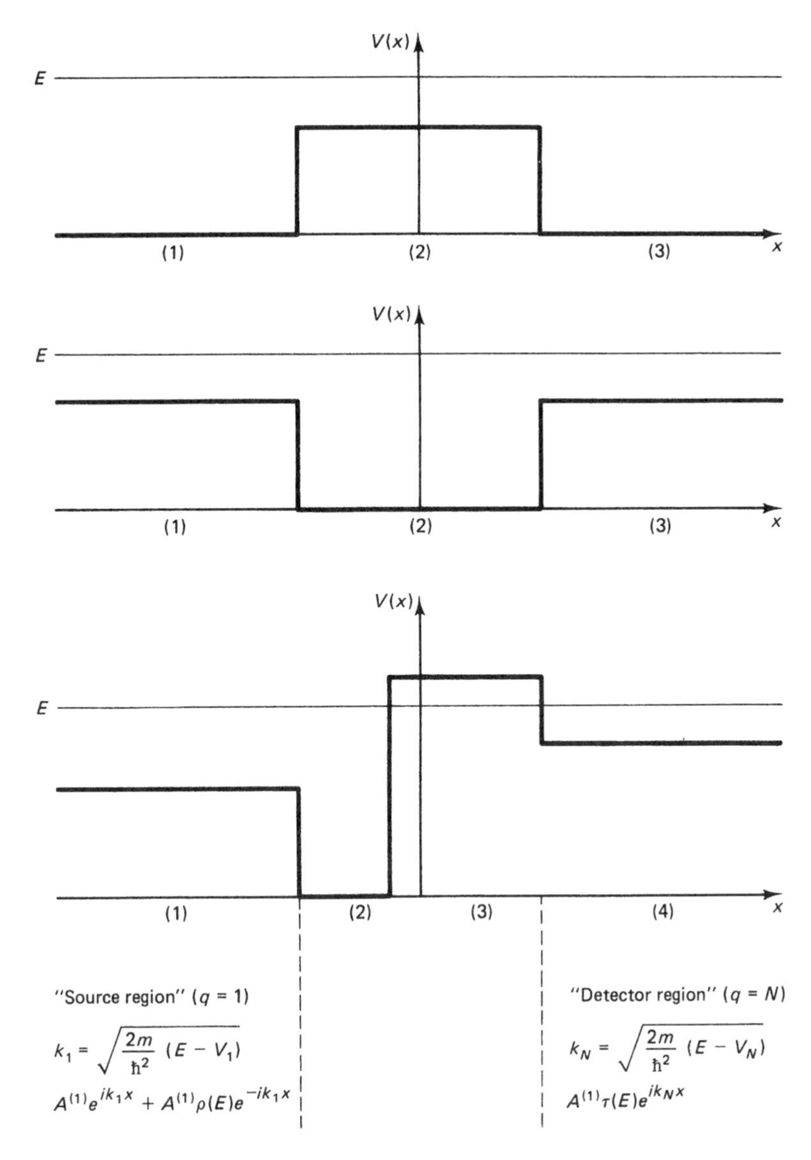

Figure 8.16 Identification of the source and detector regions for several piecewise-constant potentials.

8.6 TUNNELLING FOR FUN AND PROFIT

> "There is work,
> wild work to be done."
>
> —Professor Van Helsing
> in *Dracula*
> by Bram Stoker

We have discovered that all particles with $E < V_0$ incident on a potential step of height V_0 are reflected. So although quantum mechanics predicts *penetration* into the CF region of this potential—*i.e.*, the probability density in this region is non-zero—we can't demonstrate this phenomenon in the laboratory.

But we can demonstrate penetration into a CF region if we "chop off" the potential step, changing it into the *barrier* of width L in Fig. 8.17,[20]

$$V(x) = \begin{cases} 0 & x < 0 \\ V_0 & 0 \le x \le L \\ 0 & x > L \end{cases}. \tag{8.65}$$

According to quantum physics, particles with $E < V_0$ incident on the barrier from the left will penetrate the potential and be transmitted into the classically-allowed region beyond, where we can detect them. This is the amazing, wholly non-classical phenomenon of **tunnelling**—the topic of this section.

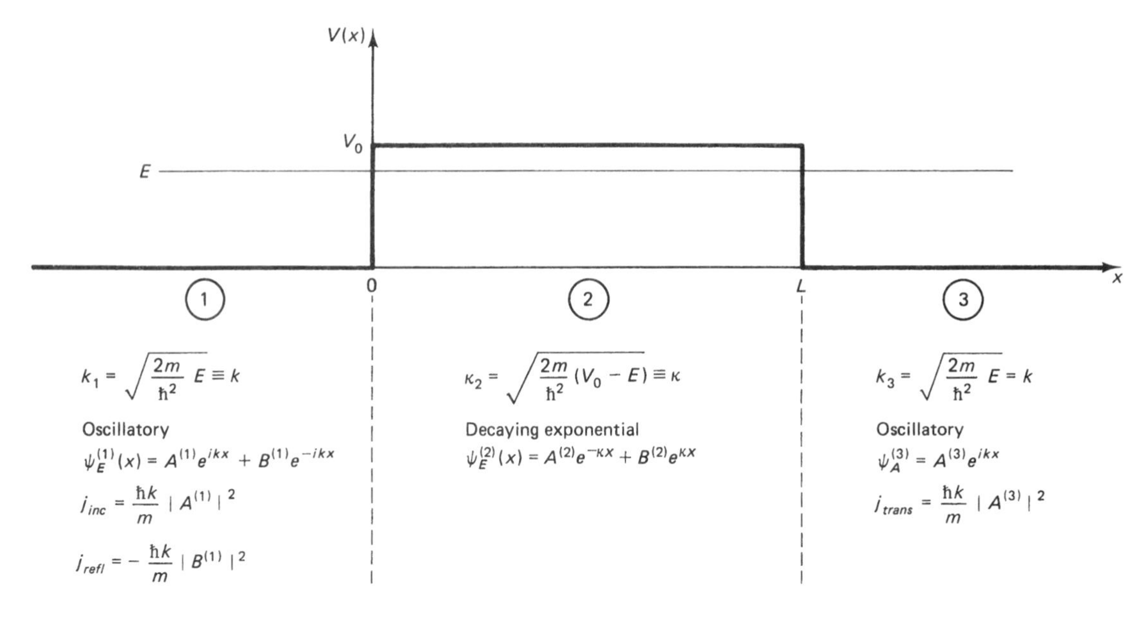

Figure 8.17 The barrier and its regions for a particle incident from the left with energy $E < V_0$.

Tunnelling is a fascinating phenomenon, both in its own right and for its many applications. For example, tunnelling explains the existence of doublets in the vibrational spectra of the ammonia molecule (NH_3) due to inversion of its nuclear geometry, a property that is important in the design of atomic clocks and the ammonia beam maser;[21] the emission of alpha particles from a radioactive nucleus, which is relevant to reactor design;[22] and the field emission of electrons.[23] And barrier penetration is important

[20]Ordinarily, the symmetry of a potential such as the one in Fig. 8.17 would argue for choosing $x = 0$ at the midpoint of the barrier. But in this problem, we can simplify the algebra slightly by choosing the origin at one of the region boundaries, the points at which we must apply continuity conditions. And in this problem, we need all the help we can get.

[21]See § 12-2 of *Atomic and Space Physics* by A. E. S. Greene and P. J. Wyatt (Reading, Mass.: Addison-Wesley, 1965).

[22]See, for example, § 4-5 of *Elements of Nuclear Physics* by W. E. Meyerhoff (New York: McGraw-Hill, 1967).

[23]See T. Ando, A. B. Fowler, and F. Stern, *Rev. Mod. Phys.* **54**, 437 (1982).

in such engineering devices as tunnel diodes, which are used as switching units in fast electronic circuits.[24]

Perhaps the most remarkable application of tunnelling is the **scanning tunnelling microscope** that has been developed by scientists at the IBM Research Laboratory in Zurich. This amazing device is based on the penetration by electrons near the surface of a solid sample through the barrier at the surface. These electrons form a "cloud" of probability outside the sample. Although the probability of detecting one of these electrons decays exponentially with distance from the surface, one can induce and measure a *current* of these electrons, and thereby attain a magnification factor of 100 million—large enough to permit resolution of a few hundredths the size of an atom! For this development, Drs. Gerd Binning and Heinrich Rohrer won half of the 1986 Nobel Prize in Physics.[25]

In this section, we're going to solve for the *transmission coefficient* for particles incident on a potential barrier. I've devoted a full section to tunnelling because I want to raise your algebraic consciousness. This is (algebraically) the most complicated problem we have yet encountered, and it offers a chance to learn some nifty organizational and problem-solving tricks that are vital to the mastery of the art and craft of quantum physics. So get out paper and pencil and follow along.[26]

The Set-Up

The regions of the barrier potential and their classification for a state with $E < V_0$ appear in Fig. 8.17. For this potential, there is but a single wavenumber (for Regions 1 and 3) and a single decay constant (for Region 2), so to simplify the ensuing algebra, I've dropped the region labels from these quantities. I've also applied the asymptotic condition to $\psi_E^{(3)}(x)$ to make the composite eigenfunction conform to an experiment in which particles are incident *from the left*.

Our next step is to apply the continuity conditions (8.40) at the boundaries $x = 0$ and $x = L$. These conditions and a little algebra lead directly to the ratio of the transmitted to incident flux—the very quantity we need to determine the transmission coefficient $T(E)$ [see Eq. (8.59)]. So we'll first evaluate the *transmission probability amplitude* $\tau(E)$, the ratio of the amplitude of the transmitted wave $A^{(3)}$ to that of the incident wave, $A^{(1)}$.[27]

From Continuity Conditions to Scattering Probabilities

The results of applying continuity conditions to the pieces of the Hamiltonian eigenfunction in Fig. 8.17 appear in Table 8.5. Using these equations, we can relate $A^{(3)}$ to $A^{(1)}$. One fairly efficient route to this relationship is to move from Region 1 "through

[24] For a practical look at this application, see Chap. 9 of *Experiments in Modern Physics* by A. C. Melissinos (New York: Academic Press, 1966). You can find a great deal more about tunnelling in *Quantum Mechanical Tunnelling and its Applications* by D. Roy (Philadelphia: Taylor and Francis, 1987).

[25] See G. Binning and H. Rohrer, *Sci. Amer.* **253**, 50, (1985) and C. F. Quate, *Physics Today* **39**, 26 (1986).

[26] Readers interested in an alternate approach that bypasses some of the algebraic machinery I want to introduce should peruse the two useful papers by J. E. Draper, *Amer. Jour. Phys.* **47**, 525 (1979) and ibid., **48**, 749 (1980).

[27] Similarly, we can evaluate the reflection coefficient $R(E)$ from the ratio of $B^{(1)}$ to $A^{(1)}$. I'll do only the transmission coefficient, leaving reflection for you to ponder as a problem.

TABLE 8.5 CONTINUITY CONDITIONS FOR THE POTENTIAL BARRIER IN EQ. (8.65).

Continuity Conditions at $x = 0$		
$\psi_E^{(1)}(0) = \psi_E^{(2)}(0) \implies$	$A^{(1)} + B^{(1)} = A^{(2)} + B^{(2)}$	(8.66)
$\left.\frac{d}{dx}\psi_E^{(1)}(x)\right\|_{x=0} = \left.\frac{d}{dx}\psi_E^{(2)}(x)\right\|_{x=0} \implies$	$ikA^{(1)} - ikB^{(1)} = -\kappa A^{(2)} + \kappa B^{(2)}$	(8.67)
Continuity Conditions at $x = L$		
$\psi_E^{(2)}(L) = \psi_E^{(3)}(L) \implies$	$A^{(2)}e^{-\kappa L} + B^{(2)}e^{\kappa L} = A^{(3)}e^{ikL}$	(8.68)
$\left.\frac{d}{dx}\psi_E^{(2)}(x)\right\|_{x=L} = \left.\frac{d}{dx}\psi_E^{(3)}(x)\right\|_{x=L} \implies$	$-\kappa A^{(2)}e^{-\kappa L} + \kappa B^{(2)}e^{\kappa L} = ikA^{(3)}e^{ikL}$	(8.69)

Region 2" into Region 3 by systematically eliminating all coefficients except the two we need.

We don't need $B^{(1)}$, so let's get rid of it:

Step 1. eliminate $B^{(1)}$ by deriving an equation relating $A^{(1)}$, $A^{(2)}$, and $B^{(2)}$.

Step 2. Next we must relate $A^{(3)}$ to the ratio $A^{(2)}/B^{(2)}$: derive expressions for $A^{(2)}$ and for $B^{(2)}$ in terms of $A^{(3)}$.

Step 3. This sets us up to complete the problem: combine the results of steps 1 and 2 to obtain an expression for the ratio $A^{(3)}/A^{(1)}$.

Step 1 is easy: we just multiply Eq. (8.66) by ik and add it to Eq. (8.67), obtaining

$$2ikA^{(1)} = (ik - \kappa)A^{(2)} + (ik + \kappa)B^{(2)}. \tag{8.70}$$

Step 2 entails two manipulations. Multiplying Eq. (8.68) by $-\kappa$ and adding the result to Eq. (8.69) gets us an expression for $A^{(2)}$, and multiplying (8.68) by κ and adding to (8.69) gets us $B^{(2)}$:

$$A^{(2)} = \frac{\kappa - ik}{2\kappa} e^{ikL} e^{\kappa L} A^{(3)} \tag{8.71}$$

$$B^{(2)} = \frac{\kappa + ik}{2\kappa} e^{ikL} e^{-\kappa L} A^{(3)}. \tag{8.72}$$

Last, we carry out Step 3 by inserting Eqs. (8.71) and (8.72) into (8.70) and rearranging slightly. The result of these machinations is the desired ratio,

$$\frac{A^{(3)}}{A^{(1)}} = \frac{4i\kappa k}{\left[(\kappa + ik)^2 e^{-\kappa L} - (\kappa - ik)^2 e^{\kappa L}\right]} e^{-ikL} \qquad (E < V_0). \tag{8.73}$$

This step doesn't complete the problem, but Eq. (8.73) is an important intermediate result. This ratio is useful, for example, in the analysis of limiting cases such as a very fat barrier or a high-energy state (see below). But before investigating these special cases, we'll use Eq. (8.73) to derive the transmission coefficient. To make this task a bit easier, I want to first simplify the form of this ratio.

Crafty Simplifications

The grail of our quest is the transmission coefficient

$$T(E) = \left|\frac{A^{(3)}}{A^{(1)}}\right|^2 = \left(\frac{A^{(3)}}{A^{(1)}}\right)\left(\frac{A^{(3)}}{A^{(1)}}\right)^* . \tag{8.74}$$

But working with an expression as complicated as (8.73) is fraught with peril. If your imagination is in good working order, you may be able to visualize the algebraic swamp we would enter were we to just plug Eq. (8.73) into (8.74). As we perform the necessary algebraic manipulations to obtain a workable expression for $T(E)$, it's very easy to omit one or another of the multitude of symbols in this expression, to switch $+$ and $-$ signs, or to omit a power, or .... One way to minimize the likelihood of such blunders is to introduce *intermediate quantities* that simplify the *form* of the equation we're manipulating.[28]

For example, it usually helps to write our result in terms of *dimensionless constants*. One such constant is the ratio of the wave number k to the decay constant κ, a quantity I'll call β:

$$\beta \equiv \frac{k}{\kappa} . \tag{8.75}$$

To write the ratio (8.73) in terms of β, let's multiply its numerator and denominator by $1/(4\kappa^2)$:

$$\frac{A^{(3)}}{A^{(1)}} = \frac{i\left(\frac{k}{\kappa}\right)}{\left[\left(\frac{1}{2} + \frac{i}{2}\frac{k}{\kappa}\right)^2 e^{-\kappa L} - \left(\frac{1}{2} - \frac{i}{2}\frac{k}{\kappa}\right)^2 e^{\kappa L}\right]} e^{-ikL} . \tag{8.76}$$

The form of the denominator in (8.76) suggests the definition of a second intermediate quantity,

$$\alpha \equiv \left(\frac{1}{2} + \frac{i}{2}\frac{k}{\kappa}\right) . \tag{8.77}$$

Notice that α and its complex conjugate appear in the denominator of (8.76).

With these definitions in hand, we can write the amplitude ratio (8.76) in the more malleable form

$$\frac{A^{(3)}}{A^{(1)}} = \frac{i\beta}{\alpha^2 e^{-\kappa L} - \alpha^{*2} e^{\kappa L}} e^{-ikL} . \tag{8.78}$$

Inserting this form into equation (8.74) for $T(E)$ we *easily* obtain

$$T(E) = \frac{\beta^2}{\alpha^2 \alpha^{*2}\left(e^{-2\kappa L} + e^{2\kappa L}\right) - \left(\alpha^{*4} + \alpha^4\right)} . \tag{8.79}$$

This formula is still far from elegant, and exploring its mathematical features, graphing it, examining it in various limiting cases, and programming it efficiently on a hand calculator or personal computer are likely to be awkward chores. So let's try to simplify it further.

[28]The gambits I'm going to illustrate are standard operating procedure for problem solvers. So please don't think of them as restricted to just this example; instead, remember them as tricks of wide applicability.

Gazing at Eq. (8.79), we notice in its denominator a pair of exponentials that differ only in their sign: $e^{-2\kappa L}$ and $e^{2\kappa L}$. We can write the difference of these functions as a *hyperbolic sine function*, *viz.*,

$$e^{-2\kappa L} + e^{2\kappa L} = \left(e^{-\kappa L} - e^{\kappa L}\right)^2 + 2 = 4\sinh^2 \kappa L + 2. \tag{8.80a}$$

And we can write the sum of α^{*4} and α^4 in terms of a *completed square*, as

$$\alpha^{*4} + \alpha^4 = \left[\left(\alpha^2 - \alpha^{*2}\right)^2 + 2\alpha^{*2}\alpha^2\right] = \left[-\beta^2 + 2\alpha^{*2}\alpha^2\right]. \tag{8.80b}$$

Now, inserting Eqs. (8.80) in (8.79), we obtain a more tractable form for the transmission coefficient for a potential barrier, our final result

$$\boxed{T(E) = \frac{\beta^2}{\frac{1}{4}\left(1+\beta^2\right)^2 \sinh^2 \kappa L + \beta^2}} \qquad \text{for } E < V_0 \tag{8.81a}$$

Investigating the Transmission Coefficient

Having gone to all this work, let's look at the transmission coefficient in several alternate forms:

$$T(E) = \left[1 + \frac{1}{4}\frac{\left(1+\beta^2\right)^2}{\beta^2}\sinh^2 \kappa L\right]^{-1} \tag{8.81b}$$

$$= \frac{4\kappa^2 k^2}{\left(k^2+\kappa^2\right)^2 \sinh^2 \kappa L + 4\kappa^2 k^2} \qquad \text{E} < \text{V}_0 \tag{8.81c}$$

$$= \left[1 + \frac{\sinh^2 \kappa L}{4\left(\frac{E}{V_0}\right)\left(1 - \frac{E}{V_0}\right)}\right]^{-1}. \tag{8.81d}$$

Each of these forms displays a different *kind* of information. For example, Eq. (8.81*c*) highlights the dependence of $T(E)$ on k and κ, and (8.81*d*) shows its dependence on the energy E.

Question 8–17

Derive Eqs. (8.81).

Question 8–18

Show that the reflection coefficient for $E < V_0$ is given by

$$R(E) = \frac{\sinh^2 \kappa L}{4\left(\frac{E}{V_0}\right)\left(1 - \frac{E}{V_0}\right)}\left[1 + \frac{\sinh^2 \kappa L}{4\left(\frac{E}{V_0}\right)\left(1 - \frac{E}{V_0}\right)}\right]^{-1}. \tag{8.82}$$

[There are two ways to do this exercise: easy and hard. Think: what condition must $T(E)$ and $R(E)$ satisfy? Now, what is the easy way?] Although this result is easily derived, it's inelegant. How many useful alternate forms for $R(E)$ can you devise?

Question 8–19

Consider a state with $E > V_0$. There are two ways to derive the transmission coefficient: the hard and the easy. The hard way is to go through algebraic contortions like those that led us to Eq. (8.81*a*) (doing so is, however, useful practice). The easy way is to set up the problem—as I did for $E < V_0$ in Fig. 8.17—and then to think carefully about how the present problem differs mathematically from the example we just solved. Solve this exercise the easy way, showing that in terms of the wave number in the region $0 \leq x \leq L$,

$$k_2 = \sqrt{\frac{2m}{\hbar^2}\left(E - V_0\right)}, \tag{8.83}$$

the transmission coefficient is

$$T(E) = \left[1 - \frac{\sin^2 k_2 L}{4\left(\frac{E}{V_0}\right)\left(1 - \frac{E}{V_0}\right)}\right]^{-1}. \qquad E > V_0 \tag{8.84}$$

The transmission coefficient of Eqs. (8.81) is on display in Fig. 8.18. This figure shows two "views" of barrier penetration. For completeness, it also includes the coefficient for the continuum states of Question 8–19, *i.e.*, those whose energies are greater than the barrier height.

The results in this figure are nothing less than remarkable. Look at Fig. 8.18a and compare its implications to your classical expectations. Classical physics would predict that no particles with $E < V_0$ are transmitted; quantum mechanics reveals that the probability of transmission of such particles *increases hyperbolically* with increasing energy. The other case is equally confounding: classical physics would predict that *all* incident particles with $E > V_0$ are transmitted; quantum mechanics shows that this condition—called **total transmission**—occurs *only at a few discrete energies.* An incident particle with an energy $E > V_0$ that lies between these special values may be reflected. The probability that such a reflection will occur diminishes rapidly with increasing E.

For another perspective on transmission and reflection by a barrier, look at Fig. 8.18b. In this graph, the energy E and the barrier height V_0 are fixed and $T(E)$ is plotted *as a function of the barrier width* L. This figure shows another bizarre result: for a given energy E, only barriers of certain widths will transmit all particles of this energy. (Such barriers are said to be "transparent.") But there is no value of the width such that a barrier of this width *reflects* all incident particles, because for all values of L, the reflection coefficient $R(E)$ is less than one.

Question 8–20

Denote by λ_2 the de Broglie wavelength in the barrier region, $0 \leq x \leq L$. How is λ_2 related to the special widths L at which total transmission occurs?

Question 8–21

The transmission and reflection coefficients tell only part of the story. Derive the ratios of $B^{(1)}$, $A^{(2)}$, and $B^{(2)}$ to $A^{(1)}$, and use your results to write down an explicit form for $\psi_E(x)$. Consider both cases: $E > V_0$ and $E < V_0$.

The probabilities of Fig. 8.18 are one step removed from the eigenfunction of Question 8–21. In Fig. 8.19, you'll find a graph of the real part of this (complex) function for an energy $E < V_0$. Notice the *penetration of the barrier* and the *oscillatory*

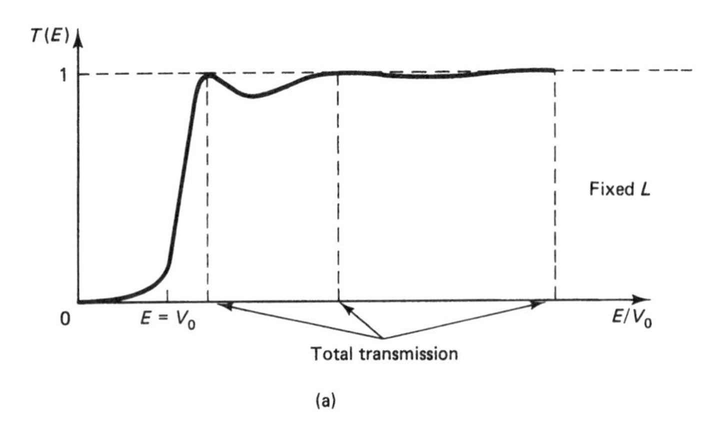

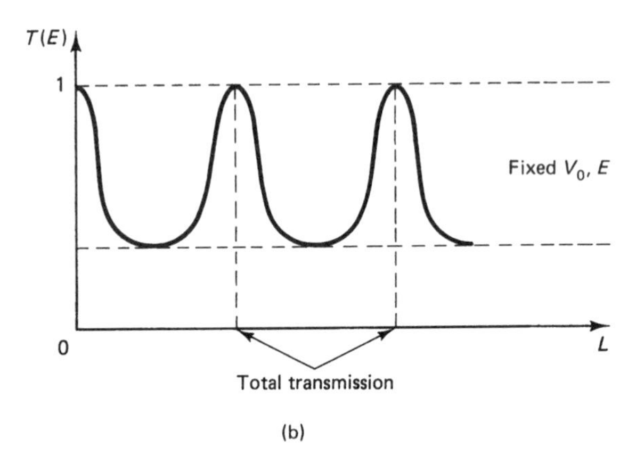

Figure 8.18 The transmission coefficient for barrier penetration (a) as a function of E for fixed V_0 and L, and (b) as a function of L for fixed V_0 and $E > V_0$. The parameters of the barrier in (a) were chosen so that $L^2 V_0 = 10\hbar^2/m$.

behavior in the two CA regions. Because of the hyperbolic decay of the eigenfunction in the CF region, the amplitude of the eigenfunction in the detector region, $x \geq L$ (and hence its probability density in this region), is reduced from its value in the source region. (Don't be misled by the oscillations in the region $x \geq L$; the only wave in this region is the transmitted wave, so the probability density there is constant.)

Question 8–22

Suppose we change the sign of V_0 so that the barrier in Fig. 8.17 becomes a potential well (see Fig. 8.16). Consider a continuum stationary state of this potential well ($E > 0$). How can you *trivially* derive from Eq. (8.81c) an expression for $T(E)$ for this case?

Aside: Scattering Theory in Three Dimensions. We now leave scattering problems and continuum states for a while. So I would like to close this section by relating what we have learned about scattering problems in the real (three-dimensional) world. In three dimensions, forward scattering (transmission) and backward scattering (reflection) are two of an infinite number of possibilities. The "detection probability" for scattering is the **differential cross section** $d\sigma/d\Omega$. The value of this quantity at angles θ and φ (see Fig. 8.6) is the probability that a particle will be detected at these angles. The corresponding *probability amplitude* is called the **scattering amplitude**;

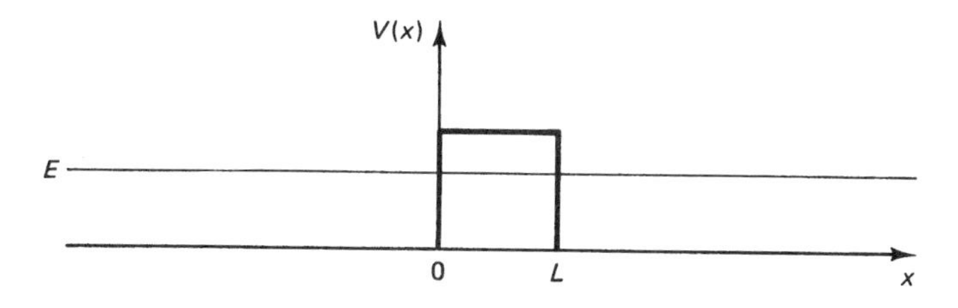

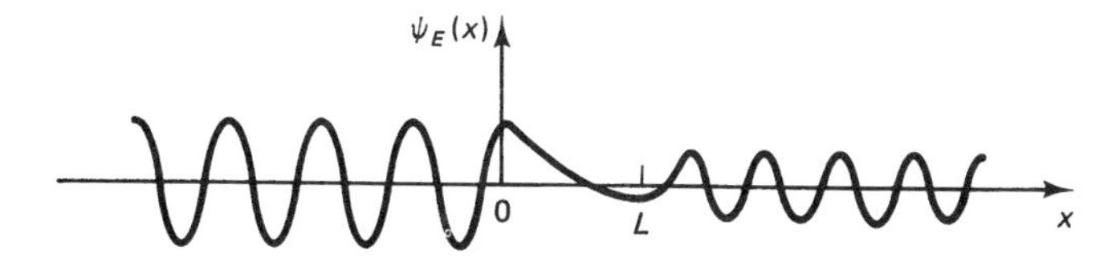

Figure 8.19 The probability density $|\psi_E(x)|^2$ for a continuum state of a particle incident on a barrier. The energy of the state is $E < V_0$, so this particle tunnels through the barrier, behavior that is mirrored in its transmission coefficient (see Fig. 8.18).

it is denoted $f(E;\theta,\varphi)$. The relationship of amplitude to probability is the familiar one,

$$\frac{d\sigma}{d\Omega}(\theta,\varphi) = |f(E;\theta,\varphi)|^2 .$$

In one dimension, we have $\rho(E) = f(E;\theta = 0)$ and $\tau(E) = f(E;\theta = \pi)$.

To finish our look at tunnelling, here's a historical note on the impact its discovery had on the developing field of quantum physics. Emilio Segré has discussed the importance of this discovery in his splendid book *From Atoms to Quarks* (Berkeley, Ca.: University of California Press, 1980). Some of you may be able to identify with Lord Rutherford in the following remark:

> George Gamow and Edward U. Condon with R. W. Gurney independently discovered the transparency of potential barriers and applied it to the explanation of the apparent paradoxes of nuclear alpha decay. This result deeply impressed Rutherford, who had been somewhat skeptical about a theory involving so much mathematics and so little intuition.

8.7 BOUND STATES OF THE FINITE SQUARE WELL

Having probed the mysteries of scattering states, we now turn to bound states. The bound-state counterpart to the extended continuum-state example of § 8.6 is the *symmetric finite square well* of Fig. 8.20b,

$$V(x) = \begin{cases} V_0 & -\infty < x \le -\frac{L}{2} \\ 0 & -\frac{L}{2} < x < \frac{L}{2} \\ V_0 & \frac{L}{2} \le x < \infty \end{cases} . \tag{8.85}$$

In this section and the next, we'll solve for the eigenfunctions (and quantized energies) of this system.

A significant difference between the bound- and continuum-state problems follows from the central contrasts between these two types of states: their boundary conditions [Eq. (8.7) *versus* Eq. (8.8)] and their allowed energies. A major difficulty in solving the TISE for bound states is that *we don't know the allowed, quantized energies in advance.* We must find these values by solving for Hamiltonian eigenfunctions that obey the boundary conditions $\psi_E(x) \to 0$ as $x \to \pm\infty$. Only such solutions can be normalized,

$$\int_{-\infty}^{\infty} \psi_E^*(x)\psi_E(x)\, dx = 1, \tag{8.86}$$

and only normalized eigenfunctions yield a normalized wave function $\Psi_E(x,t)$.

Qualitative Deductions

Rather than solve this problem by grinding out a "cookbook" implementation of the general procedure at the beginning of § 8.4, I would like to show you some shortcuts—tricks of the trade that may dramatically reduce the algebraic complexity of this problem. Our first guideline for determining *qualitative* solutions to quantum problems is

> ***Rule***
>
> Try to deduce as much as possible about the stationary-state eigenfunctions and energies by drawing analogies to states of a system you have already solved.

The first cousin of the finite square well is the *infinite square well* (Chap. 7). *(I strongly recommend that you stop reading now, go back, and review this example.)* In Example 7.2, we showed that this potential supports an infinite number of bound stationary states, with energies [Eq. (7.49)]

$$E_n^{(\infty)} = n^2 \frac{\pi^2\hbar^2}{2mL^2} \qquad n = 1, 2, 3, \ldots, \tag{8.87}$$

and eigenfunctions[29]

$$\psi_n^{(\infty)}(x) = \begin{cases} \sqrt{\dfrac{2}{L}} \cos\left(n\pi\dfrac{x}{L}\right) & n = 1, 3, 5, \ldots \\ \sqrt{\dfrac{2}{L}} \sin\left(n\pi\dfrac{x}{L}\right) & n = 2, 4, 6, \ldots \end{cases} \tag{8.88}$$

The first few spatial functions $\psi_n^{(\infty)}(x)$ are shown in Fig. 7.4, a glance at which reminds us that *each eigenfunction (8.88) has definite parity*; *i.e.*, each is either even or odd under inversion through the midpoint of the well. In this system, the parity of the Hamiltonian eigenfunctions mirrors the parity of the trigonometric functions:

$\sin kx \xrightarrow[x\to -x]{} \sin(-kx) = -\sin kx$	odd
$\cos kx \xrightarrow[x\to -x]{} \cos(-kx) = +\cos kx$	even

[29] I have introduced the label (∞) on the spatial functions and energies to distinguish them from the solutions for the finite well, which we'll determine in this section.

Another important property of the infinite-well eigenfunctions is that $\psi_n^{(\infty)}(x)$ has $n-1$ *nodes*, finite values of x at which $\psi_n^{(\infty)}(x) = 0$. Now let's use these properties to sketch the bound-state eigenfunctions of the *finite* well (8.85).

Here's what we already know about these functions. In Example 8.2, we classified the regions of the finite square well for bound states (states with energy $E < V_0$). From the arguments of § 8.4, we expect the eigenfunctions in the CA region (Region 2) to be oscillatory and those in the CF regions (Regions 1 and 3) to be decaying exponentials. (Remember that in the latter regions we must apply asymptotic conditions to guard against exploding exponentials.)

Consider the *ground state*, $\psi_1(x)$. The ground state of the *infinite* well, a simple cosine function, appears in Fig. 8.20a. To draw its counterpart for the finite well, we must somehow distort this cosine function so as to obtain a continuous and smoothly-varying function that is oscillatory for $-L/2 \leq x \leq L/2$ and that decays exponentially for $x < -L/2$ and $x > L/2$.

We do so by letting the cosine function "relax" so that at the "walls" this function goes to a *finite* value, rather than to zero (see Fig. 8.20b). Because of this change, the wavelength of $\psi_1(x)$ is slightly *larger* than that of $\psi_1^{(\infty)}(x)$. This increase in the wavelength allows the cosine function to "spill over" into the CF regions. One further change is necessary to preserve the normalization condition (8.86), which requires that the area under $|\psi_E(x)|^2$ be equal to one: the *amplitude* of $\psi_1(x)$ in the CA region is slightly smaller than the amplitude of $\psi_1^{(\infty)}(x)$ in this region.

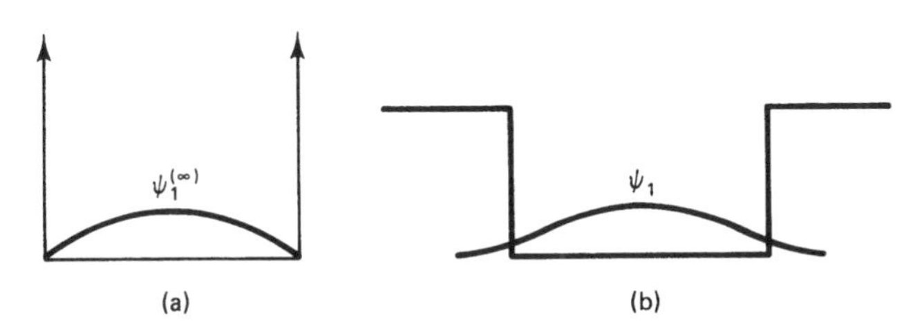

Figure 8.20 Sketches of the eigenfunctions for the ground states of (a) an infinite square well of width L, and (b) a finite square well of the same width and height V_0.

Notice that this sketch of $\psi_1(x)$ preserves two crucial properties of the corresponding infinite-well function: its *symmetry* and its *number of nodes*. When we distort an eigenfunction of the infinite well to produce an eigenfunction of the *symmetric* finite well, we must take care to preserve these properties—because in the limit $V_0 \to \infty$ any bound-state eigenfunction of the finite well must transform smoothly into the corresponding eigenfunction of the infinite well.

What can we deduce about the ground-state *energy* of the finite well? Well, this energy is related to the de Broglie wavelength in Region 2 by

$$E = \frac{\hbar^2 k^2}{2m} = \frac{4\pi^2\hbar^2}{2m\lambda^2}. \tag{8.89}$$

So an increase in λ results in a decrease in E. That is, because the de Broglie wavelength of $\psi_1(x)$ in Region 2 is slightly *larger* than that of its infinite counterpart, E_1 must be slightly *smaller* than $E_1^{(\infty)}$.

Question 8–23

Sketch the Hamiltonian eigenfunctions for the first and second excited states of the finite square well. On an energy-level diagram like Fig. 7.3, locate (approximately) the energies of the first three stationary states, showing their proximity to the bound-state energies of the infinite well.

This process of deducing qualitative information about complicated physical systems is not magic. You can do it. The principles I have applied are simple and familiar:

1. the uniqueness of $\psi_E(x)$ for a bound-state of a one-dimensional potential;
2. the normalization requirement (8.86);
3. the fact that a symmetric potential implies eigenfunctions of definite parity.

The latter property is so important that I'll summarize as follows:[30]

$$V(x) \xrightarrow[x \to -x]{} V(-x) = V(x)$$

$$\Downarrow$$

$$\psi_n(x) \xrightarrow[x \to -x]{} \psi_n(-x) = \begin{cases} +\psi_n(x) & n = 1, 3, 5, \ldots \quad \text{even parity} \\ -\psi_n(x) & n = 2, 4, 6, \ldots \quad \text{odd parity} \end{cases}$$

Thus, as we saw in Chap. 7, the ground state eigenfunction of the infinite well is even, and the excited-state eigenfunctions alternate parity with increasing n. In Chap. 9 we'll prove that this property is characteristic of *all symmetric potentials*:

$n =$	1	2	3	4	...
parity =	even	odd	even	odd	...

The moral—words to live by when solving problems—is

Rule

THINK BEFORE YOU DERIVE—and THINK SYMMETRY

These principles aid us not only in our quest for *qualitative* understanding of quantum systems, but also in the quantitative solution of the TISE, as in the next example.

Example 8.7. Solving the TISE for the Finite Square Well (with the Least Possible Work)

Let's return to the procedure of § 8.4. So far, we've identified and classified the regions of the finite square well, defined the necessary wave numbers and decay constants, and written down the pieces of the eigenfunction in the CA and CF regions (see Fig. 8.13). [Because of the parity condition, the piece of each eigenfunction in Region 2 will be a sine or a cosine

[30] Be wary of the quantum numbers we use to label Hamiltonian eigenfunctions. For most one-dimensional systems, it's conventional to choose $n = 1$ to label the ground state. But for some problems, such as the simple harmonic oscillator of Chap. 9, we pick $n = 0$ to label the ground state. The point to remember is that the parity of a state is a basic property of its eigenfunction, not a consequence of the index we use to label it.

function; so it makes sense to choose the real trigonometric form for $\psi_E^{(2)}(x)$.] The next step is applying the continuity conditions at the region boundaries.

To reduce the algebra involved in this step, let's take into account the *symmetry* of the functions we're trying to find. The parity of these functions enables us to separate the problem into two simpler problems, one for even-parity states and one for odd-parity states. That is, the general forms of the Hamiltonian eigenfunctions are[31]

$$\psi_n(x) = \begin{cases} Be^{\kappa x} \\ C\cos kx \\ Ae^{-\kappa x} \end{cases} \qquad \psi_n(x) = \begin{cases} Be^{\kappa x} & -\infty < x \le -\frac{L}{2} \\ D\sin kx & -\frac{L}{2} \le x \le \frac{L}{2} \\ Ae^{-\kappa x} & \frac{L}{2} \le x < \infty \end{cases} \tag{8.90}$$

for $n = 1, 3, 5, \ldots$ (even parity)   for $n = 2, 4, 5, \ldots$ (odd parity)

Question 8–24

Suppose we were to use the form of $\psi_E^{(2)}(x)$ (8.28), in which the exponentials e^{ikx} and e^{-ikx} appear. Write down this form for the finite square well and show that the only linear combination of these exponentials that have the necessary parity are, in fact, $\cos kx$ and $\sin kx$.

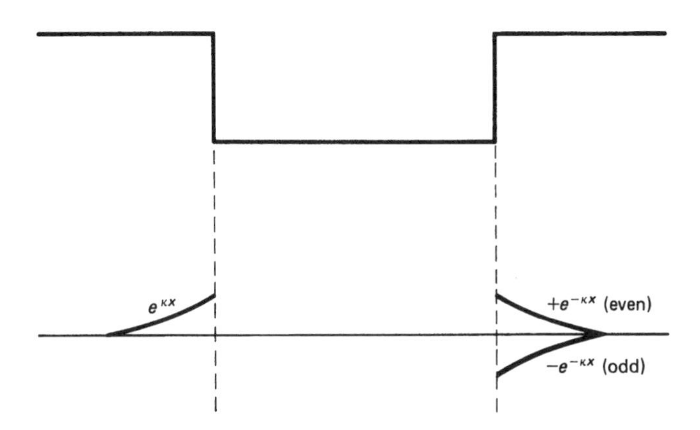

Figure 8.21 The pieces of a bound stationary-state eigenfunction of the finite square well in the classically-forbidden regions. Note the *symmetry* of these pieces of even- and odd-parity states.

Now let's apply the continuity conditions (8.40) at the region boundaries, $x = -L/2$ and $x = L/2$. We can use the power of parity to simplify this step dramatically by relating the piece of $\psi_E(x)$ in the CF Region 1 to the piece in the other CF Region, 2. Look carefully at Fig. 8.21: if $\bar{x}$ is any point in Region 3, then the parity conditions (8.89) imply

$$\psi_E^{(1)}(-\bar{x}) = \pm\psi_E^{(3)}(\bar{x}). \qquad \frac{L}{2} \le \bar{x} < \infty \tag{8.91}$$

[31] In this section, I'll write equations such as (8.90), which pertain to two cases, with the even-parity case on the left and the odd-parity case on the right. So if you wish, you can read along the left column first, then go back and read the right column. The index n will, as we'll soon see, label the allowed (quantized) values of κ and k.

This condition constrains the values of the coefficients A and B in Eq. (8.90), for Eq. (8.91) can be satisfied only if $B = \pm A$. So we can immediately eliminate one multiplicative constant from the Hamiltonian eigenfunction, leaving

$$\psi_n(x) = \begin{cases} Ae^{\kappa x} \\ C\cos kx \\ Ae^{-\kappa x} \end{cases} \qquad \psi_n(x) = \begin{cases} Ae^{\kappa x} & -\infty < x \le -\frac{L}{2} \\ C\sin kx & -\frac{L}{2} \le x \le \frac{L}{2} \\ -Ae^{-\kappa x} & \frac{L}{2} \le x < \infty \end{cases}. \tag{8.92}$$

I dropped D from (8.92), because we don't need two coefficients (C and D) in Region 2. The piece of $\psi_E(x)$ in this region is either a cosine function (if the state has even parity) or a sine function (if it has odd parity).

The simple reduction of Eq. (8.90) to Eq. (8.92) simplifies the ensuing algebra. Now we can determine $\psi_E(x)$ by solving at most *two* simultaneous algebraic equations, rather than the *four* we'd have faced had we ignored parity. [The (arduous) solution of those four equations would have led us to $B = \pm A$ anyway. But by using parity arguments, we have *deduced* a conclusion that we would otherwise have had to derive.] *Note carefully that this simplification does not apply to a potential that is not symmetric—e.g.*, to an asymmetric square well ($V_3 \neq V_1$).

This success leads us to seek additional simplifications due to symmetry. Consider the points where we must impose continuity conditions, the region boundaries $x_1 = -L/2$ and $x_2 = L/2$. These points are equidistant from the midpoint of the (symmetric) well, $x = 0$, so we can use the parity of the eigenfunction (8.92) to write

$$\psi_n(-\tfrac{L}{2}) = \pm\psi_n(\tfrac{L}{2}) \qquad \begin{cases} \text{even parity} \\ \text{odd parity} \end{cases}. \tag{8.93}$$

$$\frac{d}{dx}\psi_n(x)\bigg|_{x=-\frac{L}{2}} = \pm\,\frac{d}{dx}\psi_n(x)\bigg|_{x=\frac{L}{2}} \qquad \begin{cases} \text{odd parity} \\ \text{even parity} \end{cases}. \tag{8.94}$$

So *if we force $\psi_E(x)$ and its first derivative to be continuous at one of the boundaries, then continuity at the other boundary follows automatically from Eqs. (8.93) and (8.94), and we need not impose the conditions at the second boundary.* This wonderful property further reduces the number of equations we must solve! To convince yourself of the validity of this argument, examine Table 8.6, where I've imposed continuity conditions for the even-parity states at both boundaries. Notice that two sets of equations in this table are identical.

Now, onward to the continuity conditions. Let's do the *odd-parity eigenfunctions* (8.92). Applying Eqs. (8.93) and (8.94) to these eigenfunctions, we get a pair of *simultaneous algebraic equations*:

$$C\sin\left(\frac{1}{2}kL\right) = -Ae^{-\kappa L/2} \tag{8.95a}$$

$$Ck\cos\left(\frac{1}{2}kL\right) = \kappa Ae^{\kappa L/2}. \tag{8.95b}$$

The unknowns in these equations are A, C, *and the energy* E_n, which is buried in the wave number and decay constant

$$k \equiv \sqrt{\frac{2m}{\hbar^2}E_n} \qquad \kappa \equiv \sqrt{\frac{2m}{\hbar^2}(V_0 - E_n)}. \tag{8.96}$$

TABLE 8.6 CONTINUITY CONDITIONS FOR EVEN-PARITY STATES OF THE FINITE SQUARE WELL.

Continuity Conditions at $x = -\frac{L}{2}$		
$\psi_E^{(1)}(-\frac{L}{2}) = \psi_E^{(2)}(-\frac{L}{2})$ $\frac{d}{dx}\psi_E^{(1)}(x)\Big\vert_{x=-\frac{L}{2}} = \frac{d}{dx}\psi_E^{(2)}(x)\Big\vert_{x=-\frac{L}{2}}$	$\Longrightarrow$	$Ae^{-\kappa L/2} = C\cos\left(\frac{1}{2}kL\right)$ $\kappa Ae^{-\kappa L/2} = -kC\sin\left(-\frac{1}{2}kL\right)$
Continuity Conditions at $x = \frac{L}{2}$		
$\psi_E^{(2)}(\frac{L}{2}) = \psi_E^{(3)}(\frac{L}{2})$ $\frac{d}{dx}\psi_E^{(2)}(x)\Big\vert_{x=\frac{L}{2}} = \frac{d}{dx}\psi_E^{(3)}(x)\Big\vert_{x=\frac{L}{2}}$	$\Longrightarrow$	$C\cos\left(\frac{1}{2}kL\right) = Ae^{-\kappa L/2}$ $-Ck\sin\left(\frac{1}{2}kL\right) = -A\kappa e^{-\kappa L/2}$

Solutions to these equations exist only for certain discrete values of k and κ—*i.e.*, for *quantized* energies E_n.[32]

From Eqs. (8.95) we can easily derive a single equation that, when solved, will yield the quantized energies of the odd-parity states. To achieve this reduction for odd-parity states, we just eliminate the constants C and A, which are irrelevant to the energies, by dividing (8.95*b*) by (8.95*a*):[33]

$$\boxed{\frac{\kappa}{k} = \tan\left(\frac{1}{2}kL\right)} \qquad \boxed{\frac{\kappa}{k} = -\cot\left(\frac{1}{2}kL\right)} \tag{8.97}$$

n = 1, 3, 5, ... (even-parity states)   n = 2, 4, 6, ... (odd-parity states)

Question 8–25

Using the continuity conditions in Table 8.6, derive the equation in (8.97) for the even-parity states.

Equations (8.97) look quite different from the energy equations we've seen heretofore. The reason is that the allowed values of the energy are *implicit* in these equations. Mathematically speaking, Eqs. (8.97) are *transcendental equations.* The solution of such equations requires special tools, so I'll defer this matter until the next section. In the meantime, you can reassure yourself that they're correct by examining them in limiting cases (see Pblm. 8.11).

[32] There is a sound mathematical reason why we must leave the energy unspecified when we solve the TISE for bound stationary states. The normalization requirement (8.86) imposes an additional equation that the pieces $\psi_E(x)$ must satisfy, a condition not present for a continuum state. Were we to treat the energy of the bound state as given, we would have more equations than unknowns, and could not solve Eqs. (8.95). For more on this point, see p. 103ff in *Elementary Quantum Mechanics* by P. Fong (Reading, Mass.: Addison-Wesley, 1960).

[33] An alternate way to solve for the bound-state energies of the finite (and infinite) square wells—a method that is based on a sophisticated use of operators (Chaps. 10 and 11)—is described in the article by S. Kais and R. D. Levine, *Phys. Rev. A.*, **34**, 4615 (1987).

Aside: Still More Simplifications (the Logarithmic Derivative). If, as is often the case, we don't need the full form of $\psi_E(x)$, but rather only the equations for the bound-state energies, we can bypass most of the algebra in this example by transforming the continuity conditions (8.40) on $\psi_E(x)$ and its first derivative into a single condition on the **logarithmic derivative**. This quantity is defined as

$$\mathcal{L}[\psi_E(x)] \equiv \frac{d}{dx}\left[\ln \psi_E(x)\right] = \frac{1}{\psi_E(x)}\frac{d}{dx}\psi_E(x).$$

In terms of the logarithmic derivative, the continuity conditions (8.40) become the single equation

$$\left.\frac{1}{\psi_E^{(q-1)}(x)}\frac{d}{dx}\psi_E^{(q-1)}(x)\right|_{x=x_q} = \left.\frac{1}{\psi_E^{(q)}(x)}\frac{d}{dx}\psi_E^{(q)}(x)\right|_{x=x_q}.$$

You can easily derive Eqs. (8.97) by applying this equation to the even- and odd-parity eigenfunctions.

The last step in the solution of the TISE for the finite square well is to determine the eigenfunction. This entails using Eqs. (8.95) and (8.96) to relate the constants A and C, *viz.*,

$$A = Ce^{\kappa L/2}\cos\left(\frac{1}{2}kL\right) \quad \text{(even-parity states)} \qquad A = -Ce^{\kappa L/2}\sin\left(\frac{1}{2}kL\right) \quad \text{(odd-parity states)}. \tag{8.98}$$

These relations reduce the eigenfunctions (8.92) to

$$\psi_n(x) = \begin{cases} \left[C\cos\left(\frac{1}{2}kL\right)e^{\kappa L/2}\right]e^{\kappa x} & \\ C\cos kx & \text{for } n = 1, 3, 5, \ldots \\ \left[C\cos\left(\frac{1}{2}kL\right)e^{\kappa L/2}\right]e^{-\kappa x} & \text{(even parity)} \end{cases}$$

$$\psi_n(x) = \begin{cases} -\left[C\sin\left(\frac{1}{2}kL\right)e^{\kappa L/2}\right]e^{\kappa x} & \\ C\sin kx & \text{for } n = 2, 4, 5, \ldots \\ \left[C\sin\left(\frac{1}{2}kL\right)e^{\kappa L/2}\right]e^{-\kappa x} & \text{(odd parity)} \end{cases} \tag{8.99}$$

One lone constant, C, remains undetermined. This is the normalization constant for these eigenfunctions, and we could determine its value by imposing the normalization condition. But, because of the complexity of the finite-well eigenfunctions, we must do so by numerical integration. Since the result adds nothing to our understanding of the physics of the stationary states of this system, I won't bother.

Question 8–26

Some authors choose the zero of energy at the top of the finite square well rather than at the bottom, and let $-V_0$ denote the well depth. What changes would this choice bring about in the definitions of the wave number and decay constant (8.96) for this problem? How—if at all—would this choice change Eqs. (8.97) for the energies and (8.99) for the spatial functions? Can you see an advantage to this choice over the one I have made in this section?

What to Do if $V(x)$ Isn't Symmetric

Nature is not always so accommodating as in this example: not all potentials are symmetric. Suppose we're confronted with the *asymmetric square well*

$$V(x) = \begin{cases} V_1 & -\infty < x \leq x_1 \\ 0 & x_1 < x < x_2 \\ V_3 & x_2 \leq x < \infty \end{cases} . \tag{8.100}$$

In this problem, few of the wonderful simplifications we invoked for the symmetric case apply. Since $V_1 \neq V_3$, the decay constants in Regions 1 and 3 are not equal ($\kappa_1 \neq \kappa_3$), the eigenfunctions $\psi_n(x)$ do not have definite parity, and the coefficients of $\psi_E^{(1)}(x)$ and $\psi_E^{(3)}(x)$ are not equal [$A^{(3)} \neq B^{(1)}$].

So we know considerably less (in advance) about the eigenfunctions of the asymmetric well than we knew about those of the symmetric well. We can, however, write down the form

$$\psi_E(x) = \begin{cases} B^{(1)} e^{\kappa_1 x} & -\infty < x \leq x_1 \\ C^{(2)} \cos k_2 x + D^{(2)} \sin k_2 x & x_1 \leq x \leq x_2 \\ A^{(3)} e^{-\kappa_3 x} & x_2 \leq x < \infty \end{cases} . \tag{8.101}$$

To determine the coefficients in Eq. (8.101) we must apply continuity conditions at *both* region boundaries, x_1 and x_2. Doing so is algebraically messier than the example we have walked through here, but it's valuable practice with our new problem-solving tools. If you're game, try Pblm. 8.6.

8.8 THE FINITE SQUARE WELL: THE ANSWERS

We could use a number of strategies to conquer the transcendental equations for the finite square well, (8.97): graphical, iterative, and numerical (*e.g.*, via power-series). In this section I'll sketch two of these approaches: *graphical solution*, which highlights important mathematical features of these equations and which is useful if you can't get at a hand calculator or a computer, and *iterative solution*, which is quite accurate and easily programmable.[34]

Rewriting the Equations

To illuminate the role of the energy in the transcendental equations (8.97), let's write them in terms of the quantized values E_n. Actually, I want to use not E_n, but rather the binding energy ϵ_n. Crudely speaking, the **binding energy** of the nth bound stationary state measures how far E_n is from the top of the well. For $E = 0$ at the bottom of the well, the binding energy is defined as $\epsilon_n \equiv V_0 - E_n$. Physically, the binding energy is a measure of how "tightly bound" the particle is—the greater the binding energy, the deeper in the well is the energy eigenvalue. (You may know the binding energy from atomic physics, where it's often called the **ionization potential**.) In practical terms, the

[34] Although I don't advise skipping this section, you could omit one of these two methods. Graphical solution of transcendental equations is treated first; iterative solution begins with the subsection entitled "A Transcendental Iteration."

binding energy is the amount of energy you would have to supply to the particle to remove it from the well (with zero kinetic energy).

With this definition and Eqs. (8.96) for the wave number and decay constant, we can rewrite the transcendental equations (8.97) as

$$\tan\left[\frac{L}{2}\frac{\sqrt{2m(V_0-\epsilon_n)}}{\hbar}\right] = \left(\frac{V_0}{\epsilon_n}-1\right)^{-1/2} \qquad n = 1, 3, 5, \ldots \text{ (even-parity states)}$$

$$\cot\left[\frac{L}{2}\frac{\sqrt{2m(V_0-\epsilon_n)}}{\hbar}\right] = -\left(\frac{V_0}{\epsilon_n}-1\right)^{-1/2} \qquad n = 2, 4, 6, \ldots \text{ (odd-parity states)} \tag{8.102}$$

Only values of ϵ_n that satisfy one of these equations correspond to an eigenvalue E_n that yields an eigenfunction that is normalizable.

Question 8–27

Prove from the mathematical form of Eqs. (8.102) that the only allowed values of ϵ_n occur in the interval $0 \le \epsilon_n \le V_0$. Discuss briefly the nature of the Hamiltonian eigenfunctions for $E > V_0$.

Reducing Clutter in the Transcendental Equations

To minimize the chance that we'll make algebraic blunders while solving these equations, we should write them in terms of two *intermediate quantities*: a **potential-strength parameter**,[35]

$$\zeta \equiv \sqrt{\frac{mL^2}{2\hbar^2}V_0}, \tag{8.103a}$$

and the argument of the trigonometric functions (tangent and co-tangent),

$$\eta \equiv \sqrt{\zeta^2 - \frac{mL^2}{2\hbar^2}\epsilon_n}. \tag{8.103b}$$

With these quantities, the transcendental equations (8.102) become

$$\tan\eta = \frac{\sqrt{\zeta^2-\eta^2}}{\eta} = \sqrt{\left(\frac{\zeta}{\eta}\right)^2 - 1} \quad n = 1,3,5,\ldots \text{ (even-parity states)} \qquad \cot\eta = -\frac{\sqrt{\zeta^2-\eta^2}}{\eta} = -\sqrt{\left(\frac{\zeta}{\eta}\right)^2 - 1} \quad n = 2,4,6,\ldots \text{ (odd-parity states)} \tag{8.104}$$

[35] If we increase V_0 or L, deepening or widening the well and hence increasing its strength, we increase ζ. That's why this quantity is called a *potential strength parameter*.

We'll label the **roots** of these equations, the values of η that solve them, by the index n that labels the even- and odd-parity eigenfunctions of the Hamiltonian, *e.g.*, η_n.

Question 8–28

Derive Eqs. (8.104).

Question 8–29

Derive the following useful relationship between the roots of Eqs. (8.104) and the binding energies ϵ_n:

$$\epsilon_n = V_0 \left[1 - \left(\frac{\eta_n}{\zeta}\right)^2\right] = V_0 - \frac{2\hbar^2}{mL^2}\eta_n^2. \tag{8.105}$$

Transcendental Curves

Each of Eqs. (8.104) has the standard form of a transcendental equation (in the variable η): an equality between two functions of the variable, where at least one of the functions is a trigonometric, hyperbolic, exponential, or logarithmic function:

Even-Parity Roots

$$f_1(\eta) = f_2(\eta) \tag{8.106a}$$

$$f_1(\eta) = \tan\eta \qquad f_2(\eta) = \sqrt{\left(\frac{\zeta}{\eta}\right)^2 - 1} \tag{8.106b}$$

Odd-Parity Roots

$$f_3(\eta) = f_4(\eta) \tag{8.106c}$$

$$f_3(\eta) = \cot\eta \qquad f_4(\eta) = -\sqrt{\left(\frac{\zeta}{\eta}\right)^2 - 1} \tag{8.106d}$$

Here are some other transcendental equations you may have encountered in your studies:

$$\sin\eta = \eta, \tag{8.107a}$$

$$\ln\eta = e^{-\eta}. \tag{8.107b}$$

Notice that each transcendental equation defines *two curves*. So the roots of such an equation are the points where these curves intersect. This way of looking at a transcendental equation is the key to the simplest (though least accurate) solution technique: **graphical solution**. Just graph each curve and look for the points of intersection.

A graphical solution of the even- and odd-parity equations (8.104) for the finite square well is shown in Fig. 8.22. I've chosen the potential strength parameter ζ to produce several bound states, the first five of which are circled and labeled: three of these states are even, the other two are odd. [In Example 8.8 we'll look in more detail at the solution of Eqs. (8.104); at this point I just want you to get the general idea.] Notice that the principal branches of $f_1(\eta) = \tan\eta$ and $f_3(\eta) = \cot\eta$ are repeated for all values of η. One *even* root comes from each branch of the tangent function:

$$-\frac{\pi}{2} < \eta < \frac{\pi}{2},\ \frac{\pi}{2} < \eta < \frac{3\pi}{2},\ \frac{3\pi}{2} < \eta < \frac{5\pi}{2} \qquad \text{branches of } \tan\eta \tag{8.108a}$$

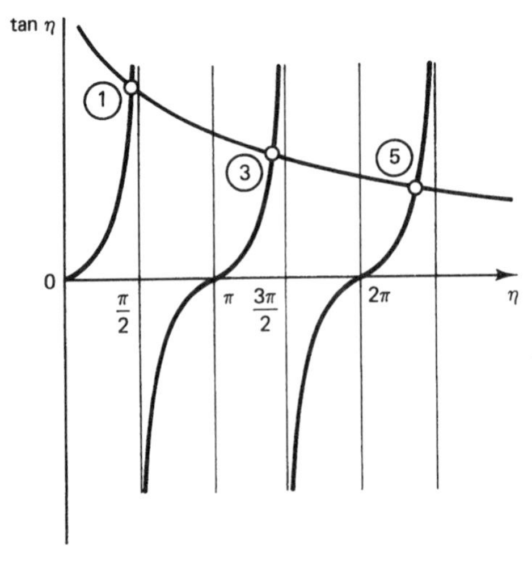

(a) Even parity

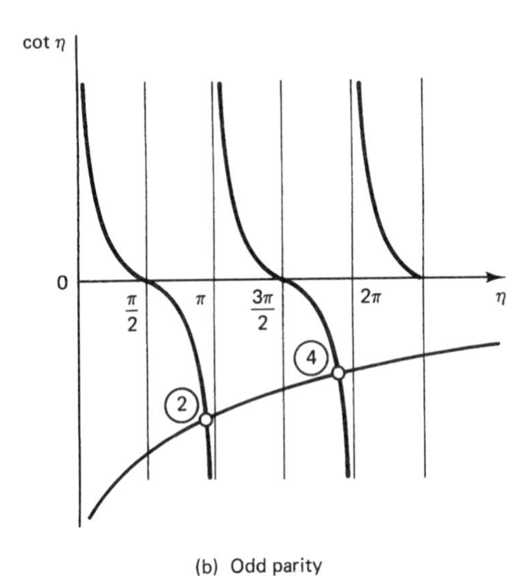

(b) Odd parity

Figure 8.22 Graphical solution of Eqs. (8.106) for three even- and two odd-parity states of a finite square well. The roots of these equations are the (circled) points of intersection of the two sets of curves. [Note that $f_2(\eta)$ and $f_4(\eta)$ are defined only up to $\eta = \zeta$; that's why these curves define only a finite number of roots.] [Adapted from *An Introduction to Quantum Physics* by A. P. French and E. F. Taylor (New York: Norton, 1978).]

And one *odd* root comes from each branch of the co-tangent function:[36]

$$0 < \eta < \pi, \quad \pi < \eta < 2\pi, \quad 2\pi < \eta < 3\pi \qquad \text{branches of } \cot\eta \tag{8.108b}$$

Example 8.8. A Transcendental Example

Consider an electron of mass m_e in a bound stationary state of a hydrogen atom. Suppose the range of the Coulomb potential binding the electron is (very roughly) $1\,a_0$. We'll model this system by a finite square well of width $L = 1\,a_0$ and depth $V_0 = 49\pi^2\hbar^2/(8mL^2)$. Like most atomic physicists, we'll solve this problem in **atomic units**:[37]

[36]Actually, only *part* of each branch contributes a root of each parity. For example, because $f_2(\eta)$ is a positive function, the even-parity roots show up in the intervals $0 < \eta < \pi/2$, $\pi < \eta < 3\pi/2$, etc,—*i.e.*, in the *first and third quadrants* of the unit circle. Similarly, the odd-parity roots show up in the *second and fourth quadrants*.

[37]If you're unfamiliar with atomic units, see Appendix F.

$$\hbar = m_e = a_0 = 1 \qquad \text{Atomic Units.} \tag{8.109}$$

In these units, the (dimensionless) potential strength parameter is $\zeta = 5.498$ and the depth of the well is $V_0 = 60.451 E_h$. (The atomic unit of energy is the **hartree**, written E_h; $1\,E_h = 27.201\,\text{eV}$.)

This potential supports *four* bound states, two of even parity and two of odd parity. Graphical solution of Eqs. (8.106) yields the roots η_n shown in Table 8.7. Also shown in this table are the binding energies ϵ_n and energy eigenvalues E_n, and the latter are compared to the bound-state energies for an *infinite* square well of the same width, $E_n^{(\infty)} = n^2\pi^2/2\,E_h$.

TABLE 8.7 BOUND-STATE ENERGIES FOR A FINITE SQUARE WELL.

		$V_0 = \frac{49\pi^2}{8}$		
n	η_n	$\epsilon_n\,(E_h)$	$E_n\,(E_h)$	$E_n^{(\infty)}\,(E_h)$
1 (e)	1.32	56.97	3.48	4.93
2 (o)	2.64	46.51	13.94	19.74
3 (e)	3.92	29.72	30.73	44.41
4 (o)	5.10	8.43	52.02	78.96

The Hamiltonian eigenfunctions obtained by inserting the finite-well energies E_n into Eqs. (8.96) and normalizing the resulting functions appear in Fig. 8.23. Finally, Table 8.8 contains, for each bound state, the wave number, decay constant, characteristic length, and wave length (in Region 2).

TABLE 8.8 WAVE NUMBERS, ETC., FOR A FINITE SQUARE WELL (IN ATOMIC UNITS).

			$V_0 = \frac{49\pi^2}{8}$		
n	k_n	λ_n	$\lambda_n^{(\infty)}$	κ_n	$1/\kappa_n$
1	2.64	2.38	2.00	10.67	0.09
2	5.28	1.19	1.00	9.64	0.10
3	7.84	0.80	0.66	7.71	0.13
4	10.20	0.62	0.50	4.11	0.24

Many of the features in these figures and tables are found in real microscopic systems. All of these features conform to our qualitative predictions:

1. The stationary-state energies are not equally spaced; rather, the spacing between adjacent bound-state energies increases with n;
2. The Hamiltonian eigenfunctions have definite, alternating parity (*c.f.*, the *qualitative* sketch of the ground-state eigenfunction in Fig. 8.20);
3. The de Broglie wavelength in the CA region is slightly *larger* than that of the infinite well, which allows for "spillage" into the CF regions;

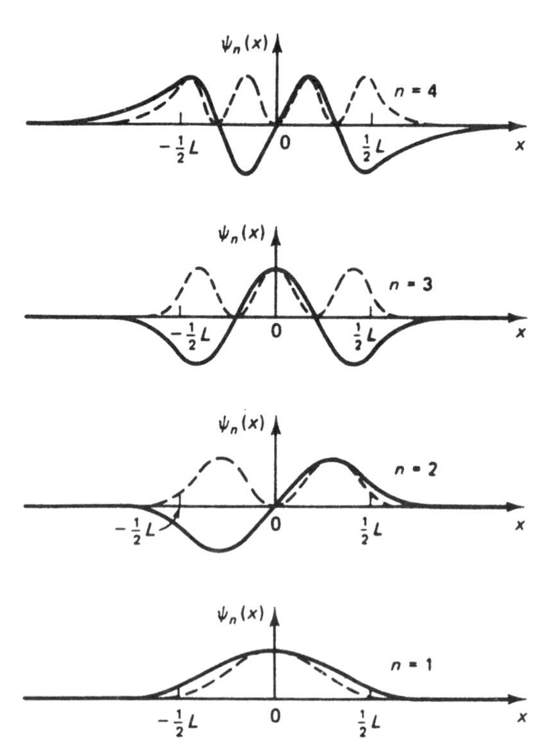

Figure 8.23 Hamiltonian eigenfunctions $\psi_n(x)$ (solid curves) and the corresponding probability densities $|\psi_n(x)|^2$ (dashed curves) for the first four stationary bound states of a finite square well. [Adapted from *An Introduction to Quantum Physics* by A. P. French and E. F. Taylor (New York: Norton, 1978).]

4. The characteristic length $1/\kappa_n$ increases with n, reflecting increased penetration of weakly-bound eigenfunctions into the forbidden zone.

Variation of the Potential Strength

Equation (8.105) shows the dependence of the binding energies ϵ_n on the depth and width of the potential well. But it may surprise you to learn that these quantities also control the *number of bound stationary states* that the well supports. To illustrate this point, Fig. 8.24 shows the graphical solution of the even-parity equation for three different values of the well depth V_0. As V_0 (and hence ζ) changes, the values η_n at which the curves (8.106) cross change and hence so do the eigenvalues E_n. From this figure, we deduce the following rule (see also Pblm. 8.2):

$$\text{increase well depth} \implies \begin{cases} \text{increase number of bound stationary states} \\ \text{increase binding energy of each state} \end{cases} \tag{8.110}$$

To illustrate this effect quantitatively, I repeated the calculations of Example 8.8 for a slightly *weaker* square well: instead of the values in that example, $\zeta = 7\pi/4$ and $V_0 = 49\pi^2/8$, I used $\zeta = 4$ and $V_0 = 32$. This well supports *three* bound stationary states, with binding energies (in atomic units)

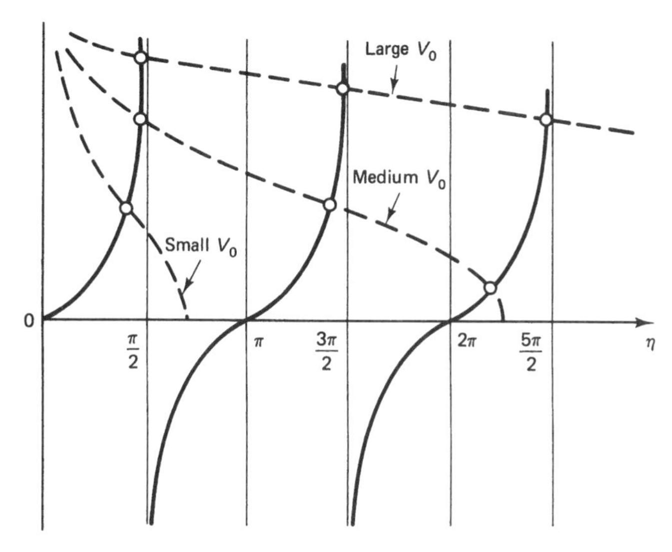

Figure 8.24 Solution of the energy equations of the finite square well for even-parity states. Three wells are considered, of varying heights as indicated. Note that both the number of bound stationary states and the energies of these states are affected by the strength of the well. [Adapted from *An Introduction to Quantum Physics* by A. P. French and E. F. Taylor (New York: Norton, 1978).]

$$\begin{aligned} \epsilon_1 &= 28.86\, E_h \qquad \text{even parity} \\ \epsilon_2 &= 19.74\, E_h \qquad \text{odd parity} \\ \epsilon_3 &= \ \ 6.14\, E_h \qquad \text{even parity} \end{aligned} \tag{8.111}$$

Compare these energies with those of the stronger well (Table 8.8). The weaker well supports one fewer bound state, and these states are *less tightly bound* than the three lowest states of the stronger well.

Figure 8.25 summarizes what we've learned about the transcendental energy equations for a finite square well. To make the curves in this figure easier to draw (and understand), I have graphed them in the equivalent form[38]

$$\begin{array}{cc} \eta \tan \eta = \sqrt{\zeta^2 - \eta^2} & -\eta \cot \eta = \sqrt{\zeta^2 - \eta^2} \\ \text{n} = 1, 3, 5, \ldots & \text{n} = 2, 4, 6, \ldots \\ \text{(even-parity states)} & \text{(odd-parity states)} \end{array} \tag{8.112}$$

[Notice that the right-hand side of each of Eqs. (8.112) is a simple quarter of a circle of radius ζ.] In Fig. 8.25, you'll find graphs of Eqs. (8.112) for several potential wells (*i.e.*, for several values of the potential strength parameter ζ). *Study this figure very carefully; it contains a great deal of useful information.*

Question 8–30

Show that the transcendental equations (8.104) can be written in the alternate forms

[38] To become a Master of Square Wells, you must be able to massage the energy equations into a variety of forms, so that you can accommodate various kinds of potentials. For extremely useful advice on graphical solution of transcendental equations, see P. H. Pitkanen, *Amer. Jour. Phys.* **23**, 111 (1955). By the way, you can buy a software package for an IBM PC that will graphically solve transcendental equations for you. It is called MATH UTILIT!ES and is available from Bridge Software, P. O. Box 118, New Town Beach, Boston, MA 02258.

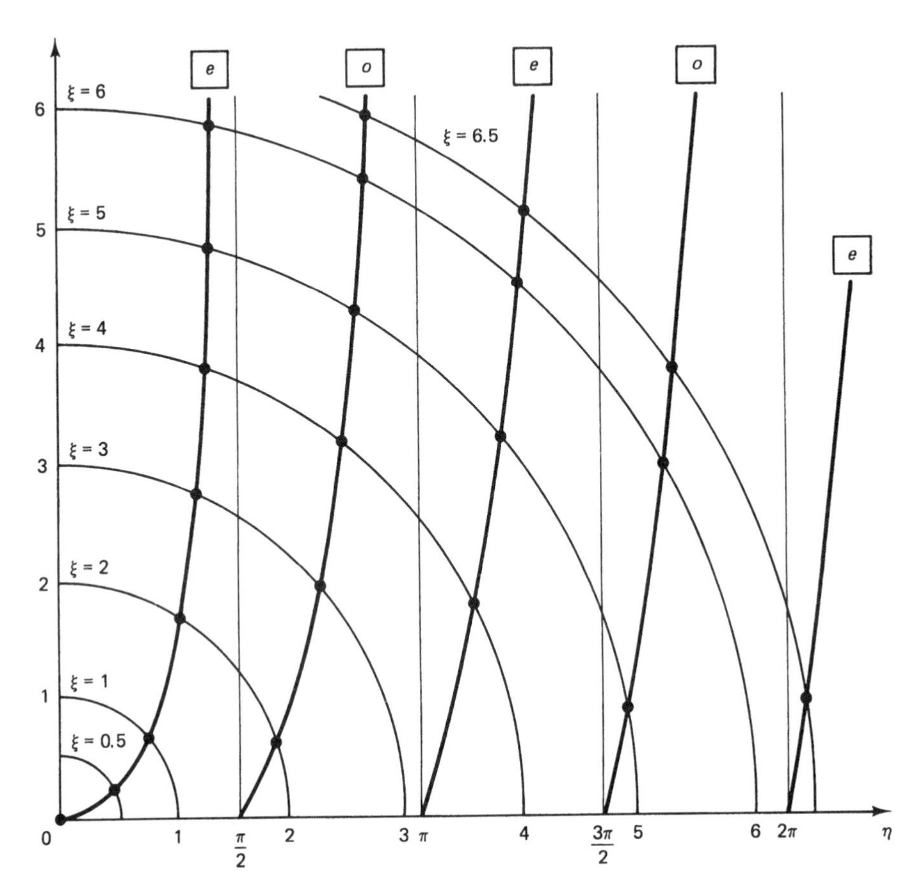

Figure 8.25 Graphical solution of the transcendental equations (8.112) of the finite square well. Even- and odd-parity states are included for several different potential strength parameters ζ. [Adapted from *Quantum Mechanics* by R. G. Winter, 2nd ed. (Davis, Calif.: Faculty Publishing, 1986).]

$$\cos\eta = \pm\frac{\eta}{\zeta} \qquad \text{even parity} \tag{8.113a}$$

$$\sin\eta = \pm\frac{\eta}{\zeta} \qquad \text{odd parity.} \tag{8.113b}$$

You should always consider using these forms when solving the energy equations graphically; Eqs. (8.113) are particularly easy to plot, because their right-hand sides are just straight lines of slope $1/\zeta$.

Question 8–31

In the above discussion, I varied the strength of the well by changing V_0. What would happen to the bound-state energies of, say, the well in Example 8.7 had I left V_0 alone and, instead, increased (or decreased) L? What would happen to the number of bound states supported by this well?

Question 8–32

Use the figures and transcendental equations of this section to prove that

$$\epsilon_n \xrightarrow[\zeta \to (n-1)\pi/2]{} 0. \tag{8.114}$$

Interpret this result physically: what does it imply about the effect on a particle in the well when the depth of the potential is varied according to the limit?

A Transcendental Iteration

Graphical solution of the transcendental energy equations for the finite square well yields a wealth of physical insights. But it is a time-consuming and not especially accurate method—at best, if you are an uncommonly good grapher, you may be able to determine ϵ_n accurate to a couple of decimal places. To calculate more accurate binding energies, you must resort to numerical methods, such as the iterative technique to be presented here.[39]

The idea behind the iterative solution of a transcendental equation is simplicity itself: construct a series of approximate solutions, called **iterates**, by repeated substitution into the equation for the root. Our starting point is the generic form (8.106) for a transcendental equation, $f_1(\eta) = f_2(\eta)$. To prepare such an equation for iterative solution, we rewrite it (yet again) as

$$\eta = F(\eta). \tag{8.115}$$

In so doing, we must carefully choose the function $F(\eta)$ to ensure that (if possible) our series of approximate solutions converges. (I'll show you how to do this below.) For example, Eq. (8.107*a*) is already in the form Eq. (8.115), $\eta = \sin\eta$, and Eq. (8.107*b*) can easily be put into this form: $\eta = e^{e^{-\eta}}$.

Once we have chosen $F(\eta)$, we develop a series of *approximate solutions* to (8.115), $\eta^{(0)}$, $\eta^{(1)}$, $\eta^{(2)}$, ... according to the simple prescription in Table 8.9. We calculate the ith iterate $\eta^{(i)}$ from the $(i-1)^{\text{th}}$ iterate by inserting the latter into the right-hand side of Eq. (8.115), *viz.*,

$$\eta^{(i)} = F\left(\eta^{(i-1)}\right). \tag{8.116}$$

Repeating this procedure, we ultimately arrive at an approximate (but very accurate) root η_n of the transcendental equation $\eta = F(\eta)$—if, that is, the sequence of iterates $\{\eta^{(0)}, \eta^{(1)}, \eta^{(2)}, \dots\}$ *converges*, *i.e.*,

$$\eta^{(i)} \xrightarrow[i\to\infty]{} \eta_n = F(\eta_n) \qquad \text{convergence of iterates} \tag{8.117}$$

Aside: On Accuracy of an Iterative Solution. Assuming that your sequence does converge, the *number* of iterates you must calculate to attain a root of a desired accuracy (say, 12 decimal places) depends on the accuracy of your initial guess $\eta^{(0)}$. So this guess should not be made in a cavalier fashion. (Indeed, whether or not your sequence of iterates converges at all may depend on how shrewd your initial guess is.) The moral is: make your initial guess an *informed* one: think carefully about the form of the transcendental equation you want to solve, perhaps evaluating it in one or more appropriate limiting cases to gain insight. Don't worry if you have to try several guesses to get smooth, rapid convergence (see Example 8.9).

[39] In this section I have drawn heavily on the valuable presentations by K. S. Krane, *Amer. Jour. Phys.* **50**, 521 (1982) and by R. D. Murphy and J. M. Phillips, ibid., **44**, 574 (1976). See also J. P. McKelvey, *Amer. Jour. Phys.* **43**, 331 (1975), J. M. Phillips and R. D. Murphy, ibid., **44**, 487 (1976), and J. A. Ball, ibid., **44**, 488 (1976).

TABLE 8.9 ITERATIVE SOLUTION OF $\eta = F(\eta)$.

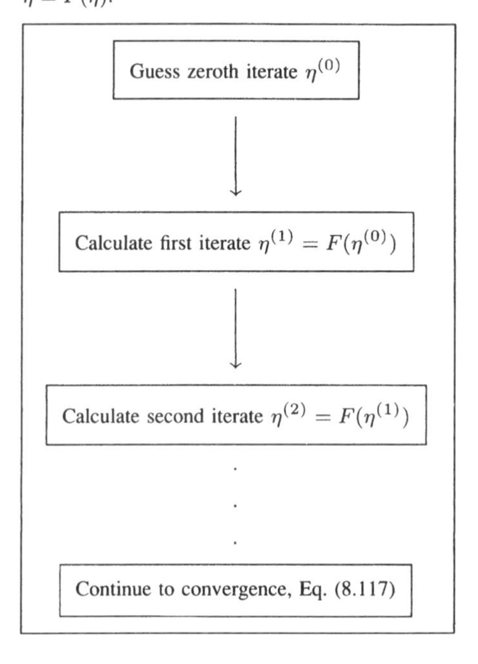

Question 8–33

Try the procedure in Table 8.9 on the simple equation $\eta = \sin\eta$. Your iterates will converge fairly quickly to a root no matter what you choose for $\eta^{(0)}$.

This book is not the place for a discourse on the convergence of iterative solutions to transcendental equations. But you need to know a bit about this topic, so you can intelligently apply this method to the transcendental equations of quantum physics. *The convergence (or lack of same) of Eq. (8.115) depends on the first derivative of $F(\eta)$, according to the following rule:*[40]

$$\text{the sequence (8.116) converges if } \left| \frac{d}{d\eta} F(\eta) \right|_{\eta=\eta_n} \leq 1. \tag{8.118}$$

This condition shows why the choice of $F(\eta)$ is critical. For example, the iterative method will fail for the simple equation $\eta = \sin\eta$ if we write this equation as $\eta = \sin^{-1}\eta$, for this form violates Eq. (8.118).

Question 8–34

Show that of the following equivalent forms of the even-parity equation,

[40]Two types of convergence are possible: If this derivative is negative, successive iterates will *oscillate* about a root η_n; if it is positive (and less than 1), successive iterates will converge monotonically to a root.

$$\eta = \cot\eta\sqrt{\zeta^2 - \eta^2} \tag{8.119a}$$

$$\eta = \tan^{-1}\sqrt{\left(\frac{\zeta}{\eta}\right)^2 - 1} \tag{8.119b}$$

$$\eta = \sqrt{\zeta^2 - \eta^2\tan^2\eta} \tag{8.119c}$$

only (8.119*b*) satisfies the condition (8.118). Write down the corresponding equation for the odd-parity roots.

Example 8.9. Iterating Our Way to the Roots of the Finite Square Well

To illustrate the iterative method of solution, let's consider (one last time) the finite square well of Example 8.8. This well supports $N_b = 4$ bound states; so we expect to find two roots of each parity (one from each quadrant of the unit circle). This knowledge guides our initial guesses for each root.[41]

I programmed Eqs. (8.119) on my trusty HP-25 hand calculator and obtained the roots in Table 8.10 (converged to the number of decimal points shown). For comparison, I have included in this table the results of graphical solution of the same problem (Example 8.8).[42]

TABLE 8.10 ACCURATE BOUND-STATE ENERGIES FOR A FINITE SQUARE WELL.

$V_0 = \frac{49\pi^2}{8}$			
n	$\eta^{(0)}$	η (converged)	η_n from Table 8.7
1 (e)	$\pi/4$	1.327016	1.32
2 (o)	$3\pi/4$	2.640594	2.64
3 (e)	$5\pi/4$	3.918889	3.92
4 (o)	$7\pi/4 - 0.001$	6.096720	5.10

8.9 FINAL THOUGHTS — AND AN APPLICATION

You can apply the procedure we used in this chapter to solve the TISE for piecewise-constant potentials to a wide range of problems. The strategy of approximating a potential near points where it varies rapidly by a discontinuous "jump" greatly simplifies the solution of many a TISE, and is widely used in research applications.

And even if we can't invoke this approximation, we may be able to use the problem-solving techniques of this chapter. For example, a powerful method for studying continuum states of atoms, molecules, and nuclei—called the "R-matrix method" (see

[41] Unless you are using a very sophisticated hand calculator or a computer, your numerical device will probably return arctangents in the principal branch of this function, so you will have to shift each root into the appropriate branch. This is easy enough, of course: just add the appropriate multiple of π [see Eqs. (8.108)].

[42] Until you work through this example yourself (which you really should do), you won't understand my choice of the initial guess for the fourth root. When I tried a point in the middle of the appropriate branch, $\eta^{(4)} = 7\pi/4$, I found that convergence was *very* slow and erratic. So I altered slightly my first guess; the choice in Table 8.10 led speedily to the converged root. Numerical methods are like guitars: you have to fool with them a bit to get them to produce what you want.

Pblm. 8.13)—is based on partitioning space into two regions and enforcing continuity (of the logarithmic derivative) at the boundary between the regions. This method is similar in spirit to our solution of the barrier problem in § 8.6.

Another rather sophisticated method for dealing with continuous potentials is the "WKB approximation." Like the approach used in this chapter, the WKB method entails constructing pieces of the Hamiltonian eigenfunction in CA and CF regions of the potential. But hooking these pieces together is rather complicated, for it requires the *exact* solution of the TISE at points very near the boundaries between adjacent CA and CF regions. The WKB method is particularly useful for calculating transmission coefficients through continuous barriers.[43]

Although we've spent a lot of time implementing the problem-solving procedure in § 8.4, the heart of this chapter is § 8.2 (on the physics of bound and continuum states) and § 8.3 and 8.5 (on the interpretation of continuum stationary states). The summaries in Tables 8.11 and 8.12 should help you to focus on the essentials of this chapter.

TABLE 8.11 BOUND AND CONTINUUM STATES FOR A PIECEWISE-CONSTANT POTENTIAL.

	Bound States	Continuum States
$\psi_E(x)$ as $x \to \pm\infty$	impose $\psi_E(x) \to 0$	$\psi_E(x)$ oscillatory (finite)
energies	only certain discrete E_n	any value of E OK
normalization	ensure $\int_{-\infty}^{\infty} \psi_E^*(x)\psi_E(x) = 1$	"normalize" to fit physics
to determine:	E_n and $\psi_n(x)$	probabilities $R(E)$ & $T(E)$
	Piecewise Constant $V(x)$	
solution in CA region	oscillatory	oscillatory
solution in CF region	decaying (exp.)	decaying (exp.)
first region	classically forbidden	classically allowed
last region	classically forbidden	classically allowed

Quantum-Well Heterostructures

Before leaving square wells, I want to introduce you to an amazing beast: the quantum-well heterostructure. This creature is a man-made, ultra-thin layer of crystalline matter,

[43]The initials stand for the last names of the inventors of this method, G. Wentzel, H. Kramers, and L. Brillouin. The WKB method straddles the nebulous line between undergraduate and graduate quantum mechanics and, with some regret, I have decided to leave it out of this book. You can read about it in most graduate and a few undergraduate texts; I particularly like the introduction in § 7.10 of *Introductory Quantum Mechanics* by R. A. Liboff (San Francisco: Holden-Day, Inc., 1980) and, at a more advanced level, the treatment in § 16.2 of *Principles of Quantum Mechanics* by R. Shankar (New York: Plenum, 1980). A clever alternative to the WKB method is the **path integral approach**, which was developed by Richard Feynman and which is explicated in the text *Quantum Mechanics and Path Integrals* by R. Feynman and A. Hibbs (New York: McGraw-Hill, 1965). To see how to apply this method to the tunnelling problem in § 8.6, see B. R. Holstein and A. R. Swift, *Amer. Jour. Phys.* **50**, 833 (1982).

TABLE 8.12 PIECES OF EIGENFUNCTIONS IN CA AND CF REGIONS. (FOR $V = 0$ AT THE BOTTOM OF THE LOWEST WELL)

Classically Allowed	Classically Forbidden
$E > V_q$	$E < V_q$
$\frac{d^2}{dx^2}\psi_E^{(q)}(x) = -k_q^2\psi_E^{(q)}(x)$	$\frac{d^2}{dx^2}\psi_E^{(q)}(x) = \kappa_q^2\psi_E^{(q)}(x)$
$k_q = \sqrt{2m(E - V_q)/\hbar^2}$	$\kappa_q = \sqrt{2m(V_q - E)/\hbar^2}$
$Ae^{ik_q x} + Be^{-ik_q x}$	$Ae^{-\kappa_q x} + Be^{\kappa_q x}$
$C\cos k_q x + D\sin k_q x$	$C\cosh \kappa_q x + D\sinh \kappa_q x$

with especially tailored electronic properties, that one can grow in a laboratory. They play an important role in the design of a variety of high-technology semiconductor devices, and you may study them someday in a solid-state physics course.

One can create very accurate models of quantum-well heterostructures using symmetric finite square wells. An example of a quantum-well heterostructure is shown in Fig. 8.26a, which shows a layer of gallium arsenide (GaAs) roughly 200 Å thick sandwiched between two thicker layers of aluminum gallium arsenide. (For reasons that need not concern us, these are labelled $Al_xGa_{1-x}As$.) A bound electron in the layer of GaAs is free to move in three dimensions. And its motion *in the plane of the sample*, which I'll denote the yz plane, is unconstrained. So the dependence of the wave function of the electron on y and z is (to a good approximation) that of a free particle.

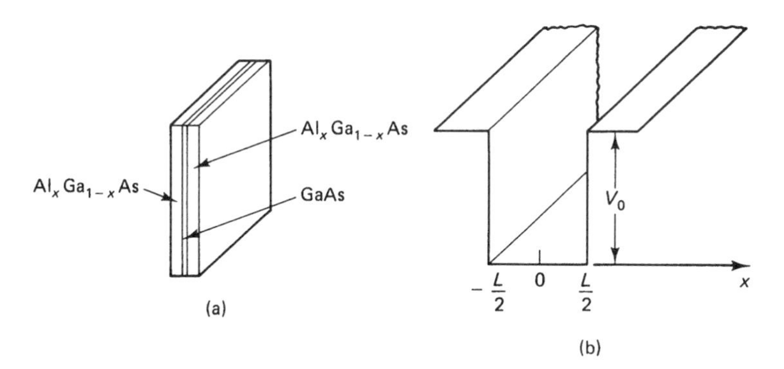

Figure 8.26 (a) A schematic of a quantum-well heterostructure. (b) The potential well experienced by an electron in the quantum-well heterostructure of (a). [After Fig. 1 of R. M. Kolbas and N. Holonyak, Jr., *Amer. Jour. Phys.* **52**, 431 (1984).]

But the potential energy perpendicular to this plane produces *a symmetric finite square well*—typically of depth $\sim 0.4\,\text{eV}$. So the x dependence of the electron's state function is given by one of the eigenfunctions in (8.99). Knowing the energies of the bound stationary states of this well, physicists can explain observations such as optical emission or luminescence spectra from a quantum-well heterostructure.[44]

[44] I don't have space here to do justice to these fascinating critters. For more, see R. M. Kolbas and N. Holonyak, Jr., *Amer. Jour. Phys.* **52**, 431 (1984) and the chapter "**Artificial Structures**" in *Solid State Physics* by Gerald Burns (New York: Academic Press, 1985).

So don't let anyone tell you that square wells are useful only to students of undergraduate quantum physics.

ANNOTATED SELECTED READINGS

Numerical Solution of the TISE

In § 8.8 I showed you two ways to solve transcendental equations: graphical and iterative. If the former is inaccurate and the latter doesn't converge, or if you want accurate eigenfunctions as well as the bound-state energies, you'll probably want to consider numerical solution of the TISE. This is a vast subject, full of powerful methods and lots of pitfalls. A good place to start learning about it—particularly if you have access to a personal computer—is

1. Killingbeck, J. P., *Microcomputer Quantum Mechanics*, 2nd ed. (Boston: Adam Hilger Ltd., 1983).
2. Merrill, J. R., *Using Computers in Physics* (Boston: Houghton Mifflin Co., 1976). Chap 12 of this book contains a nice discussion of applications of numerical methods to quantum physics.

The clearest discussion I've read of the application of numerical methods for solving differential equations to quantum problems is Appendix G of

3. Eisberg, R., and R. Resnick, *Quantum Physics of Atoms, Molecules, Solids, Nuclei, and Particles* (New York: Wiley, 1985). Like almost every discussion in this admirable book, this appendix is clear and filled with examples. It even includes a computer program.

Finally, you'll find a clever alternative method for calculating bound-state energies of a general piecewise-constant potential in the article by S. A. Shakir in *Amer. Jour. Phys.* **52**, 845 (1984).

EXERCISES AND PROBLEMS

Van Helsing: "We'll investigate the matter scientifically, without prejudice and without superstition."
Lucy Harker: "Enough of your science! I know what I have to do."

—*Nosferatu the Vampyre* by Werner Herzog

Exercises

8.1 Mysterious Transmissions

In § 8.6, we considered at length continuum states of a barrier potential for energies less than the barrier height, $E < V_0$. In Question 8.19, I asked you to derive Eq. (8.84) for the transmission coefficient for a state with $E > V_0$.

This equation shows that a particle incident from the left may be transmitted past the barrier or reflected back into Region 1, *i.e.*, $T(E) \leq 1$ and $R(E) \leq 1$. But, as shown in Fig. 8.18a, at certain discrete energies the particle will *definitely* be transmitted—for at these "resonance energies" $T(E) = 1$. Let's label the resonance energies by an index, n, as $E_n, n = 1, 2, \ldots$. We'll let ϵ_n denote their distance from the top of the well: $\epsilon_n = E_n - V_0$.

[a] **Derive** an expression for the resonance energies ϵ_n in terms of the barrier width L, the index n, and fundamental constants.

[b] Let λ_2 denote the de Broglie wavelength of the particle in the region $[-L/2 \leq x \leq L/2]$. If a transmission resonance occurs at an incident energy E_n, how must the barrier width L be related to the wavelength λ_2 for a particle with that energy?

[c] Are there transmission resonances for energies $E \leq V_0$? Why or why not?

[d] **Sketch** the transmission coefficient $T(E)$ as a function of energy for values of E from $E = 0$ to some value larger than the *third* resonance energy, $E > E_3$. On your sketch, show the location of V_0 in relation to the resonance energies.

[e] **Sketch** the Hamiltonian eigenfunction for the lowest-energy resonance, $\psi_{E_1}(x)$.

[f] In an experiment, an electron beam of current 3 nanoamperes and energy E is incident on a barrier of height $V_0 = 4\,\text{eV}$. As E is increased from zero, the first observed transmission resonance occurs at $\epsilon_1 = 2.35\,\text{eV}$. **Calculate** the width L of the barrier, expressing your result in angstroms.

8.2 One-Dimensional Scattering in the Lab

Muons are particles of charge $q = -1.602 \times 10^{-19}\,\text{C}$. The mass of a muon, m_μ, is roughly 1/9 that of a proton. Suppose a source emits a monoenergetic beam of muons in the direction of increasing x towards a target located at $x = 0$.

In this problem, we do not know the potential energy of interaction between the target and the projectile. Indeed, all we know is that this potential is constant in the source and detector regions. Far to the right of the target, there is a detector that measures a final current I_{final}.

[a] In one experiment carried out using this apparatus, the energy of the muons in the incident beam is $E_{\text{inc}} = 10.0\,\text{eV}$ and the particle density in the incident beam is 4000 muons per meter. The detector measures a current $I_{\text{final}} = 2.51 \times 10^{-11}\,\text{C-sec}^{-1}$. **Calculate** the probability that a muon in the incident beam will be reflected by its encounter with the target, expressing your result in counts per second.

[b] Suppose we modify the detector so that it can also measure the *kinetic energy* T_{final} of the muons it detects. Repeating the experiment of part [a], we find the same final current, and the final kinetic energy of $T_{\text{final}} = 6.0\,\text{eV}$. **Write down** the explicit form of the pure momentum function $\Psi_p(x,t)$ that approximately represents the state of particles in the *transmitted beam—i.e.*, determine the values of all constants, such as wavenumber and frequency, that appear in this function.

8.3 Stellar Quantum Physics

The energy of most stars results from the fusion of hydrogen nuclei to form helium nuclei. This notion was first proposed in 1920 by Eddington and was promptly rejected by his peers on the following grounds. At temperatures typical of the interior of stars, virtually *none* of the hydrogen nuclei have enough kinetic energy to overcome their mutual electrostatic repulsion and get close enough together to undergo a fusion reaction. But fusion *does* occur in stars. **Explain why.**

8.4 Quantum Mechanics of the Step Potential

A beam of electrons of wavelength 1.5833 Å is incident from the left on the step potential

$$V(x) = \begin{cases} 0 & x < 0 \\ 100\,\text{eV} & x \geq 0 \end{cases}.$$

The flux in the incident beam is 4.5941×10^{12} electrons/sec.

[a] **Write down** an explicit expression for the time-dependent pure momentum function that approximately represents the *incident* beam, fully evaluating all constants.

[b] **Derive** an explicit expression for the *spatial part* of the transmitted wave, *i.e.*, for $\psi_E(x)$ for $x > 0$, again evaluating all constants.

[c] From the definition of the probability current density and your answer to part [b], **evaluate** the transmitted current.

[d] What can you conclude from your answer to part [c] concerning the value of the transmission coefficient at this energy?

[e] **Evaluate** the position probability density at $x = 3\,\text{Å}$.

[f] **Explain** carefully why your answers to parts [d] and [e] are not mutually contradictory.

8.5 A Step-Down Scattering Problem

A beam of electrons with number density of 1×10^{15} electrons/m is incident from the left on the step potential energy

$$V(x) = \begin{cases} 0 & x < 0 \\ -V_0 & x \geq 0 \end{cases}.$$

(The constant V_0 is positive, so this is a "down step" rather than an "up step" like the potential in Fig. 8.12.)

[a] **Write down** the form of the spatial function $\psi_E^{(q)}(x)$ in each region of this potential, defining all wavenumbers and/or decay constants. Don't forget to impose asymptotic conditions for $x \to \pm\infty$.

[b] **Obtain** expressions for the transmission and reflection coefficients for this potential, expressing your results in terms of the quantities you introduced in part [a]. (This is not hard, because you don't have to derive these expressions from scratch. Instead, try to relate this problem to one you have already solved.)

[c] **Evaluate** $T(E)$ and $R(E)$ for an incident beam of kinetic energy 100 eV and a step of magnitude (V_0) 50 eV. Check your answer by verifying that probability is conserved [c.f., Example 8.6].

[d] **Evaluate** the incident and transmitted fluxes for the conditions of part [c]. Be sure to show units.

[e] **Sketch** $\psi_E(x)$, being careful to clearly show any differences in the nature of the spatial function or in its amplitude or wavelength between the regions.

8.6 Continuum States of the Finite Square Well

A beam of particles of mass m is incident from the left on the potential well in Fig. 8.13. Suppose that the energy of the particles is greater than the height of the walls, *i.e.*, $E > V_0$.

[a] **Write down** the TISE and its general solution in each region, defining appropriate wavenumbers and arbitrary constants as needed. If you can set one or more of these arbitrary constants to zero at this point, do so and explain why.

[b] **Evaluate** the reflected and transmitted *probability current densities* in each region. Be sure to indicate the direction and magnitude of each current density and label it as reflected, transmitted, etc.

[c] **List** the conditions that you would use to relate the pieces of the function $\psi_E(x)$ in each region to one another. Do you have enough conditions to *uniquely* determine all of the arbitrary constants in these solutions? If not, discuss the physical significance of the one (or ones) you cannot determine from these conditions.

[d] **Define** the probability of transmission into the region $x \geq L/2$, explaining the physical significance of this quantity and relating it to the current densities of part [c].

[e] **Show** that the transmission coefficient is given by

$$T(E) = \left[1 + \frac{1}{4}\frac{V_0^2}{E(E+V_0)}\sin^2 k_2 L\right]^{-1}, \tag{8.6.1}$$

where k_2 is the wavenumber in the region $-L/2 \leq x \leq L/2$. (This need not entail much work if you *think* about how this potential is related to one whose transmission coefficient you know.) Under what conditions on k_2 is $T(E) = 1$? Consider an energy E such that this condition is satisfied. At this energy, what is the value of the reflection coefficient $R(E)$ for this system? Why?

[f] **Compare** the descriptions given by quantum and classical physics of the motion of a (microscopic) particle with this potential.

8.7 The Mystery of the Potential Step

Here's a puzzler: Does our finding that $R(E) = 1$ for states of the potential step with $E < V_0$ contradict our finding that the probability density in Region 2 for such a state is non-zero? Suppose we were to measure the position of the electron and found it, say, in an interval in Region 2 of magnitude $1/(2\kappa_2)$. By such a measurement, we have localized the state of the particle, *i.e.*, we have changed its position uncertainty. Discuss the consequences of this localization for the momentum uncertainty of the projectile and for its energy uncertainty. Now, is it possible to detect an electron in the CF region with $T < 0$? If not, why not? If so, what sense does this make?

8.8 Applying the Correspondence Principle

Applying the correspondence principle to scattering problems can be tricky if you use the limit $\hbar \to 0$. But there is another way: taking the energy to extremes. Show that in the limits $E \gg V_0$ and $E \ll V_0$ the transmission coefficient obeys the expected classical results, *i.e.*,

$$T(E) \xrightarrow[E \gg V_0]{} 1 \tag{8.8.1}$$

$$T(E) \xrightarrow[E \ll V_0]{} 0. \tag{8.8.2}$$

Problems

8.1 A Review Problem on Stationary States of a Piecewise-Constant Potential

Consider a particle of mass m with the potential energy

$$V(x) = \begin{cases} \infty & x < 0 \\ V_1 & 0 \le x \le L/2 \, . \\ V_2 & x > L/2 \end{cases} \tag{8.1.1}$$

Were we to set $V_1 = 0$ and $V_2 = \infty$, this potential would reduce to an *infinite square well* of width $L/2$. But we've done that problem to death, so let's consider some other choices of these parameters.

Part A: Bound States

For this part, consider *only* energies in the range $V_1 < E < V_2$. Moreover, suppose that the depth of the well, $V_2 - V_1$, is great enough to support several bound states.

[A.a] Suppose that $V_2 = \infty$ but $V_1 > 0$. These choices give the potential shown in Fig. 8.1.1b. Determine the stationary-state energies and eigenfunctions for this potential.

[A.b] Suppose instead that $V_1 = 0$ but V_2 is finite and positive, *i.e.*, consider the potential of Fig. 8.1.1c.

(1) What are the boundary conditions on the bound-state eigenfunctions of this potential?

(2) Sketch the ground-state eigenfunction.

(3) Compare the energy of the function in (2) to the energy of the corresponding eigenfunction of an *infinite* square well of width $L/2$: *i.e.*, is the energy of your function less than, equal to, or greater than $E_1^{(\infty)}$? Explain your answer.

(4) The three lowest bound-state energies of a finite square well of width L and depth V_2 are $E_1 = 0.98\,V_2$, $E_2 = 0.383\,V_2$, and $E_3 = 0.808\,V_2$. Now, what is the ground-state energy of potential in Fig. 8.1.1c? Explain your answer.

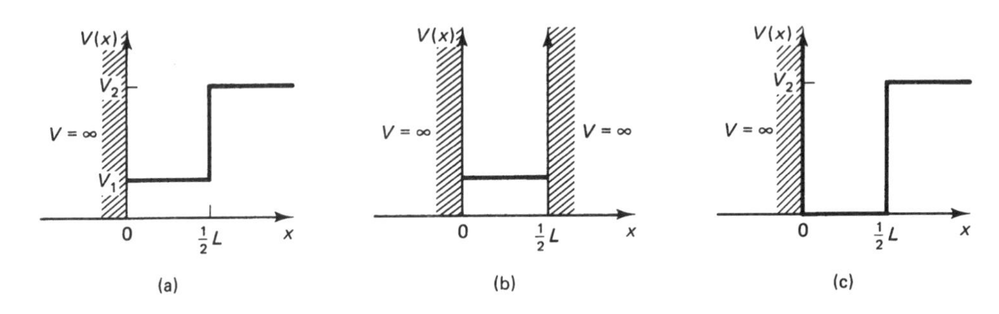

Figure 8.1.1

[A.c] Finally, consider the potential energy shown in Fig. 8.1.1a with $V_2 = 10.0\,\text{eV}$ and $V_1 = 1.0\,\text{eV}$. **Evaluate** the energy of the *ground state* of this potential.

Part B: Continuum States

In this part, consider *only* energies $E > V_2$ in Eq. (8.1.1) for $V_2 > V_1 > 0$.

[B.a] **Write down** the general solution of the TISE in each region defined by this potential. *Hint:* To simplify the rest of this problem, write your solutions in trigonometric rather than complex exponential form.

[B.b] **Apply** the boundary conditions on the eigenfunction at $x = 0$.

[B.c] **Write down** the continuity conditions that must hold at $x = L/2$. **Apply** these conditions to your general solution. (But don't try to solve the resulting equations!)

[B.d] Are the energies of the stationary states with $E > V_2$ quantized? Why or why not? Are they degenerate? Why or why not?

[B.e] Suppose $V_1 = 10.0\,\text{eV}$ and $V_2 = 20.0\,\text{eV}$. Consider a continuum state of energy $E = 30.0\,\text{eV}$.

(1) **Evaluate** the de Broglie wavelength of the electron in each region.

(2) **Draw** a sketch of the eigenfunction for this energy.

Part C: A Little Something Extra

[C] The infinity in the potential (8.1.1) at $x = 0$ has an interesting consequence for the problem of a particle beam incident on this potential *from the right* with energy E. **Evaluate** the probability that a particle in such a beam will be reflected back into Region 3. Give a derivation or (brief) justification of your answer.
Hint: Are the spatial functions for this problem travelling waves or standing waves?

8.2 How Many Bound States Are in Your Well?

From the discussion in § 8.8 on solution of the transcendental equations for bound-state energies of the finite square well, you can see why it's important to be able to determine in advance the number of bound states a particular potential supports. This problem explores some ways to do so. Unless otherwise stated, all questions pertain to a symmetric potential well of depth V_0 and width L.

[a] Have another look at the figures in § 8.8. Now, **explain** why the following rule holds:

> ***Rule***
>
> All symmetric finite square wells, however weak they may be, support at least one bound state.

Is this rule true for an *asymmetric* well, such as Eq. (8.100)? Why or why not?

[b] **Prove** that the number N_b of bound states supported by $V(x)$ is determined by the potential strength parameter ζ as follows:

$$(N_b - 1)\frac{\pi}{2} < \zeta < N_b\frac{\pi}{2} \qquad \Longrightarrow \qquad V(x) \text{ supports } N_b \text{ bound states.} \tag{8.2.1}$$

[c] Let $E_n^{(\infty)}$ denote the stationary-state energies of an *infinite* square well of width L. **Show** that the number of bound states N_b supported by the corresponding *finite* well is determined by where V_0 falls on the energy spectrum of the infinite well, according to the rule

$$E_{N_b-1}^{(\infty)} \leq V_0 \leq E_{N_b}^{(\infty)}, \tag{8.2.2}$$

with the convention that $E_0^{(\infty)} = 0$.

[d] **Determine** the number of even- and odd-parity bound stationary states in a well of depth $V_0 = 1.0\,\text{eV}$ and width $L = 1.0\,\text{Å}$. Also **determine** the number of states of each parity in a well of depth $V_0 = 100\,\text{eV}$ and width $L = 10\,\text{Å}$.

[e] Consider a quantum-well heterostructure (§ 8.9) that can be modeled by a symmetric finite square well of depth $0.361\,\text{eV}$ and width $180\,\text{Å}$. How many bound states (of an electron) does this heterostructure support?

8.3 Applications of the Finite-Square Well I: Atomic Physics

In the alkali atoms, the least tightly bound electron—the *valence electron*—experiences an attractive potential due to the nucleus and the other (so-called "core") electrons. The core electrons effectively screen the valence electron in such a way that its potential can (crudely) be modeled by a symmetric finite square well.

For example, the diameter of the potassium atom is roughly $4.324\,\text{Å}$. So we might choose as the width $L = 4.3\,\text{Å}$. Consider an electron in a symmetric finite well of this width and of depth $V_0 = 30.0\,\text{eV}$.

[a] How many bound stationary states does this potential energy support? What is the *parity* of each bound Hamiltonian eigenfunction?

[b] **Calculate** the binding energy (accurate to three decimal places) of each bound state. Express your answer in eV.

[c] **Graph** the spatial functions for each bound state. (You need not normalize your spatial functions. But in every other respect, your graphs should be *quantitatively* accurate.) **Sketch** a typical continuum-state function—*i.e.*, $\psi_E(x)$ for some energy $E > V_0$.

[d] For the ground state of this square well, **calculate** the probability for detection of a particle *outside the well*. **Discuss** the physical significance of this number.

[e] Suppose a beam of photons of wavelength λ is incident on an electron in the *ground state* of this potential well. For what values of λ could the incident photon be *absorbed*, inducing a transition of the electron into a higher-energy bound state or into the continuum? (Express your answer in angstroms.)

Remark: In actual atoms and molecules, electrons are forbidden from making certain transitions between stationary states by **selection rules**. These rules follow from the angular momentum properties of the bound stationary states of atoms. These properties are irrelevant to a one-dimensional system such as the present finite square well, but not to atoms (see Volume II).

8.4 Applications of the Finite Square Well: II (Nuclear Physics)

Guided by our understanding of classical E & M, we can write the potential energy of an atom or a molecule as the sum of Coulomb (electrostatic) interaction terms between pairs of charged particles. But in nuclear physics, there is no classical analogue from which to determine the potential energy of a nucleus. So nuclear physicists must resort to model potentials constructed from experimental data, such as the observed stationary bound-state energies of nucleons. Like much of our information about the physics of nuclei, these data are often determined in collision experiments. For example, we could model the potential of a nucleon in a nucleus by a *symmetric finite square well*.

In particular, consider a *neutron* in the ground state of such a well. From experiments in the laboratory, we find the following information:

(1) Gamma rays scattered from this nucleus can cause ejection of the neutron provided their kinetic energy is at least 8 MeV.

(2) The width of the well is $L = 3$ fermi.

How deep would a square well of this width have to be to bind a neutron by at least 8 MeV? Explain how you arrived at your answer.

8.5 Solution of the TISE by Numerical Integration

For some systems, analytical solution of the TISE (using the procedure in § 8.4) is not the easiest way to obtain the bound-state eigenfunctions and energies. For (many) others, analytic solution is impossible. In this problem, we explore an alternate way to solve for the eigenfunctions in such problems: numerical integration. (This method has the advantage that it is also applicable to continuous potentials, such as those we'll study in Chap. 9.)

For example, suppose we have a particle of mass m with the potential energy shown in Fig. 8.5.1,

$$V(x) = \begin{cases} \infty & x \leq -L/2 \\ 0 & -L/2 < x < -L/4 \\ V_0 & -L/4 \leq x \leq L/4 \\ 0 & L/4 < x < L/2 \\ \infty & x \geq L/2 \end{cases} \tag{8.5.1}$$

Let's choose $V_0 = \pi^2\hbar^2/8mL^2$ and concentrate on the *ground state*. Our goals are the energy and eigenfunction of this state, E_1 and $\psi_1(x)$.

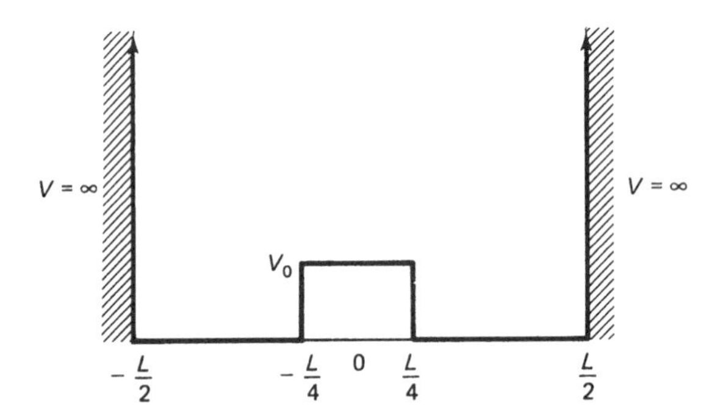

Figure 8.5.1

[a] In Chap. 7 we studied the ground state of the *infinite* square well. We can reduce the potential in Fig. 8.5.1 to such a potential by setting $V_0 = 0$. Using your understanding of the infinite well, **draw** a *qualitatively accurate* sketch of $\psi_1(x)$ for the potential shown above. Be sure to show all important changes in wavelength, etc. On the same graph, **sketch** the ground-state eigenfunction for the infinite well.

[b] Use numerical integration of the TISE to **determine** the value of E_1. On the same figure, **graph** your normalized spatial function $\psi_1(x)$ *and* the solution of the TISE for an energy near but not equal to E_1.

Hint: To start, you'll need a reasonable initial guess at E_1, which you can obtain by thinking about the ground-state energy for the infinite well. A reasonable step-size (Δx) is $L/20$. You must also decide where to start the integration. You know the symmetry of the ground-state eigenfunction and can deduce from this information the value of the first derivative (the slope) of $\psi_1(x)$ at $x = 0$. So this point is a good place to initialize your integration. Then integrate to $x = L/2$, adjusting E_1 until $\psi_1(x)$ obeys the appropriate boundary condition at $x = L/2$ (to, say, two decimal places). You need *not* normalize $\psi_1(x)$; thus you are free to pick the value of this function at the initial point equal to any convenient (real) number.

[c] Alas, sometimes numerical integration is impractical or, even worse, unstable. In such predicaments, analytic solution of the differential equation may be feasible (as in the finite square well of § 8.8). **Solve** the TISE for the potential 8.5.1. for the ground-state energy in $V(x)$ by analytic methods. That is, use the boundary and continuity conditions on $\psi_E(x)$ to derive a transcendental equation for the energies, and then solve your equation for E_1 in terms of V_0 by graphical or iterative method.

Hint: You can make your life a little easier by using real functions wherever possible.

8.6 The Dreaded Asymmetric Square Well

[a] Using the procedure of § 8.4, **evaluate** the energies of the bound stationary states of the asymmetric square well (8.100) for $V_1 = 50\,\text{eV}$ and $V_3 = 30\,\text{eV}$. Let the width of the well be $1\,\text{Å}$. (You will obtain a single transcendental equation for the binding energies in this well, which you can solve graphically or using the methods in § 8.8.)

[b] Does an asymmetric well such as the one in part [a] *necessarily* support one or more bound states, like the symmetric finite well? **Prove** your answer.

8.7 Quantum Reflection of Your Car (The Correspondence Principle)

A particle incident on a symmetric finite square well (not a barrier) of depth V_0 with energy *greater than* V_0 may be reflected from the left wall of the well. Let's see what happens in a macroscopic situation. Suppose a fast-moving automobile is hurtling towards a river bank. **Estimate** the probability that the driver will avoid drowning because his car is repelled by quantum mechanical reflection at the river bank. (There are some circumstances where you just can't count on quantum mechanics.)

8.8 Transmission in the Limit of a Very Large Barrier

Some barriers are very wide. This circumstance is a boon, for it allows us to derive a simple, physically reasonable approximation for the transmission coefficient $T(E)$, which otherwise is given by the less-than-simple result we derived in § 8.6, Eq. (8.81). (Even if your barrier isn't wide, you may want to examine its transmission and reflection coefficients in this limit. Such a study is an excellent check on your algebra.)

For an energy $E \le V_0$, consider the limiting case in which the barrier width L is so large that $\kappa L \gg 1$, where κ is the decay constant in the CF region. Since $1/\kappa$ is the characteristic length of penetration into the CF region, this condition amounts to the reasonable requirement that $L \gg 1/\kappa$.

[a] **Show** that in this limit the transmission coefficient is given by

$$T(E) \approx 16\left(\frac{E}{V_0}\right)\left(1 - \frac{E}{V_0}\right)e^{-2\kappa L}. \tag{8.8.1}$$

[b] What happens to $T(E)$ in the extreme limit $\kappa L \to \infty$? Why is this limit called *the classical limit*?

Hint: This problem is a case in point of my remark in § 8.6 that manipulating complex algebraic expressions is easier if you don't work with the final result—here Eqs. (8.81) for $T(E)$—but rather with a well-chosen *intermediate result*. In this problem, you should manipulate the expression (8.73) for the ratio $A^{(3)}/A^{(1)}$.

Remark: Transmission in the limit of a huge barrier is sometimes described in terms of a quantity called the opacity. The **opacity**, which is usually denoted α, is defined (for a barrier of width L and height V_0) as $\alpha = 2mV_0L^2/\hbar^2$.

8.9 Applications of Barrier Penetration

Let's consider some practical applications of the results we obtained in § 8.6 and in Pblm. 8.7. Suppose a beam of electrons of kinetic energy $3.8\,\text{eV}$ is incident from the left on a square potential barrier of height $14.5\,\text{eV}$ and width $5.3\,\text{Å}$.

[a] **Evaluate** the probability that a particle in the beam will be transmitted through the barrier using the exact equations in § 8.7, and with the approximate result you derived in the previous problem.

[b] Suppose $\kappa L \gg 1$. Use the result of the previous problem to **show** that for many values of E/V_0, the *order of magnitude* of the transmission coefficient is given to a good approximation by

$$T(E) \approx e^{-2\kappa L}. \tag{8.9.1}$$

What does this expression give for the order of magnitude of $T(E)$ for the physical conditions described above?

Now let's change the experiment a little. Consider two experiments, one involving a beam of protons, the other a beam of deuterons. In each experiment, the source produces a beam of incident energy 7 MeV that impinges on a barrier of height 10 MeV and width 10^{-14} m.

[c] Which beam, if either, will have a greater transmission probability? **Explain** your answer.

[d] **Verify (or contradict)** your answer to [c] by evaluating $T(E)$ for each beam.

8.10 Continuum States of a Two-Step Potential

A beam of electrons with energy E is incident from the left on the "two-step potential"

$$V(x) = \begin{cases} 0 & x < 0 \\ V_1 & 0 \le x < L \,. \\ V_2 & x \ge L \end{cases} \tag{8.10.1}$$

We'll suppose that $V_2 > V_1$ and will consider only continuum states with $E > V_2$.

[a] Derive the following expression for the probability of transmission into the region $x > L$:

$$T(E) = \frac{4k_1 k_2^2 k_3}{k_2^2 \left(k_1 + k_3\right)^2 - \left(k_1^2 - k_2^2\right)\left(k_2^2 - k_3^2\right)\sin^2 k_2 L}. \tag{8.10.2}$$

Graph $T(E)$ versus E for a range of energies $E > V_2$.

[b] Although I've given you the answer to [a], don't just *believe* it. **Show** that (8.10.2) reduces to the correct transmission coefficients for the special cases $V_1 = V_2$ and $V_2 = 0$.

[c] Let's consider this problem from a slightly different point of view. Suppose E is fixed but we can vary the width of the first step, L. **Determine** an expression for the value of L that will *maximize* the transmission coefficient $T(E)$, writing your answer in terms of the de Broglie wavelength in the region $0 \le x \le L$.

[d] In part [b] we saw that the transmission coefficient for the double step reduces to that of the single step if $V_1 = V_2$. But how is it related to the coefficients for the two-step potentials that make up the double-step in Fig. 8.10.1? Let $T_1(E)$ denote the transmission coefficient for a *single* step at $x = 0$ and $T_2(E)$ the coefficient for a *single* step at $x = L$. **Write down** expressions for $T_1(E)$ and for $T_2(E)$ and show that each is greater than or equal to $T(E)$ of part [a].

[e] Use your answer to [d] to demonstrate that $T(E) \neq T_1(E)T_2(E)$. But wait! Is this correct? Consider the following argument:

"The probability current density transmitted past the step at $x = 0$ is $T_1(E) j_{\text{inc}}$. This current is incident on the step at $x = L$. Therefore the current density transmitted into the region $x > L$ is $T_2(E)T_1(E) j_{\text{inc}}$, and we conclude that $T(E) = T_2(E)T_1(E)$."

Somewhere in this argument there lurks an incorrect assumption. What is it?

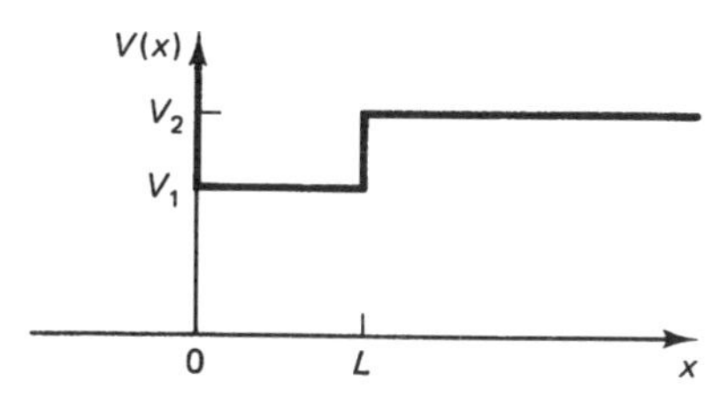

Figure 8.10.1 Two-step potential.

8.11 Square Wells: Going to Extremes

Having gone through the not-inconsiderable algebra required to determine the transcendental equations (8.97) for the bound-state energies and Eqs. (8.99) for the Hamiltonian eigenfunctions of the finite square well, how can we convince ourselves that our answers are correct? Well, one way to do so is to consider a limiting case for which we know the correct answer. For the finite square well, the obvious limit is that in which the well becomes infinite, $V_0 \to \infty$.

Show that in this limit, the finite square well *energies* and *spatial functions* reduce to Eqs. (7.49) and (7.55) for the *infinite* square well. (In considering the spatial functions, you need show only that their dependence on x reverts to that of the infinite square well spatial functions; don't worry about the normalization factor.)

8.12 A Nifty Problem-Solving Trick

You can solve lots of problems in quantum mechanics (and, for that matter, in most other fields of science and engineering) by analogy to problems that you have already worked. The key to mastering this powerful tactic is to *think carefully about how the problem you are trying to solve differs physically from problems whose solutions you know.*

For example, in some applications to solid state physics, the potential energy of an electron can be modeled rather accurately by the "semi-infinite" potential in Fig. 8.12.1,

$$V(x) = \begin{cases} \infty & x \leq 0 \\ 0 & 0 < x < L \, . \\ V_0 & x \geq L \end{cases} \tag{8.12.1}$$

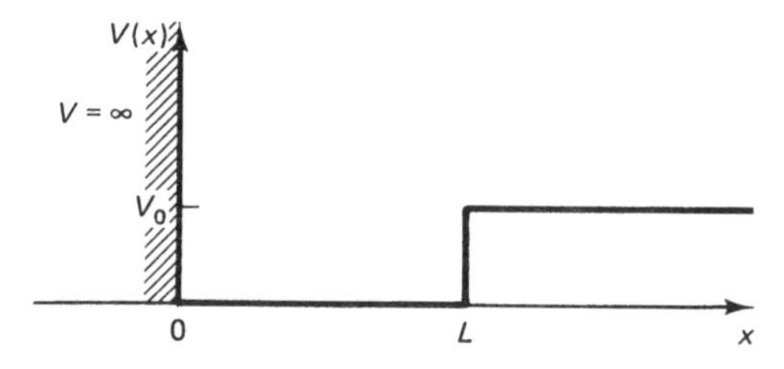

Figure 8.12.1 A "semi-infinite" square well.

[a] Suppose the well parameters V_0 and L are such that this potential energy supports *three* bound stationary states. **Sketch** the spatial functions of these states.
Hint: What are the boundary conditions on $\psi_E(x)$ at $x = 0$ and in the limit $x \to \infty$?

[b] What is the (transcendental) equation for the bound-state energies of the semi-infinite well? Answer this question in two ways:
1) by solving the TISE, imposing appropriate boundary and continuity conditions;
2) by writing down the answer from a consideration of the corresponding *finite* square well.

Which way was easier?

8.13 The R-matrix Method

In research, it's often very difficult to solve for the Hamiltonian eigenfunction of an atomic, molecular, or nuclear system, because we do not know the potential (*e.g.*, in nuclear physics) or because the TISE poses insuperable computational problems (*e.g.*, in atomic and molecular physics). One widely-used stratagem for conquering such problems is the **R-matrix method.** This method exploits the remarkable fact that you don't need to know the detailed structure of the eigenfunction in the region of the potential provided you can somehow evaluate this function (and its first derivative) at the "potential boundary"—the point in space beyond which the potential is (effectively) zero. To illustrate the R-matrix method, consider a very simple application: a symmetric, piecewise-constant one-dimensional potential.

[a] A beam of particles of energy E and unit amplitude is incident from the left on a symmetric potential $V(x)$. The potential is known to be zero except in the region $-L/2 \leq x \leq L/2$, but it is not necessarily piecewise-constant. All we know about it inside this region is that it is well-behaved (*e.g.*, continuous). **Show** that the *probability amplitudes* for reflection and transmission, $\rho(E)$ and $\tau(E)$, can be written in terms of the logarithmic derivative at $x = +L/2$,

$$\mathcal{L}_E(\tfrac{L}{2}) \equiv \frac{L}{2}\mathcal{L}[\psi_E(x)] = \frac{L}{2}\frac{d}{dx}\left[\ln \psi_E(x)\right] = \frac{L}{2}\frac{1}{\psi_E(x)}\frac{d}{dx}\psi_E(x). \tag{8.13.1}$$

[Notice that I have redefined this quantity (see § 8.7) by multiplying by $L/2$, so as to produce a conveniently dimensionless quantity.]

[b] **Derive** expressions for the reflection and transmission coefficients $R(E)$ and $T(E)$ in terms of $\mathcal{L}_E(\frac{L}{2})$.

[c] Now let's apply the result of part [a] to the barrier potential in Fig. 8.17,

$$V(x) = \begin{cases} 0 & x < -L/2 \\ V_0 & -L/2 \leq x \leq L/2\,. \\ 0 & x > L/2 \end{cases} \tag{8.13.2}$$

Show that for states with $E > V_0$, your results of part [a] reduce to Eqs. (8.81) for $T(E)$ and to the corresponding result for $R(E)$.

Hint: Since $V(x)$ is symmetric, there exist two (degenerate) eigenfunctions for each energy; these can be chosen so that each has definite parity. You can reduce the amount of algebra in this problem by defining a logarithmic derivative for each of these functions, say $\mathcal{L}_E^{(+)}(\frac{L}{2})$ for the even function and $\mathcal{L}_E^{(-)}(\frac{L}{2})$ for the odd function. You can then derive rather simple expressions for $\rho(E)$ and $\tau(E)$ in terms of these logarithmic derivatives.

Remark: In one dimension, the "R-matrix" that gives this method its name is the *inverse* of the logarithmic derivative. In applications of this method, it turns out that it is more convenient (computationally) to use this quantity rather than $\mathcal{L}[\psi_E(x)]$.

8.14 Resonances in One-Dimensional Potential Wells

One of the most tantalizing phenomena in physics is the *resonance*. Resonances are a phenomenon of scattering, and we can illustrate their basic physics in the context of a (slightly artificial) one-dimensional potential. (Actually, the creature we are studying is called a **virtual state**, because it occurs in a state with zero angular momentum; but such technicalities need not concern us. See, however, the following problem.)

Consider a particle of mass m with a "soft" square well potential, *i.e.*, a semi-infinite square well with a delta-function potential for positive x:

$$V(x) = \begin{cases} \infty & x < 0 \\ \alpha\delta(x - L) & x > 0 \end{cases}. \tag{8.14.1}$$

The constant α is called the *opacity* of the (repulsive) delta function potential, which is centered at $x = L$. We shall assume that $\alpha \gg 1$, *i.e.*, the barrier is highly opaque.

[a] Consider a continuum state of energy $E = \hbar^2 k^2/2m > 0$. **Show** that in the region *outside the well* the Hamiltonian eigenfunction for this state (normalized according to Dirac

Delta function normalization, as in Chap. 4) can be written in terms of the wavenumber k and a function of k that we'll denote $\beta(k)$ as

$$\psi_E(x) = \sin\left(kx + \beta(k)\right). \qquad (x > L) \tag{8.14.2}$$

As you may have guessed, $\beta(k)$ is the *phase* of the eigenfunction outside the "soft box." What is the form of $\psi_E(x)$ inside the well $(0 < x < L)$?

[b] **Derive** expressions for the phase $\beta(k)$ and the amplitude $A(k)$ of the eigenfunction *inside* the well.
Hint: Your expressions for $\beta(k)$ and $A(k)$ should be written in terms of k, L, and α. You will find the subsequent parts of this problem to be easier if you first transform your results from part [b] into expressions for $\tan\beta(k)$ and for $A(k)$ in terms of $\cos 2kL$ and $\sin 2kL$.

[c] **Derive** expressions for $\beta(k)$ and $A(k)$ in the limit of an infinitely opaque barrier ($\alpha \to \infty$) and thereby **show** that *except for certain special values of* k there is no penetration into the well (*i.e.*, the Hamiltonian eigenfunction inside the well is zero). *In this limit*, what are the energies corresponding to the special values of k for which penetration does occur? These special energies are the **resonance energies**, and the penetration phenomenon you have exhibited here is characteristic of scattering at a resonance energy.

[d] Analysis of the amplitude (which you need not carry out) reveals that for a finite but very large opacity ($\alpha \gg 1$), the amplitude $A(k)$ attains its maximum for values of k that satisfy the *transcendental equation*

$$\tan 2kL = -\frac{2k}{2m\alpha/\hbar^2}. \tag{8.14.3}$$

Like the transcendental equations we solved in § 8.8, this equation is satisfied only for a *discrete* set of wavenumbers. We'll label these roots $k_n; n = 1, 2, \ldots$. **Derive** expressions for the amplitude of the Hamiltonian eigenfunction inside the well at one of these "resonant wavenumbers" $[A(k_n)]$ and **show** that *for even* n, these amplitudes are huge $[A(k_n) \gg 1]$, while *for odd* n they are very small $[A(k_n) \ll 1]$. For $\alpha \gg 1$, where are the resonance energies in proximity to those of part (c)?

[e] To get a feeling for the extraordinary effect of a resonance on the eigenfunction (and hence on the position probability density), consider a specific case:

$$\alpha = \frac{\hbar^2}{2m}\frac{100\pi}{L}. \tag{8.14.4}$$

Graph $A(k)$ and $\beta(k)$ for values of k in the range $\frac{19\pi}{2L} \le k \le \frac{21\pi}{2L}$. You will find that there is one resonance in this range; discuss the effect of this resonance on $A(k)$ and on $\beta(k)$ for k in this range. What happens to the resonance energy and wavenumber as $\alpha \to \infty$?

8.15 More About Virtual States (in One Dimension)

In this problem, we'll look at virtual states from a different point of view than that adopted in Pblm. 8.13. Note: This problem does not require that you have completed Pblm. 8.13.

Consider the potential energy

$$V(x) = \begin{cases} 0 & x < -L/2 - b \\ V_0 & -L/2 - b \le x \le -L/2 \\ 0 & -L/2 < x < L/2 \\ V_0 & L/2 \le x \le L/2 + b \\ 0 & x > L/2 + b \end{cases} \tag{8.15.1}$$

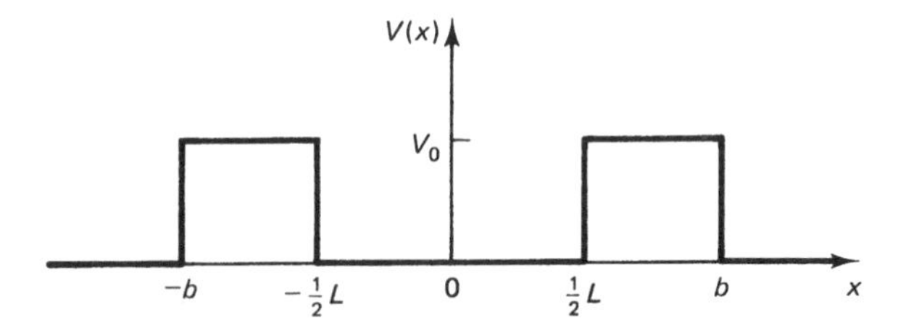

Figure 8.15.1 A piecewise-constant potential that supports a virtual state.

This potential consists of *two* barriers, each of height V_0 and equal width b, separated by a distance L (see Fig. 8.15.1). It is a repulsive potential that does not support any bound stationary states.

[a] **Discuss** qualitatively and quantitatively the stationary states of a particle with this potential energy. Consider two cases: $E \geq V_0$ and $E \leq V_0$. **Show** that in the energy range $0 \leq E \leq V_0$ there exist discrete values of the energy E_n such that an eigenfunction with energy near E_n has a much larger amplitude in the region between the barriers (*i.e.*, for $-L/2 \leq x \leq L/2$) than anywhere else in space.

At any time t_0, we can construct a *wave packet* that is highly localized in the region between the barriers by superposing a number of (continuum) stationary-state wave functions $\Psi_E(x, t_0)$ with energies in a narrow range ΔE_n about one of these "special energies" E_n. As time passes (*i.e.*, for $t \geq t_0$), the packet "leaks out" of the region between the wells. A quantum state represented by such a function is called a **virtual state** of energy E_n and energy uncertainty ΔE_n.

[b] Let Δt_n denote the time it takes the packet to "leak out" of the region between the barriers. **Show** that for a virtual state of energy E_n, the product $\Delta E_n \, \Delta t_n$ is a number of the order of Planck's constant.

[c] Consider the limit in which the width of the barriers becomes infinite: $b \to \infty$ with the condition that the width L of the well between the barriers is constant. **Show** that in this limit, the energies E_n of the virtual states of part [a] approach the *bound-state energies* of a finite square well of width L and depth V_0. Isn't that amazing?

States of a Particle in One Dimension II

Mainly Continuous Potentials

'My son, seek not
things that are too hard for thee;
and search not out
things that are above thy strength.'

—Jesus-ben-Sira
in *Between Sunset and Moonrise*
by R. H. Malden

In Chap. 8 we learned how to solve the TISE for bound and continuum stationary-state eigenfunctions of one-dimensional potentials that could be approximated by piecewise-constant functions. But most potentials are not amenable to so simple a model, for their variation with x is more complicated than a finite number of abrupt changes. Here, for example, are several such potentials, each of which plays an important role in a discipline of current research:[1]

$$V(x) = \frac{1}{2} m\omega_0^2 x^2 \qquad \text{harmonic oscillator} \tag{9.1a}$$

$$V(x) = ax \qquad \text{linear potential} \tag{9.1b}$$

$$V(x) = D\left[1 - e^{-a(x-x_e)}\right]^2 \qquad \text{Morse potential} \tag{9.1c}$$

$$V(x) = \frac{V_0}{\cos^2 ax} \qquad \text{Pöschl-Teller potential hole} \tag{9.1d}$$

[In Eqs. (9.1), everything that isn't x or a number is a parameter.] We cannot obtain eigenfunctions of Hamiltonians with these potentials using the machinery of Chap. 8, because their continuous nature prohibits us from subdividing space ($-\infty < x < \infty$) into a (finite) number of regions within which $V(x)$ is constant. Of course, the mathematical and physical *principles* of Chap. 8—boundary, asymptotic, and continuity conditions—do apply to these systems. But we need new mathematical tools for solving the TISE.

Most continuous potentials support both bound and unbound states. The eigenfunctions for *unbound states*—the continuum stationary-state functions—are not normalizable. But we can interpret them physically (as scattering probability amplitudes) because they remain finite in the asymptotic region; *i.e.*, for any E in the continuum, these eigenfunctions obey [*c.f.*, Eq. (8.8)]

$$\boxed{|\psi_E(x)| < \infty \qquad \text{Continuum state}} \tag{9.2a}$$

The eigenfunctions for bound states are normalizable—and hence physically admissible—only for certain discrete energies. All such functions must satisfy the boundary conditions [*c.f.*, Eq. (8.7)]

$$\boxed{\psi_E(x) \xrightarrow[x\to\pm\infty]{} 0 \qquad \text{Bound state}} \tag{9.2b}$$

In either case, of course, the full wave function is

$$\Psi(x,t) = \psi_E(x)e^{-iEt/\hbar}. \tag{9.3}$$

[1] For each of these potentials, *analytic* solutions of the TISE exist. In this chapter you'll see how to solve the TISE for the harmonic oscillator (§ 9.7) and the linear potential (§ 9.11). For a discussion of the Pöschl-Teller potential hole, see G. Pöschl and E. Teller, *Z. Physik*, **83**, 143 (1933).

Strategies for obtaining continuum eigenfunctions of continuous potentials differ only in detail from those for the piecewise-constant potentials of Chap. 8. So in this chapter, I won't dwell on continuum states; rather, I'll show you *how to obtain bound-state solutions of the TISE for a continuous potential*. Many and varied are the methods for doing so, and an in-depth look at several of them would take us far from the essentials of quantum physics. So I'll concentrate on a single, very powerful, widely applicable technique: *the method of power-series expansion (§ 9.7)*.

But one does not always need a *quantitative* solution of the TISE. If all we need is a rough idea of the nature of the eigenfunctions and the energies, we can use *qualitative arguments* such as those in Chap. 8. So we'll begin (in § 9.2) by generalizing these arguments from Chap. 8 and by learning a few new tricks. In subsequent sections we'll augment our qualitative understanding with quantitative results for selected systems.

So much effort goes into solving the TISE for these systems that it's easy to forget that stationary states are only a small, special subset of the quantum states of the system. So, for the principal example of this chapter (the simple harmonic oscillator), I'll supplement our study of stationary states in § 9.6–9.9 with a look at a non-stationary state (§ 9.10).

But first we must return to our scheme for analyzing stationary states according to regions defined by the potential and see what, if anything, of this procedure can be salvaged when the potential is continuous.

9.1 REGIONS, REGIONS EVERYWHERE

Identifying regions of piecewise-constant potential (Chap. 8) was trivial: we defined each region by the appropriate constant value of the potential, V_q, irrespective of the *state* we were considering. For a state of energy E, we then classified each region as classically-allowed (CA) or classically-forbidden (CF), accordingly as $E < V_q$ or $E \geq V_q$.

The difficulty posed by continuous potentials, such as those in Fig. 9.1, is that *a continuous potential does not uniquely demarcate regions of space*. Consider, for example, the three states indicated by horizontal lines in Fig. 9.1a. Two of these are continuum states; the other is a bound state. For the continuum state with the highest energy, all of space is classically allowed, because $E \geq V(x)$ for all x. But for the other continuum state, the region under the barrier is classically forbidden. And for the bound state, only the region in the well is classically allowed. Clearly, *to define the regions of a continuous potential we must take into account the energy of the state we want to investigate*. Yet, the classification scheme of Chap. 8 still applies:

$E \geq V(x)$	classically allowed
$E < V(x)$	classically forbidden

Since the regions of a continuous potential depend on the energy of the state, so do the locations of the *boundaries between regions*. Thus, as illustrated in Fig. 9.1b, the boundaries for two bound states of an attractive well occur at different values of x. Borrowing from classical physics, quantum mechanics call each boundary between a CA and a CF region a **classical turning point**. *The classical turning points are the values of x at which the total energy of a particle equals its potential energy*. (A macroscopic

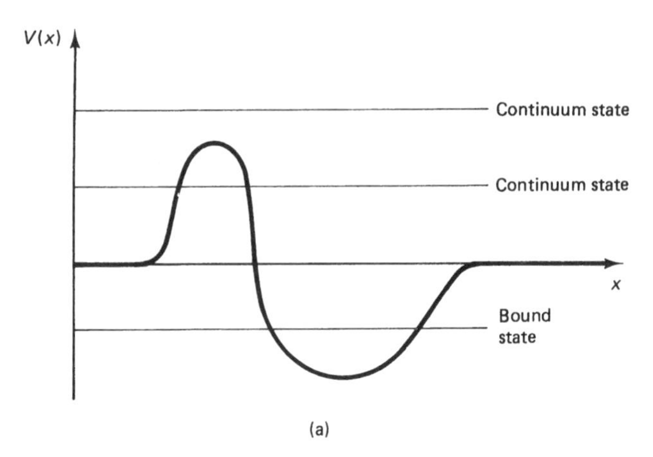

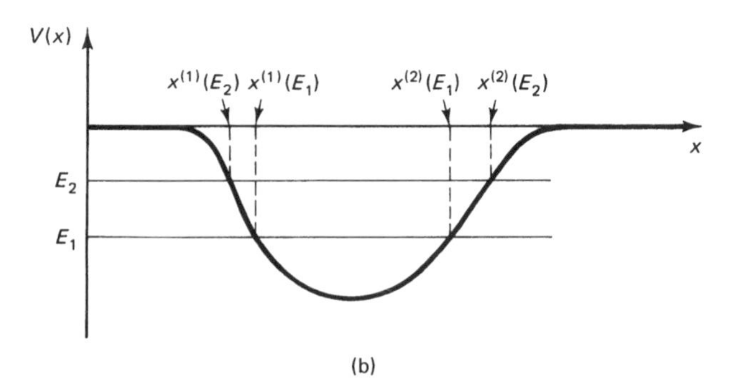

Figure 9.1 Two continuous potentials. (a) A finite-range potential with both a barrier and a well. Such a potential supports bound and continuum states, with different regions. (b) The regions and classical turning points of a continuous potential well depend on the energy E of its states, as indicated by the horizontal lines.

particle moving through a classically-allowed region will turn around when it reaches a position where its potential energy equals its total energy, because the kinetic energy of such a particle can't be negative.) To emphasize the dependence of these points on the energy, I'll denote the classical turning point between region q and region $q+1$ by the slightly ornate symbol $x^{(q)}(E)$. So, by definition,

$$V\left(x^{(q)}(E)\right) = E \qquad \text{[at a classical turning point].} \tag{9.4}$$

Having identified and classified the regions of $V(x)$ for a state, we can sketch the Hamiltonian eigenfunction. We prepare such drawings just as in Chap. 8: we consider each region separately, then join the pieces of the eigenfunction at the classical turning points in such a way that the eigenfunction and its first derivative are continuous everywhere. But when we try to implement this approach *quantitatively*, to solve the TISE as we did for a piecewise-constant potential in § 8.4, we run into big trouble.

What the Hamiltonian Eigenfunction in a Region Is Not

The potential in the TISE is the troublemaker. In the qth region, this equation is

$$\frac{d^2}{dx^2}\psi_E^{(q)}(x) = -\frac{2m}{\hbar^2}\left[E - V(x)\right]\psi_E^{(q)}(x). \qquad x^{(q-1)}(E) \le x \le x^{(q)}(E)\,. \tag{9.5}$$

Look carefully at the right-hand side. It does *not* have the simple form of a constant times the unknown function [*c.f.*, Eq. (8.22) and Eq. (8.29)]. So we cannot simply write down an analytic form for the solution in each region.

Equation (9.5) is an ordinary differential equation of the form

$$\frac{d^2}{dx^2}\psi_E(x) = (\text{function of } x)\,\psi_E(x). \tag{9.6}$$

So its solution, $\psi_E^{(q)}(x)$, is *not* a simple sum of sine and cosine functions in a CA region or of real exponentials in a CF region. Consequently, trying to work out this solution via a region-by-region analysis probably won't work. Instead, we must tackle the *full* TISE,

$$\frac{d^2}{dx^2}\psi_E(x) = -\frac{2m}{\hbar^2}\left[E - V(x)\right]\psi_E(x) \qquad -\infty < x < \infty. \tag{9.7}$$

To reinforce this point, ponder for a moment the ground-state eigenfunction of a simple harmonic oscillator [the potential (9.1a)] shown in Figure 9.2. As we'll discover in § 9.7, this eigenfunction, which for technical reasons is labelled with the index $n = 0$, is

$$\psi_0(x) = \left(\frac{m\,\omega_0}{\hbar\pi}\right)^{1/4} e^{-m\,\omega_0 x^2/2\hbar}. \tag{9.8}$$

The dependence on x of this function differs strikingly from that of an eigenfunction for a piecewise-constant potential: nowhere does $\psi_0(x)$ have the simple trigonometric or exponential dependence of the eigenfunctions of Chap. 8. So don't forget that

> ***Rule***
>
> The Hamiltonian eigenfunctions for a continuous potential are not made up of pieces each of which can be written as a sum of simple trigonometric or exponential functions. Instead, the functions must be obtained by solving the full TISE, Eq. (9.7).

Before tackling the quantitative solution of this equation, I want to show you how to deduce the *qualitative* nature of $\psi_E(x)$. Get ready to cast your mind back to high school algebra, for it is there that we shall find the key.

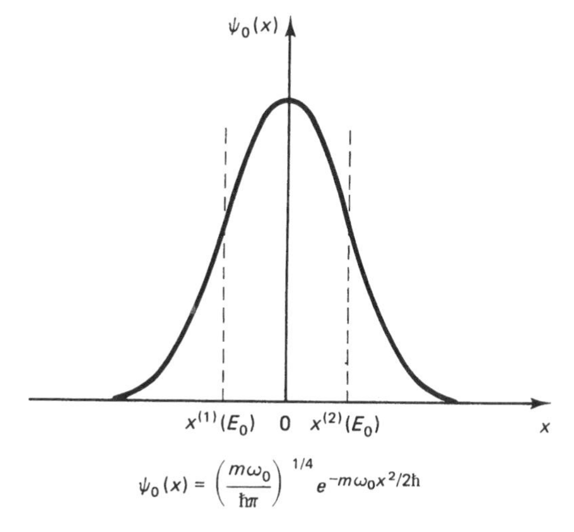

Figure 9.2 The ground-state eigenfunction of the harmonic oscillator (9.1a). The classical turning points are labelled $x^{(1)}(E_0)$ and $x^{(2)}(E_0)$.

9.2 CURVATURE IS THE KEY

You probably learned about the curvature of simple functions in your high school mathematics course. But you don't have to dig through a math book to see curvature. Curvature is the difference between a gently rolling hill and a precipitous mountain, between a narrow ravine and a wide valley. In math or in nature, the **curvature** of a function is a measure of the *rate of change of the function's slope*. The first derivative of a function is its slope, so the second derivative is its curvature. This quantity tells us "how rapidly the function turns over." (When I talk about "rapid" or "slow" variation of a function, I don't mean to imply variation in time. This colloquial usage refers to the variation of the function with x.)

A glance at the Schrödinger equation (9.7) will reveal why curvature is so important in the quantum mechanics of stationary states: the curvature of the Hamiltonian eigenfunction is $d^2\psi_E(x)/dx^2$. In this section, we'll discover what curvature can tell us about the nature of these eigenfunctions and their energies.

In Chap. 8, we learned how to stitch together spatial functions for systems with piecewise-constant potentials from oscillatory and exponential pieces. The nature of each piece—oscillatory or exponential—depended on whether its region was classically allowed or classically forbidden. More generally, *what determines the nature of any function $f(x)$ is the sign of the curvature of the function.* Examples of two types of functions are shown in Fig. 9.3. The function $f(x)$ in Fig. 9.3 is oscillatory; such a function is often said to be **concave**. The second derivative $d^2f(x)/dx^2$ is negative where $f(x)$ is positive, and $d^2f(x)/dx^2$ is positive where the function is negative. So the identifying feature of an oscillatory (concave) function is

$$\frac{1}{f(x)}\frac{d^2f}{dx^2} < 0 \qquad \Longrightarrow \qquad f(x) \text{ is oscillatory (concave).} \qquad (9.9a)$$

By contrast, the function $g(x)$ in Fig. 9.3 varies *monotonically* with x; functions with this behavior are said to be **convex**. The sign of the second derivative of this function is identical to the sign of the function itself, so the identifying feature of a monotonic (convex) function is

$$\frac{1}{f(x)}\frac{d^2f}{dx^2} > 0 \qquad \Longrightarrow \qquad f(x) \text{ is monotonic (convex)} \qquad (9.9b)$$

The simplest oscillatory and monotonic functions are, respectively, the trigonometric and (real) exponential functions of Table 8.3 (*c.f.*, Tables 8.11 and 8.12). But these functions do not solve the TISE for a continuous potential, so only their *qualitative* nature is relevant here.

Curvature and the Hamiltonian Eigenfunction

The simplest deduction we can make from the curvature of an eigenfunction is whether it is oscillatory or monotonic. The form of the TISE (9.7) illuminates this feature of the eigenfunctions:

$$\frac{d^2}{dx^2}\psi_E(x) = -\frac{2m}{\hbar^2}\left[E - V(x)\right]\psi_E(x) \qquad \text{[CA Region]} \qquad (9.10a)$$

$$\frac{d^2}{dx^2}\psi_E(x) = +\frac{2m}{\hbar^2}\left[V(x) - E\right]\psi_E(x) \qquad \text{[CF Region]} \qquad (9.10b)$$

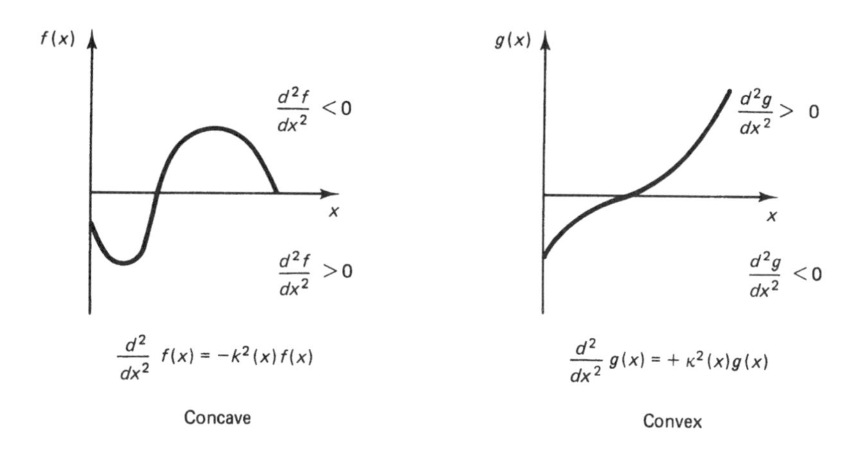

Figure 9.3 Examples of a concave and a convex function.

Clearly, the TISE determines the nature of the eigenfunction in each region defined by $V(x)$ and E. The curvature of an eigenfunction at a point x depends on the energy E, on the value of x, and, of course, on the functional dependence of $\psi_E(x)$ on x.[2]

In particular, *the sign* of the curvature of $\psi_E(x)$—and hence the nature of this eigenfunction—is determined by *the sign* of $E - V(x)$. In a CA region, the energy is greater than $V(x)$; so Eq. (9.9*a*) applies and the eigenfunction is oscillatory. The reverse argument pertains to a CF region, and we conclude[3]

$$\boxed{\begin{array}{llll} E \geq V(x) & \text{(CA region)} & \Longrightarrow & \psi_E(x) \text{ is oscillatory} \\ E < V(x) & \text{(CF region)} & \Longrightarrow & \psi_E(x) \text{ is monotonic} \end{array}} \qquad (9.10c)$$

But the nature of an eigenfunction is only a small part of what we can learn from its curvature. We can use the relationship between curvature and the energy to draw accurate sketches of eigenfunctions even for very complicated potentials. To introduce this topic, I want to compare qualitatively two eigenfunctions of the (not-very-complicated) potential in Fig. 9.4a. Superimposed on this potential are the energies of its first and second excited states. The corresponding eigenfunctions, $\psi_2(x)$ and $\psi_3(x)$, are shown in Figs. 9.4b and 9.4c.

Question 9–1

Explain why neither of the functions in Fig. 9.4 is the *ground-state* eigenfunction of the potential in Fig. 9.4a. Sketch the missing function.

Let's first deduce the shape of these eigenfunctions in the CA region. Since $E_3 - V(x) > E_2 - V(x)$, the curvature of $\psi_3(x)$ is *greater* than that of $\psi_2(x)$. Hence the de Broglie wavelength of $\psi_3(x)$ is *smaller* than that of $\psi_2(x)$ (*i.e.*, $\lambda_3 < \lambda_2$). (Notice

[2]Equations (9.10) show that the curvature of $\psi_E(x)$ depends on x. For this reason, $d^2\psi_E/dx^2$ should (strictly speaking) be called the "local curvature" of the eigenfunction.

[3]These arguments are consistent with our findings for a piecewise-constant potential. In this simple case $E - V(x) = E - V_q$. The pieces of $\psi_E(x)$ in a CA region are the simplest oscillatory functions, $\sin k_q x$ and $\cos k_q x$. Those in a CF region are the simple monotonic exponentials $e^{\pm\kappa_q x}$.

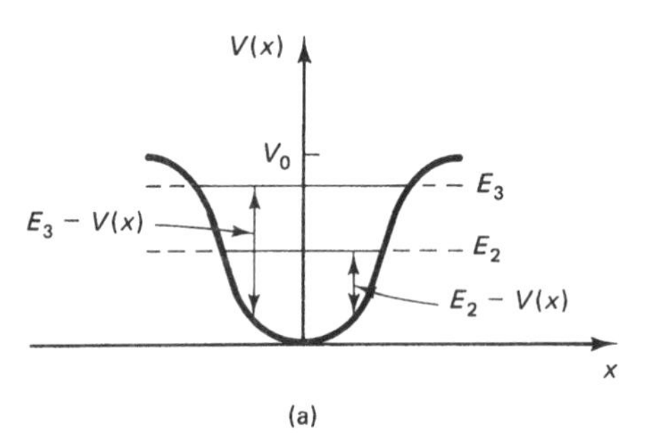

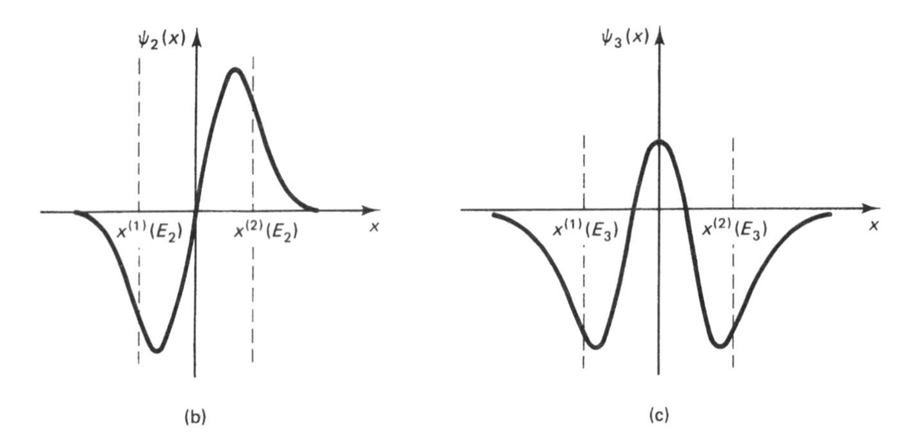

Figure 9.4 (a) A potential well and the eigenfunctions for (b) its first and (c) second excited states. The vertical dashed lines on the eigenfunction sketches are the classical turning points. [Adapted from *Introduction to Quantum Physics* by R. H. Dicke and J. P. Wittke (Reading, Mass.: Addison-Wesley, 1960).]

that the former eigenfunction has one more node than the latter.) Of course, we expect the de Broglie wavelengths to be related in this way, because $E_3 > E_2$, and $E \propto 1/\lambda^2$.

Now let's turn to the CF regions (outside the well). Here $V(x) - E_3 < V(x) - E_2$, so the curvature of $\psi_3(x)$ is less than that of $\psi_2(x)$. In these regions, both of these functions decay monotonically to zero, but the differences in their curvatures means that $\psi_3(x)$ decays to zero *more slowly* (with increasing distance from the midpoint of the well) than does $\psi_2(x)$. That is, $\psi_3(x)$ penetrates deeper into the CF regions than does $\psi_2(x)$. (If we pick $E = 0$ at the bottom of the well, as in Fig. 9.4a, we can translate this argument into a simple principle: the function with the larger energy is the most penetrating.) Mathematically, this difference appears in the decay constants for these functions, $\kappa_2 > \kappa_3$, and in the characteristic lengths, $1/\kappa_3 > 1/\kappa_2$.

Aside: Another Look at the Finite Square Well. These arguments explain differences between the various bound-state eigenfunctions of the symmetric square well of § 8.8 (*c.f.*, Fig. 8.23). Compare, for example, $\psi_1(x)$ and $\psi_2(x)$. The curvature of the latter function is roughly four times that of the former (because E_2 is roughly four times E_1). Correspondingly, the de Broglie wavelength of $\psi_2(x)$ is about half that of $\psi_1(x)$.

These arguments are also reflected in the data in Tables 8.7 and 8.8. Consider, for example, the CF regions of the finite square well. As the quantum number n (and hence the energy E_n) increases, the decay constant κ_n decreases, and $\psi_n(x)$ penetrates more deeply into the CF region.

The continuous potential in Fig. 9.4 looks rather like a "smoothed out" square well. And, indeed, the primary difference between the behavior of the eigenfunctions of this potential and those of the finite square well is that the pieces of the spatial functions for the continuous well are not simple trigonometric or exponential functions. Rather, they are complicated oscillatory or monotonic functions.

To summarize this discussion of contrasting eigenfunctions, I'll re-state our conclusions in terms of the binding energy ϵ_n (§ 8.8). *The binding energy of a particle in a well is just the energy measured from the top of the well*; *e.g.*, for the potential in Fig. 9.4a, $\epsilon_n \equiv V_0 - E_n$. So, for example, the less tightly bound of two bound states is the one with the smaller binding energy. We have argued that

> ***Rule***
>
> A weakly bound state oscillates more rapidly in a CA region and penetrates more deeply into a CF region than a tightly bound state.

The "Local Wavelength" and Other Quantities

You may have noticed that in discussing the eigenfunctions in Fig. 9.4, I used the "wavelengths" and "decay constants" of these functions. Because the potential is a continuous function of x, these quantities are not constants. Still, this usage enables us to exploit analogies between the Schrödinger Equations for continuous and piecewise-constant potentials. To acknowledge the x dependence of these quantities, we refer to them as "local" :

$$k(x) \equiv \sqrt{\frac{2m}{\hbar^2}\left[E - V(x)\right]} \qquad \text{local wave number} \tag{9.11a}$$

$$\kappa(x) \equiv \sqrt{\frac{2m}{\hbar^2}\left[V(x) - E\right]} \qquad \text{local decay constant} \tag{9.11b}$$

$$\lambda(x) \equiv \frac{2\pi}{k(x)}. \qquad \text{local wavelength} \tag{9.11c}$$

You may find it useful to write the TISE in CA and CF regions in terms of these quantities, *viz.*,

$$\frac{d^2}{dx^2}\psi_E(x) = -k^2(x)\psi_E(x) \qquad \text{CA Region} \tag{9.12a}$$

$$\frac{d^2}{dx^2}\psi_E(x) = +\kappa^2(x)\psi_E(x) \qquad \text{CF Region.} \tag{9.12b}$$

[Compare these equations to Eqs. (8.24) and (8.31).] In terms of the local wavelengths and decay constants for two eigenfunctions, $\psi_m(x)$ and $\psi_n(x)$, with $E_m > E_n$, we have:

$$\text{CA region} \quad \begin{cases} k_m(x) > k_n(x) \\ \lambda_m(x) < \lambda_n(x) \end{cases} \qquad E_m > E_n$$

$$\text{CF region} \quad \begin{cases} \kappa_m(x) < \kappa_n(x) \\ \dfrac{1}{\kappa_m(x)} > \dfrac{1}{\kappa_n(x)} \end{cases} \qquad E_m > E_n.$$

Curvature, Energy, and Boundary Conditions

The discussion of Fig. 9.4 exemplifies the intimate relationship between *the energy of a bound state* and *the curvature of its spatial function.* In Chaps. 7 and 8, we uncovered another intimate relationship, that between *the allowed (quantized) values of these energies* and *the boundary conditions the corresponding eigenfunctions must satisfy*. Indeed, in the solutions of the TISE for the infinite and finite square wells, quantization of the bound-state energies emerged as an inevitable consequence of these boundary conditions. We can use the connection between quantized energies, the curvature of eigenfunctions, and the boundary conditions on these functions to understand more deeply the phenomenon of quantization in quantum physics.

Recall the infinite square well (§ 7.4). This potential defines only one region where $\psi_E(x)$ is non-zero: within the walls. The spatial functions must equal zero at the walls, and this boundary condition restricts the *allowed energies* to $E_n = n^2\pi^2\hbar^2/\left(2mL^2\right)$ for $n = 1, 2, 3, \ldots$. This restriction, in turn, constrains the *wavelengths* of the Hamiltonian eigenfunctions in this (CA) region to the discrete values $\lambda = 2L,\ L,\ 2L/3,\ L/2, \ldots$. And this constraint controls the allowed *curvatures* of the eigenfunction. *A solution of the TISE whose curvature differs even slightly from one of the allowed values will not satisfy the boundary condition at the walls of the box and hence is inadmissible.* The principles at work in this trifling example are our guides to understanding the interplay of curvature and quantization in far more complicated systems.

Now consider a finite potential, such as the finite square well or the potential in Fig. 9.4a. Each bound-state eigenfunction of such a potential encompasses several regions. *We must impose boundary conditions in the first and last (classically forbidden) regions, where the function must monotonically decay to zero.* The rate of this decay is controlled by the local decay constant $\kappa(x)$—which, by its definition (9.11*b*), is related to the (quantized) energy of the state and which, through the TISE (9.12*b*), is related to the curvature of the function. So if the energy and curvature aren't just right, the eigenfunction won't decay—it will blow up.

Between the first and last regions are one or more intermediate regions. The pieces of $\psi_E(x)$ in these regions must hook together so that the composite spatial function is continuous and smoothly varying everywhere. But, because the pieces in the intermediate regions are determined by the values of $\psi_E(x)$ and its first derivative at the boundaries of the first and last regions (see § 9.3), the role of the intermediate regions in quantization is secondary to that of the outer regions.

To illustrate the relationship of quantization, curvature, and energy, let's return to the generic continuous potential in Fig. 9.4a and take another look at the eigenfunction for the second excited state. If we imagine ourselves travelling in from $\pm\infty$, where $\psi_3(x) = 0$, we can see that the value of the local decay constant $\kappa(x)$, which controls the decay of this function to zero in the limit $x \to \pm\infty$, determines the value of this function and its slope at the classical turning points. As we cross these points, we must ensure that the pieces in the CF regions we're leaving join continuously and smoothly to the piece in the CA region inside the well. For all these conditions to be met, the energy must be just right, $E = E_3$.

To see what can go wrong, let's change the energy just a bit—increasing it, say, to $E_3' > E_3$ (see Fig. 9.5a)—and solve the TISE anew. The resulting solution, which is sketched in Fig. 9.5a, and its first derivative are suitably continuous at the classical turning points. But in the CF regions, this function does not decay to zero; instead, it blows up as $x \to \pm\infty$.

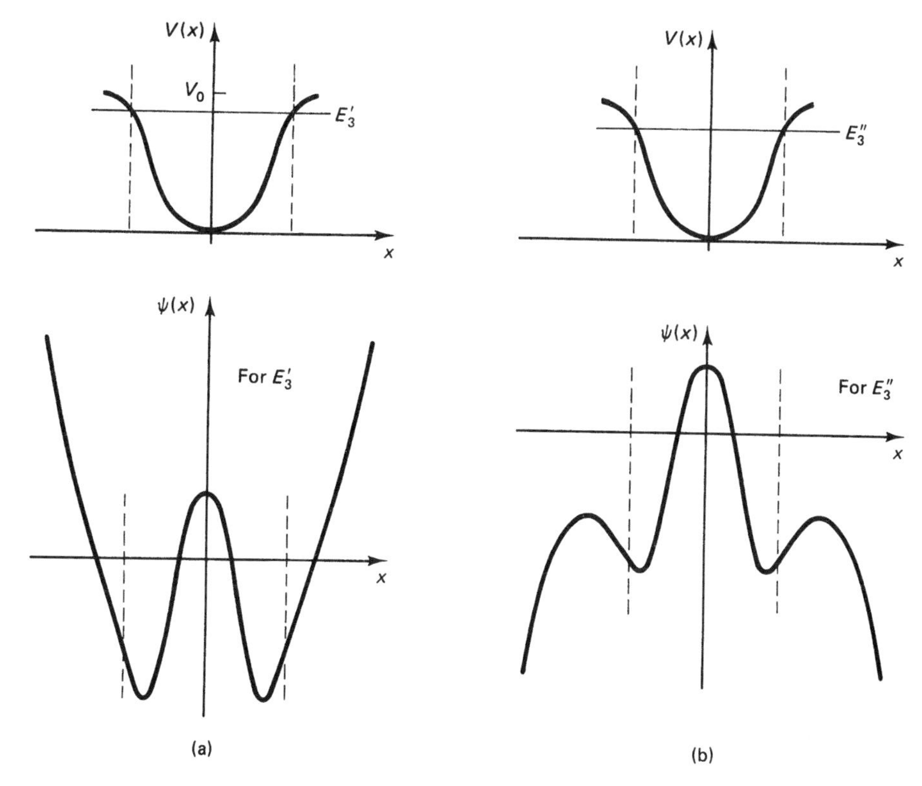

Figure 9.5 Naughty eigenfunctions: solutions of the TISE (for the potential of Fig. 9.4a) that cannot represent physically admissible bound states. The eigenfunctions shown are those for an eigenvalue slightly (a) larger and (b) smaller than E_3. [Adapted from *Introduction to Quantum Physics* by R. H. Dicke and J. P. Wittke (Reading, Mass.: Addison-Wesley, 1960).]

An equally impressive cataclysm occurs if we decrease the energy slightly, say to $E_3'' < E_3$ in Fig. 9.5b. Again, the monotonic pieces of the solution for E_3'' in the CF regions blow up in the asymptotic limits. In either case, these exploding functions are perfectly legitimate *mathematical* solutions of the TISE; but they're useless to us, for they cannot be interpreted as a probability amplitude.

The interplay of curvature, boundary conditions, and energy quantization that we have explored in one-dimensional systems characterizes all quantum systems that support bound states. We've seen that

Rule

A physically admissible bound-state eigenfunction of a continuous potential must decay to zero in two CF regions: one that includes $x \to \infty$ and one that includes $x \to -\infty$. The monotonic pieces of this function must connect to pieces in the intermediate region(s) so as to preserve the continuity of the function and its first derivative. These inviolable demands quantize the energy of the particle and constrain the curvature of the wave function that represents its state.

9.3 THE STATIONARY STATES OF AN ARBITRARY POTENTIAL

> 'My dear Watson,
> try a little analysis yourself,'
> said he, with a touch of impatience.
>
> —from *The Sign of Four*
> by Sir Arthur Conan Doyle

In § 9.2, we saw how to use the interplay of boundary conditions, energy quantization, and curvature to understand the shape of bound-state eigenfunctions of the finite square well, the harmonic oscillator, and a generic symmetric potential (Fig. 9.4a). Here I want to explore features of the *collection* of Hamiltonian eigenfunctions of a one-dimensional system; these features are invaluable guides to understanding the qualitative nature of these important functions.

A somewhat more complicated potential is the "bistable" (double-dip) potential in Fig. 9.6a. The spatial functions for the three lowest stationary states of this potential are graphed in Fig. 9.6b. The behavior of each of these functions—their oscillations in the CA region, numerous nodes, and decay in the CF regions—is explicable in terms of the principles of § 9.2.[4]

Consider, for example, the ground-state function $\psi_0(x)$. This function corresponds to the *minimum allowed energy*, so in the CA region it has *the smallest curvature and no nodes*. The local decay constant of the ground state is larger than that of any other stationary state, so in CF regions its $\psi_0(x)$ decays to zero more rapidly than any other bound-state eigenfunction.

Similar arguments explain the excited-state eigenfunctions. The energy of the first of these states, E_1, is greater than that of the ground state, E_0. The corresponding eigenfunction is the *normalizable* solution $\psi_E(x)$ of the TISE for the value of E nearest to (but greater than) E_0. Because $E_1 > E_0$, the local wavelength in the CA region of $\psi_1(x)$ is less than that of $\psi_0(x)$, and $\psi_1(x)$ has one more node than $\psi_0(x)$. Can you see how to extend these arguments to the eigenfunction for the *second* excited state?

Nodes

Physically, the nodes of a wave function correspond to minima in the position probability density (*c.f.*, Fig. 8.23). In qualitative problem solving, the number of nodes of an eigenfunction is an important clue to its shape. And in quantitative problem solving, the number of nodes is a vital check on a numerical or analytic solution of the TISE. Here's a handy summary of our deductions thus far concerning the *nodal structure* of the bound-state eigenfunctions (of a one-dimensional system):

$$\boxed{\text{number of nodes in } \psi_n(x) \text{ (if } n = 1 \text{ labels the ground state)} = n - 1} \tag{9.13}$$

[4] This potential, which is the sum of two hyperbolic cosine functions, is one of an interesting class of *bistable potentials*. One can solve the TISE of a particle in a bistable potential for low-lying bound states by *analytic* means. The strategy used to solve the particular bistable potential in Fig. 9.6 is the method of power series (see § 9.7). For the details, see M. Razavy, *Amer. Jour. Phys.* **48**, 285 (1980). Notice, by the way, that the ground state is labelled by $n = 0$, not $n = 1$.

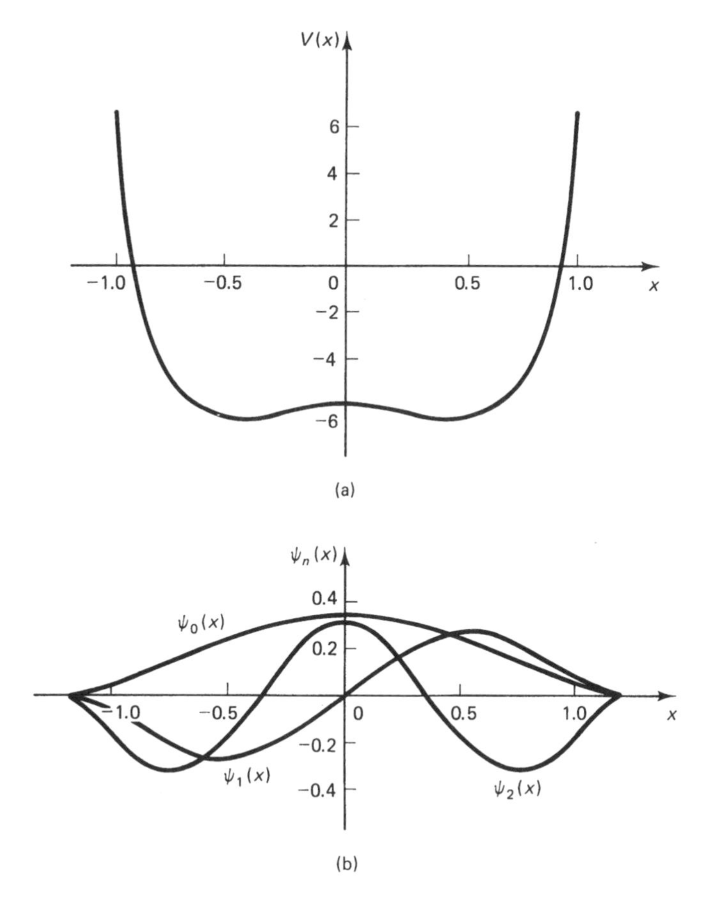

Figure 9.6 (a) The bistable potential $V(x) = \hbar^2 (\cosh 4x - 6\cosh 2x - 1)/m$ and (b) its three lowest bound-state eigenfunctions. The energies of these states, on the scale shown in (a), are $E_0 = -5.12$, $E_1 = -2.0$, and $E_2 = +3.12$. (Note that the ground state is labelled with $n = 0$.) [From M. Razavy, *Amer. Jour. Phys.* **48**, 285 (1980); used with permission.]

Unlike some of our other tools (*e.g.*, parity arguments), the nodal characteristics of bound-state eigenfunctions don't depend on the symmetry of the potential. To illustrate this point, let's consider a state of a simple, asymmetric, piecewise-constant potential.

Example 9.1. Deducing an Eigenfunction of the Asymmetric Square Well

At the end of § 8.7, I discussed briefly the asymmetric square well (see also Pblm. 8.6). Let's work out some of the *qualitative* properties of the eigenfunction for the *first excited state* of the potential (see Fig. 9.7)

$$V(x) = \begin{cases} 5 & x < 0 \\ 0 & 0 \le x \le 4 \, . \\ 2 & x > 4 \end{cases} \tag{9.14}$$

Note carefully that I've written this potential in atomic units (see Example 8.8 and Appendix F). Notice also that *since $V(x)$ is piecewise constant, the pieces of $\psi_1(x)$ consist of the simple sine, cosine, and real exponential functions of Table 8.12 [see Eq. (8.101)].*

The potential (9.14) supports two bound states. Numerical solution of the TISE for this potential reveals that the energy of the first excited state is $E_2 = 1\,E_h$. From this value,

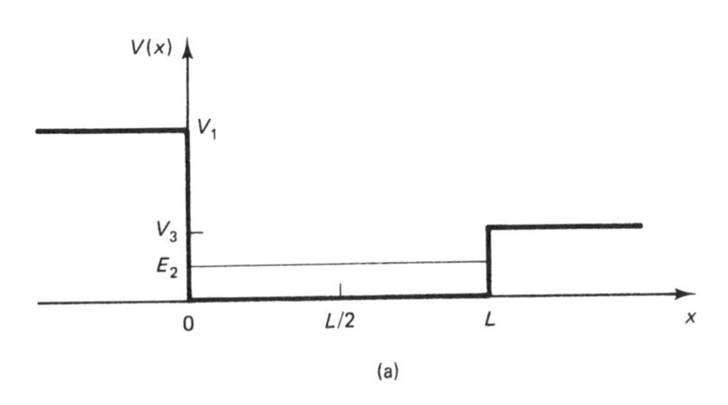

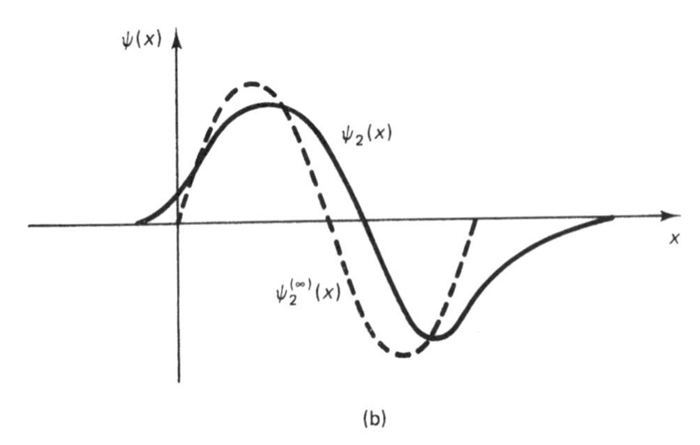

Figure 9.7 (a) An asymmetric well potential. In Example 9.1, we consider the first excited state of this potential for well parameters $V_1 = 5\,E_h$, $V_3 = 2\,E_h$, and $L = 4\,a_0$. (b) A sketch of the eigenfunction for the first excited state of this potential; also shown is the corresponding eigenfunction for an infinite square well of the same width.

we can calculate the wavelength in the CA region and the decay constants in the CF regions for this state, *viz.*,

$$\kappa^{(1)} = 2.83 \tag{9.15a}$$

$$k^{(2)} = 1.41 \tag{9.15b}$$

$$\kappa^{(3)} = 1.41. \tag{9.15c}$$

Question 9–2

What are the (atomic) units of the quantities in Eqs. (9.15)? Convert each of these quantities to SI units (see Appendix D).

One way to figure out the shape of $\psi_2(x)$ is to consider the corresponding function for an *infinite* square well of the same width. This function, $\psi_2^{(\infty)}(x)$, is sketched in Fig. 9.7b. It is a simple sine function of wavelength $\lambda_2^{(\infty)} = 4$. The wavelength of the corresponding eigenfunction $\psi_2(x)$ of the *finite well* is slightly larger than this value; as you can see from Eq. (9.15*b*), $\lambda_2 = 4.46$. This means that *almost* a full oscillation of $\psi_2(x)$ fits into the CA region of the asymmetric square well. But, unlike $\psi_2^{(\infty)}(x)$, the finite-well function is not zero at the classical turning points, where it must hook continuously to monotonically decaying functions in Regions 1 and 3.

The rates of decay of the pieces of $\psi_2(x)$ in these CF regions are not equal, because the decay constant for Region 1 is larger than that for Region 3. Hence $\psi_2(x)$ penetrates more deeply into the CF region $x > L$ than into the CF region $x < 0$. In fact, the characteristic length for decay into Region 3 is 0.71 and for Region 1 is 0.35.

Perhaps the most subtle feature of the eigenfunction $\psi_2(x)$ is the effect of this difference in the characteristic lengths upon *the location of the node in the CA region.* Because

$\psi_2(x)$ penetrates almost twice as deeply into the region $x > L$ as it does into the region $x < 0$, the value of $|\psi_2(x)|$ at $x = L$ *must be larger* than the value at $x = 0$. Consequently the position of the node in $\psi_2(x)$ must be *to the right of the midpoint* of the well, *i.e.*, at some $x > L/2$.

Putting together these Sherlockian deductions, we come up with the sketch in Fig. 9.7b. (This sketch shows all the major features of the accurate, numerical solution of the TISE for this state.) Note that even though the potential (9.14) is certainly not symmetric, $\psi_2(x)$ (like all eigenfunctions) conforms to the nodal structure (9.13).

Question 9–3

Sketch the eigenfunctions for the ground state of the potential in Fig. 9.7a.

Question 9–4

In Example 9.1, we used the value of E_2 to calculate the wavelength and decay constants of a stationary state. Doing so requires having solved the TISE, at least for the eigenvalues. But the arguments illustrated by this example don't require knowledge of the energy or, indeed, of the parameters V_0 and L that define the well. Consider an *arbitrary* asymmetric square well that supports at least *three* bound states. Sketch the eigenfunctions for these three states.

A Note on Parity

In § 8.7, I remarked that *the bound stationary-state eigenfunctions of any symmetric potential have definite parity: they are either even or odd under inversion through the midpoint of the potential, which point we usually choose as* $x = 0$. (We'll prove this property in § 9.4.) Although not a consequence of the nodal structure of these functions, this "parity structure" is clearly related to the nodal structure, for *an eigenfunction of definite parity can have a node at the midpoint of the potential only if the eigenfunction is odd.*

The relationship between *parity* and *nodes* explains why the parities of the eigenfunctions of a symmetric potential alternate with increasing n. We start with the ground-state eigenfunction, which has the minimum (allowed) curvature and hence no nodes. Consistent with this fact, the ground-state eigenfunction is even.

Going up the ladder of stationary bound states, we find that with each increase by 1 of the quantum number n, the number of nodes increases by 1. The spatial function for the first excited state has a node at the midpoint of $V(x)$, but the function for the second excited state does not. And so the parity of the eigenfunctions $\psi_n(x)$ alternates: even, odd, even, odd, ... This property is illustrated for the symmetric finite square well in Figs. 8.23, for the double-dip potential in Fig. 9.6a, and for the simple harmonic oscillator in Fig. 9.14 below.

One Last Trick: Amplitude Variation

Before turning to the mathematical solution of the TISE, I want to show you one more feature of the qualitative solution of this equation: the variation of the *amplitude* of an eigenfunction in a CA region.[5] This property follows from the continuity conditions on $\psi_E(x)$ and the variation of the local wavelength with $E - V(x)$.

[5]The development in this section draws on the discussion in A. P. French and E. F. Taylor, **39**, 961 (1971).

To deduce the principles of amplitude variation, we'll consider the first excited state of the piecewise-constant potential in Fig. 9.8a. *Notice that this potential is not symmetric.*

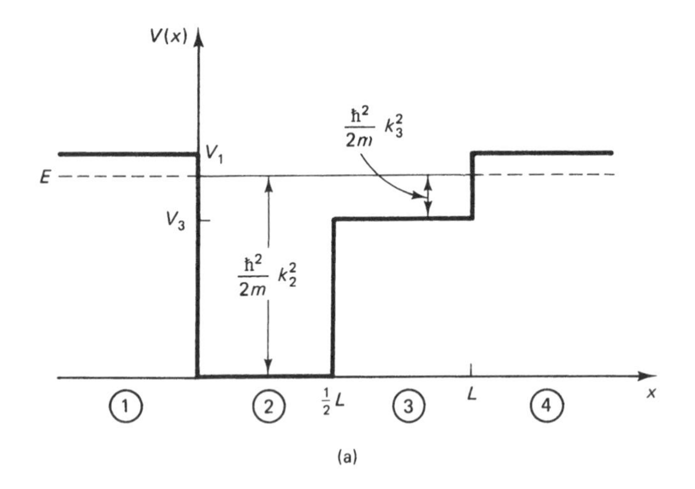

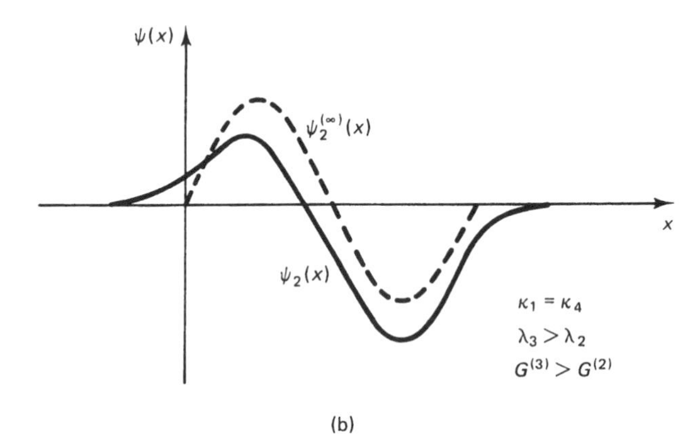

Figure 9.8 (a) An asymmetric piecewise-constant potential whose change at $x = L/2$ causes the amplitude of its bound-state eigenfunctions to change at this point. (b) A sketch of the eigenfunction for the first-excited state of this potential and its counterpart for an infinite well of width L with $V_3 = 0$.

Let's see if we can figure out how the piece of the eigenfunction $\psi_2(x)$ in the CA Region 2 differs from the piece in the CA Region 3. Since $V_3 > V_2$, the wave numbers in Regions 2 and 3 bear the relationship $k_3 < k_2$. So the de Broglie wavelengths of these pieces are

$$\lambda_2 = \sqrt{\frac{2\pi^2\hbar^2}{m}}\sqrt{\frac{1}{E}} \quad < \quad \lambda_3 = \sqrt{\frac{2\pi^2\hbar^2}{m}}\sqrt{\frac{1}{E - V_3}}. \tag{9.16}$$

Since $\lambda_2 < \lambda_3$, the curvature of $\psi_E^{(2)}(x)$, the piece in Region 2, is greater than that of $\psi_E^{(3)}(x)$, the piece in Region 3.

To show you the consequences of this change in curvature and wavelength for the *amplitude* of the eigenfunction, I want to construct a function that is continuous across the boundary between Regions 2 and 3. [We can't use the logarithmic derivative (§ 8.7) for this purpose, because when we form this quantity, by dividing $d\psi_E(x)/dx$ by $\psi_E(x)$, the amplitude of the eigenfunction cancels.] The particular function I want to construct involves the pieces of the eigenfunction in each region, the first derivatives of these pieces, and the wavelengths.

Let's start with the general form of the eigenfunction in each region. Since this potential is *piecewise-constant*, we can easily write down these forms using principles from Chap. 8, *viz.*,

$$\psi_E^{(2)}(x) = C^{(2)} \cos k_2 x + D^{(2)} \sin k_2 x \tag{9.17a}$$

$$\psi_E^{(3)}(x) = C^{(3)} \cos k_3 x + D^{(3)} \sin k_3 x. \tag{9.17b}$$

The constants $C^{(q)}$ and $D^{(q)}$ make the manipulation of these forms more complicated than necessary. So let's use elementary trigonometry to rewrite Eq. (9.17) in terms of an *amplitude* $G^{(q)}$ and a *phase constant* ϕ_q, to wit:

$$\psi_E^{(2)}(x) = G^{(2)} \sin(k_2 x + \phi_2) \tag{9.18a}$$

$$\psi_E^{(3)}(x) = G^{(3)} \sin(k_3 x + \phi_3). \tag{9.18b}$$

The first derivatives of these functions are

$$\frac{d}{dx}\psi_E^{(2)}(x) = k_2 G^{(2)} \cos(k_2 x + \phi_2) \tag{9.19a}$$

$$\frac{d}{dx}\psi_E^{(3)}(x) = k_3 G^{(3)} \cos(k_3 x + \phi_3). \tag{9.19b}$$

If we now square Eqs. (9.18) and (9.19), add the results in each region, introduce the wavelengths (9.16), and rearrange a little, we obtain

$$\left[\psi_E^{(2)}(x)\right]^2 + \frac{1}{(2\pi)^2}\lambda_2^2 \left[\frac{d}{dx}\psi_E^{(2)}(x)\right]^2 = \left[G^{(2)}\right]^2 \tag{9.20a}$$

$$\left[\psi_E^{(3)}(x)\right]^2 + \frac{1}{(2\pi)^2}\lambda_3^2 \left[\frac{d}{dx}\psi_E^{(3)}(x)\right]^2 = \left[G^{(3)}\right]^2 \tag{9.20b}$$

Look very carefully at Eqs. (9.20). They are written in terms of two quantities that *must be continuous across the boundary between the classically-allowed regions*: $\psi_E(x)$ and $d\psi_E(x)/dx$. Consider values of x extremely near this boundary. The wavelengths on the right-hand side change abruptly across the boundary, $\lambda_3 > \lambda_2$. But the eigenfunction on the left-hand side must be continuous, so to compensate for the change in the wavelength, the amplitudes on the right-hand side must be related by $G^{(2)} < G^{(3)}$.

This derivation confirms the principle used to draw the sketch in Fig. 9.8b: the amplitude of the spatial function is larger in the region where the wavelength of the function is larger.

We can generalize these findings to a continuous potential by imagining such a potential as an infinite number of piecewise-constant pieces (*c.f.*, Fig. 8.2). We must, of course, keep in mind that the amplitude of a spatial function for a continuous potential is not constant. So the *continuous* variation in $E - V(x)$ causes a *continuous* change in the curvature and the amplitude of $\psi_E(x)$ of the following sort:

Rule

The amplitude of an oscillatory piece of a bound-state spatial function increases with decreasing curvature, i.e., with increasing wavelength.

As a final illustration of this rule, consider the bound-state eigenfunctions of the harmonic oscillator potential (§ 9.8) shown in Fig. 9.16. Notice the pronounced variation of the amplitude of states of high quantum number n.

Question 9–5

Explain how to justify this generalization by representing a continuous potential as an infinite number of infinitesimally small steps, as in Fig. 8.2.

Question 9–6

Suppose that the first derivative of $\psi_E(x)$ happens to be zero at a step increase in a piecewise-constant potential like the one in Fig. 9.8a. Show that the amplitude of such an eigenfunction does not change across the region boundary.

*9.4 MATHEMATICAL PROPERTIES OF BOUND-STATE EIGENFUNCTIONS

In our inquiry into the qualitative nature of the Hamiltonian eigenfunctions for bound states of a one-dimensional potential, we've uncovered several important properties:

1. $\psi_n(x)$ must go to zero in the asymptotic limits $x \to \pm\infty$;
2. $\psi_n(x)$ and $d\psi_n(x)/dx$ must be continuous everywhere;
3. the curvature of $\psi_n(x)$ varies with $|E - V(x)|$, and influences the wave properties of this function as follows:

$$\boxed{\text{as } |E - V(x)| \text{ increases} \begin{cases} \text{in a CA region} & \begin{cases} \text{amplitude} & \text{decreases} \\ k(x) & \text{increases} \\ \lambda(x) & \text{decreases} \end{cases} \\ \text{in a CF region} & \begin{cases} \kappa(x) & \text{increases} \\ 1/\kappa(x) & \text{decreases} \end{cases} \end{cases}} \tag{9.21}$$

4. $\psi_n(x)$ has $n - 1$ nodes (if $n = 1$ denotes the ground state);
5. *for a symmetric potential*, the eigenfunctions $\psi_n(x)$ alternate parity with increasing n, beginning with an even ground-state $\psi_1(x)$.

Before turning to the mathematical solution of the TISE, I want to prove for you the last of these properties: the parity of the eigenfunctions of a symmetric potential.[6] But before we can prove this property, we must prove that the functions we're talking about are unique.

On the Uniqueness of Bound Stationary-State Eigenfunctions

The uniqueness of physically admissible solutions of the one-dimensional TISE follows from the fact that this is a *second-order ordinary differential equation*. For each value of E, this equation has two linearly-independent solutions. But, since the TISE is homogeneous, any linear combination of these functions is a third solution (for the same energy); so actually there exist an infinite number of solutions. We select one out of

[6] If you choose to omit this section, at least read the two Proclamations in it.

this infinite pile of solutions by specifying the value of the eigenfunction and its first derivative at a single point or, equivalently, the value of the eigenfunction at two points.[7] Of all these solutions, we want the one that satisfies property [1] above: the normalizable solution. We know from § 9.2 that normalizable solutions to the TISE don't exist for all values of E. But here I want to consider a related question that arises once we have found a normalizable solution: is it unique?

> ***Proclamation 9.1***
>
> Normalizable solutions of the one-dimensional TISE are unique (to within an arbitrary global phase factor).

Argument: We'll prove this assertion using the time-honored technique of **proof by contradiction**. That is, we'll pretend that Proclamation 9.1 is false—*e.g.*, that there exist two *distinct* solutions of the TISE for a single energy, say $\psi_n(x)$ and $\chi_n(x)$:

$$\frac{d^2}{dx^2}\psi_n(x) = -\frac{2m}{\hbar^2}\left[E_n - V(x)\right]\psi_n(x) \tag{9.22a}$$

$$\frac{d^2}{dx^2}\chi_n(x) = -\frac{2m}{\hbar^2}\left[E_n - V(x)\right]\chi_n(x). \tag{9.22b}$$

Then from Eqs. (9.22) we'll derive a relationship between $\psi_n(x)$ and $\chi_n(x)$ that unmasks these two functions as aliases of one another—*i.e.*, we'll prove that these functions represent the *same* physical state. Two such functions are not distinct, so our relationship will contradict the supposition, thereby proving the proclamation.[8]

To relate $\psi_n(x)$ and $\chi_n(x)$, let's combine Eqs. (9.22*a*) and (9.22*b*) by multiplying the first by $\chi_n(x)$ and the second by $\psi_n(x)$ and subtracting the results, *viz.*,

$$\chi_n(x)\frac{d^2}{dx^2}\psi_n(x) - \psi_n(x)\frac{d^2}{dx^2}\chi_n(x) = 0. \tag{9.23}$$

A clever way to simplify Eq. (9.23) is to reduce the order of differentiation by rewriting the left-hand side as a total derivative,

$$\frac{d}{dx}\left[\chi_n(x)\frac{d}{dx}\psi_n(x) - \psi_n(x)\frac{d}{dx}\chi_n(x)\right] = 0. \tag{9.24a}$$

Integrating this equation, we obtain

$$\left[\chi_n(x)\frac{d}{dx}\psi_n(x) - \psi_n(x)\frac{d}{dx}\chi_n(x)\right] = \text{constant}. \tag{9.24b}$$

Question 9–7

If you are not familiar with this step, take a few minutes to verify that Eq. (9.24*a*) is equivalent to Eq. (9.23). Then take a few more minutes to **memorize this trick**—it will serve you long and well.

[7] That is why the value of $\psi_E^{(q)}(x)$ and its first derivative at the boundary between an outer CF region and the adjacent CA region determine the function $\psi_E(x)$.

[8] The word "distinct" here is critical. Recall (Chap. 3) that two wave functions are distinct if they represent different quantum states of the systems. But two wave functions that differ *only* by a global phase factor represent the *same* physical state and so are not distinct.

Now, what is the value of the constant in Eq. (9.24*b*)? Well, the left-hand side of this equation must, obviously, be the same for all x. We know the values of $\psi_n(x)$ and $\chi_n(x)$ (even for an arbitrary potential) in the limits $x \to \infty$ and $x \to -\infty$: they are zero. So the constant in (9.24*b*) must be zero, and we can write this equation as

$$\frac{1}{\psi_n(x)} \frac{d}{dx} \psi_n(x) = \frac{1}{\chi_n(x)} \frac{d}{dx} \chi_n(x). \tag{9.25}$$

We can now integrate Eq. (9.25) to obtain the simple relationship

$$\ln \psi_n(x) = \ln \chi_n(x) + C, \tag{9.26}$$

where C is the constant of integration. Hence the solutions of Eqs. (9.22) are proportional:

$$\chi_n(x) = C\psi_n(x). \tag{9.27}$$

But the constant of proportionality in Eq. (9.27) is not arbitrary. Both $\psi_n(x)$ and $\chi_n(x)$ must be normalizable, *i.e.*,

$$\int_{-\infty}^{\infty} \psi_n^*(x)\, \psi_n(x)\, dx = \int_{-\infty}^{\infty} \chi_n^*(x)\, \chi_n(x)\, dx = 1, \tag{9.28}$$

and *these conditions can be satisfied only if the modulus of C is unity.* So C must have the form $C = e^{i\delta}$, where δ is an arbitrary real constant. Hence our two supposedly distinct solutions of the TISE are related by

$$\chi_n(x) = e^{i\delta}\, \psi_n(x). \tag{9.29}$$

Ah ha!—the constant $e^{i\delta}$ is just a *global phase factor.* Such a factor is physically meaningless, because it disappears from such measurable quantities as observables and probabilities (§ 3.5). So the wave functions

$$\psi_n(x)e^{-iE_nt/\hbar} \qquad \Longleftrightarrow \qquad e^{i\delta}\psi_n(x)e^{-iE_nt/\hbar} \tag{9.30}$$

are physically equivalent, which means that $\psi_n(x)$ and $\chi_n(x)$ are *not* distinct solutions to (9.22). This contradiction of our premise proves that each normalizable bound-state eigenfunction of a one-dimensional Hamiltonian is unique.

Q. E. D.

This proof establishes a *one-to-one correspondence* between the energies of bound stationary states and the eigenfunctions of the Hamiltonian,

$$\boxed{E_n \qquad \longleftrightarrow \qquad \psi_n(x)} \tag{9.31}$$

Physicists express this correspondence as follows:

Rule

The bound-state energies of a single particle in a one-dimensional potential are non-degenerate. (This rule does not apply to potentials in more than one dimension or to systems consisting of more than one particle.)

Question 9–8

Apply the result we just proved to the function $\chi_n(x) = \psi_n^*(x)$ and thereby prove that we are always free to choose the eigenfunctions of a single particle in one dimension to be real functions.

The Consequences of Symmetry

Now that we have established the uniqueness of the bound-state eigenfunctions of a one-dimensional potential, we can prove:

> ***Proclamation 9.2***
>
> The bound-state eigenfunctions of a symmetric potential have definite parity.

Argument: To determine the parity of a function, we invert the spatial coordinate x through the origin, *i.e.*, $x \to -x$. Since we're interested in the behavior of a solution of the TISE under this operation, let's see what happens when we invert x in this equation. The symmetric potential doesn't change,

$$V(-x) = V(x), \qquad \text{symmetric potential} \tag{9.32}$$

so under this inversion, the TISE is transformed into[9]

$$\frac{d^2}{dx^2}\psi_n(-x) = -\frac{2m}{\hbar^2}\left[E_n - V(x)\right]\psi_n(-x). \tag{9.33}$$

This equation is identical to the TISE; i.e., $\psi_n(-x)$ *satisfies the same differential equation as does* $\psi_n(x)$. And $\psi_n(-x)$ satisfies the same boundary conditions as does $\psi_n(x)$.

But for a given allowed energy, there is only *one* distinct, normalizable Hamiltonian eigenfunction (see Proclamation 9.1). So $\psi_n(x)$ and $\psi_n(-x)$ must be related by Eq. (9.29),

$$\psi_n(-x) = e^{i\delta}\psi_n(x). \tag{9.34}$$

Since we're free to choose $\psi_n(x)$ to be real (see Question 9–8), the factor in Eq. (9.34) is $e^{i\delta} = \pm 1$, *i.e.*,

$$\boxed{\psi_n(-x) = \pm\psi_n(x)} \tag{9.35}$$

Therefore each eigenfunction of the Hamiltonian for a symmetric potential is either even or odd under inversion through $x = 0$.

Q. E. D.

Proclamation 9.2 is the first of many *symmetry arguments* you'll encounter in your study of quantum physics. I cannot overstate the power of symmetry. It is little exaggeration to say that when you perceive the symmetry properties of a system (*i.e.*, of

[9]The second derivative does not change under this inversion, because

$$\frac{d}{dx} \xrightarrow[x\to -x]{} -\frac{d}{dx} \qquad \Longrightarrow \qquad \frac{d^2}{dx^2} \xrightarrow[x\to -x]{} \frac{d^2}{dx^2}.$$

its potential), you peer into the deepest level of its physics. And such arguments have a practical side, for they invariably simplify the solution of the Schrödinger Equation.

9.5 THE HARMONIC OSCILLATOR MADE SIMPLE

Yet my classic lore aggressive
(If you'll pardon the possessive)
Is exceedingly impressive
When you're passing an exam.
—from *The Grand Duke*
by Gilbert and Sullivan

Harmonic oscillators are useful models of complicated potentials in a variety of disciplines—including the quantum theory of electromagnetic radiation, the study of lattice vibrations in crystalline solids, and the analysis of infrared spectra of diatomic molecules (§ 9.6). You probably remember the classical description of oscillatory motion from a high school or freshman physics class in which you studied the **simple harmonic oscillator (SHO)**. Actually, classical systems rarely execute *simple* oscillatory motion; more often they undergo *damped* or *forced* harmonic motion.[10] But even if particles execute "complicated" harmonic motion, their *small excursions* about equilibrium can be accurately approximated by the simple harmonic oscillator. So the SHO is the starting point for study of any system whose particles oscillate about equilibrium positions.

What is "Simple" Harmonic Motion?

Consider a point moving serenely along the circumference of a circle at a constant *angular frequency* ω_0, as in Fig. 9.9. The projection of this point on the diameter of the circle, $x(t)$ in the figure, executes **simple harmonic motion**; *i.e.*, $x(t)$ changes with time according to a simple equation involving the frequency ω_0, the amplitude of motion A, and a phase constant ϕ:

$$x(t) = A \sin\left(\omega_0 t + \phi\right). \qquad \text{simple harmonic motion} \tag{9.36}$$

The angular frequency ω_0 is usually called the **natural frequency**. The **classical oscillation frequency** ν_0 is related to the natural frequency by $\nu_0 = \omega_0/(2\pi)$.

The relationship of the function (9.36) to its second derivative is also simple:

$$\frac{d^2}{dx^2} x(t) = -\,\omega_0^2 x(t). \tag{9.37}$$

This is the famous differential equation of simple harmonic motion. Notice that the amplitude and phase constant have disappeared from Eq. (9.37).

[10]This section consists primarily of a review of the physics of classical simple harmonic motion. For a more advanced treatment, see Chaps. 4 and 5 of *Classical Dynamics of Particles and Systems*, 2nd ed. by J. B. Marion (New York: Academic Press, 1970) or any other undergraduate classical mechanics text.

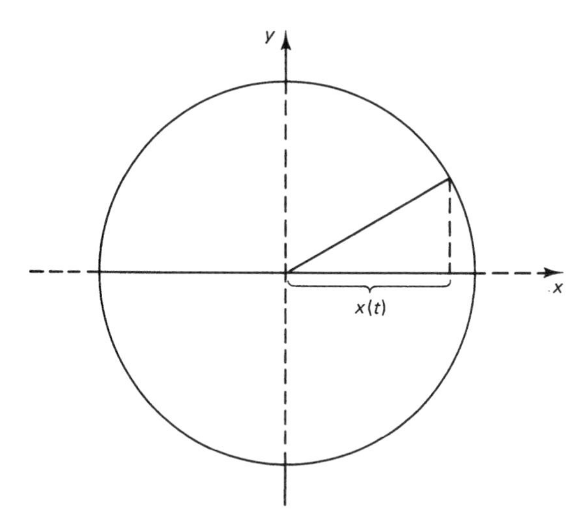

Figure 9.9 Simple harmonic motion. The projection on the diameter of a circle of a point moving along the circumference.

The Macroscopic Simple Harmonic Oscillator

The most famous simple harmonic oscillator is the mass on a spring in Fig. 9.10a.[11] If the wall to which the spring is hooked is immovable, and if the spring is "ideal" (*i.e.*, has no mass and doesn't deform), and if friction and gravity are negligible, then the force on the mass acts to restore it to its equilibrium position x_e, as illustrated in Fig. 9.10b. The force exerted by a spring with force constant k_f is linear:

$$\text{force on mass } m = -k_f(x - x_e) \qquad \text{Hooke's Law.} \tag{9.38}$$

The corresponding trajectory $x(t)$, derived from Newton's Second Law, is

$$\frac{d^2}{dx^2}x(t) = -\frac{k_f}{m}x(t). \tag{9.39}$$

Adroitly noticing the similarity between Eq. (9.39) and the mathematical equation of simple harmonic motion, (9.37), we conclude that the mass is a simple harmonic oscillator whose natural frequency is related to the physical properties of the system (the mass and the force constant of the spring) by

$$\omega_0 = \sqrt{\frac{k_f}{m}}. \tag{9.40}$$

In practice, the amplitude and phase constant in the trajectory (9.36) are determined from the initial conditions $x(t_0)$.

The potential energy of the particle follows simply from the force (9.38),[12]

$$V(x) = -\int (\text{Force on mass } m)\, dx \tag{9.41a}$$

$$= \frac{1}{2}k_f x^2 = \frac{1}{2}m\,\omega_0^2 x^2, \tag{9.41b}$$

[11]For a review of this classic example of classical simple harmonic motion, see Chap. 14 of *Fundamentals of Physics*, 2nd ed. by D. Halliday and R. Resnick (New York: Wiley, 1981) or, at a more advanced level, Chap. 2 of *Mechanics*, 3rd ed. by K. R. Symon (Reading, Mass: Addison-Wesley, 1971).

[12]Actually, the integral in (9.41*a*) determines $V(x)$ only to within an arbitrary constant, the zero of energy, which I'll choose at the bottom of the well.

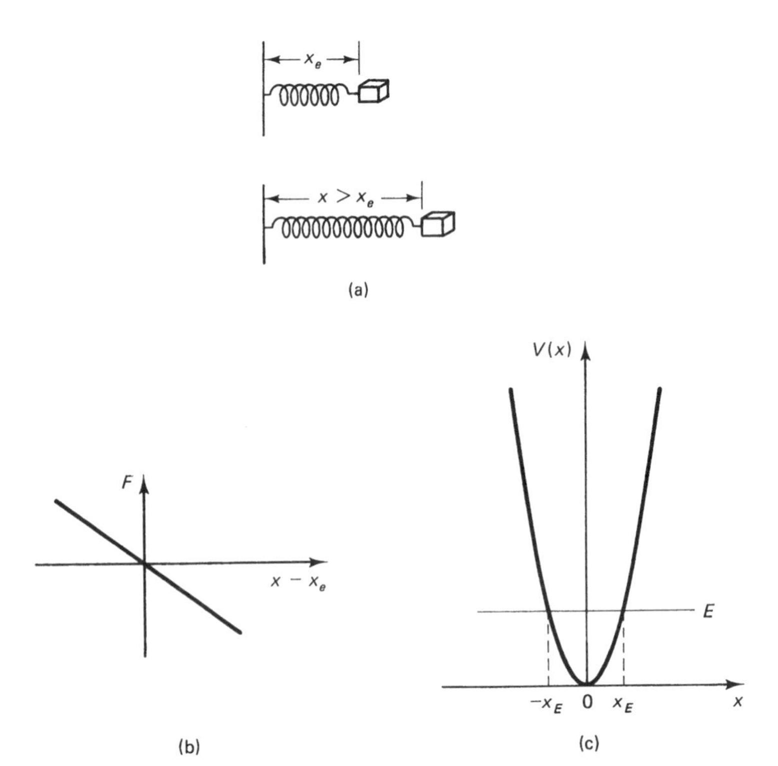

Figure 9.10 (a) A mass on a spring. The mass moves only in the x direction under the influence of the linear restoring force shown in (b). (c) The potential energy that derives from this force. Also shown are the classical turning points $\pm x_E$ for a particle with total energy E.

as shown in Fig. 9.9c. The total energy of the particle is just $V(x)$ plus its kinetic energy, *i.e.*,

$$E = \frac{p^2}{2m} + \frac{1}{2} m \omega_0^2 x^2. \tag{9.42}$$

The total energy is, of course, conserved.

As the particle oscillates back and forth, its kinetic energy and hence its speed changes because the potential energy in (9.42) changes. For example, at equilibrium, the particle's potential energy is zero. So its speed is greatest at this point. On the other hand, the particle's kinetic energy is zero at the classical turning points,

$$x_E = \pm\sqrt{\frac{2E}{m\omega_0^2}}. \qquad \text{classical turning points of the SHO} \tag{9.43}$$

So as it nears these points, the particle slows down; at them, it stops dead in its tracks and turns around.

From these observations we can predict the outcome of a measurement of the position of a classical simple harmonic oscillator. Were we to perform such a measurement at an arbitrary time during one period of oscillation, we'd most likely find the particle at positions where its speed is the smallest, *i.e.*, near the classical turning points. Later in this chapter, we'll compare this prediction with what quantum physics predicts for a microscopic oscillator. To prepare for this comparison, let's construct the classical counterpart of the quantum-mechanical probability density—the *classical position probability density* $P_{\text{cl}}(x)$—and write this quantity in terms of x and E.

The Classical Probability Density

To define a probability we must first identify an *ensemble* (§ 3.3): a large number of identical, non-interacting systems, all in the same state. But in a position measurement at a fixed time, all members of an ensemble of *macroscopic* particles would yield the same result. So to develop a classical probability function we must adapt the concept of an ensemble to a macroscopic system. *In this section only, I'll deviate from our usual use of the word "ensemble." Here we'll assume that the positions of members of the ensemble are measured at random times.*[13] In such a measurement, particles will be found at various places along the trajectory of the state, so the results will fluctuate according to the probability function we'll now derive.

By analogy with the quantum-mechanical position probability density, we define the **classical position probability density** as

$$P_{\text{cl}}(x)\,dx = \text{probability of finding the particle in the interval } x \text{ to } x+dx. \tag{9.44}$$

Although we seek an expression for $P_{\text{cl}}(x)$ in terms of x and E, our starting point is the definition (9.44), which involves time.

The time required for the particle to carry out one cycle of simple harmonic motion is the **classical period** T_0,

$$T_0 = \frac{2\pi}{\omega_0}. \qquad \text{period of a classical simple harmonic oscillator} \tag{9.45}$$

[Don't confuse the period with the kinetic energy $T = p^2/(2m)$]. So the probability function (9.44) is the fraction of one period that the particle spends in the interval dx at x, *i.e.*,

$$\boxed{P_{\text{cl}}(x)\,dx = \frac{dt}{T_0}} \qquad \text{classical probability density} \tag{9.46}$$

Now let's transform (9.46) into an expression in the variable x.

The infinitesimal dt is related to an infinitesimal of position, dx, by the velocity $v = dx/dt$. Using this fact and Eq. (9.45), we can eliminate t from $P_{\text{cl}}(x)$, *viz.*,

$$P_{\text{cl}}(x)\,dx = \frac{\omega_0}{2\pi}\frac{1}{v}\,dx. \tag{9.47}$$

Now we must express v in terms of x and E. This isn't hard for the simple harmonic oscillator, because we know the trajectory, (9.36), *viz.*,[14]

[13] There is an alternate way to define an ensemble of macroscopic systems so that the members will exhibit different values of an observable: randomize the initial conditions. But this approach is inappropriate to the present argument, for systems with different initial conditions are not in the same physical state.

[14] To avoid clutter in these equations, I've chosen the initial conditions so that the phase constant is $\phi = 0$. Convince yourself that this choice corresponds to a state in which the particle is initially at the *rightmost* classical turning point, *i.e.*, to $x(t_0) = +x_E$.

$$x(t) = x_E \sin(\omega_0 t) \tag{9.48a}$$
$$v(t) = \omega_0 x_E \cos(\omega_0 t). \tag{9.48b}$$

Since $v(t)$ involves the cosine of $\omega_0 t$ and $x(t)$ involves the sine of the same argument, we can combine Eqs. (9.48) according to the familiar trigonometric identity (see Appendix H) $\cos\theta = \sqrt{1 - \sin^2\theta}$, obtaining

$$v(x) = \omega_0 x_E \sqrt{1 - \frac{x^2}{x_E^2}} = \omega_0 \sqrt{x_E^2 - x^2}. \tag{9.49}$$

Using this result, we can express the classical probability as a function of x, to wit:

$$P_{\text{cl}}(x)\,dx = \frac{\omega_0}{2\pi} \frac{1}{\omega_0 \sqrt{x_E^2 - x^2}}\,dx = \frac{1}{2\pi} \frac{1}{\sqrt{x_E^2 - x^2}}\,dx. \tag{9.50}$$

This function is the *relative* probability that in a measurement *at a random time* of the position of a macroscopic simple harmonic oscillator with energy E, we'll obtain a value in the infinitesimal range dx at x. Notice carefully that the dependence of this quantity on *energy* is implicit in the classical turning points under the square root sign [*c.f.*, Eq. (9.43)]. Alas, we cannot meaningfully compare Eq. (9.50) to the usual quantum probability density. Can you figure out what is wrong and how to fix it?

***** Pause while reader ponders *****

Right: $P_{\text{cl}}(x)$ is a *relative* probability density. We must *normalize* this function so that, like its quantum cousin, it satisfies

$$\int_{-\infty}^{\infty} P_{\text{cl}}(x)\,dx = 1. \tag{9.51}$$

The absolute *normalized classical probability density* is

$$\boxed{P_{\text{cl}}(x) = \frac{1}{\pi\sqrt{x_E^2 - x^2}}} \qquad \text{simple harmonic oscillator} \tag{9.52}$$

Question 9–9

Derive Eq. (9.52), using the integral

$$\int_{-\infty}^{\infty} \left(x_E^2 - x^2\right)^{-1/2} dx = \pi.$$

Make explicit the energy dependence of this function, obtaining

$$P_{\text{cl}}(x) = \frac{\sqrt{m\omega_0^2}}{\pi} \frac{1}{\sqrt{2E - m\,\omega_0^2 x^2}}. \tag{9.53}$$

This function is sketched in Fig. 9.11 at three energies, which just happen to equal those of three stationary states of the microscopic simple harmonic oscillator in § 9.8. Reassuringly, these sketches are consonant with our classical expectations: $P_{\text{cl}}(x)$ *predicts that a macroscopic simple harmonic oscillator is least likely to be found at equilibrium, where it is moving most rapidly, and most likely to be found near the classical turning points, at which points it will stop and turn around.*

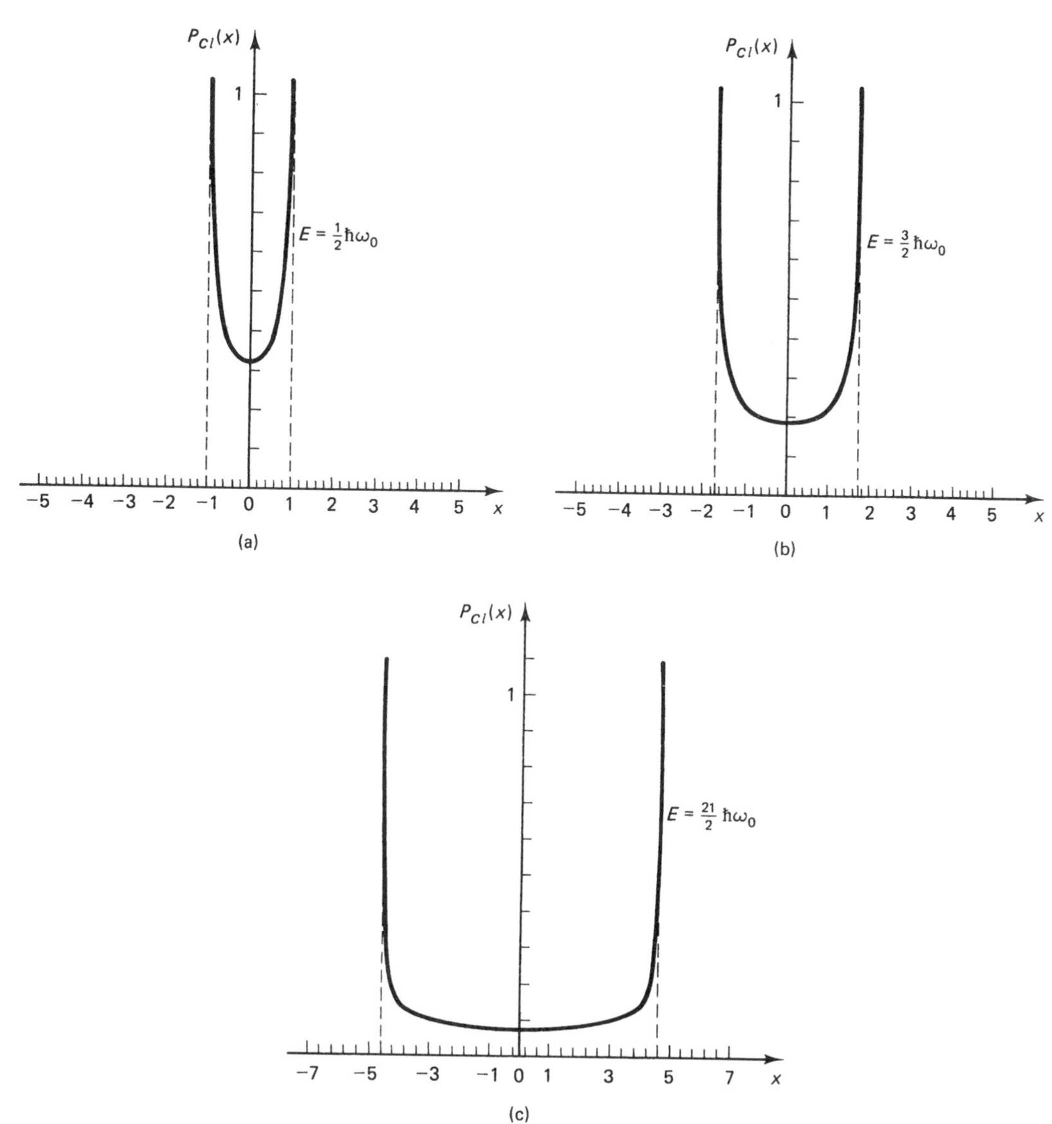

Figure 9.11 The classical probability function for three energies. These energies correspond to those of the ground state ($n = 0$), the first excited state ($n = 1$), and the tenth excited state ($n = 10$) of a microscopic oscillator with mass m and natural frequency ω_0 (compare to Fig.9.15).

*9.6 AN APPLICATION: THE VIBRATIONS OF A DIATOMIC MOLECULE

Isn't it astonishing that
all these secrets have been preserved
for so many years
just so that we could discover them!

—Orville Wright
June 7, 1903

One of the most interesting applications of the microscopic simple harmonic oscillator is to contemporary research on molecular vibrations.[15] Molecules are complicated beasties, and understanding their structure and spectra requires coping with complicated, non-spherical potentials generated by two or more force centers (the nuclei) and by lots of electrons. Fortunately, one can study the physics of the nuclei and electrons of most molecules separately.

A diatomic molecule consists of two positively-charged nuclei (of mass $\sim 10^{-28}$ kg) and a bunch of electrons (each of mass $m_e = 9.1 \times 10^{-31}$ kg). If the molecule is electrically neutral, there are enough bound electrons that their total negative charge exactly cancels the positive charge of the nuclei. In nature, bound states of molecules abound—which leads one to wonder: What holds a molecule together?

The subject of molecular binding is, to say the least, vast.[16] Some molecules (*e.g.*, NaCl) are held together by *ionic bonds* in which (crudely speaking) the molecule is bound by electrical forces between the composite positively and negatively charged ions. Others are held together by *covalent bonds*, where the reason for stability is more subtle. In such a molecule, the bound electrons, which are in a stationary "electronic state," provide a potential energy that supports several bound stationary states of the nuclei.[17]

Were the nuclei of a diatomic molecule macroscopic particles, their geometry would resemble the dumbbell in Fig. 9.12. Three kinds of nuclear motion occur:

1. *translation* through space;
2. *rotation* about the internuclear axis or about an axis perpendicular to this one;
3. *vibration* along the internuclear axis.

It is the last of these quantum "motions" that, for certain quantum states, can be approximated by simple harmonic motion.

[15]This section is just an extended example designed to convince you that the harmonic oscillator is worth the work we're about to put into it and to give you insight into how it can be used as a model. You can skip this section and still handle the rest of the chapter without difficulty.

[16]For a brief overview, see Chap. 11 of *Quantum States of Atoms, Molecules, and Solids* by M. A. Morrison, T. L. Estle, and N. F. Lane (Englewood-Cliffs, N.J.: Prentice Hall, 1977). For more detail, see *Bonding Theory* by Donald J. Royer (New York: McGraw-Hill, 1968) or nearly any chemistry text.

[17]This is a crude statement of one of the theoretical bulwarks of molecular physics: The Born-Oppenheimer approximation. This approximation is based on the relative masses of the electrons and nuclei: the former are orders of magnitude lighter than the latter, and so respond very quickly to changes in the nuclear geometry. We can conjure an (admittedly crude) analog of this approximation by imagining two hippopotami in a swamp. Surrounding the hippos is a swarm of mosquitos. As the hapless hippos lumber about, the light, carefree mosquitos whiz around, readjusting their positions in response to changes in the "dihippo force centers"—their goal being to maximize their blood intake. For a somewhat more scholarly discussion of this approximation, see Chap. 12 of *Quantum States of Atoms, Molecules, and Solids* by M. A. Morrison, T. L. Estle, and N. F. Lane (Englewood Cliffs, N.J. Prentice Hall, 1977).

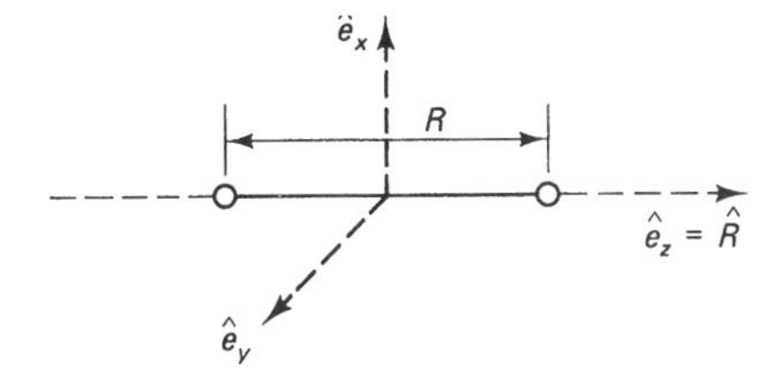

Figure 9.12 A schematic of the nuclei of a diatomic molecule.

The variable that describes the vibrations of the nuclei in a diatomic molecule is the **internuclear separation**, R in Fig. 9.12. Rather like two masses connected by a spring, the nuclei vibrate about *the equilibrium internuclear separation* $R = R_e$. Typically, the equilibrium internuclear separation of molecules is roughly an angstrom. For example, R_e for H_2 is 0.7414 Å, but for Kr_2, R_e is 4.0 Å. You will find additional data for diatomic molecules in Table 9.1.[18]

TABLE 9.1 PROPERTIES OF DIATOMIC MOLECULES (AND A COUPLE OF IONS).

molecule[a]	R_e(Å)	$k_f(\times 10^2$ N-m)	$\hbar\omega_0$ (eV)	μ(amu)
H_2^+	1.05	1.60	0.2879	0.504
H_2	0.7414	5.75	0.546	0.504
Li_2	2.673	0.25	0.0436	3.470
F_2	1.412	4.70	0.1136	9.50
N_2^+	1.1164	20.1	0.2737	7.003
N_2	1.0977	23.0	0.2924	7.003
CO	1.1283	19.0	0.2690	6.8608
HF	0.9168	9.66	0.5131	0.957
HCl	1.2746	5.17	0.3708	0.980
HBr	1.4144	4.11	0.3284	0.995
Kr_2	4.0	0.014	0.003	41.90

[a]In the ground electronic state.
Source: Data from Table 10.2 of *Reference Data for Atoms, Molecules, and Ions*, A. A. Radzig and B. M. Smirnov (New York: Springer-Verlag, 1985).

Consider, for example, the hydrogen molecule. Figure 9.13 shows the electronic potential energy in which the two protons of this molecule vibrate. It is easy to understand the large-scale features of these curves. As the nuclei get very close together ($R \to 0$) their repulsion overwhelms the attractive electron-nuclei interactions, and the potential asymptotically approaches $+\infty$. As the nuclei separate, they ultimately *dissociate* into neutral atoms or ions; this behavior is reflected in the slow decay of $V(R)$ as $R \to \infty$.[19]

The lowest-lying curve in this figure is the *ground electronic state*, which is attractive and which supports several bound vibrational states. But, although the motion of the nuclei in this state is oscillatory, it is not simple harmonic.

[18]And you can find absolutely everything you might want to know about the properties of diatomic molecules in the heroic data compilation *Constants of Diatomic Molecules* by K. P. Huber and G. Herzberg (New York: Van Nostrand Reinhold, 1979).

[19]For a more detailed introductory look at the forces responsible for such potentials, see Chap. 5 of *Spectroscopy and Molecular Structure* by G. W. King (New York: Holt, Rinehardt, and Winston, 1974).

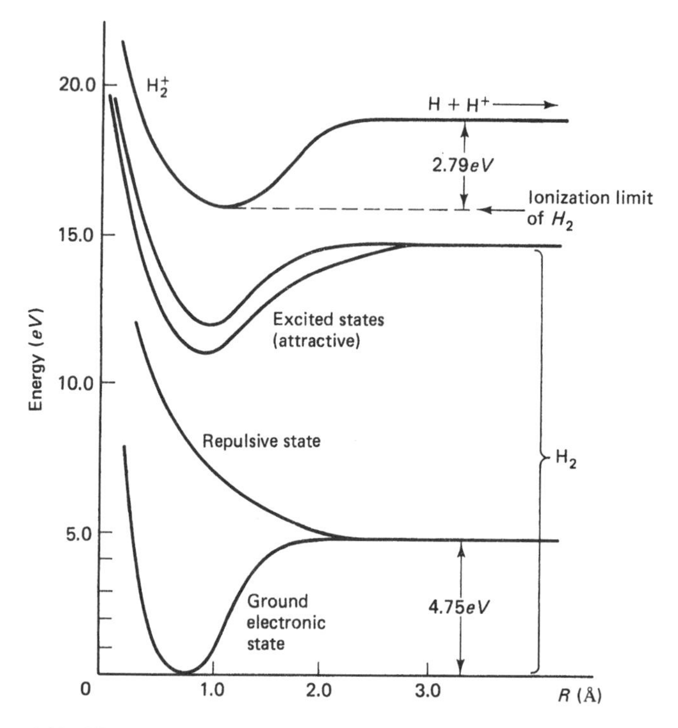

Figure 9.13 The electronic potential energies of several states of the H_2 molecule. The ground-state and three excited states (two attractive and one repulsive) are shown. The horizontal dashed line is the ionization limit of H_2; the curves above this line represent electronic states of the H_2^+ *ion*. [Adapted from the complete potential energy curve in T. Sharp, *Atomic Data*, **2**, 119 (1971).]

The function $V(R)$ deviates from that of a simple harmonic oscillator because *the restoring forces that act on the nuclei are not linear*. But consider values of R very near the minimum of the well—*i.e., consider small excursions from equilibrium,* $R \approx R_e$. Here the potential energy *is* roughly quadratic in the displacement $R - R_e$. Denoting this displacement by

$$x \equiv R - R_e, \qquad \text{displacement from equilibrium} \tag{9.54a}$$

we can write *an approximate potential energy for small-amplitude vibrations*, in terms of the force constant k_f of the molecule, as[20]

$$V(x) = k_f x^2/2. \tag{9.54b}$$

How good an approximation is the simple harmonic oscillator? In Fig. 9.14, this potential is compared to a *very* accurate model of a true electronic energy curve, the Morse potential (*c.f.*, Fig. 8.2). The Morse potential lacks some of the simplifying features of the simple harmonic oscillator: it is not symmetric and does not increase to infinity as $R \to \infty$. Consequently, the Morse potential supports only a finite number

[20]The nuclei of a *polyatomic molecule* (*i.e.*, one with more than two of the little beggars) can execute a number of different *kinds* of vibrational motion, each of which is called a normal mode of vibration. The easiest road to insight into the (vibrational) physics of such molecules is via a powerful and beautiful branch of mathematics called **group theory**. See, for example, *Chemical Applications of Group Theory*, 2nd ed. by F. A. Cotton (New York: Wiley, 1971).

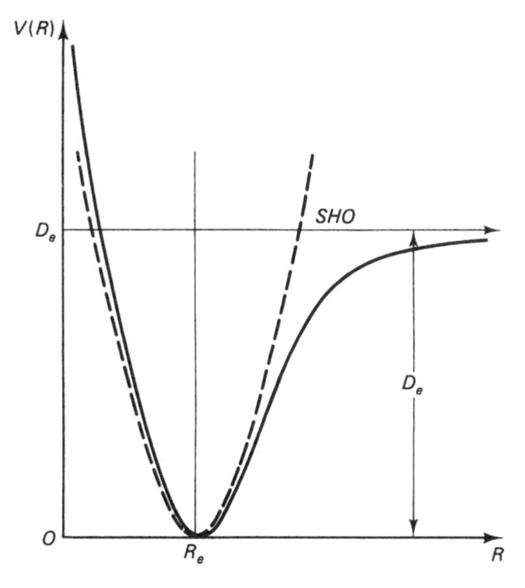

Figure 9.14 The simple harmonic oscillator (dashed curve) and a more accurate model of a molecular electronic potential energy, the Morse potential (solid curve).

of bound states, and its bound-state eigenfunctions don't have definite parity. These differences notwithstanding, the low energies and eigenfunctions of the simple harmonic oscillator are surprisingly good approximations to a more accurate solution of the TISE for nuclear vibrations.

Aside: Molecular Vibrations in the Laboratory. Experimentalists studying molecular vibrations measure such molecular properties as the natural frequencies and force constants in Table 9.1. A widely-used method for such measurements is observation of photons emitted or absorbed when a transition occurs between two stationary vibrational states. From measured spectra, physicists can also determine other quantities crucial to understanding the physics of the molecule, quantities such as the dipole moment function $D(R)$.[21]

The dipole moment of a molecule is analogous to that of a classical charge distribution. Assuming that the only motion the nuclei execute is vibrational, one finds that *the most probable transition is one in which the vibrational quantum number n changes by one unit.* These transitions are well described by the simple harmonic oscillator model. (The analysis of molecular spectra is complicated by the fact that real molecules rotate as well as vibrate. We'll learn how to represent molecular rotations in Part VI, where we'll discover that, like the vibrational energy, the rotational energy of a molecule is quantized. In the analysis of a real molecule, we must take such rotations into account.)

In the lab, one cannot measure directly the function $D(R)$, because the minimum vibrational energy of a molecule is non-zero, *i.e.*, in every state, the nuclei vibrate. So spectroscopists measure two quantities from which they construct the dipole moment function $D(R)$. From the (infrared) transition intensities they can deduce the *slope* $dD(R)/dR$ at R_e. And, by studying the response of the molecule to an external electric field (see Volume II), they can figure out the expectation values of $D(R)$ in the various stationary vibrational states. For example, the dipole moment of HCl (at equilibrium) is 1.085×10^{-18} SC-cm and its derivative (at equilibrium) is 0.86×10^{-10} SC.[22]

[21] See, for example, § 4.12 of *Molecular Structure and Dynamics* by W. H. Flygare (Englewood Cliffs, N.J. Prentice Hall, 1978).

[22] W. S. Benedict *et al.*, *J. Chem. Phys.* **26**, 1671 (1957).

Question 9–10

The measured natural frequency of the HCl molecule is $\omega_0 = 8.65 \times 10^{13}$ Hz. Calculate the force constant k_f of this molecule. (Don't forget to use the reduced mass of the nuclei in your calculation.)

9.7 THE QUANTUM SIMPLE HARMONIC OSCILLATOR—SOLVED!

The Hamiltonian of the simple harmonic oscillator is

$$\hat{\mathcal{H}} = -\frac{\hbar^2}{2m}\frac{d^2}{dx^2} + \frac{1}{2}m\,\omega_0^2 x^2. \qquad \text{simple harmonic oscillator} \tag{9.55}$$

The solutions of the TISE

$$\frac{d^2}{dx^2}\psi_n(x) = -\frac{2m}{\hbar^2}\left[E_n - \frac{1}{2}m\,\omega_0^2 x^2\right]\psi_n(x) \tag{9.56}$$

comprise the spatial parts of the stationary-state wave functions of this system,

$$\Psi_n(x,t) = \psi_n(x)e^{-iE_n t/\hbar}, \tag{9.57}$$

where E_n is one of the allowed energies of the oscillator. Because $V(x)$ increases without limit as $|x|$ increases, it supports an infinite number of bound states but no continuum states. So in any stationary state, the particle is localized in the vicinity of the well, *i.e.*,

$$P_n(x,t) \xrightarrow[x\to\pm\infty]{} 0. \tag{9.58}$$

Question 9–11

Is Eq. (9.58) correct for a *non-stationary state*? Before you answer, think carefully about how a non-stationary state is related to the stationary states of a system. (We'll study a non-stationary state of this system in § 9.10).

We know from § 9.1–9.3 that each state (9.57) is non-degenerate (unique), and that we can take each $\psi_n(x)$ to be real. We also know that, because the potential in the Hamiltonian (9.55) is symmetric, its eigenfunctions have definite parity, beginning with the even ground state. Finally, from curvature arguments we know that $\psi_n(x)$ has n nodes.[*Note carefully that, in contrast to Eq. (9.13), the ground state is labelled with* $n = 0$, *as* $\psi_0(x)$]. These qualitative deductions are all very well, but how do we *solve* (9.56) for the bound-state eigenfunctions and energies? Therein lies the tale of this section.

The Plan of Attack

Mathematicians often solve second-order differential equations with **the method of power series**. Unfortunately, for Eq. (9.56), the method doesn't work.[23] But we can transform Eq. (9.56) into an equation for which the method *does* work. Once we have

[23]Direct application of this method to Eq. (9.56) leads to a three-term recurrence relation. But the solution of this equation contains at most two arbitrary constants, which is one too few. The problem is that this equation has a singularity at $\pm\infty$.

done so, we simply expand the unknown function in our new equation in a power series, insert the series into the differential equation, and solve for the expansion coefficients. Table 9.2 contains an overview of the steps we'll carry out, so you can follow along.

TABLE 9.2 GAME PLAN FOR SOLUTION OF THE TISE OF THE SHO.

Step 1: Clean up TISE using intermediate quantities	$\psi_E(x) \longrightarrow \psi_\epsilon(\xi)$
Step 2: Find the solution in the asymptotic limit $\xi \longrightarrow \infty$	$\psi_E(x) \longrightarrow \chi_{\text{as}}(\xi)$
Step 3: Factor out the asymptotic behavior	$\psi_E(\xi) \propto AH_E(\xi)\chi_{\text{as}}(\xi)$
Step 4: Derive a DE for the unknown factor	equation for $H_E(\xi)$
Step 5: Expand the unknown function in a power series	$H_E(\xi) = \sum c_j \xi^j$
Step 6: Put series in DE & derive recurrence relation	c_j in terms of c_{j-1}
Step 7: Enforce boundary conditions	quantize E_n

Cleaning Up Our Act

To minimize the number of symbols we must manipulate, let's rewrite the Schrödinger equation in terms of *intermediate variables and parameters*. These intermediate quantities, which are summarized in Table 9.3, are suggested by the mathematical structure of the TISE we're trying to solve,

$$\left[\frac{d^2}{dx^2} - \frac{m^2\omega_0^2}{\hbar^2}x^2 + \frac{2mE}{\hbar^2}\right]\psi_E(x) = 0. \tag{9.59}$$

TABLE 9.3 INTERMEDIATE QUANTITIES FOR THE TISE OF THE SHO.

Quantity	Definition	Units
energy	$\epsilon \equiv \frac{2}{\hbar\,\omega_0}E$	dimensionless
potential strength	$\beta \equiv \sqrt{m\,\omega_0/\hbar}$	$(\text{length})^{-1}$
position	$\xi \equiv \beta x$	dimensionless

First, we scale x by the *potential strength parameter* $\beta \equiv \sqrt{m\omega_0/\hbar}$ to produce the dimensionless variable $\xi = \beta x$. Instead of solving an equation [(9.59)] in x for a function of x, $\psi_E(x)$, we'll transform (9.59) into an equation in ξ and solve for $\psi_E(\xi)$. Using the chain rule of differentiation to determine the required derivatives, we find

$$\frac{d\psi}{dx} = \frac{d\xi}{dx}\frac{d\psi}{d\xi} = \beta\frac{d\psi}{d\xi} \tag{9.60a}$$

$$\frac{d^2\psi}{dx^2} = \frac{d}{d\xi}\frac{d\xi}{dx}\left(\frac{d\psi}{dx}\right) = \beta^2\frac{d^2\psi}{d\xi^2}. \tag{9.60b}$$

Introducing the energy variable $\epsilon \equiv 2E/(\hbar\omega_0)$, we can write the transformed Schrödinger equation for the simple harmonic oscillator in the elegant form[24]

$$\boxed{\left(\frac{d^2}{d\xi^2} - \xi^2 + \epsilon\right)\psi_\epsilon(\xi) = 0} \qquad \text{Weber's equation} \tag{9.61}$$

Equation (9.61) is probably unfamiliar to you. But a practicing mathematician would recognize it as **Weber's equation.** Weber's equation is one of several differential equations whose *analytic* solutions have been exhaustively studied and tabulated. Because of their importance in applied mathematics, these are called **special functions.**[25] You will meet several special functions in this book, beginning with Hermite polynomials later in this section.

Since the solutions to Weber's equation are known, we could just write them down and quit. But we wouldn't learn much. Many and varied are the differential equations of physics that can't be solved by well-known, well-studied, well-tabulated functions; and you must know how to cope with them. So let's press on with the plan outlined in Table 9.2 and *solve* Weber's equation.

What Happens at Infinity

One of the first things you should check when confronted by an unfamiliar differential equation is its behavior in the *asymptotic limits of its variable*. Some equations blow up in these limits—in which case we must factor out the asymptotic behavior of the unknown function before we can get down to solving the equation by a method such as expansion in a power series. (Mathematicians describe such points as *singularities*. By the way, some equations blow up at finite values, such as $x = 0$, so in general it's a good idea to go on a singularity hunt before you embark on the quest for a solution.)

Weber's equation simplifies considerably in the limit of large ξ. As $\xi \to \pm\infty$, the second term in (9.61) grows, ultimately overwhelming the constant term ϵ. Letting $\chi_{\text{as}}(\xi)$ denote the limit of $\psi_\epsilon(\xi)$,

$$\psi_\epsilon(\xi) \xrightarrow[\xi\to\pm\infty]{} \chi_{\text{as}}(\xi), \tag{9.62}$$

we can write the *asymptotic form* of Weber's equation as

$$\left(\frac{d^2}{d\xi^2} - \xi^2\right)\chi_{\text{as}}(\xi) = 0. \tag{9.63}$$

[24]Don't confuse the energy variable with the binding energy of the finite square well, which in Chap. 8 we denoted by ϵ_n.

[25]Students often ask *why* such functions are "special." I know no better answer than that provided by W. H. Press, B. P. Flannery, S. A. Teukolsky, and W. T. Vetterling in their superb book *Numerical Recipes: The Art of Scientific Computing* (New York: Cambridge University Press., 1986): "There is nothing particularly special about a *special* function except that some person in authority or a textbook writer (not the same thing!) has decided to bestow the moniker." The bible of special functions (for scientists and engineers) is the massive compendium *Handbook of Mathematical Functions* edited by M. Abramowitz and I. Stegun (National Bureau of Standards Applied Mathematics Series No. 55, November 1970). I find particularly useful *Formulas and Theorems for the Special Functions of Mathematical Physics* by W. Magnus, F. Oberhettinger, and R. P. Soni (New York: Springer-Verlag, 1966), which includes some little-known relationships involving these functions.

Not surprisingly, $\chi_{\text{as}}(\xi)$ is independent of the energy variable ϵ, which disappeared from (9.61) in the limit.

Equation (9.63) is *almost* the equation for a Gaussian function,

$$\frac{d^2 f}{d\xi^2} = \left(\xi^2 \pm 1\right) f(\xi). \tag{9.64a}$$

Equation (9.64*a*) has two linearly independent solutions: the Gaussian functions $e^{-\xi^2/2}$ and $e^{+\xi^2/2}$. But we are interested only in the solution to (9.63) for *very large values* of ξ. In the asymptotic region, $\xi^2 - 1 \approx \xi^2$, and Eq. (9.64*a*) becomes identical to (9.63). So introducing arbitrary constants A and B, we can write the *general solution of Weber's equation in the asymptotic region* as

$$\chi_{\text{as}}(\xi) = Ae^{-\xi^2/2} + Be^{+\xi^2/2} \qquad \text{[solution for } \xi \to \pm\infty\text{]}. \tag{9.64b}$$

Question 9–12

What is the *width* of $e^{\xi^2/2}$?

Question 9–13

Substitute the general form (9.64*b*) into (9.63) and, using the fact that $\left|\xi^2\right|$ is large, verify that it is indeed a solution.

Do you see a problem with the asymptotic solution (9.64*b*)? As $\xi \to \infty$, one of the terms blows up: $e^{+\xi^2/2} \to \infty$. From our solutions of the TISE for a piecewise-constant potential in Chap. 8 we know how to deal with such exploding terms: we kill them off by exploiting the arbitrariness of the constants in the general form. In Eq. (9.64*b*) we choose $B = 0$, thereby ensuring that the function $\psi_\epsilon(\xi)$ decays to zero as $\xi \to \infty$ and hence is normalizable. So we have discovered how the Hamiltonian eigenfunctions of the simple harmonic oscillator behave *in the asymptotic limit*

$$\boxed{\psi_E(\xi) \xrightarrow[\xi \to \pm\infty]{} \chi_{\text{as}}(\xi) = Ae^{-\xi^2/2}} \tag{9.65}$$

Aside: More on the Asymptotic Solution of Weber's Equation. Strictly speaking, Eq. (9.65) describes the *dominant* behavior of $\psi_E(x)$ in the asymptotic limit. Closer examination of Weber's equation reveals that in this limit, it's satisfied by any function of the form $A\xi^n e^{-\xi^2/2}$, where n is an integer. As $\xi \to \pm\infty$, the decaying exponential dominates the growing power ξ^n. In fact, for any positive constant α,

$$\lim_{\xi \to \infty} \xi^n e^{-\alpha\xi^2} = 0. \qquad \alpha > 0$$

Thus we regain Eq. (9.65).

What Happens Everywhere Else

Let's write $\psi_\epsilon(\xi)$ in a way that incorporates what we know—its asymptotic behavior—and that represents what we don't know by a new function $H_\epsilon(\xi)$ and a constant A:

$$\boxed{\psi_\epsilon(\xi) = AH_\epsilon(\xi)e^{-\xi^2/2}} \tag{9.66}$$

Eventually we'll determine A via the *normalization condition*, which, expressed in terms of the dimensionless variable ξ, reads

$$\frac{1}{\beta}\int_{-\infty}^{+\infty} \psi_\epsilon^*(\xi)\,\psi_\epsilon(\xi)\,d\xi = 1. \tag{9.67}$$

But $\psi_\epsilon(\xi)$ can satisfy this condition only if $H_\epsilon(\xi)$ does not befoul the asymptotic decay to zero of the Gaussian factor in (9.66). Our eigenfunction will be normalizable if $H_\epsilon(\xi)$ has the property that

$$H_\epsilon(\xi)e^{-\xi^2/2} \xrightarrow[\xi\to\pm\infty]{} 0. \tag{9.68}$$

We already know a few things about the unknown function in (9.66). First, notice that the eigenfunction $\psi_\epsilon(\xi)$ depends on the energy variable $\epsilon = 2E/(\hbar\omega_0)$, but the asymptotic function does not. So I've labelled $H_\epsilon(\xi)$, the unknown function in (9.66), with ϵ. Evidently, this function will play a role in determining the bound-state energies.

Second, the function $H_\epsilon(\xi)$ should reflect the properties of the corresponding eigenfunction $\psi_\epsilon(x)$. One of these properties is the *definite parity* of this eigenfunction. The Gaussian $e^{-\xi^2/2}$ is even, so $H_\epsilon(\xi)$ must be responsible for the parity of $\psi_E(x)$; *i.e.*,

$$V(x) \text{ is symmetric} \quad \begin{matrix} \Longrightarrow \\ \Longrightarrow \end{matrix} \quad \begin{matrix} \psi_\epsilon(-\xi) = \pm\psi_\epsilon(\xi) \\ H_\epsilon(-\xi) = \pm H_\epsilon(\xi) \end{matrix} \quad \text{parity condition.} \tag{9.69}$$

Aside: A Clue in the Asymptotic Solution of Weber's Equation. I noted above that in the asymptotic limit Weber's equation is satisfied by any function of the form $A\xi^n e^{-\xi^2/2}$, where n is an integer. This suggests a possible form for $H_\epsilon(\xi)$: a sum of powers of ξ, *i.e.*, a polynomial, such as $c_0\xi^0 + c_2\xi^2 + c_4\xi^4 + \ldots$ (even parity) or $c_1\xi^1 + c_3\xi^3 + c_5\xi^5 + \ldots$ (odd parity). And indeed, this turns out to be the answer.

The Differential Equation to Solve

We began our assault on the TISE for the SHO by rewriting it in simplified notation and dimensionless variables (Table 9.3). Examination of the asymptotic behavior of the resulting equation revealed that its solution is dominated by a decaying Gaussian. We have factored this (known) function out of the Hamiltonian eigenfunction and must now derive an equation for the remaining, unknown function. To this end, let's substitute the form (9.66) for $\psi_\epsilon(\xi)$ into Weber's equation (9.61). After a little algebra, we obtain

$$\boxed{\frac{d^2}{d\xi^2}H_\epsilon(\xi) - 2\xi\frac{d}{d\xi}H_\epsilon(\xi) + (\epsilon - 1)H_\epsilon(\xi) = 0} \quad \text{Hermite equation} \tag{9.70}$$

This equation is called the **Hermite differential equation**, after the French mathematician Charles Hermite (1822–1901), who found and explored its solutions. These solutions are appropriately called **Hermite functions**.

Question 9–14

Derive Eq. (9.70).

Our approach to the Hermite equation is quite simple (see Table 9.2). Most functions can be represented by a *power series* in their independent variable (*e.g.*, the Taylor and Laurent series). So we'll try expanding $H_\epsilon(\xi)$ in a series with unknown coefficients c_j,

$$H_\epsilon(\xi) = c_0 + c_1\xi + c_2\xi^2 + \cdots, \tag{9.71a}$$

$$= \sum_{j=0}^{\infty} c_j \xi^j. \tag{9.71b}$$

Of course, one or more of these coefficients may be zero. To determine their values, we will substitute this series into the differential equation for $H_\epsilon(\xi)$ and, exploiting the linear independence of the functions ξ^j, will derive a relationship between various of the coefficients. This relationship is called a *recursion relation.* Last, we'll impose the boundary conditions (9.68). Doing so will *limit the series to a finite number of terms* and, as it happens, will determine the allowed energies of the oscillator.

Question 9–15

Use the relationship between the Hermite functions $H_\epsilon(\xi)$ and the Hamiltonian eigenfunctions $\psi_\epsilon(\xi)$ *and* your knowledge of the qualitative properties of these eigenfunctions to answer the following question: Which of the following describes the ξ dependence of the *lowest-order term* in the power-series expansion of the Hermite polynomial in the *ground-state eigenfunction*: a constant term, a term proportional to ξ, or a term proportional to ξ^n where $n > 1$? Next, answer this question for the *first excited state.*

Power Series: A Quick Review

Before we begin, I want to jog your memory regarding *series expansions of functions.*[26] You are probably familiar with expansions of elementary functions in power series. For example, the expansions of e^ξ and $e^{-\xi}$, which solve the differential equations

$$\left(\frac{d}{dx} \pm 1\right) f(\xi) = 0,$$

are

$$e^{\pm\xi} = 1 \pm \xi + \frac{1}{2!}\xi^2 \pm \frac{1}{3!}\xi^3 + \cdots \tag{9.72a}$$

$$= \sum_{j=0}^{\infty} b_j \xi^j, \tag{9.72b}$$

where

$$b_j = (\pm 1)^j \frac{1}{j!}. \tag{9.72c}$$

We can adapt this form to other functions, such as e^{ξ^2}:

[26]If this summary isn't sufficient, see Chap. 5 of *Mathematical Methods for Physicists*, 2nd ed. by G. Arfken (New York: Academic Press, 1970) or *Mathematical Methods in the Physical Sciences*, 2nd ed. by Mary Boas (New York: Wiley, 1963).

$$e^{\xi^2} = 1 + \xi^2 + \frac{1}{2!}\left(\xi^2\right)^2 + \frac{1}{3!}\left(\xi^2\right)^3 + \frac{1}{4!}\left(\xi^2\right)^4 + \cdots \tag{9.73a}$$

$$= 1 + \xi^2 + \frac{1}{2}\xi^4 + \frac{1}{6}\xi^6 + \frac{1}{24}\xi^8 + \cdots. \tag{9.73b}$$

To write a general form [analogous to (9.72*b*)] for the series (9.73*b*), we need only note that the *powers* in successive terms increase by 2; *viz.*,

$$e^{\xi^2} = \sum_{\substack{j=0 \\ \text{(even)}}}^{\infty} b_j \xi^j, \tag{9.74a}$$

where

$$b_j = \frac{1}{(j/2)!}. \tag{9.74b}$$

A crucial question concerning a series expansion of a function is: for what values of the independent variable does the series converge to the function? The region of ξ within which a power series $\sum_{j=0}^{\infty} b_j \xi^j$ converges to an analytic function $f(\xi)$ depends on *the behavior of the ratio of successive terms in the series as the summation index j becomes large.* According to the **d'Alembert ratio test**, such a series converges for all ξ if the ratio of successive terms, $b_{j+1}\xi^{j+1}/b_j\xi^j$, goes to zero as $j \to \infty$. For example, the ratio of successive terms in the series (9.73) is

$$\frac{\dfrac{1}{\left(\frac{j+2}{2}\right)!}\xi^{j+2}}{\dfrac{1}{\left(\frac{j}{2}\right)!}\xi^{j}} = \frac{1}{\frac{j}{2}+1}\xi^2 \sim \frac{2}{j}\xi^2 \xrightarrow[j\to\infty]{} 0. \tag{9.75}$$

[*Notice very carefully that Eq. (9.75) implies only that the series converges, not that it converges to zero.*] Of course, this series does converge, to e^{ξ^2}.

Solving the Hermite Equation the Power-Series Way

With this background, let's return to the series (9.71) for the Hermite function. What are the implications of the properties of the Hermite functions, Eqs. (9.68)–(9.69), for this series expansion? First, $H_\epsilon(\xi)$ depends on the energy variable ϵ, so at least one of the unknown coefficients c_j must also depend on ϵ; *i.e.*, $c_j = c_j(\epsilon)$. Second, the *parity condition* (9.69) means that each function $H_\epsilon(\xi)$ will include *either* even powers of ξ (if it is even) *or* odd powers (if it is odd)—but not both. This property follows from the simple parity of x^n,

$$(-x)^n = (-1)^n x^n = \pm x^n. \qquad \begin{cases} n \text{ even} \\ n \text{ odd} \end{cases} \tag{9.76}$$

That is, each Hermite function must have one of the following forms:

$$H_\epsilon(\xi) = \begin{cases} c_0 + c_2\xi^2 + c_4\xi^4 + \cdots & \text{even parity} \\ c_1\xi + c_3\xi^3 + c_5\xi^5 + \cdots & \text{odd parity.} \end{cases} \tag{9.77}$$

One consequence of (9.77) is that for each Hermite function, one of the coefficients c_0 and c_1 must be zero:[27]

$$H_\epsilon(\xi) \text{ is even} \quad \Longrightarrow \quad c_0 \neq 0 \text{ and } c_1 = 0 \tag{9.78a}$$

$$H_\epsilon(\xi) \text{ is odd} \quad \Longrightarrow \quad c_1 \neq 0 \text{ and } c_0 = 0. \tag{9.78b}$$

This requirement poses no problem. The function $H_\epsilon(\xi)$ contains two arbitrary coefficients, so two of the c_j's in (9.71) are arbitrary. One of these will appear in each even-parity function and one in each odd-parity function.

At this point, I hope you are worried by the presence in the Hermite series of an *infinite number of terms*. The worry is that this series may blow up as $\xi \to \infty$. Such a detonation would not *necessarily* be a disaster, because it's the function

$$\psi_\epsilon(\xi) = Ae^{-\xi^2/2}\left(\sum_{j=0}^{\infty} c_j \xi^j\right) \tag{9.79}$$

that must go to zero in the asymptotic limit. In fact, the blow up of these series *is* a disaster, for it *dominates* the decay of the Gaussian and renders the product of the two functions unnormalizable. Fortunately, we can avoid this disaster by making the series *finite*. This works because in the asymptotic limit, a decaying exponential dominates any *finite* polynomial.[28] So once we've derived the recurrence relation, we must find a way to terminate each series at a finite number of terms, so that it includes terms with incides from $j = 0$ to $j = j_{\max}$.

The Algebra

To substitute the series (9.71) into the Hermite differential equation, we need a couple of derivatives:

$$\frac{d}{d\xi} H_\epsilon(\xi) = \sum_{j=0}^{\infty} j c_j \xi^{j-1} \tag{9.80a}$$

$$\frac{d^2}{d\xi^2} H_\epsilon(\xi) = \sum_{j=0}^{\infty} j(j-1) c_j \xi^{j-2}. \tag{9.80b}$$

Substituting these expansions in the Hermite equation, we get

$$\sum_{j=0}^{\infty} j(j-1)c_j\xi^{j-2} - 2\sum_{j=0}^{\infty} jc_j\xi^j + (\epsilon - 1)\sum_{j=0}^{\infty} c_j\xi^j = 0. \tag{9.81}$$

To extract from (9.81) a relationship between the coefficients c_j, we use the *linear independence* of the functions $\{1, \xi, \xi^2, \xi^3, \ldots\}$. That is, the only linear combination of these functions that equals zero is the one with zero coefficients:

[27] Equations (9.78) apply to the solutions of the Hermite differential equation because, as we'll see, in the even-parity solutions of this equation the lowest-order term is constant and in the odd-parity solutions the lowest-order term is proportional to ξ. In general, we can conclude only that for even-parity solutions $c_{2n+1} = 0$ and for odd-parity solutions $c_{2n} = 0$.

[28] See, for example, § 6.8 of *Calculus and Analytic Geometry* by G. B. Thomas (Reading, Mass.: Addison-Wesley, 1960).

$$\boxed{c_0 + c_1\xi + c_2\xi^2 + \cdots = 0 \quad \Longleftrightarrow \quad c_0 = c_1 = c_2 = \cdots = 0} \qquad \text{linear independence} \tag{9.82}$$

To exploit this property, let's first manipulate Eq. (9.81) into the form

$$\sum_{j=0}^{\infty} (\text{factors involving the } c_j \text{ s})\, \xi^j = 0. \tag{9.83}$$

Then (9.82) allows us to set each factor equal to zero.

The only term in (9.81) not already in the desired form is the first. This term contains ξ^{j-2} instead of ξ^j. But look at this series:

$$\sum_{j=0}^{\infty} j(j-1)c_j\xi^{j-2} = 2c_2 + 3\cdot 2c_3\xi + 4\cdot 3c_4\xi^2 + \cdots . \tag{9.84a}$$

To write (9.84*a*) in the form (9.83), we need only reduce the value of the index j by 2; *i.e.*,

$$\sum_{j=0}^{\infty} j(j-1)c_j\xi^{j-2} = \sum_{j=0}^{\infty}(j+2)(j+1)c_{j+2}\xi^j. \tag{9.84b}$$

Question 9–16

If you aren't familiar with the re-labelling of summations, a standard gambit in the solution of differential equations, this step may seem a bit magical. The key is that the factors j and $j-1$ in (9.84*a*) kill the terms for $j=0$ and $j=1$. [a]. Write out the first few terms in (9.84*b*) explicitly and verify that it is equivalent to (9.84*a*). [b]. More formally, you can get the same effect by changing the index of summation from j to $j' = j-2$. If j runs from 2 to ∞, then j' runs from 0 to ∞. Try this procedure, then replace j' by j and regain (9.84*b*).

With this re-labelling, Eq. (9.81) assumes the form of (9.83):

$$\sum_{j=0}^{\infty} \left[(j+2)(j+1)c_{j+2} + (\epsilon - 1 - 2j)c_j\right]\xi^j = 0. \tag{9.85}$$

According to (9.82), the quantity in square brackets is zero. Therefore, successive coefficients c_{j+2} and c_j in the series for the Hermite function are related by the **two-term recurrence relation**

$$\frac{c_{j+2}}{c_j} = \frac{2j+1-\epsilon}{(j+2)(j+1)}. \qquad [j = 0, 1, 2, \cdots] \tag{9.86}$$

Notice that the energy variable appears in this relation.

"What good," you may be wondering, "is a two-term recurrence relation?" Well, we can use this equation to determine the terms in the Hermite series from the arbitrary coefficients, c_0 and c_1. Taking into account the parity of these functions [Eq. (9.69)], we can write

$$H_\epsilon(\xi) = \begin{cases} c_0\left(1 + \dfrac{c_2}{c_0}\xi^2 + \dfrac{c_4}{c_2}\dfrac{c_2}{c_0}\xi^4 + \dfrac{c_6}{c_4}\dfrac{c_4}{c_2}\dfrac{c_2}{c_0}\xi^6 + \cdots\right) & \text{even parity} \\[2ex] c_1\left(\xi + \dfrac{c_3}{c_1}\xi^3 + \dfrac{c_5}{c_3}\dfrac{c_3}{c_1}\xi^5 + \dfrac{c_7}{c_5}\dfrac{c_5}{c_3}\dfrac{c_3}{c_1}\xi^7 + \cdots\right) & \text{odd parity} \end{cases} \tag{9.87}$$

Now, for the first coupled even parity solutions we have

$$c_2 = \frac{1-\epsilon}{2} c_0$$
$$c_4 = \frac{5-\epsilon}{12} c_2 = \left(\frac{5-\epsilon}{12}\right)\left(\frac{1-\epsilon}{2}\right) c_0. \tag{9.88a}$$

Similarly, for an odd-parity function $H_\epsilon(\xi)$,

$$c_3 = \frac{3-\epsilon}{6} c_1$$
$$c_5 = \frac{7-\epsilon}{20} c_2 = \left(\frac{7-\epsilon}{20}\right)\left(\frac{3-\epsilon}{6}\right) c_1. \tag{9.88b}$$

So it looks like we're through. We have enough arbitrary coefficients to normalize the eigenfunctions, and we have developed clean forms for the Hermite functions of both parities:

$$H_\epsilon(\xi) = \begin{cases} \sum\limits_{\substack{j=0 \\ \text{(even)}}} c_j \xi^j & \text{even parity} \\ \sum\limits_{\substack{j=1 \\ \text{(odd)}}} c_j \xi^j & \text{odd parity} \end{cases} \qquad c_{j+2} = \frac{2j+1-\epsilon}{(j+2)(j+1)} c_j \tag{9.89}$$

But there is a snake in Eden.

Aside: Convergence of the Hermite Series. Is the problem *lack of convergence*? Does this series converge? Well, the ratio of successive terms is

$$\frac{c_{j+2}\xi^{j+2}}{c_j \xi^j} = \frac{2j+1-\epsilon}{(j+2)(j+1)}\xi^2.$$

This ratio does go to zero as $j \to \infty$. So the series (9.89) do converge to an analytic function of ξ. The trouble is not *whether* this series converges, but what it converges to.

Trouble at Infinity

The form of the Hamiltonian eigenfunction (9.66) for the SHO is

$$\psi_E(\xi) = (\text{infinite series in } \xi)(\text{decaying Gaussian}). \tag{9.90}$$

For this function to be physically admissible, it must go to zero as $\xi \to \infty$. The easiest way to investigate the asymptotic behavior of a series is to relate it to an *analytic function* of ξ. Mathematicians know a clever way to ferret out such relationships: *compare successive terms of the series in this limit to successive terms in the expansions of various analytic functions.* Let's apply this strategy to the even-parity Hermite series. (I'll ask you to deal with the odd-parity solutions in a question.)

For large $|\xi|$, the dominant terms in this series are those with large indices, j. In the limit $j \to \infty$, the ratio of successive terms in the Hermite series (9.89) behaves as

$$\frac{c_{j+2}\xi^{j+2}}{c_j \xi^j} = \frac{2j+1-\epsilon}{(j+2)(j+1)}\xi^2 \underset{j\to\infty}{\sim} \frac{2}{j}\xi^2. \tag{9.91}$$

Now, $2\xi^2/j$ is precisely the ratio we found in Eq. (9.75) for the power-series expansion of the function $f(\xi) = e^{\xi^2}$. Therefore, in the asymptotic limit the series for an even-parity Hermite function behaves like this function, *i.e.*,

$$\sum_{\substack{j=0 \\ (\text{even})}}^{\infty} c_j \xi^j \xrightarrow[|\xi|\to\infty]{} e^{\xi^2}. \tag{9.92}$$

We have a problem. Look at the asymptotic behavior of the *product* of this infinite series and the decaying Gaussian:

$$\left(\sum_{j=0}^{\infty} c_j \xi^j\right) e^{-\xi^2/2} \xrightarrow[|\xi|\to\infty]{} e^{\xi^2} e^{-\xi^2/2} \to \infty. \tag{9.93}$$

Since this product is not bounded, it cannot be normalized and hence isn't physically admissible.[29]

The Terminator

There is only one way out: we must terminate the series before it can blow up. That is, we must somehow force all coefficients c_j for j greater than some finite index j_{max} to be zero. This stratagem works because according to the recursion relation in (9.89), $c_{j+2} \propto c_j$, which implies that

$$\text{if } c_{j_{\text{max}}} = 0, \text{ then } c_j = 0 \text{ for all } j > j_{\text{max}}. \tag{9.94}$$

According to the recurrence relation (9.86), we can make $c_{j_{\text{max}}+2}$ equal to zero by choosing the energy variable ϵ so that the numerator in this relation is zero. So the series is terminated at j_{max} if

$$\epsilon = 2j_{\text{max}} + 1. \tag{9.95a}$$

For any value of $j_{\text{max}} > 0$, Eq. (9.95*a*) forces the Hermite series to be *finite*. The resulting function is called a **Hermite polynomial**:

$$H_\epsilon(\xi) = \sum_{j=0}^{j_{\text{max}}} c_j \xi^j \qquad \epsilon = 2j_{\text{max}} + 1. \tag{9.95b}$$

[29]This argument is heuristic, not rigorous. It can, however, be put on a rigorous foundation using a theorem concerning power series that was proved by D. F. Lawden [see his book *The Mathematical Principles of Quantum Mechanics* (London: Methuen, 1967)]: consider two power series $\sum_{j=0}^{\infty} c_j\xi^j$ and $\sum_{j=0}^{\infty} b_j\xi^j$ that, according to the d'Alembert ratio test, converge to functions $f_1(\xi)$ and $f_2(\xi)$, respectively. If there exists a value of the index j—call it j_0—such that for $j \geq j_0$ the coefficients of both series are positive and are related by $b_j/b_{j-1} > c_j/c_{j-1}$, then $f_2(\xi)$ is bounded above by $f_1(\xi)$. Applying this theorem to the series for e^{ξ^2} and for the Hermite function, one can prove [Eqs. (9.74) and (9.85)] that the Hermite series is bounded by the exponential function, *i.e.*,

$$\lim_{\xi\to\pm\infty} H_\epsilon(\xi) < e^{\xi^2}.$$

Any such function dominates the Gaussian $e^{-\xi^2/2}$ in the asymptotic limit, leading to an exponential explosion. [This connection was pointed out in M. Bowen and J. Coster, *Amer. Jour. Phys.* **48**, 307 (1980).]

When we construct the Hamiltonian eigenfunction $\psi_\epsilon(\xi) = AH_\epsilon(\xi)e^{-\xi^2/2}$ with a Hermite polynomial, we obtain a function of the form

$$\psi_\epsilon(\xi) = (\text{finite series in } \xi)\,(\text{decaying Gaussian}). \tag{9.96}$$

Such a function does go to zero in the asymptotic limit. Whew!

For example, the smallest value of $j_{\max}$ is $j_{\max} = 0$. According to (9.95*a*), this choice corresponds to $\epsilon = 1$. This value of ϵ forces all coefficients other than c_0 to be zero, *i.e.*, $c_2 = c_4 = \cdots = 0$. The resulting Hermite polynomial has only one term, $H_\epsilon(\xi) = c_0$, and is even. Here are a few more examples:

$$\begin{aligned}
j_{\max} &= 2 \quad &&\Rightarrow c_4 = c_6 = \cdots = 0 \quad &&\Rightarrow H_\epsilon(\xi) = c_0 + c_2\xi^2 \\
j_{\max} &= 3 \quad &&\Rightarrow c_5 = c_7 = \cdots = 0 \quad &&\Rightarrow H_\epsilon(\xi) = c_1\xi + c_3\xi^3 \\
j_{\max} &= 6 \quad &&\Rightarrow c_8 = c_{10} = \cdots = 0 \quad &&\Rightarrow H_\epsilon(\xi) = c_0 + c_2\xi^2 + c_4\xi^4 + c_6\xi^6.
\end{aligned} \tag{9.97}$$

The Allowed Energies

It is customary to label Hermite polynomials with the index $n = j_{\max}$. So the three polynomials in Eq. (9.97) are $H_2(\xi)$, $H_3(\xi)$, and $H_6(\xi)$. The index n is the highest power of ξ that appears in the polynomial $H_n(\xi)$, and is called the **degree of the polynomial.** We also use n to label the energy variable ϵ and write the restriction (9.95a) as

$$\epsilon_n = 2n + 1. \tag{9.98}$$

But this variable is related to the stationary state energies by $E_n = \hbar\omega_0\epsilon_n/2$ [see Table 9.3]. So *the requirement that $H_n(\xi)$ must be a finite polynomial restricts the energies of the simple harmonic oscillator* to

$$\boxed{E_n = \left(n + \frac{1}{2}\right)\hbar\omega_0. \qquad n = 0, 1, 2, \ldots} \tag{9.99}$$

For any value of E not of this form, the corresponding eigenfunction of the SHO Hamiltonian contains an *infinite* polynomial and hence blows up at infinity—socially unacceptable behavior for a function that aspires to physical significance. Once again we have uncovered a subtle kinship between mathematics and physics, the latter limiting the solutions provided by the former:

$$\boxed{\text{boundary conditions}} \quad \Longrightarrow \quad \boxed{\text{energy quantization}}$$

Question 9–17

Repeat the above argument for an *odd-parity* Hermite function, showing that this case leads to the same restriction on the series as we found for an even-parity function.

Question 9–18

Question 9–15 explains why the series Eq (9.71) is appropriate to the Hermite equation. But in general, the lowest-order term in the solution of a differential equation could be proportional to ξ to some power, say p. If we cannot deduce this power, we can let the mathematics determine it by writing the series as

$$f(\xi) = \xi^p \sum_{j=0}^{\infty} c_j \xi^j.$$

Substitute this form into the Hermite equation and show that $p = 0$ for the even-parity functions and $p = 1$ for the odd functions.

9.8 THE SHO—THE ANSWERS

The key result of § 9.7 was our discovery that only certain solutions of the Hermite equation yield physically admissible wave functions—namely, *finite* polynomials whose order n is related to the energy variable ϵ by $\epsilon = 2n + 1$. These polynomials fall into two classes according to their parity:

$$\boxed{\begin{aligned} H_n(\xi) &= \sum_{\substack{j=0 \\ \text{(even)}}}^{n} c_j \xi^j \qquad [c_0 \text{ arbitrary}] \qquad n = 0, 2, 4, \ldots \\ H_n(\xi) &= \sum_{\substack{j=1 \\ \text{(odd)}}}^{n} c_j \xi^j \qquad [c_1 \text{ arbitrary}] \qquad n = 1, 3, 5, \ldots \end{aligned}} \tag{9.100}$$

For functions of either parity, the coefficients c_j for $j > 1$ are determined from the recurrence relation (9.86), which we can write as

$$c_j = \frac{2(j - n - 2)}{j(j-1)} c_{j-2} \qquad 2 \leq j \leq n - 2. \tag{9.101}$$

With this relation, you can easily generate lots of Hermite polynomials, if necessary. (Other mathematical properties of the Hermite polynomials are explored in Pblm. 9.11.)

Example 9.2. The First Four Hermite Polynomials

The lowest-order Hermite polynomial is the even function $H_0(\xi)$, which, as we saw in § 9.7, is simply $H_0(\xi) = c_0$. It is conventional to choose $c_0 = 1$ (which we are free to do, since this coefficient is arbitrary), giving

$$H_0(\xi) = 1. \tag{9.102a}$$

Not a major thrill. [The function (9.102*a*) is not, of course, the ground-state eigenfunction of the SHO; we'll construct and normalize that function in Example 9.3.]

The Hermite polynomial for $n = 1$ is odd and consists of the single term $c_1\xi$. Following convention, we choose $c_1 = 1$, which gives

$$H_1(\xi) = \xi. \tag{9.102b}$$

The next two polynomials are a bit more interesting. The Hermite polynomial for $n = 2$ includes two terms: $c_0 + c_2\xi^2$. According to the recurrence relation, the coefficient c_2 is just -2, so[30]

[30]More conventions: In tables of Hermite polynomials, the highest power of ξ always appears first, with a plus sign (see Table 9.4). Thus, the tabulated polynomials differ by an overall factor of -1 from those calculated via the recurrence relation. In fact, we are free to multiply the Hermite polynomials by any constant we like, for they solve a *linear, homogeneous* differential equation and hence are arbitrary to within an overall multiplicative constant.

$$H_2(\xi) = 1 - 2\xi^2. \qquad (9.102c)$$

One more, then I'll stop. The form of the Hermite polynomial for $n = 3$ is $H_3(\xi) = c_1\xi + c_3\xi^3$. According to the recurrence relation, $c_3 = -2/3$, so this polynomial is

$$H_3(\xi) = \xi - \frac{2}{3}\xi^3. \qquad (9.102d)$$

Mercifully, tables of Hermite polynomials exist (see Table 9.4), so we usually don't have to go through this tiresome process. Then again, you never know when you might be separated from your trusty tables and need a Hermite polynomial. Or you might need a special function that hasn't been tabulated. So please learn the tiresome process anyway.

TABLE 9.4 A SELECTION OF HERMITE POLYNOMIALS.

$H_0(\xi) = 1$
$H_1(\xi) = 2\xi$
$H_2(\xi) = 4\xi^2 - 2$
$H_3(\xi) = 8\xi^3 - 12\xi$
$H_4(\xi) = 16\xi^4 - 48\xi^2 + 12$
$H_5(\xi) = 32\xi^5 - 160\xi^3 + 120\xi$
$H_6(\xi) = 64\xi^6 - 480\xi^4 + 720\xi^2 - 120$
$H_7(\xi) = 128\xi^7 - 1344\xi^5 + 3360\xi^3 - 1680\xi$
$H_8(\xi) = 256\xi^8 - 3584\xi^6 + 13440\xi^4 - 13440\xi^2 + 1680$
$H_9(\xi) = 512\xi^9 - 9216\xi^7 + 48384\xi^5 - 80640\xi^3 + 30240\xi$

Question 9–19

Use the recurrence relation to generate the Hermite polynomial of order 10, *i.e.*, $H_{10}(\xi)$. (Don't start from $H_0(\xi)$; use Table 9.4.)

The Normalized Hamiltonian Eigenfunctions of the SHO (At Last)

The Hamiltonian eigenfunction $\psi_n(x)$ of the SHO is a product of the Hermite polynomial $H_n(\xi)$ and a decaying Gaussian, *i.e.*,

$$\boxed{\begin{aligned} &\psi_n(x) = A_n H_n(\beta x) e^{-\beta^2 x^2/2} \\ &\beta = \sqrt{\frac{m\omega_0}{\hbar}} \qquad n = 0, 1, 2, \ldots \end{aligned}} \qquad \text{SHO eigenfunctions} \qquad (9.103a)$$

[Notice that I've used the definition of the dimensionless length variable $\xi = \beta x$ to translate $\psi_\epsilon(\xi)$ into a function of x.] The corresponding energies are

$$\boxed{E_n = \left(n + \tfrac{1}{2}\right)\hbar\omega_0. \qquad n = 0, 1, 2, \ldots} \qquad \text{SHO allowed energies} \qquad (9.103b)$$

The only parts of the eigenfunctions (9.103*a*) *that remain to be worked out are the normalization constants* A_n.

Example 9.3. The First Normalized Eigenfunction.

The ground-state eigenfunction is

$$\psi_0(\xi) = A_0 e^{-\xi^2/2}. \tag{9.104a}$$

We must choose A_0 so that the normalization condition [Eq. (9.67)] is satisfied, *i.e.*,

$$\frac{1}{\beta} A_0^2 \int_{-\infty}^{\infty} e^{-\xi^2}\, d\xi = 1. \tag{9.104b}$$

You'll find this integral—along with several others—in Table 9.5. The normalization condition (9.104*b*) is satisfied if we choose

$$A_0 = \sqrt[4]{\frac{\beta^2}{\pi}}. \tag{9.104c}$$

So the normalized ground-state wave function of the simple harmonic oscillator is

$$\Psi_0(x,t) = \sqrt[4]{\frac{\beta^2}{\pi}} e^{-\beta^2 x^2/2}\, e^{-i\,\omega_0 t/2}. \qquad \text{ground-state eigenfunction} \tag{9.105}$$

TABLE 9.5 USEFUL INTEGRALS OF GAUSSIAN FUNCTIONS.

$$\int_{-\infty}^{\infty} e^{-\alpha x^2}\, dx = \sqrt{\frac{\pi}{\alpha}}$$

$$\int_{0}^{\infty} x e^{-\alpha x^2}\, dx = \frac{1}{2}\alpha$$

$$\int_{-\infty}^{\infty} x^2 e^{-\alpha x^2}\, dx = \frac{1}{2}\sqrt{\frac{\pi}{\alpha^3}}$$

Happily, we don't have to explicitly normalize each eigenfunction. The required integrals are of a well-studied standard form. From tables of integrals involving Gaussian functions, one can show that the normalization constant for the nth stationary state is[31]

$$A_n = \left(\frac{\beta^2}{\pi}\right)^{1/4} \frac{1}{\sqrt{2^n n!}}. \tag{9.106}$$

The corresponding *wave function* is

$$\boxed{\Psi_n(x,t) = \left[\left(\frac{\beta^2}{\pi}\right)^{1/4} \frac{1}{\sqrt{2^n n!}}\right] H_n(\beta x) e^{-\beta^2 x^2/2}\, e^{-i(2n+1)\,\omega_0 t/2}} \tag{9.107}$$

[31] See Appendix J. For more information, have a look at the Appendix to *Elementary Quantum Mechanics* by D. S. Saxon (San Francisco: Holden-Day, 1968).

A handy list of the first several eigenfunctions of the SHO is provided in Table 9.6, and the first eight of these functions — and their probability densities — are graphed in Figs. 9.15 and 9.16.

TABLE 9.6 THE HAMILTONIAN EIGENFUNCTIONS OF THE SHO.

n	E_n	$\psi_n(x)$
0	$\frac{1}{2}\hbar\,\omega_0$	$\left(\frac{\beta^2}{\pi}\right)^{1/4} e^{-\beta^2 x^2/2}$
1	$\frac{3}{2}\hbar\,\omega_0$	$\left(\frac{\beta^2}{\pi}\right)^{1/4} \sqrt{\frac{1}{2}}\, 2\beta x e^{-\beta^2 x^2/2}$
2	$\frac{5}{2}\hbar\,\omega_0$	$\left(\frac{\beta^2}{\pi}\right)^{1/4} \sqrt{\frac{1}{8}}\left(4\beta^2 x^2 - 2\right) e^{-\beta^2 x^2/2}$
3	$\frac{7}{2}\hbar\,\omega_0$	$\left(\frac{\beta^2}{\pi}\right)^{1/4} \sqrt{\frac{1}{48}}\left(8\beta^3 x^3 - 12\beta x\right) e^{-\beta^2 x^2/2}$
4	$\frac{9}{2}\hbar\,\omega_0$	$\left(\frac{\beta^2}{\pi}\right)^{1/4} \sqrt{\frac{1}{384}}\left(16\beta^4 x^4 - 48\beta^2 x^2 + 12\right) e^{-\beta^2 x^2/2}$
5	$\frac{11}{2}\hbar\,\omega_0$	$\left(\frac{\beta^2}{\pi}\right)^{1/4} \sqrt{\frac{1}{3840}}\left(32\beta^5 x^5 - 160\beta^3 x^3 + 120\beta x\right) e^{-\beta^2 x^2/2}$

Question 9–20

Just for fun (?), normalize $\psi_1(x)$ (using Table 9.5 to ease the pain of integration) and show that your answer agrees with Eq. (9.107).

Question 9–21

Find the form of the normalized SHO eigenfunctions for energies of $11\hbar\,\omega_0/2$ and $23\hbar\,\omega_0/2$.

9.9 THE SHO—THE PHYSICS

Thus far we have focused on *mathematical* properties of the eigenfunctions of the simple harmonic oscillator, supporting the qualitative arguments of § 9.1–9.3 with a quantitative solution in § 9.7. But our ultimate goal is *physical insight*. So in this section we'll use the stationary-state probability densities, as calculated from these eigenfunctions, to investigate statistical properties of this system.

Question 9–22

But first, show that the classical turning points of the simple harmonic oscillator are given by

$$x_n = \pm\sqrt{\frac{2n+1}{\beta^2}}. \tag{9.108}$$

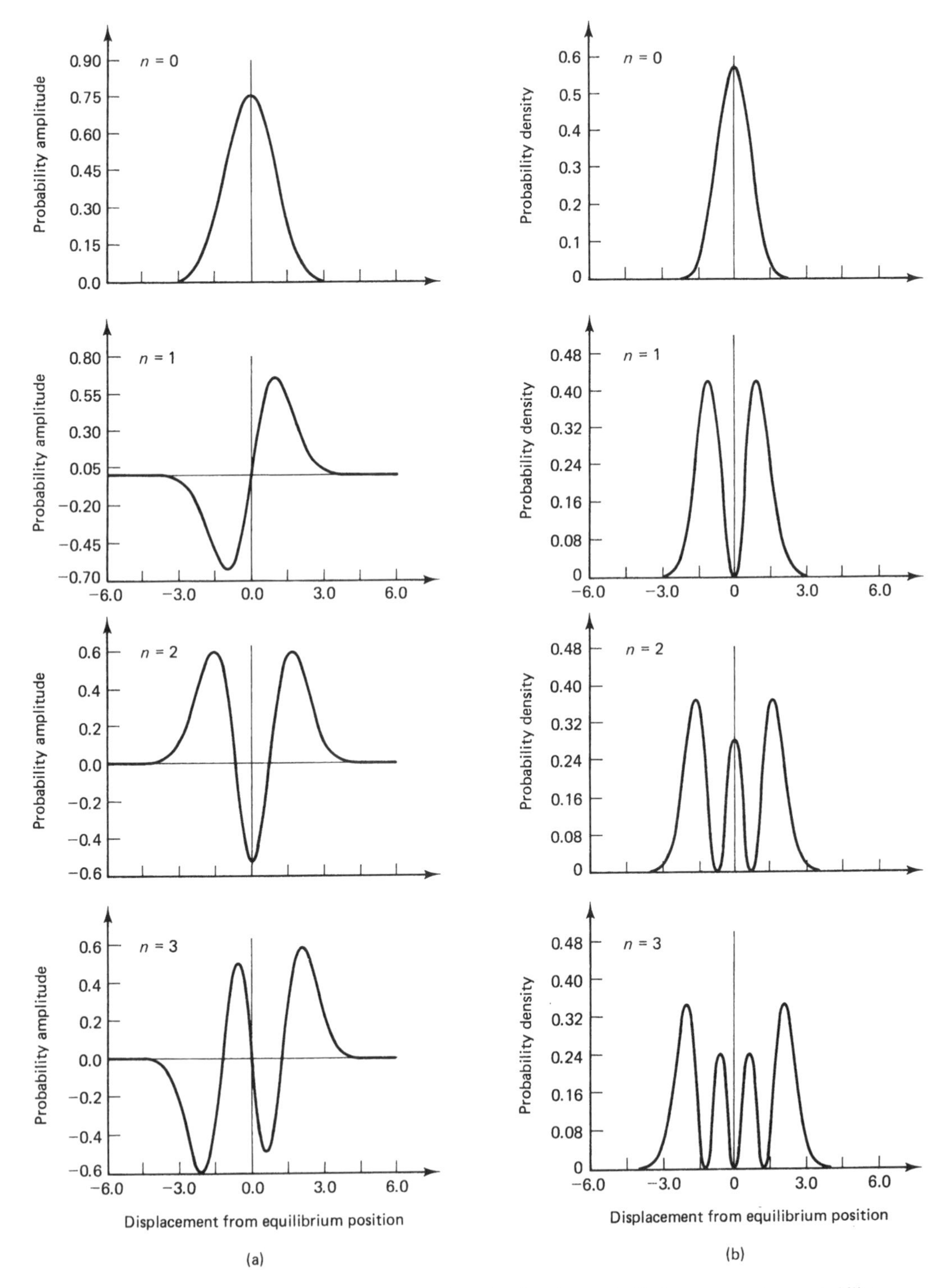

Figure 9.15 (a) Normalized Hamiltonian eigenfunctions and (b) the corresponding probability densities for the first four stationary states of the simple harmonic oscillator: $n = 0, 1, 2$, and 3. Note that these eigenfunctions are plotted as a function of the dimensionless variable $\xi = \beta x$ (see Table 9.3). [Adapted from *Quantum Mechanics* by K. Ziock (New York: Wiley, 1969).]

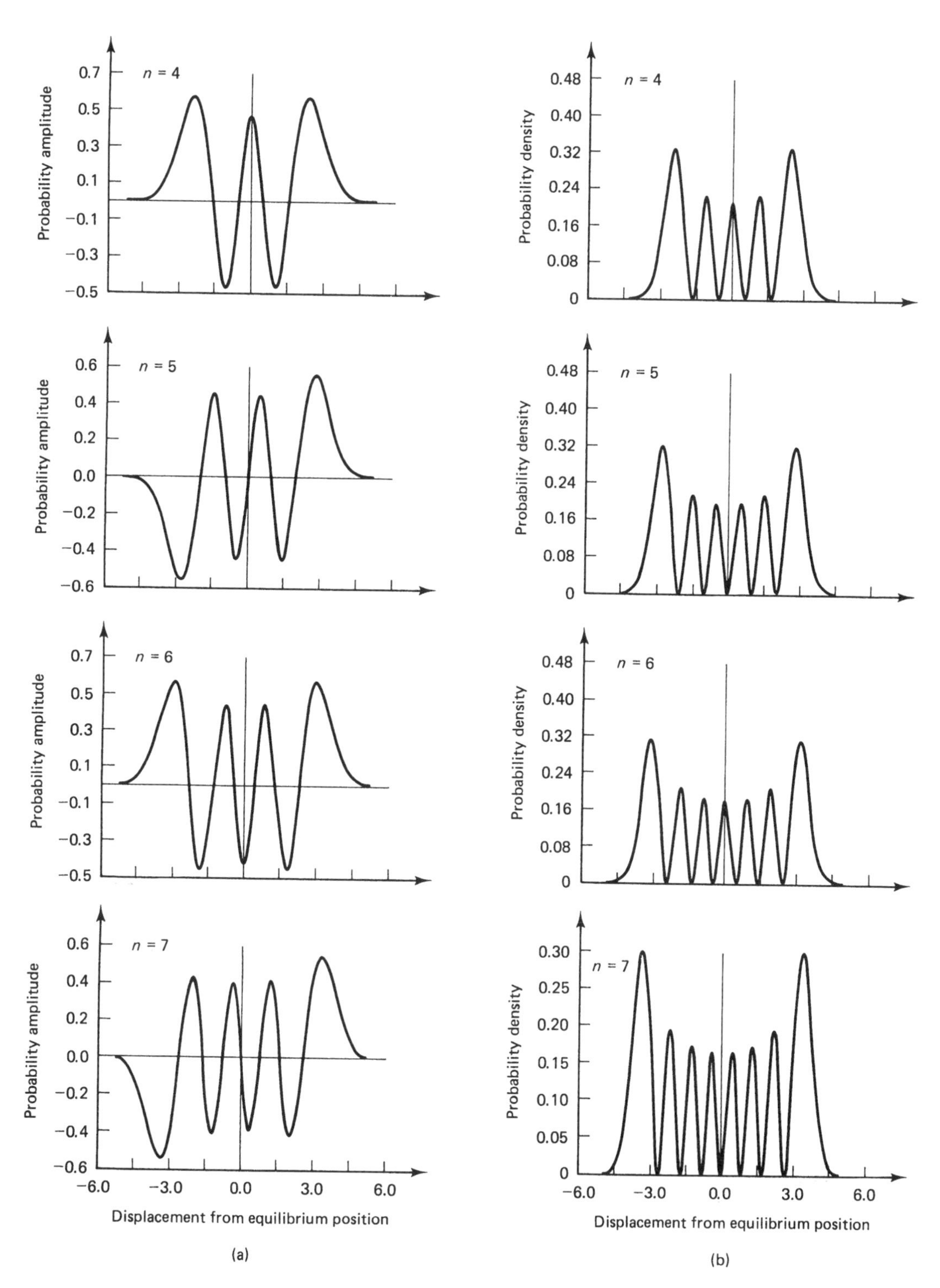

Figure 9.16 (a) Eigenfunctions and (b) probability densities for the stationary states of the SHO with $n = 4$, 5, 6, and 7. [Adapted from *Quantum Mechanics* by K. Ziock (New York: Wiley, 1969).]

The probability density $P_n(x)$ tells us what to expect in a position measurement on an ensemble of SHOs in the nth stationary state. The results predicted by these functions differ strikingly from those for a classical oscillator—as you can see by comparing $P_{\text{cl}}(x)$ in Fig. 9.11 to the corresponding $P_n(x)$ in Figs. 9.16.

The most obvious difference between these figures is that *the classical oscillator cannot penetrate into a forbidden region. But no matter what its state, a quantum oscillator does penetrate into the classically forbidden regions.* This penetration is greatest (*i.e.*, the characteristic length is largest) for low-energy states (*i.e.*, smallest n) and, as you can see in Fig. 9.16, it diminishes with increasing energy. Because of this decrease in the characteristic length as $n \to \infty$, the microscopic oscillator does behave *in this limit* somewhat like its classical counterpart.

You have probably already spotted several other differences between the quantum and classical probability densities. Compare, for example, these two functions (in Figs 9.11 and 9.15) for an energy $E = 3\hbar\omega_0/2$. (This is the first excited state of the microscopic oscillator.) If the physics of a particle with this energy were governed by classical physics, the particle would be most likely found near its classical turning points—precisely where, according to quantum physics, it is *least* likely to be found. Another striking difference is the nodes in the quantum-mechanical density functions, features that are absent from the classical densities.

But look carefully what happens to these nodes (and to the associated peaks in P_n) as the energy E_n of the microscopic oscillator increases: they appear to bunch together. Consequently the overall shape of $P_n(x)$ *for a large value of* n begins to look suspiciously like $P_{\text{cl}}(x)$. This phenomenon—an important clue to the connection between classical and quantum physics—is more evident in the quantum-mechanical density for a *very* large value of the energy, such as $P_{60}(x)$ in Fig. 9.17.

This observation strongly intimates that *the classical limit of a simple harmonic oscillator is the limit of large energy (large n).* And, as you'll see when you answer the next two Questions, it is. But be careful in taking this limit. The energy levels of the SHO $E_n = \left(n + \frac{1}{2}\right)\hbar\omega_0$ are *equally spaced*, so in the limit $n \to \infty$, these energies do not merge to form a continuum. Instead, the separation between adjacent eigenvalues of the SHO, $E_{n+1} - E_n$, is constant ($\hbar\omega_0$), and energy quantization—certainly a non-classical phenomenon—is preserved even as $n \to \infty$. To see how to take the classical limit, consider the next two questions.

Question 9–23

Show mathematically why the classical limit of the SHO is $n \to \infty$ via the following argument: Divide the CA region $[-x_n, x_n]$ into a finite number m of equally-spaced intervals of size Δx. Let $\bar{x}_m$ denote the midpoint of the mth interval. Evaluate the average over one such interval of the quantum probability density $P_n(x)$ and show that this average approaches $P_{\text{cl}}(x)$ as $n \to \infty$. In particular, show that

$$\frac{1}{\Delta x}\int_{\bar{x}_m - \frac{1}{2}\Delta x}^{\bar{x}_m + \frac{1}{2}\Delta x} P_n(x)\,dx \xrightarrow[n\to\infty]{} P_{\text{cl}}(\bar{x}_m).$$

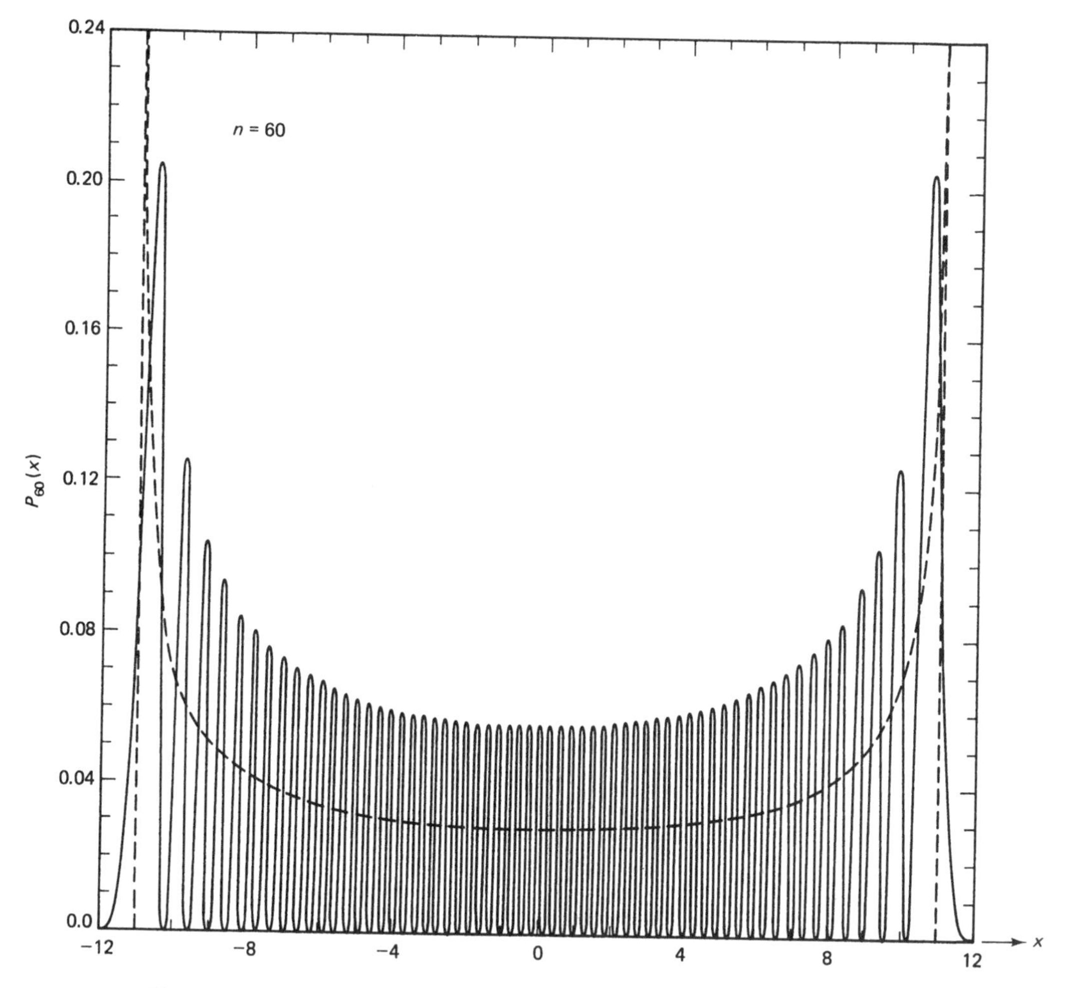

Figure 9.17 The position probability density for a simple harmonic oscillator in the stationary state $\psi_{60}(x)$. Also shown is the classical probability density for this energy. [Adapted from *Quantum Mechanics* by K. Ziock (New York: Wiley, 1969).]

Question 9–24

The correspondence principle (*c.f.*, § 4.2 and 5.7) relates classical and quantum mechanics by requiring the latter to reduce to the former in the classical limit. In many cases, we obtain the classical limit of a quantum-mechanical result by the artifice of setting $\hbar \to 0$. But this gambit won't work for the SHO energy spectrum of Fig. 9.18, because the energies of the simple harmonic oscillator are proportional to $\hbar$. But the ratio of the spacing between adjacent levels to the energy of one of these levels behaves as follows in the limit of large energy:

$$\frac{E_{n+1} - E_n}{E_n} = \frac{1}{n + \frac{1}{2}} \xrightarrow[n \to \infty]{} \frac{1}{n} \to 0.$$

Explain why this is the appropriate way to take the classical limit of an energy spectrum. (Hint: What do spectroscopists measure?)

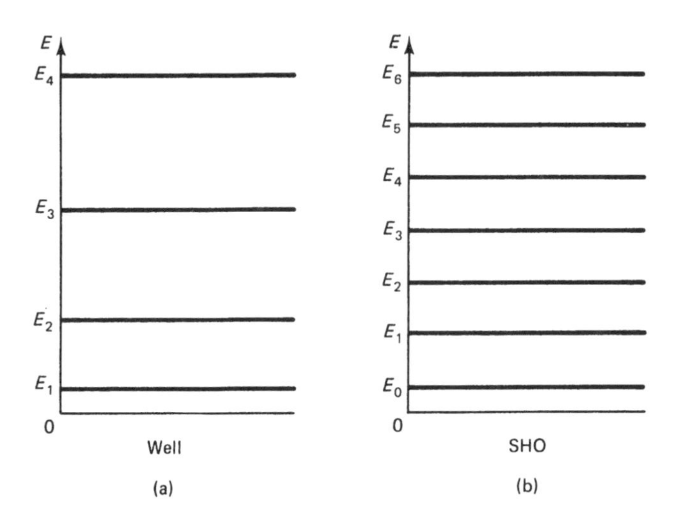

Figure 9.18 Energy level diagrams for a simple harmonic oscillator (right spectrum) and a finite square well (left spectrum). The natural frequency of the oscillator and the width V_0 and depth L of the well were chosen so that the two systems have equal ground-state energies and the square well supports several bound states.

Statistical Properties of the SHO

The results of ensemble measurements of observables such as position, momentum, and energy are described by statistical quantities—expectation values and uncertainties—that we can calculate from the eigenfunctions of the Hamiltonian. If the state is stationary, the particle's energy is sharp, and the statistical quantities for this observable are simply

$$\langle E \rangle = E_n \qquad \Delta E = 0 \qquad [n^{\text{th}} \text{ stationary state}]. \tag{9.109}$$

Only a bit more work is required to evaluate the expectation values and uncertainties for the other two key observables: position and momentum.

The position probability densities in Fig. 9.16 show vividly that the position of a particle in the nth stationary state of a SHO is uncertain:

$$0 < \Delta x < \infty \qquad [n^{\text{th}} \text{ stationary state}]. \tag{9.110a}$$

This means that the linear momentum is also uncertain, because the position and momentum uncertainties must satisfy the Heisenberg Uncertainty Principle, $\Delta x \, \Delta p \geq \hbar/2$:

$$0 < \Delta p < \infty \qquad [n^{\text{th}} \text{ stationary state}]. \tag{9.110b}$$

Thus does nature limit our knowledge of the position and momentum of a particle in a stationary state.[32]

The implications of Eq. (9.110) for experimental physics are clear: in an ensemble measurement of either position or momentum, individual results will fluctuate about the mean, $\langle x \rangle_n$ or $\langle p \rangle_n$. We can calculate these expectation values for the nth stationary state from the definitions

[32]There do exist states of the simple harmonic oscillator for which the position or the momentum are sharp *at certain times*. But these states are not stationary. It is also possible to construct state functions (by superposing stationary-state wave functions) whose expectation values $\langle x \rangle$ and $\langle p \rangle$ mimic those of a classical oscillator. These states, called "coherent quasi-classical states", are also non-stationary. Coherent states play an important role in the quantum theory of electromagnetic radiation and in quantum optics. See, for example, Complement G_V in *Quantum Mechanics*, Volume I by C. Cohen-Tannoudji, B. Diu, and F. Laloë (New York: Wiley, 1977).

$$\langle x \rangle_n = \int_{-\infty}^{\infty} \psi_n^*(x) x \psi_n(x)\, dx \tag{9.111a}$$

$$\langle p \rangle_n = -i\hbar \int_{-\infty}^{\infty} \psi_n^*(x) \frac{d}{dx} \psi_n(x)\, dx. \tag{9.111b}$$

Or we can *think* for a moment and determine these average values without doing any work. The property of the eigenfunctions that renders trivial the determination of these expectation values is (you guessed it) parity.

Look: the parity of x is odd, so *regardless of the parity of* $\psi_n(x)$, the integrand in (9.111*a*) is odd. Hence the integral is zero, and $\langle x \rangle_n = 0$. Well, that was easy enough.

Evaluating (9.111*b*) isn't much harder—if you remember that the first derivative of a function with even parity is an odd function, and vice versa:

$$\boxed{\begin{aligned} \frac{d}{dx}\,(\text{even function}) &= \text{odd function} \\ \frac{d}{dx}\,(\text{odd function}) &= \text{even function} \end{aligned}} \tag{9.112}$$

Regardless of the parity of $\psi_n(x)$, this eigenfunction times its first derivative is odd. Hence the integrand in (9.111b) is odd, the integral is zero, and $\langle p \rangle_n = 0$.

But wait a minute: *in no way did these proofs depend on the particular symmetric potential under consideration*. That is, we used only the parity of the Hamiltonian eigenfunctions, not the fact that they are eigenfunctions of the SHO. So, almost without intending to, we've proven that

$$\boxed{\langle x \rangle_n = 0 \qquad \langle p \rangle_n = 0 \qquad n^{\text{th}} \text{ stationary state of a symmetric } V(x)} \tag{9.113}$$

Knowing the expectation values of position and momentum makes evaluating their uncertainties easier. In this case, the expressions we must evaluate are [*c.f.*, Eq. (7.63)]

$$(\Delta x)_n = \sqrt{\langle x^2 \rangle_n - \langle x \rangle_n^2} = \sqrt{\langle x^2 \rangle_n} \tag{9.114a}$$

$$(\Delta p)_n = \sqrt{\langle p^2 \rangle_n - \langle p \rangle_n^2} = \sqrt{\langle p^2 \rangle_n}. \tag{9.114b}$$

So the expectation values we must evaluate are

$$\langle x^2 \rangle_n = \int_{-\infty}^{\infty} \psi_n^*(x) x^2 \psi_n(x)\, dx \tag{9.115a}$$

$$\langle p^2 \rangle_n = -\hbar^2 \int_{-\infty}^{\infty} \psi_n^*(x) \frac{d^2}{dx^2} \psi_n(x)\, dx. \tag{9.115b}$$

General forms for the remaining integrals have been derived from the properties of the Hermite polynomials in the spatial functions (see Pblm. 9.11)—the answers appear in Table 9.7.

WARNING

The results in Table 9.7 pertain only to the simple harmonic oscillator; unlike Eq. (9.113), these integrals are not general properties of a stationary state.

TABLE 9.7 HANDY INTEGRALS OF EIGENFUNCTIONS OF THE SIMPLE HARMONIC OSCILLATOR.

$$\beta = \sqrt{\frac{m\,\omega_0}{\hbar}}$$

$$\int_{-\infty}^{\infty} \psi_n^*(x) \frac{d}{dx} \psi_m(x)\,dx = \begin{cases} \beta\sqrt{\frac{n+1}{2}} & m = n+1 \\ -\beta\sqrt{\frac{n}{2}} & m = n-1 \\ 0 & \text{otherwise} \end{cases}$$

$$\int_{-\infty}^{\infty} \psi_n^*(x) x \psi_m(x)\,dx = \begin{cases} \frac{1}{\beta}\sqrt{\frac{n+1}{2}} & m = n+1 \\ \frac{1}{\beta}\sqrt{\frac{n}{2}} & m = n-1 \\ 0 & \text{otherwise} \end{cases}$$

$$\int_{-\infty}^{\infty} \psi_n^*(x) x^2 \psi_m(x)\,dx = \begin{cases} \frac{2n+1}{2\beta^2} & m = n \\ \frac{\sqrt{(n+1)(n+2)}}{2\beta^2} & m = n+2 \\ 0 & \text{otherwise} \end{cases}$$

Question 9–25

Evaluate the matrix elements $\langle \psi_0 \mid \hat{p} \mid \psi_1 \rangle$ and $\langle \psi_1 \mid \hat{p} \mid \psi_0 \rangle$ by direct integration and verify that the first two expressions in Table 9.7 are, for these cases at least, correct.

Question 9–26

Evaluate the matrix element of $\hat{p}^2$ between two eigenfunctions of the simple harmonic oscillator.

Example 9.4. The Position Uncertainty

To evaluate $(\Delta x)_n$ we need only look in Table 9.7 for the expectation value of x^2, *viz.*,

$$\left\langle x^2 \right\rangle_n = \frac{2n+1}{2\beta^2} = \left(n + \frac{1}{2}\right) \frac{\hbar}{m\,\omega_0} = \frac{E_n}{m\,\omega_0^2}, \tag{9.116}$$

where I've used the definition of β and $E_n = \left(n + \frac{1}{2}\right) \hbar\,\omega_0$ to express $\left\langle x^2 \right\rangle_n$ in a variety of handy forms. The position uncertainty is just the square root of this expectation value,

$$\begin{aligned} (\Delta x)_n &= \sqrt{\frac{1}{\beta^2}\left(n + \frac{1}{2}\right)} \\ &= \sqrt{\left(n + \frac{1}{2}\right) \frac{\hbar}{m\,\omega_0}} = \sqrt{\frac{E_n}{m\,\omega_0^2}}. \end{aligned} \qquad [n^{\text{th}} \text{ stationary state}] \tag{9.117}$$

This result expresses quantitatively a trend we saw in Fig. 9.16: *the position uncertainty increases with increasing energy (i.e., with increasing n).*

The quantity we just evaluated is a measure of the range of values of x where a *microscopic* particle with a SHO potential is likely to be found in an ensemble measurement of position. The classical counterpart of this quantity is defined by the classical turning points, Eq. (9.108). These points define the range of values $-x_n \le x \le x_n$ where a *macroscopic* oscillator with energy E_n would be found. The position uncertainty is related to x_n by

$$(\Delta x)_n = \frac{1}{\sqrt{2}}\, x_n, \tag{9.118}$$

so for this system, the position uncertainty turns out to be comparable to the classical range, *i.e.*,

$$(\Delta x)_n \approx x_n. \tag{9.119}$$

Question 9–27

Show that the average value of the potential energy of the SHO in the nth stationary state is

$$\langle V(x)\rangle_n = \tfrac{1}{2} m\,\omega_0^2 \left\langle x^2\right\rangle_n, \tag{9.120a}$$

$$= \tfrac{1}{2} E_n. \tag{9.120b}$$

We'll return momentarily to the implications of this result.

Example 9.5. The Momentum Uncertainty

At first glance, the prospect of evaluating the momentum uncertainty (9.114*b*) looks horrendous. In the position representation, $p^2 = -\hbar^2 d^2/dx^2$, so the integrand in (9.115*b*) is a product of $\psi_n(x)$ and its second derivative, which, in turn, is the product of a Hermite polynomial and a decaying Gaussian. No doubt the resulting integral is one you and I don't want to evaluate unless we absolutely have to.

We don't have to. We can get rid of the second derivative of $\psi_n(x)$ by taking advantage of its appearance in the TISE,

$$-\hbar^2 \frac{d^2}{dx^2}\psi_n(x) = 2m\left[E_n - V(x)\right]\psi_n(x). \tag{9.121}$$

Substituting this form into (9.115*b*), we obtain a much simpler expression for the expectation value,

$$\left\langle p^2\right\rangle_n = \int_{-\infty}^{\infty} \psi_n^*(x)\Big\{2m\left[E_n - V(x)\right]\Big\}\psi_n(x)\,dx, \tag{9.122a}$$

$$= 2mE_n \int_{-\infty}^{\infty} \psi_n^*(x)\psi_n(x)\,dx - 2m\,\langle V(x)\rangle_n\,. \tag{9.122b}$$

Now look carefully at the two terms in (9.122*b*). The Hamiltonian eigenfunctions are normalized, so the integral in the first term is unity. This reduces the expectation value to

$$\boxed{\left\langle p^2\right\rangle_n = 2m\left[E_n - \langle V(x)\rangle_n\right]. \quad [n^{\text{th}}\text{ stationary state of any potential } V(x)]} \tag{9.123}$$

Notice that in this derivation, $V(x)$ is arbitrary; nowhere did we use an explicit form of the simple harmonic oscillator potential or of its eigenfunctions, nor did we assume that these functions have definite parity. So Eq. (9.123) pertains to any bound stationary state of any one-dimensional potential.

We apply Eq. (9.123) to the SHO by inserting the form of the potential, *viz.*,

$$\left\langle p^2\right\rangle_n = 2m\left[E_n - \frac{1}{2}m\,\omega_0^2\left\langle x^2\right\rangle_n\right] \tag{9.124a}$$

$$= 2m\left[E_n - \frac{1}{2}m\,\omega_0^2\,\frac{E_n}{m\,\omega_0^2}\right] \tag{9.124b}$$

$$= mE_n. \tag{9.124c}$$

This simple result gives the equally simple momentum uncertainty,

$$(\Delta p)_n = \sqrt{mE_n} = \sqrt{\left(n+\tfrac{1}{2}\right)\hbar m\,\omega_0} \qquad [n^{\text{th}} \text{ stationary state of the SHO}]. \tag{9.125}$$
$$= \hbar\beta\sqrt{n+\tfrac{1}{2}}$$

The position and momentum uncertainties for several stationary states of the simple harmonic oscillator are tabulated in Table. 9.8, from which you can get an idea of their trends as n increases. For future reference, I've collected together the statistical properties of the SHO in Table 9.9.

TABLE 9.8 POSITION AND MOMENTUM UNCERTAINTIES FOR THE SIMPLE HARMONIC OSCILLATOR.

n	$\beta(\Delta x)_n$	$(\Delta p)_n/\sqrt{m\hbar\,\omega_0}$
0	0.71	0.71
1	1.22	1.22
2	1.58	1.58
3	1.87	1.87
4	2.12	2.12

TABLE 9.9 PROPERTIES OF THE nth STATIONARY STATE OF THE SHO.

Observable	$\hat{Q}$	Certainty?	$\langle Q \rangle$	$(\Delta Q)_n$
Energy	$\hat{\mathcal{H}}$	sharp	E_n	0
Position	$\hat{x}$	not sharp	0	$\sqrt{\left(n+\frac{1}{2}\right)\frac{\hbar}{m\,\omega_0}}$
Momentum	$\hat{p}$	not sharp	0	$\sqrt{\left(n+\frac{1}{2}\right)m\,\omega_0\hbar}$

Question 9–28

Evaluate the expectation value of the kinetic energy T for the nth stationary state of the SHO, using Eq. (9.125). Show that your result can be written

$$\langle T \rangle_n = \tfrac{1}{2}E_n. \tag{9.126}$$

Aside: The Virial Theorem in Quantum Physics. A most interesting feature of a *macroscopic* simple harmonic oscillator is the equal sharing of its total energy among kinetic and potential parts. Expressed mathematically, this result—called the **Virial Theorem**—reads

$$E = 2T = 2V \qquad \text{[for a macroscopic SHO]}. \tag{9.127a}$$

A similar relationship holds for a microscopic oscillator in a stationary state, as you discovered if you answered Questions 9–27 and 9–28. But the *form* of this relation reflects a fundamental difference between classical and quantum physics, for *the quantum-mechanical Virial Theorem is expressed in terms of expectation values and pertains only to stationary-states of the oscillator.* This theorem states that *on the average*, the total energy in a stationary state is equally partitioned among potential and kinetic parts, *i.e.*,

$$\langle E \rangle_n = 2\langle T \rangle_n = 2\langle V \rangle_n . \tag{9.127b}$$

Of course, the expectation value in (9.127*b*) is just $\langle E \rangle_n = E_n$.[33]

Question 9–29

Prove the following general relationship between the mean kinetic and potential energies of any stationary state,

$$\begin{aligned} \left[\hat{T} + V(x)\right] \psi_n(x) &= E_n \psi_n(x), \\ \Longrightarrow \qquad \langle T \rangle + \langle V \rangle &= E. \end{aligned}$$

Now prove that the Virial Theorem (9.127*b*) is consistent with this relationship.

The Heisenberg Uncertainty Principle for the SHO

Using our expressions for the position and momentum uncertainties for the nth stationary state of a simple harmonic oscillator, we can calculate the uncertainty product in the Heisenberg Uncertainty Principle—according to which these uncertainties *for any state—stationary or non-stationary—of any microscopic system*, are related by $(\Delta x)(\Delta p) \geq \hbar/2$. But the result, as calculated from the general forms in Table 9.9, is somewhat surprising:

$$(\Delta x)_n \, (\Delta p)_n = \left(n + \frac{1}{2}\right) \hbar. \tag{9.128}$$

The uncertainty product satisfies the uncertainty principle in a special way: for the ground state ($n = 0$) it is equal to $\hbar/2$.[34]

Question 9–30

From the data in Table 9.8, calculate the uncertainty products for the first five stationary states of the SHO.

Question 9–31

What are the limiting values of $(\Delta x)_n$ and $(\Delta p)_n$ as $n \to \infty$? Is your answer consistent with the Heisenberg Uncertainty Principle?

[33]For more on the Virial Theorem, see Chap. 14 of *Quantum Chemistry* 3rd ed. by I. N. Levine (Boston: Allyn and Bacon, 1983).

[34]This result may remind you of § 4.5, where we showed that a Gaussian function is the wave packet of minimum uncertainty [see Eq. (4.69)]. But there is a crucial difference. In Chap. 4 we were discussing *non-stationary states of a free particle*; in this chapter the same mathematical function has appeared as the ground-state eigenfunction $\psi_0(x)$ of a particle bound in a simple harmonic oscillator potential. There is nothing mysterious going on here; the feature that renders the uncertainty product for this function a minimum is not the physical significance of this function, but its mathematical form.

*9.10 THE INCREDIBLE SLOSHING OSCILLATOR

Stationary states are important but dull. Properties such as $P_n(x)$ $\langle E\rangle_n$, and $(\Delta x)_n$ are independent of time; so not even in a statistical sense does the "motion" of a microscopic oscillator in a stationary state resemble that of a macroscopic oscillator.

But the behavior of a microscopic oscillator in a *non-stationary state* is far more interesting. To investigate the motion of a particle in such a state we must calculate the expectation values of position and momentum—which, because the state is non-stationary, turn out to depend on time.[35]

Example 9.6. A Periodic Wave Packet.

Suppose the state of a system is represented by the normalized wave function

$$\Psi(x,t) = \frac{1}{\sqrt{2}}\left(\frac{\beta^2}{\pi}\right)^{1/4} e^{-\beta^2 x^2/2}\left[e^{-i\omega_0 t/2} + \sqrt{2}\beta x e^{-3i\omega_0 t/2}\right]. \tag{9.129a}$$

Pretty scary, right?

Not actually. The presence of the Gaussian $e^{-\beta^2 x^2/2}$ and factors of the form $e^{-ni\omega_0 t/2}$ are clues that this function may be a simple combination of stationary-state wave functions $\Psi_n(x,t) = \psi_n(x)e^{-iE_n t/\hbar}$. Sure enough, a few moments spent with the analytic forms in Table 9.6 will reveal that

$$\Psi(x,t) = \frac{1}{\sqrt{2}}\left[\Psi_0(x,t) + \Psi_1(x,t)\right]. \tag{9.129b}$$

To be sure, this function does not represent a stationary state, for its dependence on space and time variables can't be separated.

Question 9–32

Verify that Eqs. (9.129*a*) and (9.129*b*) are equivalent.

Evaluating statistical properties of the state is much easier if we work with the expression for $\Psi(x,t)$ in terms of stationary-state wave functions, (9.129*b*), because we know these properties for stationary states (Table 9.9). For this reason, I'll also leave the forms of all equations as general as possible for as long into the derivation as possible.

The expectation value of position at time t is

$$\langle x\rangle(t) = \langle \Psi(x,t) \mid \hat{x} \mid \Psi(x,t)\rangle = \int_{-\infty}^{\infty} \Psi^*(x,t)\hat{x}\Psi(x,t)\,dx. \tag{9.130}$$

Inserting (9.129*b*) *but not explicit forms for the eigenfunctions* into this definition, we obtain

$$\begin{aligned}\langle x\rangle(t) = \frac{1}{2}\Big(&\langle\psi_0 \mid \hat{x} \mid \psi_0\rangle + \langle\psi_1 \mid \hat{x} \mid \psi_1\rangle \\ &+ \langle\psi_0 \mid \hat{x} \mid \psi_1\rangle e^{-i\omega_0 t} + \langle\psi_1 \mid \hat{x} \mid \psi_0\rangle e^{+i\omega_0 t}\Big).\end{aligned} \tag{9.131a}$$

The first two terms in (9.131*a*) are just $\langle x\rangle_0$ and $\langle x\rangle_1$, and, according to Table 9.9, these are zero. In the remaining terms are two matrix elements that, according to Table 9.7, are both equal to $1/\sqrt{2\beta^2}$. So the expectation value of position for the non-stationary state (9.129) is simply

$$\langle x\rangle(t) = \frac{1}{2}\frac{1}{\sqrt{2}\beta}\left(e^{-i\omega_0 t} + e^{+i\omega_0 t}\right) = \frac{1}{\sqrt{2}\beta}\cos\omega_0 t. \tag{9.131b}$$

[35]We'll take up non-stationary states in their own right a few chapters from now (Chap. 12), so you can skip this extended example if you have to.

Now let's evaluate $\langle p \rangle (t)$ for this state,

$$\langle p \rangle (t) = \frac{1}{2}\Big[\langle p \rangle_0 + \langle p \rangle_1 - i\hbar \left(\int_{-\infty}^{\infty} \psi_0^*(x) \frac{d}{dx} \psi_1(x)\, dx \right) e^{-i\omega_0 t} - i\hbar \left(\int_{-\infty}^{\infty} \psi_1^*(x) \frac{d}{dx} \psi_0(x)\, dx \right) e^{+i\omega_0 t} \Big]. \tag{9.132a}$$

Calling once again on Table 9.7, we can reduce the right-hand side of (9.132*a*) to

$$\langle p \rangle (t) = \frac{1}{2} \frac{\beta}{\sqrt{2}} (-i\hbar) \left(e^{+i\omega_0 t} - e^{-i\omega_0 t} \right) = -\frac{\beta\hbar}{\sqrt{2}} \sin \omega_0 t. \tag{9.132b}$$

So, written in terms of $\beta = \sqrt{m\omega_0/\hbar}$, *the mean values of the position and momentum of a simple harmonic oscillator in the non-stationary state (9.129) are*

$$\langle x \rangle (t) = \sqrt{\frac{\hbar}{2m\omega_0}} \cos \omega_0 t \tag{9.133a}$$

$$\langle p \rangle (t) = -\sqrt{\frac{m\omega_0\hbar}{2}} \sin \omega_0 t. \tag{9.133b}$$

Well, this state is certainly non-stationary! In contrast to the stationary states in previous sections, this state is characterized by a position and momentum that, on the average, *oscillate* about zero with period $2\pi/\omega_0$, just like the classical trajectory $x(t)$ and $p(t)$ of a macroscopic oscillator.

But don't be fooled: this finding doesn't pertain to all observables—as you'll discover in the next Question.

Question 9–33

Use the TISE $\hat{\mathcal{H}}\psi_n(x) = E_n\psi_n(x)$ to show that the mean value of the energy for the state (9.130) is

$$\langle E \rangle = \left\langle \Psi(x,t) \mid \hat{\mathcal{H}} \mid \Psi(x,t) \right\rangle = \frac{1}{2}\left(E_0 + E_1\right) = \hbar\omega_0. \tag{9.134}$$

Hint: You will need to use the following property of the Hamiltonian eigenfunctions, which we'll prove in the next chapter:

$$\int_{-\infty}^{+\infty} \psi_n^*(x)\psi_m'(x)\, dx = 0 \qquad m \neq n. \tag{9.135}$$

In the non-stationary state of this example, the uncertainties in position and momentum oscillate with the classical period T_0 (see Pblm. 9.4). This is strange behavior for a wave packet—most of these critters spread; their widths increase without limit as time passes [see § 6.7 and especially Fig. 6.4]. But the motion of this packet looks something like this: according to Eq. (9.133*a*), it "sloshes" back and forth between the classical turning points (*i.e.*, its center undergoes simple harmonic motion at frequency ω_0). After one period, the packet has resumed its original shape! No spreading.

To gain further insight into this strange behavior, consider the probability density for this state:

$$P(x,t) = \Psi^*(x,t)\Psi(x,t) \tag{9.136a}$$

$$= \frac{1}{2}\left[\psi_0^*(x)\psi_0(x) + \psi_1^*(x)\psi_1(x) + \psi_0^*(x)\psi_1(x)e^{-i\omega_0 t} + \psi_1^*(x)\psi_0(x)e^{+i\omega_0 t}\right]. \tag{9.136b}$$

The time-dependent terms in $P(x,t)$ oscillate *at the natural frequency* ω_0. So at the end of a period the probability density returns to the form it had at the beginning of the period. No spreading!

A General Non-stationary State of the SHO

We can easily generalize the arguments at the end of Example 9.6 to show that *sloshing* characterizes the motion of *any wave packet of a simple harmonic oscillator*. This generalization is based on the expansion of the wave function for an arbitrary non-stationary state in Hamiltonian eigenfunctions. Any such function can be written (see Chap. 12) as a sum of stationary-state wave functions, with expansion coefficients I'll call d_n; *i.e.*,

$$\Psi(x,t) = \sum_{n=0}^{\infty} d_n \psi_n(x) e^{-iE_n t/\hbar}. \tag{9.137}$$

It is imperative that the sum include all eigenfunctions of the Hamiltonian. The coefficients d_n depend on the initial condition $\Psi(x,0)$.

To see the sloshing of such a state, let's calculate the corresponding probability density:

$$P(x,t) = \sum_{m=0}^{\infty} \sum_{n=0}^{\infty} d_m^* d_n \, \psi_m^*(x) \psi_n(x) e^{-i(n-m)\,\omega_0 t}. \tag{9.138}$$

Look at the time dependence of this function. Each term oscillates at frequency ω_0 or at an integral multiple of this frequency. So the probability density for an arbitrary state of the SHO is periodic: after a time $2\pi/\omega_0$, $P(x,t)$ returns to its original form. Every non-stationary state of the SHO is therefore periodic.

WARNING

This time dependence is peculiar to states of the simple harmonic oscillator; sloshing behavior with no long-term change in the width occurs for no other system.

Aside: Sloshing of a Coherent State. A particularly interesting periodic wave packet that is widely used in applied fields such as quantum electronics and quantum optics is one that represents a "coherent state" of the SHO. This state evolves from the initial condition

$$\Psi(x,0) = \left(\frac{\beta^2}{\pi}\right)^{1/4} e^{-\beta^2 (x-x_0)^2/2}. \tag{9.139}$$

This function has the form of the simple Gaussian that describes the ground stationary state of the simple harmonic oscillator—*c.f.*, Example 9.3 [Eq. (9.104*a*)]—but differs from $\psi_0(x)$ in that its center $\langle x \rangle (0)$ is displaced by an amount x_0.

The probability density for (9.139)—which we can obtain by expanding this function à la Eq. (9.137) and evaluating the coefficients d_n—is unexpectedly simple[36]

$$P(x,t) = \sqrt{\frac{\beta^2}{\pi}}\, e^{-\beta^2 (x - x_0 \cos \omega_0 t)^2}. \tag{9.140}$$

[36] See § 2.8 of *Quantum Mechanics: Principles and Applications* by M. Alonso and H. Valk (Reading, Mass.: Addison-Wesley, 1972).

The center of this function, which is also a Gaussian, changes with time according to the familiar formula $\langle x \rangle (t) = x_0 \cos \omega_0 t$. Evaluating the uncertainty $\Delta x(t)$, we discover that *the width of this particular wave packet is independent of time*! As illustrated in Fig. 9.19, this packet sloshes back and forth with a constant width. This state is unique: no other in nature exhibits this behavior.

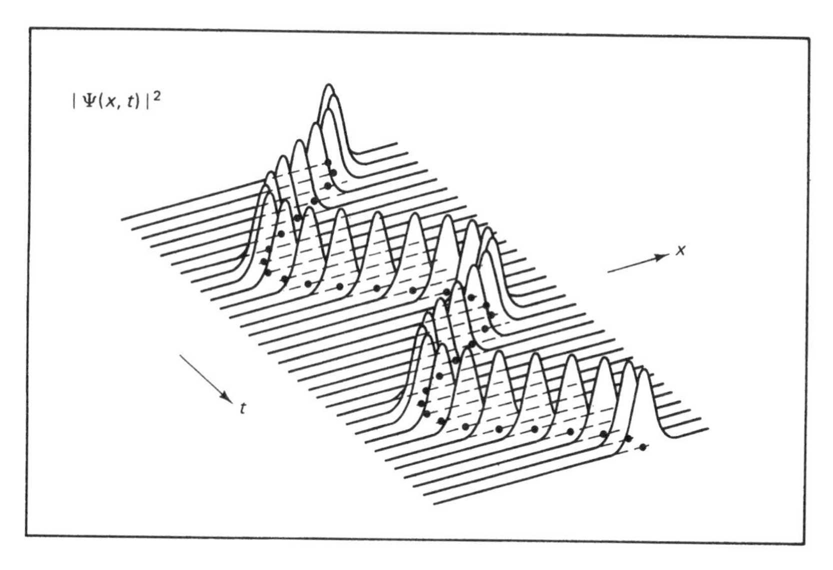

Figure 9.19 The evolution of a coherent state of a simple harmonic oscillator. The probability density (9.140) is shown as a function of time, with the motion of a classical particle indicated by the solid dots. Notice that the width of the wave packet is unchanged as it executes periodic motion under the influence of the restoring force of the harmonic potential. [From *The Picture Book of Quantum Mechanics* by S. Brandt and H. D. Dahmen (New York, John Wiley, 1985). Used with permission.]

Before leaving this section, I want to reiterate the approach we used to the derivation of the expectation values $\langle x \rangle (t)$ and $\langle p \rangle (t)$ in Example 9.6. Learn from this example! Fight the temptation—rife among novice quantum mechanics—to develop an explicit form for a state function before you carry out the required algebraic manipulations! You might get the right answer this way, but suffer far more pain than necessary. So remember

> ***Rule***
>
> When manipulating the wave function for a non-stationary state, always work with its general form as long as possible. Avoid until the last minute inserting into this form explicit dependence of the Hamiltonian eigenfunctions (or of operators) that appear in this form.

*9.11 FINAL THOUGHTS—AND AN APPLICATION OF THE LINEAR POTENTIAL

In § 9.5 I mentioned several applications of the simple harmonic oscillator. But this is just one of many continuous one-dimensional potentials used in current research to model complicated interactions. The Schrödinger equations for most such potentials can be transformed into differential equations that are solved by special functions or that can be subdued by the method of power series—although particularly recalcitrant cases must be attacked via numerical integration (*c.f.*, the Selected Readings for Chap. 8). To close our study of continuous potentials (and, for the moment, of problem-solving tactics), let's take a brief look at another important continuous example, the *linear potential* $V(x) \propto x$.

I want to introduce this model in the context of a remarkable application from solid state physics. If an ordinary metal is exposed to a magnetic field of a few hundred Gauss directed parallel to its surface, as in Fig. 9.20a, the resulting potential energy supports bound states of the electrons in the metal near the surface.[37] The potential such an electron finds itself in can be accurately approximated by a linear form.

We'll denote the distance into the metal by x and the magnetic induction of the applied field by B. The surface presents an infinite barrier to the electron (like the left wall of the infinite square well in Chap. 7), so our model is actually a **half-infinite linear potential**. For a magnetic field of strength B, the potential inside the metal is proportional to B and to a constant γ that depends on the particular solid under investigation, *i.e.*,[38]

$$V(x) = \begin{cases} \gamma B x & x > 0 \\ \infty & x \leq 0 \end{cases} \qquad \text{half-infinite linear potential.} \tag{9.141}$$

This potential, which is sketched in Fig. 9.20b, supports an *infinite number of bound stationary states*, the first four of which are shown. It can be used to model the potentials experienced by an electron in many metals in their normal phases. For example, at temperatures greater than 1.2 K, the electrons in metallic aluminum find themselves in a potential very like Eq. (9.141). (But at lower temperatures, this metal becomes a superconductor, and the model potential in Fig. 9.20c becomes more appropriate.)

Aside: Magnetospectroscopy. In the laboratory, solid-state physicists probe the mysteries of surface electron states by a technique called **resonant magnetospectroscopy**. The idea is simpler than the name: Microwave radiation of frequency ω_m is applied to the solid (in addition to the magnetic field $\mathbf{B} = B\hat{\mathbf{e}}_y$). This radiation induces transitions between bound states in the potential (9.141), provided that the applied frequency is nearly equal to the separation between the levels. In practice, the applied frequency ω_m is fixed and the field strength B varied, changing the energy levels until their separation is close enough to $\hbar\omega_m$ for a transition to be likely—a condition called **resonance**.

The TISE for the linear potential (9.141) is

$$\left[-\frac{\hbar^2}{2m}\frac{d^2}{dx^2} + \gamma B x - E\right]\psi_E(x) = 0. \qquad [\text{for } x > 0 \text{ only}] \tag{9.142}$$

The solutions of this equation that obey the boundary conditions imposed by the potential (9.141) are the bound-state eigenfunctions and energies, which I'll label by an index

[37]These states are sometimes called "skipping states," because in such a state, a classical particle, acting under the influence of the magnetic force, would skip along the surface of the metal—like a flat rock on a still pond. Technically, these states are known as "surface Landau levels." For more about these fascinating states and their physics see R. E. Doezema and J. F. Koch, *Phys. Cond. Matter*, **19**, 17 (1975) and M. Wanner, R. E. Doezema, and U. Strom, *Phys. Rev. B*. **12**, 2883 (1975).

[38]In SI units, the magnetic field strength (in non-magnetic materials) is related to the magnetic induction by $H = \mu_0 B$, where μ_0 is the constant $4\pi \times 10^{-7}$ H/m. The magnitude of the corresponding (Lorentz) force on a particle of charge q and velocity v is $F = qvB$. From this force we obtain the potential (9.141).

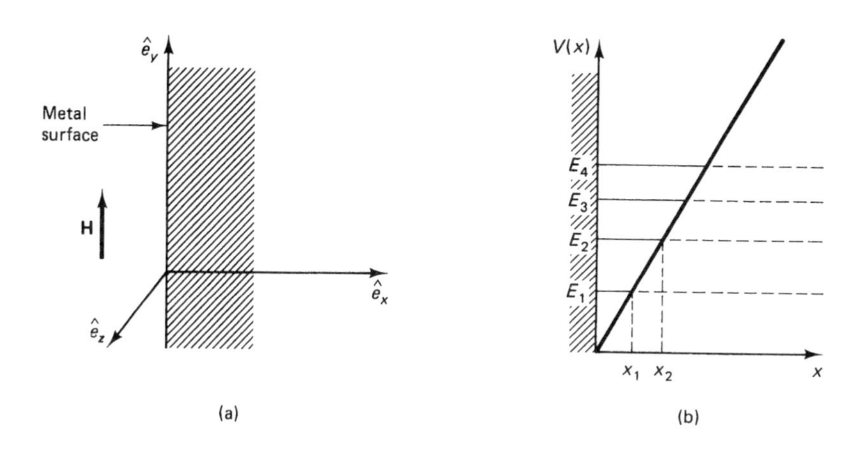

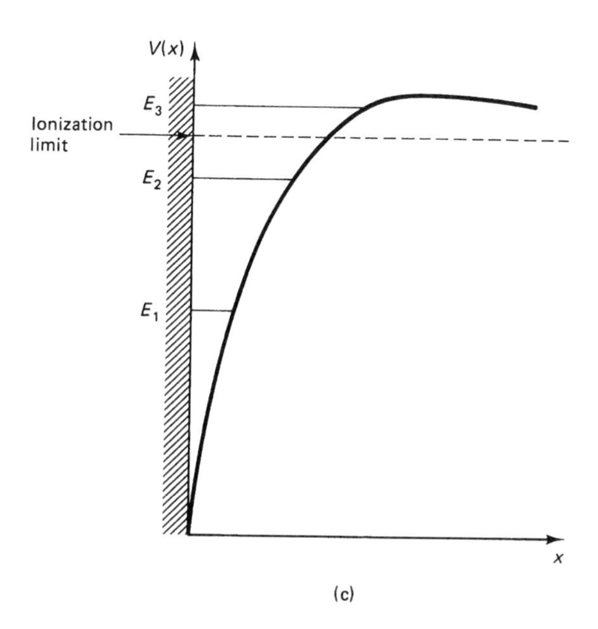

Figure 9.20 (a) The half-infinite linear model of the potential of an electron near the surface of a normal metal. The potential energy (b) supports an infinite number of bound states, four of which are shown along with the classical turning points of the first two, x_1 and x_2. For a superconductor, such as Al at temperatures below 1.2 K, the model potential (c) is used. This potential supports a finite number of bound states whose energies lie below the limit indicated by the horizontal dashed line. Note that the continuum state with energy E_3 can tunnel through the barrier (see Chap. 8). [For a discussion of this potential, see R. E. Doezema, S. C. Whitmore, and J. N. Huffaker, *Phys. Rev. B*. **34**, 4614 (1986).]

$n = 1, 2, 3, \ldots$. The boundary conditions reflect the effect of the infinite barrier at the surface of the metal:

$$\psi_E(x = 0) = 0 \tag{9.143a}$$

$$\psi_E(x) \xrightarrow[x \to +\infty]{} 0. \tag{9.143b}$$

Before launching into the mathematical solution of Eq. (9.142), let's apply our knowledge of *qualitative quantum physics* to sketch a couple of the bound-state eigenfunctions. It turns out that we can deduce quite a lot about these functions without much labor:

1. Since the potential is not symmetric, its eigenfunctions don't have definite *parity*.
2. This potential depends on x, so the *classical turning points* depend on the energy. The classical turning point x_n for the nth stationary state is (see Fig. 9.20b)

$$x_n = \frac{E_n}{\gamma B}. \qquad n = 1, 2, \ldots \tag{9.144}$$

   These points are critical: they delimit the classically allowed and classically forbidden regions and thus are instrumental in our determination of the nature of the eigenfunctions, as

$$\text{CA region } \; 0 \le x \le x_n\text{: } \psi_n(x) \text{ oscillatory}$$
$$\text{CF region } \; x_n < x < \infty\text{: } \psi_n(x) \text{ monotonically decaying}$$

3. The *amplitude* of each eigenfunction changes with x: as x increases from 0, the quantity $|E_n - V(x)|$ decreases, so the amplitude of $\psi_n(x)$ increases [see Eq. (9.21)].
4. The *wavelengths* and *decay constants* of two spatial functions, say $\psi_1(x)$ and $\psi_2(x)$, reflect the x dependence of the potential. The wavelength for $\psi_2(x)$ is smaller than that for $\psi_1(x)$, as is the local decay constant. Therefore, $\psi_2(x)$ oscillates more rapidly for $x \le x_n$ and decays to zero more slowly for $x > x_n$ (*i.e.*, is more penetrating) than does $\psi_1(x)$.
5. The function $\psi_n(x)$ has $n - 1$ nodes.

Putting these facts together, we can draw the qualitatively correct eigenfunction sketches in Fig. 9.21.

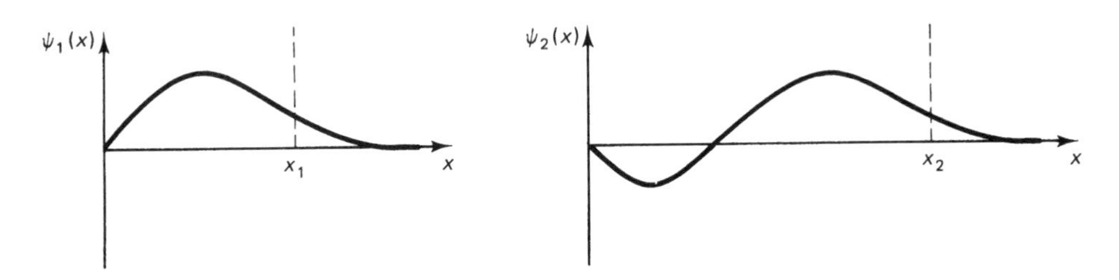

Figure 9.21 Sketches of the Hamiltonian eigenfunctions for the first two bound stationary states of the half-infinite linear potential (9.141). The vertical dashed lines are the classical turning points x_n that separate the CA and CF regions.

Question 9–34

Extend these arguments and sketch $\psi_3(x)$ and $\psi_4(x)$.

Question 9–35

Compare qualitatively the energy of the ground state of the linear potential in Fig. 9.20b with that of an infinite square well of width $L = x_1$: is the energy of the linear potential greater than, less than, or equal to that of the infinite well? Also answer this question for the first excited state, using an infinite well of width $L = x_2$.

But enough of such frivolity. How do we solve the TISE (9.142)? We could do so numerically[39] or we could transform it into a form that has already been solved by someone else. I prefer the latter approach.

The necessary transformation involves introducing a potential strength parameter, an energy variable, and a dimensionless length—just the sorts of tricks we played on the TISE for the simple harmonic oscillator in § 9.7 (see Table 9.2). First we must isolate the second derivative in the first term of (9.142); *viz.*,

$$\left[\frac{d^2}{dx^2} - \frac{2m}{\hbar^2}\gamma B\, x + \frac{2mE}{\hbar^2}\right]\psi_E(x) = 0. \tag{9.145}$$

To clean up (9.145), let's introduce the following *intermediate quantities* (*c.f.*, Table 9.3):

$$\beta \equiv \frac{2m}{\hbar^2}B\gamma \qquad \text{potential strength parameter} \tag{9.146a}$$

$$\epsilon \equiv \frac{2m}{\hbar^2}E \qquad \text{energy variable.} \tag{9.146b}$$

In this garb, the TISE looks like

$$\left(\frac{d^2}{dx^2} - \beta x + \epsilon\right)\psi_E(x) = 0. \tag{9.147}$$

According to our game plan (*c.f.*, Table 9.3), the next step is to introduce a dimensionless variable ξ. Following the example of the simple harmonic oscillator, you might try $\xi = \beta x$. Reasonable this choice is; successful it is not. After trying several reformulations of Eq. (9.147), one discovers that to reduce it to a form whose solution is known, we must change the independent variable from x to the *energy-dependent* variable

$$\xi \equiv a\left(x - \frac{E}{\gamma B}\right) \qquad \text{where } a \equiv \left(\frac{2m\gamma B}{\hbar^2}\right)^{1/3} \tag{9.148}$$

With this definition Eq. (9.147) becomes the light and frothy

$$\left(\frac{d^2}{d\xi^2} - \xi\right)\psi_\epsilon(\xi) = 0. \qquad \text{Airy equation} \tag{9.149}$$

This equation is called the Airy equation, after the man who solved it.[40]

We want only solutions of (9.149) that satisfy the boundary conditions (9.143). So we must express these conditions in terms of ξ. Applying the change of variable $x \to \xi$, we find that the condition (9.143*a*) must be imposed on $\psi_\epsilon(\xi)$ at $\xi = -aE/(\gamma B)$. The

[39]See R. D. Desko and D. J. Bord, *Amer. Jour. Phys.*, **51**, 82 (1983).

[40]Their discoverer was the English astronomer and mathematician G. B. Airy (1801–1892). Airy is better known for his work as the Astronomer Royal, as the organizer of several completely unsuccessful expeditions to study the transit of Venus in 1874 and 1882, and as an egotistical, ruthless, and much disliked astronomer. He did not, of course, introduce Airy functions in the context of quantum physics, which did not exist when he published their details (1838). Rather, he was studying rainbows. See H. M. Nussenzweig, *Sci. Amer.*, **236**, 116 (1977).

second condition is a limit, so it does not change when the variable does. We obtain the boundary conditions:[41]

$$\psi_\epsilon\left(\xi = -\frac{aE}{\gamma B}\right) = 0 \tag{9.150a}$$

$$\psi_e(\xi) \xrightarrow[\xi\to+\infty]{} 0. \tag{9.150b}$$

The two linearly-independent solutions of Eq. (9.149) are called the **Airy functions.** They are denoted by the odd notation $\text{Ai}(\xi)$ and $\text{Bi}(\xi)$. (The symbol i is not $\sqrt{-1}$; it is part of the name of the function.) Graphs of selected Airy functions appear in Fig. 9.22. Each Airy function oscillates, crossing the horizontal axis at several (discrete) values of the argument ξ. These special values of ξ, which occur only for negative values of ξ, are called the **zeroes of the Airy function**. I'll denote the zeroes of $\text{Ai}(\xi)$ by ξ_n for $n = 1, 2, \ldots$.

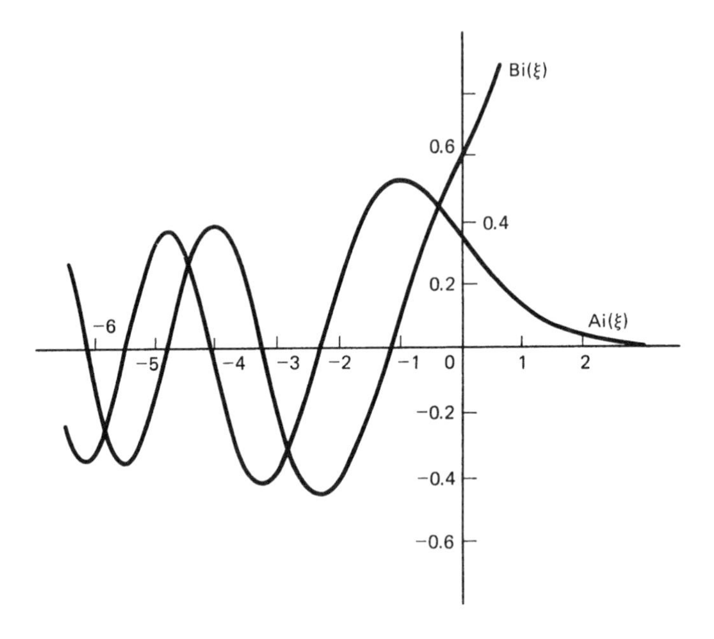

Figure 9.22 The Airy functions $\text{Ai}(\xi)$ and $\text{Bi}(\xi)$. [Adapted from Figs. 10.6 and 10.7 of *Handbook of Mathematical Functions*, ed. by M. Abramowitz and I. A. Stegun (New York: Dover, 1965).]

Using the two Airy functions, we can construct the general solution of the TISE for a linear potential. With arbitrary coefficients C and D, we have

$$\psi_\epsilon(\xi) = C\ \text{Ai}(\xi) + D\ \text{Bi}(\xi). \tag{9.151a}$$

The trouble with this form is that $\text{Bi}(\xi)$ blows up as $\xi \to \infty$, which eruption violates the boundary condition (9.150*b*). So we must set $D = 0$, leaving behind the physically admissible eigenfunction

$$\psi_\epsilon(\xi) = C\ \text{Ai}(\xi). \tag{9.151b}$$

[41]The half-infinite linear potential (9.141) admits a non-zero Hamiltonian eigenfunction only for $x > 0$. But the transformation to the variable ξ is such that $\xi = 0$ at the classical turning point $x_n = E_n/(\gamma B)$. So the CA region includes negative values of ξ; this region is $-aE/(\gamma B) \le \xi \le 0$.

We can determine the coefficient C by enforcing the normalization condition, which for this eigenfunction reads

$$\int_0^\infty \psi_E^*(x)\psi_E(x)\,dx = 1 \qquad \Longrightarrow \qquad \frac{1}{a}\int_{-aE/(\gamma B)}^{\infty} \psi_\epsilon^*(\xi)\psi_\epsilon(\xi)\,d\xi = 1. \tag{9.152}$$

As usual, I'll bypass this tedious normalization.

We took care of the boundary condition at infinity, (9.150*b*), by setting $D = 0$ in the general form (9.151*a*). But physically admissible eigenfunctions must also satisfy the condition at the left wall of the potential, Eq. (9.150*a*). As you might expect, this condition quantizes the bound-state energies. Watch.

The only way the function (9.151*b*) can obey the boundary condition (9.150*a*) is if

$$\mathrm{Ai}\left(\xi = -\frac{1}{\gamma B}E\right) = 0. \tag{9.153}$$

That is, the Airy function in the eigenfunction must have a zero at $\xi = -ae/(\gamma B)$. But we can't control where the zeroes of $\mathrm{Ai}(\xi)$ occur (see Fig. 9.22); the only way we can enforce this requirement is to choose *the energy* in ξ so that Eq. (9.153) is satisfied. This tactic becomes clear if we write explicitly the relationship between the nodes of the Airy function at ξ_1, ξ_2, ... and the stationary-state energies E_n,

$$E_n = -\frac{\gamma B}{a}\xi_n = \left(\frac{B^2\gamma^2\hbar^2}{2m}\right)^{1/3}\xi_n \qquad \text{allowed energies of the linear potential.} \tag{9.154}$$

The zeroes in Table 9.10 should help you determine the bound-state energies for any system that can be modeled by a half-infinite linear potential. You'll have a chance to try this out in the impending problems.

TABLE 9.10 THE FIRST TEN ZEROES OF THE AIRY FUNCTION $\mathrm{Ai}(\xi)$.

n	ξ_n
1	−2.33810
2	−4.08794
3	−5.52055
4	−6.78670
5	−7.94413
6	−8.02265
7	−10.04017
8	−11.00852
9	−11.93601
10	−12.82877

Example 9.7. Aluminum.

A piece of metallic Al (at a temperature above 1.2 K) is exposed to a magnetic field $B = 38.7 \times 10^{-4}$ Tesla. The velocity of an electron near the surface of the metal is 1.28×10^6 m/sec. For this system, the constant in the definition of the potential is $\gamma = 2.05 \times 10^{-13}$ C-m/sec.[42] The constant a in the definition (9.148) is $a = 5.85 \times 10^7$.

We calculate the energies of the bound stationary states from the zeroes in Table 9.10 according to

$$E_n = -1.36 \times 10^{-23} \xi_n \text{ (J)} = -8.43 \times 10^{-5} \xi_n \text{ (eV)}. \tag{9.155}$$

The ground state is very near the bottom of the well, with $E_1 = 1.97 \times 10^{-4}$ eV. The corresponding eigenfunction $\psi_1(x)$ looks like the one in Fig. 9.21a; its classical turning point $x_1 = -(E_1/\gamma B) = 4 \times 10^{-8}$ m.

Question 9–36

Suppose we are studying a system which is more accurately modeled by a half-infinite linear potential whose value at $x = 0$ is negative, *i.e.*,

$$V(x) = \begin{cases} -V_0 + \gamma B x & x > 0 \\ \infty & x \le 0 \end{cases} \tag{9.156}$$

where $V_0 > 0$. How do the stationary state energies E_n and eigenfunctions $\psi_n(x)$ of this system differ from those of Eq. (9.141)?

The half-infinite linear potential (9.141) is a useful beast. It can be used to model a "quantum bouncer" (a point mass in a uniform gravitational field that bounces elastically on a horizontal floor).[43] And, more recently, it has played a role in the theory of elementary particles. Quarks are currently thought to be the fundamental building blocks of protons, neutrons, ..., indeed, of matter itself. Theoretical physicists have speculated that the interaction between two quarks may be a constant force, for the strong binding provided by such a force would explain the curious fact that isolated quarks have never been seen. And a constant force means a linear potential.[44]

So don't let anyone tell you that continuous one-dimensional potentials are useful only to students of undergraduate quantum physics.

EXERCISES AND PROBLEMS

In order to attain the impossible
one must attempt the absurd.

—Miguel de Unamuno y Jugo

[42]The constant γ actually depends on the orientation of the metal and of the field. This is a representative value. For more data on this experiment, see T. Wegehaupt and R. E. Doezema, *Phys. Rev. D.*, **16**, 2515 (1977).

[43]See P. W. Langhoff, *Amer. Jour. Phys.*, **39**, 954 (1971); R. L. Gibbs, *Amer. Jour. Phys.*, **43**, 25 (1975), and V. C. Aguilera-Navarro, H. Iwamoto, E. Ley-Koo, and A. H. Zimerman, *Amer. Jour. Phys.*, **49**, 648 (1981).

[44]You can find a discussion of linear potentials in quark theory in § 7.4 of *Quantum Physics* by R. G. Winter (Belmont, Ca.: Wadsworth, 1979). See also Y. Nambu, *Sci. Amer.*, **235**, 48 (1976) and S. D. Drell, *Amer. Jour. Phys.*, **46**, 597 (1978). More recent investigations suggest that the nuclear force is not constant, but rather varies with the distance between quarks, growing stronger at large distances, weaker at short distances. For more information, see the article "Asymptotic Freedom" by David J. Gross in *Physics Today*, **87**, 39 (1987).

Exercises

9.1 Qualitative Behavior of Eigenfunctions

In this problem, we consider the solution of the TISE for a couple of variations on the infinite square well of Chap. 7. In each part, you'll be asked to determine how the eigenfunctions and bound-state energies change as a consequence of small changes in this potential energy.

[a] Consider a particle of mass m with potential energy (see Fig. 9.1.1):

$$V(x) = \begin{cases} \infty & -\infty < x \le 0 \\ 0 & 0 < x < L/2 \\ V_0 & L/2 \le x < L \\ \infty & L \le x < \infty \end{cases} \tag{9.1.1}$$

where V_0 is a constant $0 < V_0 < \pi^2\hbar^2/2mL^2$.

(1) For the potential in Fig. 9.1.1, is the energy E_1 *greater than or less than* the ground-state energy for the corresponding potential with $V_0 = 0$? **Explain** your answer.

(2) **Sketch** the ground-state ($n = 1$) eigenfunction for this potential. On the same figure, **sketch** the $n = 1$ eigenfunction for the case $V_0 = 0$. **Discuss** the differences between the two functions.

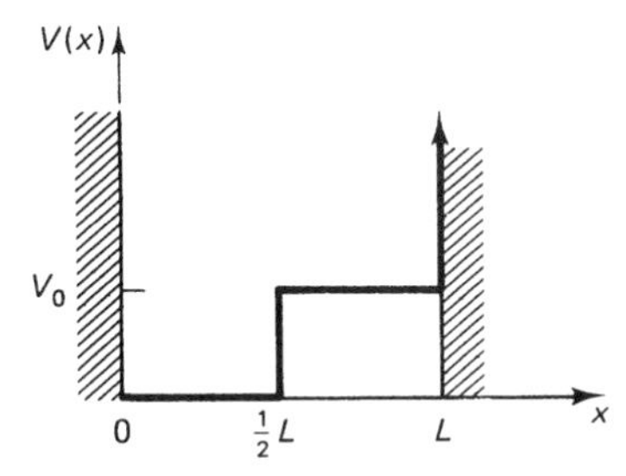

Figure 9.1.1

[b] Now consider the potential energy (Fig. 9.1.2)

$$V(x) = \begin{cases} \infty & -\infty < x \le 0 \\ 0 & 0 < x < L \\ V_0 & L \le x < 2L \\ \infty & 2L \le x < \infty \end{cases} \tag{9.1.2}$$

where $V_0 \gg \pi^2\hbar^2/2mL^2$. Notice that the width of the well is now $2L$.

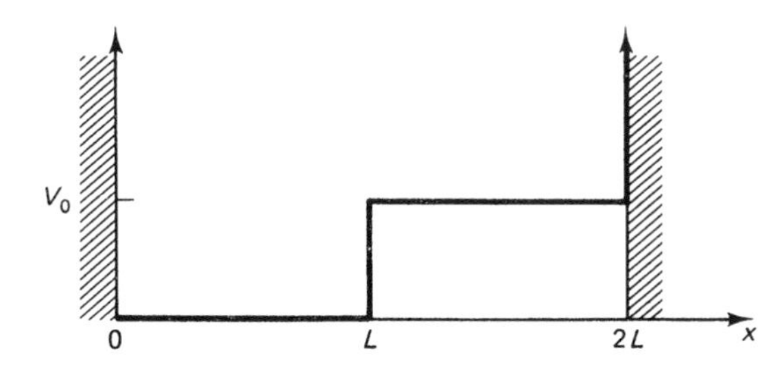

Figure 9.1.2

(1) **Sketch** $\psi_1(x)$, the ground-state eigenfunction for this potential. On the same figure, **sketch** $\psi_1(x)$ for the case $V_0 = \infty$ (*i.e.*, for a particle of mass m in an infinite well of width L) and for the case $V_0 = 0$. **Discuss** the differences between the three eigenfunctions and the reasons for these differences.

(2) Using arguments based on your sketches, **determine** upper and lower bounds on the ground-state energy for this potential.

[c] Finally, consider the potential energy (Fig. 9.1.3)

$$V(x) = \begin{cases} \infty & x \leq 0 \\ \alpha x & 0 < x < L \\ \infty & x \geq L \end{cases} \tag{9.1.3}$$

where $\alpha > 0$ is a constant. Consider the stationary state with $n = 4$.

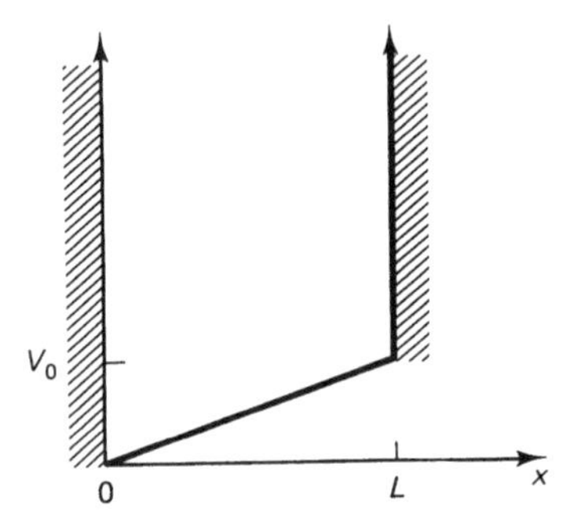

Figure 9.1.3

(1) **Sketch** $\psi_4(x)$ for this potential and for $V = 0$. **Compare** your two eigenfunctions and explain the differences between them.

(2) Is the energy E_4 for $V_0 = V(\alpha L)$ greater than or less than the corresponding energy E_4 for $V_0 = 0$? **Why**?

9.2 Application of Quantum Mechanics to a "Molecular" Potential

In § 9.6, we learned that vibrations of the nuclei in a diatomic molecule can be described by a one-dimensional potential $V(R)$, where R is the internuclear separation. Two typical molecular potentials are graphed as a function of R in Fig. 9.2.1. Assume that both potentials in this figure behave the same at the two extremes of R: *i.e.*, that $V(R) \to \infty$ as $R \to 0$ and $V(R) \to 0$ as $R \to \infty$.

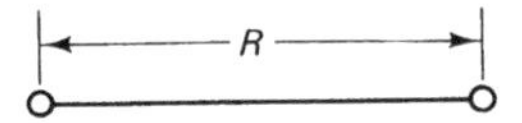

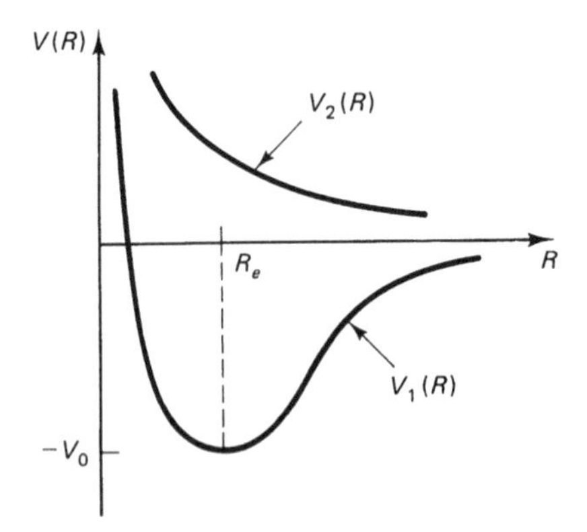

Figure 9.2.1 A model of molecular vibrations.

[a] Consider first the potential energy $V_1(R)$ in the figure. This potential has a minimum at $R = R_e$. Near the minimum, *i.e.*, for values of R such that $|R - R_e| < a$ for some

constant a, the potential energy has positive curvature $d^2V(R)/dR^2 = m\omega_0^2$, and can be approximated by the parabolic function

$$V_1(R) = \tfrac{1}{2}m\omega_0^2(R - R_e)^2. \tag{9.2.1}$$

But for $|R - R_e| > a$, the potential $V_1(R)$ deviates from this simple form (see Fig. 9.2.1).

(1) **Write** down approximate expressions for the eigenfunction $\psi_0(R)$ and energy E_0 of the *minimum energy* stationary state. Why are your results only approximate? What is the approximate expectation value $\langle R\rangle$ for this state?
Hint: Be careful to note where the origin of R and the zero of energy are defined.

(2) Let's define the value of the potential V_1 at $R = R_e \pm a$ to be $-V_a$, *i.e.*, $V_1(R_e \pm a) = -V_a$. Suppose the curvature of the potential energy at its minimum $R = R_e$ is given by

$$m\omega_0^2 = \left.\frac{d^2\,V(R)}{dR^2}\right|_{R_e} = m\frac{(V_e - V_a)^2}{36\hbar^2}. \tag{9.2.3}$$

Consider bound states with energy $E < -V_a$. **Derive** an expression for the approximate energies of these bound states. **Discuss** the qualitative nature of the corresponding bound-state eigenfunctions.

[b] Now briefly consider the "dissociating" potential energy $V_2(R)$.

(1) What are the allowed energies for this case?

(2) For what values of E (if any) are the eigenfunctions normalizable?

(3) Suppose we pick an energy E and define R_E as the corresponding classical turning point. **Discuss** the behavior of the eigenfunction $\Psi_E(x)$ in the regions $0 < R < R_E$ and $R > R_E$.

9.3 Molecular Vibrations: An Application of the SHO

The Hamiltonian that represents *small-amplitude vibrations* of a diatomic molecule is Eq. (9.55) with m replaced by μ, the reduced mass of the molecule. Suppose we use the SHO to model vibrations of the CO molecule. Spectroscopic measurements reveal that one of the stationary states of this molecule has an energy $E_n = 0.6725\,\text{eV}$. The corresponding spatial function $\psi_n(x)$ is sketched in Fig. 9.3.1.

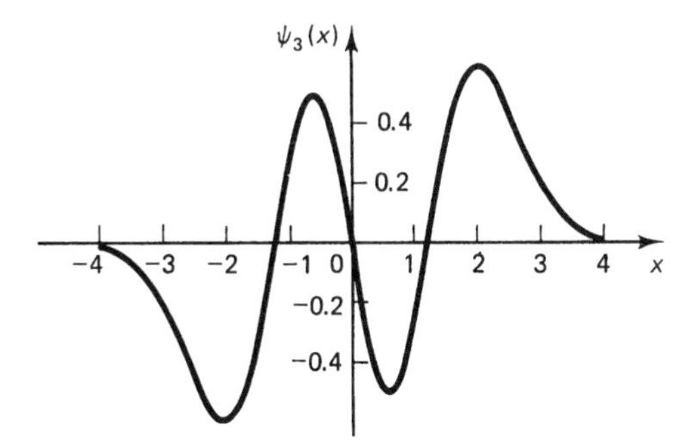

Figure 9.3.1 A stationary-state wave function of the CO molecule.

[a] What is the quantum number n for this state?

[b] **Sketch** the probability density for this state. (Don't cheat; stay away from Fig. 9.15).

[c] Suppose we perform an ensemble measurement of the internuclear separation R of CO in this stationary state.

(1) What average value of R would be obtained in the experiment?

(2) What is (are) the most probable value(s) of R for this state? You need **not** calculate these quantities! You can estimate them from the sketch of ψ_n or from your sketch in [b]. But be sure to explain how you arrived at your answer.

[d] Suppose the CO molecule undergoes a transition from the nth stationary state to the $(n-1)^{\text{th}}$ state. What is the value of the frequency ν of the photon emitted in this transition?

9.4 A "Truncated" Simple Harmonic Oscillator Potential

Consider an electron near the surface of a solid. The surface presents an infinite potential barrier to the electron. Suppose the forces inside the metal are most accurately modeled by the SHO potential energy so that the potential of the electron is

$$V(x) = \begin{cases} \infty & x < 0 \\ \frac{1}{2}m\omega_0^2 x^2 & x \geq 0 \end{cases}. \tag{9.4.1}$$

[a] What are the energies of the *three lowest stationary states* of this system? Explain how you arrived at your answer.

[b] Do the eigenfunctions of this system have definite parity? Why or why not?

[c] **Sketch** the Hamiltonian eigenfunctions for the *ground state* and the *first excited state* of this system.

9.5 Bound Stationary States: Short-Answer Questions

Consider the potential shown in Fig. 9.5.1,

$$V(x) = \begin{cases} V_1 & -\infty < x \leq -a \\ c|x| & -a < x < b \\ V_2 & b \leq x < \infty \end{cases} \tag{9.5.1}$$

where $V_1 > V_2 > 0$ and c is a positive, real constant. Also shown on the figure is the energy E_1 of the *ground stationary bound state* of this potential.

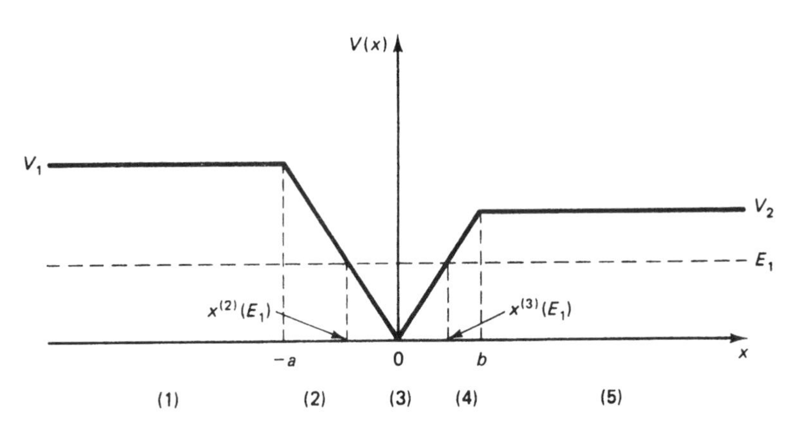

Figure 9.5.1

For this energy, we can identify five regions of x,

$$\begin{array}{ll} Region\ 1 & -\infty < x \leq -a \\ Region\ 2 & -a < x < -x^{(2)}(E_1) \\ Region\ 3 & -x^{(2)}(E_1) \leq x \leq x^{(3)}(E_1) \\ Region\ 4 & x^{(3)}(E_1) < x < b \\ Region\ 5 & b \leq x < \infty. \end{array} \tag{9.5.2}$$

Consider each of the following statements about the function $\psi_1(x)$. If the statement is correct, simply say so and go on. If, however, the statement is *in any way incorrect*, **explain** what is wrong with the statement. If *part* of an incorrect statement is correct, be sure to note that, too.

1. The spatial function $\psi_1(x)$ *in Region (5)* is a decaying, exponential-type function.
2. The spatial function $\psi_1(x)$ *in Region (2)* has the following x dependence: $e^{+\kappa x}$, where κ is a positive, real constant.
3. *In Region (3)*, $\psi_1(x)$ is a linear combination of the oscillatory functions $\sin kx$ and $\cos kx$, where k is a positive, real constant.
4. The spatial function $\psi_1(x)$ has definite parity, and, since it is the *ground state* spatial function, it is even.
5. The function $\psi_1(x)$ has one node, but this node does not occur at $x = 0$.
6. The function $\psi_1(x)$ penetrates deeper into the classically forbidden Region (5) than into Region (1).

Remark: Another way to express [6] is to say that the integrated position probability for Region (5),

$$P^{(5)} \equiv \int_b^\infty |\psi_1(x)|^2 \, dx, \tag{9.5.3}$$

is larger than the integrated position probability for Region (1),

$$P^{(1)} \equiv \int_{-\infty}^{-a} |\psi_1(x)|^2 \, dx. \tag{9.5.4}$$

Problems

9.1 Comparison of a Piecewise-Constant and a Continuous Potential

We spent most of Part III working with the *symmetric* finite square well [of height V_0 and width L (with its midpoint at $x = 0$)] and the simple harmonic oscillator [of natural frequency ω_0]. Consider the *ground-state eigenfunctions* of these two potentials: $\psi_1^{\text{SW}}(x)$ and $\psi_0^{\text{SHO}}(x)$. Choose the parameters of the well (V_0 and L) and of the SHO (ω_0) so that the following criteria are satisfied:

1. The classical turning points occur at the same values of x.
2. The local wave number at $t = 0$ of the piece of $\psi_0^{\text{SHO}}(x)$ in the CA region is equal to the wave number of the corresponding piece of $\psi_1^{\text{SW}}(x)$.
3. The local decay constant of the pieces of $\psi_0^{\text{SHO}}(x)$ in the CF regions are equal *at the classical turning points* to the decay constant of the corresponding pieces of $\psi_1^{\text{SW}}(x)$.

Now, for the resulting parameters, **prepare an accurate graph** of the ground-state eigenfunctions. **Discuss** the physical and mathematical origin and the significance of all differences between the two functions.

9.2 Selected Continuous Potentials

For each of the following systems, derive an expression for the bound-state energies in terms of fundamental constants and the zeroes of the appropriate special functions:

1. A particle of mass m bouncing elastically from a fixed floor in a gravitational field with constant acceleration g.
2. A proton moving in a uniform electric field.

9.3 Vibrations of the OH Molecule

Vibrations of the nuclei in the ground electronic state of an OH molecule can be modeled by a SHO potential with a vibrational frequency that corresponds to $\hbar\omega_0 = 9.68 \times 10^{-2}$ eV. Suppose, instead, that we try to model this system by a *finite symmetric square well* of width L and depth V_0. **Determine** values of these parameters so that the following conditions are fulfilled:

1. The ground-state energies of these systems (E_0 for the SHO and E_1 for the well) are equal.
2. The well supports *five* bound states.

Draw adjacent energy-level diagrams (on the same scale) for the two systems, showing the lowest five states of each. **Graph** the Hamiltonian eigenfunctions of these five states for each system. Which model is more realistic, and why?

9.4 A Non-Stationary state of the SHO

A one-dimensional microscopic simple harmonic oscillator is in a quantum state represented by the state function

$$\Psi(x,t) = \frac{1}{\sqrt{2}}\left[\psi_0(x)e^{-iE_0t/\hbar} + \psi_1(x)e^{-iE_1t/\hbar}\right]. \tag{9.4.1}$$

[a] **Evaluate** the position and momentum uncertainties $\Delta x(t)$ and $\Delta p(t)$ for this state and **show** that these functions oscillate at the natural frequency of the simple harmonic oscillator, ω_0.

[b] **Graph** the position probability density for this state at the following times (where T_0 is the classical period): $t = \frac{1}{4}T_0$, $\frac{1}{2}T_0$, $\frac{3}{4}T_0$, T_0, $\frac{3}{2}T_0$, and $2T_0$.

9.5 A General Non-Stationary State of the SHO

A wave function of a simple harmonic oscillator can be expanded in a linear combination of the stationary-state wave functions we have studied in this chapter, as

$$\begin{aligned}\Psi(x,t) &= \sum_{n=0}^{\infty} d_n\psi_n(x)e^{-iE_nt/\hbar} \\ &= \sum_{n=0}^{\infty} d_nA_nH_n(\beta x)\,e^{-\beta^2x^2/2}\,e^{-iE_nt/\hbar}.\end{aligned} \tag{9.5.1}$$

The expansion coefficients d_n are calculated from the initial wave function $\Psi(x,0)$ as

$$d_n = A_n\int_{-\infty}^{\infty} e^{-\beta^2x^2/2}H_n(\beta x)\Psi(x,0)\,dx. \tag{9.5.2}$$

In this problem we'll apply this expansion to a system in which the particle is initially localized at the position x_0. So the initial wave function is the Dirac Delta function (see Chap. 4)

$$\Psi(x,0) = \delta(x - x_0). \tag{9.5.3}$$

[a] **Derive** an expression for d_n for this state.

[b] **Write down** the first three terms in the expansion Eq. (9.5.1) of $\Psi(x,t)$.

[c] **Sketch** the function in the sum Eq. (9.5.1) at $t = 0$, including first one, then two, then three terms. Now consider the number of terms in the series: would you need a few terms or a large number of terms to accurately represent the initial state Eq. (9.5.3)?

9.6 Semiconductor Quantum Physics

A very thin sample of the semiconductor PbTe (lead telluride) of width L is exposed to a uniform electric field $\vec{\mathcal{E}}$ normal to the surface. The potential energy of an electron near the surface of the sample can be modeled by the one-dimensional potential

$$V(x) = \begin{cases} \infty & -\infty < x \le 0 \\ \alpha x & 0 < x < L \\ \infty & L \le x < \infty \end{cases}. \tag{9.6.1}$$

(This potential is just an infinite square well with a ramp for a floor.) We can control the width of the sample L in the process of growing it.

[a] **Write down** the general solution of the TISE for this system, introducing two arbitrary constants C and D.

[b] Now **impose** the boundary conditions appropriate to this potential and thereby **transform** the TISE into a set of two *linear homogeneous algebraic equations* for C and D. Under

what conditions does this set of equations have a non-trivial solution (*i.e.*, one other than $C = D = 0$)? Use your answer to **derive** an equation for the bound-state energies of the particle.

[c] Suppose we grow the crystal so that its thickness is equal to the difference between the two zeroes ξ_1 and ξ_6 of the Airy function $\mathrm{Ai}(\xi)$. **Solve** your equation from [b] for the bound-state energies of this potential.

9.7 Airy and Bessel Functions

[a] **Show** that the spatial functions resulting from the condition (9.153) can be expressed in terms of Bessel functions of order 1/3, as

$$\psi_\epsilon(\xi) = \begin{cases} C\frac{1}{\pi}\mathrm{K}_{\frac{1}{3}}\left(\frac{2}{3}\xi^{3/2}\right) & \xi > 0 \\ C\frac{1}{3}\sqrt{|\xi|}\left[\mathrm{J}_{\frac{1}{3}}\left(\frac{2}{3}|\xi|^{3/2}\right) + \mathrm{J}_{-\frac{1}{3}}\left(\frac{2}{3}|\xi|^{3/2}\right)\right] & \xi \le 0 \end{cases}$$

where $\mathrm{K}_{\frac{1}{3}}$ is a modified Hankel function (a Bessel function of the third kind).

[b] **Show** that the bound-state energies are the values E_n at which

$$\mathrm{J}_{\frac{1}{3}}\left(\frac{2}{3}\sqrt{\frac{2}{\gamma B\hbar^2}}\right) + \mathrm{J}_{-\frac{1}{3}}\left(\frac{2}{3}\sqrt{\frac{2}{\gamma B\hbar^2}}\right) = 0.$$

9.8 A Light and Airy Problem

The Airy function $\mathrm{Ai}(\xi)$ is actually an integral. It's defined as

$$\mathrm{Ai}(\xi) \equiv \frac{1}{\pi}\lim_{\epsilon\to 0}\int_0^\infty e^{-\epsilon u}\cos\left(\frac{u^3}{3} + \xi u\right)\,du. \tag{9.8.1}$$

(The factor $e^{-\epsilon u}$ guarantees that the integral will converge as $u \to \infty$. This factor vanishes when we take the limit, after integration.)

[a] **Prove** that the integral in Eq. (9.8.1) satisfies the Airy equation (9.149).

[b] By examining the behavior of the integrand in Eq. (9.8.1) for large values of the variable ξ, **explain** why $\mathrm{Ai}(\xi)$ goes to zero in the limit $\xi \to \infty$. This limit, you will (I hope) recall, is vital to ensuring that an eigenfunction involving an Airy function satisfies the boundary conditions for physical admissibility.

[c] **Show** that the asymptotic limit of the Airy function is

$$\mathrm{Ai}(\xi) \xrightarrow[\xi\to\infty]{} \frac{1}{\sqrt{2}\,\pi}\xi^{-1/4}e^{-2\xi^{3/2}/3}. \tag{9.8.2}$$

[d] Using Eq. (9.8.2), **compare** the behavior in the classically forbidden regions of the ground-state eigenfunctions of the half-infinite linear potential and of a simple harmonic oscillator. In particular, decide if the eigenfunction for the ground state of the linear potential decays to zero with increasing ξ *faster, slower, or at the same rate* as the ground-state eigenfunction of the SHO. Also **compare** the rate of decay of the ground-state eigenfunction of the linear potential to that of the corresponding function for a finite, symmetric square well.

[e] The other asymptotic limit is $\xi \to -\infty$. In this limit, the Airy function reduces to the form

$$\mathrm{Ai}(\xi) \xrightarrow[\xi\to-\infty]{} \frac{1}{\sqrt{\pi}}(-\xi)^{-1/4}\sin\left[\frac{2}{3}(-\xi)^{3/2} + \frac{\pi}{4}\right]. \tag{9.8.3}$$

Use this form to **determine** approximate zeroes of the Airy function ξ_n for $n = 3$, 4, and 5. Compare your answers to the exact zeroes in Table 9.10.

9.9 Bending Vibrations of Methylene

The SHO is a useful model for studying the vibrations of a diatomic molecule (see § 9.6). But more complicated systems call for more complicated models. Consider, for example, the methylene molecule CH_2. In its ground electronic state, this system is bent; the angle between the two C–H bonds is about 134°. The vibrations of this molecule are *bending vibrations* and are described as functions of this angle, which we'll denote by θ. (Other kinds of vibrations involving stretching of the C–H bonds also occur. In this problem, however, we'll keep these bonds fixed at their equilibrium values.)

A graph of this potential, shown in Fig. 9.9.1,[45] looks like a parabola with a hump in the middle. Also shown on this figure—and tabulated in Table 9.9.1—are the first seven stationary-state energies, as determined by recent, highly accurate, spectroscopic studies.[46]

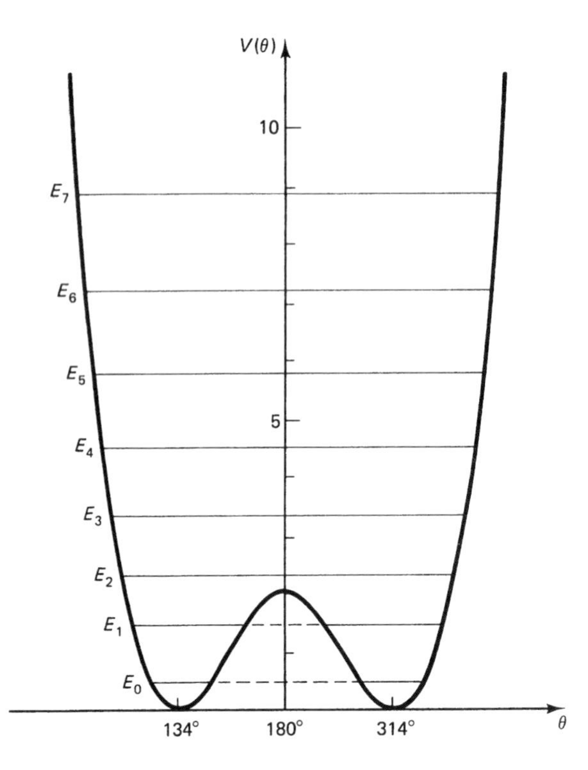

Figure 9.9.1

To study these vibrations quantum-mechanically, we must devise a model of this potential. We'll express this model in terms of the variable $x \equiv \theta - \theta_e$, *i.e.*, the displacement of the bending angle from equilibrium. Obviously, the simple SHO potential $V(x) = \frac{1}{2}m\omega_0^2 x^2$ won't do.

[a] Suppose we induce a hump in the SHO potential by adding a Gaussian term that includes two additional parameters, α and β, *i.e.*, consider

$$V(x) = \frac{1}{2}m\omega_0^2 x^2 + \alpha e^{-\beta x^2}. \tag{9.9.1}$$

(1) **Determine** values of the parameters ω_0, α, and β in Eq. (9.9.1) that fit the data in Table 9.9.1

[45] Adapted from R. B. Bunker and P. Jensen, *J. Chem. Phys.*, **79**, 1334 (1983).

[46] See D. G. Leopold, K. K. Murray, A. E. Stevens-Miller, and W. C. Lineberger, *J. Chem. Phys.*, **83**, 4849 (1985).

TABLE 9.9.1
ENERGIES OF STATIONARY (BENDING) VIBRATIONAL STATES OF METHYLENE.

n	$E_n(\text{eV})$
1	0.12
2	0.23
3	0.35
4	0.50
5	0.66
6	0.84
7	1.04

(2) **Repeat** (1) using a model that contains, in addition to the terms in Eq. (9.9.1), the *quartic* term γx^4. Which model do you think is the most accurate?

(3) Using the methods of § 9.2–9.3, **sketch** the first *four* eigenfunctions $\psi_n(x)$ of your best model potential. On these sketches, also **show** the corresponding eigenfunctions for the case $\alpha = 0$.

[b] Suppose we try the alternate model

$$V(x) = \gamma x^4 - \frac{1}{2} m\,\omega_0^2 x^2, \tag{9.9.2}$$

treating γ and ω_0 as parameters.

(1) **Graph** this equation for several choices of γ and ω_0 to convince yourself that it is possible to find parameters that give a shape similar to that in Fig. 9.9.1.

(2) Analysis of spectroscopic data for methylene reveals that the minima in the potential of the ground electronic state occur at $\theta_e = 134°$ and at $\theta_e = 314°$ and that the top of the bump is roughly $0.2\,\text{eV}$ above these minima. From these data, **determine** the values of the parameters γ and ω_0 in the model potential Eq. (9.9.2).

(3) What are the energies of the first *seven* bound states of a SHO with a natural frequency ω_0 equal to the value of this parameter you determined in (2)? Compare your answers to those in Table 9.9.1.

[c] "Double-hump" potentials can be used in a variety of molecular problems, such as the buckling motion of ring compounds in organic chemistry, and the inversion of the ammonia molecule in laser physics. In some of these applications, the minima are much deeper (compared to the hump) than in methylene. If this is the case—and if we're interested only in low-lying vibrational states—we may be able to get a (very) rough approximation to the energy levels by assuming the hump to be very large. Consider this limit and **expand** the model potential Eq. (9.9.2) in a Taylor series about equilibrium, dropping all terms above the quadratic. **Derive** an expression for the energy levels for the left and right wells in this model.

9.10 A Qualitative Solution of a TISE

Consider a particle of mass m with potential

$$V(x) = c|x|, \tag{9.10.1}$$

where c is a real, positive constant (see Fig. 9.10.1). We'll denote the ground stationary state of this system by the quantum number $n = 1$.

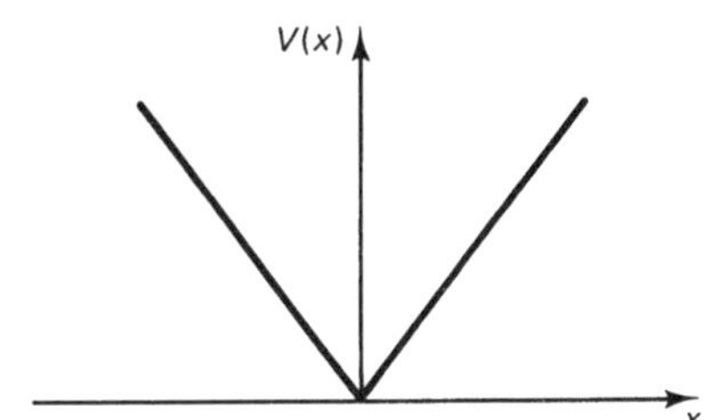

Figure 9.10.1

[a] **Sketch** $\psi_1(x)$ and $\psi_{10}(x)$. In your sketches, be careful to show accurately the changes in wavelength and amplitude as x varies. Briefly **discuss** the qualitative features of these two states, explaining them on physical grounds.

[b] **Sketch** the probability densities for these two stationary states. On your sketch of $P_{10}(x)$, also show the probability density that classical physics would predict for a particle with this potential energy.

Remark: This problem is yet another example of the *Correspondence Principle*—quantum behavior becoming classical behavior in the appropriate limit.

9.11 Hermite Polynomials: A Capsule Introduction to Their Properties

In our study of the simple harmonic oscillator, we encountered our first (but not last) special function: the Hermite polynomials of degree n, $H_n(\xi)$. From now on—in this course and in other physics, math and engineering courses—you'll be using special functions, many of which are more complicated than Hermite polynomials. The care and feeding of these functions is facilitated by the fact that they all share similar properties and satisfy similar relationships. In this problem, we'll develop the most important of these properties and relationships for the Hermite polynomials.

The Hermite polynomial of order n is defined as

$$H_n(\xi) = (-1)^n \, e^{\xi^2} \frac{d^n}{d\xi^n} e^{-\xi^2}. \tag{9.11.1}$$

The origin of this polynomial is, surprisingly, buried in an interesting property of Gaussians. Recall (from Chap. 4) that, in its simplest form, a Gaussian function has the mathematical form

$$f(\xi) = e^{-\xi^2}. \tag{9.11.2}$$

The property to which I'm referring concerns the nth derivative of $f(\xi)$. This derivative is proportional to a polynomial of degree n, to wit:

$$\frac{d^n}{d\xi^n} f(\xi) = (-1)^n \; (n^{\text{th}} \text{ order polynomial}) \; f(\xi). \tag{9.11.3}$$

[a] **Verify** Eq. (9.11.3) for $n = 0$, 1, and 2. **Show** that polynomials in this equation for these values of n are indeed the Hermite polynomials of degree 0, 1, and 2. Using Eq. (9.11.1), **derive** $H_3(\xi)$.

As the last question in part [a] shows, the definition Eq. (9.11.1) is not a particularly convenient way to determine Hermite polynomials. One alternative is to use a **recurrence relation**. A recurrence relation is an equation that relates the desired special function of order n to functions of lower order and, sometimes, to their derivatives.

[b] From the definition Eq. (9.11.1), **derive** the recurrence relation

$$H_n(\xi) = \left(2\xi - \frac{d}{d\xi}\right) H_{n-1}(\xi). \tag{9.11.4}$$

One of the most important relatives of a special function is its "generating function." The **generating function** generates the desired special function via *a Taylor series expansion.* For example, a Taylor Series expansion of

$$g(\lambda) \equiv e^{-\lambda^2+2\lambda\xi} \tag{9.11.5}$$

in the variable λ about $\lambda = 0$ "generates" the Hermite polynomials according to the following equation:

$$g(\lambda) = \sum_{n=0}^{\infty} \frac{\lambda^n}{n!} H_n(\xi). \tag{9.11.6}$$

[c] **Derive** Eq. (9.11.6).
Hint: One way to do this is to first consider the Taylor series expansion about $\lambda = 0$ of the function

$$f(\xi + \lambda) = e^{-(\xi+\lambda)^2}. \tag{9.11.7}$$

Then use Eq. (9.11.1).

You may well be wondering what earthly good Eq. (9.11.6) is. Surely it isn't likely to help us derive explicit expressions for Hermite polynomials. True enough. But this equation can be used to easily obtain some very useful recurrence relations.

[d] Use Eq. (9.11.6) to **derive** the following results:

$$\frac{d}{d\xi} H_n(\xi) = 2nH_{n-1}(\xi) \tag{9.11.8}$$

$$H_n(\xi) = 2\xi H_{n-1}(\xi) - 2(n-1)H_{n-2}(\xi). \tag{9.11.9}$$

Hint: Differentiate Eq. (9.11.6) with respect to λ and ξ. Then equate like powers of λ.

[e] Use Eq. (9.11.9) to **derive** H_9 from H_8 and H_7.

[f] Use Eq. (9.11.8) to **derive** the differential equation satisfied by $H_n(\xi)$,

$$\left[\frac{d^2}{d\xi^2} - 2\xi\frac{d}{d\xi} + 2n\right] H_n(\xi) = 0. \tag{9.11.10}$$

[g] **Prove** that the *parity* of $H_n(\xi)$ is $(-1)^n$.

9.12 Another "Truncated" SHO

An unrealistic feature of the SHO potential is that it increases without limit to infinity. To improve on this model, we might try truncating this potential as shown in Fig. 9.12.1, *i.e.*, with V_1 and V_2 as positive constants,

$$V(x) = \begin{cases} V_1 & -\infty < x \le -a \\ \frac{1}{2}m\omega_0^2x^2 & -a < x < b \\ V_2 & b \le x < \infty \end{cases}. \tag{9.12.1}$$

Rather than solve the TISE for this potential (no trivial matter!) let's explore some of qualitative features of eigenfunctions.

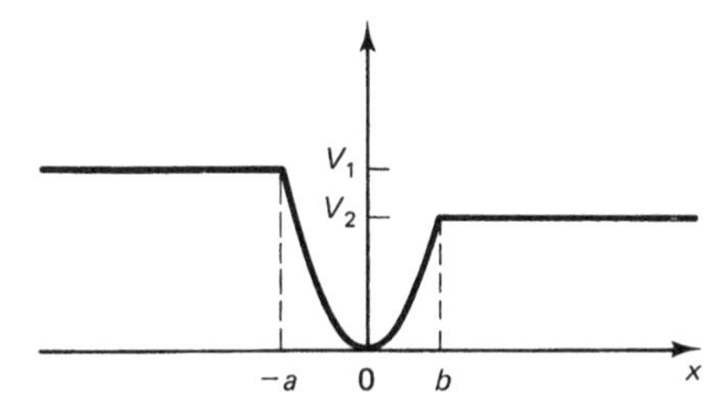

Figure 9.12.1

[a] Consider four ranges of energy:

$$\begin{array}{ll} Range\ A: & E > V_1 \\ Range\ B: & V_2 < E < V_1 \\ Range\ C: & 0 < E < V_2 \\ Range\ D: & E < 0 \end{array}$$

For each range of energy, answer the following questions:

(1) What is the nature of the eigenfunction (*i.e.*, oscillatory or monotonic) throughout space ($-\infty < x < \infty$)? (Be sure to distinguish different regions, where appropriate.)

(2) Is the energy spectrum quantized or continuous?

(3) Are the stationary states degenerate?

(4) Are the stationary states bound or continuous?

[b] Now for *Range C:* $0 < E < V_2$ *only*, do the following:

(1) **Write down** the TISE in each region of x.

(2) **Write down** the boundary and continuity conditions on $\psi_E(x)$ in terms of the pieces of this function in each region from [b.1]. **Compare** the number of boundary conditions with the number of unknowns in the general solution of the TISE [keep in mind the fact that this (second-order) differential equation has two linearly-independent solutions in each region]. Do you have sufficient flexibility to normalize the eigenfunctions?

(3) Consider the *ground state*, which we'll label with $n = 0$. Is this energy greater or less than $\hbar\omega_0/2$? **Justify** your answer. Now, **sketch** $\psi_0(x)$ and, on the same figure, the ground-state eigenfunction for a SHO [which is obtained from Eq. (9.12.1) by letting $a \to \infty$ and $b \to \infty$]. (Feel free to wildly exaggerate the differences between these functions, for clarity's sake.)

[c] Consider now the *symmetric* truncated well obtained from Eq. (9.12.1) by setting $b = a$. As in [b], we'll consider energy range C: $0 < E < V_2$ only.

(1) Do the eigenfunctions of this potential have definite parity? Why or why not? How many nodes does $\psi_0(x)$ have? What is the symmetry of this function?

(2) In designing a model potential by truncating the SHO, is there any advantage to truncating so as to give a symmetric potential?
Hint: Review your answers to question [b.2]. Think about imposing the boundary conditions.

(3) Suppose the wave-function fairy gave you the analytic form of the ground-state spatial function $\psi_0(x)$ and told you to calculate the average value of x for this state. What value would you get? Would you expect the same answer if $b < a$ (the asymmetric potential)? If so, explain why. If not, would you expect $\langle x \rangle_0$ to be greater or less than its value for $b = a$?

9.13 An Application of the SHO to Molecular Vibrations

Small-amplitude vibrations of a diatomic molecule can be studied using as a model of the true molecular vibrational potential the simple harmonic oscillator potential

$$V(x) = \frac{1}{2}k_f x^2, \tag{9.13.1}$$

where k_f is the force constant. For a typical diatomic molecule, the force constant is roughly $k_f = 1.0 \times 10^3$ J-m^{-2} (see Table 9.1).

[a] Using this value, **estimate** the value of the *zero-point* vibrational energy of a diatomic molecule (in eV).

[b] **Estimate** the energy spacing (in eV) between the ground vibrational state and the first excited vibrational state of the molecule.

[c] Suppose the molecule undergoes a transition from the first excited vibrational state to the ground state, emitting a photon in the process. **Estimate** the energy, the frequency, and the wavelength of the emitted photon. **Compare** your result with the *classical* oscillation frequency of this system. In what range of the electromagnetic spectrum is this photon found?

9.14 Quantum Oscillations in the Sky?

If you were on an airplane, staring out the window, you might notice the disquieting fact that the wings of the plane oscillate. Careful observation shows that the period of these oscillations is typically on the order of 1 sec and the amplitude is typically about 0.1 m. (If the amplitude is much greater, the plane will crash and you needn't worry about solving this problem.)

[a] **Demonstrate** that these oscillations could not be due to the quantum-mechanical zero point motion of the wings.

[b] What is the quantum number n of the observed oscillations?

[c] Suggest an alternative explanation for this phenomenon.

9.15 A One-dimensional "Hydrogen Atom"

One of the first "real systems" we'll study in Volume II is the *one-electron atom*, the paradigm of which is the hydrogen atom. The potential energy of an electron in a hydrogen atom (in MKS units) is

$$V(r) = -\frac{e^2}{4\pi\epsilon_0 r}, \tag{9.15.1}$$

where e is the charge on the electron and r is the distance of the electron from the proton. The hydrogen atom is, of course, a three-dimensional system, and r is the radial coordinate of a *spherical coordinate system* with the origin at the proton. But we need not wait until the next volume to get a feeling for the physics of the hydrogen atom. We can model this system by a simple, continuous one-dimensional potential energy whose Schrödinger Equation we can solve. The first question is, how to devise such a model.

Our first thought might be to try a potential energy of the form

$$V(x) = -\frac{e^2}{4\pi\epsilon_0 |x|}, \tag{9.15.2}$$

where we use $|x|$ to ensure that the spatial coordinate of the electron does not assume negative values. [Recall that r in Eq. (9.15.1) is defined on the domain from 0 to ∞.] The TISE for this potential can be solved, but doing so is mathematically quite difficult, because the potential Eq. (9.15.2) has a *pole* at the origin. And there is another problem with this model: since r is *defined* to be non-negative, spatial functions for the hydrogen atom must be zero for $x < 0$. But this potential does not guarantee that the corresponding bound-state spatial functions will be zero for $x < 0$.

These considerations motivate us to conjure a simpler one-dimensional model of a hydrogen atom,

$$V(x) = \begin{cases} -\dfrac{e^2}{4\pi\epsilon_0 x} & x > 0 \\ \infty & x \le 0 \end{cases}. \tag{9.15.3}$$

We can determine the bound stationary-state energies of an electron with this potential by applying the method of power series that we studied in § 9.7. Let's go through the solution together, step-by-step.

[a] Simplify the TISE for total energy E by changing the variable of differentiation from x to one that is dimensionless. Define this new variable ξ in terms of the constant

$$a_0 \equiv \frac{4\pi\epsilon_0 \hbar^2}{m_e e^2}, \tag{9.15.4}$$

which is the Bohr radius of the hydrogen atom (see Appendix B). Introduce a dimensionless energy parameter ϵ defined by

$$\epsilon^{-2} \equiv -\frac{2m_e E a_0^2}{\hbar^2}, \tag{9.15.5}$$

and the *dimensionless* **length variable** ξ

$$\xi \equiv \frac{2x}{\epsilon a_0}. \tag{9.15.6}$$

Notice that the energy E necessarily appears in the length variable.
Remark: The differential equation that results from this change of variable, which is called **Kummer's equation**, can be solved using confluent hypergeometric Wittaker functions.[47] However, we'll just plow on ahead as though we don't know this.

[b] The equation you have just derived cannot immediately be solved by, for example, inserting into it a power series expansion of $\psi_E(\xi)$. First, we must examine its *asymptotic solution*. Consider your equation in the limit $|\xi| \longrightarrow \infty$ and **determine** the asymptotic behavior of $\psi_E(\xi)$. (Don't forget to take steps to ensure that this spatial function remains physically admissible in this limit.)

[c] We now introduce a new unknown function $f(\xi)$ that is defined so the spatial function has the form of $f(\xi)$ times a function that describes the asymptotic dependence of ψ_E on ξ. **Derive** the equation that $f(\xi)$ must satisfy.

[d] Your equation from part [c] can be solved by a power series, *viz.*,

$$f(\xi) = \sum_{j=0}^{\infty} c_j \xi^j, \tag{9.15.7}$$

in which we *must* set $c_0 = 0$. **(Why?) Derive** a recurrence relation for the coefficients c_j in this series.

[e] Give an argument to show that the infinite series Eq. (9.15.7) must be truncated at some finite order, and **show** that doing so leads to the restriction that ϵ must equal a positive integer.

[f] From the result to part [e], **obtain** an expression for the allowed energies of the bound states of this model hydrogen atom. **Look up** the equation for the bound-state energies of an actual, three-dimensional hydrogen atom and compare your result to this answer. Is Eq. (9.15.3) a good model for determining the energies of this system? Can you think of any important physical effects that have been totally ignored in this model?

[g] Using the recurrence relation of part [d] **write down** unnormalized spatial functions for the *first three* bound states of this model potential energy.

9.16 The SHO in Momentum Space

In Chap. 4 we explored the *momentum probability amplitude* for a quantum state. This function, $\Phi(p)$, is the Fourier transform of the *initial* wave function in the position representation, $\Psi(x, 0)$, *i.e.*,

$$\Phi(p) = \frac{1}{\sqrt{2\pi\hbar}} \int_{-\infty}^{\infty} \Psi(x, 0) e^{-ipx/\hbar}\, dx. \tag{9.16.1}$$

If the system is in the nth stationary state, then $\Psi(x, 0) = \psi_n(x)$, and this relationship becomes

$$\Phi_n(p) = \frac{1}{\sqrt{2\pi\hbar}} \int_{-\infty}^{\infty} \psi_n(x) e^{-ipx/\hbar}\, dx. \qquad [n^{\text{th}} \text{ stationary state}] \tag{9.16.2}$$

[47] See *A Course in Modern Analysis* by Wittaker and Watson (Cambridge University Press, 1969), p. 377 ff.

In this case, the inverse **transform** relation is

$$\psi_n(x) = \frac{1}{\sqrt{2\pi\hbar}} \int_{-\infty}^{\infty} \Phi_n(p) e^{ipx/\hbar}\, dp. \qquad [n^{\text{th}} \text{ stationary state}] \tag{9.16.3}$$

Use these relationships to transform the TISE for the SHO into momentum space. To do this, substitute Eq. (9.16.3) into the TISE and derive an equation for the momentum probability amplitude $\Phi_n(p)$. Express your equation in terms of the variables in Table 9.3. Now, **solve** your equation for $\Phi_n(p)$ for the SHO.
Remark: The simple harmonic oscillator is one of the few systems (other than the free particle) for which the momentum-space Schrödinger Equation is easily solved. When the Schrödinger equations of most systems are transformed into momentum space, integral terms appear that render their solution difficult mathematically.

9.17 The Linear Potential in Momentum Space

In § 4.6 we explored the momentum representation of a quantum state. The expressions relating the momentum probability amplitude $\Phi(p)$ to the wave function in the position representation are given in Fig. 4.12. In § 5.8 we learned how to express $\hat{x}$ and $\hat{p}$ in this representation [Eq. (5.90)] and how to use these forms to evaluate expectation values.

[a] **Show** that in the momentum representation the TISE for the nth stationary state of the half-infinite linear potential (9.141) is

$$\left(\frac{p^2}{2m} + i\gamma B \frac{d}{dp}\right) \Phi_n(p) = E\Phi_n(p). \tag{9.17.1}$$

[b] **Integrate** Eq. (9.17.1), obtaining

$$\Phi_n(p) = \Phi(0) e^{-ip\left(E - p^2/6m\right)/(\gamma B\hbar)}. \tag{9.17.2}$$

[c] Now use the Fourier **transform** relationships to transform Eq. (9.17.2) into the *position representation*, thereby regaining the Hamiltonian eigenfunction $\psi_n(x)$ of § 9.11. With little difficulty, you should obtain the following integral:

$$\psi_n(x) = \sqrt{\frac{2}{\pi\hbar}}\Phi(0) \int_0^{\infty} \cos\left[\frac{p^3}{6m\gamma B} + p\left(x - \frac{E}{\gamma B}\right)\right] dp. \tag{9.17.3}$$

You can show that this is equivalent to (9.151) (to within an overall multiplicative normalization constant) by using the handy Airy relation

$$\operatorname{Ai}\left[\pm(e\alpha)^{-1/2}x\right] = \frac{\pi}{\sqrt{3\alpha}} \int_0^{\infty} \cos\left(\alpha p^3 \pm xp\right) dp. \tag{9.17.4}$$

Granted, this is not as easy a way to solve the TISE as the method of § 9.11, but it is an interesting alternative that sheds more light on the connection between the position and momentum representations of a quantum state.

CHAPTER 10

Operators in Quantum Mechanics I

The Importance of Being Hermitian

The tale of Math is a complex one,
and it resists
both a simple plot summary
and a concise statement of its meaning.

—*The Mabinogi*
by Patrick K. Ford

In Part III we solved several simple one-dimensional problems using the fundamental elements of quantum mechanics—wave functions and operators—and Postulates I–IV, which assert the existence of these elements and tell us how to interpret them. But our ultimate goal is understanding the (more complicated) systems of the microworld: atoms, molecules, solids, nuclei. Our ability to solve these systems and to extract the maximum amount of physical information from their state functions depends on our mastery of the mathematical tools of quantum mechanics.

So in Part IV we're going to shift gears again. In the next four chapters, it's back to basics for an intimate look at the elements of quantum physics—operators in this chapter and the next, wave functions in Chap. 12. Then, in Chap. 13, we'll discover how these abstract mathematical constructs describe what goes on in the real world of laboratory measurements.

Paradoxically, these chapters involve less algebra but more mathematics than those in Part III.[1] And, because this mathematics is at times rather abstract, the relevance of some of the topics I'll introduce may seem a tad remote. Trust me. All will become clear.

In this chapter we concentrate on *operators*, mathematical instructions that, in quantum physics, represent quantities we can measure in the laboratory. My goal in Chaps. 10 and 11 is to help you become familiar with operators, their mathematical properties, and the implications of these properties for the physics of microscopic systems. To this end, we'll first supplement our grab-bag of mathematical properties of operators (from Chap. 5) in § 10.1–10.4. Then, in § 10.5–10.8, we will focus on a particular type of operator equation—the eigenvalue equation—and will there discover a remarkable and beautiful example of the subtle interweaving of mathematics and physics that characterizes quantum theory. This background will prepare us to turn, in Chap. 11, to the consequences of the properties of operators for the physics of uncertainty relations, conservation laws, and constants of the motion.

10.1 OPERATORS IN REVIEW: A CAPSULE SUMMARY OF CHAPTER 5

Cast your mind back to Chap. 5. There we learned quantum physicists represent *observables* via mathematical instructions called *operators*. This essential idea was succinctly stated in The Third Postulate of Quantum Mechanics (§ 5.4), which bears repeating:

[1] The reason is that the characteristic manipulations in the branch of mathematics underlying this material, *linear algebra*, are far easier than those of the theory of differential equations. We won't study linear algebra *per se*; I have resolutely resisted the temptation to show you the mathematics of quantum mechanics in all its (considerable) glory. Still, you should be aware that this theory can be described with a high degree of mathematical rigor. For an overview of this description, see Chap. 8 of *Quantum Mechanics*, 2nd ed. by Eugen Merzbacher (New York: Wiley, 1980). More advanced discussions can be found in *Foundations of Quantum Mechanics* by J. M. Jauch (Reading, Mass.: Addison-Wesley, 1968) and in *Mathematical Methods of Quantum Mechanics* by G. Fano (New York: McGraw-Hill, 1971).

Postulate III:

In quantum mechanics, every observable is represented by an operator that is used to obtain physical information about the observable from state functions. For an observable that in classical physics is represented by the function $Q(x, p)$, the corresponding operator is $\hat{Q}(\hat{x}, \hat{p})$.

In Chap. 5 we also learned how to construct quantum-mechanical operators in the position and momentum representations, starting from the operators for $\hat{x}$ and $\hat{p}$ (see Table 10.1). Nearly all operators in quantum mechanics derive from observables that can be expressed as functions of x and p. For such a classical function $Q(x, p)$, we can design an explicit operator $\hat{Q} = \hat{Q}(\hat{x}, \hat{p})$ by inserting the instructions for $\hat{x}$ and $\hat{p}$ (in the appropriate representation) into $Q(x, p)$. Thus, all operators can be considered functions of the elemental operators $\hat{x}$ and $\hat{p}$. A compendium of the operators we have used so far appears in Table 10.2.[2]

TABLE 10.1 FORMS OF THE POSITION AND MOMENTUM OPERATORS IN THE POSITION AND MOMENTUM REPRESENTATIONS.

Fundamental Operator	Position Representation	Momentum Representation
$\hat{x}$	x	$+i\hbar\frac{\partial}{\partial p}$
$\hat{p}$	$-i\hbar\frac{\partial}{\partial x}$	p
Reference	Eq. (5.89)	Eq. (5.90)

The Algebra of Operators

When manipulating operators, we must obey their *algebra:* the collection of rules that describes how *arbitrary* operators act on functions and on each other. Particularly important are the similarities and (especially) the differences between the rules for numbers and functions and those for operators.

The definitions of the *sum* and *difference* of two operators $\hat{A}_1$ and $\hat{A}_2$ (acting on an arbitrary function f) look just like the definition of the sum and difference of two numbers or two functions:

$$(\hat{A}_1 \pm \hat{A}_2)f = \hat{A}_1 f \pm \hat{A}_2 f. \tag{10.1a}$$

But, unlike numbers and functions, *operators do not necessarily commute.* Recall that two operators $\hat{A}_1$ and $\hat{A}_2$ **commute** if *for any function* f, the function $\hat{A}_1\hat{A}_2 f$ is equal to $\hat{A}_2\hat{A}_1 f$. *This equality does not hold for two arbitrary operators*, so we say "the algebra of operators is not commutative," *i.e.*,

$$\hat{A}_1\hat{A}_2 \neq \hat{A}_2\hat{A}_1. \qquad \text{non-commutativity} \tag{10.1b}$$

[2]For a detailed look at the construction of quantum-mechanical operators, see J. R. Shewell, *Amer. Jour. Phys.*, **27**, 16 (1959).

TABLE 10.2 QUANTUM-MECHANICAL OPERATORS (IN THE POSITION REPRESENTATION).

Observable	Operator	Instructions
position	$\hat{x}$	multiply by x
momentum	$\hat{p}$	$-i\hbar\dfrac{\partial}{\partial x}$
kinetic energy	$\hat{T}$	$-\dfrac{\hbar^2}{2m}\dfrac{\partial^2}{\partial x^2}$
potential energy	$\hat{V}$	multiply by $V(x,t)$
Hamiltonian	$\hat{\mathcal{H}}$	$-\dfrac{\hbar^2}{2m}\dfrac{\partial^2}{\partial x^2} + V(x,t)$
energy	$\hat{\mathcal{E}}$	$i\hbar\dfrac{\partial}{\partial t}$

We have seen, however, that there exist *particular pairs* of operators that do commute, such as $\hat{p}$ and $\hat{T}$. These examples do not contradict the general non-commutativity rule (10.1*b*), for this rule merely states that not *all* pairs of operators commute. Clearly, whether or not two operators commute depends on the particular operators we are talking about.

Commutativity is so important a property that quantum mechanics have introduced a special operator called the commutator to express it. The **commutator** of two operators $\hat{Q}_1$ and $\hat{Q}_2$ is defined as[3]

$$\boxed{\left[\hat{Q}_1, \hat{Q}_2\right] \equiv \hat{Q}_1\hat{Q}_2 - \hat{Q}_2\hat{Q}_1} \qquad \text{commutator.} \qquad (10.2a)$$

In terms of the commutator, the general non-commutativity of arbitrary quantum-mechanical operators is written

$$\left[\hat{Q}_1, \hat{Q}_2\right] \neq 0. \qquad (10.2b)$$

Of course, the commutator of two operators that do commute equals zero (see Table 10.3). In Chap. 11 we'll investigate the commutativity—or lack of same—for various operators and the remarkable consequences of this property for measurement of the observables they represent.

The Role of Operators

Operators and their commutators play central roles in quantum physics. The key to evaluating a statistical property of an observable, such as its expectation value or uncertainty,

[3]In this chapter I will use a slightly different notational convention than in Chap. 5, for now it is crucial to distinguish properties of an arbitrary operator from properties that pertain only to operators in quantum mechanics—*i.e.*, to "Hermitian" operators (see § 10.3). Throughout the rest of this book, I will use $\hat{A}$, $\hat{B}$, $\hat{C}$, ... for an arbitrary operator and $\hat{Q}$, $\hat{R}$, $\hat{S}$, ... for a quantum-mechanical (Hermitian) operator. For example, because the commutator is always discussed for Hermitian operators, I have written Eq. (10.2) in terms of $\hat{Q}$, not $\hat{A}$.

TABLE 10.3 COMMUTING AND NON-COMMUTING OPERATORS.

$\hat{Q}_1\hat{Q}_2 = \hat{Q}_2\hat{Q}_1$	$\hat{Q}_1$ and $\hat{Q}_2$ commute	Example: $\hat{p}$ and $\hat{T}$
$\hat{Q}_1\hat{Q}_2 \neq \hat{Q}_2\hat{Q}_1$	$\hat{Q}_1$ and $\hat{Q}_2$ don't commute	Example: $\hat{p}$ and $\hat{x}$

is the operator that represents the observable (*c.f.*, § 9.9 and 9.10). For example, the expectation value at time t of an observable Q for a state $\Psi(x,t)$ of a one-dimensional system is

$$\langle Q \rangle (t) = \int_{-\infty}^{\infty} \Psi^*(x,t)\hat{Q}\Psi(x,t)\,dx\,. \tag{10.3a}$$

For a three-dimensional system, Eq. (10.3*b*) becomes the three-dimensional integral

$$\langle Q \rangle (t) = \int_{\text{all space}} \Psi^*(x,y,z,t)\hat{Q}\Psi(x,y,z,t)\,dx\,dy\,dz. \tag{10.3b}$$

In §10.2, I'll introduce a general shorthand for the expectation value that subsumes one- and three-dimensional systems. In this "Dirac notation," the expectation value of Q looks like

$$\langle Q \rangle (t) = \langle \Psi \mid \hat{Q} \mid \Psi \rangle. \tag{10.3c}$$

Operators also appear in the fundamental *laws* of quantum theory, *e.g.*, the time-dependent Schrödinger equation (TDSE), which describes the evolution of a quantum state, is the operator equation

$$\hat{\mathcal{H}}\Psi = \hat{\mathcal{E}}\Psi. \tag{10.4a}$$

We shall see in Chap. 11 that operators also play a role in the laws that describe *the evolution of a physical property*, such as position, and that these laws bear a remarkable resemblance to those of Newton's classical mechanics.

Finally, operators define their eigenvalues via *eigenvalue equations.* For example, if the state represented by Ψ in the TDSE (10.4a) is stationary, then the state function has the special form $\Psi = \psi_E e^{-iEt/\hbar}$, and this equation reduces to the time-independent Schrödinger equation (TISE),

$$\hat{\mathcal{H}}\psi_E = E\psi_E. \tag{10.4b}$$

As noted in Chap. 7, the TISE exemplifies an important class of equations called **eigenvalue equations**: ψ_E is the eigenfunction *of the operator* $\hat{\mathcal{H}}$ with eigenvalue E. All operators have eigenvalue equations, and the study of these equations reveals physical conclusions of remarkable generality—statements, for example, that limit our knowledge of the microworld. By studying operators we can probe the mysteries of this world.

10.2 MORE MATHEMATICS OF QUANTUM-MECHANICAL OPERATORS

To flesh out the skeleton of operator algebra we just reviewed, I want to show you a couple of additional features. First we'll discuss *linearity*, an important mathematical property of (nearly all) operators in quantum mechanics. We'll also find out how to manipulate functions and derivatives of operators and, last but not least, will learn a simplified notation that will save fingers and pencils as we venture deeper into the formalism of quantum physics.

Linearity (Superposition in a Mathematical Guise)

In Chap. 6 we related the *principle of superposition* to the TDSE. In particular, we showed that if Ψ_1 and Ψ_2 are (distinct) state functions of a system with Hamiltonian $\hat{\mathcal{H}}$, then for arbitrary constants c_1 and c_2, the linear combination $c_1\Psi_1 + c_2\Psi_2$ satisfies the TDSE:

$$\left(\hat{\mathcal{H}} - \hat{\mathcal{E}}\right)(c_1\Psi_1 + c_2\Psi_2) = c_1\left(\hat{\mathcal{H}} - \hat{\mathcal{E}}\right)\Psi_1 + c_2\left(\hat{\mathcal{H}} - \hat{\mathcal{E}}\right)\Psi_2 = 0. \qquad (10.5)$$

Hence this arbitrary linear combination represents a quantum state of the system. Physically speaking, this is the principle of superposition. Mathematically speaking, it is a manifestation of the *linearity of the operator* $\hat{\mathcal{H}} - \hat{\mathcal{E}}$.

Nearly all operators in quantum mechanics are linear. The definition of linearity is simplicity itself: if f_1 and f_2 are arbitrary functions and c_1 and c_2 arbitrary complex constants, then the operator $\hat{Q}$ is **linear** if

$$\hat{Q}\left(c_1 f_1 + c_2 f_2\right) = c_1\hat{Q}f_1 + c_2\hat{Q}f_2. \qquad \text{linearity} \qquad (10.6)$$

Notice that this definition is really two properties in one:

$$\hat{Q}\left(f_1 + f_2\right) = \hat{Q}f_1 + \hat{Q}f_2 \qquad (10.7a)$$

$$\hat{Q}cf = c\hat{Q}f. \qquad (10.7b)$$

It is easy to verify the linearity or lack of same of a particular operator, as is illustrated by the following little examples.

Example 10.1. Linearity of the Linear Momentum Operator

When we operate on $c_1 f_1(x) + c_2 f_2(x)$ with $\hat{p} = -i\hbar d/dx$, the constants slide right through the first derivative, unchanged. And, since differentiation is distributive, we conclude

$$-i\hbar\frac{d}{dx}\Big[c_1 f_1(x) + c_2 f_2(x)\Big] = c_1\Big[-i\hbar\frac{d}{dx}f_1(x)\Big] + c_2\Big[-i\hbar\frac{d}{dx}f_2(x)\Big], \qquad (10.8)$$

i.e., the operator $\hat{p}$ is linear.

Question 10–1

Prove linearity of $\hat{p}$ *in the momentum representation* by operating with the explicit form of this operator (see Table 10.1) on an arbitrary linear combination of two momentum-space functions $g_1(p)$ and $g_2(p)$.

Aside: Anti-linear operators. Lack of linearity does not necessarily imply that an operator is useless in quantum mechanics; "anti-linear" operators play a role in some branches of our field. An **anti-linear** operator is one that satisfies the properties

$$\hat{Q}\left(f_1 + f_2\right) = \hat{Q}f_1 + \hat{Q}f_2 \qquad (10.9a)$$

$$\hat{Q}cf = c^*\hat{Q}f. \qquad (10.9b)$$

The operator in Example 10.2 is anti-linear, as is the *time-reversal* operator, which is important in collision theory.[4] We will not, however, have occasion to use anti-linear operators in this book. By the way, an operator that is neither linear nor anti-linear is called, sensibly, a "non-linear operator."

Example 10.2. An Anti-linear Operator

Consider the operator that forms the complex conjugate of its operatee:

$$\hat{A}f = f^*. \tag{10.10}$$

This operator satisfies the first condition of linearity, Eq. (10.7*a*), because

$$\hat{A}\left(f_1 + f_2\right) = f_1^* + f_2^* = \hat{A}f_1 + \hat{A}f_2. \tag{10.11a}$$

But it fails the second test, because

$$\hat{A}cf = c^* f^* \neq cf^*. \tag{10.11b}$$

Notice that we cannot salvage this operator by requiring c to be real; the condition (10.7*b*) must hold for a *complex constant*.

Question 10–2

Consider the operator $\hat{A}$ that creates the square of the function on which it operates,

$$\hat{A}f(x) = f^2(x).$$

Is this operator linear, anti-linear, or neither? **Prove** your answer by showing which, if any, of the conditions (10.7) or (10.9) it violates.

Functions of an Operator

In our sojourns through the fields of quantum mechanics, we'll often want to manipulate *operator functions* of varying complexity. Nearly all operator functions can be written in terms of the elementary operators $\hat{x}$ and $\hat{p}$. Operator functions as simple as $\hat{p}^2/(2m)$ or $m\omega_0^2\hat{x}^2/2$ pose no problem. But more formidable concoctions, such as $e^{\hat{Q}}$, are a bit trickier.

We can subdue even the most complicated operator function via its *power-series expansion*:

Rule

Any function $\hat{f}(\hat{A})$ of an operator $\hat{A}$ can be expressed as a power series in $\hat{A}$, provided the power-series expansion of the corresponding *classical function* $f(A)$ converges.[5]

---

[4] See Chap. 6 of *Scattering Theory* by J. R. Taylor (New York: Wiley, 1972).

[5] Although the issue of convergence poses interesting mathematical questions, it is not a problem in practice. If you are interested, see, for example, § 9.1 in *Principles of Quantum Mechanics* by R. Shankar (New York: Plenum, 1980).

According to our prescription for generating operators, the coefficients in the series for the operator $\hat{f}(\hat{A})$ are the very coefficients b_j that appear in the power series expansion of the function $f(A)$,

$$f(A) = \sum_{j=0}^{\infty} b_j A^j. \tag{10.12}$$

So once we know the series expansion of the *function*, Eq. (10.12), we know the expansion of the corresponding *operator*:

$$\hat{f}(\hat{A}) = \sum_{j=0}^{\infty} b_j \hat{A}^j. \tag{10.13}$$

Note that if the function $f(A)$ is real, then the coefficients b_j in the expansion of the operator $\hat{f}(\hat{A})$ are also real.

Example 10.3. The Exponential Operator

To use the operator $e^{\hat{x}}$ we must know its effect on a function of x. To find out, we look to the series expansion of the corresponding exponential *function* e^{x} [*c.f.*, Eq. (9.72)],

$$e^{x} = 1 + x + \frac{1}{2!}x^2 + \frac{1}{3!}x^3 + \cdots \tag{10.14a}$$

$$= \sum_{j=0}^{\infty} b_j x^j \qquad b_j = \frac{1}{j!}. \tag{10.14b}$$

According to the above rule, the operator $e^{\hat{x}}$ is simply

$$e^{\hat{x}} = 1 + \hat{x} + \frac{1}{2!}\hat{x}^2 + \frac{1}{3!}\hat{x}^3 + \cdots \tag{10.15a}$$

$$= \sum_{j=0}^{\infty} b_j \hat{x}^j \qquad b_j = \frac{1}{j!}. \tag{10.15b}$$

Well, that was pretty simple.

Actually, manipulating operators as series expansions is not always this easy. But you're unlikely to go wrong if you remember the **First Golden Rule of Operator Algebra**:

> ***Rule***
>
> To evaluate the product of two or more operators, act with each operator in succession.

[For example, $\hat{Q}^3 f = \hat{Q}\hat{Q}\hat{Q}\, f$.] This rule tells us how to evaluate the terms in the power series expansion (10.13) we must retain for this series to accurately approximate the operator $\hat{f}(\hat{A})$.

The Derivative of an Operator

In Chap. 11 we'll find out how observables in quantum mechanics evolve, a topic that will lead us to expressions involving the time derivative of an operator, $d\hat{Q}(t)/dt$. Happily, such expressions are easy to deal with. The derivative of an operator, which is just another operator, is defined like the derivative of a function. And derivatives of operators obey the familiar rules of differentiation of functions, *e.g.*,

$$\frac{d}{dt}\left[\hat{A}_1(t) + \hat{A}_2(t)\right] = \frac{d}{dt}\hat{A}_1(t) + \frac{d}{dt}\hat{A}_2(t) \tag{10.16a}$$

$$\frac{d}{dt}\left[\hat{A}_1(t)\,\hat{A}_2(t)\right] = \left[\frac{d}{dt}\hat{A}_1(t)\right]\hat{A}_2(t) + \hat{A}_1(t)\left[\frac{d}{dt}\hat{A}_2(t)\right]. \tag{10.16b}$$

There is, however, a trap lurking in Eq. (10.16*b*); can you see it?

We must not blithely interchange the order of the operators in the terms on the right-hand side, for there is no guarantee that $\left[\frac{d}{dt}\hat{A}_1(t)\right]\hat{A}_2(t)$ is the same operator as $\hat{A}_2(t)\left[\frac{d}{dt}\hat{A}_1(t)\right]$. *These operators are equivalent only if they commute.* Thus, in working with derivatives of operators, don't overlook the **Second Golden Rule of Operator Algebra**:

> ***Rule***
>
> Thou shalt not interchange the order of operation of two operators unless thou art sure they commute.

Ubiquitous Integrals (Dirac Braket Notation)

Certain characteristic integrals are woven throughout the fabric of quantum mechanics. Already we've calculated *expectation values*, integrals of the form $\int_{-\infty}^{\infty} \Psi^*(x,t)\hat{Q}\Psi(x,t)\,dx$ [*c.f.*, Eq. (10.3*a*)]. Somewhat similar integrals appeared in our study of non-stationary states of the particle-in-a-box (Example 7.4) and of the simple harmonic oscillator (§ 9.10). In both of these examples we had to cope with *matrix elements*, integrals in which an operator is sandwiched between two *different* state functions, *e.g.*, $\int_{-\infty}^{\infty} \Psi_1^*(x,t)\hat{\mathcal{H}}\Psi_2(x,t)\,dx$. In these examples we also had to evaluate *overlap integrals*, which look like matrix elements but for the absence of an operator, *e.g.*, $\int_{-\infty}^{\infty} \Psi_1^*(x,t)\Psi_2(x,t)\,dx$. All these types of integrals can be easily and conveniently represented using *Dirac braket notation*.

Dirac brakets are just shorthand symbols for integrals. In its most general form, the Dirac braket is defined as

$$\boxed{\langle \Psi_1 \mid \hat{Q}\Psi_2 \rangle \equiv \int_{\text{all space}} \Psi_1^*\hat{Q}\Psi_2\,dv}\,. \tag{10.17}$$

By variously labelling the state functions in (10.17), using the appropriate volume elements and limits of integration, and retaining or omitting the operator $\hat{Q}$, we can accommodate all sorts of overlap integrals, expectation values, and matrix elements:

$$\langle Q \rangle = \langle \Psi_1 \mid \hat{Q}\Psi_1 \rangle = \int_{\text{all space}} \Psi_1^* \hat{Q} \Psi_1 \, dv \qquad \text{expectation value} \tag{10.18a}$$

$$\langle \Psi_1 \mid \hat{Q}\Psi_2 \rangle = \int_{\text{all space}} \Psi_1^* \hat{Q} \Psi_2 \, dv \qquad \text{matrix element} \tag{10.18b}$$

$$\langle \Psi_1 \mid \Psi_2 \rangle = \int_{\text{all space}} \Psi_1^* \Psi_2 \, dv. \qquad \text{overlap integral} \tag{10.18c}$$

The symbol $\langle \Psi_1 \mid \hat{Q}\Psi_2 \rangle$ *represents an integral over all the spatial coordinates in* Ψ, *so these matrix elements, expectation values, and overlap integrals are just numbers. They may be real or complex, and may depend on time.*

The Dirac braket is more "abstract" than the explicit expression for the integral it represents, and therein lies its great utility. This notation is simpler because it leaves several instructions *implicit*. To master Dirac brakets, remember these implicit instructions:

1. Take the complex conjugate of the state function to the left of the leftmost vertical bar.
2. Act with the operator $\hat{Q}$ on the function to its right.
3. Integrate the integrand $\Psi_1^* \hat{Q} \Psi_2$ over all space.

For example, the Dirac braket for the expectation value of the linear momentum of a particle in one dimension in a quantum state represented by $\Psi(x, t)$ is

$$\langle p \rangle (t) = \langle \Psi \mid \hat{p}\Psi \rangle = \int_{-\infty}^{\infty} \Psi^*(x, t) \left[\hat{p}\Psi(x, t)\right] dx. \tag{10.19a}$$

The braket that represents the energy of a particle in three dimensions in state $\Psi(x, y, z, t)$ is

$$\langle E \rangle (t) = \langle \Psi \mid \hat{\mathcal{H}}\Psi \rangle = \iiint \Psi^*(x, y, z, t) \left[\hat{\mathcal{H}}\Psi(x, y, z, t)\right] dx\, dy\, dz. \tag{10.19b}$$

Dirac braket notation is very flexible. It can even accommodate integrals of $(\hat{Q}\Psi_1)^* \Psi_2$, in which the operator acts on the *first* function, Ψ_1. By extension of the definition (10.17), we can represent such an integral by

$$\int_{\text{all space}} \left[\hat{Q}\Psi_1\right]^* \Psi_2 \, dv = \langle \hat{Q}\Psi_1 \mid \Psi_2 \rangle. \tag{10.20}$$

Note that

$$\left[\hat{Q}\Psi_1\right]^* = \hat{Q}^* \Psi_1^*, \tag{10.21}$$

so the integral in (10.20) is

$$\langle \hat{Q}\Psi_1 \mid \Psi_2 \rangle = \int_{\text{all space}} \left[\hat{Q}^* \Psi_1^*\right] \Psi_2 \, dv. \tag{10.22}$$

In practice, Dirac notation is complicated slightly by *the convention of the second vertical bar*. Usually the braket $\langle \Psi_1 \mid \hat{Q}\Psi_2 \rangle$ is written $\langle \Psi_1 \mid \hat{Q} \mid \Psi_2 \rangle$.[6] The second bar should cause no confusion provided you remember that *in general*

$$\boxed{\langle \Psi_1 \mid \hat{Q} \mid \Psi_2 \rangle = \langle \Psi_1 \mid \hat{Q}\Psi_2 \rangle \neq \langle \hat{Q}\Psi_1 \mid \Psi_2 \rangle}. \tag{10.23}$$

There is, however, an extremely important class of operators for which the second and third brakets in (10.23) are equal. If $\hat{Q}$ is a *Hermitian operator*, then the number $\langle \Psi_1 \mid \hat{Q}\Psi_2 \rangle$ is equal to $\langle \hat{Q}\Psi_1 \mid \Psi_2 \rangle$. We'll study Hermitian operators in the next section.

It may seem that the only reason for introducing braket notation is to avoid writer's cramp. Not so. This more abstract notation greatly facilitates deriving results of quantum theory, as you'll see in this and subsequent chapters. Our tools are the simple properties in Table 10.4, which follow from the definition of the Dirac braket, Eq. (10.17). These properties are easy to prove, and doing so will provide useful practice with Dirac brakets. And by memorizing these properties, you can improve your chances of becoming an ace quantum mechanic.

TABLE 10.4 HANDY PROPERTIES OF DIRAC BRAKET NOTATION.

$\langle \Psi_1 \mid c\Psi_2 \rangle = c\langle \Psi_1 \mid \Psi_2 \rangle$	(I)
$\langle c\Psi_1 \mid \Psi_2 \rangle = c^*\langle \Psi_1 \mid \Psi_2 \rangle$	(II)
$\langle \Psi_1 \mid \Psi_2 \rangle^* = \langle \Psi_2 \mid \Psi_1 \rangle$	(III)
$\langle \Psi_1 + \Psi_2 \mid \Psi_3 + \Psi_4 \rangle = \langle \Psi_1 \mid \Psi_3 \rangle + \langle \Psi_1 \mid \Psi_4 \rangle + \langle \Psi_2 \mid \Psi_3 \rangle + \langle \Psi_2 \mid \Psi_4 \rangle$	(IV)

Question 10–3

Prove properties I and III in Table 10.4. Then use these properties to prove Property IV.

10.3 OPERATORS WITH FREEDOM OF CHOICE: HERMITICITY

I hate definitions.

—Benjamin Disraeli
in *Vivian Grey*

Not every mathematical instruction can represent an observable. Just as a complex *function* can represent a quantum state only if it is normalizable, an *operator* can represent a physically measurable quantity only if it satisfies a property called *Hermiticity.*

[6] This convention originates in an abstract treatment of quantum mechanics in which the state we represent by the wave function $\Psi_2(x,t)$ is represented by an abstract beast called a *ket* and denoted $\mid \Psi_2 \rangle$. The state that results when a quantum-mechanical operator $\hat{Q}$ acts on this ket is $\hat{Q} \mid \Psi_2 \rangle$. In this formulation, the Dirac braket is actually the *scalar product* of this ket with another abstract beast that represents the state with wave function Ψ_1. Called a "bra," this critter is denoted by $\langle \Psi_1 \mid$.

Hermiticity Defined

What is Hermiticity? Well, in the last section I made a big deal out of Eq. (10.23), noting that *only for certain operators* $\hat{Q}$ is the matrix element $\langle \Psi_1 \mid \hat{Q}\Psi_2 \rangle$ equal to $\langle \hat{Q}\Psi_1 \mid \Psi_2 \rangle$. Equality of these matrix elements is, in fact, the *definition* of Hermiticity: an operator $\hat{Q}$ is **Hermitian** if *for any two physically-admissible state functions* Ψ_1 and Ψ_2,

$$\boxed{\left\langle \Psi_1 \mid \hat{Q}\Psi_2 \right\rangle = \left\langle \hat{Q}\Psi_1 \mid \Psi_2 \right\rangle = \left\langle \Psi_1 \mid \hat{Q} \mid \Psi_2 \right\rangle} \qquad \text{Hermiticity} \qquad (10.24a)$$

This definition is so important that I want to show it to you in its full integral glory:

$$\int_{\text{all space}} \Psi_1^* \left[\hat{Q}\Psi_2\right] dv = \int_{\text{all space}} \left[\hat{Q}\Psi_1\right]^* \Psi_2 \, dv. \qquad (10.24b)$$

Since the equality in Eqs. (10.24) must hold for *any* two state functions, *Hermiticity is a property of the operator itself*—as we'll see in § 10.4, where we'll liberate this definition from dependence on state functions.[7] Notice, by the way, that the double-bar notation $\langle \Psi_1 \mid \hat{Q} \mid \Psi_2 \rangle$ in (10.24a) makes eminent sense for an Hermitian operator, since such an operator has freedom of choice: it can operate on either Ψ_2 or on Ψ_1; the value of the matrix element is the same in either case.

Question 10–4

In some quantum mechanics texts, you will find this alternate definition of Hermiticity: an operator $\hat{Q}$ is Hermitian if for any physically-admissible state function Ψ,

$$\left\langle \Psi \mid \hat{Q}\Psi \right\rangle = \left\langle \hat{Q}\Psi \mid \Psi \right\rangle. \qquad (10.25)$$

Prove that this definition is equivalent to (10.24*a*) by considering a superposition state whose wave function Ψ is a linear combination of two state functions, *i.e.*,

$$\Psi = c_1\Psi_1 + c_2\Psi_2. \qquad (10.26)$$

What are the implications of Eq. (10.25) for the expectation value of $\hat{Q}$?

As you peer at Eqs. (10.24), trying to decipher the meaning of this abstract, remote definition, you may wonder why *all* operators aren't Hermitian. The answer, perhaps surprisingly, has nothing to do with the complex conjugation in this definition—as you can see from the following example.

[7] Do you remember where you first saw Hermiticity? In Chap. 6 we discovered that the Hamiltonian $\hat{\mathcal{H}}$ of a system is Hermitian. After proving (§ 6.5) that the position probability density of a particle is conserved, I noted [*c.f.*, Eq. (6.82)] that as a consequence the Hamiltonian satisfies

$$\int_{-\infty}^{\infty} \Psi^*(x,t)\hat{\mathcal{H}}\Psi(x,t)\, dx = \int_{-\infty}^{\infty} \left[\hat{\mathcal{H}}\Psi(x,t)\right]^* \Psi(x,t)\, dx,$$

i.e., the Hamiltonian is Hermitian. This finding illustrates how an abstract *mathematical* property, Hermiticity of an operator, is tied to a *physical* property, conservation of probability.

Example 10.4. A Non-Hermitian Operator

What could be simpler than a first derivative? Let's find out whether or not the operator $\hat{A} = \partial/\partial x$ is Hermitian by testing the definition (10.24*b*) for specific states. Let Ψ_1 and Ψ_2 be the wave functions of the ground and first-excited states of the simple harmonic oscillator at $t = 0$ (see Table 9.6). (Note that these functions are real.)

For this case, we can easily evaluate the left- and right-hand sides of Eq. (10.24) with the aid of Table 9.7; to wit:

$$\left\langle \Psi_1 \,\middle|\, \frac{\partial}{\partial x}\Psi_2 \right\rangle = \int_{-\infty}^{\infty} \psi_0(x)\frac{d}{dx}\psi_1(x)\,dx = \frac{\beta}{\sqrt{2}} \tag{10.27a}$$

$$\left\langle \frac{\partial}{\partial x}\Psi_1 \,\middle|\, \Psi_2 \right\rangle = \int_{-\infty}^{\infty} \left[\frac{d}{dx}\psi_0(x)\right]\psi_1(x)\,dx = -\frac{\beta}{\sqrt{2}}. \tag{10.27b}$$

Oh what a difference a minus sign can make! Since (10.27*a*) and (10.27*b*) are different, the operator $\hat{A}$ violates the definition (10.24) and hence is not Hermitian.

Notice that you can *disprove* Hermiticity by finding *a single pair* of functions that violates the definition (10.24). Novice quantum physicists sometimes mistakenly try to *prove* Hermiticity in the same way, *i.e.*, using only a single pair of functions. (For some reason, such attempts usually involve the function e^{ikx}.) So heed well this

Warning

You cannot prove that an operator is Hermitian by showing that the definition (10.24) is satisfied for a *particular* function; you must prove it for an *arbitrary* function.

Example 10.5. An Hermitian Operator

Having shown you an operator that is not Hermitian, it is incumbent upon me to show you one that is. The operator $\hat{p} = -i\hbar\partial/\partial x$ is one of the most important in quantum physics, so we certainly should check that it is legitimate. To do so, we must answer the burning question

$$\langle \Psi_1 \mid \hat{p}\Psi_2 \rangle \stackrel{?}{=} \langle \hat{p}\Psi_1 \mid \Psi_2 \rangle. \tag{10.28a}$$

For a one-dimensional system, the integral form of (10.28*a*) is

$$\int_{-\infty}^{\infty} \Psi_1^*(x,t)\left[-i\hbar\frac{\partial}{\partial x}\Psi_2(x,t)\right]dx \stackrel{?}{=} \int_{-\infty}^{\infty}\left[-i\hbar\frac{\partial}{\partial x}\Psi_1(x,t)\right]^* \Psi_2(x,t)\,dx. \tag{10.28b}$$

We can prove that $\hat{p}$ is Hermitian if we can manipulate the left-hand side of Eq. (10.28*b*) into the form of the right-hand side, *i.e.*, if we can shift the first derivative from $\Psi_2(x,t)$ to $\Psi_1(x,t)$ *without otherwise altering the integrand.* The most straightforward way to do this is to use *integration by parts*, *viz.*,[8]

[8]Remember your freshman calculus? There you learned that the definite integral $\int_a^b u\,dv$ can be written as

$$\int_a^b u\,dv = uv\Big|_a^b - \int_a^b v\,du.$$

To apply this method to Eq. (10.28*a*), we use $u = \Psi_1^*$ and $dv = \dfrac{\partial\Psi_2^*}{\partial x}\,dx$.

$$\int_{-\infty}^{\infty} \Psi_1^*(x,t)\left[-i\hbar\frac{\partial}{\partial x}\Psi_2(x,t)\right] dx$$
$$= -i\hbar\left\{\left[\Psi_1^*(x,t)\Psi_2(x,t)\right]\Big|_{-\infty}^{\infty} - \int_{-\infty}^{\infty} \Psi_2(x,t)\frac{\partial}{\partial x}\Psi_1^*(x,t)\,dx\right\}. \tag{10.29}$$

But $\Psi_1(x,t)$ and $\Psi_2(x,t)$ are normalizable and hence go to zero in the limit $x \to \pm\infty$. Therefore the first term on the right-hand side of Eq. (10.29) is zero, leaving

$$\int_{-\infty}^{\infty} \Psi_1^*(x,t)\left[-i\hbar\frac{\partial}{\partial x}\Psi_2(x,t)\right] dx = (-i\hbar)\left[-\int_{-\infty}^{\infty}\frac{\partial\Psi_1^*}{\partial x}\Psi_2(x,t)\,dx\right] \tag{10.30a}$$

$$= \int_{-\infty}^{\infty}\left[-i\hbar\frac{\partial}{\partial x}\Psi_1(x,t)\right]^* \Psi_2(x,t)\,dx \tag{10.30b}$$

$$= \langle \hat{p}\Psi_1 \mid \Psi_2 \rangle. \tag{10.30c}$$

Ergo, $\hat{p}$ is Hermitian.

Another word of advice: Readers unfamiliar with such mathematical demonstrations may be tempted to try manipulating *both* sides of Eq. (10.28*b*), seeking a common ground. This is almost invariably the *wrong* way to approach such a derivation; success is much more likely if you will try to manipulate one side into the form of the other, as in this example.

Question 10–5

Prove from the definition of Hermiticity that the position operator $\hat{x}$ is Hermitian and that $\hat{x}\hat{p}$ is not Hermitian.

Question 10–6

Prove that $\hat{A} = 3\hat{1}$ (where $\hat{1}$ is the "identity operator," which does nothing) is Hermitian and that $\hat{A} = 3i\hat{1}$ is not Hermitian.

So What? (The Consequences of Hermiticity)

Hermiticity is important in quantum mechanics because

> ***Rule***
>
> All operators in quantum mechanics are Hermitian.

Hermiticity is physically significant because it guarantees that the numbers that quantum mechanics predicts for the results of laboratory measurement will be real, not complex. This result is so important that we're now going to prove it:

> ***Proclamation 10.1***
>
> The expectation values of a Hermitian operator are real.

Argument: To prove this assertion (*c.f.*, Question 10–4), I will call upon a basic fact about complex numbers:[9]

$$\boxed{\text{a number } c \text{ is real if it is equal to its complex conjugate, } c^* = c} \tag{10.31}$$

We want to prove that for a Hermitian operator $\hat{Q}$, the expectation value

$$\langle Q \rangle = \int_{\text{all space}} \Psi^* \hat{Q} \Psi \, dv \tag{10.32}$$

is equal to its complex conjugate. This is not hard to do:

$$\langle Q \rangle^* = \int_{\text{all space}} (\Psi^*)^* \left[\hat{Q}\Psi\right]^* dv \tag{10.33a}$$

$$= \int_{\text{all space}} \Psi \left[\hat{Q}\Psi\right]^* dv \tag{10.33b}$$

$$= \int_{\text{all space}} \left[\hat{Q}\Psi\right]^* \Psi \, dv. \tag{10.33c}$$

But since $\hat{Q}$ is Hermitian, the integral in (10.33c) is equal to

$$\int_{\text{all space}} \Psi^* \hat{Q} \Psi \, dv = \langle Q \rangle . \tag{10.34}$$

Ergo

$$\boxed{\hat{Q} \text{ Hermitian} \quad \Longrightarrow \quad \langle Q \rangle^* = \langle Q \rangle \quad \textit{i.e., } \langle Q \rangle \text{ is real}} \tag{10.35}$$

Q. E. D.

Can you see why Proclamation 10.1 is so important? *The expectation value of any observable is supposed to be measurable*; it's the average of values obtained in an ensemble measurement of the observable. But numbers measured in a laboratory are real, so the expectation value must be real. According to (10.35), this requirement can be satisfied if (and only if) the operator that represents the observable is Hermitian. (In § 10.6 we'll explore further the consequences of Hermiticity.)

Before leaving this proof, I want to repeat it in Dirac notation:

$$\langle Q \rangle^* = \left\langle \Psi \mid \hat{Q}\Psi \right\rangle^* \tag{10.36a}$$

$$= \left\langle \hat{Q}\Psi \mid \Psi \right\rangle \qquad \text{[Property III: Table 10.4]} \tag{10.36b}$$

$$= \left\langle \Psi \mid \hat{Q}\Psi \right\rangle \qquad \text{[Hermiticity of } \hat{Q}\text{]} \tag{10.36c}$$

$$= \langle Q \rangle \qquad \text{[definition (10.32) of } \langle Q \rangle\text{].} \tag{10.36d}$$

[9]It's easy to verify Eq. (10.31). A complex number c can be written in terms of its real and imaginary parts, as $c = a + ib$, where a and b are real numbers. The complex conjugate of c is, by definition, $c^* = a - ib$. So $c^* = c$ only if $b = 0$, *i.e.*, if c is real. (See also Appendix K.)

Notice how much simpler and more elegant this shorthand notation is. Once you become reasonably facile with Dirac brakets, you can whip out proofs like this in your sleep.

Hermitian Operators in Combination

When presented with an operator that purports to represent an observable, the first thing you should do is verify its Hermiticity. Checking the Hermiticity of simple operators, such as the kinetic energy operator $\hat{T} = \hat{p}^2/(2m)$ or the angular momentum operator (in three dimensions) $\hat{\mathbf{L}} = \hat{\mathbf{r}} \times \hat{\mathbf{p}}$, is fairly easy. But not all operators are so simple. As the complexity of the dependence of $\hat{Q}$ on $\hat{x}$ and $\hat{p}$ increases, so does the difficulty of proving it Hermitian. Fortunately, a simple set of general rules answers the question: *what functions of one or more operators are Hermitian?* For reference, I've collected these properties together in Table 10.5. Let's see how to prove them.

TABLE 10.5 FUNCTIONS OF HERMITIAN OPERATORS: ARE THEY HERMITIAN?

For Hermitian Operators $\hat{Q}_1$ and $\hat{Q}_2$		
Operator	Hermitian	Not Hermitian
c = constant	if c is real	if c is complex
$c_1\hat{Q}_1 + c_2\hat{Q}_2$	if c_1, c_2 are real	if c_1, c_2 are complex
$\hat{Q}_1\hat{Q}_2$	if $\left[\hat{Q}_1, \hat{Q}_2\right] = 0$	if $\left[\hat{Q}_1, \hat{Q}_2\right] \neq 0$
$\hat{Q}_1^n$ (n integer)	YES	– – –
commutator[a] $\left[\hat{Q}_1, \hat{Q}_2\right]$	$i\left[\hat{Q}_1, \hat{Q}_2\right]$	$\left[\hat{Q}_1, \hat{Q}_2\right]$
anticommutator[b]	YES	– – –
a function $\hat{f}(\hat{Q}_1)$	if f is real	if f is complex

[a]The commutator is $\left[\hat{Q}_1, \hat{Q}_2\right] \equiv \hat{Q}_1\hat{Q}_2 - \hat{Q}_2\hat{Q}_1$

[b]The "anticommutator" is $[\hat{Q}_1, \hat{Q}_2]_+ \equiv \hat{Q}_1\hat{Q}_2 + \hat{Q}_2\hat{Q}_1$

The simplest operator function is *a function that involves only one operator.* This case is rarely a problem, because

> ***Rule***
>
> Any *real* function of a single Hermitian operator is Hermitian.

This rule takes care of simple operators, such as the potential energy operator of the half-infinite linear potential (§ 9.11), as well as troublemakers, such as $e^{\hat{Q}}$ of Example 10.3. You can prove it by expanding the arbitrary real operator in a power series, à la Eq. (10.13).

Question 10–7

Prove from the definition of Hermiticity that the kinetic and potential energy operators for the simple harmonic oscillator, $\hat{T}$ and $\hat{V}(x) = m\omega_0^2\hat{x}^2/2$, are Hermitian. Isn't it easier to use the properties in Table 10.5?

More interesting conundrums arise if *two or more operators* are involved, for then the Hermiticity of the operator function depends critically on the nature of the constituent operators. For example, $\hat{x}$ and $\hat{p}$ are Hermitian; but not all functions of $\hat{x}$ and $\hat{p}$ are Hermitian—a point illustrated in Table 10.6 and in the following example.

TABLE 10.6 HERMITIAN AND NON-HERMITIAN COMBINATIONS OF $\hat{x}$ AND $\hat{p}$.

Hermitian	Non-Hermitian
$\hat{x}$, $\hat{p}$	$\hat{x}\hat{p}$, $\hat{p}\hat{x}$
$i\,[\hat{x}, \hat{p}]$	$[\hat{x}, \hat{p}]$

Example 10.6. The Product of Two Hermitian Operators

Let $\hat{Q}_1$ and $\hat{Q}_2$ be Hermitian operators; *i.e.*, suppose that for arbitrary physically-admissible state functions Ψ_1 and Ψ_2,

$$\langle \Psi_1 \mid \hat{Q}_1 \mid \Psi_2 \rangle = \langle \Psi_2 \mid \hat{Q}_1 \mid \Psi_1 \rangle \tag{10.37a}$$

$$\langle \Psi_1 \mid \hat{Q}_2 \mid \Psi_2 \rangle = \langle \Psi_2 \mid \hat{Q}_2 \mid \Psi_1 \rangle. \tag{10.37b}$$

Our goal is to find the conditions (if any) under which the product operator $\hat{Q}_3 \equiv \hat{Q}_1\hat{Q}_2$ is Hermitian; *i.e.*,

$$\int_{-\infty}^{\infty} \Psi_1^*(x,t)\hat{Q}_3\Psi_2(x,t)\,dx \stackrel{?}{=} \int_{-\infty}^{\infty} \left[\hat{Q}_3\Psi_1(x,t)\right]^* \Psi_2(x,t)\,dx, \tag{10.38a}$$

or, in Dirac Notation

$$\langle \Psi_1 \mid \hat{Q}_3\Psi_2 \rangle \stackrel{?}{=} \langle \hat{Q}_3\Psi_1 \mid \Psi_2 \rangle. \tag{10.38b}$$

We can explore this question using the Hermiticity of $\hat{Q}_1$ and $\hat{Q}_2$ and the meaning of a product of two operators:

$$\int_{-\infty}^{\infty} \Psi_1^*(x,t)\hat{Q}_3\Psi_2(x,t)\,dx = \int_{-\infty}^{\infty} \Psi_1^*(x,t)\left\{\hat{Q}_1\left[\hat{Q}_2\Psi_2(x,t)\right]\right\}dx \tag{10.39a}$$

$$= \int_{-\infty}^{\infty} \left[\hat{Q}_1\Psi_1(x,t)\right]^*\hat{Q}_2\Psi_2(x,t)\,dx \tag{10.39b}$$

$$= \int_{-\infty}^{\infty} \left[\hat{Q}_2\hat{Q}_1\Psi_1(x,t)\right]^*\Psi_2(x,t)\,dx. \tag{10.39c}$$

Notice carefully that I had to put $\hat{Q}_2$ before $\hat{Q}_1$ in Eq. (10.39c). The expression in (10.39c) is *not* equal to the right-hand side of Eq. (10.38) unless $\hat{Q}_2\hat{Q}_1 = \hat{Q}_1\hat{Q}_2$, *i.e.*, unless these Hermitian operators commute.

We have discovered another rule:

> ***Rule***
>
> The product of two Hermitian operators is Hermitian if and only if the two operators commute.

It is convenient to express this finding in terms of the **commutator** of $\hat{Q}_1$ and $\hat{Q}_2$, which is defined in Eq. (10.2*a*):[10]

$$\boxed{\hat{Q}_3 = \hat{Q}_1\hat{Q}_2 \text{ is Hermitian} \iff \left[\hat{Q}_1, \hat{Q}_2\right] = 0} \tag{10.40}$$

Question 10–8

Under what conditions on three Hermitian operators $\hat{Q}_1$, $\hat{Q}_2$, and $\hat{Q}_3$ is the product $\hat{Q}_1\hat{Q}_2\hat{Q}_3$ Hermitian? Can you generalize your result to the product of N Hermitian operators?

We have already evaluated the commutator of $\hat{x}$ and $\hat{p}$ in Example 5.4,

$$\left[\hat{x}, \hat{p}\right] = \hat{x}\hat{p} - \hat{p}\hat{x} = i\hbar\hat{1}. \tag{10.41}$$

Since $\left[\hat{x}, \hat{p}\right] \neq 0$, the product $\hat{x}\hat{p}$ is not Hermitian (see Question 10–5).

Question 10–9

The **anti-commutator** of two operators $\hat{Q}_1$ and $\hat{Q}_2$ is defined as

$$[\hat{Q}_1, \hat{Q}_2]_+ \equiv \hat{Q}_1\hat{Q}_2 + \hat{Q}_2\hat{Q}_1. \qquad \text{anti-commutator} \tag{10.42}$$

Show that the *anti-commutator* of $\hat{x}$ and $\hat{p}$ is Hermitian. What quantum-mechanical operator should be used to represent the classical observable xp?

Here's what the above proof looks like in Dirac notation [see Eq. (10.38*b*)]:

$$\left\langle \Psi_1 \,\middle|\, \hat{Q}_1\hat{Q}_2\Psi_2 \right\rangle = \left\langle \hat{Q}_1\Psi_1 \,\middle|\, \hat{Q}_2\Psi_2 \right\rangle \qquad [\text{Hermiticity of } \hat{Q}_1] \tag{10.43a}$$

$$= \left\langle \hat{Q}_2\hat{Q}_1\Psi_1 \,\middle|\, \Psi_2 \right\rangle \qquad [\text{Hermiticity of } \hat{Q}_2] \tag{10.43b}$$

$$= \left\langle \hat{Q}_1\hat{Q}_2\Psi_1 \,\middle|\, \Psi_2 \right\rangle \qquad [\text{if } \left[\hat{Q}_1, \hat{Q}_2\right] = 0]. \tag{10.43c}$$

So much for the product of two Hermitian operators. But other combinations of these operators are possible, *e.g.*, their sum.

Example 10.7. Linear Combinations of Hermitian Operators

Suppose $\hat{Q}_1$ and $\hat{Q}_2$ are Hermitian operators and c_1 and c_2 are complex constants. Is $\hat{Q}_3 \equiv c_1\hat{Q}_1 + c_2\hat{Q}_2$ Hermitian? Go ahead, take a guess.

***** Pause while reader ponders *****

Now let's find the answer.

[10]The symbol $\iff$ is mathematicians' notation for "if and only if," also known in the trade as "iff."

All we need to answer this question (using Dirac notation) are the properties in Table 10.4. By Property IV, we have

$$\left\langle \Psi_1 \,\middle|\, \left(c_1\hat{Q}_1 + c_2\hat{Q}_2\right)\Psi_2\right\rangle = \left\langle \Psi_1 \,\middle|\, c_1\hat{Q}_1\Psi_2\right\rangle + \left\langle \Psi_1 \,\middle|\, c_2\hat{Q}_2\Psi_2\right\rangle. \tag{10.44}$$

Using the Hermiticity of $\hat{Q}_1$ and $\hat{Q}_2$, the right-hand side of (10.44) becomes

$$= \langle \hat{Q}_1\Psi_1 \mid c_1\Psi_2\rangle + \langle \hat{Q}_2\Psi_1 \mid c_2\Psi_2\rangle. \tag{10.45}$$

Moving the constant to the left of each bar, we obtain

$$\left\langle \Psi_1 \,\middle|\, \left(c_1\hat{Q}_1 + c_2\hat{Q}_2\right)\Psi_2\right\rangle = \left\langle c_1^*\hat{Q}_1\Psi_1 \,\middle|\, \Psi_2\right\rangle + \left\langle c_2^*\hat{Q}_2\Psi_1 \,\middle|\, \Psi_2\right\rangle \tag{10.46a}$$

$$= \left\langle \left(c_1^*\hat{Q}_1 + c_2^*\hat{Q}_2\right)\Psi_1 \,\middle|\, \Psi_2\right\rangle. \tag{10.46b}$$

[Notice carefully that I had to take the complex conjugate of the constants when I moved them. If you don't understand why, write out Eq. (10.45) in its full integral form and derive from it Eq. (10.46a).] But the operator to the left of the bar in Eq. (10.46b) is not equal to $\hat{Q}_3$ unless c_1 and c_2 are real. So only in this special case is $\hat{Q}_3$ Hermitian:

Rule

An arbitrary linear combination of two (or more) Hermitian operators is Hermitian if and only if the constants in the linear combination are real numbers,

$$\hat{Q}_3 = c_1\hat{Q}_1 + c_2\hat{Q}_2 \text{ is Hermitian} \iff c_1 \text{ and } c_2 \text{ are real} \tag{10.47}$$

Question 10–10

Repeat the proof of Eq. (10.47) using the full integral form of the matrix elements involved.

To show you how to check Hermiticity using the rules in Table 10.6, I'll work one final example.

Example 10.8. The Kinetic Energy Operator

The kinetic energy operator

$$\hat{T} = \frac{\hat{p}^2}{2m} \tag{10.48}$$

is the product of a real constant and the operator $\hat{p}^2$. Since $\hat{p}$ is Hermitian, so is any power of $\hat{p}$. And the product of a real constant times a Hermitian operator is another Hermitian operator. So, with little difficulty and without explicit recourse to the definition (*c.f.*, Question 10–7), we confidently conclude that $\hat{T}$ is a Hermitian operator.

Question 10–11

Prove that the Morse potential [see Figs. 8.2 and 9.14]

$$V(x) = D\left(1 - e^{-a(x-x_e)}\right)^2 \qquad a = \text{a real constant}$$

is Hermitian.

Question 10–12

Prove that the Hamiltonian of a particle in the Morse potential of Question 10–11, $\hat{\mathcal{H}} = \hat{T} + \hat{V}$, is Hermitian.

Question 10–13

Consider a system with a Hamiltonian $\hat{\mathcal{H}}$ in a state represented by Ψ. The probability of finding the particle anywhere in space is the *integrated probability* [see § 3.4 and Eq. (6.64)]

$$P\big([-\infty,\infty],t\big) \equiv \langle \Psi \mid \Psi \rangle. \tag{10.49}$$

Suppose the potential energy in $\hat{\mathcal{H}}$ were a complex function. Would the total probability Eq. (10.49) still be conserved? Why or why not? Can you think of any conceivable application of this exercise?

Question 10–14

Use the properties of the Dirac braket in Table 10.4 to prove that for arbitrary physically-admissible state functions Ψ_1 and Ψ_2,

$$\hat{Q}\ \text{Hermitian} \quad \Longrightarrow \quad \langle \Psi_1 \mid \hat{Q} \mid \Psi_2 \rangle = \langle \Psi_2 \mid \hat{Q} \mid \Psi_1 \rangle^*. \tag{10.50}$$

From this result prove Proclamation 10.1.

10.4 FUN WITH ADJOINTS

> Do not imagine
> that mathematics is hard and crabbed,
> and repulsive to common sense.
> It is merely
> the etherealization of common sense.
>
> —William Thompson (Lord Kelvin)
> 1824–1907

In § 10.3, I asserted that Hermiticity is a property of *operators*, unrelated to the state functions on which they operate. Yet, I defined this property in Eq. (10.24*a*) in terms of state functions Ψ_1 and Ψ_2. If Hermiticity pertains solely to an *operator*, then we should be able to define it independently of state functions. Mathematicians know how to do this, using a quaint and curious beastie called "the adjoint."

What is an adjoint? Well, consider the following whimsical puzzle: Suppose $\hat{A}$ is an arbitrary operator, not necessarily Hermitian. For any two state functions construct the matrix element

$$\langle \Psi_1 \mid \hat{A}\Psi_2 \rangle = \int_{\text{all space}} \Psi_1^* \big[\hat{A}\Psi_2\big]\, dv. \tag{10.51}$$

This matrix element is just a number. **Find** another operator, say $\hat{B}$, such that the number $\langle \hat{B}\Psi_1 \mid \Psi_2 \rangle$ equals $\langle \Psi_1 \mid \hat{A}\Psi_2 \rangle$, *i.e.*, the operator $\hat{B}$ is defined by

$$\langle \hat{B}\Psi_1 \mid \Psi_2 \rangle = \int_{\text{all space}} \big[\hat{B}\Psi_1\big]^* \Psi_2\, dv = \langle \Psi_1 | \hat{A}\Psi_2 \rangle. \tag{10.52}$$

The desired operator is called the *adjoint* of $\hat{A}$, denoted $\hat{A}^\dagger$, and I'm going to show you how to construct it.[11] That is, the **adjoint** of $\hat{A}$ is the operator $\hat{A}^\dagger$ such that

$$\boxed{\langle \hat{A}^\dagger \Psi_1 \mid \Psi_2 \rangle = \langle \Psi_1 \mid \hat{A} \Psi_2 \rangle} \qquad \text{adjoint.} \tag{10.53}$$

Example 10.9. The Adjoint of the First Derivative Operator

In Example 10.4 we demonstrated that the operator $\hat{A} = \partial/\partial x$ is not Hermitian. To find the adjoint of this operator we use integration by parts (*c.f.*, Example 10.5). For a one-dimensional system, the matrix element we must juggle is

$$\langle \Psi_1 \mid \hat{A} \Psi_2 \rangle = \int_{-\infty}^{\infty} \Psi_1^*(x,t) \frac{\partial}{\partial x} \Psi_2(x,t)\, dx \tag{10.54a}$$

$$= \left[\Psi_1^* \Psi_2 \right] \Big|_{-\infty}^{\infty} - \int_{-\infty}^{\infty} \left[\frac{\partial}{\partial x} \Psi_1^*(x,t) \right] \Psi_2(x,t)\, dx. \tag{10.54b}$$

Since the first term on the right-hand side of Eq. (10.54*b*) is zero (*why?*), we must grapple with only the second term. To identify the adjoint of $\hat{A}$, we must write this term so it looks like the left-hand side of (10.53):

$$\langle \hat{A}^\dagger \Psi_1 \mid \Psi_2 \rangle = \int_{-\infty}^{\infty} \left[-\frac{\partial}{\partial x} \Psi_1^*(x,t) \right] \Psi_2(x,t)\, dx. \tag{10.55}$$

This manipulation unmasks the adjoint:

$$\hat{A} = \frac{\partial}{\partial x} \qquad \Longrightarrow \qquad \hat{A}^\dagger = -\frac{\partial}{\partial x}, \tag{10.56}$$

which is consistent with Example 10.4.

Question 10–15

Consider $\hat{A} = \hat{c}$, where c is a *complex* number. Prove that $\hat{A}^\dagger = \hat{c}^*$.

The Adjoint of a Hermitian Operator

A Hermitian operator has a particularly simple adjoint, as you should be able to immediately deduce from the definition of Hermiticity (10.24*a*) and of the adjoint (10.53):[12]

> ***Rule***
>
> A Hermitian operator is its own adjoint, and, conversely, an operator that equals its own adjoint is Hermitian; *i.e.*,
>
> $$\hat{Q} \text{ Hermitian} \qquad \Longleftrightarrow \qquad \hat{Q}^\dagger = \hat{Q}. \tag{10.57}$$

[11] Mathematicians also refer to the adjoint of an operator as its "Hermitian conjugate" *whether or not the operator in question is Hermitian.* To avoid the obvious source of confusion, I'm going to use only the term "adjoint."

[12] Equation (10.57) explains why mathematicians often refer to a Hermitian operator as "self-adjoint."

So *one way to determine whether or not a particular operator is Hermitian is to find its adjoint and just look at it.* But, to judge from Example 10.9, this scheme is no simpler than direct application of the definition of Hermiticity. Happily, the determination of adjoints is rendered quite easy by a host of general properties, the most important of which appear in Table 10.7.

TABLE 10.7 OPERATORS AND THEIR ADJOINTS.

Operator	Adjoint
$\hat{A}$	$\hat{A}^\dagger$
$c_1\hat{A}_1 + c_2\hat{A}_2$	$c_1^*\hat{A}_1^\dagger + c_2^*\hat{A}_2^\dagger$
$\hat{A}\hat{B}$	$\hat{B}^\dagger\hat{A}^\dagger$
$\left[\hat{A},\hat{B}\right]$	$\left[\hat{A}^\dagger,\hat{B}^\dagger\right]$
$\left[\hat{A},\hat{B}\right]_+$	$\left[\hat{A},\hat{B}\right]_+$
$\hat{A}^\dagger\hat{A}$	$\hat{A}^\dagger\hat{A}$

Question 10–16

Prove that if $\hat{A}$ is linear, so is $\hat{A}^\dagger$.

Question 10–17

Prove from the definition of the adjoint that $\left(\hat{A}\hat{B}\right)^\dagger = \hat{B}^\dagger\hat{A}^\dagger$ and derive the adjoint of the anticommutator $\left[\hat{A},\hat{B}\right]_+$.

Example 10.10. The Position-Momentum Product Reconsidered

We can ascertain whether the operator $\hat{x}\hat{p}$ is Hermitian by inspecting its adjoint. With Table 10.7 at hand, evaluating the adjoint of $\hat{x}\hat{p}$ is a snap:

$$(\hat{x}\hat{p})^\dagger = \hat{p}^\dagger\,\hat{x}^\dagger \tag{10.58a}$$

$$= \hat{p}\,\hat{x}, \tag{10.58b}$$

where in the last step I used the Hermiticity of $\hat{x}$ and of $\hat{p}$. Since $\hat{x}$ and $\hat{p}$ don't commute, $\hat{p}\hat{x} \neq \hat{x}\hat{p}$, and we conclude that the operator $\hat{x}\hat{p}$ is not Hermitian.

Question 10–18

Determine the adjoint of the operator $\hat{x}\hat{T}$, where $\hat{T}$ is the kinetic energy operator (10.48). Is $\hat{x}\hat{T}$ Hermitian?

The adjoint provides a useful (albeit abstract) tool for studying Hermiticity and manipulating operators. But the real reason for introducing all these tools is to study the physics of operators in quantum mechanics. To a great extent, the importance of these operators rests in their eigenvalue equations, to which we now turn.

"Is there any other point to which
you would wish to draw my attention?"
"To the curious incident
of the dog in the night-time."
"The dog did nothing in the night-time."
"That was the curious incident,"
remarked Sherlock Holmes.

—*Silver Blaze*
by Sir Arthur Conan Doyle

An operator can wreak considerable havoc, transforming the function on which it acts into a quite different function. Consider, for example, the effect of the linear momentum operator on the initial wave function for a simple harmonic oscillator in its first excited (stationary) state (Table 9.6):

$$\hat{p}\Psi_1(x,0) = -i\hbar\frac{d}{dx}\left[2A_1\beta x e^{-\beta^2 x^2/2}\right] \tag{10.59}$$

$$= -2i\hbar A_1 \beta e^{-\beta^2 x^2/2}\left(1 - \beta^2 x^2\right), \tag{10.60}$$

where A_1 is the normalization constant of Eq. (9.106). Observe that the dependence on x of the function $\hat{p}\Psi_1(x,0)$ is quite different from that of $\Psi_1(x,0)$.

But something strange happens when the *system Hamiltonian* operates on this function: we get

$$\hat{\mathcal{H}}\Psi_1(x,0) = \hat{\mathcal{H}}\psi_1(x) = E_1\psi_1(x). \tag{10.61}$$

There are two reasons why the Hamiltonian operator gives back *the same function* times a constant: first, $\Psi_1(x,0) = \psi_1(x)$ is a stationary-state wave function; and second, Eq. (10.61) is just the time-independent Schrödinger Equation (TISE). As noted in Chap. 7, the TISE is an *eigenvalue equation*, and $\Psi_1(x,0) = \psi_1(x)$ is an *eigenfunction* of the Hamiltonian with *eigenvalue* E_1.[13]

Eigenvalue Equations on Parade

The form of the eigenvalue equation for an operator $\hat{A}$ with **eigenfunction** χ_a and **eigenvalue** a is

$$\hat{A}\chi_a = a\chi_a. \qquad \text{eigenvalue equation} \tag{10.62}$$

The collection of *all* the eigenvalues of $\hat{A}$ is called the **spectrum** of this operator. The number of eigenvalues in the spectrum may be (and usually is) infinite. And these values may be discrete, continuous, or both. For example, the spectrum of the Hamiltonian of an infinite square well—the energies obtained by solving the TISE—is discrete (§ 7.4); that of the Hamiltonian for a barrier potential (§ 8.6) is continuous; and that of the finite square well (§ 8.8) includes both discrete and continuous eigenvalues.

A notational note: when discussing *discrete* eigenvalues, we may label the eigenfunction by the eigenvalue, as in (10.62), or by an index j, *e.g.*,

[13]The word "eigenvalue" is of German origin and means "proper value," "self value," or "characteristic value." In fact, some mathematicians refer to eigenvalue equations as "characteristic equations."

$$A\chi_j = a_j\chi_j. \qquad \text{[discrete eigenvalues only]} \tag{10.63}$$

Continuum eigenvalues are their own index—*i.e.*, they are used to label their eigenfunctions, as in Eq. (10.62).

Example 10.11. A Cornucopia of Eigenvalue Equations

In our journey through quantum physics we have espied several eigenvalue equations other than the TISE. While making the acquaintance of the *linear momentum operator* (Chap. 5), we solved the equation

$$\hat{p}\chi_p = p\chi_p \tag{10.64}$$

for the eigenfunction

$$\chi_p(x) = \frac{1}{\sqrt{2\pi\hbar}}e^{ipx/\hbar}. \tag{10.65}$$

Recall that the prefactor multiplying this eigenfunction ensures that it will satisfy *Dirac Delta Function Normalization*, *i.e.*,

$$\langle \chi_{p'} \mid \chi_p \rangle = \delta(p' - p). \tag{10.66}$$

The eigenvalue equation (10.64) has only continuum eigenvalues, as does the corresponding equation for the position operator,

$$\hat{x}\chi_{x_0}(x) = x_0\chi_{x_0}(x). \tag{10.67a}$$

The Dirac Delta function shows up in a most peculiar way in the eigenvalue equation for the *position operator*: it turns out that $\delta(x - x_0)$ is the eigenfunction of this operator with eigenvalue x_0, *i.e.*,

$$\hat{x}\delta(x - x_0) = x_0\delta(x - x_0). \tag{10.67b}$$

We'll return to the ramifications of this odd result in Chap. 13. I should note, though, that it makes a screwy kind of sense: were the state function of an ensemble of particles an eigenfunction of position, then the location of the members of the ensemble would be well-defined (*i.e.*, sharp). This physical situation is reflected by the eigenfunction in (10.67*a*): out of all possible locations, the Dirac Delta function $\delta(x - x_0)$ "selects" x_0 (recall the discussion of this function in Chap. 4 and Appendix L). [Don't forget, though, that because $\delta(x - x_0)$ can't be normalized, it's not physically admissible.]

We have also glimpsed the eigenvalue equation for the *parity operator* in our studies of bound-state eigenfunctions of symmetric potentials [*c.f.*, § 8.7]. *The parity operator, which we denote by* $\hat{\Pi}$, *inverts the spatial coordinates of a function through the origin, taking* x *into* $-x$:

$$\hat{\Pi}f(x) = f(-x). \qquad \text{parity operator} \tag{10.68}$$

The simple spectrum of $\hat{\Pi}$ includes only *two discrete eigenvalues*, $+1$ (for an even function) and -1 (for an odd function):

$$\boxed{\hat{\Pi}f(x) = \begin{cases} +f(x) & \text{if } f(x) \text{ is even} \\ -f(x) & \text{if } f(x) \text{ is odd} \end{cases}} \qquad \begin{array}{c}\text{eigenfunctions} \\ \text{of the parity operator}\end{array} \tag{10.69}$$

In § 10.8 we'll prove that these are the *only* eigenvalues of the parity operator and will scrutinize the eigenfunctions of this important operator.

What to Do in Case of Degeneracy

The labelling of eigenfunctions becomes a bit more complicated for degenerate eigenvalues. If, to an eigenvalue q, there corresponds *one and only one eigenfunction* χ_q, then this eigenvalue is **non-degenerate**. This parlance is an obvious extension of our use in Chap. 9 of the word "non-degenerate" to describe energy eigenvalues. In § 9.4 (Proclamation 9.1) we proved that the bound-state eigenfunctions of a one-dimensional system are unique, so their energies are non-degenerate. In fact, this proof illustrates a general principle:

> ***Rule***
>
> The discrete eigenvalues of any operator that represents an observable of a single-particle system in one dimension are non-degenerate.

This principle does not hold for two- and three-dimensional systems; nor does it pertain to many-particle systems. The bound-state energies of the (one-electron) hydrogen atom, for example, are degenerate: to each eigenvalue E_n of the H atom Hamiltonian (where n is an integer $n \geq 1$) there correspond $2n^2$ distinct eigenfunctions. (Remember that by "distinct" we mean linearly independent.) The jargon appropriate to this situation is: the bound-state energies of the H atom are "$2n^2$-fold degenerate."

Eigenstates

In quantum mechanics, we are particularly interested in eigenvalue equations for the state function, *e.g.*,[14]

$$\hat{Q}\Psi_q = q\Psi_q. \qquad \text{eigenstate} \tag{10.70}$$

Such a state is so important we give it a special name: an *eigenstate*:

$$\boxed{\Psi \text{ an eigenfunction of } \hat{Q} \implies \text{system in an } \textbf{eigenstate} \text{ of the observable } Q} \tag{10.71}$$

There is, of course, no guarantee that the state represented by Ψ is an eigenstate of *any* operator. But for most states of most systems, we can find one (or more) operators that act on Ψ to produce this function times a constant, the eigenvalue. So remember the following:

[14] I have purposely left unspecified the spatial variables and the time in this equation (and in many others in this chapter) to emphasize the generality of abstract equations such as (10.70). In one dimension, this eigenvalue equation looks like

$$\hat{Q}\Psi(x,t) = q\Psi(x,t).$$

It is important to realize that unless $\hat{Q}$ *explicitly* contains the time or derivatives thereof, it is the *coordinate dependence* of Ψ that is important in its eigenvalue equation. For this reason, many authors write this equation as

$$\hat{Q}\Psi_t(x) = q\Psi_t(x).$$

I'm using subscripts for other things, so I won't use this convention.

Warning

Equation (10.71) must hold *at all times* if the state represented by Ψ is an eigenstate of $\hat{Q}$; the system is *not* in an eigenstate if this equation is (accidentally) satisfied at one (or more) particular times.

The most familiar example of an eigenstate is a stationary state, which is an *energy eigenstate*. To see why, we need only recall that the wave function of such a state is separable,

$$\Psi_E = \psi_E e^{-iEt/\hbar}. \tag{10.72}$$

Hence this *state function is an eigenfunction of the Hamiltonian*,

$$\hat{\mathcal{H}}\Psi_E = E\Psi_E, \tag{10.73}$$

which means that *the state is an eigenstate of the energy.*

Question 10–19

Prove that a simple harmonic oscillator in its (stationary) ground state is not in an eigenstate of linear momentum.

A system can be (and, in fact, usually is) in an eigenstate of more than one observable. All that is required is that the state function be an eigenfunction of both of the corresponding operators, *e.g.*,

$$\begin{aligned} \hat{Q}_1\Psi_{q_1,q_2} &= q_1\Psi_{q_1,q_2} \\ \hat{Q}_2\Psi_{q_1,q_2} &= q_2\Psi_{q_1,q_2}. \end{aligned} \qquad \text{simultaneous eigenstate of } Q_1 \text{ and } Q_2 \tag{10.74}$$

As the label on Eq. (10.74) attests, we call such a state a **simultaneous eigenstate** of the observables.[15]

For example, the stationary states of the *symmetric finite square well* are eigenstates of energy,

$$\hat{\mathcal{H}}\Psi_n(x,t) = E_n\Psi_n(x,t), \tag{10.75a}$$

and of parity

$$\hat{\Pi}\Psi_n(x,t) = (-1)^{n+1}\Psi_n(x,t). \tag{10.75b}$$

(Remember that the ground state of this system is even and is labelled by $n = 1$.) Thus, these states are simultaneous eigenstates of energy and parity. The subject of simultaneous eigenstates will come into its own in Chap. 11.

Question 10–20

Are the stationary states of an *asymmetric* square well simultaneous eigenstates of energy and parity? ... of energy and any other operator you can think of?

[15]There is more to this remark than meets the eye. A quantum system can be characterized, in a very abstract way, by the set of *all* operators of which the state function is an eigenfunction. Such a set is said to be "complete" (a word we'll meet in a different context in § 10.7). Since these operators all commute with one another, this set is called the "complete set of commuting operators" of the system—its CSCO. For a succinct introduction to the CSCO, see § 4.6 of *A Quantum Mechanics Primer* by Daniel T. Gillespie (London: International Textbook, 1970).

Question 10–21

In Chap. 4 we studied the pure-momentum function of the free particle, $\Psi(x,t) = Ae^{-i(px-Et)/\hbar}$. Of what two operators is this a simultaneous eigenfunction? Why is such a simultaneous eigenstate not seen in the laboratory?

Eigenstates and Measurement

Quantum mechanics care whether or not a system is in an eigenstate of an observable because of the consequences for *measurement of the observable.* For example, as we learned in Chap. 7 and saw illustrated in Chaps. 8 and 9, *in an energy eigenstate (10.72) the energy is sharp.* In statistical terms, this statement means that *the average value of energy is equal to the eigenvalue of $\hat{\mathcal{H}}$, and the uncertainty in the energy is zero:*

$$\Psi_E = \psi_E e^{-iEt/\hbar} \qquad \Longrightarrow \qquad \begin{cases} \langle E \rangle (t) = E \\ \Delta E(t) = 0 \end{cases} . \qquad \text{[energy eigenstate]} \qquad (10.76)$$

This is *not* merely an abstract mathematical result. It means that if you wander into your lab and perform an ensemble measurement of the energy of a particle that's in an energy eigenstate, you'll find that all members exhibit the same value, the eigenvalue E. Were the ensemble in a non-stationary state, however, the results of your measurement would be quite different (a situation we'll confront in Chap 12).

Equation (10.76) suggests a provocative connection between the eigenvalue equations for a state function and the results of measurement of an observable Q, a connection shown schematically in Fig. 10.1. The proof of this connection follows rather simply from the definitions of $\langle Q \rangle$ and ΔQ.

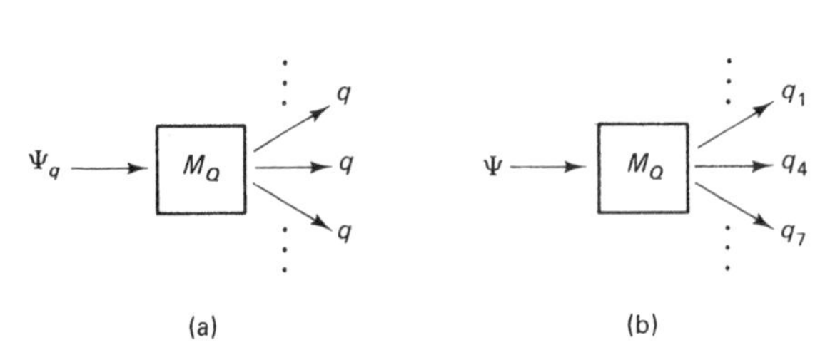

Figure 10.1 A symbolic representation of measurement M_Q of an observable Q on a quantum system initially in (a) an eigenstate of Q represented by state function Ψ_q; (b) a state Ψ that is not an eigenstate of Q. The box in each figure denotes the measurement apparatus, and the arrows to the right of the box show the results exhibited by a few of the huge number of members in the ensemble on which the measurement is performed.

> ***Proclamation 10.2***
>
> In an eigenstate of an observable, the uncertainty of this observable is zero, and its expectation value is equal to the eigenvalue of the operator that represents the observable.

Argument: In mathematical language, this conjecture reads

$$\hat{Q}\Psi_q = q\Psi_q \Longleftrightarrow \begin{cases} \Delta Q(t) = 0 \\ \langle Q \rangle (t) = q \end{cases} . \qquad (10.77)$$

We can prove both parts of (10.77) by deft use of the eigenvalue equation $\hat{Q}\Psi_q = q\Psi_q$. Thus, the expectation value

$$\langle Q \rangle (t) = \langle \Psi_q \mid \hat{Q} \mid \Psi_q \rangle = \langle \Psi_q \mid \hat{Q}\Psi_q \rangle \qquad (10.78a)$$

becomes

$$\langle Q\rangle(t) = \langle \Psi_q \mid q\Psi_q\rangle = q\langle \Psi_q \mid \Psi_q\rangle. \tag{10.78b}$$

The state function is normalizable, so $\langle \Psi_q \mid \Psi_q\rangle = 1$, and the right-hand side of (10.78*b*) is just q. Therefore

$$\boxed{\langle Q\rangle(t) = q} \qquad \text{[in an eigenstate of } Q \text{ with eigenvalue } q] \tag{10.79}$$

We'll use the same strategy to prove that $\Delta Q(t) = 0$. We begin with the square of this quantity,

$$(\Delta Q)^2 = \langle Q^2\rangle - \langle Q\rangle^2 \tag{10.80a}$$

$$= \langle Q^2\rangle - q^2. \qquad \text{[by Eq. (10.79)]} \tag{10.80b}$$

And we can easily evaluate the remaining expectation value, $\langle Q^2\rangle$, by repeated application of the eigenvalue equation and another judicious application of normalizability:

$$\langle Q^2\rangle = \langle \Psi_q \mid \hat{Q}\hat{Q}\Psi_q\rangle \tag{10.81a}$$

$$= q\langle \Psi_q \mid \hat{Q}\Psi_q\rangle \qquad \text{[by the eigenvalue equation]} \tag{10.81b}$$

$$= q^2\langle \Psi_q \mid \Psi_q\rangle \qquad \text{[by the eigenvalue equation]} \tag{10.81c}$$

$$= q^2. \qquad \text{[by normalizability of } \Psi] \tag{10.81d}$$

Substituting this result into Eq. (10.80*b*), we find, not surprisingly, that $(\Delta Q)^2 = 0$, whence

$$\boxed{\Delta Q = 0} \qquad \text{[in an eigenstate of } Q \text{ with eigenvalue } q] \tag{10.82}$$

Q. E. D.

Proclamation 10.2 assures us that we can be certain (within the limits of experimental accuracy) of the outcome of a measurement of an observable Q *on a system that is in an eigenstate of* Q: in such a measurement, all members yield the same value, the eigenvalue q.

Proclamation 10.2 also shows how to determine experimentally whether or not a system is in an eigenstate of a particular observable: we just perform an ensemble measurement of the observable and see if all members give the same value. If so, the system is in an eigenstate. If, instead, the results fluctuate about $\langle Q\rangle$, then it is not in an eigenstate of the observable we measured—although it could be in an eigenstate of some other observable.[16]

For example, the ground-state of the simple harmonic oscillator is an eigenstate of energy and parity [recall that $\psi_0(x)$ is an even function], but it's not an eigenstate of $\hat{p}$, nor of $\hat{x}$, as we proved in Examples 9.4 and 9.5. So a measurement of the energy of a SHO in such a state will yield the same result, unlike a measurement of position.

[16]Actually, this procedure is a little tricky. As we'll learn in Chap. 13, if an ensemble is not in an eigenstate of Q before measurement of this observable, it will be *after* the measurement. That is, were we to test an ensemble that originally was not in an eigenstate by measuring $\hat{Q}$, then after the measurement the ensemble *would be* in an eigenstate of Q and hence would differ from the original ensemble. So in practice we must effect our verification measurement on a sub-ensemble of the original ensemble of interest.

10.6 CONSEQUENCES OF HERMITICITY I: REAL EIGENVALUES AND ORTHONORMAL EIGENFUNCTIONS

In the last section we discussed the eigenvalue equation for an arbitrary operator. But in quantum mechanics, operators aren't arbitrary; they're Hermitian. And the properties of the eigenfunctions and eigenvalues of Hermitian operators are very special indeed.

On the Reality of Eigenvalues

We have proven (Proclamation 10.1, § 10.3) that the expectation value of a Hermitian operator $\hat{Q}$ with respect to an arbitrary quantum state Ψ,

$$\langle Q \rangle = \int_{\text{all space}} \Psi^* \hat{Q} \Psi \, dv, \tag{10.83}$$

is a real number. This property is important because we can *measure* $\langle Q \rangle$ in the laboratory. We have also proven (Proclamation 10.2, § 10.5) that for a system in an eigenstate of Q, the expectation value (10.83) reduces to the appropriate eigenvalue q of the operator $\hat{Q}$, *i.e.*,

$$\hat{Q}\Psi_q = q\Psi_q \qquad \Longrightarrow \qquad \langle Q \rangle = q. \tag{10.84}$$

The obvious implications of these two facts is that the set of all eigenvalues $\{q\}$ of *any* Hermitian operator $\hat{Q}$ consists of real numbers.

But this argument is deceptive, for it seems to hinge on the eigenfunction of $\hat{Q}$ being a state function. As we'll now prove, the eigenvalues of a Hermitian operator are real *regardless of the precise nature of the eigenfunctions.*

> ***Proclamation 10.3***
>
> The eigenvalues of a Hermitian operator are real numbers.

Argument: We seek a proof that

$$\boxed{\hat{Q} \text{ Hermitian} \qquad \Longrightarrow \qquad \text{eigenvalues } \{q\} \text{ are real}}$$

The eigenvalue equation for an arbitrary Hermitian operator is[17]

$$\hat{Q}\chi_q = q\chi_q. \tag{10.85}$$

We know one way to prove that an eigenvalue q is real: show that it is equal to its complex conjugate [*c.f.*, Eq. (10.31)], *i.e.*, prove that $q^* = q$. To this end, we first must

[17]The discussion of this section becomes needlessly complicated if the eigenvalues in question are continuous. So *in the text, I'll assume that all eigenvalues are discrete.* [But I will label eigenfunctions by their eigenvalues, not by an index, as in (10.63); students seem to find the latter notation convenient in solutions of problems but confusing in proofs of properties.] In footnotes and asides, I'll show you what these results look like for a spectrum that includes continuous eigenvalues. Feel free to skip the footnotes if you aren't interested in the more general case.

isolate q. We do so by multiplying Eq. (10.85) on the left by χ_q^* and integrating over all space, *viz.*,[18]

$$\langle \chi_q \mid \hat{Q}\chi_q \rangle = q\langle \chi_q \mid \chi_q \rangle. \tag{10.86}$$

Now, the eigenfunctions $\{\chi_q\}$ can be normalized, so we can use

$$\langle \chi_q \mid \chi_q \rangle = 1 \qquad \text{normalization of eigenfunctions} \tag{10.87}$$

to simplify Eq. (10.86), obtaining

$$q = \langle \chi_q \mid \hat{Q}\chi_q \rangle. \tag{10.88a}$$

The complex conjugate of (10.88*a*) is

$$q^* = \langle \chi_q \mid \hat{Q}\chi_q \rangle^*. \tag{10.88b}$$

The rest of the proof that $q^* = q$ is a snap:

$$q^* = \langle \hat{Q}\chi_q \mid \chi_q \rangle, \qquad \text{[Property III: Table 10.4]} \tag{10.89a}$$

$$= \langle \chi_q \mid \hat{Q}\chi_q \rangle, \qquad \text{[Hermiticity of } \hat{Q}\text{]} \tag{10.89b}$$

$$= q. \qquad \text{[by Eq. (10.88a)]} \tag{10.89c}$$

[Notice that in (10.89*b*) I moved $\hat{Q}$ to the right of the bar, so I could use the eigenvalue equation to eliminate this operator from the Dirac braket.]

Q. E. D.

The importance of this property will become evident in Chap. 13, which deals with measurement in quantum mechanics. There we'll discover the remarkable fact that *the eigenvalues of a Hermitian operator are the only values we could obtain in a measurement of the observable represented by the operator.* So these numbers must be real. And, for Hermitian operators, they are.

On the Orthogonality of Eigenfunctions

Of the many properties of the set $\{\chi_q\}$ of eigenfunctions of a Hermitian operator $\hat{Q}$, the most familiar is its *linear independence*, *i.e.*,

$$\sum_{\text{all } q} c_q\chi_q = 0 \iff c_q = 0 \qquad \text{for all } q. \qquad \text{linear independence} \tag{10.90}$$

But the functions $\{\chi_q\}$, which we obtain by solving (somehow) the eigenvalue equation, have three other properties of great importance in quantum theory, each of which has affiliated with it special jargon:

> ***Rule***
>
> The eigenfunctions of a Hermitian operator are *orthogonal*, constitute a *complete set*, and satisfy *closure*.

[18]Remember that the "integral over all space" embraces all spatial coordinates in the eigenfunction. For a single particle in one dimension, this integral is $\int_{-\infty}^{\infty} \cdots dx$; for a single particle in three dimension, it is $\int_{\text{all space}} \cdots dv$. For a many-particle system, it is a multiple integral over the spatial coordinates of all particles (see Volume II).

I now want to show you what these words mean, beginning with the one you're probably familiar with: *orthogonality*.[19]

Orthogonality is an important (and familiar) property of *vectors*. The x and y axes of a Cartesian coordinate system, for example, are orthogonal. Letting $\hat{e}_x$ and $\hat{e}_y$ denote unit vectors along these axes, we can express their orthogonality mathematically using the dot product, as $\hat{e}_x \cdot \hat{e}_y = 0$. There is, in fact, a close analogy between the mathematics of vectors and of eigenfunctions; this parallel is a consequence of the fact that both are described by the same branch of mathematics, linear algebra.

But in quantum mechanics, when we say that two (non-degenerate) eigenfunctions with different eigenvalues are **orthogonal**, we mean simply that their overlap integral is zero. To illustrate this general property in a specific case, consider the Hamiltonian eigenfunctions for the ground and first-excited states of a one-dimensional simple harmonic oscillator. In Fig. 10.2 are graphs of these eigenfunctions and of their product $\psi_0(x)\psi_1(x)$. The area under this curve, $\int_{-\infty}^{\infty} \psi_0(x)\psi_1(x)\,dx$, is the **overlap** of $\psi_0(x)$ and $\psi_1(x)$. It is clear from the figure that the positive and negative contributions to this integral are equal, so

$$\int_{-\infty}^{\infty} \psi_0(x)\psi_1(x)\,dx = 0, \qquad (10.91a)$$

or, in Dirac notation,

$$\langle \psi_0 \mid \psi_1 \rangle = 0. \qquad (10.91b)$$

Equations (10.91) are the mathematical statements of the orthogonality of $\psi_0(x)$ and $\psi_1(x)$. But this property is no accident; it occurs because these functions are eigenfunctions of an operator ($\hat{\mathcal{H}}$) that is Hermitian, as I now propose to prove.

Proclamation 10.4

Eigenfunctions of a Hermitian operator whose (non-degenerate) eigenvalues are different are orthogonal.

Argument: In Dirac notation, we express this assertion as[20]

[19]This is the only one of these three properties that I'll prove. To discuss in detail the mathematical justification for completeness and closure would take us too far afield into the formalism of quantum mechanics. The mathematically inquisitive among you can find proofs in Chap. V, § 6, and Chap. VII of *Quantum Mechanics* by A. Messiah (Amsterdam: North-Holland, 1965).

[20]If the eigenvalues are continuous, the Dirac Delta function enters the picture, and the orthogonality relation (10.92) reads

$$\left.\begin{aligned} \hat{Q}\chi_q &= q\chi_q \\ \hat{Q}\chi_{q'} &= q'\chi_{q'} \end{aligned} \quad (q' \neq q) \right\} \implies \langle \chi_{q'} \mid \chi_q \rangle = \delta(q'-q). \qquad \text{[continuous eigenvalues]}$$

For example, the eigenvalues p of the linear momentum operator $\hat{p}$ are continuous, and the orthonormality integral for its eigenfunctions (10.65) is

$$\langle \chi_{p'} \mid \chi_p \rangle = \frac{1}{2\pi\hbar} \int_{-\infty}^{\infty} e^{-ip'x/\hbar} e^{ipx/\hbar}\,dx.$$

This integral is a fairly well-known form of the Dirac Delta function (see Appendix L) $\delta(p-p')$. That is, the eigenfunctions of $\hat{p}$ satisfy the orthonormality relation

$$\langle \chi_{p'} \mid \chi_p \rangle = \delta(p-p').$$

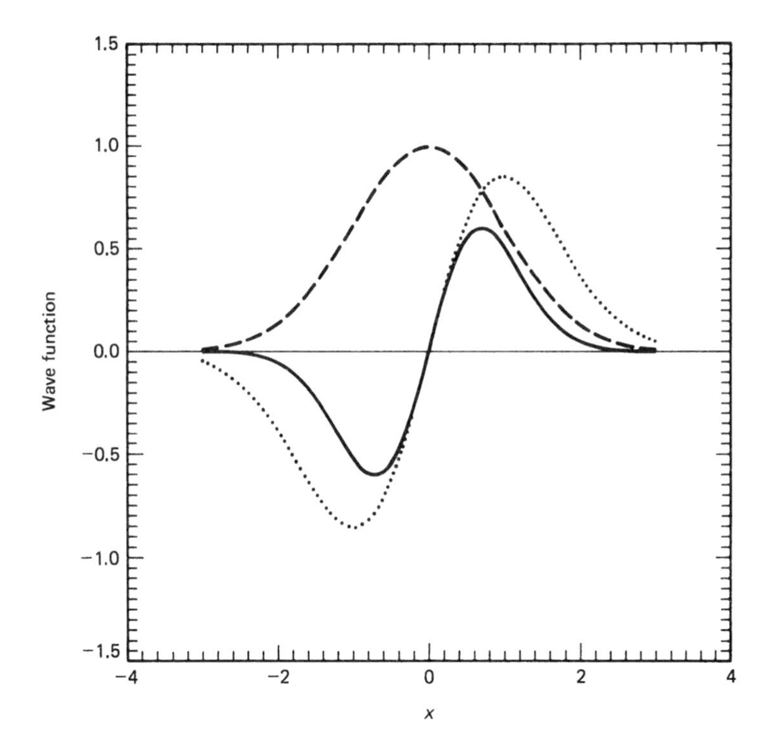

Figure 10.2 Graphic evidence that the first two Hamiltonian eigenfunctions of the simple harmonic oscillator are orthogonal. The solid curve is the wave function of the ground state, the dashed curve is the wave function of the first excited state, and the dotted curve is their overlap.

$$\left.\begin{aligned}\hat{Q}\chi_q &= q\chi_q \\ \hat{Q}\chi_{q'} &= q'\chi_{q'}\end{aligned}\quad (q' \neq q)\right\} \quad \Longrightarrow \quad \langle \chi_{q'} \mid \chi_q \rangle = 0 \qquad \text{orthogonality} \quad (10.92)$$

To prove that the overlap integral $\langle \chi_{q'} \mid \chi_q \rangle$ equals zero, we must extract from the eigenvalue equations in (10.92) an expression involving this integral. To this end, let's multiply the first of these equations by $\chi_{q'}^*$ and integrate over all space. In Dirac notation, the result is

$$\langle \chi_{q'} \mid \hat{Q}\chi_q \rangle = q \langle \chi_{q'} \mid \chi_q \rangle. \tag{10.93}$$

We can easily simplify the left-hand side of this equation using the Hermiticity of $\hat{Q}$, *viz.*,

$$\begin{aligned}
\langle \chi_{q'} \mid \hat{Q}\chi_q \rangle &= \langle \hat{Q}\chi_{q'} \mid \chi_q \rangle && \text{[Hermiticity of } \hat{Q}\text{]} && (10.94a) \\
&= \langle q'\chi_{q'} \mid \chi_q \rangle && \text{[the eigenvalue equation (10.85)]} && (10.94b) \\
&= q'^* \langle \chi_{q'} \mid \chi_q \rangle && \text{[Property II, Table 10.4]} && (10.94c) \\
&= q' \langle \chi_{q'} \mid \chi_q \rangle. && \text{[Proclamation 10.2]} && (10.94d)
\end{aligned}$$

Pondering Eqs. (10.93) and (10.94), we light on the next step: subtract (10.94*d*) from (10.93), *viz.*,

$$(q - q')\langle \chi_{q'} \mid \chi_q \rangle = 0. \tag{10.95}$$

But by hypothesis $q \neq q'$. So Eq. (10.95) can hold only if the overlap integral is zero.

Q. E. D.

Remembrances of Series Past

Orthogonality is a property familiar from our study of elementary functions. In Chap. 4, while discussing the Fourier series expansion of a function $f(x)$,

$$f(x) = a_0 + \sum_{n=1}^{\infty} \left[a_n \cos\left(2\pi n \frac{x}{L}\right) + b_n \sin\left(2\pi n \frac{x}{L}\right)\right], \tag{10.96}$$

we noted that the sine and cosine functions satisfy the relations

$$\int_{-L/2}^{L/2} \cos\left(2\pi n \frac{x}{L}\right) \sin\left(2\pi m \frac{x}{L}\right) \, dx = 0 \tag{10.97a}$$

$$\begin{aligned} \int_{-L/2}^{L/2} \cos\left(2\pi n \frac{x}{L}\right) \cos\left(2\pi m \frac{x}{L}\right) \, dx &= \int_{-L/2}^{L/2} \sin\left(2\pi n \frac{x}{L}\right) \sin\left(2\pi m \frac{x}{L}\right) \, dx \\ &= \begin{cases} 0, & \text{if } m \neq n; \\ 1, & \text{if } m = n. \end{cases} \end{aligned} \tag{10.97b}$$

Technically, Eqs. (10.97) are called "orthonormality relations," because they combine orthogonality and the normalization condition; the latter is obtained when $m = n$.

These trigonometric functions are, of course, related to the Hamiltonian eigenfunctions of the infinite square well [Eq. (7.55)], with real eigenvalues $E_n = n^2\pi^2\hbar^2/(2mL^2)$. So Eqs. (10.97) illustrate Proclamation 10.4, expressing the orthogonality of the eigenfunctions of the (Hermitian) Hamiltonian of the infinite square well.

> **Aside: Dealing with Degeneracy by Orthogonalization.** In the statement of Proclamation 10.4, I deftly slipped in the qualifier "non-degenerate," because two eigenfunctions that correspond to the same eigenvalue are not *necessarily* orthogonal. This is *the case of degenerate eigenvalues*: *i.e.*, $q = q'$ in the discussion above. Most eigenvalues of most quantum-mechanical operators are degenerate; so what do we do in this case?
>
> Fortunately, *we can always form from the eigenfunctions* $\{\chi_q\}$ *another set of eigenfunctions all of whose elements are orthogonal.* Each element of the new set is a linear combination of the original functions $\{\chi_q\}$. Several procedures exist for constructing such sets; we'll examine one of these, *the Gram-Schmidt orthogonalization procedure*, in Volume II. The point to retain is that *we can always find a set of linearly-independent eigenfunctions of a Hermitian operator whose elements are orthonormal.*

10.7 CONSEQUENCES OF HERMITICITY II: COMPLETENESS AND CLOSURE

In § 10.6 we proved two properties of the solutions of the eigenvalue equation for a Hermitian operator: the eigenvalues are real, and the eigenfunctions are orthogonal. In this section, we'll discuss (but not prove) a slightly more arcane property of these solutions. Although this property, completeness, is probably unfamiliar to you, it is actually one you've seen—in the Fourier series (10.96).

This series illustrates that the set of functions $\{\cos\left(2\pi n \frac{x}{L}\right), \sin\left(2\pi n \frac{x}{L}\right)\}$ is *complete in the variable x. The completeness of this set of functions means that any (suitably well-behaved) function $f(x)$ that is periodic on the interval $[-L/2, L/2]$ can be expanded in the elements of this set.* The coefficients of this expansion, a_n and b_n in (10.96), are determined from $f(x)$ as [*c.f.*, Eq. (4.42)],

$$a_0 = \frac{1}{L}\int_{-L/2}^{L/2} f(x)\,dx \tag{10.98a}$$

$$a_n = \frac{2}{L}\int_{-L/2}^{L/2} f(x)\cos\left(2\pi n\frac{x}{L}\right)dx \qquad (n = 1, 2, \ldots) \tag{10.98b}$$

$$b_n = \frac{2}{L}\int_{-L/2}^{L/2} f(x)\sin\left(2\pi n\frac{x}{L}\right)dx. \qquad (n = 1, 2, \ldots) \tag{10.98c}$$

In quantum mechanics, completeness is a property of the set $\{\chi_q\}$ of eigenfunctions of any Hermitian operator $\hat{Q}$. (And it is not restricted to periodic functions.) *Completeness of a set of eigenfunctions means that any well-behaved function of the variables on which the eigenfunctions depend can be expanded in the set*, *i.e.*, there exist coefficients c_q such that a function f can be written as the series $\sum_q c_q \chi_q$.[21]

Completeness is an extremely important property, for it is the key to *the expansion of an arbitrary state function in an arbitrary (complete) set of eigenfunctions*—a tactic the full value of which will come clear in the next chapter. Here's how such an expansion works mathematically:[22]

[21] By "well-behaved" I mean that f must be continuous everywhere; this adjective also usually requires that f be finite (but not necessarily normalizable) everywhere. For the mathematical details, see § 7 of *Quantum Mechanics*, Volume I by K. Gottfried (New York: Benjamin, 1966).

[22] Yet another note regarding the trouble caused by pesky continuous eigenvalues: if the spectrum of an operator is continuous, then the completeness relation cannot be written as in (10.99), because we cannot *sum* over a continuous index. Instead we must integrate, *viz.*,

$$\hat{Q}^\dagger = \hat{Q} \quad \Longrightarrow \quad \Psi = \int_{\text{all } q} c_q \chi_q \, dq. \qquad \text{[continuous spectrum]}$$

For example, the Fourier integral for a free-particle wave function [*c.f.*, Eq. (4.77)] is an expansion of this function in the complete set of eigenfunctions of the linear momentum operator:

$$\Psi(x,t) = \int_{-\infty}^{\infty}\left[\Phi(p)e^{-i\omega t/\hbar}\right]\left[\frac{1}{\sqrt{2\pi\hbar}}e^{ipx/\hbar}\right]dp = \int_{-\infty}^{\infty} c_p(t)\chi_p(x)\,dp.$$

Note how the time dependence enters the expansion coefficients $c_p(t) = \Phi(p)e^{-i\omega t/\hbar}$ in this example. [Recall that for a free particle, the dispersion relation is $\omega = E/\hbar = \hbar k^2/(2m)$.] If the spectrum includes both discrete and continuous eigenvalues, the completeness relation looks even worse:

$$\hat{Q}^\dagger = \hat{Q} \quad \Longrightarrow \quad \Psi = \sum_{\text{discrete } q} c_q\chi_q + \int_{\text{continuous } q} c_q\chi_q\,dq.$$

This form would be required, for example, to expand a function in the set of eigenfunctions of the Hamiltonian of the *finite* symmetric square well.

Rule

Any (physically-admissible) state function can be expanded in the complete set of eigenfunctions of any Hermitian operator, provided the eigenfunctions satisfy the same boundary conditions as the state function:

$$\hat{Q}^\dagger = \hat{Q} \quad \Longrightarrow \quad \Psi = \sum_{\text{all } q} c_q \chi_q \qquad \text{eigenfunction expansion} \tag{10.99}$$

In the parlance of quantum mechanics, the set of eigenfunctions we choose for the expansion is called the **basis set**. The coefficient c_q is called the **projection** of Ψ onto the **basis function** χ_q. We'll see shortly how to calculate these projections for a particular wave function—although you may be able to guess how from Eq. (10.98).

For example, we expand a wave function of a single particle in one dimension in the complete set of eigenfunctions $\{\chi_q(x)\}$ as

$$\Psi(x,t) = \sum_{\text{all } q} c_q(t)\chi_q(x), \tag{10.100a}$$

and that of a single particle in three dimensions in the set $\{\chi_q(\mathbf{r})\}$ as

$$\Psi(\mathbf{r},t) = \sum_{\text{all } q} c_q(t)\chi_q(\mathbf{r}). \tag{10.100b}$$

All such expansions have a common form:

$$\begin{pmatrix}\text{wave}\\ \text{function}\end{pmatrix} = \sum_{\substack{\text{all}\\ \text{eigenfunctions}}} \begin{pmatrix}\text{projection on}\\ \text{the eigenfunction}\end{pmatrix}\begin{pmatrix}\text{eigenfunction of}\\ \text{Hermitian operator}\end{pmatrix}. \tag{10.100c}$$

Notice that the state function depends on time, so the *expansion coefficients* c_q in (10.99) perforce depend on t. Note also that *all eigenfunctions of* $\hat{Q}$ *must be included in these expansions.* This means, in principle, that the sum in Eqs. (10.100) will contain an infinite number of terms. To be sure, for a *particular* state, some of the expansion coefficients c_q *may* happen to be zero—if, as in Example 10.12 below, the function we are expanding has a symmetry property not shared by the eigenfunctions $\{\chi_q\}$. Nevertheless, you should memorize the following rule:

Rule

In general do not omit any eigenfunctions from a complete set expansion.

Example 10.12. The Eigenfunctions of the SHO Hamiltonian

The Hamiltonian of the one-dimensional harmonic oscillator, $\hat{\mathcal{H}} = \hat{p}^2/(2m) + m\omega_0^2\hat{x}^2/2$, is Hermitian. So the set of its eigenfunctions, $\{\psi_n(x)\}$ where $n = 0, 1, 2, \ldots$, is *complete in the variable* x and can be used as the *basis set* in an expansion of any well-behaved function of x that is spatially localized. The *spectrum* of the SHO Hamiltonian consists solely of discrete energies $E_n = \left(n + \frac{1}{2}\right)\hbar\omega_0$, and all of the corresponding eigenfunctions satisfy the boundary conditions $\psi_n(x) \to 0$ as $x \to \pm\infty$.

Because this Hamiltonian is Hermitian, we can expand any physically-admissible wave function $\Psi(x,t)$ (that satisfies these boundary conditions) in the set of its eigenfunctions. According to Eq. (10.100*a*), the expansion of a function $\Psi(x,t)$ in this basis looks like

$$\Psi(x,t) = \sum_{n=0}^{\infty} c_n(t)\psi_n(x). \tag{10.101}$$

We implemented just such an expansion for a non-stationary state of the SHO in Example 9.6: the state represented by [Eq. (9.129*a*)]

$$\Psi(x,t) = \frac{1}{\sqrt{2}}\left(\frac{\beta^2}{\pi}\right)^{1/4} e^{-\beta^2 x^2/2}\left[e^{-i\omega_0 t/2} + \sqrt{2}\beta x e^{-3i\omega_0 t/2}\right] \tag{10.102a}$$

$$= \frac{1}{\sqrt{2}}\psi_0(x)e^{-iE_0 t/\hbar} + \frac{1}{\sqrt{2}}\psi_1(x)e^{-iE_1 t/\hbar}. \tag{10.102b}$$

The second line in Eq. (10.102) is the expansion of $\Psi(x,t)$ in the complete set of SHO eigenfunctions. [The state function (10.102*a*) was so simple that we could determine the coefficients in (10.102*b*) by inspection; later in this section we'll derive a straightforward procedure for calculating these coefficients when this simple gambit fails.] Note that *for this particular state function*, all but two of the expansion coefficients are zero.

We can exploit the completeness of the eigenfunctions $\{\psi_n(x)\}$ to expand more complicated SHO wave functions, such as the (initial) wave packet

$$\Psi(x,0) = \frac{1}{\sqrt{2\pi}}\frac{1}{a}x^2 e^{-x^2/(2a^2)}. \tag{10.103a}$$

This particular function has *even parity*, so although its expansion (10.101) does contain an infinite number of terms, the coefficients $c_n(t)$ for all odd n *must* be zero. [Only even-parity functions can appear in the expansion of an even function; just one odd-parity function on the right-hand side of (10.103*b*) would befoul the purity of the parity of $\Psi(x,t)$.] So *for this state*, the complete set expansion becomes

$$\Psi(x,t) = \sum_{\substack{n=0 \\ \text{(even)}}}^{\infty} c_n(t)\psi_n(x). \qquad \text{[even-parity } \Psi] \tag{10.103b}$$

Aside: On the Choice of a Complete Set for Expansion. A remarkable feature of completeness is its versatility: *the Hermitian operator that defines the basis set in which we expand a state function need not have anything to do with the system whose state function we expand.* For example, we could use the compete set of SHO eigenfunctions to expand a state function of some other system—say, a stationary-state wave function of the symmetric finite square well (§ 8.7–8.8), or a *free-particle* wave packet such as the Lorentzian

$$\Psi(x,0) = \frac{1}{\pi}\frac{a}{x^2 + a^2}. \tag{10.104a}$$

The form of the expansion of (10.104*a*) is

$$\begin{pmatrix}\text{free-particle} \\ \text{wave packet}\end{pmatrix} = \sum_{\substack{\text{all} \\ \text{eigenfunctions}}} c_n(t)\begin{pmatrix}\text{eigenfunction of} \\ \text{the SHO Hamiltonian}\end{pmatrix} \tag{10.104b}$$

This feature of completeness isn't magic; completeness is a mathematical property of eigenfunctions, not tied to the systems whose states they may represent.

Question 10–22

Could we expand a wave function of the *asymmetric finite* square well in the set of eigenfunctions of the SHO Hamiltonian? Why or why not? How about a wave function of the *infinite* square well?

A Familiar Analogue

Does completeness seem remote, abstract, too mathematical to visualize? Would you like a *picture*? In discussing orthogonality (§ 10.6), I noted an analogy between the mathematics of eigenfunctions and that of vectors. Look at Fig. 10.3, where a vector [a directed line segment in geometrical space ($\Re^3$)] is expanded in the set of unit vectors $\{\hat{e}_x, \hat{e}_y, \hat{e}_z\}$. Any (continuous) directed line segment **V** can be so expanded,

$$\mathbf{V} = c_x\hat{e}_x + c_y\hat{e}_y + c_z\hat{e}_z, \tag{10.105}$$

because the set $\{\hat{e}_x, \hat{e}_y, \hat{e}_z\}$ is complete. These three unit vectors, in fact, comprise a *basis* for the expansion of the arbitrary vector **V**. The "expansion coefficients" c_x, c_y, and c_z are called the **components** of **V** and are determined as *projections* of this vector on the unit vectors, the elements of the complete set:

$$c_x = \hat{e}_x \cdot \mathbf{V} \tag{10.106a}$$

$$c_y = \hat{e}_y \cdot \mathbf{V} \tag{10.106b}$$

$$c_z = \hat{e}_z \cdot \mathbf{V}. \tag{10.106c}$$

This analogy is the origin of the use of the word "projection" for the expansion coefficients $c_q(t)$ in the expansion of Ψ in the complete set $\{\chi_q\}$, Eq. (10.99). Drawing further on this analogy, physicists sometimes call these coefficients the **components** of Ψ on the eigenfunctions $\{\chi_q\}$.[23]

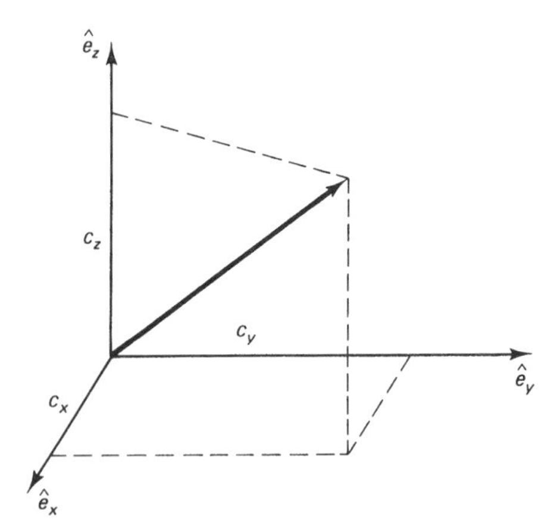

Figure 10.3 A vector and its components on the complete set of axes of a Cartesian (rectangular) coordinate system.

[23]There is more here than meets the eye. As I have remarked before, quantum states can be represented by "vectors" in an abstract "linear vector space" called *Hilbert space*. Hilbert space is *not* a three-dimensional space like the one we live in, but the analogy between the two provides a powerful geometrical interpretation of quantum mechanics. For a readable introduction to this viewpoint, see § 4.4 of *Introductory Quantum Mechanics* by R. L. Liboff (San Francisco: Holden-Day, 1980). For the whole story, see *Principles of Quantum Mechanics*, 4th ed. by P. A. M. Dirac (Cambridge: Oxford University Press, 1970).

Determination of the Expansion Coefficients

In the vector analogy to the eigenfunction expansion (10.99), the coefficients c_x, c_y, and c_z are determined from **V** and from the elements of the basis set in which the vector was expanded. In the Fourier series analogy (10.98), the coefficients a_n and b_n were determined from $f(x)$ and from the elements of the set $\{\cos\left(2\pi n\frac{x}{L}\right), \sin\left(2\pi n\frac{x}{L}\right)\}$. So it should come as no surprise that *the coefficients $c_q(t)$ in the expansion of Ψ in $\{\chi_q\}$ are determined from the state function and the eigenfunctions in this set* (see Table 10.8). In fact, the equation for these coefficients, which we'll now derive, bears a suggestive similarity to those for the components of a vector and for the Fourier coefficients.

TABLE 10.8 THREE EXPANSIONS: AN ANALOGY.

	Expand in	Components	Ref.
vector	$\{\hat{e}_x, \hat{e}_y, \hat{e}_z\}$	$c_x,\ c_y,\ c_z$	(10.105)
periodic function	$\{\cos\left(2\pi n\frac{x}{L}\right), \sin\left(2\pi n\frac{x}{L}\right)\}$	$a_n,\ b_n$	(10.96)
state function	$\{\chi_q\}$	$c_q(t)$	(10.99)

To derive from the expansion $\sum_q c_q\chi_q$ an expression for the coefficient c_q we must isolate this coefficient. We can do so by exploiting what we know about the eigenfunctions in this expansion: they are orthonormal, *i.e.*,

$$\langle\chi_{q'} \mid \chi_q\rangle = \delta_{q',q}. \tag{10.107}$$

How can we make an *orthogonality integral* appear in the eigenfunction expansion? Easy: we first multiply the expansion (on the left) by $\chi^*_{q'}$:

$$\chi^*_{q'}\Psi = \sum_{\text{all } q} c_q\chi^*_{q'}\chi_q. \tag{10.108}$$

We then integrate both sides of Eq. (10.108) over all space, and voilà!

$$\int_{\text{all space}} \chi^*_{q'}\Psi\,dv = \sum_{\text{all } q} c_q \int_{\text{all space}} \chi^*_{q'}\chi_q\,dv. \tag{10.109a}$$

In Dirac notation Eq. (10.109*a*) is

$$\langle\chi_{q'} \mid \Psi\rangle = \sum_{\text{all } q} c_q\langle\chi_{q'} \mid \chi_q\rangle. \tag{10.109b}$$

Now, the overlap integrals on the right-hand side are equal to Kronecker delta functions $\delta_{q,q'}$. But each of these Kronecker delta functions equals zero except for $\delta_{q,q} = 1$. So this beast kills all terms in the sum over q except $q = q'$, thereby isolating the coefficient $c_{q'} = \langle\chi_{q'} \mid \Psi\rangle$. Dropping the unnecessary prime on the eigenvalue, we can write our result for *the projection of Ψ on χ_q* as

$$\boxed{\text{projection of } \Psi \text{ of } \chi_q = c_q = \langle\chi_q \mid \Psi\rangle} \tag{10.110}$$

Note that this result isn't restricted to wave functions: we can apply it to any function $f(x)$ that obeys the same boundary conditions as the eigenfunctions $\{\chi_q\}$: *i.e.*, $c_q = \langle \chi_q \mid f \rangle$.

To illustrate this important result, here are the coefficients in the expansions in Eqs. (10.100) of one- and three-dimensional wave functions:

$$c_q(t) = \int_{-\infty}^{\infty} \chi_q^*(x)\Psi(x,t)\,dx \tag{10.111a}$$

$$c_q(t) = \int_0^{\infty}\int_0^{\pi}\int_0^{2\pi} \chi_q^*(\mathbf{r})\Psi(\mathbf{r},t)\,dv, \tag{10.111b}$$

where in (10.111*b*) the volume element in spherical coordinates is $dv = r^2 dr\, \sin\theta d\theta\, d\varphi$.

Question 10–23

Derive Eq. (10.102*b*) from Eq. (10.102*a*) using the general relationship (10.110).

To fully appreciate the awesome niftiness of Eq. (10.110) we'll have to wait until Chap. 12, where we'll use it to solve the time-*dependent* Schrödinger equation for non-stationary states. Since that chapter, which is entirely devoted to eigenfunction expansions, is filled with applications of this procedure, I'll eschew further examples here. We have, after all, a bit more to cover in our sojourn through the mathematics of Hermitian operators.

Closure

The final property of the eigenfunctions of a Hermitian operator is the one of the least obvious relevance: closure. Closure is intimately related to the other properties we've examined, orthogonality and completeness (see Pblm. 10.5), but it has a very odd form. One viewpoint on the closure relation is that it's a way to represent the Dirac Delta function $\delta(x - x')$ in terms of the eigenfunctions of an operator $\hat{Q}$:

$$\boxed{\sum_{\text{all } q} \chi_q(x)\chi_q(x') = \delta(x - x')} \qquad \text{closure} \tag{10.112}$$

Notice carefully that in each term in the sum in (10.112), the eigenvalue q appears twice: once as a label on the eigenfunction χ with argument x and a second time on the same eigenfunction but with the argument x'.

Question 10–24

The Dirac Delta function $\delta(x - x')$ acts in an integral $\int_{-\infty}^{\infty} \delta(x - x')f(x')\,dx'$ to "pluck out" the value $f(x)$ (see Appendix L). Prove that

$$\int_{-\infty}^{\infty} \sum_{\text{all } q} \chi_q(x)\chi_q(x')f(x')\,dx' = f(x).$$

We won't often use the closure relation. But this property is important in more advanced treatments of quantum theory, so I include it for (ahem) completeness.

Where's the Physics?

The major results of this chapter are the properties we've been discussing in the past two sections. They are summarized in Table 10.9, where I have also included the results for continuous eigenvalues so you'll have them if you need them. (If you haven't been reading the footnotes, ignore this column.) Before leaving these properties, *which you should memorize*, I must stress one point.

Way back in Chap. 3, Postulate I declared that any quantum state could be represented by a wave function $\Psi(x,t)$. It is extremely important to appreciate that *the collection of coefficients $\{c_q(t)\}$ in the expansion of a state function in any complete set is merely an alternate way to represent this physical state. These coefficients contain precisely the same physical information as the state function.*

The wave function $\Psi(x,t)$ highlights the observable *position, i.e.*, this function—the *position probability amplitude*—is especially well-suited to investigating probabilities, average values, and uncertainties of the observable x. Similarly, the set $\{c_q(t)\}$ highlights properties of the system related to the observable Q.

For example, in Chap. 4, we first encountered the notion of *multiple representations of a quantum state*. There we developed the momentum-space representation of a state, its *momentum probability amplitude* $\Phi(p)$. In that Chapter, I emphasized that $\Phi(p)$ contained no more or less physical information than the state function in the position representation, $\Psi(x,t)$. Rather, the momentum-space function is just a way to represent that information that is particularly convenient for studying the *momentum properties* of the particle.

In fact, as we'll see in Chap. 13, the coefficient $c_q(t) = \langle \chi_q \mid \Psi \rangle$ is a "probability amplitude for the observable Q," just as $\Psi(x,t)$ is a probability amplitude for position. Thus we have the following alternative ways of representing the physical information about the particle in a particular quantum state:

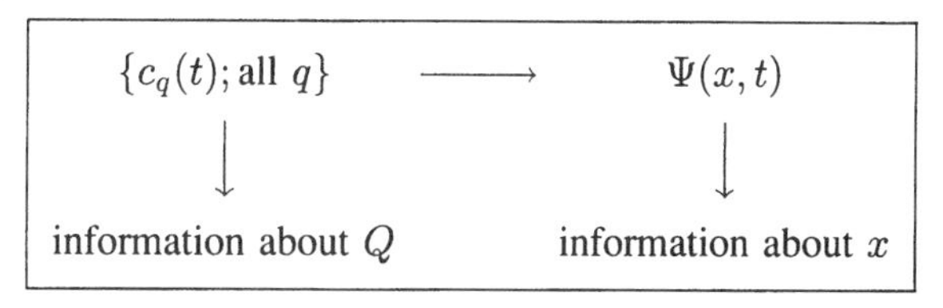

Since there are a very large number of observables Q for any system, there are a huge number of alternative representations of that information. The three most important are the position, momentum, and energy representations. The latter will be introduced in Chap. 12.

10.8 FINAL THOUGHTS—AND A FINAL EXAMPLE: THE PARITY OPERATOR

To review the many mathematical machinations I've introduced in this Chapter, I want to explore further the properties of a particularly important operator: $\hat{\Pi}$.[24] We defined the parity operator in Eq. (10.68) by its action on an arbitrary function,

[24]The parity operator is peculiar in that, unlike $\hat{p}$ or $\hat{\mathcal{H}}$, it cannot be expressed as a function of $\hat{x}$ and $\hat{p}$. Instead, the parity operator is one of a class of *symmetry operators*. Symmetry operators play an important role in quantum mechanics when this theory is formulated in terms of a branch of advanced mathematics called "group theory." (Other symmetry operators include translation, reflection and rotation; we'll study rotation operators in Volume II.) Nevertheless, the parity of a quantum state can be measured in the laboratory; in fact, the Chinese-American physicists T-D Lee and C. N. Yang shared the Nobel Prize in 1957 for demonstrating that parity is not conserved in all microscopic interactions (see § 11.9 for the saga of the fall of parity conservation).

TABLE 10.9 PROPERTIES OF THE EIGENFUNCTIONS AND EIGENVALUES OF A HERMITIAN OPERATOR Q.

	Discrete Spectrum	**Continuous Spectrum**
eigenvalues	q_n real	q real
orthonormality	$\langle \chi_{q'} \mid \chi_q \rangle = \delta_{q',q}$	$\langle \chi_{q'} \mid \chi_q \rangle = \delta(q' - q)$
completeness	$\Psi = \sum_{\text{all } q} c_q \chi_q$	$\Psi = \int_{\text{all } q} c_q \chi_q \, dq$
projection of Ψ	$c_q = \langle \chi_q \mid \Psi \rangle$	$c_q = \langle \chi_q \mid \Psi \rangle$
closure	$\sum_{\text{all } q} \chi_q(x')\chi_q(x) = \delta(x' - x)$	$\int_{\text{all } q} \chi_q(x')\chi_q(x) \, dq = \delta(x' - x)$

$$\hat{\Pi} f(x) = f(-x), \tag{10.113}$$

and noted two of its eigenvalue equations in Eqs. (10.69),

$$\hat{\Pi} f(x) = \begin{cases} +f(x) & \text{if } f(x) \text{ is even} \\ -f(x) & \text{if } f(x) \text{ is odd} \end{cases} \tag{10.114}$$

But my remarks on parity thus far have left unanswered a host of questions: are ± 1 the *only* eigenvalues of this operator? Is $\hat{\Pi}$ a valid quantum-mechanical operator, *i.e.*, is it Hermitian? And what about a *symmetric* potential allows the eigenfunctions of its Hamiltonian to have definite parity, *i.e.*, to be *simultaneously* eigenfunctions of the parity operator? We are now in a position to answer these questions.

Question 10–25

The parity operator in three dimensions inverts a vector $\mathbf{r}$ through the origin, as

$$\hat{\Pi} f(\mathbf{r}) = f(-\mathbf{r}). \tag{10.115}$$

In rectangular (Cartesian) coordinates, the effect of this operator is

$$\hat{\Pi} f(x, y, z) = f(-x, -y, -z). \tag{10.116}$$

What is the effect of $\hat{\Pi}$ in *spherical coordinates*, *i.e.*, what is $\hat{\Pi} f(r, \theta, \varphi)$?

Like most good quantum-mechanical operators, $\hat{\Pi}$ is linear:

$$\hat{\Pi}\left[c_1 f_1(x) + c_2 f_2(x)\right] = c_1 f_1(-x) + c_2 f_2(-x) \tag{10.117a}$$

$$= c_1 \hat{\Pi} f_1(x) + c_2 \hat{\Pi} f_2(x). \tag{10.117b}$$

And, like *all* good quantum-mechanical operators, it is Hermitian. The proof of the latter property is worth contemplating, for it illustrates yet another important problem-solving strategy.

Example 10.13. Hermiticity of the Parity Operator

This is one proof that is best *not* carried out in Dirac notation. The issue at hand is the question

$$\int_{-\infty}^{\infty} \Psi_1^*(x,t)\left[\hat{\Pi}\Psi_2(x,t)\right] dx \stackrel{?}{=} \int_{-\infty}^{\infty} \left[\hat{\Pi}\Psi_1(x,t)\right]^* \Psi_2(x,t)\, dx. \tag{10.118}$$

Let's go to work on the *left-hand* side of (10.118),

$$\int_{-\infty}^{\infty} \Psi_1^*(x,t)\hat{\Pi}\Psi_2(x,t)\,dx = \int_{-\infty}^{\infty} \Psi_1^*(x,t)\Psi_2(-x,t)\,dx. \tag{10.119}$$

To prove (or disprove) Hermiticity, we must transform the integrand on the right-hand side of (10.119) into a form in which the parity operator acts on $\Psi_1(x,t)$, *i.e.*, a form that contains $\Psi_1(-x,t)$ and $\Psi_2(x,t)$. Can you see how to effect this change?

Sure ... just change the dummy variable of integration from x to $-x$ (being very careful with the limits of integration when we do so):

$$\int_{-\infty}^{\infty} \Psi_1^*(x,t)\Psi_2(-x,t)\,dx = \int_{\infty}^{-\infty} \Psi_1^*(-x,t)\Psi_2(x,t)\,d(-x) \tag{10.120a}$$

$$= -\int_{\infty}^{-\infty} \left[\hat{\Pi}\Psi_1(x,t)\right]^*\Psi_2(x,t)\,dx. \tag{10.120b}$$

We now use the minus sign to interchange the limits of integration, obtaining

$$\int_{-\infty}^{\infty} \Psi_1^*(x,t)\hat{\Pi}\Psi_2(x,t)\,dx = \int_{-\infty}^{\infty} \left[\hat{\Pi}\Psi_1(x,t)\right]^*\Psi_2(x,t)\,dx, \tag{10.121}$$

which proves quite handily that $\hat{\Pi}$ is Hermitian.

So even though $\hat{\Pi}$ cannot be expressed in the form $\hat{Q}(\hat{x},\hat{p})$, it can represent a physically-measurable quantity. And it has all the accoutrements of a Hermitian operator: real eigenvalues and (as we'll see below) a complete set of orthogonal eigenfunctions that satisfy closure.

Question 10–26

What is the adjoint of the parity operator?

Question 10–27

The parity operator also has some other easily-proven properties. For one thing,

$$\hat{\Pi}^2 = \hat{1}.$$

Prove this property.

Eigenvalues and Eigenfunctions of the Parity Operator

Let's now solve the eigenvalue equation for the parity operator:

$$\hat{\Pi}\chi_q(x) = q\chi_q(x), \tag{10.122}$$

where q, the eigenvalue, must be real. Our first stab at solving (10.122) might be to let $\hat{\Pi}$ act on the eigenfunction. At first glance, the result looks unpromising:

$$\chi_q(-x) = q\chi_q(x). \tag{10.123}$$

We need an equation *both sides of which* contain $\chi_q(x)$.

We can change the sign of the argument on the left-hand side of Eq. (10.123) back to x by a second application of $\hat{\Pi}$ to *both* sides of this equation [since $\hat{\Pi}\chi_q(-x) = \chi_q(x)$], *viz.*,

$$\chi_q(x) = q\hat{\Pi}\chi_q(x). \tag{10.124a}$$

We now use the eigenvalue equation (10.122) to let Π act on the right-hand side of (10.124*a*), obtaining

$$\chi_q(x) = q^2 \chi_q(x). \tag{10.124b}$$

This means that $q^2 = 1$, which, since q must be real, means that $q = \pm 1$.

So *the parity operator has only two, discrete eigenvalues*:[25]

$$\hat{\Pi}\chi_{\pm 1}(x) = \pm 1 \, \chi_{\pm 1}(x), \tag{10.124c}$$

in agreement with Eq. (10.69). Ergo, *any function that is an eigenfunction of* $\hat{\Pi}$ *is either even or odd.*

Note carefully that this property of the eigenfunctions of the parity operator is *not* a consequence of the *definition* of that operator—although the glib way I have referred to "even- and odd-parity functions" earlier in this book may have led you to think so. Rather, it is a consequence of the Hermiticity of the parity operator. Note also that the above argument does *not* imply that all functions are eigenfunctions of $\hat{\Pi}$. There is no reason why functions $g(x)$ cannot exist such that $\hat{\Pi}g(x) \neq \pm g(x)$; indeed, we have seen several such functions in our study of one-dimensional systems with *asymmetric* potentials.

The *orthogonality* of the eigenfunctions of $\hat{\Pi}$ means that the overlap integral of an even- and an odd-parity function is zero, *i.e.*,

$$\langle \chi_{+1} \mid \chi_{-1} \rangle = 0. \tag{10.125}$$

More interestingly, *completeness* of the set of eigenfunctions $\{\chi_{+1}, \chi_{-1}\}$ means that *any function can be written as the sum of an even function and an odd function, i.e.*,

$$f(x) = c_1 \chi_{+1}(x) + c_2 \chi_{-1}(x). \qquad \text{[for arbitrary } f(x)\text{]} \tag{10.126}$$

This property is useful in a variety of contexts other than quantum physics, so it's worth investigating further.

Question 10–28

Suppose $f(x)$ does not have definite parity, and that this function and the eigenfunctions $\chi_{+1}(x)$ and $\chi_{-1}(x)$ in (10.126) are normalized. What must be the values of the expansion coefficients c_1 and c_2? Derive expressions for the eigenfunctions $\chi_{+1}(x)$ and $\chi_{-1}(x)$ in terms of $f(x)$.

Parity and the Hamiltonian

In Chaps. 8 and 9, we scrutinized several one-dimensional *symmetric* potentials—noting that their bound-state wave functions have definite parity, *i.e.*, their Hamiltonian eigenfunctions $\psi_E(x)$ are eigenfunctions of $\hat{\Pi}$. This situation is special; as we have seen, the Hamiltonian eigenfunctions of systems with asymmetric potentials, such as the half-infinite linear potential and the asymmetric square well, are neither even nor odd.

In § 9.4 we supported these observations with the proof of Proclamation 9.2, which stated that *the bound-state eigenfunctions of a symmetric potential have definite parity.*

[25]This derivation told us nothing about the eigenfunctions χ_q. In fact, it's impossible to solve the eigenvalue equation (10.122) for an analytic form of these functions (as we could, say, the Hamiltonian eigenvalue equation for the simple harmonic oscillator) because of the peculiar nature of the parity operator. To each eigenvalue of $\hat{\Pi}$ there corresponds an *infinite number of eigenfunctions*; so the eigenvalues of $\hat{\Pi}$ are "infinite-fold degenerate."

That proof appealed directly to the TISE for a symmetric Hamiltonian. But the reasons for this result are deeper even than the Schrödinger equation; they strike to the very essence of the nature of the system itself. To show you why, I want to reconsider this property from an *operator viewpoint*. In particular, I want to explore the commutation relation of the Hamiltonian and the parity operator—and its implications. To this end, we'll consider in succession the two terms in $\hat{\mathcal{H}}$, beginning with the potential operator.

One way to express the symmetry (under inversion) of a symmetric potential is in terms of the function $V(x)$, *i.e.*,

$$V(-x) = V(x). \tag{10.127a}$$

How could we express this essential feature of the system as an *operator equation*? Well, the function $V(x)$ is related to the *potential operator*, one of the two operators in the system Hamiltonian (see Chap. 5). In the position representation, $\hat{V}$ is the mathematical instruction: multiply by the *function* $V(x)$.

The statement (10.127*a*) of the symmetry of the potential involves an inversion of x through the origin, so it's not surprising that its operator expression involves the parity operator. Equation (10.127*a*) shows us that this operator commutes with $\hat{V}$: since $V(x)$ is symmetric, we can interchange $\hat{V}$ and $\hat{\Pi}$ at will, *i.e.*,[26]

$$\hat{\Pi}\hat{V} = \hat{V}\hat{\Pi} \qquad \Longrightarrow \qquad \left[\hat{\Pi}, \hat{V}\right] = 0. \qquad \text{[for a symmetric potential]} \tag{10.127b}$$

The commutator relation (10.127*b*) expresses the symmetry of the system in the operator viewpoint. But this is only half of the story: to see if the Hamiltonian commutes with $\hat{\Pi}$, we must consider next the kinetic energy operator.

Example 10.14. The Parity Operator and the Kinetic Energy Operator

Our demonstration that the parity operator commutes with a symmetric potential operator may lead you to wonder if $\hat{\Pi}$ commutes with the kinetic energy operator. The form of this operator, $\hat{T} = \hat{p}^2/(2m)$, suggests that we back up one step and ask whether $\hat{\Pi}$ commutes with $\hat{p}$. In one dimension, $\hat{p} = -i\hbar d/dx$, and it's not hard to see that the order of operation of these operators is important. Since $\hat{\Pi}$ leaves the constant factor $-i\hbar$ unchanged, we have

$$\hat{\Pi}\hat{p}f(x) = -i\hbar\hat{\Pi}\frac{df(x)}{dx} \tag{10.128a}$$

$$= -i\hbar\left[-\frac{df(x)}{dx}\right] \tag{10.128b}$$

$$= -\hat{p}\hat{\Pi}f(x). \tag{10.128c}$$

So the operators $\hat{\Pi}$ and $\hat{p}$ don't commute, *i.e.*,

$$\hat{\Pi}\hat{p} = -\hat{p}\hat{\Pi}. \tag{10.128d}$$

But $\hat{T}$ entails *two* successive applications of $\hat{p}$. From the form of Eq. (10.128*d*), you might guess that the form of $\hat{T}$ causes this operator to commute with $\hat{\Pi}$. Let's see:

[26] Be careful when working with operator equations such as (10.127*b*); 'tis safest to let them act on an arbitrary function, as

$$\hat{\Pi}\hat{V}f(x) = \hat{V}\hat{\Pi}f(x).$$

Be *sure* you understand why the left-hand side of this equation is $V(-x)f(-x)$ instead of $V(-x)f(x)$.

$$\hat{\Pi}\hat{T} = \frac{1}{2m}\hat{\Pi}\hat{p}\hat{p} \tag{10.129a}$$

$$= -\frac{1}{2m}\hat{p}\hat{\Pi}\hat{p} \tag{10.129b}$$

$$= \frac{1}{2m}\hat{p}\hat{p}\hat{\Pi} \tag{10.129c}$$

$$= \hat{T}\hat{\Pi}. \tag{10.129d}$$

You were right: $\hat{\Pi}$ and $\hat{T}$ do commute:

$$\left[\hat{\Pi}, \hat{T}\right] = 0. \tag{10.129e}$$

We can now deduce the desired commutation relation for $\hat{\Pi}$ and the Hamiltonian of a system with a symmetric potential. Combining the result of Example 10.13 with Eq. (10.127*b*), we find

$$\boxed{\hat{\Pi}\hat{\mathcal{H}} = \hat{\mathcal{H}}\hat{\Pi} \qquad \Longrightarrow \qquad \left[\hat{\Pi}, \hat{\mathcal{H}}\right] = 0} \qquad \text{[for a symmetric potential]} \tag{10.130}$$

i.e.,

Rule

The Hamiltonian of a system with a symmetric potential commutes with the parity operator.

Question 10–29

Prove that if $V(-x) \neq V(x)$, then $\left[\hat{\mathcal{H}}, \hat{\Pi}\right] \neq 0$.

The central question of this section concerns the *consequences* of this commutation relation for the eigenfunctions of the Hamiltonian. Let's act on such an eigenfunction with the operator equation (10.130):

$$\hat{\Pi}\hat{\mathcal{H}}\psi_E(x) = \hat{\mathcal{H}}\hat{\Pi}\psi_E(x). \tag{10.131}$$

The presence of $\hat{\mathcal{H}}\hat{\Pi}\psi_E(x)$ on the right-hand side of (10.131) leads us to wonder: *might the function $\hat{\Pi}\psi_E(x)$ be an eigenfunction of $\hat{\mathcal{H}}$?* To find out, look at the left-hand side of this equation. Using the TISE to simplify this side, we get

$$\hat{\Pi}\hat{\mathcal{H}}\psi_E(x) = \hat{\Pi}E\psi_E(x) = E\left[\hat{\Pi}\psi_E(x)\right], \tag{10.132}$$

with which we can write Eq. (10.131) as

$$\hat{\mathcal{H}}\left[\hat{\Pi}\psi_E(x)\right] = E\left[\hat{\Pi}\psi_E(x)\right]. \tag{10.133}$$

Egad! *Both functions $\psi_E(x)$ and $\hat{\Pi}\psi_E(x)$ are eigenfunctions of the operator $\hat{\mathcal{H}}$, with the same eigenvalue.* But we know (from Proclamation 9.1, § 9.4) that the (bound-state) eigenvalues of a one-dimensional, single-particle system are non-degenerate. Consequently, these two functions must be proportional to one another. Using q as the constant of proportionality, we can write

$$\hat{\Pi}\psi_E(x) = q\psi_E(x). \tag{10.134}$$

But this is just the eigenvalue equation of the parity operator, Eq. (10.124), and the proportionality constants are just the eigenvalues $q = \pm 1$.

This argument proves that the Hamiltonian eigenfunctions of a symmetric potential are *simultaneous eigenfunctions* of two operators, $\hat{\Pi}$ and $\hat{\mathcal{H}}$:[27]

$$\hat{\mathcal{H}}\psi_E(x) = E\psi_E(x) \qquad \text{Hamiltonian eigenvalue equation} \tag{10.135a}$$

$$\hat{\Pi}\psi_E(x) = \pm\psi_E(x). \qquad \text{parity eigenvalue equation} \tag{10.135b}$$

This alternate proof of Proclamation 9.2 lays bare the essence of why stationary-state wave functions of a symmetric Hamiltonian have definite parity. We have seen that this property is an inevitable consequence of the fact that a symmetric Hamiltonian commutes with the parity operator. This example hints at a profound connection between the commutation relations satisfied by a pair of operators and the eigenfunctions of those operators. In the next chapter, we'll explore this relationship further, and later, in Part IV, we'll discover the remarkable implications of this connection for measurements on microscopic systems.

ANNOTATED SELECTED READINGS

Operators

Don't limit your reading to my treatment of the mathematics and physics of operators. Nearly every quantum mechanics text written in the last 30 years has a chapter or so on this material; some of them are very good. One I particularly like is

1. Feynman, R. P., R. B. Leighton, and M. Sands, *The Feynman Lectures on Physics:* Volume III (Reading, Mass.: Addison-Wesley, 1970).

The entirety of Feynman's treatment of quantum physics is well worth your attention. Chap. 20 deals with operators and ranges over most of the material covered in more conventional texts.

A particularly excellent treatment of the mathematics of quantum mechanical operators is

2. Jordan, T. F., *Linear Operators in Quantum Mechanics*, (New York: Wiley, 1969).

Jordan's text emphasizes the mathematical foundations of quantum theory, but I think most of you will find it accessible and enlightening.

Eigenvalue Equations and Completeness

Many of the pivotal equations in quantum theory are eigenvalue equations. Eigenvalue equations and their solutions are a subject in themselves. Two fine sources for background reading on this (and related mathematical topics) are

3. Jackson, J. D., *Mathematics for Quantum Mechanics* (New York: W. A. Benjamin, 1962).
4. Hylleraas, E. A., *Mathematical and Theoretical Physics:* Volume I, (New York: Wiley, 1970). For specific information on eigenvalue equations from a mathematical viewpoint, see Chap. 25.

[27]You may be wondering why I didn't label these eigenfunctions with *both* eigenvalues, E and ± 1. Doing so isn't wrong, but neither is it necessary. If we know the energy, we can *deduce* the parity of the function, so the eigenvalue ± 1 is redundant. Note that the argument doesn't go the other way: knowledge of the parity of a particular Hamiltonian eigenfunction certainly does not enable us to deduce its energy.

The subject of completeness is fraught with subtleties. But complete sets of functions are at the heart of the mathematics of quantum physics, so such matters are critical. Regrettably, most serious treatments of this topic are at a very advanced level. The best treatment I know of is in

5. Cushing, J. T., *Applied Analytical Mathematics for Physical Scientists* (New York: Wiley, 1975). Chap. 4.

EXERCISES AND PROBLEMS

"... at times
the intention is puzzling
the discontinuities inexplicable."

—*The Endless Short Story*
by Ronald Sukenick

I've saved several nifty problems on operators for Chap. 11, which concludes the story begun here. Unfortunately, most of the problems in this chapter are a bit dry and "exercise-y." Sad to tell, to be a quantum physicist you must be a good *operator mechanic*, and that means *lots of practice*. Like practicing a musical instrument, practicing quantum mechanics is sometimes a little tedious, but nearly always rewarding.

Exercises

10.1 Operators: Short-Answer Questions for Review

[a] If $\hat{A}$ represents an observable in quantum mechanics, what condition must it satisfy (be explicit)?

[b] **Discuss** carefully the meaning and consequences of the statement "a quantum system is in an eigenstate of $\hat{A}$."

[c] **Write down** the eigenvalue equation for $\hat{A}$. What are the consequences for the solutions of this equation if $\hat{A}$ is Hermitian?

[d] Suppose you were given the classical expression for an observable $Q = Q(x, p)$. How would you form the quantum-mechanical operator that represents this observable?

[e] If $\langle A \rangle$ is equal to one of the eigenvalues of $\hat{A}$, is the system *necessarily* in an eigenstate of $\hat{A}$? Why or why not?

10.2 Coping with Operators: I

Consider the following operators:

1. $\hat{A}_1$, which differentiates once with respect to x and multiplies the result by $\sqrt{x}$, *i.e.*,

$$\hat{A}_1 f(x) = \sqrt{x}\,\frac{d}{dx} f(x).$$

2. $\hat{A}_2$, which inverts the function it acts on.
3. $\hat{A}_3$, which translates the function it acts on by a, *i.e.*,

$$\hat{A}_3 f(x) = f(x + a).$$

4. $\hat{A}_4 = 8d^2/dx^2$.
5. $\hat{A}_5 = -d^4/dx^4 + 6$.

Which of these operators is linear? Which is Hermitian? **Justify** your answers.

10.3 Coping with Operators II

Consider the operators in Exercise 10.2 *and* the additional operators:

1. $\hat{A}_6$, which squares the first derivative with respect to x of the function it acts on, *i.e.*,

$$\hat{A}_6 f(x) = \left[\frac{d}{dx} f(x)\right]^2 .$$

2. $\hat{A}_7$, which averages the function it acts on over an interval of magnitude L centered on $x = 0$.

Which of these operators could, in principle, represent an observable in quantum mechanics? Fully **justify** your answer.

10.4 Operators in Quantum Mechanics

[a] Let $\hat{Q}_1$ and $\hat{Q}_2$ be Hermitian Operators. **Prove** that the operator $\hat{Q}_3 \equiv i\left[\hat{Q}_1, \hat{Q}_2\right]$ is Hermitian.
Hint: Use Dirac notation and the Hermiticity of $\hat{Q}_1$ and $\hat{Q}_2$.

[b] What can you conclude from part [a] about the expectation value of $\hat{Q}_3$?

[c] What is the adjoint of $\hat{Q}_3$?

[d] Suppose $\hat{Q}_3$, the operator defined in [a], is a *non-zero* operator. Consider the second operator $\hat{Q}_4 = \hat{Q}_1\hat{Q}_2$. Is $\hat{Q}_4$ Hermitian? **Prove** your answer, preferably by examining the adjoint of $\hat{Q}_4$.

[e] Suppose $\hat{Q}_1$ is the Hamiltonian of a conservative system,

$$\hat{Q}_1 = \hat{\mathcal{H}} = \hat{T} + \hat{V}(x).$$

Suppose further that $\hat{Q}_2 = \hat{p}$. **Determine** $\hat{Q}_3$, as defined in part [a].
Hint: Be careful; $\hat{Q}_3$ is non-zero.

10.5 Are you positive? Are you definite?

A Hermitian Operator is said to be **positive-definite** if its expectation value is positive *for any state function, i.e.*,

$$\langle Q \rangle = \left\langle \Psi \mid \hat{Q} \mid \Psi \right\rangle > 0. \qquad \text{(all } \Psi\text{)}$$

[a] **Prove** that the eigenvalues of a positive-definite operator are positive, real numbers.

[b] Which of the operators $\hat{x}^2$ and $\hat{p}^2$ are positive-definite? **Prove** your answer.

10.6 Working with Dirac Notation

Write the following integrals in full integral form and in Dirac braket notation:

[a] the expectation value *in spherical coordinates* of the energy of a three-dimensional system with wave function $\Psi(\mathbf{r}, t)$;

[b] the matrix element $\langle \Psi_1 \mid \hat{Q} \mid \Psi_2 \rangle$, where the operator $\hat{Q}$ is defined in terms of a two-variable function $K(x, x')$ as

$$\hat{Q} f(x) \equiv \int_{-\infty}^{\infty} K(x, x') f(x')\, dx'.$$

10.7 Anti-Hermiticity

Just as some operators are anti-linear, others are *anti-Hermitian*. Although such operators play little role in quantum theory, they do provide practice with operator mathematics. An **anti-Hermitian** operator is one that satisfies

$$\left\langle \Psi_1 \mid \hat{A}\Psi_2 \right\rangle = -\left\langle \hat{A}\Psi_1 \mid \Psi_2 \right\rangle. \qquad \text{anti-Hermitian operator}$$

[a] What is the adjoint of an anti-Hermitian operator?

[b] Is the expectation value of an anti-Hermitian operator real, complex, or imaginary? **Prove** your answer.

[c] Under what conditions on the number c is the operator $\hat{A} = c\hat{1}$ anti-Hermitian?

[d] Prove that $\hat{A} = \partial/\partial x$ is anti-Hermitian.

[e] Is $\left[\hat{x}, \hat{p}\right]$ anti-Hermitian? Why or why not?

10.8 Eigenvalue Equations

Consider a system with the Hamiltonian

$$\hat{\mathcal{H}} = \frac{d^2}{dx^2} + x^2.$$

[a] Is $\hat{\mathcal{H}}$ Hermitian?

[b] **Determine** whether or not the function $f(x) = Axe^{-x^2/2}$ is an eigenfunction of $\hat{\mathcal{H}}$ and, if so, determine its eigenvalue.

[c] **Normalize** $f(x)$ by determining the value of the constant A.

[d] Conjure a system with this Hamiltonian. What state does $f(x)$ represent?

Problems

10.1 The Exponential Operator I

The exponential operator of Example 10.3 affords a wonderful example for practicing operator manipulations, because for this operator most such manipulations are pretty straightforward. In this problem and the next, we'll consider various aspects of operators of the form $e^{\hat{Q}t}$, *where $\hat{Q}$ is Hermitian.*

[a] Use the series expansions of two such operators, $e^{\hat{Q}_1 t}$ and $e^{\hat{Q}_2 t}$, to **prove** that in general

$$e^{\hat{Q}_1 t} e^{\hat{Q}_2 t} \neq e^{\hat{Q}_2 t} e^{\hat{Q}_1 t}. \tag{10.1.1}$$

[b] Under what conditions (if any) on $\hat{Q}_1$ and $\hat{Q}_2$ is

$$e^{\hat{Q}_1 t} e^{\hat{Q}_2 t} \stackrel{?}{=} e^{(\hat{Q}_1 + \hat{Q}_2)t}. \tag{10.1.2}$$

10.2 The Exponential Operator II: The Derivative

The exponential operator you worked with in Pblm. 10.1 is also useful for practicing with operator derivatives. Try your hand at the following exercises:

[a] **Determine** whether the operators $\hat{Q}$ and $e^{\hat{Q}t}$ commute.

[b] Now, **derive** an expression for the operator $de^{\hat{Q}t}/dt$, assuming that $\hat{Q}$ does not explicitly depend on t.

[c] **Validate or refute** the following proposition:

$$\frac{d}{dt} e^{\hat{Q}_1 t} e^{\hat{Q}_2 t} \stackrel{?}{=} (\hat{Q}_1 + \hat{Q}_2) e^{\hat{Q}_1 t} e^{\hat{Q}_2 t}. \tag{10.2.1}$$

[d] Consider a Hermitian operator that explicitly depends on t. **Validate or refute** the following proposition

$$\frac{d}{dt} e^{\hat{Q}(t)} \stackrel{?}{=} \frac{d\hat{Q}(t)}{dt} e^{\hat{Q}(t)}. \tag{10.2.2}$$

10.3 The Parity Operator

Prove that the *anti-commutator* of the parity operator and the linear momentum operator is zero, *i.e.*,

$$\left[\hat{\Pi}, \hat{p}\right]_+ = 0.$$

10.4 Operators and Eigenfunctions

A particle of mass m is in the ground state of a finite symmetric square well of width L.

[a] What is the expectation value of $\hat{p}$ for this state?

[b] **Show** that the Hamiltonian eigenfunction of this state is not an eigenfunction of $\hat{p}$.

[c] **Write** the Hamiltonian eigenfunction as a linear combination of eigenfunctions of $\hat{p}$ and evaluate the *ratio* of the coefficients in your expansion. **Discuss** your results physically, commenting on their implications for the probability current density in this state.

[d] Is the Hamiltonian eigenfunction also an eigenfunction of the kinetic-energy operator T? **Justify** your answer.
Hint: Consider the TISE for $\psi_1(x)$.

[e] Is the linear momentum in this state sharp? Why or why not?

[f] Answer parts [d] and [e] for the ground state of the *infinite* square well.

10.5 An Incomplete Treatment of Completeness

In § 10.7 we discussed the mathematical property called *completeness*. There we learned that the set of eigenfunctions $\{\chi_q(x)\}$ of an Hermitian operator $\hat{Q}$ is **complete**; *i.e.*, any (suitably well-behaved) function $f(x)$ can be expanded in this set [*c.f.*, Eq. (10.99)], as

$$f(x) = \sum_{\text{all } q} c_q \chi_q(x). \tag{10.5.1}$$

The coefficients in this expansion are evaluated à la Eq. (10.110) as

$$c_q = \int_{-\infty}^{\infty} \chi_q^*(x') f(x')\, dx'. \tag{10.5.2}$$

Equation (10.5.1) is useful in practice once we know that a given set of functions is complete, but it's awkward if we want to *determine* whether a particular set of functions is complete. For that purpose, we'd prefer a relationship expressed *entirely in terms of the eigenfunctions*, not in terms of a nameless "arbitrary function." Happily, we can derive just such a relationship by inserting Eq. (10.5.2) into the expansion (10.5.1), to wit:

$$f(x) = \sum_{\text{all } q} \int_{-\infty}^{\infty} \chi_q^*(x') \chi_q(x) f(x')\ dx'. \tag{10.5.3}$$

Now, since $f(x)$ is arbitrary, Eq. (10.5.3) can hold if and only if the eigenfunctions $\{\chi_q(x)\}$ satisfy the closure relationship (10.112),

$$\sum_{\text{all } q} \chi_q^*(x) \chi_q(x') = \delta(x - x'), \tag{10.5.4}$$

where $\delta(x - x')$ is the Dirac Delta function. So *one way to check completeness of a set of functions is to see if they satisfy Eq.* (10.5.4).

Prove that the set of eigenfunctions of the Hamiltonian *of a simple harmonic oscillator* is complete, by showing that they satisfy (10.5.4).

Hint: First, use the integral

$$\int_{-\infty}^{\infty} e^{-a^2x^2+bx}\, dx = \frac{\sqrt{\pi}}{a} e^{b^2/(4a^2)} \qquad [\text{for Re } a^2 > 0] \tag{10.5.5}$$

to show that the normalized spatial functions for the SHO can be represented as definite integrals, *viz.*,

$$\psi_n(x) = \frac{(-i)^n}{2\sqrt{(2^n \pi n!)}} \left(\frac{m\omega_0}{\pi\hbar}\right)^{1/4} e^{\beta^2 x^2/2} \int_{-\infty}^{\infty} e^{-t^2/4+it\beta x}\, t^n\, dt. \tag{10.5.6}$$

10.6 A Strange Way to Represent Operators

Some operators are very complicated in their effect on functions. Perhaps the most extreme examples are the *non-local operators*, whose action on a function entails integrating the function over all space.

Any operator $\hat{A}$, whether local or non-local, can be written as an *integral* over the product of the function it acts on and a two-variable function called the **kernel**. This is called the *integral representation* of $\hat{A}$ and looks like

$$\hat{A}f(x) = \int_{-\infty}^{\infty} K(x,x')f(x')\,dx', \tag{10.6.1}$$

where $K(x,x')$ is the kernel of $\hat{A}$. We have seen one example of an integral representation of an operator: the Fourier integral relations (4.48) and (4.49) are integral representations of the operators that determine the Fourier transform and inverse Fourier transforms of the functions on which they act.

[a] What is the kernel of the operator that creates the Fourier transform of a (normalized) function $f(x)$?

[b] What is the kernel in the integral representation of the unit operator $\hat{1}$?

[c] **Show** that the integral representation of the linear momentum operator (for a one-dimensional system) is $i\hbar\frac{d}{dx'}\delta(x-x')$.

10.7 The Time Reversal Operator

In the quantum theory of scattering of elementary particles, an important role is played by a most peculiar operator. The **time reversal** operator $\hat{\mathcal{T}}$ is akin to the parity operator of § 10.8 in that it acts on the variables in coordinate- and momentum-space wave functions. In particular, this operator changes x into $+x$ and p into $-p$; these changes correspond to a "time reversal" $t \to -t$:

$$\hat{\mathcal{T}} \;:\; \begin{cases} x \to x \\ p \to -p \end{cases}. \tag{10.7.1}$$

In high-energy physics, an important property of scattering processes is *invariance of the physical properties of a system under time-reversal.* The effect of $\hat{\mathcal{T}}$ on a state function is to produce a function that represents the same state evolving *backward in time.* For example, the mean value of the linear momentum in a "time reversed" state $\hat{\mathcal{T}}\Psi$ is

$$\langle \hat{\mathcal{T}}\Psi \mid \hat{p} \mid \hat{\mathcal{T}}\Psi \rangle = -\langle \Psi \mid \hat{p} \mid \Psi \rangle. \tag{10.7.2}$$

[a] **Prove** that the effect of $\hat{\mathcal{T}}$ on the state function of a system $\Psi(x,t)$ is

$$\hat{\mathcal{T}}\Psi(x,t) = \Psi^*(x,t). \tag{10.7.3}$$

[b] What does $\hat{\mathcal{T}}$ do to the *momentum-space* wave function $\Phi(p) = \mathcal{F}[\Psi]$?

[c] **Prove** that $\hat{\mathcal{T}}$ is anti-linear.

CHAPTER 11

Operators in Quantum Mechanics II

To Commute or Not to Commute

At the present moment
you thrill with the glamour of the situation
and the anticipation of the hunt.
Where would that thrill be
if I had been as definite as a timetable?
I ask only a little patience, Mr. Mac,
and all will be clear to you.

—Sherlock Holmes
in *The Valley of Fear*
by Sir Arthur Conan Doyle

Like a weevil in a tin of flour, the commutator lurked throughout Chap. 10; every time we considered more than one observable, commutators appeared. Thus, in Example 10.6 (§ 10.3) we learned that whether the commutator of two Hermitian operators is nonzero determines whether their product is Hermitian. And, most strikingly, in § 10.8 we discovered that for a symmetric one-dimensional system, the existence of bound-state spatial functions with definite parity is intimately tied to the commutator of the Hamiltonian and the parity operator.

In this chapter the commutator comes into its own. First, we'll investigate simultaneous eigenfunctions and their origin in the commutativity of certain Hermitian operators (§ 11.1), generalizing the remarks at the end of the last chapter. Next, we'll master the manipulation of commutators in algebraic expressions (§ 11.2–11.3). Then, for the *pièce de résistance*, we'll return to uncertainty principles. These principles are among the most powerful and general results of physical theory, for they express fundamental limitations on our ability to acquire through measurement knowledge about two or more observables. As we'll discover in § 11.4–11.5, these physical limitations are a direct consequence of a rather abstract mathematical question: do the operators that represent the observables being measured commute?

The commutator does more than limit our knowledge of the microworld; it appears in conservation laws (§ 11.5–11.6). The form and structure of conservation laws in quantum physics differs strikingly from those of classical physics, as you might expect since we abandoned trajectories in Chap. 2. Yet, there is a connection between these theories: the Correspondence Principle.

Last seen in § 5.7, the Correspondence Principle (as we now know it) tells us that we relate measurable quantities of microscopic systems to their analogs in the macroscope via expectation values. But physics is the study not just of physical quantities but also of laws and relationships between these quantities. In this chapter the commutator will lead us to the connection between the laws of quantum and classical physics. This result, called Ehrenfest's Theorem (§ 11.7), is one of the most satisfying strands in the beautiful tapestry of quantum mechanics.

To bring to a close our exploration of operators in quantum physics we'll look at one final uncertainty relation. Not all uncertainty relations are alike, and in § 11.8 we'll meet a most peculiar one, the energy-time uncertainty relation. We now know enough to understand the origin of this uncertainty principle and its role in physical phenomena such as the decay of excited states of atoms, a physical example I'll bring on stage in § 11.8.

11.1 SIMULTANEOUS EIGENSTATES AND THE COMMUTATOR

In § 10.5 I introduced you to *eigenstates*: special states whose wave functions are eigenfunctions of one (or more) operator(s), *e.g.*,

$$Q\Psi_q = q\Psi_q. \tag{11.1}$$

What is special about an eigenstate of the observable Q is that in this state Q is sharp. That is, an ensemble measurement of Q will yield the value q and only that value; for this state, $\Delta Q = 0$ and $\langle Q \rangle = q$.

Chapter 10 ended on a somewhat enigmatic note, with a remark concerning state functions that satisfy eigenvalue equations like (11.1) *for two operators*. In § 10.8 we proved that such eigenfunctions exist for the operators $\hat{\mathcal{H}}$ and $\hat{\Pi}$ if and only if the potential operator $\hat{V}$ in $\hat{\mathcal{H}} = \hat{T} + \hat{V}$ commutes with $\hat{\Pi}$. This, in turn, happens if and only if $V(x)$ is symmetric.

Generalizing this example, we can define a **simultaneous eigenstate** as one whose state function is an eigenfunction of two (or more) operators. For observables Q and R, we could label such a state function by the eigenvalues q and r in the eigenvalue equations and write these equations as

$$\begin{aligned} \hat{Q}\Psi_{q,r} &= q\Psi_{q,r} \\ \hat{R}\Psi_{q,r} &= r\Psi_{q,r}. \end{aligned} \qquad \text{simultaneous eigenstate of } Q \text{ and } R \tag{11.2}$$

Don't forget that the state function $\Psi_{q,r}$ in Eqs. (11.2) is (implicitly) a function of time. For example, the (time-dependent) wave function that represents a stationary state of a symmetric one-dimensional potential satisfies the simultaneous eigenvalue equations

$$\hat{\mathcal{H}}\Psi(x,t) = E\Psi(x,t) \tag{11.3a}$$

$$\hat{\Pi}\Psi(x,t) = \pm\Psi(x,t). \tag{11.3b}$$

In a simultaneous eigenstate, both observables Q and R are sharp, so

$$\langle Q \rangle = q \qquad \Delta Q = 0 \tag{11.4a}$$

$$\langle R \rangle = r \qquad \Delta R = 0. \tag{11.4b}$$

In general, of course, the statistical quantities $\langle Q \rangle$, $\langle R \rangle$, ΔQ, and ΔR depend on time. But if, as in Eqs. (11.4), the state function used to calculate them represents a simultaneous eigenstate of both observables, then these quantities are independent of time. [Nonetheless, we *formally* evaluate these quantities from the state function Ψ at the same time. This seemingly pedantic point is important because of the implications of Eqs. (11.4) for measurement.]

Question 11–1

Prove that Eqs. (11.4) follow from (11.2) for a simultaneous eigenstate of Q and R.

The essential point of Eqs. (11.4) is that a "simultaneous measurement" of Q and R—one in which both properties are measured at the same time—yields *uniquely* the values q and r. To appreciate the physical significance of this remark, you'll have to wait until Chap. 13, where we confront the problem of measurement. But for a preview, look at Fig. 11.1, which is a generalization of our schematic for measurement of a single observable (Fig. 10.1).

Question 11–2

Consider a state of the infinite square well that is represented by the state function

$$\Psi(x,t) = \sqrt{\frac{2}{L}} \cos(2\pi \frac{x}{L}).$$

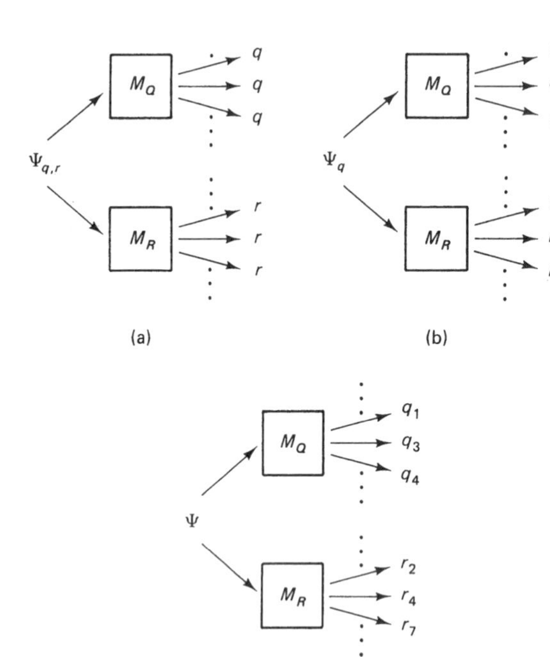

Figure 11.1 A schematic representation of *simultaneous* measurement of two observables, Q and R, on a system. Three quantum states are considered. The state in (a) is a simultaneous eigenstate of these two observables, so all members of the ensemble yield the same value of Q and the same value of R. The state in (b) is an eigenstate of Q but not of R, so measurement of the latter observable yields a distribution of values, *i.e.*, the value of R is uncertain. In the state in (c), which is not an eigenstate of either observable, both measurements yield a distribution of results.

Is this state a simultaneous eigenfunction of $\hat{\mathcal{H}}$ and $\hat{p}$? Of any two other observables? Explain your reasoning.

Do All Pairs of Operators Define Simultaneous Eigenfunctions?

The definition (11.2) of a simultaneous eigenstate raises two questions about any pair of observables Q and R:

1. Is *a particular state* a simultaneous eigenstate of Q and R?
2. Do *any* simultaneous eigenstates of Q and R exist?

In an important sense, the second of these questions is more general than the first, which can easily be answered by checking whether or not the particular function satisfies Eqs. (11.2) [*c.f.*, Question 11–2].

The answer to the second question depends on the operators $\hat{Q}$ and $\hat{R}$. We know, for example, that there exist simultaneous eigenfunctions of $\hat{\Pi}$ and $\hat{\mathcal{H}}$ (for a symmetric potential) [see Eqs. (11.3)]. But there do *not* exist simultaneous eigenfunctions of $\hat{x}$ and $\hat{p}$. According to the Heisenberg Uncertainty Principle, *for any state of any system* the uncertainty product of x and p is positive:

$$\Delta x(t)\, \Delta p(t) \geq \tfrac{1}{2}\hbar > 0. \tag{11.5}$$

This principle explicitly prohibits the existence in nature of states in which x or p (or both) is sharp; *i.e.*, a simultaneous eigenfunction of $\hat{x}$ and $\hat{p}$ cannot represent a physically realizable quantum state. Clearly, not all pairs of observables have simultaneous eigenstates.

This observation raises several questions. What property of $\hat{x}$ and $\hat{p}$ proscribes the existence of simultaneous eigenfunctions of these operators? And what property of other pairs of operators allows the existence of simultaneous eigenfunctions? Given arbitrary observables Q and R, how can we ascertain whether or not simultaneous eigenstates of Q and R exist?

We already have several clues. We know (from § 10.8) that the existence of simultaneous eigenstates of $\hat{\mathcal{H}}$ and $\hat{\Pi}$ hinges on whether or not $[\hat{\mathcal{H}}, \hat{\Pi}] = 0$; this commutator is zero if $V(x)$ is symmetric, in which case such eigenstates do exist, but it's nonzero otherwise. We also know (from Chap. 5) that $\hat{x}$ and $\hat{p}$, which don't support simultaneous eigenfunctions, don't commute, for $[\hat{x}, \hat{p}] = i\hbar\hat{1}$. Could it be that only operators that commute support simultaneous eigenfunctions? You bet it could.

The Commutator is the Key

One of the most important theorems in quantum mechanics states conditions for the existence of simultaneous eigenfunctions:

> ***Proclamation 11.1***
>
> A complete set of simultaneous eigenfunctions of two Hermitian operators exists if and only if the operators commute.

Note carefully that this statement says *two* things:

1. Operators that commute define a complete set of simultaneous eigenfunctions.
2. Two operators that share a complete set of simultaneous eigenfunctions commute.

Before discussing this proclamation, let's write down its mathematical form. If Hermitian operators $\hat{Q}$ and $\hat{R}$ obey the commutation relation

$$[\hat{Q}, \hat{R}] = 0, \tag{11.6}$$

then they define a set of functions $\{\chi_{q,r}\}$, each element of which satisfies the equations

$$\hat{Q}\chi_{q,r} = q\chi_{q,r} \tag{11.7a}$$

$$\hat{R}\chi_{q,r} = r\chi_{q,r}. \tag{11.7b}$$

Moreover, the set $\{\chi_{q,r}\}$ is *complete*, so any well-behaved function Ψ can be expanded as[1]

$$\boxed{\Psi = \sum_{\text{all } q} \sum_{\text{all } r} c_{q,r}\chi_{q,r}} \qquad \left[\begin{array}{c}\text{expansion in complete set}\\ \text{of simultaneous eigenfunctions}\end{array}\right] \tag{11.8}$$

[1] In writing such expansions, be careful not to drop any variables on the right-hand side. Any independent variables on which Ψ depends that are not variables of the eigenfunctions $\chi_{q,r}$ must appear in the coefficients $c_{q,r}$. Usually, this means that these coefficients depend on time, as in Eq. (11.10). Another common situation will surface in Volume II, where we'll study quantum states of a particle in three dimensions. In spherical coordinates, the wave functions for such states depend on r, θ, and ϕ, and we often expand these wave functions in sets that are complete in θ and ϕ but not r. In such expansions, the coefficients depend on r and on t.

The coefficient $c_{q,r}$ is the **projection** of Ψ on the eigenfunction $\chi_{q,r}$.

Aside: The Importance of Completeness. The *completeness* of the set of simultaneous eigenfunctions defined by two commuting operators is a crucial feature of Proclamation 11.1 and is often overlooked or misunderstood by novice quantum mechanics. *The existence of a single simultaneous eigenfunction of two operators is no guarantee that the operators commute.* The most famous such counterexample occurs in the study of the *orbital angular momentum operator* $\hat{\mathbf{L}}$. This is a vector operator, *i.e.*, it has three component operators, $\hat{L}_x$, $\hat{L}_y$, and $\hat{L}_z$. No two of these commute; *e.g.*, $\left[\hat{L}_x, \hat{L}_y\right] \neq 0$. Yet, there exists *one* quantum state in which L_x, L_y, and L_z are sharp: the state in which the orbital angular momentum is zero. But a *complete set* of simultaneous eigenfunctions of any two component operators—$\hat{L}_x$ and $\hat{L}_y$, say—does not exist. We will find out why such counterexamples exist in § 11.4, where we will connect the commutator to uncertainty relations. And we will investigate $\hat{\mathbf{L}}$ and its properties in Volume II.

We can evaluate the coefficients in the expansion (11.8) from the state function via our usual expansion (10.109*b*),

$$c_{q,r} = \langle \chi_{q,r} \mid \Psi \rangle. \tag{11.9}$$

These coefficients depend on time. For example, the wave function of a single particle in three dimensions depends on three spatial variables $\mathbf{r}$ (x, y, and z in rectangular coordinates; r, θ, and ϕ in spherical coordinates) and on time. We can expand such a function, $\Psi(\mathbf{r}, t)$, in a complete set $\{\chi_{q,r}(\mathbf{r})\}$ of eigenfunctions of (commuting) operators $\hat{Q}$ and $\hat{R}$ that act on functions of $\mathbf{r}$, *i.e.*,

$$\Psi(\mathbf{r}, t) = \sum_{\text{all } q} \sum_{\text{all } r} c_{q,r}(t) \chi_{q,r}(\mathbf{r}). \tag{11.10}$$

The projection $c_{q,r}(t)$ of $\Psi(\mathbf{r}, t)$ on $\chi_{q,r}$ is, according to Eq. (11.9),

$$c_{q,r}(t) = \int_{\text{all space}} \chi^*_{q,r}(\mathbf{r}) \Psi(\mathbf{r}, t)\, dv. \tag{11.11}$$

In the next chapter we'll explore the uses and interpretation of Eq. (11.11). For now, just note the time dependence of the projections, $c_{q,r}(t)$, and the simple extension of Eq. (10.110) to get (11.9).

Argument: The proof that commuting operators define a complete set of simultaneous eigenfunctions is rather abstract and convoluted; rather than reproduce it here, I'll refer the curious to more advanced texts.[2] But I do want to show you the proof of the other side of this theorem, that *two operators that share a complete set of simultaneous eigenfunctions necessarily commute.* This proof illustrates techniques of operator algebra of great power and applicability.

This demonstration hinges on the meaning of completeness and commutativity and on a simple consequence of the equations satisfied by the functions $\chi_{q,r}$. Consider what

[2]Actually, this proof is not too hard if we are dealing only with non-degenerate eigenvalues. If we operate on the eigenvalue equation $\hat{Q}\chi_q = q\chi_q$ with $\hat{R}$ and use the (assumed) commutativity of the two operators, we obtain $\hat{Q}[\hat{R}\chi_q] = q[\hat{R}\chi_q]$. Now, if q is non-degenerate, then the function $\hat{R}\chi_q$ must be proportional to χ_q, and the proportionality constant is just the eigenvalue of $\hat{R}$, *i.e.*, $\hat{R}\chi_q = r\chi_q$. This proves that χ_q is also an eigenfunction of $\hat{R}$. No such simple argument, however, can cope with the common case of degenerate eigenvalues. For details, see § V.7 of *Elementary Quantum Mechanics* by D. Saxon (San Francisco: Holden-Day, 1968).

happens when we apply the operator $[\hat{Q}, \hat{R}]$ to one of the simultaneous eigenfunctions of $\hat{Q}$ and $\hat{R}$:

$$[\hat{Q}, \hat{R}]\chi_{q,r} = \hat{Q}\hat{R}\chi_{q,r} - \hat{R}\hat{Q}\chi_{q,r}. \tag{11.12}$$

Because $\chi_{q,r}$ satisfies the eigenvalue equations (11.2), the operators on the right-hand side of (11.12) yield

$$[\hat{Q}, \hat{R}]\chi_{q,r} = qr\chi_{q,r} - rq\chi_{q,r}, \tag{11.13a}$$

$$= (qr - rq)\chi_{q,r}. \tag{11.13b}$$

But *numbers commute.* So the commutator of two commuting observables "annihilates" any simultaneous eigenfunction of its constituent operators,

$$[\hat{Q}, \hat{R}]\chi_{q,r} = 0. \tag{11.14}$$

I hope you can see that *Equation (11.14) does not, in itself, prove that the operators* $\hat{Q}$ *and* $\hat{R}$ *commute. Remember that to ascertain whether or not two operators commute, we must let their commutator act on an arbitrary (physically-admissible) state function* Ψ *and see if the result is zero*:

$$[\hat{Q}, \hat{R}]\Psi \stackrel{?}{=} 0. \tag{11.15}$$

It is at this point that the *completeness* of the set of functions $\{\chi_{q,r}\}$ comes in handy: we can expand the state function Ψ à la Eq. (11.8) and insert the result into (11.15), *viz.*,

$$[\hat{Q}, \hat{R}] \sum_{\text{all } q} \sum_{\text{all } r} c_{q,r}\chi_{q,r} \stackrel{?}{=} 0. \tag{11.16}$$

The commutator doesn't affect the expansion coefficients—they're just complex numbers–so this operator slides through to the eigenfunction, giving

$$\sum_{\text{all } q} \sum_{\text{all } r} c_{q,r}[\hat{Q}, \hat{R}]\chi_{q,r} \stackrel{?}{=} 0. \tag{11.17}$$

But according to Eq. (11.14), each term on the left-hand side of (11.17) is zero. Hence the sum is zero and we conclude that

$$[\hat{Q}, \hat{R}] = 0, \tag{11.18}$$

i.e., $\hat{Q}$ and $\hat{R}$ commute.

Q. E. D.

Examples and Counterexamples

We will encounter *lots* of examples of commuting operators and their simultaneous eigenfunctions for two- and three-dimensional systems in Volume II. For instance, the Hamiltonian of the hydrogen atom, which consists of one electron and one proton, commutes with the operator $\hat{L}^2$ that represents the *square of the orbital angular momentum of the electron with respect to the proton.* Consequently there exists a complete set of stationary states—eigenfunctions of the Hamiltonian—that are also eigenfunctions of $\hat{L}^2$. The wave functions of these simultaneous eigenstates represent the familiar atomic orbitals ($1s$, $2s$, $2p_0$, etc.) that you probably studied in high school chemistry. These eigenfunctions, and their counterparts for orbitals of other atoms, are the building blocks of the quantum mechanics of atomic and molecular physics.

For now, though, our principal example of a pair of commuting operators is $\hat{\Pi}$ and the Hamiltonian of a symmetric potential. In § 10.8 we showed that $\hat{\Pi}$ and the

Hamiltonian of a SHO commute, and we discussed the corresponding complete set of simultaneous eigenfunctions, $\{\psi_n(x); n = 0, 1, 2, \dots\}$. The form of the expansion of an arbitrary wave function $\Psi(x,t)$ in this set is slightly simpler than Eq. (11.8) because it requires only one summation, *i.e.*

$$\Psi(x,t) = \sum_{n=0}^{\infty} c_n(t)\psi_n(x). \tag{11.19}$$

Formally, the general result (11.8) prescribes a second summation over all allowed eigenvalues of $\hat{\Pi}$: -1 and $+1$. But *for this system* such a summation would be redundant, because each Hamiltonian eigenfunction $\psi_n(x)$ has definite parity, even or odd.[3]

Be careful not to generalize too hastily. Simplifications such as the elimination of a summation in (11.19) are rare, and one must be wary of blithely throwing away summations in an expansion of a wave function in a set of simultaneous eigenfunctions. In general, one should live by the following rule:

Rule

When expanding a function in a complete set of simultaneous eigenfunctions of two (or more) commuting operators, sum over all eigenvalues of all observables.

In our investigation into continuum states we encountered another pair of commuting operators: $\hat{T}$ and $\hat{p}$. For example, the potential of a free particle is zero, so $\hat{\mathcal{H}} = \hat{T}$. Hence the Hamiltonian commutes with $\hat{p}$, and these commuting operators define the simultaneous eigenfunctions

$$\psi_{E,p}(x) = \frac{1}{\sqrt{2\pi\hbar}} e^{-ipx/\hbar} \tag{11.20}$$

by the eigenvalue equations[4]

[3] I could, of course, be tiresomely pedantic and write Eq. (11.19) as

$$\Psi(x,t) = \sum_{\text{all } E} \sum_{q=-1}^{+1} c_{E,q}(t)\psi_E(x).$$

I would then note that for even-parity Hamiltonian eigenfunctions the coefficient $c_{E,-1} = 0$ and for odd-parity eigenfunctions, $c_{E,+1} = 0$, which would lead me back to a single summation, over n.

[4] If you read the footnotes in Chap. 10 concerning expansions in a complete set of eigenfunctions with continuous eigenvalues, you know how to write such an expansion for $\{\psi_{E,p}(x)\}$. We integrate rather than sum over the eigenvalues p. We need *not* integrate over E because the free-particle equation $E = p^2/2m$ establishes a correspondence between the values of p and the energies E. (This correspondence is not, however, one-to-one.) Consequently if we integrate over p from $-\infty$ to $+\infty$ we are assured of picking up all the eigenfunctions in the set $\{\psi_{E,p}(x)\}$. (That is to say, the index E on these eigenfunctions is, for this system, redundant.) The complete set expansion,

$$\Psi(x,t) = \int_{-\infty}^{\infty} \left[\Phi(p)e^{-i\omega t/\hbar}\right] \left[\frac{1}{\sqrt{2\pi\hbar}} e^{ipx/\hbar}\right] dp = \int_{-\infty}^{\infty} c_p(t)\psi_{E,p}(x)\, dp,$$

is, of course, just the Fourier integral for a wave packet in terms of the momentum-space amplitude $\Phi(p)$. As I noted in Chap. 10, this momentum-space state function appears in the projection $c_p(t)$ of $\Psi(x,t)$ on $\psi_{E,p}(x)$.

$$\hat{p}\psi_{E,p}(x) = p\psi_{E,p}(x) \tag{11.21a}$$

$$\hat{\mathcal{H}}\psi_{E,p}(x) = E\psi_{E,p}(x) \tag{11.21b}$$

where

$$E = \frac{p^2}{2m}. \tag{11.21c}$$

What Commutativity Does Not Imply

Just as important as mastering the implications of the commutativity of two operators is knowing what this property does not imply. Readers new to quantum mechanics often *incorrectly* conclude that if $\hat{Q}$ and $\hat{R}$ commute, then all eigenfunctions of one operator—say, $\hat{Q}$—are necessarily eigenfunctions of the other. But only under very special circumstances is this conclusion correct.[5] To see a simple, familiar situation where it is wrong, consider Table 11.1, which contains three quantum states of a one-dimensional simple harmonic oscillator. State 1 is stationary, so its wave function is a simultaneous eigenfunction of the energy and parity operators. The wave function for State 2 is a linear combination of two *even-parity* stationary-state functions: the wave functions for the ground state and for the second excited state. State 2 is not, of course, an eigenstate of energy, but because of the particular stationary states that make it up, it *is* an eigenstate of parity. So the wave function that represents State 2 is an eigenfunction of $\hat{\Pi}$ but not of $\hat{\mathcal{H}}$, even though these two operators do commute. Finally, the superposition of the ground and first excited states, State 3, is not an eigenfunction of either operator.

My point is that the commutativity of two operators certifies *only* that simultaneous eigenfunctions of the two operators are there if you need them. This property says nothing about a particular state of the system; such a state may or may not be an eigenstate of one or both of the commuting operators. Finally, and in this regard, it's vital that you keep straight in your mind the difference between *simultaneous eigenfunctions* and a *simultaneous eigenstate*. The former need not be (and, in general, are not) state functions. The latter is a special quantum state whose wave function Ψ is a simultaneous eigenfunction of two operators.

TABLE 11.1 THREE STATES OF A SYMMETRIC ONE-DIMENSIONAL SYSTEM.

State	Wave Function	$\hat{\mathcal{H}}$	$\hat{\Pi}$
1	$\Psi_n(x,t) = \psi_n(x)e^{-iE_nt/\hbar}$	$\Delta E = 0$	$\Delta\Pi = 0$
2	$\Psi(x,t) = \frac{1}{\sqrt{2}}\left[\Psi_0(x,t) + \Psi_2(x,t)\right]$	$\Delta E > 0$	$\Delta\Pi = 0$
3	$\Psi(x,t) = \frac{1}{\sqrt{2}}\left[\Psi_0(x,t) + \Psi_1(x,t)\right]$	$\Delta E > 0$	$\Delta\Pi > 0$

[5]The simplest case in which all eigenfunctions of $\hat{Q}$ are necessarily eigenfunctions of $\hat{R}$ is when the eigenvalue q is non-degenerate; in this case, χ_q is necessarily an eigenfunction of $\hat{R}$. For more on this matter, see Chap. II.D of *Quantum Mechanics* by C. Cohen-Tannoudji, B. Diu, and F. Laloë (New York: Wiley, 1977).

Clearly, the commutator plays a crucial role in determining the nature of possible quantum states of a microscopic system. (Any residual doubts on this matter can be dispelled by peeking ahead to § 11.4.) To become more agile manipulators of commutators, let's take a (brief) detour into the algebra of these beasties.

11.2 EVALUATING COMMUTATORS: THE RIGHT WAY, THE WRONG WAY, AND THE EASY WAY

> A singer can shatter a glass
> with the proper high note. . . .
> But the simplest way for anyone
> to break a glass is
> simply to drop it on the floor.
>
> —Marius to Lestat in
> *The Vampire Lestat*
> by Anne Rice

Commutators are ubiquitous; in our voyage through the microworld, we will meet them at every turn. So far, we've encountered only rather simple commutators. But to navigate the deeper, more complex waters to come, we must be able to "evaluate" a commutator—*i.e.*, to reduce it to the simplest (operator) form. We might, for example, have to contend with the likes of $[\hat{p}, \hat{x}^2]$, $[\hat{T}, \hat{x}]$, or even $[\hat{Q}, e^{\hat{R}}]$.

The most challenging commutators are those involving *operator functions*, such as $\hat{x}^2$, $\hat{p}^2$, or $e^{\hat{R}}$. We know from § 10.2 how to subdue such operators: expand 'em in their power series [see Eqs. (10.12)–(10.13)]. And we know from § 5.5 the rule for manipulating commutators, which, after all, are just operators:

> ***Rule***
>
> Evaluate an operator expression by letting the expression act on a function of the variables in the constituent operators.

If, for example, we want to prove, for arbitrary operators $\hat{A}$ and $\hat{B}$ the relation

$$[\hat{A}, \hat{B}] = -[\hat{B}, \hat{A}], \tag{11.22}$$

we may be tempted to expand the left-hand side, then switch the order of the resulting products, *viz.*,

$$[\hat{A}, \hat{B}] = \hat{A}\hat{B} - \hat{B}\hat{A} \tag{11.23a}$$

$$= -(\hat{B}\hat{A} - \hat{A}\hat{B}) \tag{11.23b}$$

$$= -[\hat{B}, \hat{A}]. \tag{11.23c}$$

This approach isn't *wrong*, but it fosters bad habits.

In this section I'll show you safe and efficient ways to evaluate commutators, point out a couple of traps for the unwary, and provide useful tables of commutator relations. As you work through this section, please obey the above rule (even when, for purposes of illustration, I violate it). After you have practiced (a lot) with commutator algebra, you will be able to confidently and correctly perform algebraic feats of legerdemain in the style of (11.23).

Basic Commutator Relations

We'll consider first commutator relations satisfied by arbitrary operators that may or may not be Hermitian. (So I'll denote these operators by $\hat{A}$, $\hat{B}$, $\hat{C}$, ...) Then we'll focus on commutator relations for Hermitian operators, *i.e.*, those that involve $\hat{x}$, $\hat{p}$, and functions thereof.

Table 11.2 contains most commutator relations you're likely to need. You can prove each of these operator equations by following the guidelines suggested above.

TABLE 11.2 HANDY COMMUTATOR RELATIONS.

Relation	
$\left[\hat{A}, \hat{A}^n\right] = 0$	(I)
$\left[\hat{A}, \hat{B}\right] = -\left[\hat{B}, \hat{A}\right]$	(II)
$\left[\hat{A}, c\hat{B}\right] = c\left[\hat{A}, \hat{B}\right]$	(III)
$\left[\hat{A}, (\hat{B}+\hat{C})\right] = \left[\hat{A}, \hat{B}\right] + \left[\hat{A}, \hat{C}\right]$	(IV)
$\left[\hat{A}, \hat{B}\hat{C}\right] = \left[\hat{A}, \hat{B}\right]\hat{C} + \hat{B}\left[\hat{A}, \hat{C}\right]$	(V)
$\left[\hat{A}, \hat{f}(\hat{A})\right] = 0$	(VI)

Question 11–3

The *order* of operators is, as we know, crucial and must not be cavalierly changed. To illustrate, prove that the relation obtained by interchanging the order of operators in the second term on the right-hand side of Property V is not, in general, true. That is, prove that

$$\left[\hat{A}, \hat{B}\hat{C}\right] \neq \left[\hat{A}, \hat{B}\right]\hat{C} + \left[\hat{A}, \hat{C}\right]\hat{B}. \tag{11.24}$$

You can prove this either by general argument or by counterexamples. Under what conditions, if any, are the left- and right-hand sides of Eq. (11.24) equal?

Question 11–4

Prove that

$$\left[\hat{A}, \left[\hat{B}, \hat{C}\right]\right] + \left[\hat{B}, \left[\hat{C}, \hat{A}\right]\right] + \left[\hat{C}, \left[\hat{A}, \hat{B}\right]\right] = 0. \tag{11.25}$$

As I noted above, the ringers in commutator algebra are relations involving operator *functions*, but we can conquer these using the power series expansions of the functions. For example, to prove Property VI, $\left[\hat{A}, \hat{f}(\hat{A})\right] = 0$, we need only expand the operator $\hat{f}(\hat{A})$ à la Eq. (10.13), using the coefficients b_j of the series expansion of the *function* $f(A)$:

$$\hat{f}(\hat{A}) = \sum_{j=0}^{\infty} b_j \hat{A}^j. \tag{11.26}$$

With this expansion, the commutator on the left-hand side of Property VI, when acting on an arbitrary function $g(x)$, gives[6]

$$\left[\hat{A}, \hat{f}(\hat{A})\right] g(x) = \sum_{j=0}^{\infty} b_j \left[\hat{A}, \hat{A}^j\right] g(x). \tag{11.27}$$

According to Property I, $\left[\hat{A}, \hat{A}^n\right] = 0$ for any power n, so each term in (11.27) is zero. This proves Property VI.

And so forth.

Question 11–5

Prove that

$$\left[\hat{A}, \hat{B}\right] = 0 \quad \Longrightarrow \quad \left[\hat{B}, \hat{f}(\hat{A})\right] = 0. \tag{11.28}$$

Commutation Relations Involving $\hat{x}$ *and* $\hat{p}$

Most operators in quantum mechanics can be written as functions of $\hat{x}$ and $\hat{p}$. So the most important commutation relations are those involving these operators and their powers. We can evaluate all such commutators using the properties in Table 11.2 and the fundamental relation

$$\boxed{\left[\hat{x}, \hat{p}\right] = i\hbar\hat{1}} \tag{11.29}$$

which we proved in Example 5.4.

Example 11.1. A Commutator for the Linear Potential

In the text and problems for Chap. 9 we saw the usefulness of the linear potential $V(x) = ax$, where a is a constant that depends on the particular system. If, in the course of applying this model to a physical system, we run into the commutator $\left[\hat{\mathcal{H}}, \hat{T}\right]$, how do we proceed?

Well, since the Hamiltonian is $\hat{\mathcal{H}} = \hat{T} + \hat{V}$ and the kinetic energy operator $\hat{T} = \hat{p}^2/(2m)$ commutes with itself, the commutator of these two operators trivially reduces to

$$\left[\hat{\mathcal{H}}, \hat{T}\right] = \left[\hat{V}, \hat{T}\right] = a\left[\hat{x}, \hat{T}\right] = \frac{a}{2m}\left[\hat{x}, \hat{p}^2\right]. \tag{11.30}$$

One way to simplify the remaining commutator $\left[\hat{x}, \hat{p}^2\right]$ is to expand $\hat{p}^2$ and use Property V in Table 11.2, *viz.*,

$$\left[\hat{x}, \hat{p}^2\right] f(x) = \left[\hat{x}, \hat{p}\hat{p}\right] f(x) \tag{11.31a}$$

$$= \left[\hat{x}, \hat{p}\right]\hat{p} f(x) + \hat{p}\left[\hat{x}, \hat{p}\right] f(x) \tag{11.31b}$$

$$= 2i\hbar\hat{p} f(x). \tag{11.31c}$$

So the commutator of $\hat{\mathcal{H}}$ and $\hat{T}$ is actually a much simpler operator than its initial form suggests:

$$\left[\hat{\mathcal{H}}, \hat{T}\right] = \frac{i\hbar}{m} a\hat{p}. \tag{11.32}$$

[6] For simplicity, I have assumed in this section that all arbitrary operators act on functions of x. But the relations we are proving all pertain to operators that act on functions of more variables, as we'll see in Volume II.

Always try to simplify complicated commutators as soon as they appear in problems. The operator on the right-hand side of (11.32) is much easier to work with than the one on the left-hand side.

Machinations such as those of Example 11.1 are so common in quantum mechanics that a table of commutation relations involving these fundamental operators is a valuable time saver. For future reference, I've compiled such a compendium in Table 11.3.[7]

TABLE 11.3 COMMUTATOR RELATIONS INVOLVING THE POSITION AND MOMENTUM OPERATORS

Relation	
$\left[\hat{x}^2, \hat{p}\right] = 2i\hbar\hat{x}$	(I)
$\left[\hat{x}, \hat{p}^2\right] = 2i\hbar\hat{p}$	(II)
$\left[\hat{x}^2, \hat{p}^2\right] = 2i\hbar\left(2\hat{x}\hat{p} - i\hbar\hat{1}\right)$	(III)
for arbitrary operator functions $\hat{f}(\hat{p})$ and $\hat{g}(\hat{x})$:	
$\left[\hat{x}, \hat{f}(\hat{p})\right] = i\hbar\frac{d}{d\hat{p}}\hat{f}(\hat{p})$	(IV)
$\left[\hat{p}, \hat{g}(\hat{x})\right] = -i\hbar\frac{d}{d\hat{x}}\hat{g}(\hat{x})$	(V)
for a Hamiltonian $\hat{\mathcal{H}} = \hat{T} + \hat{V}(x)$:	
$\left[\hat{\mathcal{H}}, \hat{x}\right] = -\frac{i\hbar}{m}\hat{p}$	(VI)
$\left[\hat{\mathcal{H}}, \hat{p}\right] = i\hbar\frac{\partial V}{\partial x}\hat{1}$	(VII)

Question 11–6

Use Property IV in Table 11.3 to prove Property VI. Use Property IV to prove Property II. Can you prove Property IV?

Question 11–7

Prove that for any power n,

$$\left[\hat{x}, \hat{p}^n\right] = i\hbar n\hat{p}^{n-1}. \tag{11.33}$$

The easiest way to prove this relation is via a technique called **mathematical induction**. First prove that it is true for the lowest power, $n = 1$. Then *assume* it to be true for power n and prove that it therefore holds for power $n + 1$. (Since the relation is true for $n = 1$, it must be true for $n = 2$ and hence for all higher values of n, by the same "stair-step" argument.)

[7]Note carefully that the arbitrary functions in Properties IV and V are functions of one or the other of the operators $\hat{x}$ and $\hat{p}$. Under some circumstances these properties apply to functions of both operators, such as $\hat{h}(\hat{x}, \hat{p})$. But because $\hat{x}$ and $\hat{p}$ don't commute, the applicability of these properties depends on the nature of the function $h(x, p)$. So be very careful if you ever have to work out a commutator involving such a function and want to use Properties IV and V.

Trouble with Commutators

Mulling over Example 11.1, you may have wondered, "Why on earth does he persist in cluttering up derivations such as Eqs. (11.31) with arbitrary functions? Surely so simple a derivation can safely be performed without lugging along $f(x)$."

Right you are. We could work through this derivation using operator equations with minimal chance of error if, as in (11.31), we work at a rather high "level of abstraction." That is, if at no point do we introduce an *explicit form* for the instructions $\hat{x}$ or $\hat{p}$—a form such as $\hat{p}$ in the position representation ($\hat{p} = -i\hbar d/dx$) or $\hat{x}$ in the momentum representation ($\hat{x} = i\hbar d/dp$).

This example suggests a useful guideline for all operator manipulations, whether or not commutators are involved:

> ***Rule***
>
> Whenever possible, manipulate operators as abstract entities rather than as explicit instructions in a particular representation.

To see this rule in action and to appreciate how ignoring it can be hazardous to the health of your solution, let's consider another example.

Example 11.2. A Commutator for the Simple Harmonic Oscillator

Suppose we need to evaluate $\left[\hat{\mathcal{H}}, \hat{p}\right]$ for the SHO Hamiltonian $\hat{\mathcal{H}} = \hat{T} + \frac{1}{2}m\omega_0^2\hat{x}^2$. [Such an evaluation is a critical step in determining the constants of the motion of this system (see § 11.5).]

Let's first express the operators in this commutator in the position representation and let them act on a function $f(x)$. Noting that $\hat{T}$ commutes with $\hat{p}$, we begin with

$$\left[\hat{\mathcal{H}}, \hat{p}\right] f(x) = \left[\hat{V}, \hat{p}\right] f(x) = \frac{1}{2}m\omega_0^2\left[\hat{x}^2, \hat{p}\right] f(x). \tag{11.34}$$

Getting explicit, we write the commutator on the right-hand side of (11.34) as

$$\left[\hat{x}^2, \hat{p}\right] f(x) = \hat{x}^2\hat{p}f(x) - \hat{p}\hat{x}^2 f(x) \tag{11.35a}$$

$$= -i\hbar\left\{x^2\frac{d}{dx}f(x) - \frac{d}{dx}\left[x^2 f(x)\right]\right\}. \tag{11.35b}$$

The derivatives in (11.35*b*) are easy to evaluate, *e.g.*,

$$\frac{d}{dx}\left[x^2 f(x)\right] = x^2\frac{df}{dx} + 2xf(x). \tag{11.36}$$

We now have only to insert (11.36) into (11.35*b*) to simplify the right-hand side of the latter to

$$\left[\hat{x}^2, \hat{p}\right] f(x) = -i\hbar x^2\frac{df}{dx} + i\hbar x^2\frac{df}{dx} + 2i\hbar x f(x) \tag{11.37}$$

$$\left[\hat{x}^2, \hat{p}\right] f(x) = 2i\hbar x f(x). \tag{11.38}$$

Substituting this simple result into the right-hand side of Eq. (11.34), we conclude (*finally*) that

$$\left[\hat{\mathcal{H}}, \hat{p}\right] = i\hbar m\omega_0^2\hat{x}. \qquad \text{[for the SHO]} \tag{11.39}$$

Is all this work really necessary? I would argue that it is. Here (with trenchant commentary) is an example of the kind of trouble one might get into working this problem with operators in the position representation but no functions:

$$\left[\hat{\mathcal{H}}, \hat{p}\right] = \left[\frac{1}{2}m\omega_0^2 x^2, -i\hbar\frac{d}{dx}\right] \qquad \text{okay} \qquad (11.40a)$$

$$= -\frac{i\hbar}{2}m\omega_0^2\left(x^2\frac{d}{dx} - \frac{d}{dx}x^2\right) \qquad \text{Bad judgment, but okay} \qquad (11.40b)$$

$$= -\frac{i\hbar}{2}m\omega_0^2\left(x^2\frac{d}{dx} - 2x\right) \qquad \text{oops!} \qquad (11.40c)$$

$$\Longrightarrow \qquad \left[\hat{\mathcal{H}}, \hat{p}\right] = \frac{1}{2}m\omega_0^2\left(\hat{x}^2\hat{p} - 2i\hbar\hat{x}\right). \qquad \text{nope} \qquad (11.40d)$$

We got into trouble at Eq. (11.40*c*) because we were trying to manipulate derivatives as operators rather than manipulating derivatives of functions *as functions*. This is, obviously, the *wrong* way to evaluate a commutator.

The *easiest* way to evaluate $\left[\hat{\mathcal{H}}, \hat{p}\right]$ for the SHO is the most abstract way :

$$\left[\hat{\mathcal{H}}, \hat{p}\right] = \left[\hat{T}, \hat{p}\right] + \left[\hat{V}, \hat{p}\right] \qquad (11.41a)$$

$$= \left[\hat{V}, \hat{p}\right] \qquad \text{[Table 11.2: Property I]} \qquad (11.41b)$$

$$= \frac{1}{2}m\omega_0^2\left[\hat{x}^2, \hat{p}\right] \qquad \text{[Table 11.2: Property III]} \qquad (11.41c)$$

$$= \frac{1}{2}m\omega_0^2\left(2i\hbar\hat{x}\right) = i\hbar m\omega_0^2\hat{x}. \qquad \text{[Table 11.3: Property I]} \qquad (11.41d)$$

Notice how this evaluation proceeds without mishap *if we avoid an explicit representation.*

Question 11–8

Evaluate the commutator of $\hat{x}$ and $\hat{\mathcal{H}}$ for the SHO two ways (the "right" and "easy" ways).

The bottom line in operator manipulations—as in all problem solving—is the theme I keep harping on: *never proceed through algebra mindlessly, on autopilot. Instead, think at every step about what you are doing and the alternate paths open to you.* If you solve problems this way you will learn more, have more fun, and be far less likely to wind up with the wrong answer.

11.3 OPERATOR MANIPULATIONS AND MATRIX REPRESENTATIONS

Many and varied are the theorems of quantum mechanics. But few are the theorems you'll find in this book. Our focus is not on mathematical proofs but on quantum physics. Still, some theorems are so powerful, so useful, that they cannot be ignored. The one we'll prove in this section illustrates the interplay of various aspects of the mathematics of operators that we've examined in Chap. 10 and § 11.1–11.2.[8]

[8]The usefulness of this theorem will not be apparent until Volume II, where we'll learn how to solve problems so complicated that their Schrödinger equation is not amenable to exact solution. Implementing these *approximation methods* is much easier if full use is made of Proclamation 11.2. Still, it's a short proof, a useful exercise, so I recommend that you don't skip it.

Consider two *commuting Hermitian* operators: Q and R such that

$$\hat{Q}^\dagger = \hat{Q} \text{ and } \hat{R}^\dagger = \hat{R} \tag{11.42a}$$

$$\left[\hat{Q}, \hat{R}\right] = 0. \tag{11.42b}$$

Suppose that χ_q and $\chi_{q'}$ are two *distinct* eigenfunctions of $\hat{Q}$ [*i.e.*, with different eigenvalues $(q \neq q')$] but that these functions are *not eigenfunctions of* $\hat{R}$. Therefore, the only eigenvalue equations these functions satisfy are

$$\hat{Q}\chi_q = q\chi_q \tag{11.43a}$$

$$\hat{Q}\chi_{q'} = q'\chi_{q'}. \qquad (q' \neq q) \tag{11.43b}$$

We'll now prove that the matrix element of $\hat{R}$ between χ_q and $\chi_{q'}$ is zero:

$$\langle \chi_q \mid \hat{R} \mid \chi_{q'} \rangle = 0. \tag{11.44}$$

Question 11–9

Prove that the matrix element of $\hat{Q}$ between these eigenfunctions, $\langle \chi_q \mid \hat{Q} \mid \chi_{q'} \rangle$ for $q \neq q'$, is zero. Justify carefully and in detail each step of your argument.

Equation (11.44) may seem magical, for $\hat{Q}$, the operator that defines the *functions* in the matrix element $\langle \chi_q \mid \hat{R} \mid \chi_{q'} \rangle$, has only the most ephemeral relationship to $\hat{R}$, the operator that appears in this matrix element. Novices trying to prove that (11.44) follows from (11.42) sometimes muddy the waters by introducing a complete set of simultaneous eigenfunctions of $\hat{Q}$ and $\hat{R}$. Although not a foolish approach to the proof, this gambit isn't necessary, as the argument to follow will attest. First let's state our conjecture in words.

Proclamation 11.2

The matrix element of an operator with respect to eigenfunctions of a second operator that commutes with the first is zero if the eigenvalues of the two eigenfunctions are different.

Argument: To prove this claim, we must somehow develop an expression involving the matrix element $\langle \chi_q \mid \hat{R} \mid \chi_{q'} \rangle$. The only connection between $\hat{R}$ and the functions in this braket is the commutation relation Eq. (11.42*b*). So a sensible first step is to sandwich the operator $\left[\hat{Q}, \hat{R}\right]$ between the eigenfunctions χ_q and $\chi_{q'}$. According to (11.42) the resulting Dirac braket is zero:

$$\left\langle \chi_q \,\middle|\, \left[\hat{Q}, \hat{R}\right] \,\middle|\, \chi_{q'} \right\rangle = \langle \chi_q \mid \hat{Q}\hat{R} - \hat{R}\hat{Q} \mid \chi_{q'} \rangle = 0. \tag{11.45}$$

The matrix element in (11.45) consists of two terms,

$$\langle \chi_q \mid \hat{Q}\hat{R} - \hat{R}\hat{Q} \mid \chi_{q'} \rangle = \langle \chi_q \mid \hat{Q}\hat{R} \mid \chi_{q'} \rangle - \langle \chi_q \mid \hat{R}\hat{Q} \mid \chi_{q'} \rangle, \tag{11.46}$$

each of which looks like it might be persuaded to yield a matrix element of $\hat{R}$ only.

To get such a result we must eliminate $\hat{Q}$ from each braket in Eq. (11.46). This chore is rather simple, because the functions in these brakets are eigenfunctions of $\hat{Q}$. To position this operator so it can act on its eigenfunctions, we use its Hermiticity, *viz.*,

$$\langle \hat{Q}\chi_q \mid \hat{R}\chi_{q'} \rangle - \langle \chi_q \mid \hat{R}\hat{Q}\chi_{q'} \rangle = 0. \tag{11.47}$$

Now, in the first term, $Q\chi_q = q\chi_q$. We must be careful in removing the constant from this braket, for $\hat{Q}\chi_q$ appears to the *left* of the bar in this Dirac braket. Were q complex, its complex conjugate q^* would emerge from the braket. But q is an eigenvalue of a Hermitian operator and hence is real. So this number emerges from the Dirac braket without change (*see Property II of Table 10.4*). In the second term, no such questions arise in dealing with $\hat{Q}\chi_{q'} = q'\chi_{q'}$. Therefore, Eq. (11.47) becomes

$$q\langle \chi_q \mid \hat{R} \mid \chi_{q'} \rangle - q'\langle \chi_q \mid \hat{R} \mid \chi_{q'} \rangle = 0. \tag{11.48a}$$

Factoring out the matrix element common to both terms in (11.48*a*), we have

$$(q - q')\langle \chi_q \mid \hat{R} \mid \chi_{q'} \rangle = 0. \tag{11.48b}$$

But since the eigenvalues q and q' are different, their difference $q - q'$ is not zero. This means that the matrix element must be zero:

$$\langle \chi_q \mid \hat{R} \mid \chi_{q'} \rangle = 0. \tag{11.48c}$$

Q. E. D.

Question 11–10

Do you understand how to use this result? Suppose you need to evaluate the matrix element of the parity operator between two distinct simple harmonic oscillator eigenfunctions, $\langle \psi_n \mid \hat{\Pi} \mid \psi_{n'} \rangle$. Explain in detail how to apply Proclamation 11.2 to this problem.

Matrix Representations of Operators

Another way to express Proclamation 11.2 is to say "the off-diagonal matrix elements of $\hat{R}$ with respect to eigenfunctions of $\hat{Q}$, with which it commutes, are zero." Implicit in this statement is the notion of representing an operator by a *matrix*, a particular arrangement of the Dirac brakets $\langle \chi_q \mid \hat{R} \mid \chi_{q'} \rangle$.

When we say a matrix "represents" an operator we mean that the matrix contains *the same information* as the operator, albeit expressed in a different way. The idea of matrix representations of operators originated with Heisenberg, who formulated quantum theory as *matrix mechanics*.[9] It achieved full flowering with Dirac, who formulated quantum theory using linear algebra in a vector space, an approach that leads inexorably to matrices.

It is easy to see the genesis of this idea. Each Dirac braket $\langle \chi_q \mid \hat{R} \mid \chi_{q'} \rangle$ is a (complex) number that can be uniquely labelled by the quantum numbers q and q'. Let's call this number $R_{q,q'}$. Collect together all such matrix elements—by constructing all possible Dirac brakets using all elements of the complete set of eigenfunctions $\{\chi_q\}$—and arrange these numbers into a matrix:

$$\mathsf{R} = \begin{pmatrix} R_{q_1,q_1} & R_{q_1,q_2} & R_{q_1,q_3} & \cdots \\ R_{q_2,q_1} & R_{q_2,q_2} & R_{q_2,q_3} & \cdots \\ R_{q_3,q_1} & R_{q_3,q_2} & R_{q_3,q_3} & \cdots \\ \vdots & \vdots & \vdots & \vdots \end{pmatrix}. \tag{11.49}$$

[9] For accessible treatments of matrix mechanics, see, for example, Chap. 2 of *Quantum Mechanics* by H. Hameka (New York: Wiley, 1981) and Chap. 11 of *Introduction to Quantum Mechanics* by B. Dicke and D. Wittke (Reading, Mass.: Addison-Wesley, 1960).

Notice that *the eigenvalue to the left of the operator is the row index of the matrix and the eigenvalue to the right is the column index.* The matrix $\mathbf{R}$ is called **the matrix representation of $\hat{R}$ with respect to the eigenfunctions of $\hat{Q}$.**[10] By the way, there is nothing special about the set of eigenfunctions with respect to which we define the matrix $\mathbf{R}$; a huge number of different matrix representations of a given operator can be constructed, by simply choosing different sets of eigenfunctions.

In terms of the matrix representation (11.49), Proclamation 11.2 takes on its most elegant form:

$$\boxed{\left[\hat{Q}, \hat{R}\right] = 0 \qquad \Longrightarrow \qquad \mathbf{R} \text{ is diagonal}} \tag{11.50}$$

To see a typical matrix representation, let's write down the representation $\mathbf{x}$ of the position operator $\hat{x}$ for the simple harmonic oscillator with respect to the eigenfunctions of $\hat{\mathcal{H}}$, the functions $\psi_n(x)$. Each element of this matrix is a Dirac braket of the form

$$\langle \psi_n \mid \hat{x} \mid \psi_{n'} \rangle = \int_{-\infty}^{\infty} \psi_n^*(x) x \psi_{n'}(x)\, dx. \tag{11.51}$$

So the matrix representation of $\hat{x}$ with respect to the eigenfunctions of $\hat{\mathcal{H}}$ looks like[11]

$$\mathbf{x} = \begin{pmatrix} \langle \psi_0 \mid \hat{x} \mid \psi_0 \rangle & \langle \psi_0 \mid \hat{x} \mid \psi_1 \rangle & \langle \psi_0 \mid \hat{x} \mid \psi_2 \rangle & \cdots \\ \langle \psi_1 \mid \hat{x} \mid \psi_0 \rangle & \langle \psi_1 \mid \hat{x} \mid \psi_1 \rangle & \langle \psi_1 \mid \hat{x} \mid \psi_2 \rangle & \cdots \\ \langle \psi_2 \mid \hat{x} \mid \psi_0 \rangle & \langle \psi_2 \mid \hat{x} \mid \psi_1 \rangle & \langle \psi_2 \mid \hat{x} \mid \psi_2 \rangle & \cdots \\ \vdots & \vdots & \vdots & \vdots \end{pmatrix} \tag{11.52}$$

Question 11–11

Use Table 9.7 to evaluate the integrals in $\mathbf{x}$. Is this matrix diagonal?

Question 11–12

Write down and evaluate the matrix representation of $\hat{x}^2$ for the SHO with respect to the Hamiltonian eigenfunctions.

Introducing this topic, I said that *the matrix representation of an operator contains the same information as does the operator.* That information concerns the observable represented by the operator; in effect, this operator and its various matrix representations (with respect to different sets of eigenfunctions) are just different ways to represent the observable. But operators act on functions to yield, in general, other functions,

[10] If we refer to the eigenfunctions of a Hermitian operator as a *basis*, then we call the matrix R *the matrix representation of $\hat{R}$ in the basis defined by $\hat{Q}$.*

[11] *Caveat emptor.* In the matrix x, the quantum number for row 1 and column 1 is $n = 0$; but according to matrix notation, this element is labelled $x_{1,1}$, *i.e.*,

$$x_{1,1} = \langle \psi_0 \mid \hat{x} \mid \psi_0 \rangle.$$

So in this case the *indices* on each element of the matrix are not the same as the *quantum numbers* that define the element being so labelled. Such complications often arise, and it is important to keep separate in your mind the indices on the matrix elements and the (physical) quantum numbers that label the eigenfunctions in the Dirac brakets. (In the quantum physics of a single particle in three dimensions or of a many-particle system, the situation becomes even more complicated, for a single index usually represents the values of many quantum numbers.)

while matrices multiply other matrices to yield still other matrices. So clearly, the way we solve a problem in quantum physics using the matrix representations of the relevant observables is very different than the way we solve the same problem using the operator representations of these observables. The most obvious parallel between operator and matrix mechanics is the eigenvalue problem. For an observable Q, the operator eigenvalue equation, which we have discussed at length in this and the previous chapter, is

$$\hat{Q}\chi_q = q\chi_q.$$

For the *matrix* **q** the matrix eigenvalue equation is

$$\mathbf{q}\underline{\chi}_q = q\underline{\chi}_q,$$

where $\underline{\chi}_q$ is a *column vector*—a matrix with only one column—that is the **eigenvector** of the matrix **q** with eigenvalue q.

The Dirac formulation of quantum mechanics is powerful and beautiful, but it is also quite abstract. So we won't pursue matrix representations further in this book.

11.4 COMMUTATORS AND UNCERTAINTY PRINCIPLES

In § 11.2 we proved (Proclamation 11.1) that operators that commute share simultaneous eigenfunctions. If the state function Ψ of a system is one such eigenfunction, then the system is in an eigenstate of the observables represented by the commuting operators, and Ψ satisfies the corresponding eigenvalue equations (11.2). In such a state, both observables are sharp, so their uncertainties are simultaneously zero [Eq. (11.4)]. Therefore, in this state simultaneous ensemble measurements of these observables yield uniquely the eigenvalues of (11.2).

This means that if operators $\hat{Q}$ and $\hat{R}$ commute, then the only uncertainty relation satisfied by the product of their uncertainties is the obvious one

$$\Delta Q(t)\,\Delta R(t) \geq 0, \qquad \text{[any state } \Psi \text{ of any system]} \tag{11.53}$$

which follows from the definition of the uncertainty of an observable.

But we have seen pairs of operators whose commutator is a *non-zero constant*. Our paradigm for this case is the pair $\hat{x}$ and $\hat{p}$, whose commutator $[\hat{x}, \hat{p}]$ equals $i\hbar\hat{1}$. There exist no eigenstates of either operator, for the Heisenberg Uncertainty Principle

$$\Delta x\,\Delta p \geq \tfrac{1}{2}\hbar \qquad \text{[any state } \Psi \text{ of any system]} \tag{11.54}$$

prohibits the existence of such a state.[12]

The Generalized Uncertainty Principle

These examples hint at a connection between the *commutator* of two quantum-mechanical operators and the *uncertainty relation* for the corresponding observables—a connection

[12] The Dirac Delta function $\delta(x - x_0)$ is an eigenfunction of $\hat{x}$ and the pure momentum function $\chi_p(x)$ is an eigenfunction of $\hat{p}$. For the former $\Delta x = 0$ and for the latter $\Delta p = 0$. But since neither function is physically admissible, neither can represent a quantum state.

with profound and far-reaching consequences for laboratory measurements. Mathematically, this connection is expressed by the **Generalized Uncertainty Principle** (GUP), which we'll discuss and illustrate (but won't prove) in this section:[13]

$$\boxed{\Delta Q\,\Delta R \geq \frac{1}{2}\left|\left\langle i\big[\hat{Q},\hat{R}\big]\right\rangle\right|} \qquad \text{Generalized Uncertainty Principle} \tag{11.55}$$

This rather forbidding inequality holds *for any state of any system.* Its truth hinges on only the definitions of the uncertainty and the expectation value and on properties of inequalities.

The GUP says a mouthful. Its language is that of statistical quantities. On the left, we have the uncertainties

$$\Delta Q(t) = \sqrt{\langle Q^2\rangle(t) - \langle Q\rangle^2(t)} \tag{11.56a}$$

$$\Delta R(t) = \sqrt{\langle R^2\rangle(t) - \langle R\rangle^2(t)}. \tag{11.56b}$$

And on the right, we have the expectation value of the operator $i\big[\hat{Q},\hat{R}\big]$,[14]

$$\left\langle i\big[\hat{Q},\hat{R}\big]\right\rangle(t) = \left\langle \Psi \left| i\big[\hat{Q},\hat{R}\big]\right| \Psi\right\rangle, \tag{11.57a}$$

which, for a one-dimensional wave function, is

$$\left\langle i\big[\hat{Q},\hat{R}\big]\right\rangle(t) = i\int_{-\infty}^{\infty} \Psi^*(x,t)\big(\hat{Q}\hat{R} - \hat{R}\hat{Q}\big)\Psi(x,t)\,dx. \tag{11.57b}$$

These statistical quantities are information concerning the (aggregate) results of simultaneous measurements at time t of Q and R on a system in the state $\Psi(x,t)$. The uncertainty product $\Delta Q(t)\,\Delta R(t)$ on the left-hand side of (11.55) must be positive or zero; the absolute value bars on the right side ensure that it, too, is non-negative. To further probe the generalized uncertainty principle, let's watch it in action.

Example 11.3. The GUP Implies the HUP

Suppose $\hat{Q}$ and $\hat{R}$ in (11.55) are the familiar duo $\hat{x}$ and $\hat{p}$. Then the operator in the expectation value on the right-hand side of the GUP is just

$$i\big[\hat{Q},\hat{R}\big] = i\big[\hat{x},\hat{p}\big] = i(i\hbar\hat{1}) = -\hbar\hat{1}. \tag{11.58}$$

So for these operators, the generalized uncertainty principle is

$$\Delta x\,\Delta p \geq \frac{1}{2}\left|-\hbar\langle\Psi \mid \Psi\rangle\right|. \tag{11.59}$$

Since the state function Ψ is normalized, $\langle\Psi \mid \Psi\rangle = 1$, and Eq. (11.59) becomes the Heisenberg Uncertainty Principle,

$$\Delta x\,\Delta p \geq \tfrac{1}{2}\hbar. \tag{11.60}$$

[13] You can find proofs in, for example, § 6-12 of *Quantum Mechanics* by G. Powell and B. Crasemann (Reading, Mass.: Addison-Wesley, 1961) or in § V.8 of *Elementary Quantum Mechanics* by D. Saxon (San Francisco: Holden-Day, 1968). For a critique of the Generalized Uncertainty Principle, see D. Deutsch, *Phys. Rev. Lett.*, **50**, 63 (1983).

[14] The cryptic i multiplying this commutator in the expectation value on the right-hand side of the GUP (11.55) is a formal nicety that ensures that the operator in this expectation value is Hermitian. (In Table 10.5 I noted that the commutator of two Hermitian operators $\big[\hat{Q},\hat{R}\big]$ is not Hermitian, but that $i\big[\hat{Q},\hat{R}\big]$ is.)

Notice that nowhere in this example did I specify a particular *state system*. This simple derivation therefore *proves* that the Heisenberg Uncertainty Principle, one of the three great ideas of quantum physics discussed in Chap. 2, is an inevitable consequence of the postulates of quantum mechanics. Just as uncertainty in a laboratory measurement on a microscopic system is a consequence of the wave nature of particles, so are uncertainty relations a consequence of the wave mechanics that describes their physics; both features of the microworld are intrinsic to nature.

State-Dependent Uncertainty Relations

The result of applying the GUP to $\hat{x}$ and $\hat{p}$ was a clean, quite general uncertainty relation. But for other operators the GUP may yield more complicated relations, expressions that depend explicitly on a particular quantum state. *If* $\left[\hat{Q}, \hat{R}\right]$ *is a non-zero operator not of the form (constant)* $\hat{1}$, *then the right-hand side of (11.55) may be zero even though* $\left[\hat{Q}, \hat{R}\right] \neq 0$. *This happens if for the state* Ψ, *the matrix element* $\langle \Psi \mid i\left[\hat{Q}, \hat{R}\right] \mid \Psi \rangle$ *happens to be zero.* This situation is a little tricky and warrants examination via a specific example.

Question 11–13

Show that if Ψ is an eigenfunction of the operator $\hat{C} = i\left[\hat{Q}, \hat{R}\right]$, then

$$\langle \Psi \mid i\left[\hat{Q}, \hat{R}\right] \mid \Psi \rangle \neq 0.$$

Example 11.4. Uncertainties in Energy and Position for the SHO

In Question 11.8 you discovered that the Hamiltonian $\hat{\mathcal{H}}$ of a one-dimensional simple harmonic oscillator does not commute with the position operator $\hat{x}$. But unlike $\left[\hat{x}, \hat{p}\right]$, the commutator of $\hat{\mathcal{H}}$ and $\hat{x}$ is not a constant times the identity operator $\hat{1}$; rather, it's proportional to the linear momentum operator:

$$\left[\hat{\mathcal{H}}, \hat{x}\right] = -\frac{i\hbar}{m}\hat{p}. \tag{11.61}$$

At first glance, you might guess that the GUP (11.55) for $\hat{Q} = \hat{\mathcal{H}}$ and $\hat{R} = \hat{x}$ predicts a non-zero uncertainty product $\Delta E\, \Delta x$. But it ain't necessarily so.

To obtain an expression for this product, let's substitute (11.61) into the GUP, *viz.*,

$$\Delta E\, \Delta x \geq \frac{\hbar}{2m} \left| \langle p \rangle \right|. \tag{11.62a}$$

For a state represented by $\Psi(x,t)$, this uncertainty relation reads

$$\Delta E(t)\, \Delta x(t) \geq \frac{\hbar}{2m} \left| \int_{-\infty}^{\infty} \Psi^*(x,t) \hat{p} \Psi(x,t)\, dx \right|. \tag{11.62b}$$

Clearly, if the mean value of p for the state Ψ is zero, then the uncertainty relation (11.62) reduces to the trivial statement that the uncertainty product $\Delta E(t)\, \Delta x(t)$ is not negative [Eq. (11.53)].

This happens, for example, if the state is stationary. In § 9.9 we proved that *for the* n^{th} *stationary state of any symmetric potential,* $\langle p \rangle_n = 0$ [Eq. (9.113)]. So if we have $\Psi(x,t) = \psi_n(x) e^{-iE_n t/\hbar}$, then $\langle p \rangle$ on the right-hand side of Eqs. (11.62) is zero, and this inequality reduces to the uninformative utterance

$$(\Delta E)_n (\Delta x)_n \geq 0. \qquad \text{[any stationary state of the SHO]} \tag{11.63}$$

But what if the state is not stationary? Consider, for example, a state of the SHO that is a superposition of two stationary states:

$$\Psi(x,t) = \frac{1}{\sqrt{50}}\left[\psi_1(x)e^{-iE_1 t/\hbar} + 7\psi_2(x)e^{-iE_2 t/\hbar}\right]. \tag{11.64}$$

[You can find the eigenfunctions $\psi_1(x)$ and $\psi_2(x)$ in Table. 9.6.] To determine the uncertainty product $\Delta E(t)\,\Delta x(t)$ for this state we must, according to (11.62), first evaluate $\langle p\rangle(t)$. [There is nothing new here; in Example 9.6 we calculated the mean momentum for a (different) non-stationary state of the SHO.] For the state (11.64), we find

$$\langle p\rangle(t) = -\frac{14}{50}\,\hbar\beta\sin\omega_0 t, \tag{11.65}$$

where, as usual, ω_0 is the natural frequency of the oscillator and $\beta = \sqrt{m\omega_0/\hbar}$. The uncertainty product for energy and position at time t is therefore governed by the relation

$$\Delta E(t)\,\Delta x(t) \geq \frac{7\hbar^2\beta}{50m}\sin\omega_0 t \tag{11.66a}$$

i.e.,

$$\Delta E(t)\,\Delta x(t) \geq 0.1400\,\frac{\hbar^2\beta}{m}\sin\omega_0 t. \tag{11.66b}$$

How interesting. For this state, the mean value of p—and hence the energy-position uncertainty product—is not zero *except at special times* t, those for which $\omega_0 t$ is an integral multiple of π, *i.e.*,

$$\Delta E(t)\,\Delta x(t) \geq 0 \qquad \text{for } t = \frac{m\pi}{\omega_0} \qquad m = 0, 1, 2, \ldots \tag{11.67}$$

At these special discrete times, therefore, E and x *could be* sharp. [But even at these times, Eq. (11.66b) does not necessarily reduce to an equality.] We cannot tell, however, from the uncertainty relation alone whether they *are* sharp at these times.

Let's verify that the uncertainty principle (11.66) is correct for this state. Doing so requires evaluating the left- and right-hand sides of this relation for the state (11.64)—by no means an impossible chore, but one that entails some algebraic contortions. Since the required manipulations are familiar, I won't fill in the details; you can if you want to.

The energy uncertainty is the square root of

$$(\Delta E)^2 = \left\langle E^2\right\rangle - \langle E\rangle^2. \tag{11.68}$$

We evaluate the inevitable expectation values from the wave function (11.64) using the TISE, obtaining

$$\langle E\rangle = \frac{62}{25}\,\hbar\omega_0 = 2.48\hbar\omega_0. \tag{11.69a}$$

Notice that although the state is non-stationary, the mean value of the particle's energy is independent of time. The same is true of the mean value of the square of the energy, which turns out to be

$$\left\langle E^2\right\rangle = 6.17\,\hbar^2\omega_0^2. \tag{11.69b}$$

Using Eqs. (11.69) in (11.68) we find that for this state of the SHO,

$$(\Delta E)^2 = 0.0196\,\hbar^2\omega_0^2 \qquad \Longrightarrow \qquad \Delta E = 0.1400\,\hbar\omega_0. \tag{11.70}$$

Similar manipulations for the expectation value of x and x^2 lead us to the position uncertainty, the other factor in the product in (11.66):

$$\langle x \rangle (t) = \frac{14}{50\beta} \cos\omega_0 t \tag{11.71a}$$

$$\left\langle x^2 \right\rangle = \frac{62}{25\beta^2} \tag{11.71b}$$

$$\Longrightarrow \quad (\Delta x)^2(t) = \frac{1}{25\beta^2}\left[62 - 1.96\cos^2\omega_0 t\right] \tag{11.71c}$$

$$\Longrightarrow \quad \Delta x(t) = \frac{1}{50\beta}\sqrt{62 - 1.96\cos^2\omega_0 t}. \tag{11.71d}$$

The result of our explicit evaluation of the uncertainty product is

$$\Delta E(t)\,\Delta x(t) = (0.0280)\frac{\hbar^2\beta}{m}\sqrt{62 - 1.96\cos^2\omega_0 t}. \tag{11.72}$$

To illustrate the uncertainty relation (11.62), let's consider a particular time—say, $t = \pi/(2\omega_0)$. For this time, $\sin\omega_0 t = 1$ and $\cos\omega_0 t = 0$, and the uncertainty product (11.72) becomes

$$\Delta E\left(t = \tfrac{\pi}{2\omega_0}\right)\,\Delta x\left(t = \tfrac{\pi}{2\omega_0}\right) = 0.2205\,\frac{\hbar^2\beta}{m}. \tag{11.73a}$$

For this time the right-hand side of the uncertainty relation (11.66) is

$$\frac{\hbar}{2m}\left|\langle p \rangle \left(t = \frac{\pi}{2\omega_0}\right)\right| = 0.1400\frac{\hbar^2\beta}{m}. \tag{11.73b}$$

As expected, (11.73*a*) is greater than (11.73*b*), and the uncertainty relation is verified—for this particular time. (This verification does not, of course, constitute a proof of the generalized uncertainty principle.)

Question 11–14

(To be done only if you feel you need practice evaluating uncertainties and expectation values.) Fill in the steps leading to Eqs. (11.73). [In Chap. 12 we'll learn a very easy way to evaluate expectation values of E and its powers; you might want to wait to fill in the steps of this example until then.]

Aside: The Implications for Measurement. Our explicit results for the energy and position uncertainties shed some light on the question of what would happen were we to measure these observables at one of the special times $t = m\pi/\omega_0, m = 0, 1, 2, \ldots$, when the uncertainty relation reduces to Eq. (11.67). At these times, $\cos\omega_0 t = 1$, so the *position uncertainty* equals the constant $1.55/\beta$. The *energy uncertainty* is independent of time and so is equal to $0.1400\,\hbar\omega_0$. Both uncertainties are non-zero, so even at these special times neither energy nor position are sharp—a fact we could not have deduced from the uncertainty relation (11.67).

Question 11–15

Argue from the form of the state function that at no time could the position and energy uncertainties be zero.

Compatible and Incompatible Observables

The moral of Example 11.4 is that there is more to uncertainty relations than one might first expect. Any pair of observables Q and R falls into one of three categories, depending on the commutator of the corresponding operators $\hat{Q}$ and $\hat{R}$. These cases and their consequences are summarized in Table 11.4.

TABLE 11.4 SUMMARY OF CONSEQUENCES OF THE GUP.

	Compatible	Incompatible	
	$\left[\hat{Q}, \hat{R}\right] = 0$	$\left[\hat{Q}, \hat{R}\right] = c\hat{1}$	$\left[\hat{Q}, \hat{R}\right] = \hat{S}$
simultaneous eigenfunctions	exist	don't exist	may exist*
complete set of simultaneous eigenfunctions	exists	doesn't exist	doesn't exist
restrictions on uncertainties	none	$\Delta Q > 0$ & $\Delta R > 0$	special states**
simultaneous measurement of Q and R	possible	not possible	special states†

*Complete set of simultaneous eigenfunctions prohibited.

**Those for which $\langle \Psi \mid i\left[\hat{Q}, \hat{R}\right] \mid \Psi \rangle \neq 0$.

†A single (exceptional) simultaneous eigenstate may exist.

If $\hat{Q}$ and $\hat{R}$ commute, then the observables Q and R are said to be **compatible**; otherwise they are **incompatible**. Simultaneous eigenstates of two or more observables exist if (and only if) the observables are compatible. The uncertainty relation for two compatible observables tells us nothing; this relation is the trivial result (11.53). The uncertainty relation for incompatible observables, however, *may* limit the precision of simultaneous measurements of the observables; if the commutator of $\hat{Q}$ and $\hat{R}$ is a constant operator, then (rigorously) no states exist in which either observable is sharp. But if $\left[\hat{Q}, \hat{R}\right]$ is a non-constant, non-zero operator, then the limits imposed by the GUP depend on the state with respect to which the uncertainties are evaluated. In this case there does not exist a complete set of simultaneous eigenfunctions of $\hat{Q}$ and $\hat{R}$, but there may exist one such state.

Granted, this whole business is rather complicated. So spend a little time pondering Table 11.4, the examples of this section, and the marathon summary figure at the end of the chapter. Particularly important things to remember are:

1. Only if two observables are compatible could we (*in principle*) measure their values simultaneously to infinite precision.
2. Even if two observables are compatible, the system may be in a state in which one or both is not sharp.

11.5 CONSTANTS OF THE MOTION IN CLASSICAL AND QUANTUM PHYSICS

In several pairs of observables that we've examined, one was the total energy E. Any observable that is compatible with E has a special status in quantum mechanics and is called a **constant of the motion**. By definition, an observable Q is a **constant of the motion** if the operator that represents it, $\hat{Q}$, commutes with the operator that represents the energy, $\hat{\mathcal{H}}$:

$$\boxed{[\hat{\mathcal{H}}, \hat{Q}] = 0 \quad \Longrightarrow \quad Q \text{ is a constant of the motion}} \tag{11.74}$$

You have learned (or, depending on your curriculum, you will learn) about constants of the motion of a macroscopic system in a classical mechanics course, where these quantities are sometimes called "first integrals of the motion." But a constant of the motion for a microscopic system is a very different sort of creature, and the difference highlights a fundamental contrast between classical and quantum physics. This contrast, which will now become the dominant theme of this chapter, will ultimately return us to the Correspondence Principle, which connects classical and quantum physics.

Constants of the Motion of a Macroscopic System

In dealing with a complicated many-particle *macroscopic* system (twelve billiard balls on a table, for example) we usually cannot solve the classical equations of motion for the trajectories $\{\mathbf{r}_i(t), \mathbf{p}_i(t), i = 1, 2, 3, \ldots, N\}$ of all N particles. But we can still investigate the physics of the system, by focusing on *particular observables, functions of* $\mathbf{r}_i$ *and* $\mathbf{p}_i$*, whose values do not change as the system evolves. These special observables are the classical constants of the motion of the system.*

The most familiar constant of the motion is the total energy. For a single macroscopic particle in one dimension (one billiard ball on a wire), the energy is a function of x and p and is sometimes written as the classical Hamiltonian $H(x,p)$—*i.e.*, $E = H(x,p) = p^2/(2m) + V(x)$. As time passes, the energy doesn't change. In fact, we can prove (from Newton's Second Law) that

$$\frac{\partial V}{\partial t} = 0 \quad \Longrightarrow \quad \frac{dE}{dt} = 0. \qquad \text{[classical physics]} \tag{11.75}$$

Consequently $E = H(x,p)$ is a constant of the motion—*i.e.*, as has been known since 1847, the total energy of a conservative system is conserved.[15] So in classical physics knowledge of the constants of the motion leads to *conservation laws*. Not unexpectedly, the same is true in quantum physics (§ 11.6).

Clearly the classical definition of a constant of the motion won't wash in quantum physics, because at no time can we precisely specify the position and the linear momentum (the trajectory) of a microscopic system. To clarify this difficulty, let's consider a specific state—say, a non-stationary state of a simple harmonic oscillator [*e.g.*, (11.64)].

[15]Notice that I specified that V does not depend on t, thereby limiting this discussion to conservative systems. In classical as in quantum physics, the energy of a non-conservative system in an arbitrary state is not conserved. For a discussion of constants of the motion and conservation laws in classical physics, see Chap. 1 and § 2.6 of *Classical Mechanics*, 2nd ed. by H. Goldstein (Reading, Mass.: Addison-Wesley, 1980) or Chap. 2 of *Classical Dynamics of Particles and Systems*, 2nd ed. by J. B. Marion (New York: Academic Press, 1970.)

Such a state *does not have* an energy, a position, or a linear momentum, for none of these observables are sharp. So the classical description of one or another of these quantities as a "constant of the motion" is meaningless. To use this concept, we must re-define it in a way that is appropriate to the microworld and, yet, that remains consistent with the classical definition in the classical limit. As you'll see, Eq. (11.74) is the definition we need.

Constants of the Motion of a Microscopic System

The first thing to notice about the definition (11.74) is that it makes sense. If $\hat{\mathcal{H}}$ and $\hat{Q}$ commute, then there exist simultaneous eigenstates of energy and Q. In such a state, the energy and Q are sharp; *i.e.*, their uncertainties, as calculated from $\Psi(x,t)$ at any time, are zero. Hence a measurement of Q will yield the eigenvalue q of $\Psi(x,t)$ *regardless of when the measurement is performed*; *i.e.*, Q has a well-defined value that does not change as the system evolves. It is certainly reasonable to call such an observable a "constant of the motion."

If you'll flip back through this chapter, you'll find a couple of examples of constants of the motion for various systems: the linear momentum of a free particle, and the parity of a particle with a symmetric potential. And you'll see why the momentum of the simple harmonic oscillator is not a constant of the motion.

Question 11–16

Is position a constant of the motion for a SHO? ... for a symmetric square well? ... for an asymmetric square well? Explain your answer in each case.

Question 11–17

Prove that the energy of any conservative system is a constant of the motion of the system. (This exercise is easy and very important. Do it.)

Question 11–18

Consider an electron in a hydrogen atom that is exposed to an external electromagnetic wave of frequency ω_0 and electric field strength $\mathcal{E}_0$. One term in the potential energy of this electron is

$$V'(x,y,z) = e\mathcal{E}_0 z \cos\omega_0 t.$$

Is the total energy of the electron a constant of the motion? Why or why not?

The Importance of Constants of the Motion

What are the consequences for the wave functions of the fact that constants of the motion exist? *Because $\hat{\mathcal{H}}$ commutes with the operator $\hat{Q}$ that represents a constant of the motion, there exists a complete set of simultaneous eigenfunctions of these operators, $\{\psi_{E,q}(x)\}$. These simultaneous eigenfunctions describe the spatial dependence of stationary-state wave functions of the system, viz.,*

$$\Psi_{E,q}(x,t) = \psi_{E,q}(x)\, e^{-iEt/\hbar}. \tag{11.76}$$

Such a wave function represents a simultaneous eigenstate of E and Q; it is separable because it is an eigenfunction of $\hat{\mathcal{H}}$.

Constants of the motion are important in quantum mechanics primarily because they define the *quantum numbers* that we use to label stationary states of systems such as atoms, molecules, solids, and nuclei. These are the states between which transitions occur in spectroscopic experiments. And they are the initial and final states of targets in collision experiments. To determine these vital constants of the motion we must find a set of operators that commute with one another and with the Hamiltonian. The operators must commute with $\hat{\mathcal{H}}$ so that in stationary states, the observables they represent could be sharp. They must commute with one another so that in these states, they *all* could simultaneously be sharp. So a facility with commutators and simultaneous eigenfunctions is an essential skill of any physicist.

Aside: Constants of the Motion for Three-Dimensional Systems. Finding the constants of the motion is not an urgent matter in the study of one-dimensional microscopic systems, for we can uniquely identify the (non-degenerate) bound-states of such systems by the single quantum number that labels the energies. Additional labels are redundant.

But for three-dimensional systems, where most bound-state energies are degenerate, the energy quantum number alone is not sufficient to *uniquely* specify a quantum state. One or more additional labels are needed. These labels correspond to observables that are sharp in the stationary states so labelled and hence to constants of the motion.

For example, one of the simplest realistic three-dimensional systems is the *hydrogen atom* (one electron and one proton). All bound-state energies except the ground-state energy are degenerate, and the stationary-state wave functions for the bound states are usually (but not necessarily) labelled by quantum numbers corresponding to the energy, the square of the orbital angular momentum of the electron with respect to the proton, and the projection of this angular momentum along one space-fixed axis (usually $\hat{e}_z$). The latter two observables correspond to the operators $\hat{\mathbf{L}}^2$ and $\hat{L}_z$; and using their quantum numbers to label stationary states is legitimate *only* because these observables commute with $\hat{\mathcal{H}}$. (To *uniquely* specify a state of an electron in a hydrogen atom, one must also specify the orientation of another, purely quantum-mechanical angular momentum called the *electron spin*. We'll study spin in Volume II.)

To close this section, let's summarize the results we expect in a simultaneous measurement of E and a constant of the motion Q. If the system is in a simultaneous eigenstate of these observables, then

$$\left.\begin{aligned}\Delta E(t) &= 0\\ \langle E\rangle(t) &= E\end{aligned}\right\} \quad \text{at any time } t \quad \Longleftrightarrow \quad \text{state is stationary} \tag{11.77a}$$

$$\left.\begin{aligned}\Delta Q(t) &= 0\\ \langle Q\rangle(t) &= q\end{aligned}\right\} \quad \text{at any time } t \quad \Longleftrightarrow \quad Q \text{ is a constant of the motion.} \tag{11.77b}$$

The statement that an observable is constant in time sounds like a classical conservation law. But Eqs. (11.77) certainly don't look much like such laws, which are usually expressed in terms of time derivatives, as in Eq. (11.75). In fact, quantum mechanics does yield laws of this form, and they are equivalent to Eqs. (11.77). To see how to derive them, we must first consider the evolution of an arbitrary observable, which is the topic of the next section. So don't stop reading yet; you've only heard part of the tale of constants of the motion in quantum physics.

11.6 CONSERVATION LAWS IN QUANTUM PHYSICS

You have probably noticed that in quantum mechanics we often find ourselves questioning concepts we have lived with for years, re-defining familiar notions so they can be applied to the uncertain microscopic universe.

Consider, for example, the statement "the observable Q is conserved." In classical physics, this declaration means simply that the value of Q does not change with time. Conservation of Q is a property *of the system*; it holds for any state (*i.e.*, any trajectory). For a single macroscopic particle in one dimension, the simplest classical system, the definition of a conserved quantity is

$$\boxed{\frac{d}{dt}Q[x(t),p(t)] = 0 \qquad \text{for any state } \{x(t),p(t)\} \iff \begin{array}{c} Q \text{ is conserved} \\ \text{(classical physics)} \end{array}} \tag{11.78}$$

This is the form of a conservation law in classical physics.

The energy of an isolated *macroscopic* system, for example, does not change as the system evolves, regardless of the trajectories of the constituent particles. A measurement of the energy—or of any other conserved observable—yields the same value no matter when it is performed.

But for a *microscopic* system we cannot define a conserved observable as one whose value in any state does not change with time, because for *any* observable Q there exist quantum states in which the observable *does not have a value*—namely, those in which $\Delta Q > 0$.

The difficulty with the classical definition (11.78) is vividly illustrated by the simplest microscopic system: a free particle in one dimension (Chap. 4). For any physically admissible wave function, any wave packet, the *energy*, *position*, and *momentum* of the particle are uncertain, so in no state does a free particle *have* a value of E, of x, or of p. It is clearly nonsense, then, to ask if "the value of E or x or p" is independent of time. Yet, we expect the energy of a free particle to be conserved, for no external influences act on the particle to induce a change in this observable. We need, clearly, a new definition of *a conserved quantity*.

Conserved Quantities for Microscopic Systems

Our definition should express the time independence of some *measurable* quantity, so we won't consider definitions expressed directly in terms of the operator, such as $d\hat{Q}/dt = 0$. And our definition must not depend on a particular state of the system. A sensible way to define a conserved quantity is: one whose *mean value in any state* does not change as the system evolves, *i.e.*,

$$\boxed{\frac{d}{dt}\langle Q\rangle(t) = 0 \qquad \text{for any state } \Psi \iff \begin{array}{c} Q \text{ is conserved} \\ \text{(quantum physics)} \end{array}} \tag{11.79}$$

Note that we must add the qualifier "for any state Ψ" to this definition, because the expectation value $\langle Q \rangle$ is state dependent. For a single particle in one dimension, this definition reads[16]

$$\frac{d}{dt}\langle Q \rangle (t) = \frac{d}{dt}\langle \Psi \mid \hat{Q} \mid \Psi \rangle = \frac{d}{dt}\int_{-\infty}^{\infty} \Psi^*(x,t)\hat{Q}\Psi(x,t)\,dx = 0. \tag{11.80}$$

Note carefully that to prove that an observable Q is *not* conserved, you need find only one state that violates (11.79).

Aside: Conservation Laws and the Correspondence Principle. Using $\langle Q \rangle$ to define a conserved observable is also suggested by the Correspondence Principle. According to § 5.7, the transition between quantum and classical physics is made via expectation values [*go look at Eqs. (5.86)*]. So by defining a conserved observable for a quantum system by (11.79) we set ourselves up to regain the corresponding classical conservation law (11.78) in the classical limit—a satisfying connection we'll explore in § 11.7.

Conserved Quantities and Constants of the Motion

In practice, the definition (11.79) suffers from its state dependence. Were we to apply it literally to an observable Q, we'd have to evaluate $\langle Q \rangle = \langle \Psi \mid \hat{Q} \mid \Psi \rangle$ for all states Ψ, checking each expectation value for time dependence. This definition is not wrong, it's just awkward to apply. But there is a deeper reason than mere inconvenience why it is less than ideal.

Whether or not an observable is conserved, in quantum as in classical physics, is a property of the observable and, via the Hamiltonian, of the system dynamics. So we should be able to define this property without reference to states of the system. [Recall that in Chap. 10 we freed the definition of Hermiticity, another property of *operators*, from its dependence on state functions in Eq. (10.24), by introducing the adjoint (§ 10.4) in Eq. (10.57).]

In classical physics the conserved observables of a system are its constants of the motion. Similarly, as we'll now prove, *a conserved observable in quantum mechanics is a constant of the motion, i.e., is one whose operator commutes with $\hat{\mathcal{H}}$:*

$$\boxed{\left[\hat{\mathcal{H}}, \hat{Q}\right] = 0 \quad \Longleftrightarrow \quad \frac{d}{dt}\langle Q \rangle (t) = 0 \qquad \text{for all } \Psi} \tag{11.81}$$

The commutation relation $\left[\hat{\mathcal{H}}, \hat{Q}\right] = 0$ is the state-independent expression of the status of Q as a conserved observable. To prove (11.81) we must back off for a moment from conserved observables and consider the evolution of the mean value of an *arbitrary Hermitian operator.*

[16] You may be wondering if we could define a conserved observable in terms of some other measurable quantity, such as the probability for obtaining a particular value in a measurement of the observable. The trouble with such a definition is that it would be unwieldy, because to encompass all possible outcomes we would have to require constancy of the probabilities for all values that could be obtained in the measurement. But from these probabilities we can always calculate $\langle Q \rangle$ (see § 3.3 and 3.6), so the more convenient definition (11.79) is preferable. Moreover, as noted, this definition facilitates the transition to classical physics. [In the next chapter, we'll discover that if Q is conserved according to (11.79), then the probabilities for all possible outcomes of measurement of Q are independent of time. So all this hangs together.]

The Time Development of an Expectation Value

Let $\hat{Q}$ be an arbitrary Hermitian operator—it may even depend on time [*c.f.*, the potential energy in Question 11.18]. The first derivative of the expectation value of any such operator is given by Eq. (11.80). Our first step in deriving a more convenient expression for this quantity is to move the time derivative inside the integral.[17] When we pull this switch, we must change the total derivative to a partial derivative, because this derivative now acts on functions of time and spatial variables, *e.g.*,

$$\frac{d}{dt}\left\langle Q(t)\right\rangle = \int_{-\infty}^{\infty} \frac{\partial}{\partial t}\Big[\Psi^*(x,t)\hat{Q}(t)\Psi(x,t)\Big]\,dx. \tag{11.82}$$

Expanding this partial derivative, we obtain

$$\frac{d}{dt}\left\langle Q(t)\right\rangle = \int_{-\infty}^{\infty} \left[\frac{\partial \Psi^*}{\partial t}\hat{Q}\Psi + \Psi^*\left(\frac{\partial \hat{Q}}{\partial t}\Psi\right) + \Psi^*\hat{Q}\frac{\partial \Psi}{\partial t}\right] dx. \tag{11.83}$$

Note the presence in the second term on the right-hand side of Eq. (11.83) of the partial derivative of $\hat{Q}$ with respect to time. This term allows for the possibility that $\hat{Q}$ depends *explicitly* on time.[18]

I want next to focus on the first and third terms on the right-hand side of (11.83), so I'll separate out the term involving $\partial\hat{Q}/\partial T$, writing it as an expectation value, and will temporarily move it to the left-hand side of the equation, *viz.*,

$$\frac{d}{dt}\left\langle Q(t)\right\rangle - \left\langle \frac{\partial \hat{Q}}{\partial t}\right\rangle = \int_{-\infty}^{\infty} \left[\frac{\partial \Psi^*}{\partial t}\hat{Q}\Psi + \Psi^*\hat{Q}\frac{\partial \Psi}{\partial t}\right] dx. \tag{11.84}$$

Now, to derive a useful expression from (11.84) we must get rid of the partial derivatives in it. We faced a similar labor in Chap. 6 when deriving the rate equation for probability flow, Eq. (6.113) (§ 6.8). There we used the Time-Dependent Schrödinger Equation to eliminate the first partial time derivative of $\Psi(x,t)$ by writing

$$\frac{\partial \Psi}{\partial t} = -\frac{i}{\hbar}\hat{\mathcal{H}}\Psi \tag{11.85a}$$

$$\frac{\partial \Psi^*}{\partial t} = \frac{i}{\hbar}\left(\hat{\mathcal{H}}\Psi\right)^*. \tag{11.85b}$$

Using the same trick on (11.84), we obtain

$$\frac{d}{dt}\left\langle Q(t)\right\rangle - \left\langle \frac{\partial \hat{Q}}{\partial t}\right\rangle$$

$$= \frac{i}{\hbar}\int_{-\infty}^{\infty} \left\{\left[\hat{\mathcal{H}}\Psi(x,t)\right]^*\hat{Q}\Psi(x,t) - \Psi^*(x,t)\hat{Q}\hat{\mathcal{H}}\Psi(x,t)\right\} dx \tag{11.86a}$$

$$= \frac{i}{\hbar}\Big[\langle\hat{\mathcal{H}}\Psi \mid \hat{Q}\Psi\rangle - \langle\Psi \mid \hat{Q}\hat{\mathcal{H}}\Psi\rangle\Big]. \tag{11.86b}$$

[17] I haven't been picky about mathematical rigor in this book, but I should note here that this step is legitimate because of the continuity of Ψ and its first derivative, and because the limits of integration in (11.80) are constant.

[18] Actually, this possibility is unlikely; such explicit time dependence occurs rarely in quantum-mechanical operators. But it does show up in the response of microscopic systems to applied external fields, a matter we'll leave for Volume II.

We can now use the Hermiticity of $\mathcal{H}$ to prepare the *first* Dirac braket in (11.86*b*) to be combined with the second, *viz.*,

$$\frac{d}{dt}\Big\langle Q(t)\Big\rangle - \Big\langle\frac{\partial\hat{Q}}{\partial t}\Big\rangle = \frac{i}{\hbar}\Big[\langle\Psi \mid \hat{\mathcal{H}}\hat{Q}\Psi\rangle - \langle\Psi \mid \hat{Q}\hat{\mathcal{H}}\Psi\rangle\Big] \qquad (11.87a)$$

$$= \frac{i}{\hbar}\langle\Psi \mid \hat{\mathcal{H}}\hat{Q} - \hat{Q}\hat{\mathcal{H}} \mid \Psi\rangle. \qquad (11.87b)$$

We instantly recognize the right-hand side of Eq. (11.87*b*) as the commutator $[\hat{\mathcal{H}}, \hat{Q}]$. So the rate of change of the expectation value of an arbitrary Hermitian operator is

$$\boxed{\frac{d}{dt}\langle Q(t)\rangle = \Big\langle\frac{\partial\hat{Q}}{\partial t}\Big\rangle + \frac{1}{\hbar}\langle i[\hat{\mathcal{H}}, \hat{Q}]\rangle} \qquad (11.88)$$

Question 11–19

Use the general result (11.88) to prove that the normalization of an arbitrary state function is independent of time, *i.e.*,

$$\frac{d}{dt}\Big\langle\Psi \mid \Psi\Big\rangle = 0.$$

This is, of course, not a new result—we proved it in Chap. 6 using a different stratagem.

Summary and Return to Classical Conservation Laws

If, as is usually the case, the operator $\hat{Q}$ does not explicitly depend on time, then

$$\Big\langle\frac{\partial\hat{Q}}{\partial t}\Big\rangle = 0, \qquad (11.89)$$

in which case the equation that governs the evolution of the expectation value of Q is simply

$$\boxed{\frac{d}{dt}\langle Q\rangle = \frac{1}{\hbar}\Big\langle i\left[\hat{\mathcal{H}}, \hat{Q}\right]\Big\rangle} \qquad [\hat{Q} \text{ independent of } t] \qquad (11.90)$$

This equation pertains to all the operators we have studied thus far (to $\hat{\mathcal{H}}$, $\hat{x}$, $\hat{p}$, $\hat{\Pi}$) and to most of those whose acquaintance we'll make in Volume II.[19]

[19]There is a rather advanced formulation of classical mechanics that leads to an equation of motion strikingly analogous to the quantum-mechanical result (11.88). This formulation involves a classical quantity called a *Poisson bracket*. For a classical observable $Q[x_i(t), p_i(t); i = 1, 2, \ldots, N] = Q(t)$ that pertains to a (one-dimensional) system of N macroscopic particles whose classical Hamiltonian is H, the *Poisson bracket* is defined as

$$\{Q, H\} \equiv \sum_{i=1}^{N}\left(\frac{\partial Q}{\partial x_i}\frac{\partial H}{\partial p_i} - \frac{\partial Q}{\partial p_i}\frac{\partial H}{\partial x_i}\right). \qquad \text{Poisson bracket of } Q \text{ and } H$$

The rate of change of the observable $Q(t)$ can be expressed in terms of the Poisson bracket of Q and H as

$$\frac{d}{dt}Q(t) = \frac{\partial Q}{\partial t} + \{Q, H\}.$$

This equation looks very like (11.88); but all quantities in it are functions, not operators.

Now *suppose Q is a constant of the motion of the system with Hamiltonian $\mathcal{H}$.* Then, by the definition (11.74), $\left[\hat{\mathcal{H}}, \hat{Q}\right] = 0$. Ergo the right-hand side of (11.90) is zero and, according to the definition (11.79), Q is conserved. This little argument proves Eq. (11.81). Its point is that *you'll find it much easier to determine whether or not a given observable is conserved by evaluating the commutator of its operator and the Hamiltonian rather than by checking all possible expectation values for time dependence.*

Example 11.5. The Parity Operator

We have already demonstrated that the parity operator $\hat{\Pi}$ commutes with the Hamiltonian of a system whose potential energy is symmetric. Hence parity is a constant of the motion and, according to (11.81), is conserved, *i.e.*,

$$\frac{d}{dt}\langle \Pi \rangle = 0 \text{ for all } \Psi \qquad \text{[conservation of parity]} \tag{11.91}$$

But if $V(x)$ is not symmetric (*e.g.*, for the asymmetric square well of Fig. 9.7), then $\left[\hat{\mathcal{H}}, \hat{\Pi}\right] \neq 0$, and parity is not conserved.

The conservation of parity for submicroscopic physical systems was an issue that, a few years ago, generated a lot of controversy (and a Nobel prize); we'll look briefly at its history in the last section of this chapter.

Example 11.6. Conservation of Energy

One of the most important laws in physics is conservation of energy. According to the definition (11.79), the observable E is conserved if

$$\boxed{\frac{d}{dt}\langle \mathcal{H} \rangle = 0 \qquad \text{for any state } \Psi} \qquad \text{conservation of energy} \tag{11.92}$$

Evidently, Eq. (11.92) holds *for a stationary state*, since for such a state the expectation value

$$\langle \mathcal{H} \rangle = \left\langle \Psi \mid \hat{\mathcal{H}} \mid \Psi \right\rangle = \left\langle \Psi \mid \hat{\mathcal{H}}\Psi \right\rangle \tag{11.93}$$

is simply equal to the sharp energy of the state, E. But that's not good enough; to prove that energy is conserved we must verify (11.92) *for any state, stationary or non-stationary.*

But how much easier to use the alternate formulation in (11.81): any operator commutes with itself, so $\left[\hat{\mathcal{H}}, \hat{\mathcal{H}}\right] = 0$. This implies (11.92), *i.e.*, that the energy is conserved.

Still, there's a subtle feature to this proof. In it, we assumed that $\hat{\mathcal{H}}$ did not explicitly depend on time. What happens if $V = V(x,t)$? Using our general result (11.88) we find

$$\frac{d}{dt}\langle \mathcal{H} \rangle(t) = \left\langle \frac{\partial V}{\partial t} \right\rangle. \qquad \text{[for a time-dependent potential]} \tag{11.94}$$

As expected, the energy of a non-conservative microscopic system (one whose potential depends on t) is not conserved.

In Chap. 12 we'll investigate further the consequences of an observable being conserved; but we've already discovered the essential features. The crucial mathematical and physics statements of §11.5–11.6 are summarized in the following handy diagram:

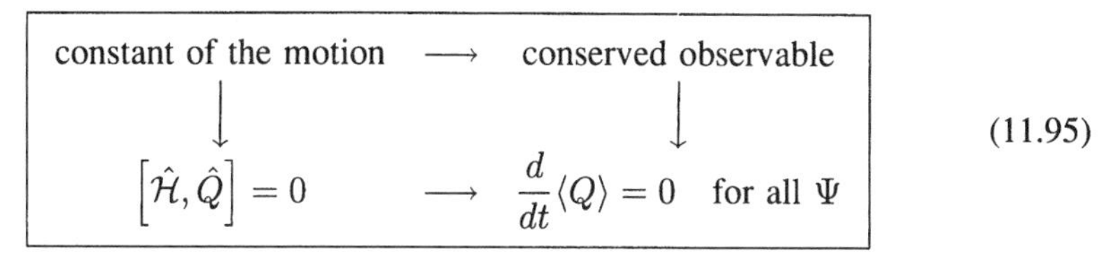

$$\begin{array}{ccc} \text{constant of the motion} & \longrightarrow & \text{conserved observable} \\ \downarrow & & \downarrow \\ \left[\hat{\mathcal{H}}, \hat{Q}\right] = 0 & \longrightarrow & \frac{d}{dt}\langle Q \rangle = 0 \quad \text{for all } \Psi \end{array} \tag{11.95}$$

Question 11–20

Supose the initial state of a quantum system (Ψ at $t = 0$) is an eigenstate of a constant of the motion Q. Time passes, and the system evolves (but no measurements are performed on it). Is $\Psi(x, t)$ for $t > 0$ an eigenfunction of $\hat{Q}$? Why or why not?

I began this section by discussing conservation laws in classical physics; we write these laws in terms of observables that are represented by functions of time. We've now seen the nature of the corresponding laws of quantum physics: we write these laws in terms of the expectation values of observables that are represented by operators. The transition between a quantum-mechanical conservation law and its classical counterpart is fairly straightforward, since according to Eqs. (5.85)

$$\begin{array}{ccc} \langle Q \rangle(t) & & Q(t) \\ & \xrightarrow[\text{in the classical limit}]{} & \\ \Delta Q(t) & & 0 \end{array} \qquad \text{Correspondence Principle} \tag{11.96}$$

But the Correspondence Principle is not limited to conservation laws. Perhaps *all* laws in quantum mechanics, if expressed in terms of expectation values, can be related through the Correspondence Principle, to their classical counterparts. Indeed they can, as we'll see in the next section.

11.7 PHYSICAL LAWS AND THE CORRESPONDENCE PRINCIPLE

> ... listen. The sound of physics.
> The soft, breathless whir of Now.
> Just listen.
> Close your eyes, pay attention:
> Murder, wouldn't you say?
> A purring electron? Photons, protons?
> Yes, and the steady hum
> of a balanced equation.
>
> — from *The Nuclear Age (A Novel)*
> by Tim O'Brian

The bedrock of classical mechanics is Newton's Second Law of Motion, popularly known (for a single particle in one dimension) as

$$F = ma. \tag{11.97}$$

You have probably seen an alternate form of this law, in which it is written in terms of the linear momentum,

$$p(t) = m\frac{d}{dt}x(t), \tag{11.98}$$

and the potential energy $V(x,t)$, which is related to the force by

$$F(x,t) = -\frac{\partial}{\partial x} V(x,t). \tag{11.99}$$

Using these equations and $a = dp(t)/dt$ in (11.97), we can easily derive this alternate form of Newton's Second Law, an expression for the *rate of change of the momentum function* $p(t)$:

$$\boxed{\frac{d}{dt}p(t) = -\frac{\partial}{\partial x}V(x,t) = F(x,t)} \qquad \text{[classical physics]} \tag{11.100}$$

We can also easily derive the corresponding equation for the *rate of change of the position function* $x(t)$, Eq. (11.98), *viz.*,[20]

$$\boxed{\frac{d}{dt}x(t) = \frac{1}{m}p(t)} \qquad \text{[classical physics]} \tag{11.101}$$

According to the Correspondence Principle (11.96), we retrieve classical observables, such as $x(t)$ and $p(t)$, in the classical limit from the expectation values of the corresponding operators, such as $\langle x \rangle (t)$ and $\langle p \rangle (t)$. The classical laws (11.100) and (11.101) describe the evolution of the *observables* in this limit, while quantum laws derived from the general evolution equation (11.88) describe changes in the *expectation values*. What are these quantum laws? And do they indeed translate into the corresponding classical laws in the classical limit?

Ehrenfest's Theorem

These provocative questions were answered in 1927 by the Austrian physicist Paul Ehrenfest (1880–1933). Ehrenfest was a member of the "second generation" of founders of quantum mechanics. He was a friend of Albert Einstein (a leader of the first generation) and shared Einstein's interests in the foundations of quantum theory and its relationship to thermodynamics and to statistical physics.

In a paper published in a 1927 issue of the distinguished German journal *Zeitschrift für Physik* [**45**, 455 (1927)], Ehrenfest proved "by a short elementary calculation without approximations" that for a single particle (in one dimension)

$$\frac{d}{dt}\langle p \rangle (t) = \left\langle -\frac{\partial V}{\partial x} \right\rangle. \qquad \text{Ehrenfest's Theorem (one dimension)} \tag{11.102}$$

[20]One reason for rewriting Newton's law in the form of Eqs. (11.100) and (11.101) is that these equations highlight the functions that make up the trajectory, $x(t)$ and $p(t)$. Equations (11.100) and (11.101) are called *Hamilton's equations* (yes, the same Hamilton) and are often written in terms of the *classical Hamiltonian* $H(x,p,t)$ as

$$\begin{aligned} \frac{d}{dt}x(t) &= \frac{\partial}{\partial p}H(x,p,t) \\ \frac{d}{dt}p(t) &= -\frac{\partial}{\partial x}H(x,p,t). \end{aligned}$$

If you're interested, take a minute to convince yourself that these are identical to (11.100) and (11.101) for a classical Hamiltonian $H(x,p,t) = T(p) + V(x,t)$.

This is the quantum-mechanical analog of Newton's Second Law, Eq. (11.100): it relates the rate of change of the momentum to the slope of the potential energy (*i.e.*, to the force) for a state of the system. Ehrenfest went on to show that in the classical limit, this result reduces to its classical counterpart.[21]

The impact of Ehrenfest's research on the physics community in the early thirties was enormous. Physicists thought, quite reasonably, that if this upstart new quantum theory was correct, then not only its *physical quantities* but also its *physical laws* must reduce to those of classical physics in the appropriate limit. Ehrenfest's work therefore contributed greatly to the acceptance of Schrödinger's theory.

Newton's Second Law Regained (Almost)

We'll prove Ehrenfest's Theorem in two stages. First, we will develop the quantum-mechanical counterpart of the classical equation for the *rate of change of position*, Eq. (11.101). Then we'll tackle the rate of change of momentum, which will lead us to Ehrenfest's analog of Newton's Second Law.

The rate of change of $\langle x\rangle(t)$ is given by Eq. (11.88) as

$$\frac{d}{dt}\langle x\rangle = \frac{1}{\hbar}\Big\langle i\big[\hat{\mathcal{H}},\hat{x}\,\big]\Big\rangle, \tag{11.103}$$

since $\partial x/\partial t = 0$. Conveniently, the commutator we need appears in Table 11.3:

$$\big[\hat{\mathcal{H}},\hat{x}\,\big] = -\frac{i\hbar}{m}\hat{p}. \tag{11.104}$$

Substituting this commutator into (11.103) we easily obtain

$$\boxed{\frac{d}{dt}\langle x\rangle(t) = \frac{1}{m}\langle p\rangle(t)}\,, \tag{11.105}$$

whose resemblance to (11.101) I need hardly mention.

To evaluate the rate of change of $\langle p\rangle$ we pull the same trick. The evolution equation for this observable is

$$\frac{d}{dt}\langle p\rangle = \frac{1}{\hbar}\Big\langle i\big[\hat{\mathcal{H}},\hat{p}\big]\Big\rangle. \tag{11.106}$$

In Table 11.3 we find the commutator

$$\big[\hat{\mathcal{H}},\hat{p}\big] = i\hbar\frac{\partial V}{\partial x}\hat{1}. \tag{11.107}$$

This result leads us immediately to

$$\boxed{\frac{d}{dt}\langle p\rangle(t) = \Big\langle -\frac{\partial V}{\partial x}\Big\rangle} \tag{11.108}$$

[21]But Ehrenfest did not formulate his theorem in terms of expectation values. Rather he discussed the evolution of the mean value of the linear momentum in terms of averages over wave packets. Equation (11.101) is a "translation" of his finding into the lingo of modern quantum mechanics. The year following publication of Ehrenfest's paper, A. E. Ruark showed how to extend it to N-particle systems [*Phys. Rev.*, **31**, 533 (1928)].

Now, by identifying the first derivative of the potential with the force [Eq. (11.100)] and taking the derivative of $\langle p \rangle$ via Eq. (11.105), we can turn (11.108) into

$$m\frac{d^2}{dt^2}\langle x \rangle (t) = \langle F(x,t) \rangle . \qquad (11.109)$$

At first glance, this looks very much like Newton's Second Law, Eq. (11.100). But they're not the same. Not yet.

Can you see the difference? Ponder carefully the *expectation value of the force*. For a state Ψ this quantity is

$$\int_{-\infty}^{\infty} \Psi^*(x,t)F(x,t)\Psi(x,t)\,dx = -\int_{-\infty}^{\infty} \Psi^*(x,t)\left[\frac{\partial}{\partial x}V(x,t)\right]\Psi(x,t)\,dx. \qquad (11.110)$$

The difference is that $\langle F \rangle (t)$ on the right-hand side of (11.109) is not *necessarily* equal to $F\big(\langle x \rangle (t)\big)$. To appreciate the importance of this difference, consider what would happen to Eq. (11.108) if, as is the case *for some systems*, these quantities happened to be equal:

$$\langle F \rangle (t) = F\big(\langle x \rangle (t)\big) \qquad \Longrightarrow \qquad \frac{d}{dt}\langle p \rangle (t) = F\big(\langle x \rangle (t)\big). \qquad (11.111)$$

In this special case, the quantum-mechanical equations that govern the evolution of the expectation values $\langle x \rangle (t)$ and $\langle p \rangle (t)$ have precisely the same form as the classical laws for $x(t)$ and $p(t)$.

"But," you ask, "isn't the relevant question whether $\langle x \rangle (t)$ and $\langle p \rangle (t)$ evolve according to the laws of classical physics *in the classical limit?*"

Right you are. And in this limit, all is well, for[22]

$$\int_{-\infty}^{\infty} \Psi^*(x,t)F(x,t)\Psi(x,t)\,dx \xrightarrow[\langle x \rangle (t) \to x(t)]{\Delta x \to 0} F\big(\langle x \rangle (t)\big). \qquad (11.112)$$

This is the essence of Ehrenfest's theorem:

> ***Rule***
>
> In the classical limit, the equations that describe the time development of $\langle x \rangle (t)$ and $\langle p \rangle (t)$ are identical to the equations of Newtonian mechanics.

Notice that the mathematical form of Eq. (11.105) for $d\langle x \rangle (t)/dt$ is identical to that of Eq. (11.101) for $dx(t)/dt$. And, as I remarked above, *for certain systems* (such as the simple harmonic oscillator), Eq. (11.108) for $d\langle p \rangle (t)/dt$ is identical in form to (11.100) for $dp(t)/dt$. But the peculiar structure of Eq. (11.109) for an arbitrary potential throws a monkey wrench into the works, which we can get rid of only by taking the classical limit.

[22]Remember that in the classical limit the spatial extent of $P(x,t) = \Psi^*(x,t)\Psi(x,t)$ is negligible (for the duration of any measurement). Mathematically speaking, this is a statement of *spatial localization*. For example, if the center of $P(x,t)$ occurs at x_0, then in the classical limit this function can be replaced by the Dirac Delta function $\delta(x - x_0)$ with no loss of accuracy.

Question 11–21

Demonstrate that the form of the classical and quantum equations for evolution of position and its expectation value are identical for a simple harmonic oscillator, *i.e.*, that

$$\langle F(x,t)\rangle = F\left(\langle x\rangle(t)\right) \qquad \text{for } V(x) = \tfrac{1}{2}m\omega_0^2 x^2. \tag{11.113}$$

Question 11–22

Demonstrate that (11.112) follows in the classical limit $\Delta x(t) \to 0$.

Question 11–23

Show why we run into trouble if the potential depends on a power of x higher than the second by demonstrating that

$$\langle F(x,t)\rangle \neq F\left(\langle x\rangle(t)\right) \qquad \text{for } V(x) \propto x^3. \tag{11.114}$$

Just What IS the Relationship of Quantum to Classical Physics?

The equations we've derived for the time development of the mean values of position and momentum add a remarkable new dimension to the Correspondence Principle, revealing that the connection of classical and quantum physics is deeper than we heretofore perceived. From Chap. 5 we knew that statistical properties, which must be used to describe an observable in quantum mechanics, go over correctly in the classical limit to functions of time, which describe the observable in classical physics. But now we have discovered that *the relationships between these statistical properties*—in particular between the expectation values of the fundamental observables, position and momentum—also go over into the corresponding relationships of classical physics. So the *laws of nature* in quantum mechanics are, in this sense, consonant with those of classical physics.[23]

Still, as satisfying as this consonance is, it is not quite as profound as you might think. There is a sense in which we implicitly *assumed* Ehrenfest's theorem when we set up our postulates in Part II. The operator postulate (Postulate III, Chap. 5) prescribes that quantum-mechanical operators, such as $\hat{\mathcal{H}}$, shall be constructed *from classical observables*, such as $T + V(x,t)$. This assumption leads inexorably (if circuitously) to Ehrenfest's theorem.

So the Correspondence Principle leaves certain aspects of the relationship between quantum mechanics and Newtonian mechanics a bit vague. The implications of this

[23]There is another way to retrieve the laws of classical physics written in a form that looks like the commutator-infested equations of this chapter. In an earlier footnote I introduced the Poisson brackets for two classical observables Q and R. We can recast the laws of classical physics in terms of these brackets, arriving at results such as the following expressions for the derivatives of an arbitrary observable $Q[x(t), p(t)]$:

$$\frac{\partial Q}{\partial p} = -\{Q, x\} \qquad \frac{\partial Q}{\partial x} = \{Q, p\}.$$

Regardless of the particular classical law you are working with, you can obtain from it the corresponding quantum-mechanical equation by merely replacing the Poisson brackets by the commutator of the relevant observables, appropriately scaled:

$$\text{replace } \{Q, R\} \text{ by } -\frac{i}{\hbar}\left[\hat{Q}, \hat{R}\right]$$

For more on Poisson brackets, see § 9.4ff of *Classical Mechanics*, 2nd ed. by H. Goldstein (Reading, Mass.: Addison-Wesley, 1980).

ambiguity have been well expressed by Daniel T. Gillespie in his fine book *A Quantum Mechanics Primer* (London: International Textbook, 1973) (p.113):

> ... it is not altogether clear just what the precise logical relationship is between classical and quantum mechanics. ... *we ourselves* as the ultimate observers of any physical system, are essentially classical objects, in that our sense can directly perceive only *macroscopic phenomena* (*e.g.*, dial readings, instrument settings, etc). This fact probably places severe restrictions not only on what things we can perceive about a *microscopic* system, but also on how we interpret what we perceive.

We shall return to this issue in Chap. 13, where we directly confront the Problem of Measurement.

Aside: A Philosophical Note. Before we leave the Correspondence Principle, I want to add a note on "quantum epistemology." Discussions of this principle seduce some readers into the impression that quantum mechanics somehow subsumes classical physics, that it is more fundamental and obviates the need for classical mechanics in physical theory. This is wrong. Although some physicists have tried to conjure up an interpretation of quantum mechanics that does describe everything,[24] the prevailing view is that a dual description is unavoidable, that neither quantum mechanics nor classical mechanics *alone* describes the physical universe. The reasons for this necessary dualism require an appreciation of subtle features of the Correspondence Principle that we haven't the space to discuss;[25] suffice it to say here that the many possibilities inherent in the (Copenhagen) description of a microscopic system do not always go over smoothly into a single actuality, a necessary characteristic of a classical description.

*11.8 THE STRANGEST "UNCERTAINTY RELATION" OF THEM ALL

> "The tragedy of science is
> the heartless murder of beautiful theories
> by ugly facts."
>
> —John Bishop
> in *Artifact*
> by Gregory Benford

All of the elements of quantum mechanics that we have investigated in this chapter—commutators, simultaneous eigenfunctions, uncertainty relations, conservation laws—interlock via the theorems and equations we have discussed. In the flow chart in Fig. 11.2, I have attempted to summarize and systematize this maze of connections. Many of these elements come together in the definition and derivation of the "energy-time uncertainty relation," which for reasons I'll explain in a minute, I prefer to call **the energy-time inequality**:

$$\Delta E \, \Delta t \geq \frac{\hbar}{2}. \tag{11.115}$$

[24] Such as the many-world interpretation of H. Evertt and the hidden variable theories of D. Bohm and others. For a low-key, accessible introduction to these matters, see Chaps. 9, 12, and 13 of *Quantum Reality: Beyond the New Physics* by N. Herbert (New York: Anchor Press, 1985).

[25] A useful overview is provided by C. H. Woo in a paper in *Amer. Jour. Phys.*, **54**, 923 (1986).

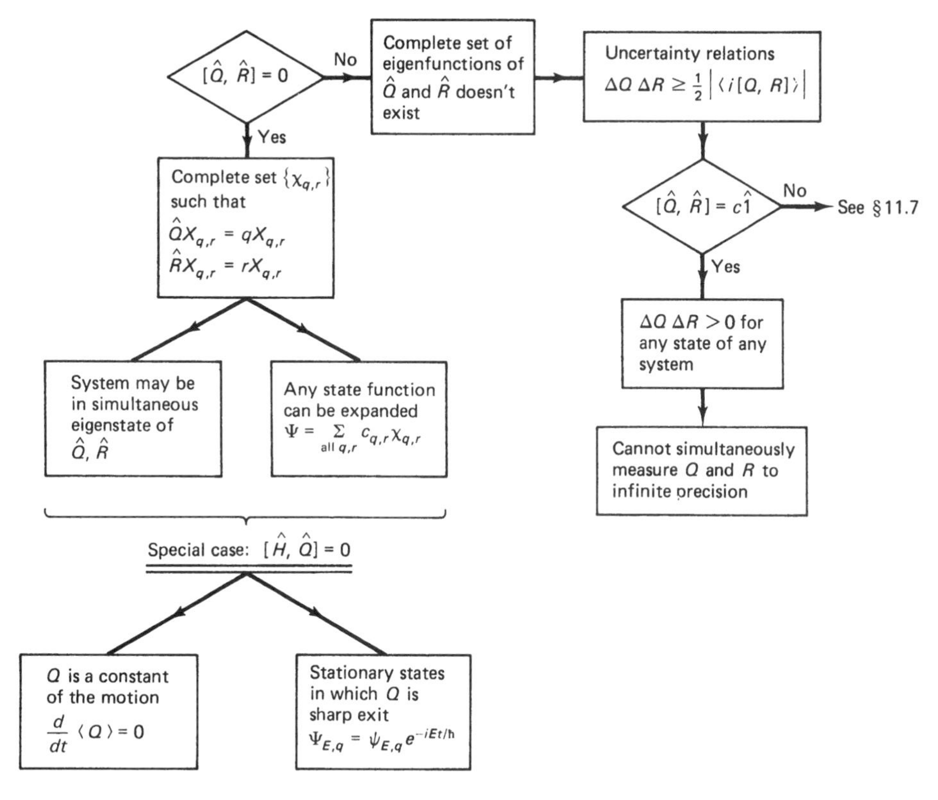

Figure 11.2 A marathon summary of the relationships between the elements of quantum mechanics examined in Chap. 11.

I commented on this relation at various places in Part II, and if you worked Pblm. 4.2 you should have a fairly good intuitive sense of its origin and meaning. This inequality is subtle, even for an uncertainty relation, but we now have the background to define its terms and understand its content. That content is of considerable practical import, for, as you'll see at the end of this section, the energy-time inequality finds application in describing the stationary states of atoms.

The first thing to notice about Eq. (11.115) is that it's not an *uncertainty relation*—at least not in the sense that we have been using the phrase. Think about time. *In quantum mechanics, time is not an observable; it is a parameter.* We may choose to investigate a quantum state at one or another time, but we cannot *measure* time as we can energy or position. And there is no operator $\hat{t}$ in quantum mechanics. So what is meant by "the uncertainty in time Δt"? If by Δt we mean to imply use of the usual definition of uncertainty, $\Delta Q = [\langle Q\rangle^2 - \langle Q^2\rangle]^{1/2}$, then this symbol means nothing.

Time for a Change

The question of the meaning of Δt in the inequality (11.115) takes us once again to the heart of quantum physics: measurement. This inequality has to do with *change* in a quantum state. The issue it addresses is: *how much time must elapse before we can say that a system has changed?* As I have remarked several times, in quantum mechanics we deal with such matters in a very pragmatic context—restricting ourselves to statements that can be quantified via experiment:

> ***Rule***
>
> To measure a change in a system, we must detect in an experiment a change in an observable of the system.

For example, one way to observe that there has been a change in the state of an ensemble that was initially in, say, a stationary state with energy E_n is to detect the photons emitted when the members undergo *transitions* to another, lower-lying state. (The transition may occur of its own accord, as in the spontaneous decay of an atom from an excited state, or by absorption of a photon, or in response to a collision with another microscopic particle, such as an electron.)

Defining a "measurable change" in a property of a microscopic system is a bit tricky, because such a definition must accommodate states in which the property is uncertain. But regardless of the state, the results of measurement of the observable can be described *en masse* by the expectation value $\langle Q \rangle$. So *we would be confident that an uncertain observable Q has actually changed were we to observe that its mean value $\langle Q \rangle$ has changed by (at least) the uncertainty ΔQ. The time required for a change of this magnitude to occur is the quantity we call Δt.*[26] So only after a time interval Δt *could* we (in principle) say whether or not a property of the system has changed. (Actual times involved in measuring changes in the laboratory are much larger than the minimum allowed by the energy-time inequality.)

Let us now put this definition on a quantitative footing. The rate of change of $\langle Q \rangle$ is $d\langle Q \rangle / dt$. So the time required for $\langle Q \rangle$ to change from its initial value (at $t = 0$) by an amount $\Delta Q(0)$ (the uncertainty evaluated at $t = 0$) is[27]

$$\boxed{\Delta t \equiv \left. \frac{\Delta Q}{\left| \frac{d}{dt} \langle Q \rangle \right|} \right|_{t=0}} \tag{11.116}$$

According to the energy-time inequality, the minimum time required to measure a change in *any observable* is related to the uncertainty in the energy of the particle by Eq. (11.115). On the face of it, this claim seems quite remarkable: why should the uncertainty in the *energy* control how long it takes to measure a change in some other observable? Let's see.

Derivation of the Energy-Time Inequality

To develop an expression for the product of the energy uncertainty and the "change time" (11.116), we start with a more conventional uncertainty relation: the inequality for the

[26] As you can probably tell, I consider use of the symbol Δt and the name "energy-time uncertainty relation" a little misleading. As defined here, Δt is in no sense an uncertainty; rather it is a *time interval*—in more precise quantum-mechanical terms, it is the amount of time that must pass before we can distinguish $\Psi(x, \Delta t)$ from its progenitor $\Psi(x, 0)$. This time interval is sometimes called "the measurement time" and is sometimes labelled by the particular observable being measured, as Δt_Q.

[27] The absolute value bars in the denominator of (11.116) are there just to ensure that $\Delta t \geq 0$, as it must for this definition to make sense.

product of the uncertainty in the energy, ΔE, and the uncertainty in the observable we're measuring, ΔQ. We'll then introduce into this relation the rate of change of $\langle Q \rangle$ and thereby transform it into Eq. (11.115).[28]

According to the Generalized Uncertainty Principle (11.55) the uncertainty product $\Delta E\, \Delta Q$ is bounded from below as follows:

$$\Delta E\, \Delta Q \geq \frac{1}{2} \left| \left\langle i [\hat{\mathcal{H}}, \hat{Q}\,] \right\rangle \right|. \tag{11.117}$$

The commutator $\left[\hat{\mathcal{H}}, \hat{Q}\,\right]$ is related to the rate of change of $\langle Q \rangle$ by the general evolution equation (11.90), *i.e.*,

$$\left\langle i [\hat{\mathcal{H}}, \hat{Q}\,] \right\rangle = \hbar \frac{d}{dt} \langle Q \rangle \,. \tag{11.118}$$

If we replace the commutator in (11.117) by (11.118), we wind up with

$$\Delta E\, \Delta Q \geq \frac{\hbar}{2} \left| \frac{d}{dt} \langle Q \rangle \right|. \tag{11.119}$$

We're almost there. Dividing both sides of (11.119) by the rate of change of $\langle Q \rangle$ gives

$$\Delta E \frac{\Delta Q}{\left| \frac{d}{dt} \langle Q \rangle \right|} \geq \frac{\hbar}{2}. \tag{11.120}$$

According to the definition of the change time Δt, this is just the energy-time inequality,

$$\boxed{\Delta E\, \Delta t \geq \frac{\hbar}{2}} \qquad \text{energy-time inequality} \tag{11.121}$$

An Application: Natural Widths and Lifetimes of Atoms

When an atom initially in an excited stationary bound state decays, it emits a photon whose energy, according to conservation of energy, should equal the difference between the energies of the initial and final states of the atom. Spectroscopists, who measure the energy of the emitted photons from decay of an ensemble of atoms, can deduce a great deal about the properties of the atom.

But high-precision spectroscopic measurements reveal that not all photons emitted from an ensemble (all members of which are in the same quantum state) have the same energy, even if all correspond to decay to the *same final state*. (Usually, the final state is the ground state.) Instead, the frequencies of individual emitted photons fluctuate around the expected value, $1/\hbar$ times the difference between the initial- and final-state energies. From this observation, physicists conclude that associated with the energy of an excited stationary state of an atom is a **width** ΔE.

[28]The derivation draws on one in *Quantum Mechanics* by A. Messiah (Amsterdam: North-Holland, 1965), p. 319–320. See also pp. 152ff in *Quantum Physics* by R. G. Winter (New York: Wadsworth, 1979).

On the face of it, this conclusion seems nonsensical, for haven't we said innumerable times that the energy of a stationary state is sharp? Yes, we have.[29] But we're no longer discussing "a stationary state" in the abstract; we're measuring the energy of a stationary state in the laboratory. And the precision of such a measurement is limited by the energy-time inequality.

In this case, the observable we're measuring is the energy of the nth excited state, E_n.[30] So Δt is the time interval that must elapse before we can ascertain that a transition from this state to the ground state has occurred. Atomic physicists call this interval the **natural lifetime** of the state and often denote it by τ_n. With this convention, we write the energy-time inequality as

$$(\Delta E)_n \tau_n \geq \frac{\hbar}{2}, \tag{11.122}$$

where $(\Delta E)_n$ is the **width** of the nth stationary state. (An atom in the ground state cannot decay to any other state, so the width of the ground state of an atom is zero.)

One can calculate the natural lifetime τ_n by measuring the intensity of the beam of emitted photons as a function of distance along this beam.[31] Typically, the lifetimes of excited states of atoms range from 10^{-7} sec to 10^{-9} sec—not very long on a macroscopic scale, but not zero either. Consider, for example, the natural lifetimes in Table 11.5 of the first five excited states of the hydrogen atom.

Question 11–24

Calculate the natural widths of each of the excited states of the hydrogen atom whose lifetimes are shown in Table 11.5. Compare your answers to the energies of these stationary states, which are given (in eV) by $E = -13.6\,n^2$, where $n = 1$ labels the ground state.

Not all excited states are so short-lived. Most atoms have a few special stationary states called **metastable states** whose lifetimes are typically around 10^{-3} sec. (Metastability of states, where it occurs, arises for reasons other than the energy-time inequality.) In Table 11.6 are shown lifetimes of several excited states of several atoms; the point of this table is to suggest the range of lifetimes one finds in nature.

Question 11–25

Comment on the truth or falsity of this statement:
"Quantum mechanics lets you violate the law of conservation of energy if you do so quickly enough."

[29] I did qualify this blanket statement in § 7.7 by alluding to the interaction of systems with the vacuum electromagnetic field, which interactions eventually induce a decay from any excited stationary state to the ground state. This qualification points out why stationary states as rigorously defined in quantum mechanics do not actually exist in nature. But this fact rarely influences experiments, for the time required for the vacuum field to induce decay is long compared to the duration of most measurements. In this section we are concerned with a quite different phenomenon: limitations imposed by nature on the precision of measurement of the energy of a stationary state.

[30] Except for the hydrogen atom (in a simple approximation that neglects certain weak interactions), more than one quantum number is required to uniquely label an energy level. So in an actual application to an actual atom, n here must be replaced by several quantum numbers, however many are needed to uniquely identify a level. Such quibbles are irrelevant, however, to our immediate concerns.

[31] For details of such an experiment, see § 9.7 of *Modern Atomic Physics: Quantum Theory and Its Applications* by B. Cagnac and J. C. Pebay-Peyroula (New York: Wiley, 1975).

TABLE 11.5 NATURAL LIFETIMES OF THE FIRST FIVE EXCITED STATES OF THE HYDROGEN ATOM.[a]

n	τ_n (10^{-8}sec)
2	0.21
3	1.02
4	3.35
5	8.8
6	19.6

[a] Mean lifetimes for principal quantum number n.

TABLE 11.6 LIFETIMES OF SOME (NON-METASTABLE) EXCITED STATES OF SOME ATOMS.

Atom	State[b]	Energy [eV]	Lifetime [ns][a]
He	$3p\,(^3P^o)^c$	23.01	94.7
	$3p\,(^1P^o)$	23.09	1.724
	$4s\,(^3S)$	23.59	64.0
Li	$2p\,(^2P^o)$	1.85	27.3
	$3p\,(^2P^o)$	3.83	200.0
	$4p\,(^2P^o)$	4.52	450.0
F	$3s\,(^4P)$	12.72	7.0
	$3s\,(^2P)$	13.0	3.5
Cl	$4s\,(^2P)$	9.23	1.0
Ga	$5p\,(^2P^o)$	4.106	50.0
Cs	$6p\,(^2P^o)$	.432	31.0
Hg	$6p\,(^3P^o)$	4.887	118.0
	$7p\,(^3P^o)$	8.829	170.0

[a] $1\,ns = 1 \times 10^{-9}$ sec.

[b] For details of state labels, see Volume II.

[c] Post-superscript indicates parity.

Source: [Data from Table 7.4 of *Reference Data for Atoms, Molecules, and Ions*, A. A. Radzig and B. M. Smirnov (New York: Springer-Verlag, 1985).]

The implications of the energy-time inequality may be unsettling. Must we abandon stationary states, with their convenient wave functions and sharp energies? Certainly not. We can continue to probe the mysteries of quantum systems using these states as theoretical constructs, knowing that we are working within the model of the microscopic universe that is defined by quantum mechanics. But this inequality does mean that when we seek to measure a property of such a state in the laboratory, we must accept the limitations imposed by nature, which prohibits us from pinning down the properties of these systems as precisely as we might like.[32]

*11.9 FINAL THOUGHTS: CONSERVATION "LAWS," DOGMA, AND THE FALL OF PARITY

Throughout this chapter, we have dignified a number of results by the name "laws." To so describe a mathematical relationship or a physical principle is to make it sound like absolute truth. But, of course, all such "laws" are no more than expressions of how natural scientists (at the moment) understand the workings of the part of the physical universe they study. Some laws, such as Einstein's statement that the speed of light is independent of the motion of the observer, still hold; others, such as Newton's "Laws of Motion" have required considerable modification.[33]

Sometimes the elevation of a physical result into a physical law comes about because of non-scientific factors, such as human nature and aesthetics. Which brings us to *conservation of parity*. What is important about this physical law is that it is not true. What is interesting about it is how its falsity was discovered. As Emilio Segre remarks in his popularization *From X-Rays to Quarks* (San Francisco: Freeman, 1980):

> The violation of parity conservation is perhaps the greatest theoretical discovery of the postwar era. It has extinguished a prejudice that had been transformed into a principle on the basis of insufficient tests.

Parity Reviewed

In Chap. 9 we proved that the bound-state Hamiltonian eigenfunctions of a single particle in one dimension *with a symmetric potential* are eigenfunctions of the parity operator $\hat{\Pi}$. This discovery achieved its most elegant statement in this chapter, where we showed that the commutativity of a symmetric Hamiltonian and $\hat{\Pi}$ was responsible for the existence of such simultaneous eigenfunctions and for the conservation of parity as expressed by Eq. (11.91), $d\langle\Pi\rangle/dt = 0$. From this equation it follows that a state that initially has definite parity has the same parity at all subsequent times. Of course, if $\hat{\mathcal{H}}$ is not symmetric, then $\left[\hat{\mathcal{H}}, \hat{\Pi}\right] \neq 0$ and Eq. (11.91) does not hold for all states.

But if the potential is symmetric, then a bound stationary state wave function $\Psi_n(x,t)$ is transformed under inversion ($x \to -x$) into either itself or minus itself; consequently the probability density $|\Psi_n(x,t)|^2$ is unchanged by inversion, and we cannot tell by any measurement that inversion has taken place. A colorful, metaphoric way to describe this fact is to say "the state is (experimentally) indistinguishable from its mirror image." This, in a nutshell, is the meaning of *parity invariance*.

[32]For more on uncertainty relations in general and on the energy-time inequality in particular, see the articles in *The Uncertainty Principle and the Foundations of Quantum Mechanics*, edited by W. C. Price and S. S. Chissick (New York: Wiley, 1977). I particularly recommend the article by J. Rayski and J. M. Rayski, which begins on p. 13.

[33]See *How the Laws of Physics Lie* by N. Cartwright (New York: Oxford University Press, 1983).

This description translates with but slight modification to systems with one or more particles in three dimensions—systems such as atoms, molecules, or elementary particles. In three dimensions, the parity operator transforms $\mathbf{r}$ into $-\mathbf{r}$, and parity invariance is colloquially referred to as "reflection symmetry." The situation in three dimensions is, in some respects, unlike the one-dimensional case. For example, the bound Hamiltonian eigenfunctions of a three-dimensional system whose Hamiltonian commutes with $\hat{\Pi}$ do not *necessarily* have definite parity. But commutativity of $\hat{\mathcal{H}}$ and $\hat{\Pi}$ does guarantee the existence of simultaneous eigenstates, *i.e.*, stationary states with definite parity.

If the system contains more than one particle, then *reactions* among the particles are possible; the particles can, for example, collide or break apart, forming new particles. According to the "law" of conservation of parity, the parity of a state of the system is unchanged by such a physical process.[34]

One can measure the intrinsic parity of various subatomic particles by studying reactions among them, but the deduction of the parities of such a particle hinges on the equality of the parities of the state before and after the reaction. That is, parity must be conserved. But is it?

Sometimes a Great Notion

From the 1930's until 1956, physicists assumed that the answer was yes. In 1924 Laporte discovered evidence of the parity of states of atoms in his studies of atomic transitions, and soon thereafter the theoretician Eugene Wigner showed parity conservation in such transitions to follow via quite beautiful symmetry arguments (rather like those of § 9.4). What then happened is revealing that physicists are, after all, only human. As Segre puts it:

> Whenever a regularity is valid in many cases, there is a tendency to generalize it to other untested circumstances and perhaps even to make it into a "principle." ... Such occurred with the principle of conservation of parity.

Then, in 1956, while trying to explain the decay of a particular subatomic particle, the Chinese physicists Tsung-Dao Lee and Chen Ning Yang reviewed the hard evidence for conservation of parity and pointed out that no such evidence existed to substantiate the belief that parity is conserved *in all interactions*. Lee and Yang discovered that only for the hadronic and electromagnetic interactions had parity conservation been *demonstrated in the laboratory*.[35] Subsequent experiments by Chien Shiung Wu and others showed that, sure enough, parity conservation had somehow left the rarefied realm of scientific

[34]Discussing parity conservation in reactions is complicated by the multi-particle nature of the system; at least *two* particles must, obviously, be present. The state of such a system must be described by a multi-particle state function, a beast we won't meet until Volume II. Suffice it to say that parity is a multiplicative property, *e.g.*, that the parity of a state of two particles is equal to the product of the intrinsic parity of the first particle, the parity of the second particle, and the parity of their relative motion. For an accessible introduction to these matters, see Chap. 9 of *Subatomic Physics* by H. Frauenfelder and E. M. Henley (Englewood Cliffs, N. J.: Prentice Hall, 1974), which also includes detailed discussions of the experimental determination of the intrinsic parities of particles.

[35]All interactions in nature fall into four categories: the nuclear (also called "strong" or "hadronic") interaction, which holds the nucleus together; the electromagnetic interaction, which dominates in atoms, molecules, and solids; the weak interaction, which controls some non-electromagnetic decay reactions; and our friend, gravity. For a fine introduction to such matters, see *The Forces of Nature*, 2nd ed. by Paul Davies (New York: Simon and Schuster, 1986.)

fact and entered the dubious domain of dogma: systems in which the weak interaction plays a role are not invariant under inversion. The parity of states of such a system is not conserved.

Aside: The Puzzle At Hand. Lee and Yang were trying to explain a mystery known as "the θ—τ puzzle." (Only physicists would come up with such a name for a mystery.) Briefly, the problem was this: A particular subatomic particle (called a K-meson, or kaon) was observed to decay into two different states, one a state of even parity, the other of odd parity. (These were called the "θ decay mode" and the "τ decay mode," respectively. Hence the name.) The state of the kaon before decay had definite parity, so it appeared that unless two different particles were involved (a hypothesis refuted by other experiments), a parity-changing reaction could occur. The existance of such a reaction violated the tenet of conservation of parity.[36]

I bring up this historical sidelight hoping it will help you keep the "laws" of quantum physics in perspective. There are no true laws in physical theory; there are only the fruits of man's efforts to understand the workings of nature. The results of those efforts are accomplishment enough.

EXERCISES AND PROBLEMS

"How in the hell
do we hunt that kind of country?"
asked Fric.
"Slowly," said Ratnose,
"and on foot."

— from *Blood Sport:*
A Journey up the Hassayampa
by Robert Jones

Exercises

11.1 A Little More about the Parity Operator

Prove that

$$\left[\hat{\Pi}, \hat{p}\right]_{+} = 0, \tag{11.1.1}$$

where $\left[\hat{Q}, \hat{R}\right]_{+}$ is the *anticommutator* of $\hat{Q}$ and $\hat{R}$, defined as

$$\left[\hat{Q}, \hat{R}\right]_{+} \equiv \hat{Q}\hat{R} + \hat{R}\hat{Q}.$$

11.2 Operators and Measurement

[a] **Calculate** the commutator of the position operator and the Hamiltonian for a particle in an infinite square well.

[b] In a simultaneous measurement of energy and position for such a particle, is it possible to determine both observables to infinite precision? Is it possible to determine one or the other to infinite precision? If only one, which one? **Explain** each of your answers.

[36] This is a greatly simplified sketch of the drama and controversy involved in overthrowing the conservation of parity. I strongly recommend that you seek out the wonderful, vivid account of this adventure, written largely by the participants, in *Adventures in Experimental Physics: γ volume* (Princeton, N. J.: World Science Education, 1973). Another valuable account is C. N. Yang, *Elementary Particles* (Princeton, N. J.: Princeton University Press, 1962).

11.3 Measurement

A particle of mass m is in a constant force field of magnitude $-g$, where g is a positive constant.

[a] Do the eigenfunctions of the Hamiltonian of this system have definite parity? **Prove your answer.**

[b] **Evaluate** the commutator of $\hat{\mathcal{H}}$ and $\hat{p}$ for this system.

[c] Does a complete set of simultaneous eigenfunctions of $\hat{\mathcal{H}}$ and $\hat{p}$ exist? Why or why not?

[d] Consider a *simultaneous* measurement of the energy and linear momentum of such a particle. Let ΔE and Δp denote the uncertianties for these observables. **Evaluate** a lower bound on the product of these uncertianties.

[e] Suppose that in the energy measurement in [d] we discover that the particular state of our ensemble is *stationary*. Will a (simultaneous) measurement of the linear momentum show that in this state the momentum is sharp (*i.e.*, that $\Delta p = 0$)? **Explain your answer carefully and fully.**

11.4 Some Statements to Critique

Consider each of the following statements. **If the statement is correct, say so and go on to the next one. If, however, the statement is in any way incorrect, write a brief, concise, clear explanation of what is wrong with it. Moreover, if you decide that a statement is wrong, give a specific example to support your judgment.**

[a] The wave function of a system is always an eigenfunction of the system Hamiltonian.

[b] If two Hermitian operators $\hat{Q}$ and $\hat{R}$ commute, then every eigenfunction of $\hat{Q}$ is necessarily an eigenfunction of $\hat{R}$.

[c] An operator that is equal to its own adjoint is necessarily Hermitian.

[d] If two operators $\hat{Q}$ and $\hat{R}$ do *not* commute, then the product of the uncertainties ΔQ and ΔR is necessarily positive—*i.e.*, $\Delta Q\, \Delta R > 0$.

11.5 More Little Bitty Questions about Operators

[a] Let $\hat{\mathcal{H}}$ denote the Hamiltonian of a one-dimensional simple harmonic oscillator (SHO) of mass m and natural frequency ω_0. **Evaluate** the commutator of $\hat{\mathcal{H}}$ of this system and the momentum operator.

[b] Is the observable p a *constant of the motion* of the microscopic simple harmonic oscillator? Why or why not?

[c] Could there exist stationary states of the SHO in which the momentum is sharp? **Provide** a *thorough* justification of your answer.

[d] **Derive** an expression for the rate of change of $\langle p \rangle$ *for an arbitrary state* of the SHO. What is the rate of change of $\langle p \rangle$ for a *stationary state* of the SHO? Is the momentum of the simple harmonic oscillator a conserved quantity? Why or why not?

[e] Suppose $\left[\hat{\mathcal{H}}, \hat{p}\right] \neq 0$. Does it necessarily follow that $\Delta E\, \Delta p > 0$ for all states of a conservative system? Why or why not?

[f] Consider a system with a constant potential energy, $V(x) = V_0$. What is $\left[\hat{\mathcal{H}}, \hat{p}\right]$ for such a system? Is it possible (in principle) to measure the energy and momentum of such a system to infinite precision? Briefly **explain** your answer.

11.6 Conserved Observables

[a] There are two distinct circumstances in which the expectation value of an observable can be independent of time, one involving a property of the *state*, the other involving a property of the *system*. **Discuss** these two circumstances.

[b] Which of the properties mentioned in [a] must be fulfilled if the observable is conserved?

[c] Suppose the condition in [b] is satisfied. What can we then conclude about the observable for the special case that the system is in a stationary state? **Justify** your answer.

[d] Consider an observable Q such that

$$\left[\hat{\mathcal{H}}, \hat{Q}\right] = c\hat{1}. \tag{11.6.1}$$

Suppose the system is initially in an *eigenstate of* Q corresponding to the eigenvalue q. **Derive** an expression for $\langle Q \rangle (t)$.

Problems

11.1 A Little Commutator Conundrum

Suppose operators $\hat{A}$ and $\hat{B}$ each commute with a third operator, $\hat{C}$.

[a] Do $\hat{A}$ and $\hat{B}$ *necessarily* commute—*i.e.*, is $\left[\hat{A}, \hat{B}\right]$ necessarily zero?

[b] *Could* $\hat{A}$ and $\hat{B}$ commute? If so, under what circumstances? If not, why not?

[c] Suppose $\hat{C} = \left[\hat{A}, \hat{B}\right]$. **Prove** that

$$\left[\hat{A}, \hat{f}(\hat{B})\right] = \left[\hat{A}, \hat{B}\right] \frac{d}{d\hat{B}} \hat{f}(\hat{B}) \tag{11.1.1}$$

[d] Use Eq. (11.1.1) to **evaluate** $\left[\hat{p}, \hat{g}(\hat{x})\right]$.

11.2 More Classical Physics Regained!

In § 11.7 we derived the equation for the rate of change of the expectation value of an arbitrary operator and showed how these equations for $\hat{x}$ and $\hat{p}$ reduce to their classical counterparts in the classical limit (Ehrenfest's Theorem). Here we shall explore a few ramifications of this theorem.

[a] Consider two Hermitian operators, $\hat{Q}$ and $\hat{R}$, that may depend explicitly on the time. **Show** that

$$\frac{d}{dt}\langle \hat{Q}\,\hat{R}\rangle = \left\langle \frac{\partial \hat{Q}}{\partial t}\,\hat{R}\right\rangle + \left\langle \hat{Q}\,\frac{\partial \hat{R}}{\partial t}\right\rangle + \frac{1}{\hbar}\left\langle i\left[\hat{\mathcal{H}}, \hat{Q}\right]\hat{R}\right\rangle + \frac{1}{\hbar}\left\langle i\hat{Q}\left[\hat{\mathcal{H}}, \hat{R}\right]\right\rangle. \tag{11.2.1}$$

[b] Use this result and the appropriate commutator from Table 11.3 to **evaluate** the rate of change of the mean kinetic energy: $d\langle T\rangle/dt$, where $\hat{T}$ is the kinetic energy operator. **Discuss** the relationship of your result to *the classical work-energy theorem*,

$$\frac{d}{dt}T = -\mathbf{v}\cdot\Delta V(\mathbf{r}, t). \qquad \text{[classical physics]} \tag{11.2.2}$$

[c] **Derive** an expression for $d\langle px\rangle/dt$. Now, consider this quantity *for a stationary state.* **Show** that for this case

$$\langle T\rangle = \frac{1}{2}\left\langle x\,\frac{dV}{dx}\right\rangle. \qquad \text{[for a stationary state]} \tag{11.2.3}$$

This result is called **the Virial Theorem.**

[d] Use the Correspondence Principle to **evaluate** this result in the classical limit, translating it into an expression in terms of classical functions of time and obtaining

$$T = \frac{1}{2}x\frac{d}{dx}V(x) \qquad \text{[in the classical limit]} \tag{11.2.4a}$$

Derive from this equation the more familiar form

$$T = -\tfrac{1}{2}xF \qquad \text{classical Virial Theorem} \tag{11.2.4b}$$

11.3 Measurements on a Free Particle

A free particle in one dimension is initially in a state represented by the wave function

$$\Psi(x, 0) = \left(\frac{\beta}{\sqrt{\pi}}\right)^{1/2} e^{ik_0 x}\, e^{-\beta^2 x^2/2}, \tag{11.3.1}$$

where β and k_0 are constants.

[a] Is the linear momentum of the particle sharp at $t = 0$? If not, **evaluate** its uncertainty.

[b] Do you think that $\langle x \rangle$ and/or $\langle p \rangle$ for this state depend on time? Why or why not?

[c] **Evaluate** $\langle x \rangle (0)$ and $\langle p \rangle (0)$ for this state.

[d] Now **evaluate** the rate of change of the expectation values $\langle x \rangle (t)$ and $\langle p \rangle (t)$ using the general evolution equation (11.88). Was your answer to [b] correct?

[e] In Chaps. 4 and 6 we studied extensively the behavior of a wave packet such as Eq. (11.3.1). Let Δt be the time required for the center of the packet to move a distance Δx, *i.e.*,

$$\Delta t \equiv \frac{\Delta x}{\left| \frac{d\langle x \rangle}{dt} \right|}. \tag{11.3.2}$$

Is this a reasonable definition of the "measurement time" for this system? **Discuss** its relation to the "change time" defined in § 11.8.

[f] **Evaluate** ΔE and Δt for the state in Eq. 11.3.1. and **show** that the uncertainty product is

$$\Delta E \, \Delta t = \frac{\hbar}{2} \left[1 + \frac{\beta^2}{4k_0^2} \right]^{1/2} \tag{11.3.3}$$

Show that this product satisfies the energy-time inequality.

11.4 A Problem with Correspondence

Consider a macroscopic particle—say a ball of mass $m = 10\,\text{gm}$—that undergoes simple harmonic motion with a natural frequency of 2 cycles/sec.

[a] Suppose the particle is in its ground state. **Evaluate** the uncertainty in its position and momentum. Do your results satisfy the Heisenberg Uncertainty Principle?

[b] Suppose now that we set the particle into motion, giving it an initial amplitude of 10^{-3} m. **Calculate** its energy. What is the order of magnitude of the quantum number appropriate to a stationary state of this energy? **Explain** why this system does not manifest quantum behavior.

11.5 Another Slant on Uncertainty

Consider a particle of mass m in the ground state of a simple harmonic oscillator potential.

[a] **Write down** the wave function for this state. In the tables of Chap. 9, look up the expectation values and uncertainties of position and momentum for this state.

[b] **Determine** the momentum-space wave function for this state and **deduce** from it an expression for the probability that in a measurement of the momentum you would obtain a value between p and $p + dp$.

[c] **Evaluate** the uncertainty product for position and momentum. Why is this state called "the minimum uncertainty wave packet"?

11.6 Constants of the Motion and Certainty

Prove that the uncertainty ΔQ in a constant of the motion Q is independent of time, *i.e.*, that

$$\frac{d}{dt} \Delta Q(t) = 0 \qquad \text{for any state } \Psi. \tag{11.6.1}$$

11.7 Evolution of Eigenstates

Suppose that a single particle in one dimension is initially (*i.e.*, at $t = 0$) in an eigenstate of an observable Q *that is a constant of the motion.*

[a] **Prove** that the system remains in that eigenstate as the state evolves.

[b] **Explore** what happens as the state evolves if the observable Q is not a constant of the motion. At the least, prove that even though the initial state is an eigenstate of Q, the state at $t > 0$ is not *necessarily* an eigenstate of this observable. (You can, if you are clever, prove that it cannot be an eigenstate of Q.)

CHAPTER 12

Eigenfunction Expansions

How to Solve Any Problem in Quantum Mechanics

The thing that most exasperates you
is to find yourself at the mercy of
the fortuitous, the aleatory, the random,
in things and in human actions —
carelessness, approximation, imprecision,
whether your own or others.

—*If on a Winter's Night a Traveller*
by Italo Calvino

We are almost through. We need only tie up a few loose ends to complete the first leg of our journey through quantum physics. In particular, we must remove two apparent limitations to quantum mechanics, limitations on our ability to access knowledge about observables and to solve the TDSE. In the process, we'll gain further physical insight into the advanced mathematical tools of Chaps. 10 and 11. Then, in Chap. 13, we'll confront a major mystery.

The first limitation is that we don't know how to extract from a state function the full measure of allowed information about any observable except position. Second, we don't know a practical way to calculate the wave function at $t > 0$ from its initial form, unless the state is stationary.

The apparent exalted status of position is illusory, a side-effect of our decision to represent quantum states by *functions of position variables (and time)*. According to Postulate II, these functions are to be understood as *position probability amplitudes*; this is the Born interpretation (Chap. 3). If we know *only* about such functions, we're in good shape if asked to predict the results of a *position measurement*, but not if faced with a measurement of any other observable. Knowing the wave function at t, we can calculate the position probability density—from which we can, in turn, calculate the probability of detecting the particle in an infinitesimal volume element about a point in space. From this density we can also calculate statistical quantities relating to position: its mean value and uncertainty.

But if we want to predict the results of a measurement at t of, say, the *energy*, we are in trouble. To be sure, we can evaluate from the wave function at t the statistical quantities $\langle E \rangle (t)$ and $\Delta E(t)$—with no small amount of work—but we don't know how to predict *the probabilities of obtaining various values of energy in the measurement.* Not yet, anyway. In this sense, quantum theory seems incomplete.

Happily, we can easily remedy this defect. In § 12.2 we'll learn how to calculate probabilities for an arbitrary observable, reasoning from the expansion of the wave function in a complete set of eigenfunctions of the operator that represents the observable. Viewed in this light, the eigenfunction expansions of Chaps. 10 and 11 stand revealed as tools for "translating" the information buried in the wave function into the "languages" of observables other than position.

The second deficiency in our quantum theory derives from our emphasis on stationary states. In Part III, for example, we learned several deft mathematical gambits for solving the *time-independent Schrödinger equation* for the Hamiltonian eigenfunctions of various (conservative) systems. These functions appear as the space-dependent factors in the wave function of any state that is stationary. But, as I have noted repeatedly, stationary states are special, uncommon in nature. *Most quantum states are non-stationary.* To solve for the wave function of such a state at $t > 0$, we must solve the time-*dependent* Schrödinger equation, no easy task.

Happily, the machinery of Chaps. 10 and 11 has, almost as a by-product, afforded us a crafty way to bypass this sinister partial differential equation: a surprisingly simple way to solve the TDSE without actually solving the TDSE. We'll discuss and illustrate this method in § 12.6 and will explore some of its consequences in §12.7–12.9.

You may be relieved to learn that we won't need to introduce any new theoretical constructs in this chapter. We can remedy both deficiencies mentioned above by combining the mathematical devices of Chaps. 10 and 11 with concepts and methods from previous chapters.

A final note. Because this is (almost) the last chapter, I have (none too surreptitiously) folded into it a review of most of the high points of this volume. So among the discussions and examples of this chapter,[1] you'll renew the acquaintance of many old friends. I hope the review aspect of this chapter will help prepare you to start Volume II, where we'll plunge vertiginously into new and strange territory.

12.1 THE FUNDAMENTAL PROBLEM OF QUANTUM PHYSICS

To elaborate a bit on the problems we'll tackle in this chapter, I want to consider briefly a question of knowledge. Every field of physics has an archetypal problem. In classical mechanics, for example, it is to calculate the trajectories of the particles comprising a system, knowing their initial positions and momenta and the forces acting on them. From these trajectories we can calculate any desired property of the system to any desired precision.

In quantum mechanics, the nature of the archetypal problem is influenced by the limitations nature imposes on the questions we can ask and by the peculiar language in which we answer. Our ultimate objective is qualitative and quantitative understanding of observed microscopic phenomena and their time development. But in pursuit of this grail we are allowed only a special kind of description of these phenomena, for we can access only probabilistic information about a quantum system.

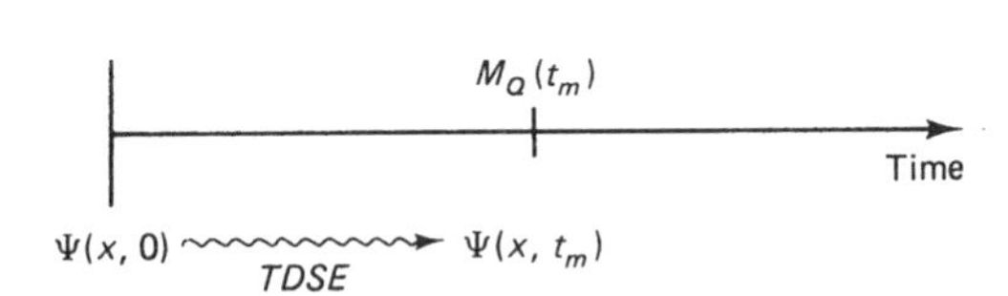

Figure 12.1 A schematic representation of the quintessential conundrum of quantum physics: to predict the results of an ensemble measurement of an observable at measurement time t knowing the Hamiltonian of the system and the wave function that represents the initial state.

The fundamental problem of quantum mechanics is illustrated schematically in Fig. 12.1. We are given the Hamiltonian $\hat{\mathcal{H}}$ of a system, which operator includes a potential energy derived from the forces acting on and within the system, and the wave function that represents the initial state of the system, Ψ at $t = 0$. At some time $t > 0$, a measurement of an observable Q is performed on an ensemble of systems in the state that evolved from the initial condition. The problem is: What happens? That is, what are the possible outcomes of this measurement?

[1] Students often complain, with some justification, that textbook authors work only easy examples, leaving all the hard stuff for the homework problems. This chapter is, in part, a review, and we have sufficient machinery and sophistication to deal with examples that are more complicated than we have hitherto faced. You'll find a few here, beginning in § 12.4.

This question is, of course, loaded. Unless the state happens to be an eigenstate of Q, nature forbids a precise answer. But an imprecise answer is allowed, and quantum mechanics should enable us to determine the values that could be obtained in the measurement and the probabilities that a member of the ensemble will exhibit each of these values. Yet, quantum mechanics, as we have developed it thus far, does not do this.

Let's review just what we do know how to do. Suppose that by hook or crook we have solved the time-dependent Schrödinger Equation,

$$\hat{\mathcal{H}}\Psi = i\hbar\frac{\partial}{\partial t}\Psi, \tag{12.1}$$

for the wave function at t; *i.e.*, we have "propagated" this function from $t = 0$ to the measurement time (see Fig. 12.1). We then can calculate the aforementioned statistical quantities from

$$\langle Q \rangle (t) = \langle \Psi(t) \mid \hat{Q} \mid \Psi(t) \rangle \tag{12.2}$$

and

$$\Delta Q(t) = \sqrt{\langle Q^2 \rangle (t) - [\langle Q \rangle (t)]^2}. \tag{12.3}$$

But how do we determine the probabilities $P(q, t)$? Unless $Q = x$, or $Q = E$ (and the system is in an *energy eigenstate*), we're up the proverbial estuary without the proper form of locomotion.[2]

Only for the special case of an *energy eigenstate* are solving the TDSE and determining probabilistic information about energy trivial tasks. If $\Psi(x, 0)$ is an eigenfunction of $\hat{\mathcal{H}}$, then at the measurement time the wave function is

$$\Psi(x,t) = \psi_E(x)\, e^{-iEt/\hbar}. \tag{12.4}$$

The probabilities of obtaining various values of the energy are equal to 1 if the energy in question is E in (12.4), and equal to 0 otherwise. The mean energy is E, the energy uncertainty is zero. Moreover, for *any* observable Q, we can calculate the mean value and uncertainty from the eigenfunction $\psi_E(x)$. As we know from Chap. 7, these quantities are independent of time.

But even in the special case of a stationary state, we don't know how to calculate the probabilities for an observable other than E. Here's a hint: consider, for example, a simple harmonic oscillator in its ground state. The probabilities of obtaining various values of position and energy for such a state are

$$P(x) = |\psi_0(x)|^2, \qquad \text{[position]}. \tag{12.5}$$

$$P(E) = \begin{cases} 1 & E = E_0 = \frac{1}{2}\hbar\omega_0 \\ 0 & \text{otherwise} \end{cases} \qquad \text{[energy]}. \tag{12.6}$$

But how would we calculate, say, the probability of getting $+1$ in a measurement of parity?

You already have, somewhere in memory, a clue to the answer to such questions. Consider linear momentum. What values could be obtained in a measurement of this observable, and what are the probabilities of obtaining each? If you remember how to

[2] Up a creek without a paddle.

answer this question, then you know where we must go next: back to one of the bulwarks of quantum theory, the Born Interpretation.

12.2 THE GENERALIZED BORN INTERPRETATION

> On fire that glows with heat intense
> We turn the host of common sense
> And out it goes at a small expense!
>
> —*Iolanthe*
> by Gilbert & Sullivan

Let's defer for the moment the problem of solving the TDSE and suppose that we know the wave function for a state at time $t > 0$ of a single particle in one dimension. In this section, we'll find out how to extract from this function the full panoply of accessible information about a measurement at t of an arbitrary observable Q. For simplicity, suppose that the spectrum of Q (the eigenvalues of its operator $\hat{Q}$) is discrete and non-degenerate. We'll label the eigenvalues with an index n in *ascending* order, as $q_1 < q_2 < q_3 < \cdots$. (We'll consider continuous and mixed spectra in § 12.5.) Finally, suppose the state at this time is not an eigenstate of Q.

Easy Calculation of the Mean

To begin our generalization of quantum theory, let's consider the average of the results of this measurement, $\langle Q \rangle (t)$. Calculating this quantity directly from $\Psi(x,t)$ à la

$$\langle Q \rangle (t) = \int_{-\infty}^{\infty} \Psi^*(x,t)\, \hat{Q}\Psi(x,t)\, dx \tag{12.7}$$

might involve considerable complicated algebra—as we've seen in earlier examples, exercises, and problems. Indeed, just working out the effect of $\hat{Q}$ on $\Psi(x,t)$ could be a mess, for, by hypothesis, the wave function is not an eigenfunction of that operator. There is, I am happy to report, an easier way.

The considerations of Chaps. 10 and 11 suggest the following alternative: expand $\Psi(x,t)$ in eigenfunctions of $\hat{Q}$ and develop an expression for $\langle Q \rangle$ in terms of the coefficients in this expansion, the projections of Ψ on the eigenfunctions of $\hat{Q}$. Denoting the eigenfunction for eigenvalue q_n by χ_n and the corresponding coefficient by $c_n(t)$, we can write

$$\Psi(x,t) = \sum_n c_n(t)\chi_n(x). \qquad \text{[expansion observable: } Q]. \tag{12.8}$$

The projection of Ψ on χ_n, which carries the time dependence, is[3]

$$c_n(t) = \langle \chi_n \mid \Psi(t) \rangle = \int_{-\infty}^{\infty} \chi_n^*(x)\Psi(x,t)\, dx. \tag{12.9}$$

[3]In this chapter and the next, the time at which these coefficients are evaluated is critical. So I am introducing a slight modification to the familiar bracket notation: labelling the state function Ψ *within a bracket* by t. Of course, this function also depends on spatial variables, but these variables are suppressed in the conventions of Dirac braket notation. It is our job to remember that they're there.

In general, there is one such projection for each eigenvalue, although in particular cases one or more projections may be zero. Such an expansion is possible for any $\Psi(x,t)$, because $\hat{Q}$ is Hermitian, so the set of its eigenfunctions is complete. Let's see how this strategem simplifies the evaluation of $\langle Q \rangle (t)$.

To evaluate the expectation value we require the expansion (12.8) of $\Psi(x,t)$ and that of its complex conjugate,[4]

$$\Psi^*(x,t) = \sum_{n'} c^*_{n'}(t)\chi^*_{n'}(x). \tag{12.10}$$

Substituting these expansions into Eq. (12.7) (and grouping terms suggestively), we obtain

$$\langle Q \rangle (t) = \int_{-\infty}^{\infty} \Big[\sum_{n'} c^*_{n'}(t)\chi^*_{n'}(x)\Big] \Big[\hat{Q}\sum_{n} c_n(t)\chi_n(x)\Big]\, dx. \tag{12.11}$$

The operator $\hat{Q}$ in the second set of square brackets slides through the coefficient in each term in the sum over n and acts on its eigenfunctions χ_n to give $q_n\chi_n$. So (12.11) becomes

$$\langle Q \rangle (t) = \int_{-\infty}^{\infty} \Big[\sum_{n'} c^*_{n'}(t)\chi^*_{n'}(x)\Big] \Big[\sum_{n} c_n(t)q_n\chi_n(x)\Big]\, dx \tag{12.12a}$$

$$= \sum_{n'}\sum_{n} c^*_{n'}(t)c_n(t)q_n\big\langle \chi_{n'} \mid \chi_n \big\rangle. \tag{12.12b}$$

In writing Eq. (12.12*b*) I lapsed back into Dirac notation in the hopes that doing so would clue you in to our next maneuver. Since $\hat{Q}$ is Hermitian, its eigenfunctions are orthonormal, *i.e.*,

$$\big\langle \chi_{n'} \mid \chi_n \big\rangle = \delta_{n',n} = \begin{cases} 0 & n' \neq n \\ 1 & n' = n \end{cases}. \tag{12.13}$$

We can use this Kroneker delta to eliminate one of the sums in Eq. (12.12*b*), *e.g.*, by reducing the sum over n' to only one term, $n' = n$. The expectation value then assumes the simple form

$$\boxed{\langle Q \rangle (t) = \sum_{n} c^*_n(t)c_n(t)q_n = \sum_{n} \left|c_n(t)\right|^2 q_n} \tag{12.14}$$

Notice that the remaining summation in this expression must include all eigenvalues of $\hat{Q}$. Of course, terms for which the projections $c_n(t) = 0$ need not explicitly be included, but whether or not any of these projections are zero depends on the state in question [see Eq. (12.9)]. In general, you must not leave any eigenvalues out of this sum.

Equation (12.14) is our alternate method for computing the expectation value. Rather than insert the wave function into the definition (12.7), then act on it with $\hat{Q}$ and integrate over x, we first determine the *projections* on the eigenfunctions of the observable, then simply evaluate the sum in (12.14). Although this method may require

[4]Can you figure out why in this equation I used n' as the summation index? There is method in my madness: Soon we're going to substitute this expansion and Eq. (12.8) into (12.7). Were we to sum over n in both summations, we would have the devil of a time keeping the two summations distinct. Moral: *Always use different dummy summation indices when combining several sums in a single expression.*

evaluating several integrals [the expansion coefficients (12.9)], it is in many cases still easier to implement than the definition (12.7).

Question 12–1

Suppose $\Psi(x,t)$ is an eigenfunction of $\hat{Q}$. Use Eqs. (12.9) and (12.14) to evaluate $\langle Q \rangle$ in this case.

Example 12.1. The Average Energy of a Non-Stationary State

In Example 11.4 we investigated a non-stationary state of the simple harmonic oscillator represented by the wave function [*c.f.*, Eq. (11.64)]

$$\Psi(x,t) = \frac{1}{\sqrt{50}} \left[\psi_1(x) e^{-iE_1 t/\hbar} + 7\psi_2(x) e^{-iE_2 t/\hbar} \right]. \tag{12.15}$$

At one point in this example, we needed the expectation value of the energy, and I claimed that this quantity is

$$\langle E \rangle = \frac{62}{25}\, \hbar\omega_0 = 2.48\hbar\omega_0. \tag{12.16}$$

Let's see if I was right.

Conveniently, the wave function (12.15) is already written as an expansion in eigenfunctions $\{\psi_n(x),\ n = 0, 1, 2, \ldots\}$ of $\hat{\mathcal{H}}$. So we need not compute the projections $c_n(t)$ as such.[5] Comparing (12.15) with the general form (12.8), we find

$$\begin{aligned} c_1(t) &= \frac{1}{\sqrt{50}} e^{-iE_1 t/\hbar} \\ c_2(t) &= \frac{7}{\sqrt{50}} e^{-iE_2 t/\hbar} \\ c_n(t) &= 0 \qquad n = 0 \text{ and } n > 2 \end{aligned} \tag{12.17}$$

To calculate the mean value of E, we need only insert these projections in the general result Eq. (12.14), which for $Q = E$ takes on the following form:[6]

$$\langle E \rangle (t) = \sum_{n=0}^{\infty} c_n^*(t) c_n(t) E_n = \sum_{n=0}^{\infty} |c_n(t)|^2 E_n. \tag{12.18}$$

Squaring the moduli of the non-zero coefficients in (12.17) and inserting the result into this equation, we obtain

$$\langle E \rangle = \frac{1}{50}\frac{3}{2}\hbar\omega_0 + \frac{49}{50}\frac{5}{2}\hbar\omega_0 = \frac{62}{25}\,\hbar\omega_0, \tag{12.19}$$

in agreement with (12.16). *Notice that the mean value does not depend on time. Although the projections in (12.17) do depend on time, this dependence vanishes because it is restricted to complex exponential factors* $e^{-iE_n t/\hbar}$, *which disappear when multiplied by their complex conjugates as prescribed by Eq. (12.18).* Notice also how easy this calculation was compared to direct substitution of $\Psi(x,t)$ into the integral form of $\langle E \rangle$.

[5] Throughout this chapter, you need to keep careful track of the "expansion observable." In this example we chose $\hat{Q} = \hat{\mathcal{H}}$, for the observable we're interested in is the energy. I'll rely on you to remember that $c_n(t)$ here is a coefficient in an expansion in *energy* eigenfunctions, not, as earlier in this section, in eigenfunctions of an arbitrary observable.

[6] *Do not use this expression unless the spectrum of the Hamiltonian of the system is solely discrete.* It does not apply, for example, to a finite potential, such as the Morse potential (Chap. 9) or the finite square well (Chap. 8). The energy spectra of these systems are a mixture of discrete and continuous portions, and methods to be introduced in § 12.4 are required.

We could also use the expansion of Ψ in eigenfunctions of Q to evaluate the expectation value of an operator function $\hat{f}(\hat{Q})$. But first let's explore further the expression (12.14) for $\langle Q \rangle$, which is a mathematical clue to the physical significance of the projections $c_n(t)$.

The Born Interpretation of the State Projections

I want to recall an argument we made in Chap. 3 concerning the expectation value of *position*,[7]

$$\langle x \rangle (t) = \int_{-\infty}^{\infty} |\Psi(x,t)|^2 \, x \, dx. \tag{12.20}$$

Our understanding of $\langle x \rangle$ as the average of results obtained in a position measurement follows from the Born interpretation of $\Psi(x,t)$ as a *position probability amplitude.* According to Postulate II, $P(x,t) = |\Psi(x,t)|^2$ is the probability that in an ensemble measurement at time t the particle will be detected in the infinitesimal volume element dx at x. Thus $|\Psi(x,t)|^2$ serves as the "weighting factor" for the value x in Eq. (12.20) for the mean value. This quantity is calculated as an integral rather than a discrete sum because the allowed values of this particular observable encompass the *continuum* from $-\infty$ to $+\infty$.

Now compare Eq. (12.20) for $\langle x \rangle$ to Eq. (12.14) for $\langle Q \rangle$. In the latter, the squared modulus $|c_n|^2$ of each projection c_n serves as a weighting factor for the value q_n of the observable Q. That is, the structure of Eq. (12.14) is

$$\begin{matrix}\text{average} \\ \text{value}\end{matrix} = \sum_{\substack{\text{possible} \\ \text{results}}} \left[\begin{pmatrix}\text{value of} \\ \text{observable}\end{pmatrix} \begin{pmatrix}\text{probability of} \\ \text{getting that value}\end{pmatrix} \right]. \tag{12.21}$$

This form suggests that we interpret $|c_n(t)|^2$ as a probability for obtaining the value q_n in a measurement of Q at time t. In terms of the ensemble on which the measurement is performed, $|c_n(t)|^2$ is the fraction of members that exhibit the value q_n [provided the wave function is normalized such that $\sum_n |c_n(t)|^2 = 1$, as in Eq. (12.33) below]. This is the **Generalized Born Interpretation.**[8]

To summarize this interpretation, we write Eq. (12.14) as

$$\langle Q \rangle (t) = \sum_n P(q_n, t) q_n, \tag{12.22}$$

[7] A word of advice: Before proceeding further, go back and re-read § 3.4 and 3.6. Doing so will make the analogy I am about to wheel in more striking.

[8] Born introduced this interpretation in the second of two papers he published in 1926 [*Z. für Phys.*, **38**, 803]. In the first of these papers [*Goett. Nachr.*, p. 14], Born introduced the idea of interpreting quantum-mechanical quantities as probabilities. (Interestingly, this suggestion—the key to the entire interpretative structure of quantum physics—appeared only in a footnote Born added to the paper after it had been set in type, a "note added in proof.") Both papers dealt with electron scattering, but Born's innovation was in no way limited to such processes. Deftly summarizing the structure of the new theory, Born writes: "The motion of particles follows probability laws but the probability itself propagates according to the law of causality." If you are curious about *how* Born came up with his idea, have a look at Chap. 12 of A. Pais' remarkable book on "matter and forces in the physical world," *Inward Bound* (New York: Oxford University Press, 1986).

where

$$P(q_n, t) = |c_n(t)|^2 = \text{probability that in a measurement of } Q \text{ at time } t \text{ a member of the ensemble will exhibit the value } q_n \tag{12.23}$$

Example 12.2. Probabilities for Energy Measurement on a Non-Stationary State

Let's return briefly to the non-stationary state (12.15) of the simple harmonic oscillator and use the Generalized Born Interpretation to predict the outcome of a measurement of *energy*. According to Eq. (12.17), only two values will be obtained: $E_1 = 3\hbar\omega_0/2$ and $E_2 = 5\hbar\omega_0/2$. So no members will exhibit, say, the ground-state energy $E = \hbar\omega_0/2$; nor will any exhibit an energy intermediate between the two observed values, such as $2\hbar\omega_0$. (There is a more profound reason why such an energy isn't seen, as we'll discover in the next section.) The probabilities that a member of the ensemble will actually exhibit one of these two energies are

$$P(E_1) = |c_1(t)|^2 = \frac{1}{50} = 0.02 \tag{12.24}$$

$$P(E_2) = |c_2(t)|^2 = \frac{49}{50} = 0.98. \tag{12.25}$$

So we predict that 2% of the member systems will exhibit the lower of the two energies, $3\hbar\omega_0/2$, and that the other 98% will exhibit the higher, $5\hbar\omega_0/2$. As expected, these results don't depend on when we perform the measurement, and the sum of the probabilities (12.24) and (12.25) is 1.

Simple Calculation of Complicated Expectation Values

Let's now generalize to *functions* of $\hat{Q}$ our new method for calculating expectation values from the projections of the state function. Consider, for example, $\langle Q^2 \rangle$, which we need to calculate the uncertainty ΔQ [see Eq. (12.3)],

$$\langle Q^2 \rangle (t) = \int_{-\infty}^{\infty} \Psi^*(x,t)\, \hat{Q}^2 \Psi(x,t)\, dx. \tag{12.26}$$

Now watch carefully. Insertion of the expansions (12.8) and (12.10) into (12.26) leads to an equation just like (12.11)—except, of course, that in place of $\hat{Q}$ in (12.11) we have $\hat{Q}^2$. The operator $\hat{Q}$ acting twice on χ_n yields q_n^2, instead of q_n as in Eq. (12.12*a*). In short order, we obtain the following result:

$$\boxed{\langle Q^2 \rangle (t) = \sum_n c_n^*(t) c_n(t) q_n^2 = \sum_n |c_n(t)|^2 q_n^2} \tag{12.27}$$

Question 12–2

Fill in the missing steps in the derivation of (12.27).

Example 12.3. The Energy Uncertainty of a Non-Stationary State

To really appreciate how useful (12.27) is you should reflect on the awkward algebraic difficulties that would arise were we to calculate the uncertainty of a non-stationary state using the integral form for $\langle Q^2 \rangle$. In contrast, watch how easy Eq. (12.27) makes this chore.

In Example 12.2 we calculated $\langle E \rangle$ for the non-stationary state (12.15) from the projections of this wave function on the eigenfunctions of $\hat{\mathcal{H}}$. These projections also appear in the expression for $\langle E^2 \rangle$,

$$\left\langle E^2 \right\rangle (t) = \sum_{n=0}^{\infty} |c_n(t)|^2 \, E_n^2. \tag{12.28}$$

Using Eqs. (12.17), we can calculate this expectation value by trivial arithmetic, obtaining

$$\left\langle E^2 \right\rangle = \frac{1}{50}\left(\frac{3}{2}\hbar\omega_0\right)^2 + \frac{49}{50}\left(\frac{5}{2}\hbar\omega_0\right)^2 = \frac{1234}{200}\,\hbar^2\omega_0^2 = 6.17\,\hbar^2\omega_0^2, \tag{12.29}$$

which agrees with Eq. (11.69*b*). This result, together with (12.19), leads us immediately to the uncertainty $\Delta E = 0.14\,\hbar\omega_0$, in accord with Eq. (11.70).

Question 12–3

Further generalize the derivations of this section to show that

$$\langle f(Q) \rangle = \sum_n |c_n(t)|^2 \, f(q_n). \tag{12.30}$$

Discuss how you would evaluate this mean value from the coefficients of the power series expansion of the *function* $f(q)$ [see Eq. (10.12) and surrounding discussion].

Evidently, the mathematical technique of eigenfunction expansion can render calculations such as evaluating uncertainties comparatively easy. One wonders what other problems this technique might help us subdue. For example, might it assist in solving the time-dependent Schrödinger equation? No textbook author would ask such a question unless the answer is yes, as you'll see in § 12.5.

The Languages of Quantum Mechanics

The central result of this section is the Generalized Born Interpretation (12.23). Although we have thus far applied this interpretation only to the energy, it can be used to extract from a state function probabilistic information about any observable. In Chap. 4, for example, we used a similar technique to extract momentum information via the Fourier Transform of a wave function (§ 4.7; see especially Table 4.2); we'll take another look at momentum in § 12.4. Underlying the Generalized Born Interpretation is the mathematical technique of *eigenfunction expansion*.

This technique provides a variety of ways to look at a state. When we consider the particular representation of a state afforded by the *wave function* $\Psi(x,t)$, we are "viewing the state through position glasses." That is, this function is the most convenient representation of the state for study of position.

But the wave function is not particularly convenient if we are interested in another observable—say, energy. By expanding the wave function in the complete set of eigenfunctions of the operator that represents that observable—the eigenfunctions of $\hat{\mathcal{H}}$—we develop a representation of the state that lets us view it "through energy glasses." *The collection of all projections of* Ψ *on the energy eigenfunctions is an alternate representation of this state.* This representation consists of the numbers

$$\{c_n(t) = \langle \psi_n \mid \Psi(t) \rangle; \quad \text{all } E_n\}. \qquad \text{energy representation} \tag{12.31}$$

This set may be infinite in number and may be complex. The energy representation contains no more or less information than does the position representation, the wave function $\Psi(x,t)$. But, as the examples of this section attest, the energy representation is as convenient for the study of energy as $\Psi(x,t)$ is for the study of position.

I have spoken of "viewing a state through various glasses." If you don't like this optical metaphor, you might prefer to think of an eigenfunction expansion as a device for *translating* physical information concerning the state from its expression in $\Psi(x,t)$, in the "language of position," to its expression in (12.31), in the language of energy. Thus, the expression for $\langle E \rangle$ we used in the examples of this section is the translation

$$\langle E \rangle (t) = \int_{-\infty}^{\infty} \Psi^*(x,t)\hat{\mathcal{H}}\Psi(x,t)\,dx \quad \longleftrightarrow \quad \langle E \rangle (t) = \sum_n |c_n(t)|^2 E_n. \tag{12.32}$$

Using the power and flexibility of eigenfunction expansion, we can translate all of the fundamental equations of quantum physics into the language of any observable (see Fig. 12.2). Here, for example, is the translation of the *normalization requirement* $\langle \Psi \mid \Psi \rangle = 1$ from the language of position into that of energy:

$$\int_{-\infty}^{\infty} \Psi^*(x,t)\Psi(x,t)\,dx = 1 \quad \longleftrightarrow \quad \sum_n |c_n(t)|^2 = 1. \tag{12.33}$$

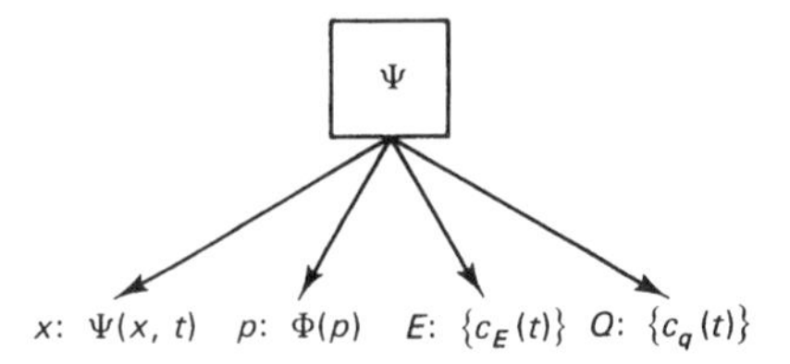

Figure 12.2 Several representations of a quantum state Ψ. To translate the wave function into the desired representation, we merely expand it in the complete set of eigenfunctions of the operator that represents the observable of interest; the resulting projections are an alternate, often more convenient, representation of the state.

Question 12–4

Write down the translation of the normalization requirement into the language of an arbitrary observable $\hat{Q}$ with a discrete spectrum. How could you use this form to normalize an initial wave function that was written as a linear combination of eigenfunctions of $\hat{Q}$?

12.3 ALL THAT HEAVEN ALLOWS: POSSIBLE OUTCOMES OF MEASUREMENT

Surely our profession, Mr. Mac,
would be a drab and sordid one
if we did not sometimes
set the scene so as to glorify our result.
... the quick inference, the subtle trap,
the clever forecast of coming events,
the triumphant vindication of bold theories
—are these not the pride and
the justification of our life's work?

—Sherlock Holmes,
in *The Valley of Fear*
by Sir Arthur Conan Doyle

In the previous section we considered an energy measurement on a simple harmonic oscillator in the non-stationary state (12.15). In discussing the possible outcomes of this measurement (Example 12.2), I remarked in passing that no values other than $E_1 = 3\hbar\omega_0/2$ and $E_2 = 5\hbar\omega_0/2$ would be seen; no other eigenvalues (E_n for $n \neq 1, 2$), nor any energies not equal to an eigenvalue. The probabilities (12.24) and (12.25) support this remark: the sum of $P(E_1) = 0.02$ and $P(E_2) = 0.98$ is 1.0, so E_1 and E_2 exhaust the possibilities inherent in this state function. This crucial feature of this state is mirrored in its wave function, which is revealed in (12.15) as a superposition of the first and second excited states *only*.

We can generalize these findings about the *possible outcomes of measurement* to an energy measurement on an arbitrary state of an arbitrary quantum system. Assuming, for simplicity, that the (energy) spectrum of the system is discrete, the expansion of the wave function of the state in eigenfunctions $\{\psi_n(x)\}$ looks like

$$\Psi(x,t) = \sum_{n=1}^{\infty} c_n(t)\psi_n(x). \tag{12.34}$$

The interpretation of the projections $c_n(t)$ as energy probability amplitudes admits the *possibility* that a system *could* exhibit any of the eigenvalues E_n. The *probability* for obtaining the nth eigenvalue, $P(E_n) = |c_n(t)|^2$, is non-zero only if the projection $c_n(t) = \langle \psi_n \mid \Psi \rangle$ happens to be non-zero for the state Ψ. *But no other energies are possible; the list of eigenvalues of the Hamiltonian,* $\{E_n, \text{all } n\}$ *exhausts the possibilities in any state function* Ψ.

This seems a radical conclusion, and you may be wondering, nervously, "How can I be sure *all* possibilities are included in (12.34)?" Let's add the probabilities for each of the eigenvalues and see if the result is equal to 1. If this sum is *less than* 1, then a member might exhibit an energy other than one of the eigenvalues of $\hat{\mathcal{H}}$. But if the sum is *equal* to one, then each member will exhibit one of the eigenvalues. (If it is greater than one, we've made a mistake.)

The sum of the probabilities of obtaining one of the eigenvalues of $\hat{\mathcal{H}}$ is

$$\sum_{n=1}^{\infty} P(E_n) = \sum_{n=1}^{\infty} |c_n(t)|^2 . \tag{12.35a}$$

Do you recognize this sum?

*** * * Pause while reader ponders * * ***

Sure; it is the *normalization integral* $\langle \Psi \mid \Psi \rangle$ expressed in the language of energy [*c.f.*, (12.33)]. And for any state Ψ, the normalization integral is unity, so

$$\sum_{n=1}^{\infty} P(E_n) = 1. \tag{12.35b}$$

Consequently, in an energy measurement of a microscopic system (whose Hamiltonian has a discrete spectrum), each member exhibits *either* E_1 *or* E_2 *or* E_3 *or* any other eigenvalue E_n. But no other value. This phenomenon is, of course, energy quantization.

Example 12.4. Energy Quantization in the Laboratory

Quantization of energy was stunningly demonstrated in famous energy-loss experiments conducted by James Franck and Gustav Hertz in 1914. Today, energy-loss experiments play an important role in the study of electron scattering from atoms and molecules. The goal of a modern energy-loss experiment is usually to determine accurate *cross sections*—the probabilities that an incident electron will scatter in various directions, with (or without) an attendant loss of energy. (Recall § 8.3.)

Were the (bound-state) energies of the target not quantized, a projectile with incident kinetic energy T could lose any amount of energy from 0 to T, and the observed energy-loss spectrum—the energies of the scattered electrons, as measured by the detector—would be smooth, revealing this continuum. Instead, it shows spikes (see Fig. 12.3a).

We now know the reason. *No matter what the state of a target particle is prior to the collision, an incident electron can induce transitions only between the stationary states of the target, for the only energies available to the target are the eigenvalues of its Hamiltonian.* That is, the state of the target before and after the collision must either be a stationary state or a superposition of stationary states; neither possibility admits energies other than the Hamiltonian eigenvalues.

Consider, for example, a gas of molecules that are initially in their *ground vibrational state* ($n = 0$ in the simple harmonic oscillator model). An electron incident on such a gas can lose energy only in discrete multiples of $\hbar$ times the fundamental vibrational frequency of the molecule. (Of course, the simple harmonic oscillator model is only an approximation, and observed energy-loss spectra reveal deviations from the equal spacing predicted by this model. The point, however, is not the size of the spacings but the fact that they are invariably seen.) This behavior is illustrated by the energy-loss spectrum for e-F_2 scattering in Fig. 12.3.[9] In this experiment, a beam of nearly mono-energetic electrons is fired out of an electron gun at an incident kinetic energy of 1.6 eV. This electron beam crosses a beam of F_2 molecules, and the resulting scattered electrons are picked up by a detector that is capable of measuring energy within an *experimental* uncertainty of about 0.03 eV.

The energy-loss spectrum in Fig. 12.3a shows the *intensity* of detected electrons that have lost less than 0.8 eV. At each energy loss, the intensity is proportional to the fraction of electrons in the incident beam that lost that amount of energy, *i.e.*, to the quantum-mechanical probability of an incident electron losing that amount. This spectrum consists of *regularly-spaced peaks* separated by about 0.11 eV. (We note from Table 9.1 that for this molecule, $\hbar\omega_0 = 0.1136\,\text{eV}$.) The large peak at zero energy loss corresponds to elastic scattering, but each of the other peaks corresponds to excitation of F_2 molecules from the ground vibrational state ($n = 0$) to excited vibrational levels with $n = 1, 2, \ldots, 7$. (Excitations to states with $n > 7$ are so unlikely that they aren't seen in this experiment.)

The levels involved in these excitations are superimposed on the potential energy curve for F_2 in Fig. 12.3b [compare to Fig. 9.13, the potential energy curve for H_2]. You can see that the potential well of the ground electronic state deviates somewhat from the purely parabolic shape of a true SHO potential. These deviations are responsible for the barely perceptible differences in the spacings between adjacent peaks in the spectrum.[10]

This argument, that the normalization requirement proves that only eigenvalues of the Hamiltonian can be obtained in a measurement of energy, generalizes easily to an

[9] Y. Fujita, S. Yagi, S. S. Kano, H. Takuma, T. Ajiro, T. Takayanagi, K. Wakiya, and H. Suzuki, *Phys. Rev. A.*, **34**, 1568 (1986).

[10] The width of each spike in the spectrum is due primarily to the finite "energy resolutions" of the detector and of the source. That is, an experimentalist can determine the energy only to within the experimental uncertainty of these devices. (This uncertainty is much larger than the natural linewidth, the quantum-mechanical uncertainty imposed by nature.) This imprecision manifests itself as a width in the measured spectrum.

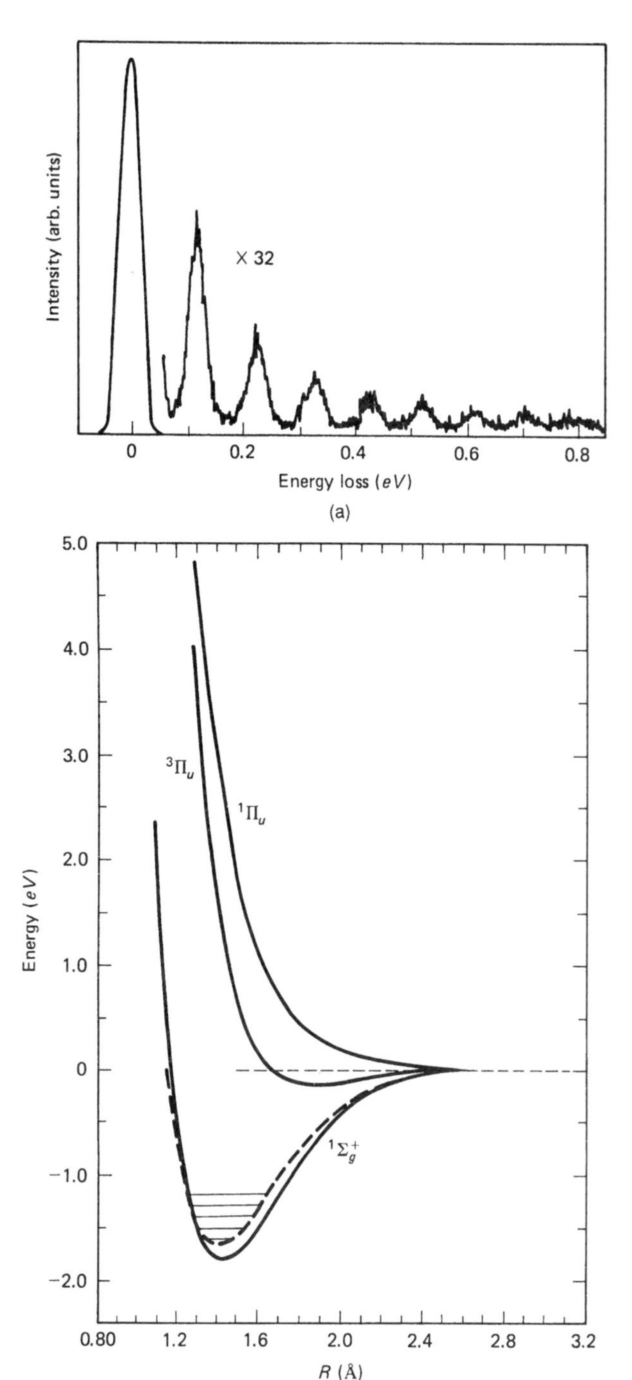

Figure 12.3 (a) An energy-loss spectrum for e-F_2 collisions showing excitation of bound vibrational states from the ground state, $n = 0$. The incident energy is 1.6 eV and the detector is placed at an angle of 30° with respect to the axis defined by the incident electron beam. [From Y. Fujita, S. Yagi, S. S. Kano, H. Takuma, T. Ajiro, T. Takayanagi, K. Wakiya, and H. Suzuki, *Phys. Rev. A.*, **34**, 1568 (1986); used with permission]. (b) Potential energy curves for the ground state of F_2 and for two excited states. In each case, the potential determined in a theoretical study (in which the TISE for the molecule was solved numerically) is represented by solid curves, and the potential deduced from experimental data is shown as a dashed curve. [From D. C. Cartwright and P. J. Hay, *J. Chem. Phys.*, **70**, 3191 (1979); used with permission.]

arbitrary observable. We can expand $\Psi(x, t)$ in the complete set of eigenfunctions of any Hermitian operator, à la Eq. (12.7). No matter what the observable, the normalization condition translates into a mathematical statement that the sum of the probabilities for all the eigenvalues is unity. Hence no other possibilities are allowed:

Rule

The only possible outcomes of measurement of an observable on any quantum state are the eigenvalues of the operator that represents the observable.

Aside: Thinking about the Continuum. In thinking about this rule, it is very important to remind yourself that *these eigenvalues need not be discrete.* I picked energy quantization of the SHO to illustrate this rule in a dramatic context that would lead us back to ideas we encountered early in this book. But the spectra of eigenvalues of most operators consists of discrete and continuous parts; *any value in the continuum of such an operator is allowed.* But *no* value is allowed from the range encompassing the discrete part except one of the eigenvalues. We'll discuss continuous and mixed spectra further in the next section.

The Pause That Refreshes

Stop and think for a moment about how utterly remarkable this rule is. It is a pinnacle of our exploration of the microworld, the point at which all the basic ideas we discussed in Chap. 2—uncertainty, duality and quantization—have emerged naturally and without additional hocus pocus from the postulates of quantum mechanics. Our study of the (admittedly abstract) machinery of quantum mechanics has thus led us to a deep, profound understanding of the most remarkable phenomena in the microverse, phenomena that mark striking departures in that world from classical behavior. Such insight is no mean accomplishment, and you should give yourself a little time to savor it. Then read on.

*12.4 THE MANY USES OF EIGENFUNCTION EXPANSIONS: TWO EXAMPLES

To see how to use the Generalized Born Interpretation and the technique of eigenfunction expansion, let's work a couple of examples. The first is a more realistic (and intricate) version of the examples in § 12.3. The second turns this interpretation on its head to show how we can deduce information about an unknown state function from data obtained in an experiment.

Example 12.5. A Non-Stationary State of an Organic Molecule

In the problems at the end of Chap. 7, we learned that even the simplest (non-trivial) quantum system—a single particle in a one-dimensional infinite square well—finds application in the study of nuclei, atoms, and molecules. For example, in Pblm. 7.5 we used such a well to model *the hexatriene molecule*, a complicated organic molecule that can be approximated by an infinite square well of length $L = 7.3\,\text{Å}$.

Suppose that from the set-up of our experiment, we deduce that the initial state of the electron can be represented by the wave function (with $x = 0$ at the left wall)

$$\Psi(x, 0) = \sqrt{\frac{30}{L^5}}\, x(x - L). \tag{12.36}$$

In the laboratory, an ensemble measurement of the energy of the particle is performed at $t = 0$. What does the experimentalist find?

Question 12–5

Verify that this state function is normalized. [You will find it easier to choose the left wall at $x = 0$—see Eq. (12.42) below.]

The first thing to notice about this state is that it is not stationary. (*Why not?*) So the energy of the electron is not sharp ($\Delta E > 0$), and various members of the ensemble could exhibit one or another of the eigenvalues [Eq. (7.56)]

$$E_n = n^2 \frac{\pi^2 \hbar^2}{2mL^2} \qquad n = 1, 2, \ldots \tag{12.37}$$

Quantization is at hand. Without even analyzing the state, we proclaim that *no energies other than the discrete values (12.37) will be seen in the measurement*.

Since the state is non-stationary, the mean of the observed values is not equal to any of the eigenvalues (12.37) (except by remarkable accident). We can, of course, calculate $\langle E \rangle (0)$ from the wave function (12.36), *viz.*,[11]

$$\langle E \rangle (0) = \langle \Psi(0) \mid \hat{\mathcal{H}} \mid \Psi(0) \rangle \tag{12.38a}$$

$$= \int_0^L \Psi^*(x,0)\, \hat{T}\, \Psi(x,0)\, dx \tag{12.38b}$$

$$= -\frac{\hbar^2}{2m} \frac{30}{L^5} \int_0^L x(x-L) \frac{d^2}{dx^2} \Big[x(x-L) \Big]\, dx \tag{12.38c}$$

$$= -\frac{\hbar^2}{m} \frac{30}{L^5} \int_0^L x(x-L)\, dx \tag{12.38d}$$

$$= -\frac{30\hbar^2}{mL^5} \left[\frac{1}{3} x^3 \Big|_0^L - \frac{1}{2} L x^2 \Big|_0^L \right]. \tag{12.38e}$$

So the average of the results of the energy measurement will be

$$\langle E \rangle (0) = \frac{5\hbar^2}{mL^2}. \tag{12.39}$$

Using $L = 7.3 \times 10^{-10}$ m, the mass $m_e = 9.1095 \times 10^{-31}$ kg of the electron, and Planck's constant $\hbar = 1.0546 \times 10^{-34}$ J-sec, we calculate the value $\langle E \rangle (0) = 1.1455 \times 10^{-19}$ J $=$ 0.7150 eV. [Note that 1 J $= 6.2415 \times 10^{18}$ eV (see Appendix D).]

Fine. But suppose we want the *probabilities* of obtaining one or another of the eigenvalues (12.37)? According to the Generalized Born Interpretation (12.23), these probabilities are the squared moduli of the projections of $\Psi(x, 0)$ on the eigenfunctions of the energy operator, $\hat{\mathcal{H}}$—which brings us back to the eigenfunction expansion

$$\Psi(x,0) = \sum_{n=1}^{\infty} c_n(0) \psi_n(x). \tag{12.40}$$

We need to evaluate the *energy probability amplitudes*

$$c_n(0) = \langle \psi_n \mid \Psi(0) \rangle = \int_0^L \psi_n^*(x) \Psi(x,0)\, dx. \tag{12.41}$$

[11]Notice that in (12.38*b*) I integrate from 0 to L, not from $-L/2$ to $L/2$. Earlier in this book I argued that when dealing with a symmetric potential we should put the origin of coordinates at the midpoint of the potential. But for this state, the presence of the factors x and $x - L$ in $\Psi(x, 0)$ argue for the unusual choice of $x = 0$ for the left wall. In Question 7.7, you converted the eigenfunctions (7.55) to such a coordinate system, finding the functions in Eq. (12.42) below. (Didn't you?)

With $x = 0$ at the left wall of the potential, the Hamiltonian eigenfunctions are [see Question 7.7]

$$\psi_n(x) = \sqrt{\frac{2}{L}} \sin\left(n\pi\frac{x}{L}\right) \qquad n = 1, 2, \ldots. \tag{12.42}$$

So the nth projection of the wave function (12.36) is

$$c_n(0) = \sqrt{\frac{2}{L}}\sqrt{\frac{30}{L^5}} \int_0^L \sin\left(n\pi\frac{x}{L}\right) x(x-L)\,dx \tag{12.43a}$$

$$= \frac{\sqrt{60}}{L^3}\left\{\frac{L^3}{n^3\pi^3}\left[(-1)^{n+1}(n^2\pi^2 - 2) - 2 - (-1)^{n+1}n^2\pi^2\right]\right\} \tag{12.43b}$$

$$= \frac{\sqrt{60}}{L^3} \cdot \begin{cases} 0 & n \text{ even} \\ -\dfrac{4L^3}{n^3\pi^3} & n \text{ odd} \end{cases}. \tag{12.43c}$$

Aside: Filling in the Gaps. "Whoa there!" I hear you cry. "About those steps between (12.43*a*) and (12.43*b*) ... " Well, okay. This integral is rather more complicated than ones we've done so far, so here's a quick sketch of its evaluation. The first thing to do is to get rid of the unsightly factors of π and L parading through the integrand; we accomplish this by changing the variable of integration from x to $u \equiv n\pi x/L$. This transforms the integral in (12.34*a*), which I'll call

$$I_n \equiv \int_0^L \sin\left(n\pi\frac{x}{L}\right) x(x-L)\,dx, \tag{12.44}$$

into

$$I_n = \left(\frac{L}{n\pi}\right)^3 \int_0^{n\pi} u(u - n\pi)\sin u\,du \tag{12.45a}$$

$$= \left(\frac{L}{n\pi}\right)^3 \left[\int_0^{n\pi} u^2 \sin u\,du - n\pi\int_0^{n\pi} u \sin u\right] du. \tag{12.45b}$$

(Don't overlook the dependence of the upper limit of integration on n.) Stalking these integrals in various tables, I found

$$\int_0^{n\pi} u^2 \sin u\,du = (-1)^{n+1}(n^2\pi^2 - 2) - 2 = \begin{cases} -n^2\pi^2 & n \text{ even} \\ n^2\pi^2 - 4 & n \text{ odd} \end{cases} \tag{12.46a}$$

$$\int_0^{n\pi} u \sin u\,du = (-1)^{n+1}n\pi = \begin{cases} -n\pi & n \text{ even} \\ n\pi & n \text{ odd} \end{cases} \tag{12.46b}$$

Substituting these expressions into (12.45*b*), I came up with (12.43*b*).

So the energy probability amplitudes (12.43*c*) are

$$c_n(0) = \begin{cases} 0 & n \text{ even} \\ -\dfrac{1}{n^3}\dfrac{8\sqrt{15}}{\pi^3} & n \text{ odd} \end{cases} \tag{12.47}$$

For this state, only energies that correspond to odd quantum numbers n have non-zero probabilities. So *none* of the systems in the ensemble will exhibit the energies E_2, E_4, E_6, ... Furthermore, the probability of obtaining E_n (for n odd, of course) is proportional to n^{-6} and so drops off very rapidly: *e.g.*,

$$\begin{aligned} P(E_1) &= |c_1|^2 = 0.9986 \\ P(E_3) &= |c_3|^2 = 0.0014 \end{aligned} \tag{12.48}$$

So most members will exhibit the ground-state energy $E_1 = 0.7057\,\text{eV}$, which is why the mean value, $\langle E \rangle = 0.7150\,\text{eV}$, is so close to this energy. But the mean isn't *equal* to E_1, because a few members exhibit the energy of the second excited state, $E_3 = 6.3509\,\text{eV}$.

As a consistency check, let's calculate the mean energy (12.39) from the probability amplitudes, *viz.*,

$$\langle E \rangle (0) = \sum_{\substack{n=1 \\ (n \text{ odd})}}^{\infty} |c_n(0)|^2 E_n \tag{12.49a}$$

$$= \sum_{\substack{n=1 \\ (n \text{ odd})}}^{\infty} \frac{1}{n^6} \left(\frac{8\sqrt{15}}{\pi^3} \right)^2 \cdot n^2 \frac{\pi^2 \hbar^2}{2mL^2}. \tag{12.49b}$$

Referring to a table of infinite series, we find the series in (12.49*b*),[12]

$$\sum_{\substack{n=1 \\ n \text{ odd}}}^{\infty} \frac{1}{n^4} = 1 + \frac{1}{3^4} + \frac{1}{5^4} + \cdots = \frac{\pi^4}{96}. \tag{12.50}$$

Hence the ensemble average of the energy at $t = 0$ is

$$\langle E \rangle (0) = \left(\frac{8\sqrt{15}}{\pi^3} \right)^2 \frac{\pi^2 \hbar^2}{2mL^2} \frac{\pi^4}{96} = \frac{5\hbar^2}{mL^2}, \tag{12.51}$$

in happy agreement with (12.39).[13]

Question 12–6

Use the energy probability amplitudes (12.47) to evaluate the energy uncertainty for the state represented by (12.36).

The above example was the first in which we had to explicitly evaluate the projections of a state function on the eigenfunctions of the observable being measured. Previously, the projections were handed us on a silver platter, the wave function being already expressed as the appropriate linear combination [*c.f.*, Eq. (12.15)]. As well you might imagine, nature is rarely so kind; algebraic contortions like those of this example are typically required to determine the amplitudes.

Actually, such problems are usually done backwards. The most common situation in the laboratory is the following: we have in hand the results of one or more ensemble measurements and need to deduce the state function. Let's see how to do this for a simple case.

Example 12.6. Deducing the State Function from Measurement Results

Consider the following challenge: You are trying to determine the form of the wave function that represents the state of an ensemble of CO molecules. In an energy measurement at $t = 0$, 70% of the members exhibit $E = 0.4035$ eV, while 30% exhibit the value $E = 0.1345\,\text{eV}$. From this information, what can you deduce about the wave function?

[12]My favorite such compendium is *Table of Integrals and Other Mathematical Data*, 4th ed. by H. B. Dwight (New York: Macmillan, 1961), where you'll find this series as Eq. (48.14).

[13]This example may seem a poor argument for using eigenfunction expansion techniques for evaluating expectation values. To be sure, in this case one could argue, especially if one is unfamiliar with series expansions such as (12.50), that evaluating the *integral* (12.38) is the easier approach. Perhaps. (I, however, loath evaluating integrals and will do almost anything to avoid them.) If you still think the integral approach might be easier, try evaluating the uncertainty $\Delta E(0)$ using the integral form of $\left\langle E^2 \right\rangle (0)$.

Since we have in hand data about *energy*, we ought to work in the language of this observable—*i.e.*, in the energy representation of Eq. (12.31). Let's model the CO molecules by a simple harmonic oscillator.[14] We know the Hamiltonian eigenfunctions (Table 9.6) and the molecular data which we need to determine $\beta = \sqrt{m\omega_0/\hbar}$.

Using $\hbar\omega_0 = 0.2690\,\text{eV}$ for CO (see Table 9.1) in the general form of an energy eigenvalue, $E_n = (n + \frac{1}{2})\hbar\omega_0$, we deduce that the observed energies are, in fact, E_0 and E_1. Therefore the initial wave function is a linear combination of the ground- and first-excited state eigenfunctions *only*,

$$\Psi(x,0) = c_0(0)\psi_0(x) + c_1(0)\psi_1(x). \tag{12.52}$$

We can determine the initial coefficients c_0 and c_1 in this expansion from the energy data, since the measured probabilities are just the squares of these coefficients:

$$\begin{aligned} P(E_0) &= |c_0|^2 = 0.30 \\ P(E_1) &= |c_1|^2 = 0.70. \end{aligned} \tag{12.53}$$

Comes now the tricky part. *It is incorrect to conclude from this data, as many do, that* $c_0 = \sqrt{0.30} = 0.5477$ *and* $c_1 = \sqrt{0.70} = 0.8367$. Think for a moment; can you figure out why this conclusion is not warranted?

The answer is that *the projections c_n are not necessarily real numbers*; they could be, but we would be wrong to deduce from the given data that they are. From the measured energy probabilities we can deduce only *the magnitude* of these projections, *i.e.*,

$$\text{Equations (12.53)} \qquad \Longrightarrow \qquad \begin{aligned} |c_0(0)| &= \sqrt{\frac{3}{10}} = 0.5477 \\ |c_1(0)| &= \sqrt{\frac{7}{10}} = 0.8367 \end{aligned} \; . \tag{12.54}$$

This means that *we do not fully know the coefficients in the wave function* (12.40). In particular, we don't have enough information to determine the *phase* of either projection. To introduce the phase, I'll write each of these complex numbers in a form containing two as-yet-unknown **phase constants** which I'll denote by δ_0 and δ_1:

$$c_0(0) = |c_0(0)|\, e^{i\delta_0} \tag{12.55a}$$

$$c_1(0) = |c_1(0)|\, e^{i\delta_1}. \tag{12.55b}$$

Using (12.54) we write the initial wave function (12.52) as[15]

[14]The SHO is a good approximation for low-lying vibrational states of this molecule. In Chap. 9 we learned that deep within the potential well of a molecule [see Figs. 9.13 and 9.14] the function $V(R)$ is accurately approximated by the simple harmonic oscillator potential defined with respect to the equilibrium internuclear separation R_e, *i.e.*,

$$V(R) \approx \frac{1}{2} m\omega_0^2 (R - R_e)^2.$$

From the energy data we deduce that in the state in this example, only low-lying vibrational levels are involved. Were this not true, were there a significant probability of obtaining a high-lying energy, we might need more arcane models, such as the Morse potential.

[15]Actually, only the *relative phase* of the constituents of $\Psi(x,0)$ is known. Like any other wave function, this one is indeterminate to within a global phase factor [recall § 3.5], which I'll call δ_{glob}. That is, we can write (12.56) as

$$\Psi(x,0) = e^{i\delta_{\text{glob}}} \left[\sqrt{\frac{3}{10}}\, \psi_0(x) + e^{i\delta_{\text{rel}}} \sqrt{\frac{7}{10}}\, \psi_1(x) \right],$$

$$\Psi(x,0) = \sqrt{\frac{3}{10}}\, e^{i\delta_0}\psi_0(x) + \sqrt{\frac{7}{10}}\, e^{i\delta_1}\psi_1(x). \tag{12.56}$$

[Note carefully that Eq. (12.56) represents the state of the ensemble of systems *before an energy measurement has been performed.* Said measurement *changes* that state (and hence the wave function) in a way we'll discuss in Chap. 13.] In § 12.5, we'll learn how to calculate from this initial function the form of the wave function for any $t > 0$.

Question 12–7

Try to guess the state function for $t > 0$ that evolves from the initial state represented by (12.56). You can probably do this without actually knowing why you are doing what you're doing.

Question 12–8

Evaluate the average value of and the uncertainty in the energy from the state function (12.56). Repeat the evaluation, working directly with the energy probability amplitudes. In either case, do the phase factors appear in your final result? Can you ascribe any physical significance to these factors?

Additional experiments, in which another observable is measured, are required to determine more about this wave function. We'll consider multiple measurements in Chap. 13.

*12.5 COPING WITH THE CONTINUUM

According to the Generalized Born Interpretation, *the projection of a state function on an eigenfunction of a Hermitian operator is the probability amplitude for obtaining the corresponding eigenvalue in a measurement of the observable represented by that operator.*[16] We can apply this interpretation to continuous as well as to discrete spectra, provided we take account of the peculiar nature of the continuum.

Recall that some operators, such as $\hat{\Pi}$, define *purely discrete spectra*; others, such as $\hat{x}$, define *purely continuous spectra*; and still others, such as $\hat{\mathcal{H}}$ for a (sufficiently deep) finite potential well, define *mixed spectra*, which consist of discrete and continuous eigenvalues. An example of the latter is the symmetric finite square well. Such a well of width $L = 1\,a_0$, and depth $V_0 = 49\pi^2/8$ [the parameters of Example 8.8] supports four bound stationary states (see Table 8.7), *i.e.*, four of the eigenvalues of its Hamiltonian are discrete. The rest, the eigenvalues for $E > V_0$, comprise the continuum, as suggested by Fig. 12.4.

Into the Continuum

To adapt the Born Interpretation to continuous eigenvalues we have only to take into account the fact that they are continuous. For contrast, consider first an observable

where the relative phase constant δ_{rel} is just the difference between δ_1 and δ_0 in Eq. (12.56). The global phase factor vanishes in the calculation of any measurable quantity and so is irrelevant to the physics of the state represented by this function.

[16]Be sure you understand all the jargon in, and the meaning of, this intentionally lingo-laden restatement. The words are important. Incidentally, in Chaps. 10 and 11, I introduced some of the material in this section (without justification) in footnotes. You need not have read those footnotes to understand this section, but if you skipped them, you probably ought to go back and read them after studying it.

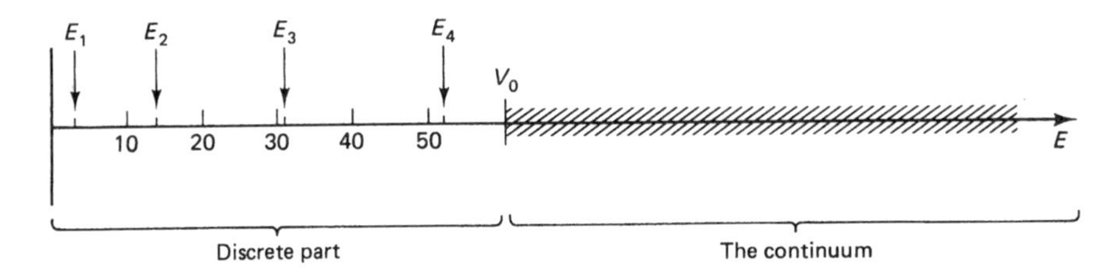

Figure 12.4 The energy spectrum of a finite square well. The energies for the well that we analyzed in Example 8.8 are shown as the discrete part of the spectrum. The rest is the continuum.

whose spectrum is *purely discrete*. To evaluate, say, the expectation value of such an observable, we *sum* [Eq. (12.14)],

$$\langle Q \rangle (t) = \sum_n |c_n(t)|^2 \, q_n, \qquad \text{[purely discrete spectrum]} \tag{12.57}$$

being careful to remember that the sum over n must include *all* discrete eigenvalues of $\hat{Q}$. Now, what if the spectrum is *purely continuous*?

Sure. We *integrate*:

$$\langle Q \rangle (t) = \int_{\text{all } q} P(q,t)\, q \, dq. \qquad \text{[purely continuous spectrum]} \tag{12.58}$$

The integral includes *all* (continuous) eigenvalues of $\hat{Q}$. This generalization comes, I hope, as no surprise; it is precisely the method we've been using to evaluate $\langle x \rangle (t)$ from $\Psi(x,t)$ [see Eq. (12.20)].

But there is more to the Born Interpretation for continuous eigenvalues than replacing a sum by an integral. We must also reconsider the meaning of the *probability amplitude* in an eigenfunction expansion. The quantity $|c_n(t)|^2$ in Eq. (12.57), which applies only to a discrete spectrum, is a *probability*, the probability of obtaining q_n in a measurement of Q at t. But the quantity $P(q,t)$ in (12.58) is a *probability density* for the observable Q, just as $P(x,t) = |\Psi(x,t)|^2$ in $\langle x \rangle (t)$ is a probability density for position.

To write down this interpretation explicitly, we need the projections of a state function Ψ on the eigenfunctions of $\hat{Q}$, which we'll now take to be an operator with a *purely continuous spectrum*. By analogy with the discrete case [Eq. (12.8); see also Table 12.1], I'll denote these projections by $c(q,t)$ and write the eigenfunction expansion as

$$\Psi(x,t) = \int_{\text{all } q} c(q,t)\chi_q(x)\, dq. \qquad \text{[expansion observable: } Q\text{]} \tag{12.59}$$

The projection $c(q,t)$ is

$$c(q,t) = \langle \chi_q \mid \Psi(t) \rangle = \int_{-\infty}^{\infty} \chi_q^*(x)\Psi(x,t)\, dx, \tag{12.60}$$

as in the discrete case [see Eq. (12.9)].

TABLE 12.1 SUMMARY OF FUNDAMENTAL QM EQUATIONS FOR DISCRETE AND CONTINUOUS OBSERVABLES.

	Discrete	Continuous
eigenvalue equation	$\hat{Q}\chi_n = q_n \chi_n$	$\hat{Q}\chi_q = q\chi_q$
projection of Ψ	$c_n(t) = \langle \chi_n \mid \Psi(t) \rangle$	$c(q,t) = \langle \chi_q \mid \Psi(t) \rangle$
probability	$\lvert c_n(t) \rvert^2$	$\lvert c(q,t) \rvert^2 \, dq$
mean value $\langle f(Q) \rangle (t)$	$\sum_n \lvert c_n(t) \rvert^2 f(q_n)$	$\int \lvert c(q,t) \rvert^2 f(q) \, dq$

Question 12–9

Substitute Eq. (12.59) into (12.60) and thereby verify that the two are consistent. Use the fact that eigenfunctions corresponding to continuous eigenvalues satisfy the normalization condition

$$\left\langle \chi_{q'} \mid \chi_q \right\rangle = \int_{-\infty}^{\infty} \chi_{q'}^*(x) \, \chi_q(x) \, dx = \delta(q' - q) \qquad \text{Dirac Delta function normalization} \tag{12.61}$$

where $\delta(q' - q)$ is the Dirac Delta function.

Now, $c(q,t)$ in (12.59) is a *probability amplitude* for Q, but since the eigenvalues of this operator are continuous, the corresponding *probability* is $\lvert c(q,t) \rvert^2 \, dq$ (*c.f.*, § 3.3 and 3.5). That is,

$$P(q,t) \, dq = \left| c(q,t) \right|^2 \, dq = \begin{array}{l} \text{probability that in a measurement of } Q \text{ at} \\ \text{time } t \text{ a member of the ensemble will exhibit} \\ \text{a value between } q \text{ and } q + dq \end{array} \tag{12.62}$$

That is what we mean when we say $\lvert c(q,t) \rvert^2$ is a *probability density*.

Aside: Concerning the Position Operator. It may not be obvious how to apply the expansion (12.59) to our most familiar operator, $\hat{x}$. The arguments in this section suggest that, for each value x_0 of position, the value of the wave function $\Psi(x_0, t)$ should be the projection of the state function on an eigenfunction of the Hermitian operator $\hat{x}$. But is it? In § 10.5, I noted that the eigenfunctions of $\hat{x}$ are the Dirac Delta functions [see Eq. (10.67*b*)], *i.e.*,

$$\hat{x}\delta(x - x_0) = x_0 \delta(x - x_0). \tag{12.63a}$$

If we use these eigenfunctions for $\chi_q(x)$ in Eq. (12.60), we get for the projections

$$c(x_0, t) = \int_{-\infty}^{\infty} \delta(x - x_0) \Psi(x, t) \, dx. \tag{12.63b}$$

The effect of the Dirac Delta function in the integral on the right-hand side of this equation is to select from the continuum of possible positions the value $x = x_0$, which gives for the position probability amplitude

$$c(x_0, t) = \Psi(x_0, t). \tag{12.63c}$$

The corresponding probability density is, therefore,

$$P(x_0, t) = \left| c(x_0, t) \right|^2 = \left| \Psi(x_0, t) \right|^2. \tag{12.63d}$$

Equations (12.63) reveal the Born Interpretation of Chap. 3 (for position) to be a special case of the Generalized Born Interpretation.

Momentum Reconsidered

One of the most important observables with a purely continuous spectrum is the linear momentum. The eigenvalue equation for the momentum operator $\hat{p}$,

$$\hat{p}\chi_p = p\chi_p, \tag{12.64}$$

defines only continuous eigenvalues, the spectrum $-\infty < p < \infty$. Its eigenfunctions [normalized according to Eq. (12.61)] are

$$\chi_p(x) = \frac{1}{\sqrt{2\pi\hbar}} e^{ipx/\hbar}. \tag{12.65}$$

Let's see what happens when we expand an arbitrary wave function in the complete set of eigenfunctions (12.65).

This expansion, according to (12.59), is an integral over p:

$$\Psi(x,t) = \int_{-\infty}^{\infty} c(p,t)\,\chi_p(x)\,dp = \frac{1}{\sqrt{2\pi\hbar}} \int_{-\infty}^{\infty} e^{ipx/\hbar} c(p,t)\,dp. \tag{12.66}$$

Note well that in (12.66) we integrate with respect to the eigenvalue p, not with respect to x. That is how the eigenfunction expansion (12.59) works.

Equation (12.60) prescribes how to calculate the projection $c(p,t)$ of Ψ on the eigenfunction χ_p:

$$c(p,t) = \langle \chi_p \mid \Psi(t) \rangle = \int_{-\infty}^{\infty} \chi_p^*(x)\Psi(x,t)\,dx. \tag{12.67}$$

(*Here* is where we integrate with respect to x.) Using the explicit form of the eigenfunction, this projection becomes suggestively familiar:

$$c(p,t) = \frac{1}{\sqrt{2\pi\hbar}} \int_{-\infty}^{\infty} e^{-ipx/\hbar}\,\Psi(x,t)\,dx. \tag{12.68}$$

Equations (12.66) and (12.68) look suspiciously like the Fourier transform relations we used in Chap. 4 to go between the position and momentum representations of a state function for a free particle. Compare, for example, Eq. (12.66) to Eq. (4.77*a*)

$$\Psi(x,t) = \frac{1}{\sqrt{2\pi\hbar}} \int_{-\infty}^{\infty} \Phi(p)\,e^{-iEt/\hbar}\,e^{ipx/\hbar}\,dp. \tag{12.69}$$

The *momentum probability amplitude for a free particle* is therefore related to the *momentum wave function* $\Phi(p)$ of Chap. 4 by[17]

$$c(p,t) = \langle \chi_p \mid \Psi(t) \rangle = \Phi(p)\,e^{-iEt/\hbar}. \qquad \text{[for a free particle]} \tag{12.70a}$$

The momentum wave function itself is just the momentum amplitude at $t = 0$, *i.e.*,

$$\Phi(p) = c(p,0). \tag{12.70b}$$

[17] We'll explore further the time-dependence of $c(p,t)$ in § 12.8. And in § 12.9 we'll discover that the simple, exponential time dependence in (12.70*a*) is not a general result; it applies only to systems for which p is a constant of the motion. One such system, as we learned in Chap. 10, is the free particle.

Question 12–10

Consider a *free particle* in one dimension in a state represented by a wave packet $\Psi(x,t)$. Substitute the expansion of this state function in momentum eigenfunctions into the time-dependent Schrödinger equation and derive a differential equation for the coefficients $c(p,t)$. Show that your equation is solved by

$$c(p,t) = c(p,0)\, e^{-iEt/\hbar}. \tag{12.71}$$

We now see that the interpretation of $\Phi(p)$ as the momentum probability amplitude, which we introduced rather tentatively in Chap. 4, is just an application of the Generalized Born Interpretation (12.62). The *probability density for momentum* is

$$P(p,t) = |c(p,t)|^2, \tag{12.72}$$

so *for the free particle*, the probability of getting a value between p and $p + dp$ in a momentum measurement *at any time* is

$$P(p,t)\, dp = |c(p,t)|^2\, dp \tag{12.73a}$$

$$= \left|\Phi(p)\, e^{-iEt/\hbar}\right|^2\, dp \tag{12.73b}$$

$$= |\Phi(p)|^2\, dp. \qquad \text{[for a free particle]} \tag{12.73c}$$

From this result we can easily regain expressions such as (4.78) for the ensemble average of momentum,

$$\langle p \rangle = \int_{-\infty}^{\infty} |c(p,t)|^2\, p\, dp = \int_{-\infty}^{\infty} |\Phi(p)|^2\, p\, dp. \qquad \text{[for a free particle]} \tag{12.74}$$

Mixed Spectra

Mixed spectra, consisting of both discrete and continuous eigenvalues, are quite common in the real (microscopic) world, so I want to discuss briefly the extension of the Generalized Born Interpretation to an operator with a mixed spectrum. This case is very straightforward; all of the basic formulae are just blends of those in Table 12.1.

Consider, for example, the finite square well of Fig. 12.4. In an investigation of the *energy* of an arbitrary state of this system, our first step would be to expand the wave function $\Psi(x,t)$ in eigenfunctions of the Hamiltonian. This expansion entails a "sum" over *all the eigenvalues of* $\hat{\mathcal{H}}$: a true sum over the discrete part and an integral over the continuous part, *i.e.*,

$$\boxed{\Psi(x,t) = \sum_{\substack{\text{all discrete}\\ \text{eigenvalues}}} c_n(t)\psi_n(x) + \int_{\substack{\text{all continuous}\\ \text{eigenvalues}}} c(E,t)\psi_E(x)\, dE} \tag{12.75}$$

For the spectrum in Fig. 12.4, this looks like

$$\begin{aligned}\Psi(x,t) &= c_1(t)\psi_1(x) + c_2(t)\psi_2(x) + c_3(t)\psi_3(x) \\ &\qquad + c_4(t)\psi_4(x) + \int_{V_0}^{\infty} c(E,t)\psi_E(x)\, dE.\end{aligned} \tag{12.76}$$

From this expansion we can easily write down expressions for other relevant quantum-mechanical quantities. For example, the probability that in an energy measurement at time t a member system will exhibit the value E_n (for $n = 1, 2, 3,$ or 4) is $|c_n(t)|^2$. But the probability that a member will exhibit an energy in the continuum is, for a value between E and $E + dE$, equal to $|c(E,t)|^2\, dE$. The mean energy obtained in such a measurement would be

$$\langle E \rangle (t) = |c_1(t)|^2 E_1 + |c_2(t)|^2 E_2 + |c_3(t)|^2 E_3 + |c_4(t)|^2 E_4 + \int_{V_0}^{\infty} |c(E,t)|^2 E\, dE. \tag{12.77}$$

And so forth.

Aha! Insight! (Into Momentum)

Before we turn to our next trick, solving the time-dependent Schrödinger equation, let's reflect for a moment on what we have done with momentum. Once again we have knitted into the growing fabric of quantum mechanics a topic encountered earlier in this book. Now, with heightened insight, we know what we are doing.

By applying the techniques of Chaps. 10 and 11 to momentum, we have put the arguments of Chap. 4—which there were based on expediency (the need to introduce a localized wave packet to represent the state of a free particle)—on the rock-solid foundation of quantum theory. When we used the method of eigenfunction expansion to translate the wave function into the language of momentum, we obtained Eq. (12.66), the Fourier transform relationship that we used (in an *ad hoc* way) in Chap. 4. The reason this equation is the Fourier transform relation is simply that the momentum eigenfunctions (12.65) are precisely the functions that appear in that relation. Ruminate further on the connections between this chapter and Chap. 4 as you ponder the handy summary in Table 12.2.[18]

TABLE 12.2 TWO VIEWS OF MOMENTUM INFORMATION IN QUANTUM MECHANICS.

Chapter 4 (Free Particle)	Chapter 12 (General State)
$\Psi(x,t) = \dfrac{1}{\sqrt{2\pi\hbar}} \int_{-\infty}^{\infty} \Phi(p)\, e^{i(px-Et)/\hbar}\, dp$	$\Psi(x,t) = \int_{-\infty}^{\infty} c(p,t)\, \chi_p(x)\, dp$
$\Phi(p) = \dfrac{1}{\sqrt{2\pi\hbar}} \int_{-\infty}^{\infty} \Psi(x,0)\, e^{-ipx/\hbar}\, dx$	$c(p,t) = \int_{-\infty}^{\infty} \chi_p^*(x) \Psi(x,t)\, dx$
$P(p,0) = \|\Phi(p)\|^2$	$P(p,t) = \|c(p,t)\|^2$
$\langle p \rangle (0) = \int_{-\infty}^{\infty} \Phi^*(p)\, p\, \Phi(p)\, dp$	$\langle p \rangle (t) = \int_{-\infty}^{\infty} \|c(p,t)\|^2\, p\, dp$

[18]In Chap. 4 we focused (primarily) on a single system: the free particle. But nowhere in this section did we specify a particular system. All our results here apply whether or not the particle is free. But if the particle is free, *i.e.*, if $V = 0$, then Eq. (12.70*a*) takes on a particularly simple form, since *for the free particle only*, the energy is related to the momentum by the simple dispersion relation $E = p^2/(2m)$. If, on the other hand, the particle is not free, then Eq. (12.70*a*) still applies, but the relationship of E to p is, to say the least, more complicated.

There is nothing magical or surprising about the appearance here of the formulae of Chap. 4. And it is really quite beautiful how all the elements of quantum mechanics hang together, how the threads of quantum physics from Part II are woven together by the mathematical machinery of Chaps. 10 and 11 to unify quantum theory. Isn't it? (Still, Chap. 13 looms, ominously, on the horizon.)

12.6 THE ARTFUL DODGE: SOLVING THE TDSE THE EASY WAY

Wolfe,
who had moved around the desk
and into his chair,
put up a palm at him:
"Please, Mr. Hobart.
I think it is always advisable
to take a short-cut
when it is feasible."

—*The Rubber Band*
by Rex Stout

Using the method of eigenfunction expansion and interpreting the expansion coefficients via the Generalized Born Interpretation, we can extract and understand information about measurement of any observable from a wave function at the time the measurement takes place ... *provided we know the wave function.* The trouble is, usually we don't.

Usually we know at most the Hamiltonian of the system and the *initial* wave function $\Psi(x, 0)$ (see Example 12.6). Without a way to determine $\Psi(x, t)$ from this initial function, the whole beautiful edifice of the last five sections becomes all but useless.

Since Chap. 6, we have known how to obtain $\Psi(x, t)$ *formally*: just solve the time-dependent Schrödinger equation

$$\hat{\mathcal{H}}\Psi = i\hbar\frac{\partial}{\partial t}\Psi \tag{12.78}$$

subject to the known initial condition. But it is facile to claim that this solves our predicament, for the TDSE is a very nasty equation indeed. Even for merely one particle in merely one dimension, it is a second-order partial differential equation in two variables:[19]

$$-\frac{\hbar^2}{2m}\frac{\partial^2}{\partial x^2}\Psi(x,t) + V(x)\Psi(x,t) = i\hbar\frac{\partial}{\partial t}\Psi(x,t), \tag{12.79}$$

the solving of which is assuredly not trivial.

The one case in which solving (12.79) *is* trivial is that of a stationary state. If $\Psi(x, 0)$ is an eigenfunction of $\hat{\mathcal{H}}$ with eigenvalue E, then solutions of the TDSE ex-

[19]I have assumed that we are dealing with a conservative system, so the potential energy in (12.79) doesn't depend on t. [In Volume II we'll implement a similar procedure, albeit in a more subtle way, to cope with non-conservative systems. This step will lead us to a powerful method for calculating approximate solutions to complicated time-dependent Schrödinger equations called *time-dependent perturbation theory.*] Somewhat less realistically, I'll also assume we have solved the time-*independent* Schrödinger equation for the system, so we know the eigenfunctions $\{\psi_n(x)\}$ and their eigenvalues E_n.

ist that are separable into a product of a space-dependent function (the Hamiltonian eigenfunction) and a time-dependent function (the usual exponential time factor):

$$\Psi_n(x,t) = \psi_n(x)e^{-iE_nt/\hbar}. \tag{12.80}$$

But if $\Psi(x,0)$ *is not a Hamiltonian eigenfunction, then* $\Psi(x,t)$ *is not separable.* Understanding this statement is the key to discovering how to conquer non-stationary states. To begin, let's recall that we can expand any wave function at any time in the complete set of eigenfunctions of $\hat{\mathcal{H}}$. Denoting the projections of this state by $c_n(t)$, we have[20]

$$\boxed{\Psi(x,t) = \sum_n c_n(t)\psi_n(x)} \qquad \text{[expansion observable: } E\text{]} \tag{12.81}$$

Although each term in (12.81) is a product of a space-dependent function and a time-dependent function, we must, in general, sum several such terms—maybe an infinite number of terms—in order to obtain a representation of $\Psi(x,t)$. So this wave function is certainly *not separable.*

To see if we could use this eigenfunction expansion to solve the TDSE, consider the projection $c_n(t)$,

$$c_n(t) = \langle \psi_n \mid \Psi(t) \rangle = \int_{-\infty}^{\infty} \psi_n^*(x)\Psi(x,t)\,dx. \tag{12.82}$$

At first glance, the prospects look grim. To determine $\Psi(x,t)$ from (12.81), we must know the projections at time t. But according to (12.82), to determine these projections we must know the very function we seek, $\Psi(x,t)$.

On further reflection, we realize that Eq. (12.81) is a boon. Think about what happens if we *substitute this expansion for* $\Psi(x,t)$ *in the TDSE (12.79).* The Hamiltonian acts on each eigenfunction in this sum, returning $E_n\psi_n(x)$ *and thereby eliminating from the Schrödinger equation the derivatives with respect to* x. What results is a set of equations *in the time variable only* for the projections $c_n(t)$, equations that are very easy to solve. Once we know the projections as functions of time, we know the state function (12.81). We have solved the TDSE without actually solving the TDSE! Let's now work out this wizardry in detail.

Equations for the Energy Amplitudes

First we substitute the expansion (12.81) into the TDSE (12.78), *viz.*,

$$\hat{\mathcal{H}} \sum_{n'} c_{n'}(t)\psi_{n'}(x) = i\hbar\frac{\partial}{\partial t} \sum_{n'} c_{n'}(t)\psi_{n'}(x). \tag{12.83}$$

[So my final result, Eq. (12.88), will look pretty I'm using n' for the (dummy) summation index in this derivation.] The Hamiltonian does not alter functions of t only, just as $\partial/\partial t$ does not alter functions of x only, so we can write (12.83) as

[20]To simplify the notation, I'll assume that the energy spectrum is purely discrete, so the eigenvalues and eigenfunctions can be labelled by $n = 1, 2, \ldots$. Results analogous to those of this section can be written down for continuous and mixed spectra using the generalizations of § 12.5.

$$\sum_{n'} c_{n'}(t)\hat{\mathcal{H}}\psi_{n'}(x) = i\hbar \sum_{n'} \psi_{n'}(x)\frac{d}{dt}c_{n'}(t). \tag{12.84a}$$

When the Hamiltonian acts on its eigenfunctions, this equation simplifies to

$$\sum_{n'} c_{n'}(t)E_{n'}\psi_{n'}(x) = i\hbar \sum_{n'} \psi_{n'}(x)\frac{dc_{n'}}{dt}. \tag{12.84b}$$

Fine. How can we extract from (12.84*b*) an equation for *a single projection* $c_n(t)$? We must somehow evaluate the sums over n' so as to select $n' = n$, and we must get rid of the x-dependent functions, the Hamiltonian eigenfunctions. Evidently, we need a Dirac Delta function. And I'll bet you know where we can get one.

Orthonormality. The x-dependent functions in this equation are eigenfunctions of a Hermitian operator and hence are orthonormal, *i.e.*,

$$\langle \psi_n \mid \psi_{n'} \rangle = \int_{-\infty}^{\infty} \psi_n^*(x)\psi_{n'}(x)\, dx = \delta_{n,n'}. \tag{12.85}$$

To exploit this handy fact, we multiply both sides of (12.84*b*) by $\psi_n^*(x)$ and integrate the resulting equation over all values of x.

From this point, I'm sure you can follow the rest of the derivation:

$$\sum_{n'} c_{n'}(t)E_{n'}\psi_n^*(x)\psi_{n'}(x) = i\hbar \sum_{n'} \psi_n^*(x)\psi_{n'}(x)\frac{dc_{n'}}{dt} \tag{12.86a}$$

$$\sum_{n'} c_{n'}(t)E_{n'} \int_{-\infty}^{\infty} \psi_n^*(x)\psi_{n'}(x)\, dx = i\hbar \sum_{n'} \frac{dc_{n'}}{dt} \int_{-\infty}^{\infty} \psi_n^*(x)\psi_{n'}(x)\, dx \tag{12.86b}$$

$$E_n c_n(t) = i\hbar \frac{dc_n}{dt}. \tag{12.86c}$$

Equation (12.86*c*) is a *first-order, ordinary differential equation* for the projection $c_n(t)$ of the state Ψ at time t on the Hamiltonian eigenfunction $\psi_n(x)$:

$$\frac{d}{dt}c_n(t) = -\frac{i}{\hbar}\, E_n c_n(t). \qquad n = 1, 2, \ldots \tag{12.87}$$

There is one such equation for each eigenfunction $\psi_n(x)$.

Integrating both sides of Eqs. (12.87) from the initial time $t = 0$, we obtain their solutions

$$\boxed{c_n(t) = c_n(0)\, e^{-iE_n t/\hbar} \qquad n = 0, 1, 2, \ldots} \tag{12.88}$$

Well, that result certainly is simple. And it leads straight to the solution of the TDSE for time t. Here's why. *We can calculate the projections* c_n *at* $t = 0$ *from the initial wave function, which (by hypothesis) we know:*

$$c_n(0) = \langle \psi_n \mid \Psi(0) \rangle = \int_{-\infty}^{\infty} \psi_n^*(x)\Psi(x, 0)\, dx. \qquad n = 1, 2, \ldots \tag{12.89}$$

Having done so, we merely multiply each projection by the familiar time factor $e^{-iE_n t/\hbar}$ and *voilà!* The expansion (12.81) is the desired wave function.

Before going on, let's cogitate a bit. *The coefficients* $\{c_n(0), n = 1, 2, 3, \dots\}$ *of (12.89) represent the state* Ψ *(in the energy representation).* The simple equation (12.88) tells us how to "propagate" the state to any time $t > 0$. From the energy representation of the state at this time we can easily determine the position representation, the wave function $\Psi(x,t)$, via our original expansion, Eq. (12.81).

So by "translating" the problem of finding the wave function from its initial condition into the language of energy, we have dodged entirely the onerous chore of solving the TDSE! We now have two ways to solve the fundamental problem of quantum mechanics, one direct, the other roundabout:

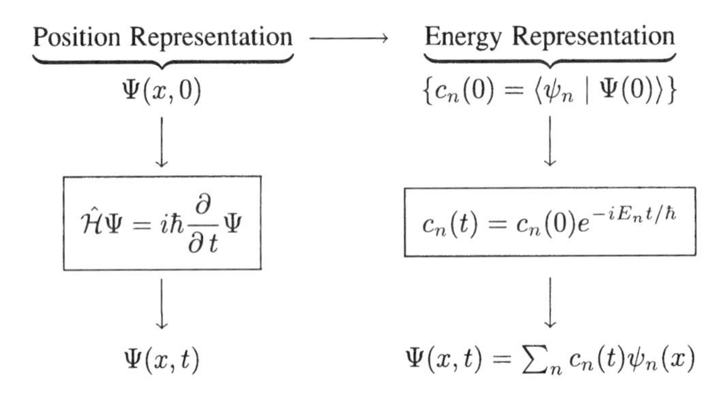

Example 12.7. Easy Evaluation of State Functions

In § 12.4 we left unfinished two applications of the Generalized Born Interpretation. In Example 12.5 we considered a hexatrine molecule, modeled by an infinite square well, in a non-stationary state. To predict the possible outcomes of an energy measurement (at $t = 0$) on this system, we expanded the initial wave function in Hamiltonian eigenfunctions, obtaining

$$\Psi(x,0) = \sqrt{\frac{30}{L^5}}\, x(x-L) = \sum_{\substack{n=1 \\ n \text{ odd}}}^{\infty} \left(-\frac{1}{n^3}\frac{8\sqrt{15}}{\pi^3}\right) \psi_n(x). \tag{12.90}$$

From that example, we have all the information needed to *write down* the solution of the TDSE for any time $t > 0$. We have determined the projections of this state at $t = 0$, and, from Chap. 7, we know the eigenvalues $E_n = n^2\pi^2\hbar^2/(2mL^2)$. Ergo the wave function is

$$\Psi(x,t) = \sum_{\substack{n=1 \\ n \text{ odd}}}^{\infty} \left(-\frac{1}{n^3}\frac{8\sqrt{15}}{\pi^3}\right) \psi_n(x)\, e^{-iE_n t/\hbar}. \tag{12.91}$$

In the second example of § 12.4, Example 12.6, we deduced from energy measurements on an ensemble of CO molecules that the (non-stationary) initial state could be represented by the wave function [Eq. (12.56)]

$$\Psi(x,0) = \sqrt{\frac{3}{10}} e^{i\delta_0} \psi_0(x) + \sqrt{\frac{7}{10}} e^{i\delta_1} \psi_1(x), \tag{12.92}$$

where δ_0 and δ_1 are (unknown) phase constants. I then asked you to guess the form of the wave function for $t > 0$. I hope you guessed

$$\Psi(x,0) = \sqrt{\frac{3}{10}} e^{i\delta_0} \psi_0(x)\, e^{-i\omega_0 t/2} + \sqrt{\frac{7}{10}} e^{i\delta_1} \psi_1(x)\, e^{-i3\omega_0 t/2}, \tag{12.93}$$

for, as we now know, this is the correct answer.

Almost embarrassingly easy, isn't it?

The Physics of Expansion in Hamiltonian Eigenfunctions

The solution of the TDSE by expansion in Hamiltonian eigenfunctions is, as you might imagine, a wonderful problem-solving tool. But it is also rich with important physics. For one thing, it clearly reveals *the explicit connection between the initial condition and the state function:*

$$\boxed{\Psi(x,t) = \sum_n \langle \psi_n \mid \Psi(0) \rangle\, \psi_n(x).e^{-iE_n t/\hbar}.} \tag{12.94}$$

Equation 12.94 is a succinct mathematical statement of *causality in quantum mechanics* (*c.f.*, Chap. 6). Recall that in the microscopic world, cause and effect take on a strange new form: the only true causality is between complex functions at different times, functions that do not themselves have any physical reality. We now see that the wave function depends on its progenitor, the initial state, in a very special way: through the initial *energy probability amplitudes* $\langle \psi_n \mid \Psi(0) \rangle$.

For another thing, the expansion (12.94) returns us to one of the most important underpinnings of quantum physics: the **Principle of Superposition**. To highlight this feature, let's write Eq. (12.94) in terms of the stationary-state wave functions (12.80),

$$\boxed{\Psi(x,t) = \sum_n c_n(0)\, \Psi_n(x,t).} \tag{12.95}$$

From this vantage point, we view the projections $c_n(0)$ as "mixture coefficients" in the *superposition of stationary states* that is the non-stationary state represented by $\Psi(x,t)$.

Question 12–11

Redo the above derivation for the special case that $\Psi(x,0)$ is an eigenfunction of $\hat{\mathcal{H}}$—say, with eigenvalue E_m. Explain briefly why, as claimed earlier in this section, separability of $\Psi(x,t)$ follows *only* in this special case.

I have boxed Eqs. (12.81), (12.94), and (12.95) to emphasize the importance of thinking about and remembering all of these perspectives on eigenfunction expansions. They tie the findings of this chapter to the rest of quantum mechanics. We have known since first we laid out the postulates of quantum physics (Part II) that inherent in a (general) wave function are many possible outcomes of an energy measurement—not, as we know now, *arbitrary* outcomes, but rather many eigenvalues of $\hat{\mathcal{H}}$. This is the essence of the concept of a superposition state. The probability that upon measurement at t, a member system will actually exhibit one of these values is also buried in the wave function—and we now know how to get at it.

Aside: The Energy and Other Representations. It is very important to understand the critical role played by *the energy representation* in our simplification of the TDSE. We could have expanded $\Psi(x,t)$ in the complete set of eigenfunctions of *any* Hermitian operator—say, $\{\chi_q(x)\}$ of some operator $\hat{Q}$. But substitution of the resulting expansion into the TDSE would lead to a far more complicated set of equations for the projections $c_n(t)$. In § 12.9 we'll pursue this matter further, deriving the following equations for the projections of Ψ on eigenfunctions of an *arbitrary observable* Q:

$$\frac{d}{dt}c_n(t) = -\frac{i}{\hbar}\sum_{n'}\left\langle \chi_n \mid \hat{\mathcal{H}} \mid \chi_{n'}\right\rangle c_{n'}(t). \qquad n = 1, 2, 3, \ldots \tag{12.96}$$

See the problem? *The equation we must solve for a particular projection, say* $c_n(t)$, *is coupled by matrix elements of* $\hat{\mathcal{H}}$ *to equations for other projections,* $c_{n'}(t)$. Such sets of coupled equations can be solved, as we'll learn in Volume II, but only with considerably more exertion than is necessary if we choose $\hat{Q} = \hat{\mathcal{H}}$ for our expansion.

Question 12–12

Derive Equations (12.96). What simplifications to these equations, if any, follow in the special case that the observable Q is a constant of the motion?

Implications for Energy Measurements

In this section I have emphasized the role of the projections $c_n(t) = \langle \psi_n \mid \Psi(t)\rangle$ as tools for solving the Schrödinger equation. But, of course, these quantities are also *energy probability amplitudes*, and their special form, $c_n(t) = c_n(0)\, e^{-iE_nt/\hbar}$, tells us something important about energy measurements: their results don't depend on time.

In such a measurement, the probability of obtaining an eigenvalue E_n is

$$P(E_n) = |c_n(t)|^2 \tag{12.97a}$$

$$= \left|c_n(0)\, e^{-iE_nt/\hbar}\right|^2 \tag{12.97b}$$

$$= |c_n(0)|^2 . \tag{12.97c}$$

It's clear what happened to the time. *Because of the special, complex exponential form of the time factor in the energy probability amplitudes, the squared moduli of these amplitudes—the energy probabilities—are independent of the measurement time. Be careful: this result does not pertain to probabilities for an arbitrary observable.* A convenient summary contrasting the energy-related properties of stationary and non-stationary states appears in Table 12.3.

Question 12–13

Use the form of the energy probability amplitudes and Eq. (12.18) for the ensemble average of energy to prove that the energy of a conservative microscopic system is conserved.

An Exhortation

I hope the discussion in this section—and the example in § 12.6—will convince you of the power of the method of eigenfunction expansion as a tool for solving the TDSE. The good news is that this method can be applied to a wide variety of differential equations.

Look carefully at the steps. We expanded a function $[\Psi(x,t)]$ of two variables (x and t) in a complete set in one of the variables (x). We then substituted the expansion into the multi-variable partial differential equation for the function (the TDSE) and used orthonormality of the complete set $[\{\psi_n(x)\}]$ to eliminate entirely one variable (x). The result was *a set* of simpler equations in a single variable.

TABLE 12.3 ENERGY-RELATED INFORMATION CONCERNING STATIONARY AND NON-STATIONARY STATES.

	Stationary $\Psi_n = \psi_n e^{-iE_n t/\hbar}$	Non-Stationary $\Psi(x,t)$
wave function	$\hat{\mathcal{H}}\Psi_n = E_n\Psi_n$	$\hat{\mathcal{H}}\Psi = i\hbar\frac{\partial}{\partial t}\Psi$
energy probability*	$P(E_{n'}) = \delta_{n',n}$	$P(E_{n'},t) = \lvert\langle\psi_{n'} \mid \Psi(t)\rvert^2$
mean energy*	$\langle E\rangle = E_n$	$\langle E\rangle(t) = \sum_n \lvert\langle\psi_n \mid \Psi(t)\rvert^2 E_n$
energy uncertainty*	$\Delta E(t) = 0$	$\Delta E > 0$

*Independent of time.

These steps can easily be extended to other partial differential equations, in more than two variables, and to complete sets of observables other than $\hat{\mathcal{H}}$. The only requirement is that the operator that defines the complete set be Hermitian, so that the eigenfunctions in the set are orthogonal.

In the interests of honesty, I should note that we have greatly simplified the mathematical problem, but at a certain cost. Rather than face a *single multi-variable partial differential equation*, we must cope with *a set of single-variable ordinary differential equations*. But solving the latter is (almost) always simpler than solving the former—which is what makes this method so powerful.

Anyway, my message is: *the technique of subduing partial differential equations via expansion in a complete set of orthonormal eigenfunctions is one of the two or three most powerful weapons in a physicist's armament. It will get you out of tight spots time and time again, not just in quantum physics, but in almost every physics course you take. Do not fail to commit it to memory.*

*12.7 THE AMAZING EXPANDING SQUARE WELL (AN EXAMPLE)

> Science isn't certainty, Sergo, it's probability.
>
> —John Bishop
> in *Artifact*
> by Gregory Benford

I now want to introduce one of the most remarkable, confounding examples in quantum physics: the effect on an observable—here, the energy—of a sudden change in the potential of a system. The system we'll consider, a particle in an infinite square well, is very simple; but nonetheless, the behavior it manifests is a paradigm for complicated physical processes such as beta decay. Stay alert. This one will test your quantum intuition to the fullest.

First, let's apply our *classical* intuition to a *macroscopic* counterpart of this system. Suppose a billiard ball (of mass m) bounces elastically (in one dimension) between the fixed walls of a box of width L, as in Fig. 12.5a.[21] The particle, being macroscopic, *has* an energy, which depends on its velocity v as $E = mv^2/2$.

At $t = 0$ the right wall *suddenly* moves to the right, causing the box to expand—to be specific, suppose it moves from $x = L$ to $x = 3L$, as in Fig. 12.5b. (By "sudden," I mean effectively instantaneous—*i.e.*, the change in the potential occurs during so short a time interval that, in effect, the position of the particle can be considered unchanged during the expansion.) What happens to the energy of the particle? Right—absolutely nothing. The particle's energy is independent of the size of the box, so the expansion has no effect on it. The particle just traverses a larger orbit as it bounces back and forth in the expanded box.

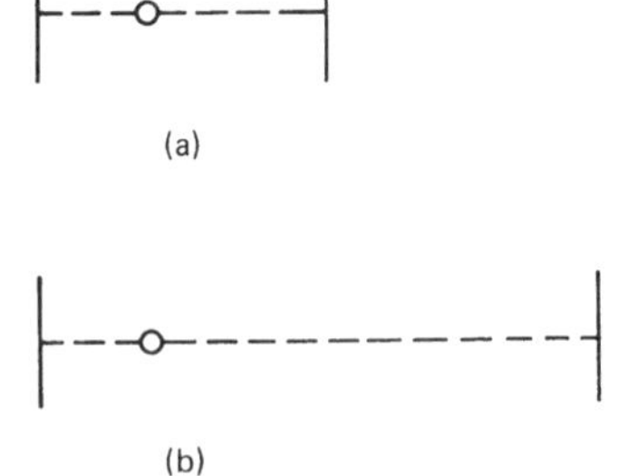

Figure 12.5 A bouncing macroscopic ball in one dimension (a) before and (b) after a *sudden* change at $t = 0$ in the position of the right wall. As a consequence of this change in the potential, the trajectory of the particle changes (in a predictable way), but its total energy does not change.

An Expansion in the Microverse

Consider now a microscopic box, the infinite square well of width L shown in Fig. 12.6a.[22] A particle in an energy eigenstate at $t = 0$ has a well-defined energy $E_n = n^2\pi^2\hbar^2/(2mL^2)$. If nothing happens to the system, the state evolves just like any other stationary state: at $t > 0$ the wave function is $\Psi_n(x,t) = \psi_n(x)e^{-iE_nt/\hbar}$. At any time, the state is an energy eigenstate, and an energy measurement will yield the value E_n with zero uncertainty. The mean of the results of such a measurement is $\langle E \rangle (t) = E_n$. Simple enough.

But suppose the system changes (instantaneously) at $t = 0$. To keep the analysis simple, let's set up the *initial condition* (*before* the change) so that the particle is in the ground stationary state of the well of width L; *i.e.*,[23]

$$\Psi(x,0) = \psi_1^{(L)}(x) = \sqrt{\frac{2}{L}}\,\sin\left(\pi\frac{x}{L}\right). \qquad [0 \leq x \leq L] \tag{12.98}$$

Outside the box, of course, this wave function is zero.

At $t = 0$ the box *suddenly* expands, as in Fig. 12.6b, the right wall abruptly moving to $x = 3L$. What is the state of the particle at some time after the sudden expansion? Is it still an energy eigenstate? What is its wave function?

[21]In one of those idealizations so beloved of textbook authors, we could imagine the ball sliding *without friction* along a *perfectly straight* wire *in the absence of gravity*. The perfectly elastic collision, too, is an idealization.

[22]For consistency with Example 12.5, I'll put $x = 0$ at the left wall of the box. Obviously, the physics of the states of the system doesn't depend on where I put the origin of coordinates.

[23]Because we will consider two square wells of different widths, I'll label the Hamiltonian eigenfunctions and eigenvalues, the stationary state energies, with a superscript that indicates the width of the well.

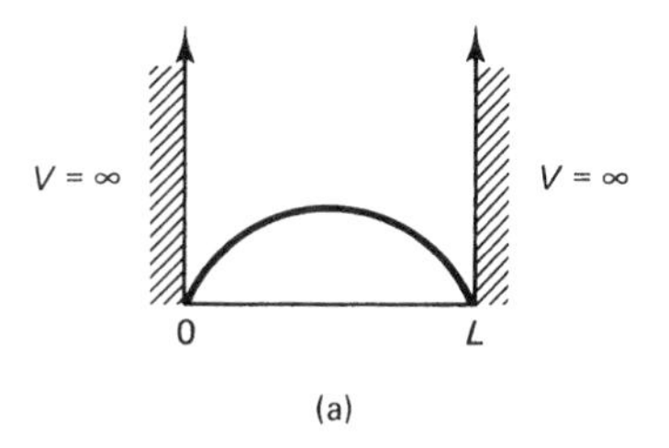

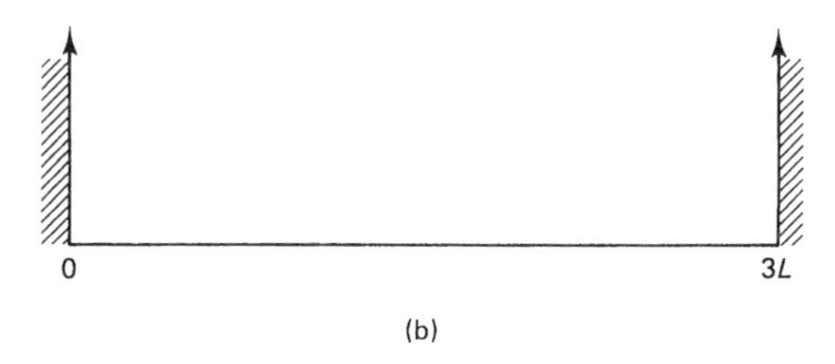

Figure 12.6 The infinite square well: (a) before and (b) after a *sudden* expansion at $t = 0$. As a consequence of this change in the potential energy, the state of the particle (not shown) changes in strange and remarkable ways.

Erroneous Answers

Some novice quantum mechanics answer the last question with the wave function for the original well, $\psi_1^{(L)}(x)\, e^{-iE_1^{(L)}t/\hbar}$, arguing that since the initial state was an energy eigenstate of the "unexpanded well," so should be the subsequent state.

This answer should make you very nervous. The equation that governs evolution of the state from the moment of expansion to time t is the time-dependent Schrödinger equation *of the expanded well, i.e.*, the Hamiltonian of this system is that of the expanded well, not of the initial well.[24] The point is that $\psi_1^{(L)}(x)$ is not an eigenfunction of *this* Hamiltonian, so the wave function $\psi_1^{(L)}(x)\, e^{-iE_1^{(L)}t/\hbar}$ is not appropriate to the expanded well.

Abandoning this guess, we might look for guidance at the allowed energies of the particle *after expansion.* From § 12.4 we know that these values are just the *eigenvalues of a well of width* $3L$,

$$E_j^{(3L)} = j^2\frac{\pi^2\hbar^2}{2m(3L)^2} = j^2\frac{\pi^2\hbar^2}{18mL^2}. \qquad j = 1, 2, 3, \ldots \tag{12.99}$$

[24]For the expanding square well, the difference between the systems before and after expansion is the boundary conditions on the eigenfunctions, *e.g.*,

$$\text{for a well of width } L: \quad \begin{cases} \psi_n^{(L)}(x=0) = 0 \\ \psi_n^{(L)}(x=L) = 0 \end{cases}$$

$$\text{for a well of width } 3L: \quad \begin{cases} \psi_n^{(L)}(x=0) = 0 \\ \psi_n^{(L)}(x=3L) = 0 \end{cases}$$

This is, I admit, an unfortunate (although minor) complicating feature of this example. It does not arise if we consider, say, a simple harmonic oscillator whose natural frequency suddenly changes at $t = 0$.

The corresponding eigenfunctions are

$$\psi_j^{(3L)}(x) = \sqrt{\frac{2}{3L}} \sin\left(j\pi\frac{x}{3L}\right). \qquad j = 1, 2, 3, \ldots \tag{12.100}$$

Energy level diagrams for wells of width L and $3L$ are shown side by side in Fig. 12.7. Lo and behold, the energy of the particle *before expansion* — the ground state energy $E_1^{(L)} = \pi^2\hbar^2/(2mL^2)$ of the well of width L—is equal to the energy of the *second excited state* ($j = 3$) *of the expanded well*, $E_3^{(3L)}$. From this observation, one might guess that the system at $t > 0$ is in the second excited state of the expanded well and that consequently its state function is $\psi_3^{(3L)}(x)\, e^{-iE_3^{(3L)}t/\hbar}$. What do you think? Does that sound right?

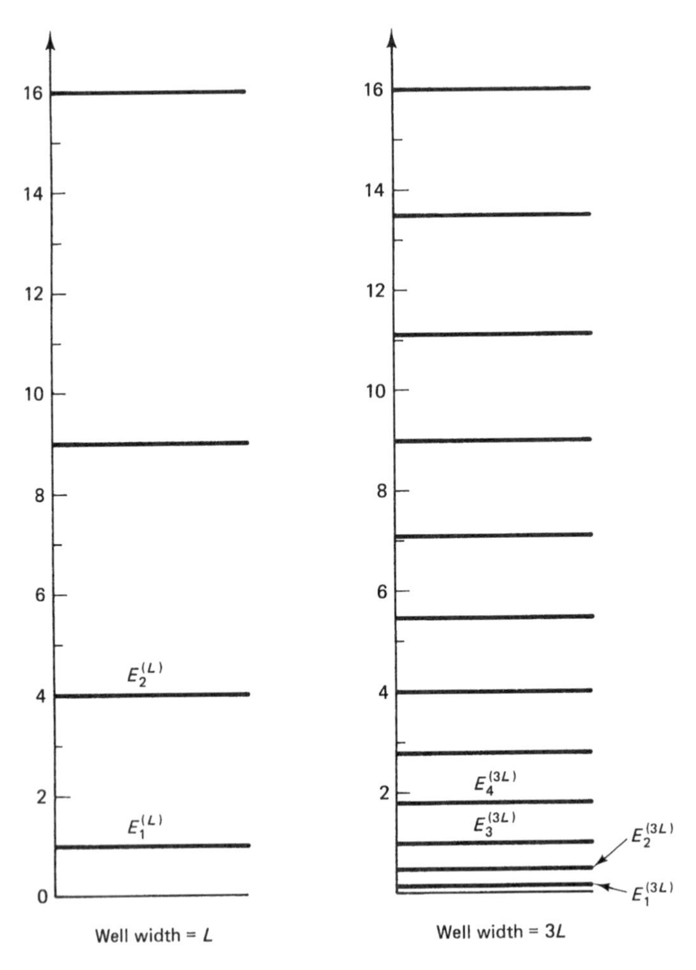

Figure 12.7 The first few stationary-state energies of infinite square wells of width L (left diagram) and $3L$ (right diagram). The *purely accidental* coincidence of $E_1^{(L)}$ and $E_3^{(3L)}$ can lead the unwary astray.

Question 12–14

> *Before you read further, work this very simple exercise.* Suppose the well expands at $t = 0$ to a width of $3L/2$. Write down the equation for the allowed energies $E_j^{(3L/2)}$. For what value of j does $E_1^{(L)}$ coincide with one of these allowed energies? Now answer the question I posed above.

This guess should trouble you. Underlying it is the assumption that the state of the particle at $t > 0$ that evolves from the initial condition $\Psi(x, 0) = \psi_1^{(L)}(x)$ is *identical* to the state that would have evolved from the initial condition $\Psi(x, 0) = \psi_3^{(3L)}(x)$. The latter initial wave function would be appropriate only if the width of the well had always been $3L$. But that's not right: those aren't the circumstances of the problem. Initially our particle wasn't in an expanded well; it was in a well of width L. Surely this fact will affect the state after expansion.

The Correct Answer

Evidently something rather subtle is going on. Simple intuitive arguments are leading us astray. We would be well advised, then, to analyze this problem systematically, sticking rigorously to the procedure for solving the TDSE we laid out in § 12.6. The analysis isn't mysterious, but it does demand care and thought at each step.

At the heart of our procedure is *the expansion of the state function in eigenfunctions of the Hamiltonian*. The coefficients in that expansion, the projections of Ψ on these eigenfunctions, have a simple exponential time dependence that is determined from the initial wave function and the eigenvalues of the Hamiltonian. The only tricky part is determining which eigenfunctions and eigenvalues to use. Let's see how this goes.

After expansion, the width of the box is $3L$, so we should expand $\Psi(x, t)$ in eigenfunctions (12.100), *i.e.*,

$$\Psi(x, t) = \sum_{j=1}^{\infty} c_j(t) \psi_j^{(3L)}(x). \qquad \text{[expansion observable: } E \text{]} \tag{12.101}$$

The projections $c_j(t)$ are proportional to an exponential *that includes the eigenvalues of the expanded well*[25]

$$c_j(t) = c_j(0)\, e^{-iE_j^{(3L)} t/\hbar}. \qquad j = 1, 2, 3, \ldots \tag{12.102}$$

Comes now the tough question: what is the expression for the projections $c_j(0)$? According to Eq. (12.89) the jth projection is the integral over all space of the complex conjugate of the jth eigenfunction *of the expanded well* times the initial wave function (12.98):

$$c_j(0) = \langle \psi_j^{(3L)} \mid \Psi(0) \rangle \tag{12.103a}$$

$$= \int_{-\infty}^{\infty} \psi_j^{(3L)*}(x)\, \Psi(x, 0)\, dx \tag{12.103b}$$

$$= \int_0^L \psi_j^{(3L)*}(x)\, \psi_1^{(L)}(x)\, dx. \tag{12.103c}$$

[25]If it is not clear as brook water why we use $E_j^{(3L)}$ rather than $E_j^{(L)}$ in this expression, go back and study carefully the derivation of (12.88). Do not pass this point until you understand it.

There are two possible traps in Eqs. (12.103). First, we must carefully follow the prescription of § 12.6 and insert for the initial wave function in (12.103*b*) the *ground-state eigenfunction of the unexpanded well* (not an eigenfunction of the expanded well). Second, the limits of integration on the integral in (12.103*b*) are changed in (12.103*c*), for this integral is cut off at $x = 0$ and $x = L$ *by the initial wave function* (12.98), which is zero outside the box. (Even fairly adept quantum mechanics sometimes goof up the limits of integration in problems like this one.) The rest is algebra.

Now we must evaluate the projections Eq. (12.103*c*),

$$c_j(0) = \sqrt{\frac{2}{3L}}\sqrt{\frac{2}{L}}\int_0^L \sin\left(\frac{j}{3}\pi\frac{x}{L}\right)\sin\left(\pi\frac{x}{L}\right)dx. \tag{12.104}$$

The obvious thing to do at this point is to change the variable of integration from x to $u \equiv \pi x/L$, *viz.*,

$$c_j(0) = \sqrt{\frac{2}{3L}}\sqrt{\frac{2}{L}}\frac{L}{\pi}\int_0^\pi \sin\left(\frac{ju}{3}\right)\sin u\,du. \tag{12.105}$$

(Notice that I have changed the upper limit of integration accordingly.) A trip to your friendly neighborhood integral table yields

$$c_j(0) = \begin{cases} \dfrac{1}{\sqrt{3}} & j = 3 \\ -\dfrac{18}{\pi\sqrt{3}}\dfrac{\sin\left(\frac{j}{3}\pi\right)}{j^2 - 9} & j \neq 3 \end{cases}. \tag{12.106}$$

Aside: Filling in the Gaps. Want to see the details? There's not much to them. First we combine the sine functions, using a familiar trig identity, to obtain for *the integral* in (12.105)

$$I_j \equiv \int_0^\pi \sin\left(\frac{ju}{3}\right)\sin u\,du \tag{12.107a}$$

$$= \frac{1}{2}\left[\int_0^\pi \cos\left(\frac{ju}{3} - u\right)du - \int_0^\pi \cos\left(\frac{ju}{3} + u\right)du\right] \tag{12.107b}$$

$$= \frac{3}{2}\left[\frac{\sin\left(\frac{j}{3} - 1\right)\pi}{j - 3} - \frac{\sin\left(\frac{j}{3} + 1\right)\pi}{j + 3}\right]. \qquad j \neq 3 \tag{12.107c}$$

The integral for $j = 3$ is particularly simple:

$$I_3 = \int_0^\pi \sin^2 u\,du = \frac{\pi}{2}. \tag{12.107d}$$

The integral for $j \neq 3$ is not particularly simple, so let's simplify it. Manipulating the sine functions in (12.107*c*) (more trig identities) and putting both terms over $(j-3)(j+3) = j^2 - 9$, we obtain

$$I_j = -9\frac{\sin\left(\frac{j}{3}\pi\right)}{j^2 - 9}. \qquad j \neq 3 \tag{12.108}$$

Equations (12.107*d*) and (12.108) lead immediately to (12.106).

The projections (12.106), when multiplied by the exponential time factors and the appropriate Hamiltonian eigenfunctions, determine the wave function $\Psi(x,t)$ at any $t > 0$ [Eq. (12.101)]. Clearly, more than one of these projections are non-zero (in fact, an infinite number of them are non-zero). So the state represented by this function is not stationary, *i.e.*, it is not an energy eigenstate.

Question 12–15

Write out the first five terms in the expansion for $\Psi(x,t)$, using explicit expressions for the energies and eigenfunctions.

The projections (12.106) are also the representation of the state Ψ in the language of energy—*i.e.*, they are the *energy probability amplitudes*. The number $|c_j(0)|^2$ is the probability that in an energy measurement at $t > 0$ *on this state of the particle in the expanded well*, we'll obtain the eigenvalue $E_j^{(3L)}$. I've tabulated some of these energies and their respective probabilities in Table 12.4.

TABLE 12.4 ENERGIES, PROJECTIONS, AND PROBABILITIES FOR A STATE OF THE EXPANDED INFINITE WELL.[a]

j	$E_j^{(L)}$	$E_j^{(3L)}$	$c_j(0)$	$P(E_j^{(3L)})$
1	1.0	0.11	0.358	0.128
2	4.0	0.44	0.573	0.328
3	9.0	1.00	0.577	0.333
4	16.0	1.78	0.409	0.167
5	25.0	2.78	0.179	0.032
6	36.0	4.00	0.0	0.0
7	49.0	5.44	0.071	0.005
8	64.0	7.11	0.055	0.003
9	81.0	9.00	0.0	0.0
10	100.0	11.11	-0.031	0.001
11	121.0	13.44	-0.026	0.0007
12	144.0	16.00	0.0	0.0

[a]Energies in units of $\pi^2\hbar^2/(2mL^2)$.

Let's return now to a question I posed at the beginning of this section. Before the well expanded, the particle had a (sharp) energy $E_1^{(L)} = \pi^2\hbar^2/(2mL^2)$. What is the probability of a member particle exhibiting this value in an energy measurement at $t > 0$, *after expansion*?

This energy is $E_3^{(3L)}$, and according to Table 12.4 the corresponding probability is $P(E_3^{(3L)}) = 0.333$. Hmmm ... not even 50% of the systems in our ensemble exhibit this energy.[26] Some energies, such as $E_6^{(3L)}$, although "allowed," do not appear in an

[26]Note carefully: a member that exhibits this energy in a measurement at $t > 0$ was not (immediately prior to the measurement) in an energy eigenstate with this energy. We know from the distribution of energies in Table 12.4 that the state of the ensemble immediately before we measure the energy is not an energy eigenstate (recall Table 12.3): the energy uncertainty is positive [$\Delta E(t) > 0$] for $t > 0$. (We'll consider the energy uncertainty, which exhibits some disquieting behavior, in a minute.)

energy measurement on this state, because the projections of the state function on the corresponding eigenfunctions are zero. Still, rather a lot of energies do show up in the measurement. The probabilities $P(E_j^{(3L)})$ appear to drop off rapidly with increasing j. Nonetheless—and contrary to our classical intuition—there is a small chance that a member of the ensemble will exhibit an energy much *larger* than the particle's value prior to expansion of the well.

Statistical Quantities of the Expanded Well

The sudden change in the potential at $t = 0$ brought about a radical change in the quantum state of the particle in the well. To gain further insight into this change, let's compute the average value and uncertainty in the energy at $t > 0$. Their initial values, of course, are

$$\begin{aligned} \langle E \rangle (0) &= E_1^{(L)} \\ \Delta E(0) &= 0 \end{aligned} \qquad \text{[before expansion]}. \tag{12.109}$$

Unfortunately, this calculation is complicated whether we use the *integral* or *series* forms of the required expectation values, $\langle E \rangle$ and $\langle E^2 \rangle$. In such cases, a physicist will sometimes abandon the quest for the exact values of these quantities and settle for numerical approximations to them. One could compute such approximations either by integrating numerically the integral forms (quadrature) or by summing a finite number of terms in each series. In the latter case, we must include enough terms to identify the value to which the sum is tending with increasing number of terms. Because we have not considered this gambit before, I want to show you how to evaluate the mean value and uncertainty in the energy by summing truncated series in the energy amplitudes $c_j(0)$.

Question 12–16

Write down explicit expressions for the integral forms of $\langle E \rangle (0)$ and $\langle E^2 \rangle (0)$ for the state of the system *after expansion*. Why are the values of these quantities not equal to Eqs. (12.109)?

The series for $\langle E \rangle$ and $\langle E^2 \rangle$ include an infinite number of terms [see Table 12.3] unless the projections $c_j(0)$ have a very special behavior: unless there exists a "maximum value" j_{max} beyond which all coefficients $c_j(0)$ are zero. Examining (12.106) we find that, unhappily, this is not the case for the problem at hand.

We could, of course, try to sum the resulting expressions exactly using the theory of infinite series, and in this simple case that strategem would work. But suppose it didn't. Or suppose we can't figure out the answer. Or suppose we have an answer and want to check it. In any case, we would like to estimate the values of $\langle E \rangle$ and ΔE. To get an idea of these values, we'll "truncate" the infinite series for $\langle E \rangle$ and $\langle E^2 \rangle$ at a fairly small number of terms and see what we get. That is, we'll *introduce* a terminal value j_{max} and calculate

$$\langle E \rangle = \sum_{j=1}^{j_{\text{max}}} |c_j(0)|^2 \, E_j \tag{12.110a}$$

$$\langle E^2 \rangle = \sum_{j=1}^{j_{\text{max}}} |c_j(0)|^2 \, E_j^2. \tag{12.110b}$$

Then the uncertainty follows from the usual formula

$$\Delta E = \sqrt{\langle E^2 \rangle - \langle E \rangle^2}. \tag{12.110c}$$

The results of these calculations for increasing values of $j_{\max}$ appear in Fig. 12.8.

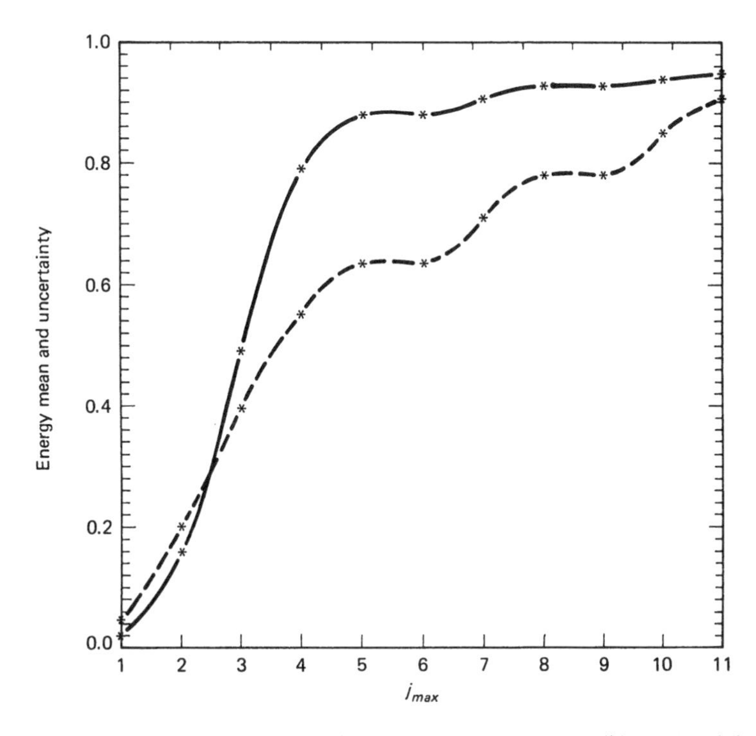

Figure 12.8 Finite-series approximations to the average energy (solid curve) and the energy uncertainty (dashed curve) for the state represented by the energy amplitudes (12.106). The horizontal axis shows the *maximum value of j included in the series for* $\langle E \rangle$ and $\left\langle E^2 \right\rangle$—see Eqs. (12.110). *Note that the units of the vertical axis are* $\pi^2\hbar^2/(2mL^2)$. Therefore 1.0 on the vertical axis is the value $E_1^{(L)}$ of the energy of the particle before the well expands.

Let's look first at the mean energy. The trend is clear. As the number of terms in (12.110*a*) increases, the average energy approaches the value $\pi^2\hbar^2/(2mL^2)$, *i.e.*,

$$\langle E \rangle \xrightarrow[j_{\max}\to\infty]{} \frac{\pi^2\hbar^2}{2mL^2}. \tag{12.111a}$$

Remarkably, *this is the energy of the particle before the well expanded at* $t = 0$, *i.e.*, the eigenvalue of the Hamiltonian of the well of width L that corresponds to the initial condition $\Psi(x, 0) = \psi_1^{(L)}(x)$.[27]

[27]This result is not a fluke. Had the particle initially been in *any* eigenstate of the well of width L, with energy $E_n^{(L)}$, its mean value at $t > 0$, after expansion, would be

$$\langle E \rangle = E_n^{(L)} \qquad \text{[for the expanded well].}$$

This result holds no matter what the width of the expanded well. [See J. G. Cordes, D. Kiang, and J. Nakajima, *Amer. Jour. Phys.*, **52**, 155 (1984).]

Equally remarkable—and far more alarming—is the behavior displayed in Fig. 12.8 by the uncertainty ΔE: as the number of terms included in $\langle E \rangle$ and $\langle E^2 \rangle$ increases, the uncertainty seems to diverge! Indeed, a detailed numerical analysis shows that

$$\Delta E \xrightarrow[j_{\max}\to\infty]{} \infty. \tag{12.111b}$$

[The same distressing behavior is found if the particle is initially in *any* eigenstate of the unexpanded well—*i.e.*, for $\Psi(x,0) = \psi_n^{(L)}(x)$. The energy uncertainty of the state after expansion is finite only if the initial state is non-stationary. (See Pblm. 12.4.)] As a consequence of the expansion, the particle has wound up in a state of maximal energy uncertainty.

Question 12–17

Verify Eqs. (12.110) and the appalling Eqs. (12.111) by continuing the analysis in Fig. 12.8 to *at least* $j_{\max} = 20$.

Question 12–18

1. Look at Fig. 12.7. Where, in relation to the energy levels of the system before expansion, is the energy of the *most probable* state after expansion?
2. Does the mean value of the energy change as a consequence of the expansion? Can you figure out why or why not?
3. We've seen that some members of the ensemble will exhibit an energy after expansion that is greater than the energy of the system before expansion. Discuss the implications of this finding for the law of conservation of energy.

This example will repay your careful perusal and considered contemplation. In addition to rather elaborately illustrating the powerhouse method of eigenfunction expansion, it illustrates the nest of counter-intuitive phenomena into which quantum mechanics sometimes leads us. And it reminds us of the marvels and wonders that await us in the microworld.

*12.8 MOMENTUM REDUX: THE CASE OF THE CONTINUUM CONTINUED

Earlier in this chapter we learned how to use the Generalized Born Interpretation to extract probabilistic information from state functions (§ 12.2–12.5). We also learned how to use the method of eigenfunction expansion to obtain these functions from the TDSE (§ 12.6–12.7) and how to translate them into the languages of various observables. These strategies are among the most important in this book. Here I want to tie them together with an example in which we contend with a more general problem and a more general quantum state than that of the previous section.

Consider a simple harmonic oscillator that is initially in a *non-stationary state* represented by the (initial) wave function $\Psi(x,0)$.[28] Suppose that at time $t > 0$, the *linear momentum* of the oscillator is measured. We want to predict the probability that a member of an ensemble of such oscillators will give a value between p and $p + dp$. How do we proceed? Ultimately we must get our hands on the *momentum probability*

[28] In § 9.10 we treated one such state—a "sloshing" state—and discovered the periodic behavior of its position probability and properties. For deep insight into sloshing states, see Pblm. 12.5.

amplitude $c(p,t)$, for the squared modulus of this quantity is the probability density we need. But this problem is complicated by the difficulty in determining the wave function at time t from which we calculate this amplitude. That is, first we must determine the wave function $\Psi(x,t)$ at the measurement time, from the initial condition $\Psi(x,0)$. Different languages are appropriate to each of these steps:

Step 1. Solve the TDSE: use the *energy* representation (§ 12.5).
Step 2. Predict results of momentum measurement: use the *momentum* representation (§ 12.4).

Let's take up with the second task first. According to the Generalized Born Interpretation, we can calculate the momentum probability density from the representation of the state Ψ *in the language of momentum, i.e.*,

$$P(p,t) = |c(p,t)|^2. \tag{12.112}$$

The projection $c(p,t)$ of the state Ψ on a momentum eigenfunction χ_p with eigenvalue p—the momentum probability amplitude—appears in the eigenfunction expansion

$$\Psi(x,t) = \int_{-\infty}^{\infty} c(p,t)\chi_p(x)\,dp \qquad \text{[expansion observable: } p\text{]}, \tag{12.113a}$$

and is related to the wave function at time t by

$$c(p,t) = \langle \chi_p \mid \Psi(t)\rangle. \tag{12.113b}$$

We know the momentum eigenfunctions in (12.113*a*); they are the familiar pure-momentum-state functions of Eq. (12.65). But what about the state function at $t > 0$?

The easiest way to determine this function is to work in the language of energy (§ 12.6). We can easily write down the solution of the TDSE using the energy representation of the state, *viz.*,

$$\Psi(x,t) = \sum_{n=0}^{\infty} c_n(0)\, e^{-iE_n t/\hbar}\psi_n(x) \qquad \text{[expansion observable: } E\text{]}. \tag{12.114}$$

The energies in the exponential time factor are

$$E_n = \left(n + \frac{1}{2}\right)\hbar\omega_0 \qquad n = 0,1,2,\ldots \tag{12.115}$$

The Hamiltonian eigenfunctions $\psi_n(x)$ are those of Table 9.6. And the projections of the initial state *on these energy eigenfunctions* are

$$c_n(0) = \langle \psi_n \mid \Psi(0)\rangle \qquad n = 0,1,2,\ldots. \tag{12.116}$$

One important feature of Eqs. (12.114)–(12.116) may seem obvious when I state it but is a frequent source of confusion: the *energy* probability amplitudes $c_n(t) = c_n(0)e^{-iE_n t/\hbar}$ are not the same as the *momentum* probability amplitudes $c(p,t)$. The two sets $\{c_n(0); n = 1,2,\ldots\}$ and $\{c(p,t); -\infty < p < \infty\}$ represent the *the same state* in different languages. Just as words that represent the concept "love" in French (*amour*) and in English are different, so the quantities that represent the state Ψ in the languages of momentum and energy are different.

But it's easy to derive a relationship between the two representations by substituting the expansion in energy eigenfunctions, Eq. (12.114), into the expression for the momentum amplitude, Eq. (12.113b); to wit:

$$\boxed{\underbrace{c(p,t)}_{\substack{\text{language of}\\\text{momentum}}} = \sum_{n=0}^{\infty} \underbrace{c_n(0)\, e^{-iE_nt/\hbar}}_{\substack{\text{language of}\\\text{energy}}} \langle \chi_p \mid \psi_n \rangle}. \tag{12.117}$$

Notice that $c(p,t)$ does not have the simple form of a constant times $e^{-iEt/\hbar}$. This simple form obtains only in special cases, such as a stationary state.

Question 12–19

Show that *if the system is initially in the m*th *energy eigenstate*, then the momentum probability amplitude is

$$c(p,t) = \langle \chi_p \mid \psi_m \rangle e^{-iE_mt/\hbar}. \qquad \text{[system in } m^{\text{th}} \text{ stationary state]} \tag{12.118}$$

Example 12.8. Momentum Measurements on a SHO

Let's get specific. To make life easier, suppose we are handed the expansion of the wave packet $\Psi(x,0)$ in Hamiltonian eigenfunctions,

$$\Psi(x,0) = \frac{1}{\sqrt{2}}\Big[\psi_0(x) + \psi_1(x)\Big]. \tag{12.119}$$

This function represents the state in the position representation. Its representation in the language of energy, which we'll use to solve the TDSE, is therefore

$$c_0(0) = \frac{1}{\sqrt{2}} \tag{12.120a}$$

$$c_1(0) = \frac{1}{\sqrt{2}} \tag{12.120b}$$

$$c_n(0) = 0. \qquad n > 1 \tag{12.120c}$$

Question 12–20

Write down the wave function (in the position representation) that represents this state at times $t > 0$.

Question 12–21

In an energy measurement at $t > 0$, what values would be found, and with what probabilities?

Now, to predict the results of a measurement of *momentum*, we insert the energy amplitudes (12.120) into Eq. (12.117) for the momentum amplitude $c(p,t)$, which reduces to just two terms,

$$c(p,t) = c_0(0)e^{-iE_0t/\hbar}\langle \chi_p \mid \psi_0 \rangle + c_1(0)e^{-iE_1t/\hbar}\langle \chi_p \mid \psi_1 \rangle \tag{12.121a}$$

$$= \frac{1}{\sqrt{2}}e^{-i\omega_0t/2}\langle \chi_p \mid \psi_0 \rangle + \frac{1}{\sqrt{2}}e^{-i\omega_1t/2}\langle \chi_p \mid \psi_1 \rangle. \tag{12.121b}$$

There remains only to evaluate the integrals

$$\langle \chi_p \mid \psi_n \rangle = \frac{1}{\sqrt{2\pi\hbar}} \int_{-\infty}^{\infty} e^{-ipx/\hbar}\, \psi_n(x)\, dx \tag{12.122}$$

for $n = 0$ and 1. This explicit expression reveals these integrals to be *the inverse Fourier transforms of the Hamiltonian eigenfunctions* $\psi_n(x)$, a result we could have anticipated from Chap. 4. For $n = 0$ and $n = 1$, the particular cases that appear in (12.121), these integrals are easily evaluated, by recourse to a Table of Fourier Transforms, as[29]

$$\langle \chi_p \mid \psi_0 \rangle = \frac{1}{2}\left(\frac{1}{\pi\beta^2\hbar^2}\right)^{1/4} e^{-p^2/(2\beta^2\hbar^2)} \tag{12.123a}$$

$$\langle \chi_p \mid \psi_1 \rangle = \frac{1}{\sqrt{2}}\left(\frac{1}{\pi\beta^2\hbar^2}\right)^{1/4} \frac{p}{\hbar}\, e^{-p^2/(2\beta^2\hbar^2)}. \tag{12.123b}$$

The constant β is (Table 9.3) $\beta = \sqrt{m\omega_0/\hbar}$.

We can now calculate the probability of obtaining a value of momentum between p and $p + dp$. We merely insert the integrals (12.123) into our expression for the momentum probability amplitude, Eq. (12.121*b*); multiply the resulting (complex) quantity $c(p,t)$ by its complex conjugate; and then multiply the result by dp. You should be able to see without going through the algebra that $P(p,t)$ does depend on time. When $c(p,t)$ is multiplied by $c^*(p,t)$, the resulting probability density contains "cross terms" that oscillate with time. This time-dependent momentum probability density contrasts strikingly with the momentum density for a stationary state, which quantity is independent of t.

Question 12–22

For the state (12.119), derive an expression for $P(p,t)$ (in terms of $\langle \chi_p \mid \chi_n \rangle$, the natural frequency of the oscillator ω_0, the particle mass m, and fundamental constants) that explicitly displays the time dependence of this probability density.

Question 12–23

Do the mean momentum $\langle p \rangle$ and the uncertainty Δp for a non-stationary state depend on time? Why or why not? How does this behavior differ from that for a stationary state?

The general expression (12.117) can be applied to any non-stationary state, so the arguments of Example 12.8 imply that the momentum probability density for such a state does depend on time.

A final thought: In problem solving we get a lot of mileage by translating the TDSE into the energy representation. And you may have wondered if it is possible to translate the TDSE into the language of an arbitrary observable. Of course it is; the

[29] In this evaluation, the parity of the Hamiltonian eigenfunctions simplifies the algebra. The general form of the Fourier transform of a function $f(x)$ is $g(k) = \int_{-\infty}^{\infty} f(x)\, e^{ikx}\, dx$. If $f(x)$ has definite parity, this expression simplifies, *i.e.*,

$$g(k) = \begin{cases} \int_0^\infty f(x)\cos kx\, dx & f(x) \text{ even} \\ \int_0^\infty f(x)\sin kx\, dx & f(x) \text{ odd} \end{cases}.$$

You can find a gathering of Fourier Transforms in any good book of mathematical formulae and tables. A handy compendium (in an inexpensive book) is in § 1.17 of *Formulas, Facts, and Constants for Students and Professionals in Engineering, Chemistry, and Physics* by H. J. Fischbeck and K. H. Fischbeck (New York: Springer-Verlag, 1982). (Of the multitude of books of tables I've accumulated, this is the cheapest and most useful.)

momentum representation is a common example.[30] But in practice such translations are rare because—as you learned in Question 12.9—the resulting equations for the projections of the state on eigenfunctions of an observable other than energy do not simplify as did the equations for the energy projections (§ 12.6).

Nevertheless, the procedure of translating quantum-mechanical equations and problems into various representations is a powerful one. First described in its full mathematical glory by P. A. M. Dirac (who in 1926 earned the first PhD ever awarded for research in quantum mechanics), this method has been codified using the mathematics of matrices, transformations, and linear algebra into the predominant problem-solving technique of quantum physics.[31] You will, I hope, have a chance to study it in your next quantum course.

12.9 FINAL THOUGHTS: CONSTANTS OF THE MOTION RECONSIDERED

> You is feeling like
> you was lost in the bush, boy?
> You says: It is a puling sample jungle
> of woods. You most shouts out:
> Bethicket me for
> a stump of a beech
> if I have the poultriest notions
> what the farest he all means.
> Gee up, girly!
>
> —*Finnegan's Wake*
> by James Joyce

In § 12.8 we discovered that the momentum probability density for a non-stationary state changes with time. Earlier in this chapter we ascertained that the energy probabilities for such a state do not depend on time. The obvious question raised by these findings is: *Is energy the only observable whose probabilities (for an arbitrary state) are time-independent?* The answer is no. As we're about to prove, *the probabilities for the energy and all other constants of the motion of a system are independent of time*. Once again, the proof uses the method of eigenfunction expansion.

First, a brief review. In Chap. 10 we defined a constant of the motion as *an observable that is compatible with the energy.* Such an observable is represented by an operator that commutes with $\hat{\mathcal{H}}$ [Eq. (11.74)]. We then proved (§ 11.6) that a constant of the motion in quantum mechanics is conserved, *i.e.*, its *mean value* in any state doesn't depend on time:

$$\left[\hat{\mathcal{H}}, \hat{Q}\,\right] = 0 \qquad \Longrightarrow \qquad \frac{d}{dt}\langle Q\rangle(t) = 0. \qquad \text{[any state } \Psi] \tag{12.124}$$

[30]In Pblm. 9.16 I asked you to consider the solution of this equation for a stationary state in the momentum representation. Similarly, in Pblm. 4.6 we solved for the momentum amplitude and density for the infinite square well.

[31]Dirac wrote what many consider to be the most elegant, clear presentation of this theory, in his landmark book *Principles of Quantum Mechanics* (Oxford: Clarendon Press, 1958), now in its fourth edition (and in paperback!).

Now I want to generalize this result by proving that

> ***Proclamation 12.1***
>
> All probabilistic and statistical quantities that refer to a constant of the motion of a system are independent of time.

In light of (12.124), this claim certainly looks reasonable. The probability for obtaining a (discrete) eigenvalue q_n in a measurement of Q is $c_n^{(Q)}(t)$.[32] These quantities appear in the expression for $\langle Q \rangle$,

$$\langle Q \rangle (t) = \sum_n \left| c_n^{(Q)}(t) \right|^2 q_n. \tag{12.125}$$

(For simplicity, I've assumed that the spectrum of $\hat{Q}$ is purely discrete. Proclamation 12.1 and its proof generalize easily to constants of the motion with continuous and mixed spectra.) It is unlikely indeed that $\langle Q \rangle$ would be *independent of time* unless each probability $|c_n^{(Q)}(t)|^2$ were also independent of time. [If correct, this conclusion does not, of course, mean that $c_n^{(Q)}(t)$ is itself independent of time.] But this plausibility argument doesn't constitute a proof.

Suppose, however, that we can prove that all *individual probabilities* are independent of time,

$$\frac{d}{dt} P(q_n, t) = \frac{d}{dt} |c_n^{(Q)}(t)|^2 \stackrel{?}{=} 0 \qquad \text{all } n. \tag{12.126}$$

Then we can conclude that all expectation values $\langle f(Q) \rangle$ are also independent of time, and therefore that the uncertainty, which depends only on $\langle Q \rangle$ and $\langle Q^2 \rangle$, is time independent. So let's focus our attention on Eq. (12.126).

Argument: The obvious way to proceed, I suppose, is to try to derive an equation for the time dependence of $c_n^{(Q)}(t)$—an equation analogous to (12.88) for the energy amplitudes $c_n^{(E)}(t)$—in the hopes that when we multiply $c_n^{(Q)}(t)$ by its complex conjugate in (12.126), the time dependence will just go away. At first, this road seems to lead to a dead end.[33]

To explore the problem, we start with the expansion of $\Psi(x,t)$ in the complete set $\{\chi_n^{(Q)}(x)\}$ of eigenfunctions of $\hat{Q}$,

$$\Psi(x,t) = \sum_n c_n^{(Q)}(t)\, \chi_n^{(Q)}(x). \tag{12.127}$$

Substituting this expansion into the TDSE for $\Psi(x,t)$, Eq. (12.79), we get

$$\sum_{n'} c_{n'}^{(Q)}(t)\, \hat{\mathcal{H}}\, \chi_{n'}^{(Q)} = i\hbar \sum_{n'} \chi_{n'}^{(Q)} \frac{d}{dt} c_{n'}^{(Q)}(t), \tag{12.128}$$

[32] In this section we'll use projections of a state on eigenfunctions of $\hat{\mathcal{H}}$ and on eigenfunctions of $\hat{Q}$. To keep straight these two types of projections, I'll use a superscript to indicate the observable in which the wave function was expanded.

[33] We shall follow here very closely the derivation of (12.88). The present derivation will be much easier to follow if you'll take a minute to refresh your memory about the earlier one.

which is very similar to Eq. (12.84*a*). *But we cannot now proceed with the rest of this derivation as we did in getting to Eq. (12.88), for the functions $\chi_{n'}^{(Q)}$ are not eigenfunctions of $\hat{\mathcal{H}}$.*

We must leave $\hat{\mathcal{H}}$ where it sits in (12.128). Its lurking presence means, as we shall see, that the equation for the nth projection $c_n^{(Q)}(t)$ is not independent of the other projections. To derive this equation, we invoke orthogonality of the eigenfunctions $\{\chi_n^{(Q)}\}$ of the Hermitian operator $\hat{Q}$,

$$\int_{-\infty}^{\infty} \chi_n^{(Q)*}(x)\, \chi_{n'}^{(Q)}(x)\, dx = \delta_{n,n'}. \tag{12.129}$$

Multiplying Eq. (12.128) from the left by $\chi_n^{(Q)*}(x)$ and integrating the result over all space, we obtain an equation for each $c_n^{(Q)}(t)$:

$$\sum_{n'} c_{n'}^{(Q)}(t) \int_{-\infty}^{\infty} \chi_n^{(Q)*}(x)\, \hat{\mathcal{H}}\, \chi_{n'}^{(Q)}(x)\, dx = i\hbar \frac{d}{dt} c_n^{(Q)}(t). \tag{12.130}$$

These equations look a bit like (12.86*b*), with one crucial difference. *The equations (12.130) for the various $c_n^{(Q)}(t)$ are coupled to one another.* [Notice that I used orthonormality of the eigenfunctions of the (Hermitian) operator $\hat{Q}$ to eliminate the sum over n' on the right-hand side. The trouble is, we can't pull the same trick on the left-hand side to make Eq. (12.129) look like (12.86c).] To solve the equation for, say, $c_1^{(Q)}(t)$, you must know all the other projections $c_n^{(Q)}(t), n \neq 1$, because they appear on the left-hand side. But to solve the equation for, say, $c_2^{(Q)}(t)$, you must know $c_1^{(Q)}(t)$, for it appears on the left-hand side of *that* equation. The dead end I mentioned above looks like a cul-de-sac. Oh well. You can't win 'em all.

But wait! We have not used all we know about $\hat{Q}$. So far we've been treating this little guy like an arbitrary operator; but $\hat{Q}$ is not arbitrary at all. *The operator $\hat{Q}$ represents a constant of the motion and therefore commutes with $\hat{\mathcal{H}}$.* To see the consequences of this fact, let's clarify the structure of (12.130) by introducing Dirac notation for the integral:

$$\sum_{n'} c_{n'}^{(Q)}(t) \langle \chi_n^{(Q)} \mid \hat{\mathcal{H}} \mid \chi_{n'}^{(Q)} \rangle = i\hbar \frac{d}{dt} c_n^{(Q)}(t). \tag{12.131}$$

Now, look very carefully at the matrix element in (12.131). *It is the matrix element of an operator ($\hat{\mathcal{H}}$) with respect to eigenfunctions of a second operator ($\hat{Q}$) that commutes with the first.* Do you recall the handy theorem we proved in § 11.3? According to Proclamation 11.3, such a matrix element is zero if the eigenvalues of the two eigenfunctions are different, *i.e.*,

$$\left[\hat{\mathcal{H}}, \hat{Q}\right] = 0 \qquad \Longrightarrow \qquad \langle \chi_n^{(Q)} \mid \hat{\mathcal{H}} \mid \chi_{n'}^{(Q)} \rangle = 0 \quad \text{if } n' \neq n. \tag{12.132}$$

So things aren't as bad as they seemed: only one of the matrix elements on the left-hand side of the equation for each $c_n^{(Q)}(t)$ is non-zero. The sum goes away. We are left with (uncoupled) equations for each projection,

$$\frac{d}{dt} c_n^{(Q)}(t) = -\frac{i}{\hbar} \langle \chi_n^{(Q)} \mid \hat{\mathcal{H}} \mid \chi_n^{(Q)} \rangle\, c_n^{(Q)}(t). \tag{12.133}$$

The mathematical structure of this equation is identical to that of (12.87). Its solution is

$$c_n^{(Q)}(t) = c_n^{(Q)}(0)\, e^{-i\langle \chi_n^{(Q)} | \hat{\mathcal{H}} | \chi_n^{(Q)} \rangle t/\hbar}. \tag{12.134}$$

The rather complicated matrix element in the exponential factor in (12.134) causes no difficulty, for when we multiply $c_n^{(Q)}(t)$ by its complex conjugate, the offending exponential goes away. We are left with

$$\left| c_n^{(Q)}(t) \right|^2 = \left| c_n^{(Q)}(0) \right|^2, \tag{12.135}$$

which handily proves (12.126) and Proclamation 12.1.

Q. E. D.

Our results here and in Chap. 11 concerning constants of the motion reveal that in quantum physics, as in classical mechanics, there is about these observables a kind of stasis. But their constancy takes on a special form in quantum physics, a form dictated by the special, restricted nature of accessible knowledge about a microscopic system.

EXERCISES AND PROBLEMS

> Unearned knowledge is perilous.
> Only by the seeking
> and gaining of it
> may its uses be understood,
> its true worth measured.
>
> —Pitchwife
> in *White Gold Wielder*
> by Stephen Donaldson

The problems for this chapter focus primarily on its contents. But because this chapter is, secondarily, a "wrap up" of the whole book, I have included some exercises that draw on earlier material.

Exercises

12.1 States of the Simple Harmonic Oscillator—Stationary and Otherwise

All questions in this exercise concern a one-dimensional simple harmonic oscillator with natural frequency ω_0. Those of Part A pertain to the basic quantum physics of a one-dimensional conservative system, those of Part B to the physical properties of the oscillator, those of Part C to its stationary and non-stationary states, and those of Part D to measurements.

Part A:

[a] **Write down** the *time-dependent* Schrödinger equation for this system, including the explicit expression for the Hamiltonian.

[b] Using the mathematical technique of *separation of variables*, **derive** an expression for the wave function $\Psi(x,t)$ of a *stationary state* of the SHO, showing explicitly the time-dependence of this function. Also **derive** the equation for the Hamiltonian eigenfunction $\psi_n(x)$. (Do not, however, solve the TISE.)

[c] **Prove** that the Hamiltonian you wrote down in part [a] is Hermitian. (You do not have to go back to first principles to do this; *i.e.*, you can use the fact that $\hat{x}$ and $\hat{p}$ are Hermitian. But justify each step in your argument.)

[d] From the Hermiticity of the Hamiltonian, there follow several important quantum-mechanical properties. For each of the following items, state the appropriate property:
(1) the expectation value of the energy, $\langle E \rangle$;
(2) the eigenvalues of the Hamiltonian, E_n;
(3) the set of Hamiltonian eigenfunctions, $\{\psi_n(x)\}$;
(4) the "overlap integral" of two Hamiltonian eigenfunctions, $\langle \psi_n(x) \mid \psi_m(x) \rangle$, where $n \neq m$.

Part B:

[a] For each pair of observables listed below, answer each of the following questions:
(1) are the observables compatible?
(2) is it possible *in principle* to measure the two observables simultaneously to infinite precision?
(3) does there exist a complete set of simultaneous eigenfunctions of the operators that represent these observables?
(4) is the product of the uncertainties of the two observables in each pair, when evaluated *for an arbitrary state*, necessarily positive?

1. energy and parity;
2. energy and position;
3. parity and momentum;
4. position and momentum

[b] From this list, pick *one* pair that *does* have a complete set of simultaneous eigenfunctions and **write down** the form of these eigenfunctions.
[c] Is the *parity* of this system conserved? **Why or why not?**

Part C:

In this part we shall consider two states of the SHO. The initial conditions that define these states are represented by the following (normalized) wave functions:

$$\Psi(x,0) = \frac{1}{\sqrt{13}}\left[2\psi_0(x) + 3\psi_1(x)\right], \tag{12.1.1}$$

where $\psi_n(x)$ are Hamiltonian eigenfunctions of the SHO, and

$$\Psi'(x,0) = \frac{1}{\sqrt{2}}\left(\frac{\beta^2}{\pi}\right)^{1/4}\left(2\beta^2x^2 - 1\right)e^{-\beta^2x^2/2}. \tag{12.1.2}$$

For *each* of these states, answer *each* of the following questions:

[a] Is either of these states stationary? If so, which one(s)?
[b] **Write down** the wave functions $\Psi(x,t)$ and $\Psi'(x,t)$ that represent the states at times $t > 0$ that evolve from these initial states.
[c] At $t = 0$ the *parity* of each state is measured. What value (or values) will be obtained in the measurements, and with what probabilities? Is the parity of either state sharp? **Why or why not?**
[d] **Evaluate** the *expectation value* of the parity of each state. What is the physical significance of these quantities?
[e] Now, suppose that at $t = 0$ we measure the *energy* of each of these states instead of the parity. **Evaluate** the probability that a member of the ensemble will exhibit each of the following values. Do this for the state represented by $\Psi(x,t)$ and for the state represented by $\Psi'(x,t)$. **In each case, fully justify your answer to each question.**

1. $E = 0$
2. $E = \frac{1}{2}\hbar\omega_0$
3. $E = \frac{7}{2}\hbar\omega_0$
4. $E = \hbar\omega_0$

[f] **Calculate** the *average value* and the *uncertainty* of the energy of the SHO that would be obtained in this measurement.
[g] Are $\langle E \rangle$ and ΔE independent of time for these states?

Part D:

For the following questions, assume that the SHO is initially in its ground state.
[a] Is this state a momentum eigenstate? **Prove your answer.**
[b] At time $t = 10\,\text{sec}$ we measure the *momentum* of the particle. **Write down** an expression for the probability that we will obtain in this measurement a value between $\hbar k_0$ and $\hbar k_0 + dk$. (You do not have to fully evaluate your expression, but write it completely enough that the only thing left to do is to evaluate an integral.)
[c] How would your answer to question [b] differ if the momentum measurement were carried out at $t = 40\,\text{sec}$ instead of at $t = 10\,\text{sec}$? **Justify your answer.**

12.2 The Physics of a Non-Stationary State

Consider a one-dimensional simple harmonic oscillator with natural frequency ω_0 that is initially in a quantum state represented by the *unnormalized* wave function

$$\Psi(x,0) = [2\psi_0(x) + 4\psi_1(x) - 3\psi_3(x)], \tag{12.2.1}$$

where, of course, the functions $\psi_n(x)$'s are the Hamiltonian eigenfunctions for this system.
[a] **Write down** the solution to the time-dependent Schrödinger Equation subject to this initial condition, *i.e.*, the wave function $\Psi(x,t)$ that represents the state for times $t > 0$.
[b] Is this state an energy eigenstate? Why or why not?
[c] **Evaluate** the mean energy for this state.
[d] In a measurement of the energy of this system at some time $t > 0$, **determine** the probability of obtaining each of the following values:
(1) $\frac{1}{2}\hbar\omega_0$;
(2) $\frac{5}{2}\hbar\omega_0$;
(3) $8\hbar\omega_0$.
[e] Is parity a constant of the motion for this system? Why or why not?
[f] Is this state an eigenstate of parity? **Prove** your answer.
[g] **Evaluate** the average momentum at $t = 0$.
[h] Do simultaneous eigenstates of parity and energy exist for this system? If so, tell why and give an example. If not, tell why not.

Problems

12.1 Energy Measurements on a Non-Stationary State

Suppose a particle in the well of Fig. 12.4 is in a state represented by the function

$$\begin{aligned}\Psi(x,t) = &\frac{1}{\sqrt{30}}\left[4\psi_1(x)e^{-iE_1t/\hbar} + \psi_2(x)e^{-iE_2t/\hbar}\right.\\ &\left.-2i\psi_3(x)e^{-iE_3t/\hbar} + 3\psi_4(x)e^{-iE_4t/\hbar}\right].\end{aligned} \tag{12.1.1}$$

In an energy measurement at time t, what value is most likely to be obtained? ... least likely to be obtained? What are the probabilities of a member system exhibiting each of these values? To describe the aggregate of results of the measurement, **evaluate** the expectation value and the uncertainty at time t.

12.2 Average Values for Non-Stationary States

Suppose we have in hand the energy probability amplitudes $c_n(0)$ for a state Ψ of a one-dimensional quantum system. Suppose the energy spectrum of the system is discrete, with eigenvalues $E_n, n = 1, 2, \ldots$ and eigenfunctions $\psi_n(x)$. A measurement at time $t > 0$ is performed on an observable Q (not the energy). **Derive** the following expression for the mean value of Q obtained in this state:

$$\begin{aligned}\langle Q \rangle (t) =& 2\sum_n \sum_{n'>n} c_{n'}^*(0) c_n(0) \langle \psi_{n'} \mid \hat{Q} \mid \psi_n \rangle \cos \frac{(E_{n'} - E_n)t}{\hbar} \\ &+ \sum_n |c_n(0)|^2 \langle \psi_n \mid \hat{Q} \mid \psi_n \rangle.\end{aligned} \tag{12.2.1}$$

12.3 Return of the Amazing Expanding Square Well

At $t = 0$ an infinite square well suddenly expands, its width changing from L to $2L$. Before expansion, the microscopic particle in the well was contentedly minding its own business in the ground stationary state, with energy $E_1^{(L)}$. At $t = 10\,\text{sec}$, the particle's energy is measured.

[a] What is the *most probable* value obtained in the measurement? **Calculate** the probability of a member system exhibiting this value.

[b] What is the probability that a member will exhibit the value $\pi^2\hbar^2/(2mL^2)$? **Discuss** briefly the implications of your answer for the state of the system after expansion.

[c] What is the probability that a member will exhibit a value greater than $8\pi^2\hbar^2/(2mL^2)$?

[d] What is the ensemble average obtained in the measurement?

[e] How does $P(E_j^{(2L)})$ depend on j in the limit $j \to \infty$? What does your answer imply about the energy uncertainty for this state?

12.4 Son of the Amazing Expanding Square Well

Consider again the expanding square well of Pblm. 12.3. Suppose the initial state of the particle *before expansion* is a non-stationary state of the well of width L, *i.e.*,

$$\Psi(x, 0) = \frac{1}{\sqrt{10}} \left[3\psi_1^{(L)}(x) + \psi_3^{(L)}(x) \right]. \tag{12.4.1}$$

[a] Let's first analyze the state at $t = 0$ *before expansion.*

(1) Were we to measure the energy of this state, what values would we find, and with what probabilities?

(2) In such a measurement, what value would we get for the average energy?

(3) What value would we get for the energy uncertainty?

[b] Now the well expands, its width changing from L to $2L$. At $t > 0$ we measure the energy of the particle in the expanded well.

(1) **Thought Question:** Is the wave function at $t > 0$

$$\Psi(x, t) \stackrel{?}{=} \frac{1}{\sqrt{10}} \left[3\psi_1^{(L)}(x)\, e^{-iE_1^{(L)}t/\hbar} + \psi_3^{(L)}(x)\, e^{-iE_3^{(L)}t/\hbar} \right]? \tag{12.4.2a}$$

If so, tell why and go to part [b (2)]. If not, is the wave function

$$\Psi(x, t) \stackrel{?}{=} \frac{1}{\sqrt{10}} \left[3\psi_1^{(2L)}(x)\, e^{-iE_1^{(2L)}t/\hbar} + \psi_3^{(2L)}(x)\, e^{-iE_3^{(2L)}t/\hbar} \right]? \tag{12.4.2b}$$

If so, explain why. If not, tell why not.

(2) In the measurement at $t = 10\,\text{sec}$, what values of the energy are allowed? **Derive** an expression for their probabilities.

(3) Of the first ten eigenvalues $E_j^{(2L)}$, what values are allowed but do not appear in the measurement for this state? Why do these values not appear?

(4) What is the most probable value for this state? What is its probability?

(5) **Calculate** the mean value of the results obtained in the energy measurement.

(6) **Calculate** the energy uncertainty for this state.
Hint: The uncertainty for this state is *not* infinite. To help you find out what it is, here's a handy series you may not know about:

$$\sum_{j=0}^{\infty} \frac{1}{(2j+1)x^2 - y^2} = \frac{\pi}{4xy} \tan\left(\frac{\pi}{2}\frac{y}{x}\right). \tag{12.4.3}$$

[c] What property of the non-stationary state (12.4.1) causes the uncertainty ΔE after expansion of the well to be finite? Would a finite uncertainty follow for *any* non-stationary state? Why, or why not?

12.5 The Incredible Sloshing Oscillator—Explained!

In § 9.10 we investigated the behavior of a non-stationary state of a particle in a simple harmonic oscillator potential $V(x) = m\omega_0^2 x^2/2$, *i.e.*, a wave packet [see Example 9.6]. We found that such a packet exhibits *periodic behavior*, sloshing back and forth, as illustrated in Fig. 9.20. In the discussion of Example 9.6, I remarked on the eigenfunction expansion of a non-stationary state wave function [Eq. (9.137)], and now I want to illuminate that discussion with what we have learned in this chapter.

Consider, then, an *arbitrary* wave packet $\Psi(x,t)$ of a SHO with natural frequency ω_0.

[a] **Write down** the expansion of this function in eigenfunctions of the SHO Hamiltonian. **Show** in your expansion the explicit time dependence of the projections $c_n(t)$.

[b] The *classical* period of such an oscillator is $T = 2\pi/\omega_0$. Based on the expansion of part [a], **figure out** the wave function at this time, $\Psi(x,T)$. How is this function related to $\Psi(x,0)$? How are the position probability densities at the two times related?

[c] **Discuss** the relationship of the properties of *all observables in this state at* $t = T$—energy, position, momentum, etc.—to these observables in the state at $t = 0$. **Discuss** how the probabilistic and statistical properties of all observables vary with time.

[d] The classical limit of the simple harmonic oscillator is $n \to \infty$, the limit in which the spacing between adjacent (quantized) energies becomes very small compared to the values of those energies (§ 9.8). Use the Correspondence Principle to **explain** how your answers to part [c] go over in this limit smoothly to the classical description of the time-dependence of observables for a macroscopic oscillator, such as a pendulum.

12.6 Inducing Non-Stationary State Behavior

In the extended example of § 12.6 we saw one way to coerce a system into a non-stationary state: make a sudden change in some parameter of the system, thereby changing its potential. In that example, we altered the size of an infinite square well that enclosed a microscopic particle. Here we'll consider an alternate way to create a non-stationary state, working with our other omnipresent example: the simple harmonic oscillator.

Suppose the potential energy of the oscillator is

$$V(x) = \frac{1}{2} m\omega_0^2 (x - x_0)^2. \tag{12.6.1}$$

This potential differs from that of a conventional oscillator (Chap. 9) in that its midpoint is at x_0 rather than at $x = 0$.

Suppose further that the initial state of the particle with this potential is the ground stationary state, with energy $E = \hbar\omega_0/2$.

Now, imagine that a *sudden change* in the system is effected at $t = 0$ so that the center of the oscillator is shifted from x_0 to $x = 0$. We shall suppose the change to be so sudden that, to a good approximation, we can consider it *instantaneous*. The Hamiltonian of the system after the change is the familiar one

$$\hat{\mathcal{H}} = \hat{T} + \frac{1}{2} m\omega_0^2 x^2. \tag{12.6.2}$$

Let's explore the consequences of this sudden change on the state of the system.

[a] **Write down** an explicit expression for the initial wave function before the change.

[b] Suppose we measure the *energy* at some $t > 0$. **Determine** the probability that we will obtain each of the following values:

(1) $\frac{1}{2}\hbar\omega_0$;

(2) $\frac{3}{2}\hbar\omega_0$;

(3) a value greater than $\frac{3}{2}\hbar\omega_0$.

[c] **Write down** the wave function $\Psi(x, t)$ for $t > 0$. (You need not evaluate the general expansion coefficient in your expression, but explain how you would do so if you had to.)

12.7 In Search of State Functions

As noted in § 12.6, in applying quantum mechanics we must often *construct* the wave function that represents a quantum state from bits and pieces of information gleaned from various experiments. Usually it is not possible to infer from experimental data all the details of the state function, but, given sufficient experimental data, we can often determine enough about the function to use it in predictive calculations. We considered an instance of this in Example 12.6. Here is another.

Consider a particle of mass m in an infinite square well with "walls" at $x = -L/2$ and $x = L/2$. In our first of two experiments, we perform an ensemble measurement at some time t of the *energy* of the particle. We find that the members of the ensemble exhibit *only* the values E_1 and E_2, and that the average value of energy is

$$\langle E \rangle (t) = \frac{3\pi^2\hbar^2}{2mL^2}. \tag{12.7.1}$$

[a] **Write down** an explicit expression for the most general wave function $\Psi(x, t)$ that is consistent with this information. (Be careful in how you deal with the *phase* of each term in your function.)

[b] In our second experiment, we measure the *position* of the same particle, ensuring that it is in the same state as *before* we measured the energy. In the position measurement, we learn that at $t = 0$ the particle is more likely to be found in the right half of the box (*i.e.*, at some value of x between 0 and $L/2$) than in the left half. Use this additional information to refine your wave function from part [a], *i.e.*, come up with a wave function that is consistent with the results of both the energy and the position measurements. Is your wave function now *completely* specified, or do you need more experimental data to fully determine it?

Remark: The curious caveat in the first sentence of part [b] is necessary because the energy measurement will change the state of the system, and I don't want you to worry about this effect yet. Were we actually going to measure two observables, such as position and energy, in the laboratory we could proceed as follows: Sub-divide the initial ensemble into two identical sub-ensembles and perform the energy measurement on one sub-ensemble and the position measurement on the other. This ensures that the energy measurement will not alter the position characteristics of the state and vice versa. We'll look at the remarkable effects of measurement on a quantum state in Chap. 13.

12.8 A Tricky Question

Derive the momentum uncertainty $(\Delta p)_n$ for an arbitrary stationary state of a particle of mass m in an infinite square well of width L.

CHAPTER 13

Altered States

The Great Measurement Mystery

"Careful, Dorothy,"
he cautioned her.
"Nothing that actually happens
is or can be preposterous."

—from *Skylark of Valeron*
by E. E. "Doc" Smith

We now confront a major mystery: *what happens to the state of a system when we measure an observable*? So far, we have avoided this question, learning instead how to describe the evolution of a quantum state *in isolation, i.e.*, in the absence of measurement. We know how to follow the development of a state from its initial condition to the time of a measurement and how to predict the probabilities for the various results that might be obtained in the measurement. But until now we haven't faced the question: what is the state of the system *after* the measurement?

The answer, although strictly part of the formalism of quantum mechanics, is not to be found in the Schrödinger equation. As we shall see (§ 13.2), *a measurement of an observable causes the state to change in a way that is inherently uncontrollable and unpredictable, via a mechanism the nature of which we do not know.* That such a change takes place is not at issue; as the distinguished mathematician John von Neumann (1903–1957) showed in 1932, this phenomenon is built into quantum theory.[1] What is at issue is why and how the state function changes. This conundrum is called **the problem of measurement.** In a recent review of this problem,[2] John G. Cramer remarked that the problem of measurement "remains the most puzzling and counterintuitive aspect of the interpretation of quantum mechanics."

Why, in an introduction to quantum mechanics, should we worry about the problem of measurement? For one thing, *physics is the science of measurement.* Our understanding of the physical universe in its macroscopic and microscopic realms ultimately rests on experiments, on measurements. Indeed, physicists believe in quantum theory because it continues to be verified in a huge number of experiments on a variety of systems under a range of physical conditions. But to understand the behavior of the microworld, as manifested in these experiments, we must understand the effect on an ensemble of human observation. This effect is the subject of this chapter.

Were we studying classical physics, we could ignore the experimental dimension of our subject (although it would be unwise to do so), because in principle, an experiment on a macroscopic system does not alter its state; *i.e.*, a measurement *can* leave unchanged all physical properties of a classical system. But that's not true for a microscopic system. Moreover, investigating a system whose states are represented by wave functions is quite different from studying one whose states are described by trajectories.

The reason for this difference is that, unlike a trajectory, a state function has no real physical existence. Rather, this function is the mathematical embodiment of accessible knowledge about the state it represents. Only in the laboratory can we seek such information about the wave function as nature allows. (In Example 12.6, for instance, we used data from an energy measurement to deduce information about the state function of an ensemble of CO molecules.) But when we perform measurements,

[1] Von Neumann's trail-blazing work *Mathematical Foundations of Quantum Mechanics* (Princeton, N. J.: Princeton University Press, 1955) does not make for easy reading, but I nonetheless recommend its early chapters to readers who are determined to understand in depth the problem of measurement.

[2] *Rev. Mod. Phys.*, **58**, 647 (1986).

the state function changes as a consequence.[3] So to plan and execute experiments and to apprehend their results, we must understand their effect on the state of the system on which they are performed.

If two or more observables are measured, the problem of measurement becomes a problem of "knowability" (§ 13.3). If the observables are incompatible, then their uncertainties satisfy an inequality—the Generalized Uncertainty Principle (§ 11.4)—that may prohibit our knowing their precise, simultaneous values. We shall examine the measurement of two incompatible observables in § 13.4 and, as a happy by-product of our inquiry into the effects of measurement on a state, will gain insight into how uncertainty works, a rare glimpse into the innermost secrets of the microcosm.

13.1 THE ROLE OF MEASUREMENT IN QUANTUM PHYSICS

> When you cannot measure it,
> when you cannot express it in numbers,
> your knowledge of it
> is of a meagre and unsatisfactory kind.
>
> —Lord Kelvin (1824–1907)

Anyone who has worked in a laboratory knows the *pragmatic* definition of a measurement: *a measurement is the experimental determination of information about the physical properties of a system.* In this chapter I want to focus on the *conceptual* definition of a measurement, and to this end will consider only "idealized measurements"—idealized in a sense that will become clear as we proceed.

Our first step towards idealizing the process of measurement is to assume that a measurement can be performed *instantaneously.*[4] Accordingly, we will speak of a "measurement time." I'll denote the measurement time for an unspecified observable (or observables) by t_m and for a specific observable Q by t_Q.

Limitations on Measurement

Both nature and human fallibility limit the accuracy of experiments. The latter source of uncertainty is the most familiar. Experimental error and the inadequacies of man-made apparatus limit the precision of measured data. For example, in the energy-loss spectrum for electron–F_2 scattering in Fig. 12.3a each peak has a width which results from the inability of the source and detector to determine *precisely* the energy of a microscopic particle. Similar *experimental uncertainties* plague physicists, regardless of the size of the systems they study.

More insidious (though less a problem in practice) are *quantum uncertainties*, which arise from the intrinsic "unknowability" of precise information about microscopic systems, as codified in the Generalized Uncertainty Principle of Chap. 11. Uncertainty may arise in a measurement of a single observable if the state being investigated happens not

[3]Like other statements in these introductory remarks, this is true in general. If, however, the state before measurement happens to be an eigenstate of the observable being measured, then the state function is unchanged by the measurement. (See § 13.2.)

[4]Like practically every aspect of measurement theory, defining the time at which an experiment is performed is a subtle business. For a provocative discussion of this question see "When is a Quantum Measurement" by Asher Peres, *Amer. Jour. Phys.*, **54**, 688 (1986).

to be an eigenstate of that observable.[5] Or it may arise (in a more complicated way) in attempts to measure simultaneously two (or more) observables. We know from § 11.4 that if observables Q and R are incompatible, then for any state their uncertainties obey the inequality [the Generalized Uncertainty Principle (11.55)]

$$\Delta Q \, \Delta R \geq \frac{1}{2} \left| \left\langle i[\hat{Q}, \hat{R}] \right\rangle \right| . \tag{13.1}$$

Whether or not this principle actually limits the "knowability" of precise information about Q and R depends on the state—more precisely, on whether the expectation value on the right-hand side of (13.1) is zero (*c.f.*, Table 11.4). If not, then we cannot simultaneously measure Q and R to infinite precision.

In the introduction to this chapter, I noted the role of measurement as an *exploratory tool* for probing the nature of a state function, which is itself inaccessible. Measurement also plays a part in the *predictive* function of physics. A primary objective of an experiment is to determine sufficient information about the state of a system at one time to enable prediction of the system's subsequent behavior. That is, we'd like to measure enough observables at t_m to be able to predict the evolution of the system for $t > t_m$.

But in quantum mechanics, such predictions are impossible unless we know what the measurement does to the state function. For, unlike classical physicists, quantum physicists must live with the unavoidable **observer-observed interaction** I first mentioned in Chap. 1:[6]

> ***Rule***
>
> It is impossible to perform a measurement on a microscopic system that does not disturb the system in a significant, unpredictable, and uncontrollable way.

Concerning this remarkable feature of quantum mechanics, John Von Neumann said:[7]

> ... in the atom we are at the boundary of the physical world, where each measurement is an interference of the same order of magnitude as the object measured and therefore affects it basically.

Now there's a sobering thought.

The Meaning of "Measurement" (A Review)

In Chap. 3 (§ 3.4) we thoroughly examined the conceptual definition of a **measurement** on a microscopic system. This topic is so crucial to our present concerns that I want to take a moment to review it. Central to the interpretation of quantum mechanics is

[5]Recall (Chap. 10) that a system in an eigenstate of Q *has a value* of this observable—namely, the eigenvalue obtained when $\hat{Q}$ acts on the state function, as $\hat{Q}\Psi(x,t) = q\Psi(x,t)$. So for such a state $\Delta Q = 0$, and the only limits to precision are experimental uncertainties.

[6]Among other disquieting implications, this rule prohibits us from inferring from measurements at t_m the nature of the state for $t < t_m$. In general, the most we can hope to determine uniquely is the state *at the instant before* we made the measurement. See § 7.1 of *Introduction to Quantum Mechanics* by R. H. Dicke and D. Wittke (Reading, Mass.: Addison-Wesley, 1960).

[7]Quoted in *The Philosophy of Quantum Mechanics* by M. Jammer (New York: Wiley, 1974), p. 266. This fascinating volume contains lengthy historical and philosophical discussions concerning the problem of measurement.

the concept of an **ensemble of microscopic systems**. Recall (Fig. 3.5) that *an ensemble is a collection of a huge number of identical, non-interacting systems in the same quantum state.* Individual systems in the ensemble are called **members**. And according to Postulate I, any state of the ensemble can be represented by a state function Ψ.[8]

When we speak of "a measurement at t_Q of Q" we mean simultaneous measurements of the observable Q on all members of the ensemble.[9] I will use the symbol $\mathsf{M}_Q[t_Q]$ as shorthand for such a measurement.

And Then a Miracle Occurs

It is surprisingly easy to see that an experiment must change the state of the system. Imagine a measurement of an observable Q. Suppose the experimental set-up is such that the state before measurement is not an eigenstate of Q—*i.e.*, that the uncertainty ΔQ is positive. In terms of an ensemble, this condition means that *at the instant before* $\mathsf{M}_Q[t_Q]$ *the members of the ensemble do not have a value of* Q; rather, their state allows for several *possible* values of Q, any one of which may appear in the measurement. These values, we now know, are among the eigenvalues of $\hat{Q}$.

Now, think about what happens to the ensemble when we measure Q: each member exhibits one or another of the eigenvalues of $\hat{Q}$. *After* $\mathsf{M}_Q[t_Q]$, *each member has a value of* Q*—the value it exhibited in the measurement—and so is in an eigenstate of* Q.[10]

So after the measurement each member is in a state in which Q is sharp. But not all members are in the same state: some exhibited one value of Q in $\mathsf{M}_Q[t_Q]$, others exhibited a different value. Clearly, *the measurement changed the nature of the ensemble*—in fact, what was an ensemble is no longer an ensemble, for the various members are in different quantum states. The measurement changed the state! It is this marvel that we will now look at in detail.

13.2 THE EFFECT OF A MEASUREMENT ON A STATE FUNCTION

I don't know,
we may be into
some kind of quantum thing here,
or some exotic relativistic physics,
in which consciousness snapped me into
some new sharp state.
My physics is lousy.

—Dr. Edward Jessup
in *Altered States* (A Novel)
by Paddy Chayefsky

[8]In this book we consider only ensembles that *can* be represented by a state function; such ensembles are said to be in *pure states*. Someday you may encounter a *mixed state* of an ensemble, a state in which the available information is insufficient to specify a state function. Such an ensemble must be described by more arcane mathematical beasts called *density matrices*.

[9]Some physicists prefer to think of a measurement as a huge number of successive measurements on a single system. This is OK provided one remembers that *the state of the system prior to each measurement must be the same*. I find this definition confusing and so won't use it in this book.

[10]We are in treacherous semantic waters. Even the word "after" has to be carefully defined. By "after" I mean *an infinitesimal time after—i.e.*, at $t_Q + \delta t$. The reason for this qualification is that once the measurement is over, the members of the ensemble continue to evolve according to the time-dependent Schrödinger Equation. This subsequent time development may *further* change the state, in ways we have discussed in Chaps. 6 and 12. But at the moment we are not concerned with this evolution. (See § 13.2)

At the end of the previous section, we deduced that measuring an observable Q on an ensemble changes the state of the members so that afterward, each is in an eigenstate of Q. *Which* eigenstate a member winds up in depends on which eigenvalue of $\hat{Q}$ it exhibited in the measurement. In fact, what is left after the measurement is not a single ensemble, but rather several **sub-ensembles**. *Each sub-ensemble consists of all members of the original ensemble that exhibited the same eigenvalue of $\hat{Q}$ in* $\mathbf{M}_Q[t_Q]$. So each sub-ensemble is a collection of (identical, non-interacting) member systems each of which is in a particular eigenstate. That is, each sub-ensemble is itself an ensemble. Here is a handy summary of this argument:

MEASUREMENT OF Q AT TIME t_Q:

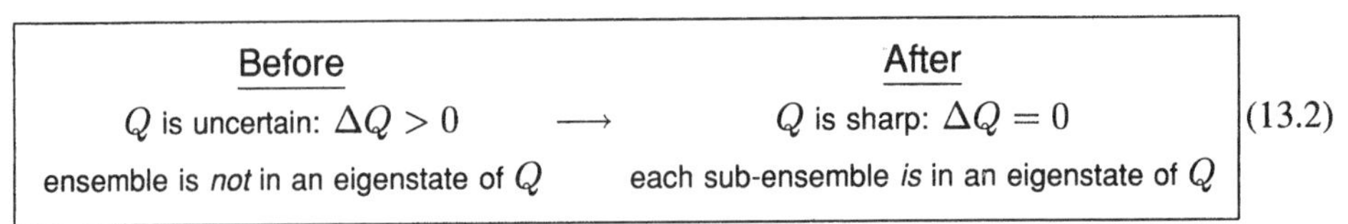

(13.2)

Clearly, the measurement wreaks a dramatic change on the ensemble on which it is performed.

In this section and the next I want to illustrate the effects of measurement by reexamining a familiar experiment and a couple of familiar examples.

Another Look at the Single-Slit Experiment

In Chap. 2 we considered the determination of *the position of electrons* in the double-slit experiment (§ 2.5). In this classic demonstration, the electrons seem to behave like waves in their propagation through space and like particles in their conduct at the detector.

The dual nature of sub-atomic particles is also evident in the **single-slit experiment** sketched in Fig. 13.1. The source emits a beam of (non-interacting, nearly mono-energetic) electrons; this beam is our ensemble. The electrons in the beam propagate through the apparatus until they reach the detector, where they form spots. Of course, we now know that the state of the electrons in the beam is represented by a wave function, the squared modulus of which is their position probability density. In this figure, I have sketched this density at various times during the journey from source to detector.

The question I want to pose now is: *what is the significance of the appearance of spots on the detector*? That is, what do these events imply about the state of the electrons after measurement? To find out, let's consider the wave function as the electrons travel to the detector.

Drawing on our knowledge of wave packets (Chaps. 4 and 6), we can explain the behavior of the electrons. They emerge from the source in a state represented by a *free-particle wave packet*—a state in which the position uncertainty Δx is positive. As time passes, this packet propagates, according to the time-dependent Schrödinger Equation with $V = 0$, to the diaphragm. Passage through the slit changes the state of the electrons, which emerge in a state represented by a different wave packet (see Fig. 13.1). But *until they arrive at the detector, the electrons do not have a position* (*i.e.*, after traversing the slit, the position uncertainty Δx is still positive). This state is characterized by several *possible positions*, as is evident in the probability densities in Fig. 13.1.

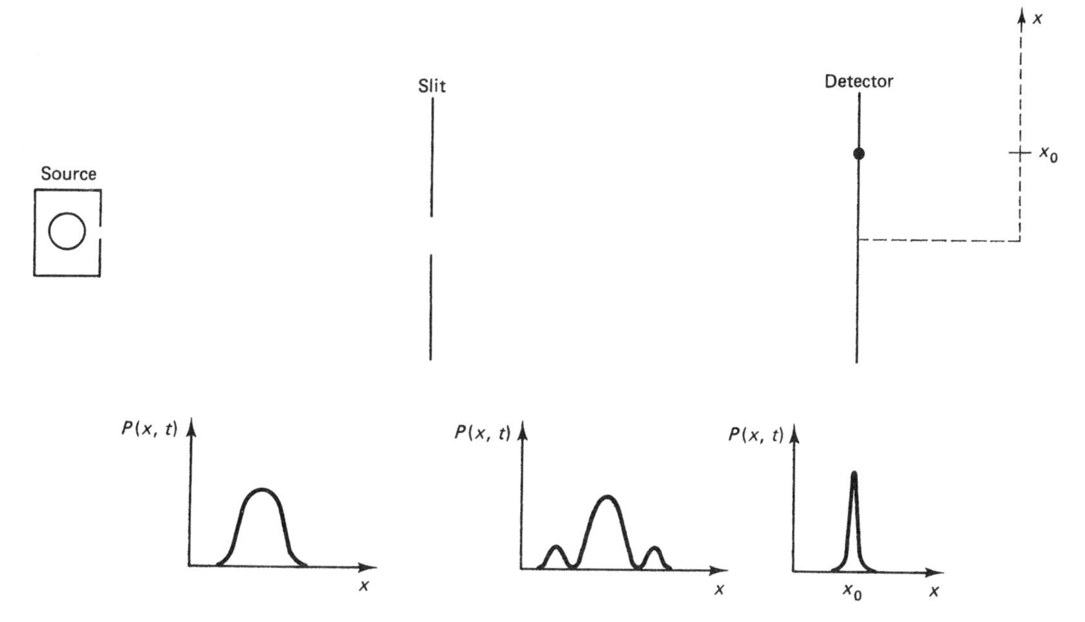

Figure 13.1 The single-slit diffraction experiment. The position probability densities $P(x, t) = |\Psi(x, t)|^2$ shown below the apparatus are *crude representations only*—that is, they are supposed to suggest changes that occur in $P(x, t)$ as the electrons saunter from the source to the detector, not to represent accurately this density function. Note that the last probability density, which corresponds to the state that represents (the sub-ensemble of) electrons that exhibited a spot at x_0, is *not* a consequence of evolution à la the time-dependent Schrödinger Equation.

At some time t_x, an electron arrives at the detector, a phosphor that measures position. It is here that we see evidence of particle-like behavior, for *on the detector, spots form.*[11] This means that at t_x the state function changed, for at this time each electron suddenly "acquired" a definite position—the value x_0 at which the spot appears (see Fig. 13.1). *Evidently the measurement "actualized" one of the multitude of possible positions that were implicit in the wave function before measurement.* That is, $\mathbf{M}_x[t_x]$ altered the state of the ensemble, transforming the wave function from one for which $\Delta x > 0$ to ones for which $\Delta x = 0$. (Position is sharp only at the instant after measurement. Afterward, the states of the electrons evolve, according to the TDSE, into ones for which Δx is positive.)

In this discussion we have identified the two radically different ways that a quantum state can change:

1. In the absence of measurement (no interaction with an observer), a state undergoes "natural Schrödinger evolution," changing with time according to the TDSE.
2. When an observable is measured, the state undergoes an "induced mutation," a change that is not described by the TDSE.

[11] What we actually see in the laboratory are photons emitted when molecules in the phosphor decay out of the state into which they were excited by the electrons. As discussed in Chap. 2, the distribution of these photons and the physics of the interaction of the electron with the molecules of the phosphor are analogous to the classical behavior of particles, not waves.

As I noted in § 13.1, to understand these "induced mutations," it is crucial that we fully understand the change in *the ensemble* to which the electrons belong. To clarify this aspect of measurement theory, let's return to an example from § 12.4, one that concerns an *energy measurement.*

Example 13.1. Converting a Non-Stationary State into a Stationary State

In Example 12.5 we analyzed a non-stationary state of a particle in a one-dimensional infinite square well of width L. There we sought to predict the outcome of an energy measurement at $t = 0$ for a state that *prior to the measurement* was represented by the wave function (with $x = 0$ at the left wall)

$$\Psi(x, 0) = \sqrt{\frac{30}{L^5}}\, x(x - L). \tag{13.3}$$

To this end, we "translated" this wave function into the *energy representation*, obtaining the energy probability amplitudes[12]

$$c_n^{(E)}(0) = \begin{cases} 0 & n = 2, 4, 6, \ldots \\ -\dfrac{1}{n^3}\dfrac{8\sqrt{15}}{\pi^3} & n = 1, 3, 5, \ldots \end{cases}. \tag{13.4}$$

The square modulus of $c_n^{(E)}(0)$ is the probability that a member of an ensemble of systems in this state will exhibit the eigenvalue $E_n = n^2\pi^2\hbar^2/(2mL^2)$ in the measurement. We evaluated some of these probabilities in Eq. (12.48)—finding for example that $P(E_1) = 0.9986$ and $P(E_3) = 0.0014$.

To understand the implications of these findings, consider an ensemble of 10,000 particle-in-a-box systems initially in the state (13.3). In an average measurement, 9986 of the members would yield the ground-state energy E_1, while 14 would yield the energy of the second excited state, E_3. [We might obtain one of the higher excited-state energies, because any member *could* exhibit any E_n (for odd n). But $P(E_n)$ for $n > 3$ is so small that a larger ensemble would be required for even one member to have a fighting chance of showing one of these values.]

But the measurement also "forces" each member into the energy eigenstate corresponding to the value it exhibited. So *the number* $|c_n(0)|^2$ *is also the probability that immediately after* $\mathbf{M}_E[t_E = 0]$ *the system will be in the* n^{th} *stationary state.*

We could verify this prediction by measuring the energy a second time, as in Fig. 13.2. Suppose that in this additional experiment, we measure the energy *of the sub-ensemble of systems that exhibited* E_1 *in the first energy measurement.* This sub-ensemble is in an energy eigenstate, so in the second measurement all its member systems would exhibit the value E_1.

It's easy to write down the wave function that represents the state of each sub-ensemble immediately after the first energy measurement. This measurement altered the wave function of those member systems that exhibited E_1 from $\Psi(x, 0)$ of Eq. (13.3) to that of the ground stationary state: $\Psi_1(x, 0) = \psi_1(x)$. Similarly, it changed the wave function of those systems that exhibited E_3 from $\Psi(x, 0)$ to $\Psi_3(x, 0) = \psi_3(x)$.

The nature of the change of state in this example is clear: the initial wave function $\Psi(x, 0)$ represented a state with an uncertain energy, *i.e.*, one in which the particle has no energy. The (first) energy measurement "actualized" one of the *possible energies* inherent in this wave function, thereby transforming the non-stationary state into one of two stationary states.

[12] In this chapter I will label projections of a state function with its expansion observable whenever doing so might avoid confusion.

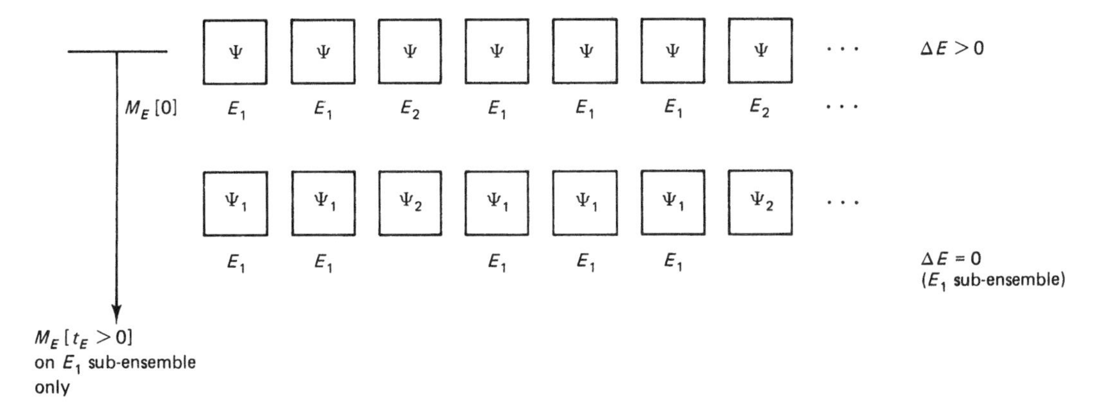

Figure 13.2 Two energy measurements on an ensemble of infinite square wells whose particles are in the state (13.3) prior to the first measurement. After the first measurement, we select those systems that gave E_1 and re-measure the energy of this sub-ensemble. These systems exhibit *only* the value E_1, for they are in a state in which the energy is sharp.

One important use of such measurements is to "prepare" a system in a particular stationary state—say, for subsequent measurements of other properties. To prepare a system in this way, simply take an ensemble of systems, measure the energy, and create a sub-ensemble consisting of all members that exhibited the energy you want. Such preparations yield the initial conditions we have seen in previous examples in this book.

Question 13–1

Suppose the second energy measurement on the E_1 sub-ensemble were carried out at, say, $t = 10\,\text{sec}$. Would the results differ from those obtained in a second energy measurement immediately after the first measurement? Why or why not?

Collapse of the State Function

Physicists have a quaint name for the remarkable transformation of a wave function that occurs when a measurement is performed: **the collapse of the state function** (also known as **reduction of the wave packet**). This name is supposed to conjure up an image of the measurement causing the state function to "collapse" onto one of the eigenfunctions of the corresponding operator. A convenient way to represent collapse of the wave function is via a "branch diagram" like the one in Fig. 13.3.

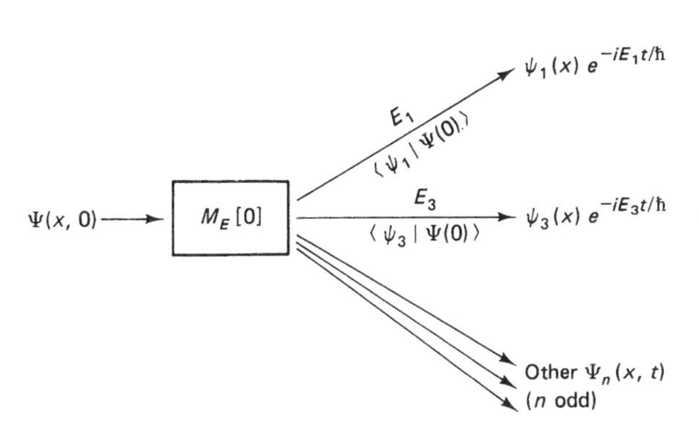

Figure 13.3 A branch diagram for collapse of the state function, here illustrated for an energy measurement on an ensemble of infinite square wells initially in the state (13.3). Each arrow to the right of the box that represents the measurement $\mathbf{M}_E[t_E]$ corresponds to one of the *possibilities* implicit in the state function prior to measurement. Each arrow is therefore labelled with one of the eigenvalues and with the projection of $\Psi(x, 0)$, the squared modulus of which is the *probability* that the measurement will cause the state of the system to "collapse" into this possible eigenstate.

13.3 THE EVOLUTION OF A STATE AFTER COLLAPSE

What happens after the collapse of a state function into an eigenstate? Well, if no further measurements are made, "natural Schrödinger evolution" takes over, and the state evolves according to the time-dependent Schrödinger Equation. In general, this development may be quite complicated and should be tackled using the machinery of Chap. 12. *But if the observable measured was energy, the subsequent evolution of the state is simple, because the resulting state is an energy eigenstate.* The time development of the wave function for an energy eigenstate was the subject of Chap. 7.

To clarify the evolution of an eigenstate after collapse, we might think about the energy measurement as having "re-initialized" the state of each member in the ensemble. For example, the members of an ensemble that exhibit the eigenvalue E_1 in an energy measurement at $t_E = 0$ (as in Example 13.2) have, *immediately after the measurement*, the new initial wave function

$$\Psi_1(x, 0) = \psi_1(x). \tag{13.5}$$

So at subsequent times their wave function is simply

$$\Psi(x, t) = \psi_1(x) e^{-iE_1 t/\hbar} \qquad t > t_E = 0. \tag{13.6}$$

Be careful. If the measured observable is not energy, then the wave function after measurement does not necessarily have the simple time dependence (13.6). After perusing another example, we'll reckon with measurements of arbitrary observables.

Example 13.2. Molecular Measurements

In Example 12.6 we deduced the state function of an ensemble of CO molecules from data obtained in an energy measurement at $t = 0$. In the measurement, 30% of the molecules exhibited the ground-state energy E_0, and 70% exhibited the first excited state energy E_1. From these facts, we reasoned that the initial wave function was a linear combination of the first two stationary states, as in Eq. (12.56). In that example, we chose (for simplicity) to model the vibrating CO molecule by a simple harmonic oscillator. So we wrote the initial wave function in terms of the eigenfunctions $\psi_n(x)$ of the SHO Hamiltonian, as

$$\Psi(x, 0) = \sqrt{\frac{3}{10}}\, e^{i\delta_0} \psi_0(x) + \sqrt{\frac{7}{10}}\, e^{i\delta_1} \psi_1(x). \tag{13.7a}$$

Equation (13.7*a*) represents the state of the molecules *before* the measurement, when their energy was not sharp. But *after* this measurement, the energy *is* sharp. So the measurement transformed the original ensemble into two distinct sub-ensembles, one consisting of particles with (sharp) energy E_0, the other of particles with (sharp) energy E_1.

Correspondingly, the wave functions that represent the state of the molecules in each sub-ensemble *after the energy measurement* are one of two stationary-state wave functions, *i.e.*,

$$\Psi_0(x, t) = \psi_0(x)\, e^{-iE_0 t/\hbar} \tag{13.7b}$$

$$\Psi_1(x, t) = \psi_1(x)\, e^{-iE_1 t/\hbar}. \tag{13.7c}$$

We're familiar (from Chap. 9) with the detailed nature of the Hamiltonian eigenfunctions and energies in these functions.

Question 13–2

Describe the results of a second energy measurement on each sub-ensemble.

Measurement of Observables Other Than Energy

We can easily extend the reasoning of this section to measurements of an arbitrary observable. The (partial) branch diagram in Fig. 13.4 shows what happens if we measure an observable Q represented by operator $\hat{Q}$ with eigenvalue equation $\hat{Q}\chi_n^{(Q)} = q_n\chi_n^{(Q)}$. Here's the rule:[13]

> ***Rule***
>
> A measurement of Q will collapse the state into one of the eigenstates of Q. A member that exhibits the (non-degenerate) eigenvalue q_n in $\mathsf{M}_Q[t_Q]$ will, immediately after the measurement, be in a state represented by the corresponding eigenfunction $\chi_n^{(Q)}$. The probability that a member will be in this eigenstate is the squared modulus of the projection $c_Q^{(n)}(t)$ of Ψ on $\chi_n^{(Q)}$.

The evolution of the state after measurement of an arbitrary observable differs in one very important respect from that of the state that results from a measurement of the energy. *The eigenfunction of $\hat{Q}$ into which Ψ collapses due to $\mathsf{M}_Q[t_Q]$ represents the state only in the instant after the measurement.* As time passes, the state will evolve according to the TDSE (unless additional measurements are performed—a complication we'll confront in the next section). Unlike the time development of the stationary state that results from an energy measurement, the evolution of this state, which is almost always non-stationary, could be very complicated.

Question 13–3

Can you think of any observable (or class of observables) other than $\hat{\mathcal{H}}$ whose measurement yields sub-ensembles whose subsequent evolution is described by a simple separable wave function? (Hint: See § 11.5.)

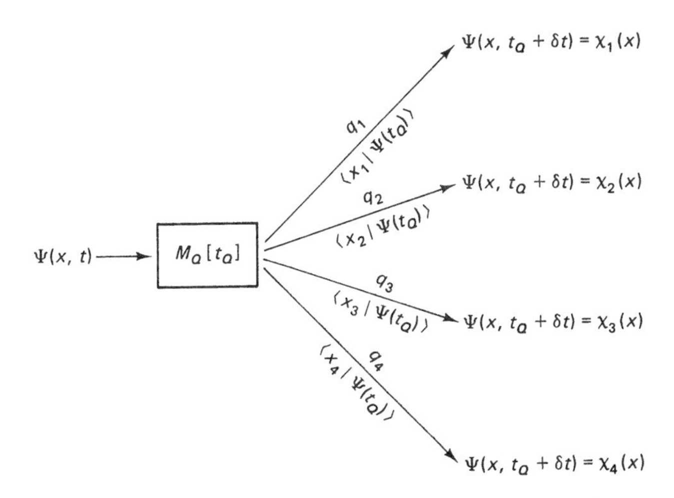

Figure 13.4 A branch diagram for the collapse of the state function due to the measurement of an arbitrary observable Q.

[13]As usual, I'll consider only operators with *discrete spectra*. This rule, and others we shall deduce, applies with the usual modifications to operators whose spectra are continuous or mixed.

Let's review our two methods for obtaining the wave function for $t > t_Q$. The direct way is to solve the TDSE

$$\hat{\mathcal{H}}\Psi(x,t) = i\hbar\frac{\partial}{\partial t}\Psi(x,t) \tag{13.8}$$

subject to the (new) initial condition $\Psi(x,t_Q) = \chi_n^{(Q)}(x)$. The easier (indirect) way is to expand $\Psi(x,t)$ in the complete set of Hamiltonian eigenfunctions, as

$$\Psi(x,t_Q) = \sum_n c_n^{(E)}(t_Q)\psi_n(x). \tag{13.9a}$$

Once we have determined the projections

$$c_n^{(E)}(t_Q) = \langle \psi_n \mid \Psi(t_Q)\rangle, \tag{13.9b}$$

we can easily write down the wave function that evolves from this initial function as

$$\Psi(x,t) = \sum_n c_n^{(E)}(t_Q)\psi_n(x)e^{-iE_n(t-t_Q)/\hbar}. \tag{13.9c}$$

Equation (13.9*c*) makes clear the point I noted above: in general the system will not remain in the eigenstate of Q into which it was thrown by $\mathbf{M}_Q[t_Q]$. Its "natural Schrödinger evolution" causes the state to evolve into the one represented by this wave function—a state in which Q is not sharp.

Question 13–4

Explain the funny-looking time dependence in the exponential in (13.9*c*).

Question 13–5

Consider the special case that the observable measured is a constant of the motion, *i.e.*, that $[\hat{Q},\hat{\mathcal{H}}] = 0$, and that its spectrum is purely discrete (and non-degenerate). Argue (or prove) that *in this special case*, the complicated time dependence of Eq. (13.9*c*) reduces to the familiar form

$$\Psi(x,t) = \psi_n(x)e^{-iE_n(t-t_Q)/\hbar}. \qquad t > t_Q \tag{13.10}$$

[*Hint*: Recall (§ 12.9: Proclamation 12.1) that all probabilistic and statistical quantities that refer to a constant of the motion of a system are independent of time.]

*13.4 MULTIPLE MEASUREMENTS OF COMPATIBLE OBSERVABLES I

Things then did not delay
in turning curious . . .
there were revelations
in progress all around her.

—*The Crying of Lot 49*
by Thomas Pynchon

We have seen that a measurement causes an uncontrollable, unavoidable *change* in a quantum state. This change comes about because the measurement necessarily involves an interaction between the observer and the system. This interaction causes the state to "collapse" into an eigenstate of the measured observable, so that afterward, this observable is sharp. According to the Born Interpretation of the state function, the probability of collapse into a *particular* eigenstate is the squared modulus of the projection of the state function (before measurement) on the corresponding eigenfunction of the operator that represents the measured observable.

This summary suggests an interesting question: what happens to the state if we measure *two observables simultaneously*? Each measurement can transform the state into an eigenstate of the measured observable; so we might guess that after both measurements the system would be in an eigenstate of both observables. But such an outcome may be impossible: *if the observables are incompatible, then a simultaneous eigenstate is prohibited* (§ 11.1).[14] But if so, then what *does* the simultaneous measurement do to the state?

In the next three sections we'll discover the answer. We begin, after a couple of important definitions and a quick review of compatibility, with a look at the simplest case: compatible observables. Then we take up the more arresting case of incompatible observables. And there we'll learn how uncertainty principles work.

A Simultaneous Measurement Defined

Our starting point is the eigenvalue equations for two observables, Q and R,

$$\hat{Q}\chi_n^{(Q)} = q_n \chi_n^{(Q)} \tag{13.11a}$$

$$\hat{R}\chi_m^{(R)} = r_m \chi_m^{(R)}. \tag{13.11b}$$

For the moment I want to leave unspecified the commutator of these operators; *i.e.*, $[\hat{Q}, \hat{R}]$ may or may not be zero. My notation for the eigenvalues, q_n and r_m, assumes, as usual, that the spectra of $\hat{Q}$ and $\hat{R}$ are purely discrete. (For instance, the examples in this section deal with measurement of the parity and total energy of a simple harmonic oscillator; the spectra for this $\hat{\mathcal{H}}$ and for $\hat{\Pi}$ are purely discrete.) I'll denote the simultaneous eigenfunctions of two compatible observables Q and R by the obvious extension of this notation, $\chi_{n,m}^{(QR)}$.

Now a little review. In Chap. 11 we learned that if two observables are *compatible* (*i.e.*, if $[\hat{Q}, \hat{R}] = 0$), then their operators define a complete set of simultaneous eigenfunctions. In this case the Generalized Uncertainty Principle (13.1) allows us (in principle) to measure both simultaneously to infinite precision. If, on the other hand, the observables are *incompatible* (*i.e.*, if $[\hat{Q}, \hat{R}] \neq 0$), then such a set does not exist, and the precision of a simultaneous measurement of Q and R may be constrained by their uncertainty relation.

Now, by a **simultaneous measurement** of Q and R we mean simply two measurements $\mathsf{M}_Q[t_Q]$ and $\mathsf{M}_R[t_R]$ that are carried out *one right after the other*. That is, the time interval between the measurements, $t_R - t_Q$, is arbitrarily small.[15]

[14] To keep the discussion in this section as simple as possible, I want to exclude the possibility of a *single* simultaneous eigenstate of Q and R. Although rare, such exceptions do occur—for example, the zero-angular-momentum state of a particle in three dimensions, which I noted in § 11.1. Such states merely muddy already opaque waters and are not directly relevant to our immediate concerns.

[15] Actually, the measurements we'll consider differ from real measurements in that they induce the least possible change in the state of the system. I'll tell you more about these "ideal" measurements later in this section.

Simultaneous Measurement on an Eigenstate

Suppose the measured observables are compatible, *i.e.*, that $\left[\hat{Q}, \hat{R}\right] = 0$. In principle we can measure simultaneously and precisely the values of two compatible observables for any state. But the particular outcome of such a measurement depends on whether or not the state is a simultaneous eigenstate of Q and R.

For example, suppose that prior to either measurement the ensemble is the simultaneous eigenstate $\Psi = \chi_{n,m}^{(QR)}$. Then $\mathsf{M}_Q[t_Q]$ and $\mathsf{M}_R[t_R]$ leave the state unchanged. To understand why, look at the measurement diagram for this case, Fig. 13.5. In the initial state Q is sharp, with value q_n. So in the first measurement, $\mathsf{M}_Q[t_Q]$, all members exhibit the same value of Q, the eigenvalue q_n. This measurement need not collapse the state into an eigenstate of Q. In fact, it need not change the state function at all.

The same reasoning applies to the second measurement, $\mathsf{M}_R[t_R]$. In the state before (and after) $\mathsf{M}_Q[t_Q]$, R is sharp, with value r_m. So in $\mathsf{M}_R[t_R]$, each member exhibits the same value, r_m, and this measurement, too, need not change the state. After all is said and done, the state function is the same as before either measurements, $\chi_{n,m}^{(QR)}$. To summarize our conclusions:

> ***Rule***
>
> A simultaneous measurement of two compatible observables on an eigenstate of both observables does not change the state of the system.

Example 13.3. Parity and Energy Measurements on a Stationary State

In the next several examples, we'll perform (in the laboratory of our imaginations) simultaneous measurement of *parity* and *energy* on an ensemble of simple harmonic oscillators in various quantum states. We begin with *the ground state*—which is an eigenstate of both observables, with eigenvalues $\Pi = +1$ and $E = E_0$. The measurement diagram for this experiment appears in Fig. 13.6. Compare this diagram carefully to the general one in Fig. 13.5.

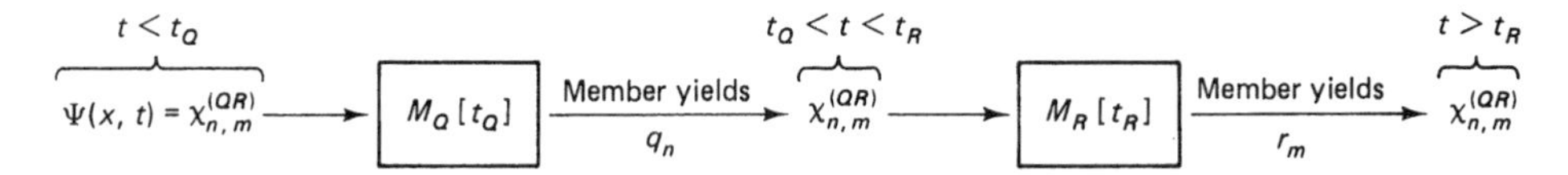

Figure 13.5 A measurement diagram for simultaneous measurement of two *compatible* observables Q and R on a simultaneous eigenstate of these observables. In the state prior to measurement, both observables are sharp, so the measurement does not change the state or the wave function.

On Ideal Measurements

You may be wondering: *could* the parity measurement $\mathsf{M}_\Pi[t_\Pi]$ in Fig. 13.6 change properties related to the energy? This is a thoughtful question: the laws of quantum physics guarantee only that the parity measurement *need not* change the energy properties of the system. That is different from saying that the measurement *does not* change these properties.

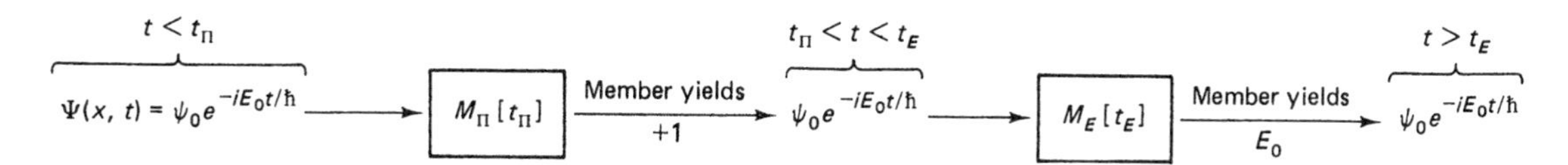

Figure 13.6 A measurement diagram for simultaneous measurement of the parity and energy of a simple harmonic oscillator in the ground (stationary) state. Because this state is a simultaneous eigenstate of both observables, nothing remarkable happens to the wave function. The measurement simply confirms the (sharp) values $\Pi = +1$ (even parity) and $E_0 = \hbar\omega/2$.

In this subtle issue lies a potential complication of the theory of measurement, for a *real* measurement of one of two compatible observables *could* change the value of the other observable. The change could be extensive; for example, if the other observable were sharp before measurement, it could be rendered uncertain after. So a real parity measurement *might* leave the system in any state with definite parity. For example, the state of a simple harmonic oscillator after a measurement $\mathbf{M}_\Pi[t_\Pi]$ that yielded $+1$ could be *any even-parity* eigenstate: the wave function could be Ψ_n for any even n or a linear combination of these even-n stationary states.[16] So a real measurement of the parity of the simple harmonic oscillator could change the energy of the system, even though energy and parity are compatible. But such an outcome is not possible in an *ideal* measurement.

An **ideal measurement** *is one that produces the minimum allowable disturbance in the physical properties of the system.* This definition is a bit tricky, so read it again. Think about it a minute. *An ideal measurement of an observable won't change properties related to any observable that is compatible with the measured one unless such a change is mandated by the laws of quantum physics.*[17] Thus, if we perform an ideal measurement of Q on an ensemble in a state in which an observable R (that is *compatible* with Q) is sharp, then after $\mathbf{M}_Q[t_Q]$, the observable R will still be sharp, its value unchanged by the measurement.

Simultaneous (Ideal) Measurements on an Arbitrary State

The effect of measuring compatible observables on a simultaneous eigenstate is none too thrilling; nothing happens to the state. More interesting things happen when we *measure two compatible observables on a system that is not in an eigenstate of one, the other, or both observables.* To be specific, let's consider the "worst case scenario": a state in which before the measurement *both* Q and R are uncertain.

In a simultaneous measurement on an ensemble initially in a state Ψ with $\Delta Q > 0$ and $\Delta R > 0$, different members may exhibit different eigenvalues of $\hat{Q}$ and $\hat{R}$. For either $\mathbf{M}_Q[t_Q]$ or $\mathbf{M}_R[t_R]$, we can calculate *the probability that a member will exhibit a particular eigenvalue* by squaring the modulus of the projection of Ψ on the corresponding eigenfunction immediately before the measurement. This much we know

[16]Of course, an experiment in which the parity was found to be even could not throw the system into an odd-parity state, for that change would contradict the result of the measurement.

[17]Such a measurement will, however, transform the state if quantum mechanics says it must. If, for example, the state prior to measurement is *not* an eigenstate of the measured observable, then the measurement—ideal or otherwise—must transform the state function into an eigenfunction of that observable. The distinction between an ideal and a real measurements will come more sharply into focus when we consider measurements of incompatible observables (§ 13.6), for a measurement of one such observable *must* change properties of the system related to the other observable.

from Chaps. 11 and 12. From the present chapter we know that each measurement will collapse the state into an eigenstate of the measured observable. The aforementioned probabilities tell us the likelihood of finding a member in one or another eigenstate after the measurement. To see this phenomenon in action, follow Fig. 13.7 as we analyze a successive measurement of compatible observables Q and R.

$0 < t < t_Q$ $\quad$ $t_Q < t < t_R$ $\quad$ $t > t_R$

$\Psi(x,t) \longrightarrow$ $M_Q[t_Q]$ $\xrightarrow{\text{Member yields } q_n}$ $\chi_n^{(Q)} \longrightarrow$ $M_R[t_R]$ $\xrightarrow{\text{Member yields } r_m}$ $\chi_{n,m}^{(QR)}$

$P(q_n, t_Q) = |c_n^{(Q)}(t_Q)|^2$ $\quad$ $P(r_m, t_R) = |c_m^{(R)}(t_R)|^2$

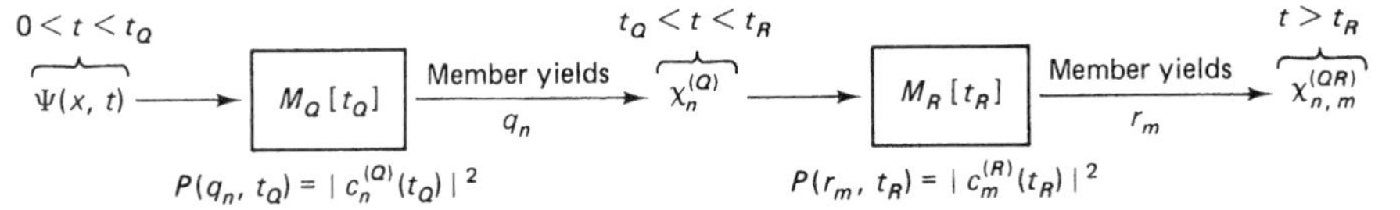

Figure 13.7 A measurement diagram for simultaneous measurement of two *compatible* observables Q and R on a state that is initially not an eigenstate of either observable. Each measurement causes the state function to collapse into one of the eigenstates of the measured observable. The state after both measurements is a simultaneous eigenstate of both observables, a possibility allowed by the compatibility of Q and R.

From § 12.2 we know that to predict the outcome of the first measurement, $\mathsf{M}_Q[t_Q]$, we must discuss this measurement in "the language of Q." To translate the state function into this language, we expand $\Psi(x,t)$ in the complete set of eigenfunctions of $\hat{Q}$:

$$\Psi(x, t_Q) = \sum_n c_Q^{(n)}(t)\chi_n^{(Q)}(x). \tag{13.12}$$

The number $|c_Q^{(n)}(t)|^2$ is the probability that the eigenvalue q_n will be obtained in $\mathsf{M}_Q[t_Q]$. A member that exhibits this value is thrown by the measurement into the nth eigenstate, so its eigenfunction *immediately after* $\mathsf{M}_Q[t_Q]$ is $\chi_n^{(Q)}$. Hence $|c_Q^{(n)}(t)|^2$ *is also the probability of a member system winding up in this eigenstate after* $\mathsf{M}_Q[t_Q]$.

Now turn to the second measurement, of the compatible observable R. We'll perform $\mathsf{M}_R[t_R]$ *on the sub-ensemble of members that exhibited* q_n *in* $\mathsf{M}_Q[t_Q]$ (look carefully at Fig. 13.7). To predict the outcome of $\mathsf{M}_R[t_R]$, we translate the problem into "the language of R" by expanding the wave function after $\mathsf{M}_Q[t_Q]$, which we just said was $\chi_n^{(Q)}$, in the set of eigenfunctions of $\hat{R}$:[18]

$$\chi_n^{(Q)}(x) = \sum_m c_R^{(m)}(t)\chi_m^{(R)}(x). \tag{13.13}$$

The probability of obtaining the eigenvalue r_m in $\mathsf{M}_R[t_R]$ is just $|c_R^{(m)}(t_R)|^2$, which is also the probability that immediately after this measurement, the system will be in the mth eigenstate of R.

What is the ultimate effect of our simultaneous measurement? The first measurement collapsed each member of the original ensemble into an eigenstate of Q. The second further collapsed each member of each sub-ensemble into one of the eigenstates of R. So *after both measurements, each member will be in one of the simultaneous eigenstates of both observables.*

We can *prepare* a state that is an eigenstate of several compatible observables by measuring one right after the other. (If, as is often the case, the observables are *constants*

[18]It may happen that $\chi_n^{(Q)}$ is an eigenfunction of $\hat{R}$—*i.e.*, $\mathsf{M}_Q[t_Q]$ could cause Ψ to collapse into one of the *simultaneous eigenfunctions of* Q *and* R (whose existence is guaranteed by the commutativity of $\hat{Q}$ and $\hat{R}$). In this special case, all the projections $c_R^{(m)}(t)$ but one will be zero. (See Example 13.5.)

of the motion, the simultaneity of their measurement is not important, because the system won't evolve out of an eigenstate of a constant of the motion while it waits for you to perform the next measurement.) For example, to prepare an ensemble of simple harmonic oscillators in a simultaneous eigenstate of parity and energy, you could first measure Π and then E. Or, if more convenient, you could measure E first, then Π.

Example 13.4. Parity and Energy Measurements on a Non-Stationary State

Return now to the ensemble of simple harmonic oscillators of Example 13.3 and suppose that their state prior to measurement of parity and energy is the following *non-stationary* initial condition:

$$\Psi(x,0) = \sqrt{\frac{1}{20}}\left[2\underbrace{\psi_0(x)}_{\text{even}} + 4\underbrace{\psi_1(x)}_{\text{odd}}\right]. \tag{13.14}$$

This state is a superposition of one even-parity state (the ground state) and one odd-parity state (the first excited state). So *for this state neither parity nor energy are sharp.* We can analyze each measurement using the rules of § 13.2, as shown in Fig. 13.8.

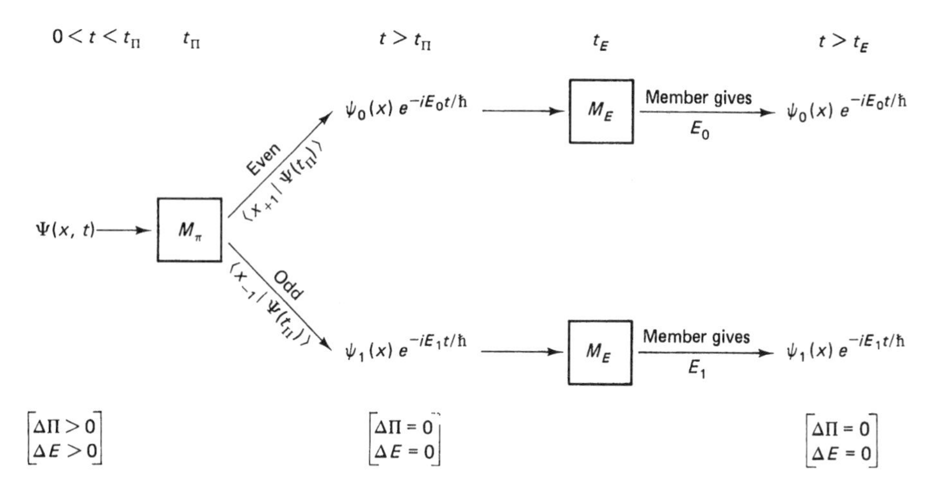

Figure 13.8 A measurement diagram for simultaneous measurement of the parity and energy of a simple harmonic oscillator in the non-stationary state (13.14). Because this state is not an eigenstate of parity, the first measurement causes collapse into one of the parity eigenstates represented in the superposition. The resulting states are also energy eigenstates, so no further change of state occurs when energy is measured.

The parity measurement causes each member to collapse into one of two parity eigenstates, with probabilities determined by the projections in Eq. (13.14): 1/5 of the systems wind up in the even-parity ground state, and 4/5 in the odd-parity first excited state. But *each of the resulting states is also an energy eigenstate.* So the second measurement, $\mathbf{M}_E[t_E]$, need not (and, in an ideal measurement, does not) further alter the state. After all is said and done, the system is in a simultaneous eigenstate of parity and energy.

Question 13–6

Equation (13.14) gives the wave function at the initial time $t = 0$. Write down the form of the state function for times between $t = 0$ and t_Π.

Question 13–7

I don't have space here to discuss in detail the calculation of joint probabilities for measurements of multiple observables. However, to see how one puts such probabilities together, **explain** in detail why the probability of obtaining q_n and r_m in a simultaneous measurement of compatible observables Q and R on a system not in an eigenstate of either observable is

$$P(q_n, r_m) = |\langle \chi_n^{(Q)} \mid \Psi(t_Q)\rangle|^2 |\langle \chi_m^{(R)} \mid \chi_n^{(Q)}\rangle|^2. \tag{13.15}$$

What would the joint probability be if $\chi_m^{(R)}$ happened to be orthogonal to $\chi_n^{(Q)}$? What would it be if the state prior to either measurement were an eigenstate of Q but not of R? ... of R but not of Q?

Analyzing the non-stationary state in Example 13.4 was particularly simple, because this superposition included only one even-parity and one odd-parity term. As a consequence of this simplicity, the second measurement $\mathsf{M}_E[t_E]$ played no significant role in determining the final state. This situation is the exception, not the rule, as we'll see in the next section.

*13.5 MULTIPLE MEASUREMENTS II: AN EXAMPLE

To show you how to cope with measurements on a more complicated state, I want to return to the ensemble of simple harmonic oscillators, this time considering the following initial condition:[19]

$$\Psi(x,0) = \sqrt{\frac{1}{38}}\left[2\underbrace{\psi_0(x)}_{\text{even}} + 3\underbrace{\psi_1(x)}_{\text{odd}} - 5\underbrace{\psi_2(x)}_{\text{even}}\right]. \tag{13.16}$$

In the absence of a measurement, this state would evolve according to the TDSE, as

$$\Psi(x,t) = \sqrt{\frac{1}{38}}\left[2\psi_0(x)\, e^{-iE_0t/\hbar} + 3\psi_1(x)\, e^{-iE_1t/\hbar} - 5\psi_2(x)\, e^{-iE_2t/\hbar}\right]. \tag{13.17}$$

But suppose that at time $t_\Pi = 0$, we measure the *parity* of the system.[20] This state is clearly not an eigenstate of parity, for it contains one term whose parity is odd and two whose parities are even. So to determine the outcome and effect of $\mathsf{M}_\Pi[t_\Pi]$, we turn to the method of eigenfunction expansion. Fortunately, we don't have to explicitly translate the function in (13.16) into the "language" of another observable to implement this method, because each energy eigenfunction of this system is also a parity eigenfunction. [That is, the expansion of $\Psi(x,t)$ we use to predict the outcome of a parity measurement looks identical to the one we would use to analyze an energy measurement.] But we do need to think carefully about the meaning of the coefficients in Eq. (13.16).

Let's look first at the *odd-parity* possibility. It's a trivial matter to calculate from Eq. (13.16) the probability that in $\mathsf{M}_\Pi[t_\Pi]$ a member will exhibit the value -1. Since the coefficient of the odd-parity term in (13.16) is $c_1^{(\Pi)}(0) = 3/\sqrt{38}$, we have

[19]This section contains the most complicated example yet of multiple measurements. If you are running out of time, you can omit it and press on to the next section.

[20]I choose $t_\Pi = 0$ solely to simplify the form of some of the results in this example. This choice in no way affects the physics or problem-solving strategy.

$$P(-1, t_\Pi) = \left| \frac{3}{\sqrt{38}} \right|^2 = 0.2368. \tag{13.18}$$

Analyzing the other possibility, that $\mathbf{M}_\Pi[t_\Pi]$ will collapse the oscillator into an *even-parity* state, is a bit trickier. There are *two* terms in (13.16) that correspond to this possibility: the ground state, with probability amplitude $2/\sqrt{38}$, and the second excited state, with amplitude $-5/\sqrt{38}$. So the overall probability of getting $\Pi = +1$ in $\mathbf{M}_\Pi[t_\Pi]$ is the *sum* of the probabilities for each of these terms:

$$P(+1, t_\Pi) = \left| \frac{2}{\sqrt{38}} \right|^2 + \left| -\frac{5}{\sqrt{38}} \right|^2 = 0.7632. \tag{13.19}$$

Notice that, satisfyingly, the probabilities for the two possible outcomes of the parity measurement sum to one.

It may help to think about this analysis in the following way: it's as if there were two "routes" to even parity: one corresponding to the ground state, the other to the second excited state. In probability theory we *add probabilities for mutually-exclusive possibilities*, so Eq. (13.19) is the correct way to calculate the probability of (one way or the other) getting even parity in $\mathbf{M}_\Pi[t_\Pi]$. But be careful with this analogy; the parity measurement does not throw members that yield $+1$ into one or the other of the stationary states $\psi_0(x)$ or $\psi_2(x)$; rather, as we'll see in a moment, this measurement forces them into an (unknown) superposition of these states.

Warning

Note that we add the *probabilities* for the two (or more) terms that correspond to the same value of a measurement observable, not the probability amplitudes.

Question 13–8

How would Eqs. (13.18) and (13.19) change if $t_\Pi > 0$?

Question 13–9

There are only two possible outcomes of $\mathbf{M}_\Pi[t_\Pi]$: $+1$ and -1. The sum of the probabilities for these outcomes, Eqs. (13.18) and (13.19) is, happily, equal to 1. Suppose we computed $P(+1, t_\Pi)$ by adding the probability *amplitudes* rather than the probabilities. Show that this gambit does not lead to a total probability of 1.

The states of the two sub-ensembles created by the parity measurement are indicated in the diagram in Fig. 13.9. Understanding the *odd-parity* outcome is easy. Since the only odd-parity term in the wave function prior to $\mathbf{M}_\Pi[t_\Pi]$, Eq. (13.16), is the one for the first excited state, each member that exhibits $\Pi = -1$ collapses into this eigenstate. The wave function of such a member after $\mathbf{M}_\Pi[t_\Pi]$ is therefore $\psi_1(x)\, e^{-iE_1 t/\hbar}$. Were we to perform a subsequent energy measurement on the sub-ensemble of odd-parity oscillators, we would find only the value $E_1 = 3\hbar\omega_0/2$. So the odd-parity sub-ensemble behaves just like the one in Example 13.4.

Not so the *even-parity* sub-ensemble. The rules of quantum mechanics demand that the state of this sub-ensemble after $\mathbf{M}_\Pi[t_\Pi]$ be an eigenstate of $\hat{\Pi}$ with eigenvalue $+1$. But *these rules do not specify which even-parity state will result.* In a real measurement,

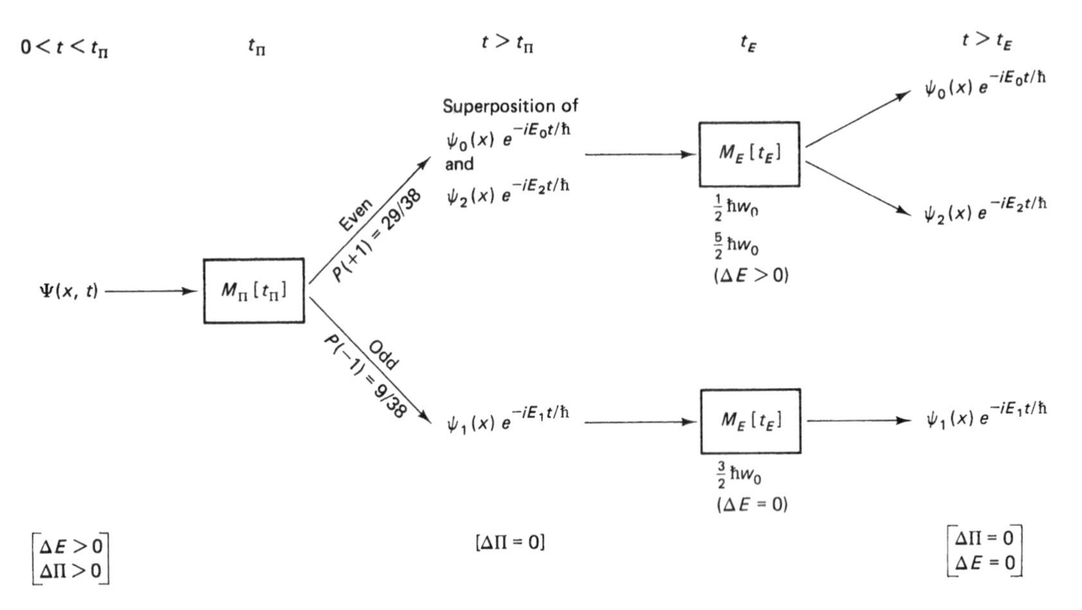

Figure 13.9 Measurement diagram for a simultaneous measurement of parity and energy on the initial SHO state represented by Eq. (13.16). The superposition referred to as the even-parity outcome of $\mathbf{M}_\Pi[t_\Pi]$ is given in Eq. (13.20).

the resulting state could be the ground state, the second excited state, or a superposition of the two, *i.e.*,

$$\Psi(x,t) = d_0^{(E)}(t)\,\psi_0(x)\,e^{-iE_0t\hbar} + d_2^{(E)}(t)\,\psi_2(x)\,e^{-iE_2t/\hbar}. \tag{13.20}$$

But we are here considering only *ideal* measurements, and for such a measurement we can determine the *ratio* of $d_0^{(E)}(t)$ to $d_2^{(E)}(t)$ in (13.20). Let's see how.

An ideal measurement of parity introduces the *least possible disturbance* in the energy-related properties of the system. This does not mean that $\mathbf{M}_\Pi[t_\Pi]$ doesn't change these properties at all. For example, the state for $t < t_\Pi$, Eq. (13.16), allows for the possibility that a member will exhibit E_1 in a subsequent energy measurement. But this possibility is not present in the (even-parity) superposition (13.20). What, then, does "least possible disturbance" mean in this case?

Well, one important energy-related property of the system prior to $\mathbf{M}_\Pi[t_\Pi]$ is that *the energy of a particle is not sharp.* Since the parity measurement *need not* collapse members that exhibit $\Pi = +1$ into an energy eigenstate, it won't do so—and the energy of the even-parity ensemble after an ideal $\mathbf{M}_\Pi[t_\Pi]$ will still not be sharp.[21] That is, the disturbance in the energy-related properties of the state (13.20) will be less if both coefficients $d_0^{(E)}(t)$ and $d_2^{(E)}(t)$ are non-zero than if one of them is zero.

To pin down the coefficients in Eq. (13.20), we must think more deeply about this matter of a "minimal disturbance."[22] The squared moduli of these coefficients are probabilities: *e.g.*, $|d_0^{(E)}(t)|^2$ is the probability that an energy measurement on the sub-

[21]Careful now. It should be evident from Eqs. (13.16) and (13.20) that the *value* of ΔE will be changed by the parity measurement. If not, glance ahead to Table 13.1.

[22]See H. Araki and M. M. Yanese, *Phys. Rev. A*, **120**, 622 (1960) and J. Cohn, *Amer. Jour. Phys.*, **36**, 749 (1968).

ensemble in this state will yield the value $E_0 = \hbar\omega_0/2$. That is, these coefficients contain information about the *distribution of possible energies that is implicit in the wave function Eq. (13.20).* The least possible disturbance in the energy-related properties of the oscillator is one that doesn't change this distribution for the allowed energies E_0 and E_2 from what it was before $\mathbf{M}_\Pi[t_\Pi]$.

For these two energies, the probability amplitudes in the state *before the parity measurement* are to be found in Eq. (13.16). These amplitudes correspond to a distribution in which the probability for E_0 is $4/38 = 0.1053$ and for E_2 is $25/38 = 0.6579$. The *ratio* of the probability for E_2 to that for E_0 is 6.25. *We can preserve the energy distribution for E_0 and E_2 by choosing the coefficients in the state function after $\mathbf{M}_\Pi[t_\Pi]$ [in Eq. (13.20)] to maintain this ratio*:

$$\frac{|d_2^{(E)}(t)|^2}{|d_0^{(E)}(t)|^2} = \frac{|c_2^{(E)}(t)|^2}{|c_0^{(E)}(t)|^2} = 6.25. \tag{13.21}$$

This reasoning may seem complicated, so here's a summary. To minimize the effect of a *parity measurement* on the *energy properties* of the oscillator we must ensure that there is no change due to $\mathbf{M}_\Pi[t_\Pi]$ in the *relative probabilities* for obtaining the energies of the ground state and the second-excited state, the only two possibilities for the even-parity sub-ensemble. The probabilities for these energies are just the squared moduli of the energy probability amplitudes in the state functions before and after the parity measurement, so by equating the ratio of these squared moduli as in (13.21) we can preserve the relative probabilities.

We cannot preserve the *absolute energy probabilities*, because the wave function before $\mathbf{M}_\Pi[t_\Pi]$ contains a possibility that is not present for the even-parity sub-ensemble that results from this measurement—namely, the possibility of the (odd-parity) first excited state. The presence in Eq. (13.16) of this possibility, represented by the term proportional to $\psi_1(x)\, e^{-iE_1t/\hbar}$, affects the *values* of the coefficients $c_n^{(E)}(t)$ in this function, because these coefficients must be normalized so that

$$\langle \Psi \mid \Psi \rangle = \left|c_0^{(E)}\right|^2 + \left|c_1^{(E)}\right|^2 + \left|c_2^{(E)}\right|^2 = 1 \qquad t < t_\Pi. \tag{13.22}$$

[This requirement is reflected in the prefactor $1/\sqrt{38}$ in Eq. (13.16).] The corresponding normalization condition for the state (13.20) (after the parity measurement) is

$$\langle \Psi \mid \Psi \rangle = \left|d_0^{(E)}\right|^2 + \left|d_2^{(E)}\right|^2 = 1 \qquad t > t_\Pi. \tag{13.23}$$

Now, if you're still with me, all we need do is choose $d_0^{(E)}(t)$ and $d_2^{(E)}(t)$ to conform to Eqs. (13.21) and (13.23). Doing so leads to a wave function that is *almost* completely determined. The remaining indeterminacy is a familiar one: we cannot specify the *phase* of each coefficient in (13.20). Denoting the phase constants for the E_0 and E_2 terms by δ_0 and δ_2, respectively, we can write (finally!) the form of the state function after an ideal parity measurement:

$$\Psi(x,t) = \sqrt{\frac{1}{29}}\left[2e^{i\delta_0}\,\psi_0(x)\, e^{-iE_0t/\hbar} + 5e^{i\delta_2}\,\psi_2(x)\, e^{-iE_2t/\hbar}\right]. \tag{13.24}$$

Whew!

Question 13–10

Verify that the wave function (13.24) is correctly normalized. Now calculate the probabilities that would be obtained in an *energy* measurement using this normalized wave function and verify that (13.24) predicts the same relative distribution for E_0 and E_2 as does (13.16).

Let's examine the state function (13.24) to see what changes the parity measurement has wrought. Recall that before $\mathbf{M}_{\Pi}[t_{\Pi}]$, energy and parity were uncertain. The superposition state after this measurement has the following properties:

1. $\Delta\Pi = 0$
2. $\Delta E > 0$
3. the relative distribution of energy probabilities for E_2 and E_0 is the same as before $\mathbf{M}_{\Pi}[t_{\Pi}]$ [Eq. (13.21)].

A closer look reveals that the *values* of the *statistical* properties related to energy were changed by the parity measurement. This change is evident in Table 13.1, which contains the values of the expectation value and uncertainty in the energy before and after $\mathbf{M}_{\Pi}[t_{\Pi}]$. The "after values" in this table correspond to the even-parity sub-ensemble. This change should come as no great surprise, because the mixture of energy eigenstates in this sub-ensemble is different from that of the original ensemble.

TABLE 13.1 STATISTICAL ENERGY-RELATED PROPERTIES BEFORE AND AFTER THE PARITY MEASUREMENT

	Before	After
$\langle E \rangle$	$2.0526\,\hbar\omega_0$	$2.2241\,\hbar\omega_0$
$\langle E^2 \rangle$	$4.6711\,\hbar^2\omega_0^2$	$5.4224\,\hbar^2\omega_0^2$
ΔE	$0.6767\,\hbar\omega_0$	$0.6898\,\hbar\omega_0$

Question 13–11

Verify each of the values in Table 13.1. (This exercise is easy and provides valuable practice.)

One reason for going through the analysis leading to Eq. (13.24) is that we need this wave function to predict the outcome of the *second* measurement, which you may have forgotten. At the beginning of this section, we proposed to follow the parity measurement by *an energy measurement*. The analysis of this second experiment is quite simple, as you can see by another look at Fig. 13.9.

An energy measurement has no effect on the state of the odd-parity sub-ensemble. It does, however, collapse the members of the even-parity sub-ensemble into one of two energy eigenstates—either the ground state or the second excited state. The probabilities for these outcomes are easy to calculate from (13.24), to wit:

$$\begin{aligned} P(E_0) &= 0.14 \\ P(E_2) &= 0.86. \end{aligned} \tag{13.25}$$

Question 13–12

According to (13.18) the probability of a member of the original ensemble winding up with sharp energy E_1 is $9/38 = 0.24$. The sum of this and the probabilities in (13.25) is 1.24. But the probability of a member winding up with energy E_0 *or* E_1 *or* E_2 should be 1. What went wrong?

Summary

Simultaneous measurement of compatible observables is not forbidden by the laws of quantum mechanics. If the state prior to the experiment happens to be a simultaneous eigenstate of the measured observables, then the measurement need not—and in an ideal measurement will not—change the state. But if the initial state is not an eigenstate of either observable, then the measurement forces the members into a state in which *both* observables are sharp. These conclusions, however, depend on the compatibility of the measured observables. What happens if the little beggars are incompatible? Stay tuned.

*13.6 DUELING OBSERVABLES: MULTIPLE MEASUREMENTS OF INCOMPATIBLE OBSERVABLES

Ah! my dear Watson, there we come
into those realms of conjecture,
where the most logical mind
may be at fault.
Each may form his own hypothesis
upon the present evidence,
and yours is as likely to be correct
as mine.

—Sherlock Holmes,
in *The Adventure of the Empty House*
by Sir Arthur Conan Doyle

The easiest part of analyzing a multiple measurement is determining the effect of the first experiment. Whether or not the observables are compatible, the first measurement collapses the system into a state in which the measured observable is sharp (unless, of course, the system was already in such a state). The fun begins when we turn to the second measurement, which forces the system into an eigenstate of the other observable. If this observable is *incompatible* with the first, then the state after the second experiment is one in which the *first* observable is no longer sharp. That is, *measuring one observable can change the properties of the system relating to another.* As we'll see, this seemingly magical alteration of the state is an unavoidable consequence of the non-commutativity of the operators that represent the measured observables.

Inescapable Changes of State

You must remember this: non-commuting operators do not define a complete set of simultaneous eigenfunctions. In general, no states exist in which the uncertainties of two non-commuting operators are zero. (Recall that we agreed to exclude the exceptional case of a single simultaneous eigenfunction of two non-commuting operators.) For example, if $\hat{Q}$ and $\hat{R}$ do not commute, then in any eigenstate of Q the incompatible observable R *must* be uncertain, and vice versa. This familiar property is the key to understanding multiple measurements of incompatible observables.

Let's consider only the "second" measurement. That is, let's pretend that we are given an ensemble of systems that are initially in an eigenstate of Q,[23]

$$\Psi(x,0) = \chi_n^{(Q)}(x) \qquad \text{[state prior to } \mathbf{M}_R[t_R]\text{]} \tag{13.26}$$

On this ensemble we are going to measure the incompatible observable R at t_R. In the state (13.26) the observable Q is sharp, *i.e.*

$$\Delta Q(0) = 0 \qquad \text{[state prior to } \mathbf{M}_R[t_R]\text{]} \tag{13.27}$$

In the measurement $\mathbf{M}_R[t_R]$ each member of the ensemble could exhibit any of the eigenvalues of $\hat{R}$ [see Eq. (13.11b)]. But not all members will exhibit the same value, for R is *necessarily* uncertain:

$$\Delta R(0) > 0 \qquad \text{[state prior to } \mathbf{M}_R[t_R]\text{]} \tag{13.28}$$

To learn more about the outcome of $\mathbf{M}_R[t_R]$, we could use the machinery of Chap. 12 to predict the probabilities for each allowed value. We translate the state function (13.26) into the language of R via expansion in the complete set $\{\chi_m^{(R)}\}$, *viz.*,

$$\Psi(x,0) = \sum_m c_m^{(R)}(0)\chi_m^{(R)}(x). \tag{13.29}$$

The probability amplitude for eigenvalue r_m is

$$c_m^{(R)}(0) = \langle \chi_m^{(R)} \mid \Psi(0) \rangle = \langle \chi_m^{(R)} \mid \chi_n^{(Q)} \rangle. \tag{13.30}$$

In the initial state $\chi_n^{(Q)}$, the observable R cannot be sharp, so the probabilities $P(r_m,0) = |c_m^{(R)}(0)|^2$ will be non-zero for at least two of the eigenvalues r_m.

There is nothing particularly unusual in the analysis of Eqs. (13.28)–(13.30). But now let's turn to the "Q-related properties" of the state *after measurement of R*. This measurement changed the state function into an eigenfunction of $\hat{R}$. For example, the sub-ensemble of members that exhibited r_m in $\mathbf{M}_R[t_R]$ winds up in a state represented by the corresponding eigenfunction:

$$\Psi(x,0) = \chi_m^{(R)}(x) \qquad \text{[an instant after } \mathbf{M}_R[t_R]\text{]} \tag{13.31}$$

If we calculate ΔQ for this state, we get a positive number; *i.e.*, in the state after $\mathbf{M}_R[t_R]$, the observable Q must be uncertain. The reason, as argued above, is simply that the state function that results from $\mathbf{M}_R[t_R]$ *cannot* be one in which both Q and R are sharp, for such simultaneous eigenstates (of incompatible observables) do not exist.

[23]For convenience, we'll initialize our clock ($t = 0$) at the instant prior to t_R. In practice, we could have prepared the system in such a state by measuring Q *immediately before the measurement of* R, *i.e.*, at $t_R - \delta t$. From the resulting members, we select a sub-ensemble of systems that exhibited the same eigenvalue of $\hat{Q}$. [If Q is a constant of the motion, we could have performed $\mathbf{M}_Q[t_Q]$ a finite time before $\mathbf{M}_R[t_R]$, for once $\mathbf{M}_Q[t_Q]$ has collapsed the system into an eigenstate of a constant of the motion, it will remain in that eigenstate, evolving according to the TDSE, until a second measurement (of an incompatible observable) is performed.]

So a multiple measurement of incompatible observables causes the state to "flip-flop" between eigenstates of each observable,[24]

$$\left.\begin{array}{c}\underline{\underline{t < t_R}} \\ \Delta Q = 0 \\ \Delta R > 0\end{array}\right\} \longrightarrow \boxed{\mathbf{M}_R[t_R]} \longrightarrow \left\{\begin{array}{c}\underline{\underline{t > t_R}} \\ \Delta Q > 0 \\ \Delta R = 0\end{array}\right. \tag{13.32}$$

So here is the rule for this case:

> ***Rule***
>
> Measurement of one of two incompatible observables (on a state in which the other observable is sharp) induces an unavoidable, uncontrollable uncertainty in the unmeasured observable.

Example 13.5. The Anharmonic Oscillator

In Chap. 9 we studied at length the stationary states of the simple harmonic oscillator. These energy eigenstates are also parity eigenstates, because the potential energy

$$V(x) = \tfrac{1}{2}m\omega_0^2 x^2 \tag{13.33}$$

is symmetric with respect to $x = 0$, *i.e.*, $V(-x) = V(x)$.

The SHO is a valuable model of low-energy vibrational states of diatomic molecules. To use it thus, we let x in (13.33) represent the deviation of the internuclear separation R from its equilibrium value R_e, *i.e.*, $x \equiv R - R_e$. But this model is crude, and to represent accurately the vibrational states of some molecules (particularly high-lying states) we must resort to model potentials that more accurately reflect the shape of the molecular potential curve [*c.f.*, Fig. 9.14]. One prominent feature of such actual molecular potentials is their *lack of symmetry about equilibrium, the point* $R = R_e$. So one criterion we would build into a more accurate model potential $V^{(M)}(x)$ is this lack of symmetry with respect to $x = 0$,

$$V^{(M)}(-x) \neq V^{(M)}(x). \tag{13.34}$$

Because of this condition, the Hamiltonian $\hat{\mathcal{H}}^{(M)}$ of a particle of mass m with such a model potential,

$$\hat{\mathcal{H}}^{(M)} = \frac{\hat{p}^2}{2m} + \hat{V}^{(M)}(\hat{x}) \tag{13.35}$$

does not commute with the parity operator. So for this model, parity and energy are incompatible, and there are no simultaneous eigenstates of these observables. Unlike the stationary states of the SHO, those of the more realistic Hamiltonian

$$\hat{\mathcal{H}}^{(M)}\psi_n^{(M)} = E_n^{(M)}\psi_n^{(M)} \tag{13.36}$$

are not also eigenstates of parity.

[24]Note carefully that this change occurs even though the measurement is "ideal." An ideal measurement of one of two *compatible* observables need not—and therefore does not—change the properties of the other observable. But any measurement of one of two *incompatible* observables may have to change the properties of the other observable. If such a measurement is ideal, however, these changes are minimal (consistent with the laws of quantum physics).

Question 13–13

> Devise an asymmetric potential $V^{(M)}(x)$ and for it, evaluate the commutator of $\hat{\mathcal{H}}^{(M)}$ and the parity operator. Write down the Generalized Uncertainty Principle for energy and parity. Discuss the implications of your answer for measurement of the energy on various kinds of states of this system.

With this background, let's consider a state of a molecule whose vibrational motion we represent with an asymmetric potential. For my own nefarious reasons, I want to consider what may seem to you a capricious choice of initial condition: *suppose the state of interest is initially in the ground state of the simple harmonic oscillator potential.* That's right: even though we are *not* using the SHO to model the molecule, we'll consider the initial wave function

$$\Psi(x,0) = \psi_0(x) = \left(\frac{\beta^2}{\pi}\right)^{1/4} e^{-\beta^2 x^2/2}, \tag{13.37}$$

where, of course, our old friend β is just $\sqrt{m\omega_0/\hbar}$. *This wave function is an eigenfunction of* Π, *so at the instant* $t = 0$ *the particle is in a parity eigenstate.* What does this mean about the energy at this instant?

The energy must be uncertain, for there are no simultaneous eigenfunctions of $\hat{\mathcal{H}}^{(M)}$ and $\hat{\Pi}$. The function (13.37) is an eigenfunction of the Hamiltonian of the simple harmonic oscillator, *not* of the asymmetric Hamiltonian (13.36). So if (13.37) is the initial condition of a particle in such a potential, it represents a state for which $\Delta E(0) > 0$.

Now let's imagine measuring the energy at $t_E = 0$. Of course, each member of the ensemble will yield one of the eigenvalues $E_n^{(M)}$ of Eq. (13.36). The *energy probability amplitudes* are just the projections of the state function immediately before $\mathbf{M}_E[t_E]$ on the eigenfunctions $\psi_n^{(M)}$ of (13.36). These amplitudes appear in the expansion

$$\Psi(x,0) = \sum_{n=0}^{\infty} c_n^{(E)}(0)\psi_n^{(M)}(x). \tag{13.38}$$

For the initial state (13.37), they are[25]

$$c_n^{(E)}(0) = \langle \psi_n^{(M)} \mid \Psi(0) \rangle = \langle \psi_n^{(M)} \mid \psi_0 \rangle. \tag{13.39}$$

The probabilities $|c_n^{(E)}(0)|^2$ describe the likelihood that each member of the original ensemble will wind up in one of the sharp-energy sub-ensembles that results from $\mathbf{M}_E[t_E]$.

Our energy measurement has dramatically changed the energy properties of the original ensemble. But it has also changed the parity properties, even though we did not measure this observable. The parity of the initial state (13.37) was sharp; the parity of each sub-ensemble created by $\mathbf{M}_E[t_E]$ is not sharp. Because $\hat{\Pi}$ does not commute with $\hat{\mathcal{H}}^{(M)}$, an energy measurement collapses the system into a state in which parity is uncertain—another flip-flop:

[25]Notice that the integral represented by the Dirac braket $\left\langle \psi_n^{(M)} \mid \psi_0 \right\rangle$ is not equal to $\delta_{n,0}$. The function $\psi_0(x)$ is an eigenfunction of the SHO Hamiltonian, while $\psi_n^{(M)}$ is an eigenfunction of our model Hamiltonian.

$$\left.\begin{array}{l}\underline{\underline{t < t_E}} \\ \Delta\Pi = 0 \\ \Delta E > 0\end{array}\right\} \implies \boxed{\mathbf{M}_R[t_R]} \implies \left\{\begin{array}{l}\underline{\underline{t > t_E}} \\ \Delta\Pi > 0 \\ \Delta E = 0\end{array}\right. \tag{13.40}$$

Question 13–14

Draw a branch diagram like the one in Fig. 13.8 for the energy measurement in this example. Then add to your diagram additional branches to show the outcome of a *subsequent* measurement of the parity on each sub-ensemble that resulted from the initial energy measurement. (To simplify your diagram, show only three branches resulting from each measurement.) Discuss briefly how the system evolves during the time interval after $\mathbf{M}_E[t_E]$ and before $\mathbf{M}_\Pi[t_\Pi]$.

Example 13.6. The Energy and Linear Momentum of a SHO

In § 13.3 we explored simultaneous measurements of the energy and parity of a one-dimensional simple harmonic oscillator, two compatible observables. Two observables of this system we know to be incompatible are the energy and linear momentum. In Example 11.2 we derived the commutator of their operators,

$$\left[\hat{\mathcal{H}}, \hat{p}\right] = i\hbar m\omega_0^2 \hat{x}. \tag{13.41}$$

The non-commutativity of $\hat{\mathcal{H}}$ and $\hat{p}$ implies that in an eigenstate of one of these operators, the other operator must be uncertain. The uncertainties and eigenfunctions are summarized in Table 13.2.

TABLE 13.2 ENERGY AND MOMENTUM EIGENSTATES OF THE SIMPLE HARMONIC OSCILLATOR.

Energy Eigenstates	Momentum Eigenstates
$\Delta E = 0$	$\Delta p = 0$
$\psi_n(x)$ (Table 9.6)	$\chi_p(x) = (2\pi\hbar)^{-1/2} e^{-ipx/\hbar}$
$\Delta p = \sqrt{(n + \frac{1}{2}) m\omega_0\hbar} > 0$	$\Delta E > 0$

Aside: Momentum and the Energy. It may strike you as curious that the momentum eigenfunction $\chi_p(x)$, which is an energy eigenstate *of the free particle*, represents a state of the simple harmonic oscillator in which the energy is not sharp. We can easily convince ourselves that the energy in this state is indeed uncertain. The energy uncertainty is

$$\Delta E = \sqrt{\langle E^2 \rangle - \langle E \rangle^2}. \tag{13.42}$$

For a momentum eigenstate, both $\langle T \rangle$ and $\left\langle T^2 \right\rangle$ are zero. (Why?) So the energy uncertainty (13.42) simplifies to

$$\Delta E = \sqrt{\langle V^2 \rangle - \langle V \rangle^2} \tag{13.43a}$$

$$= \frac{1}{2} m\omega_0^2 \sqrt{\langle x^2 \rangle - \langle x \rangle^2}. \tag{13.43b}$$

Now, the mean value of position for a momentum eigenstate, $\langle x \rangle$, is zero; but $\langle x^2 \rangle > 0$ (see the next Question). This is why the energy of a SHO in a momentum eigenstate is uncertain: $\Delta E > 0$.

Question 13–15

It is not necessary to evaluate $\langle x^2 \rangle$ to convince yourself that this quantity must be positive for the SHO. Argue the case from the Heisenberg Uncertainty Principle and the expression for the position uncertainty that is analogous to (13.42).

Suppose we measure the linear momentum and select from the results the sub-ensemble corresponding to the eigenvalue p_0.[26] Immediately after $\mathbf{M}_p[t_p]$, at time t_E, we measure the energy. The branch diagram for this measurement is shown in Fig. 13.10.

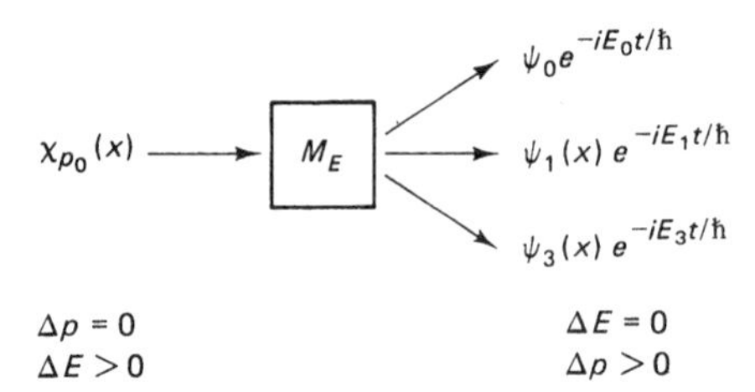

Figure 13.10 A branch diagram for an energy measurement on a simple harmonic oscillator that is initially in a momentum eigenstate. For simplicity, only three of the infinity of branches are shown.

The energy measurement causes the system to collapse into an energy eigenstate. The probability of a member winding up in the nth stationary state of the SHO is, as usual,

$$P(E_n, t_E) = \left|\langle \psi_n \mid \chi_{p_0} \rangle\right|^2 \qquad n = 0, 1, 2, \ldots. \tag{13.44}$$

The first few probabilities are easy to evaluate; we worked out the integrals $\langle \psi_n \mid \chi_{p_0} \rangle$ for $n = 0$ and $n = 1$ in Chap. 12, while toiling in Example 12.8. For instance, from Eq. (12.123*a*) we find $\langle \psi_0 \mid \chi_{p_0} \rangle$, which we can use in (13.44) to determine the probability that a member will collapse into the ground state:

$$P(E_0, t_E) = \frac{1}{4\beta\hbar\sqrt{\pi}}\, e^{-p_0^2/(\beta^2\hbar^2)}. \tag{13.45}$$

Question 13–16

Evaluate the probability that a member will exhibit the value $E_1 = 3\hbar\omega_0/2$ in the energy measurement.

Fine. But the central question here is *what has the energy measurement done to the momentum properties of the oscillator*? Remember that the state prior to $\mathbf{M}_E[t_E]$ was characterized by a sharp momentum, $p = p_0$. But after $\mathbf{M}_E[t_E]$, the system is in an energy eigenstate, so its linear momentum is uncertain (see Table 13.2). For example, the momentum uncertainty for the sub-ensemble with energy E_0 is $\Delta p(t_E + \delta t) = \sqrt{m\hbar\omega_0/2}$. Once again a measurement of one of two incompatible observables (here the energy) has changed characteristics of the system that relate to the other, unmeasured observable (here, the momentum).

[26] As we know from Chap. 8, momentum eigenstates are not normalizable and hence do not truly exist in nature. So to be precise in this analysis, we should speak of a sub-ensemble of systems in a state represented by a very fat wave packet whose momentum probability amplitude is very sharply peaked about p_0.

Uncertainty in Action

We have seen that after measurement of one of two incompatible observables, the system is in a state in which the unmeasured observable is necessarily uncertain. Imagine a *series* of successive measurements of two incompatible observables: first $\mathbf{M}_Q[t_Q]$, then, immediately thereafter, $\mathbf{M}_R[t_R]$, then $\mathbf{M}_Q[t_Q]$ again, ... These measurements cause the system to "flip-flop" between states in which one or the other of the observables (but not both) is sharp.

Uncertainty originates in the commutator properties of operators and in the formal relationship between commutativity and the possible existence of simultaneous eigenstates. The genesis and role of uncertainty in the mathematics of quantum physics is one of the most profound examples of a theme I have emphasized throughout this book: *an understanding of the physics of microscopic systems, at the experimental level of measurement and prediction of their properties, is inextricably linked to an understanding of the abstract mathematics of quantum mechanics.* This deep interrelationship, which goes far beyond the mere assertion that "mathematics is the language of physics," is one of the great and abiding beauties of quantum physics.

13.7 FINAL THOUGHTS: MYSTERIES ABOUND

> ... every branch of human knowledge
> if traced up to its source
> and final principles
> vanishes into mystery.
>
> —*The Novel of the White Powder*
> by Arthur Machen

In this chapter, we have delved into one of the great mysteries of nature: the effect of measurement on a quantum state. Actually, the effect itself is not mysterious—quantum mechanics provides a representation of the various states of a system that could result from a measurement and a way to calculate the probabilities that the system will wind up in each of these states. The mystery is, *how does this work?* By what magic does the interaction of a microscopic system and a measuring apparatus bring about this change? This is *the problem of measurement*, and I don't have a satisfactory solution to it.

One response to this problem is to ignore it, *i.e.*, just accept quantum mechanics as a collection of (remarkably successful) tools for the quantitative analysis of physical systems. I suspect that this is the attitude of most practicing physicists. It is the stance that the Nobel laureate Richard P. Feynman recommends in his wonderful book *The Character of Physical Law* (Chicago: University of Chicago Press, 1967):

> There was a time when the newspapers said that only twelve men understood the theory of relativity. I do not believe that there ever was such a time. ... On the other hand, I think it is safe to say that no one understands quantum mechanics. ... Do not keep saying to yourself, if you can possibly avoid it, "But how can it be like that?" because you will get "down the drain" into a blind alley from which nobody has yet escaped. Nobody knows how it can be like that.

Underlying the problem of measurement there is a deeper question. As a consequence of an ensemble measurement of an observable Q, the original state collapses into one of the eigenstates of Q. The question is: *what mechanism determines which eigenstate a particular member collapses into?* According to the conventional epistemology of quantum mechanics, the answer is that *random chance* governs what happens to each member of an ensemble. Many (your author included) consider this no answer at all.

This question lurks just beneath the surface of most of the examples in this chapter. In Example 13.2, for example, we considered a system that prior to measurement was in a superposition of two eigenstates of the observable being measured. Upon measurement, each member collapses into one of these eigenstates.[27] This much quantum mechanics can tell us. It can also tell us the probability that each member will collapse into one of these eigenstates. What it cannot tell us is precisely what will happen to a *particular* member of the ensemble. That information is forever shrouded in the fundamental unknowability of nature.

No less a figure than Albert Einstein found this situation unsatisfactory. In a letter to Max Born in 1926, Einstein wrote[28]

> Quantum mechanics is very impressive. But an inner voice tells me that it is not yet the real thing. The theory produces a good deal but hardly brings us closer to the secret of the Old One. I am at all events convinced that *He* does not play dice.

Always More Questions

Since the early days of quantum theory, physicists have tried to understand the collapse of the state function. Perhaps the most remarkable of these efforts is that of John Von Neumann. In his research on measurement theory, Von Neumann used quantum mechanics to describe both the (microscopic) systems on which a measurement is performed and the (macroscopic) measuring apparatus. The central question he addressed is: *at what step in the measurement process does the collapse of the wave function occur*? For example, in the single slit experiment (§ 13.2), does the wave packet collapse into an eigenstate of position when the electron interacts with phosphor molecules at the detector? ... when these molecules emit a photon, enabling us to see that the electron has arrived? ... when an observer sees one of the photons? ... or at some other point in the experiment?

Von Neumann arrived by a mathematical analysis at a remarkable answer: *the collapse of the wave function of a quantum system occurs when a physical signal from a measurement device is registered by a human consciousness.* From this conclusion, some physicists have come to an even more astounding inference: *microscopic particles do not have well-defined physical attributes until a human mind perceives that they do.*

Such conclusions seem extreme. Yet, they are logically consistent with the Copenhagen interpretation of quantum mechanics. As you might imagine, this state of affairs

[27] A state function cannot collapse into an eigenstate that is not represented in its expansion in eigenfunctions of the operator that represents the measured observable. That is, if the projection of the state function on a particular eigenfunction is zero, then the *probability* of finding the system in that eigenstate after measurement is $|0|^2 = 0$. Obviously, a measurement of the observable cannot collapse the state function into such an eigenfunction.

[28] Quoted in A. Pais's scientific biography of Einstein, *"Subtle Is the Lord . . . " The Science and Life of Albert Einstein* (Oxford: Clarendon Press, 1980). If you have even the slightest interest in Einstein as a scientist or as a man, you should rush out right now and buy a copy of this splendid book. Einstein's correspondence with Born sheds considerable light on quantum physics; it has been collected in *The Born-Einstein Letters*, edited by M. Born (New York: Walker, 1971).

has led many physicists to search for alternative interpretations. Their fascinating theories are, alas, outside the scope of this book.[29]

The End of All Our Exploring

So to what place have we come? We have learned the underlying concepts, basic tools, and fundamental laws of quantum mechanics, a powerful and successful mathematical theory for predicting the behavior of microscopic systems. We have seen how quantum theory reduces to classical physics in the appropriate limit. And we have learned how to express its concepts in terms of state functions and operators and how to manipulate these elements to solve practical problems. (You will see much more of this in Volume II of this book.) Yet, quantum theory remains fraught with mysteries and marvels.

Einstein went to his grave convinced that in some fundamental way, quantum mechanics is incomplete. Even now, the issue is unresolved. From the standpoint in this book, which presents only the "standard" (Copenhagen) interpretation of quantum mechanics, we must accept the troublesome questions that lurk at the frontiers of our understanding of the physical universe: Just how *does* a measurement wreak its changes on a state function? Do physicists study reality or a self-generated chimera of reality, and is there a difference between the two? What powers actually govern what happens to an individual member of an ensemble? And on and on and on. These are the great unanswered questions of quantum physics.

ANNOTATED SELECTED READINGS

The literature of the measurement theory and its implications for the nature of reality is vast and fascinating. A recent review that will take you deep into this theory is

1. Bohm, D., B. J. Hiley, and P. N. Kaloyerou, "An Ontological Basis for the Quantum Theory," *Phys. Repts.*, **144**, 321 (1987).

There are no true textbooks on this subject, but you'll find a wealth of information clearly explained in

2. d'Espagnat, B., *Conceptual Foundations of Quantum Mechanics*, 2nd ed. (Reading, Mass: W. A. Benjamin, 1976).

Maybe you want something a little less advanced. If so, I strongly encourage you to seek out one of the popularizations listed in the Selected Readings for Chap. 2, to which I would add

3. Polkinghorne, J. C., *The Quantum World* (London: Longmann, 1984).
4. Mermin, N. David, "Quantum Mysteries for Everyone," *Jour. Phil.*, **78**, 397 (1981).

And, for historical background,

5. Crease, R. P., and Mann, C. C., *The Second Creation: Makers of the Revolution in 20th-Century Physics* (New York: Macmillan, 1987).

[29]But I don't want to leave you with the impression that none of the alternative formulations of quantum mechanics are viable. Several are. Alternatives to the Copenhagen interpretation include hidden variable theories, semiclassical interpretations, the many-worlds interpretation, the advanced-action interpretation, and a nonlocal model of reality. The most readable introduction to these interpretations is *Quantum Reality: Beyond the New Physics* by Nick Herbert (New York: Anchor Press, 1985). A useful overview of hidden-variable theories is given in *A Survey of Hidden-Variables* by F. J. Belinfante (New York: Pergamon, 1973). And you can find a more advanced, up-to-date discussion of these issues in *Open Questions in Quantum Physics*, edited by G. Tarozzi and A. A. van der Merwe (Dordrecht, The Netherlands: Reidel, 1985). Happy reading.

EXERCISES AND PROBLEMS

It's a beginning.
It's an end.
I leave to you the problem
of ordering your perceptions
and making the journey
from one to the other.

—from *Empire Star*
by Samuel R. Delaney

Exercises

13.1 Interchangeable Measurements.

Consider two states of a one-dimensional simple harmonic oscillator. The state Ψ_1 is a superposition of the first two stationary states; the (unnormalized) initial condition for this state is

$$\Psi_1(x,0) = 2\psi_0(x) + 3\psi_1(x). \tag{13.1.1}$$

The second state, Ψ_2, is the stationary state with energy $5\hbar\omega_0/2$.

[a] At $t = 0$ the parity of each state is measured. **Draw** branch diagrams for each state showing the effect on them of $\mathbf{M}_\Pi[t_\Pi]$.

[b] A short time after $\mathbf{M}_\Pi[t_\Pi]$, we measure the energy of each sub-ensemble that was created in the measurements of part [a]. **Complete** your branch diagram, showing the effect of $\mathbf{M}_E[t_E]$ on each state.

[c] Suppose that instead of the measurements in [a] and [b], we first measure E, *then* measure Π. **Draw** branch diagrams for this case. **Discuss** the differences (if any) between your diagram for this sequence and that of part [b].

Problems

13.1 Measurements on an Infinite Square Well

Consider a neutron of mass m_n in an infinite square well of width $L = 15\,\text{fermi}$. In an ensemble energy measurement at $t = 0$, we find that 1/10 of the members exhibit the value $2\pi^2\hbar^2/m_nL^2$ and 9/10 exhibit the value $225\pi^2\hbar^2/2m_nL^2$.

[a] **Write down** the wave function of this system prior to the energy measurement.

[b] **Draw** a branch diagram showing the effect of this measurement on the state.

[c] Suppose that at $t = 0$ we measure the parity *instead of the energy*. What is the probability that $\mathbf{M}_\Pi[t_\Pi]$ will collapse the state in part [a] into an even-parity eigenstate? . . . an odd-parity eigenstate?

[d] What is the mean value of Π that would be obtained in the measurement of part [c]?

[e] Do your answers to parts [c] and [d] depend on $\mathbf{M}_\Pi[t_\Pi]$? **Explain** your answer.

[f] Suppose the ensemble suffers an infinitesimal perturbation at $t = 0$ causing it to decay to the ground state, emitting a photon in the process. **Find** the frequencies of the photons that will be emitted by various members of the ensemble.

13.2 Reviewing Measurement Theory

Suppose that the initial state of a particle of mass m in a SHO potential is represented by the (unnormalized) wave function

$$\Psi(x,0) = \psi_1(x) + 4\psi_3(x) + 2\psi_5(x) - i\psi_6(x), \tag{13.2.1}$$

where $\psi_n(x)$ is the nth Hamiltonian eigenfunction for the SHO potential. At time $t_\Pi > 0$, we perform an ensemble measurement of the parity. A finite time after this measurement, at t_E, we measure the energy of each sub-ensemble that resulted from $\mathsf{M}_\Pi[t_\Pi]$.

[a] **Write down** an expression for the wave function of the original ensemble for times $0 < t < t_\Pi$.

[b] **Calculate** the average value and uncertainty of the parity for the state in part [a].

[c] In $\mathsf{M}_\Pi[t_\Pi]$, what is the probability that a member of the original ensemble will exhibit the eigenvalue $+1$? ... the value -1?

[d] Consider the sub-ensemble of members that exhibited odd parity. **Write down** the *normalized* state function of this sub-ensemble for $t_\Pi < t < t_E$.

[e] Now turn to the energy measurement at t_E on the sub-ensemble of part [d]. In this measurement, will all members exhibit the same value of E? Why or why not?

[f] If your answer to part [e] is yes, give the value of the (sharp) energy of the state. If your answer is no, **list** the values that will be exhibited by various members of the ensemble and evaluate the corresponding energy probabilities.

[g] Will $\mathsf{M}_E[t_E]$ of part [e] *necessarily* change the state of the system? If your answer is no, **explain** why not. If it is yes, **write down** the wave function(s) after the energy measurement.

13.3 The Final Problem: More Measurements

A particle of mass m in an infinite square well of width L whose left wall is at $x = 0$. At $t = 10\,\text{sec}$, an infinitesimal perturbation causes the particle to emit a photon of (angular) frequency $\omega = 3\pi^2\hbar/(2mL^2)$.

[a] From the above information only, a student concludes: "Prior to the photon emission, the particle was in the first excited state ($n = 2$) of the infinite square well." **Critique** this statement: is it correct? Why or why not? If not, what is the faulty assumption that underlies it?

[b] Suppose we are given the additional information that *all members* of the ensemble emitted a photon of the frequency ω at $t = 10\,\text{sec}$. Is this sufficient to draw the above quoted conclusion? **Explain** your answer fully.

13.4 Measuring a Constant of the Motion

A particle of mass m with a *symmetric* potential **that supports only bound states** is initially in a non-stationary state. In terms of the eigenfunctions $\psi_n(x)$ of the Hamiltonian of the system, this state is the superposition

$$\Psi(x,0) = N\left[3\psi_1(x) + \psi_3(x) + 2\psi_4(x)\right]. \tag{13.4.1}$$

At time $t > 0$, the *parity* of this system is measured.

[a] Does the initial state have definite parity? What is $\Psi(-x,0)$?

[b] **Write down** the *normalized* wave function for $t > 0$ (before the parity measurement is performed).

[c] Is parity a constant of the motion of this system?

[d] What is the probability of obtaining the value $+1$ (even parity) in the parity measurement?

[e] What is the probability of obtaining -1 (odd parity)?

Hint: Be careful in working part [e]. For the state (13.4.1) there are two "routes" to the odd-parity result. Remember that we add the probabilities for alternate routes to a given result.

Appendixes

APPENDIX A

SI Units (The International System of Units)

THE "BASE UNITS"

Physical quantity	SI unit		Dimension
	Name	Symbol	
length	metre	m	L
mass	kilogram	kg	M
time	second	s	T
electric current	ampere	A	A
thermodynamic temperature	kelvin	K	
amount of substance	mole	mol	

SUPPLEMENTARY UNITS

Physical quantity	SI Unit	
	Name	Symbol
plane angle	radian	rad
solid angle	sterradian	sr

DERIVED UNITS

Physical quantity	SI unit			Dimension
	Name	Symbol	Definition	
electric charge	coulomb	C	s A	TA
energy	joule	J	kg m^2 s^{-2}	ML^2T^{-2}
force	newton	N	kg m s^{-2}	MLT^{-2}
frequency	hertz	Hz	s^{-1}	T^{-1}
pressure	pascal	Pa	kg m^{-1} s^{-2}	$ML^{-1}T^{-2}$
power	watt	W	kg m^2 s^{-3}	ML^2T^{-3}
electric potential	volt	V	kg m^2 s^{-3} A^{-1}	$ML^2T^{-3}A^{-1}$
magnetic flux	weber	Wb	kg m^2 s^{-2} A^{-1}	$ML^2T^{-2}A^{-1}$

DERIVED UNITS RELATED TO THE ELECTROMAGNETIC FIELD

Physical quantity	Symbol for SI Unit	Dimension
electric current density **j**	A m^{-2}	$L^{-2}A$
electric dipole moment	C m	LTA
magnetic dipole moment	J T^{-1}	L^2A
electric field strength **E**	V m^{-1}	$MLT^{-3}A^{-1}$
electric displacement **D**	C m^{-2}	$L^{-2}TA$
electric potential ϕ	V	$ML^2T^{-3}A^{-1}$
electric charge density ρ	C m^{-3}	$L^{-3}TA$
magnetic field strength **H**	A m^{-1}	$L^{-1}A$
magnetic flux density **B**	T	$MT^{-2}A^{-1}$
vector potential **A**	T m	$MLT^{-2}A^{-1}$
permeability μ	H m^{-1}	$MLT^{-2}A^{-1}$

Additional Notes

1 atomic mass unit = 1 a.u. = $1.6605402 \times 10^{-27}$ kg

1 electron volt = 1 eV = 1.602177×10^{-19} J

COMMON UNITS AND THEIR DEFINITION IN SI UNITS

Quantity	Name	Symbol	Definition
Length	inch	in	0.0254 m
	foot	ft	0.3048 m
Mass	slug	slug	14.594 kg
Energy	British thermal unit	Btu	1055.05585252 J
Temperature	degree Celsius	°C	$t[°C] = T[K] - 273.15$
	degree Fahrenheit	°F	$t[°F] = \frac{9}{5}T[K] - 459.67$
Density	standard density of fluid Hg	ρ_{Hg}	13595.1 kg–m^{-3}
Pressure	standard atmosphere	atm	101325 Pa
	torr	Torr (mm Hg)	133.322368 Pa
	conventional mm of Hg	mmHg	133.322387 Pa

APPENDIX B

Fundamental Constants in SI (and CGS) Units

The values in this table were selected from B. N. Taylor, W. H. Parker, and D. N. Langenberg, *Rev. Mod. Phys.* **41**, 375 (1969), from *Reference Data on Atoms, Molecules, and Ions* by A. A. Radzig and B. M. Smirnov (New York: Springer-Verlag, 1980) and from E. R. Cohen and B. N. Taylor, *Rev. Mod. Phys.*, **59**, 1121 (1987). Because of the widespread use of CGS units in advanced quantum mechanics courses (and applications), I have also given appropriate powers of 10 for these units.

Note: $\epsilon_0 = 10^7/(4\pi c^2) \approx 8.854 \times 10^{-12}$ F–m^{-1}

Quantity	Symbol	Mantissa	CGS exponent	SI exponent
Speed of light in vacuum	c	2.99792458	10^{10} cm $\cdot$ s^{-1}	10^8 m $\cdot$ s^{-1}
Elementary charge	e	1.602177	10^{-20} e.m.u.	10^{-19} C
Planck constant	h	6.62607	10^{-27} erg $\cdot$ s	10^{-34} J $\cdot$ s
Rationalized Planck constant	$\hbar$	1.054573	10^{-27} erg $\cdot$ s	10^{-34} J $\cdot$ s
Rydberg constant	R_∞	1.097373153	10^5 cm^{-1}	10^7 m^{-1}
Fine structure constant	α	7.297353	10^{-3}	10^{-3}
	$1/\alpha$	1.3703599	10^2	10^2
Bohr radius	a_0	5.2917725	10^{-9} cm	10^{-11} m
Electron Compton wavelength	λ_C	2.4263106	10^{-10} cm	10^{-12} m
Classical electron radius	r_e	2.8179409	10^{-13} cm	10^{-15} m
Avogadro's constant	N_A	6.02214	10^{23} mol^{-1}	10^{23} mol^{-1}
Atomic Mass Unit	$1u$	1.66054	10^{-24} g	10^{-27} kg
Energy equivalent of 1 u	$E(u)$	1.49243	10^{-3} erg	10^{-10} J

Relevant Equations for the Above Table

$$R_\infty = \frac{m_e e^4}{4\pi\hbar^3 c}$$

$$\alpha = \frac{e^2}{\hbar c}$$

$$a_0 = \frac{\alpha}{4\pi R_\infty}$$

$$\lambda_C = \frac{\alpha^2}{2R_\infty}$$

$$r_e = \frac{\alpha\lambda_C}{2\pi}$$

$$1u = 10^{-3} N_A^{-1}\,\text{kg}\cdot\text{mol}^{-1}$$

$$E(u) = m_u c^2$$

Quantity	Symbol	Mantissa	CGS exponent	SI exponent
Electron rest mass	m_e	9.1094	10^{-28} g	10^{-31} kg
Proton rest mass	m_p	1.672623	10^{-24} g	10^{-27} kg
		1.0072765	a.m.u.	a.m.u.
Ratio	m_p/m_e	1.8361527	10^3	10^3
Neutron rest mass	m_n	1.674929	10^{-24} gm	10^{-27} kg
		1.0086649	a.m.u.	a.m.u.
Permeability of vacuum	μ_0	1.2566371	—	$10^{-6}\,\text{H}\cdot\text{m}^{-1}$
Permittivity of vacuum	ϵ_0	8.8541878	—	$10^{-12}\,\text{F}\cdot\text{m}^{-1}$
Gravitational constant	G	6.67260	$10^{-8}\,\text{dyn}\cdot\text{cm}^2\cdot\text{g}^{-2}$	$10^{-11}\,\text{N}\cdot\text{m}^2\cdot\text{kg}^{-2}$
Boltzmann constant	k_B	1.380658	$10^{-16}\,\text{erg}\cdot\text{K}^{-1}$	$10^{-23}\,\text{J}\cdot\text{K}^{-1}$
Stefan-Boltzmann constant	σ	5.67051	$10^{-5}\,\text{erg}\cdot\text{s}^{-1}\cdot\text{cm}^{-2}\cdot\text{K}^{-4}$	$10^{-8}\,\text{W}\cdot\text{m}^{-2}\cdot\text{K}^{-4}$

$m_e = 0.511 MeV/c^2$
$m_p = 938.28 MeV/c^2$
$m_n = 937.57 MeV/c^2$

Relevant Equations for the Above Table

$$\epsilon_0 = \frac{1}{\mu_0 c^2}$$

$$\sigma = \frac{\pi^2}{60}\frac{k^4}{\hbar^3 c^2}$$

APPENDIX C

Order-of-Magnitude Estimates

Constants

$$c = 3 \times 10^8 \text{ m–s}^{-1}$$

$$\hbar = 10^{-34} \text{ J–s}$$

$$e = 1.6 \times 10^{-19} \text{ C}$$

$$R_\infty = 1 \times 10^7 \text{ m}^{-1}$$

$$\alpha = 1/137$$

$$a_0 = 5 \times 10^{-11} \text{ m}$$

$$m_e = 1 \times 10^{-31} \text{ kg}$$

$$m_p = 1 \times 10^{-27} \text{ kg}$$

$$m_n = 1.7 \times 10^{-27} \text{ kg}$$

$$\mu_0 = 4\pi \times 10^{-7} \text{ H–m}^{-1}$$

$$\epsilon_0 = 8.8 \times 10^{-12} \text{ F–m}^{-1}$$

$$k_B = 1.4 \times 10^{-23} \text{ J–K}^{-1}$$

$$k_b T_{\text{room}} = 0.025 \text{ eV}$$

APPENDIX D

Conversion Tables

Convert from units in the first column to units in the first row by multiplying by the conversion factor in the corresponding row-column position, as

$$1\,\text{eV} = 1.6022 \times 10^{-19}\,\text{J}$$

ENERGY

Unit	eV	J[a]	kcal/mol	cm^{-1}
1 eV	1	1.6022×10^{-19}	23.045	8065.48
1 J	6.2415×10^{18}	1	1.4384×10^{20}	5.0340×10^{22}
1 kcal/mol	4.3393×10^{-2}	6.9524×10^{-21}	1	3.4999×10^{2}
1 cm^{-1}	1.23985×10^{-4}	1.9865×10^{-23}	2.8574×10^{-3}	1
1 K	8.617×10^{-5}	1.3807×10^{-23}	1.9859×10^{-3}	6.9501×10^{-1}
1 a.u.	27.2116	4.3597×10^{-18}	6.2709×10^{2}	2.1947×10^{5}
1 Ry	13.606	2.1799×10^{-18}	3.1355×10^{2}	1.0974×10^{5}
1 MHz	4.1357×10^{-9}	6.6262×10^{-28}	9.5308×10^{-8}	3.3356×10^{-5}

[a] $1\,\text{J} = 10^{7}$ erg.

Unit	K	a.u.	Ry	MHz
1 eV	11604	3.6749×10^{-2}	7.3498×10^{-2}	2.4180×10^{8}
1 J	7.2430×10^{22}	2.2937×10^{17}	4.5873×10^{17}	1.5092×10^{27}
1 kcal/mole	5.0356×10^{2}	1.5946×10^{-3}	3.1893×10^{-3}	1.0492×10^{7}
1 cm^{-1}	1.4388	4.5563×10^{-6}	9.1127×10^{-6}	2.9979×10^{4}
1 K	1	3.1688×10^{-6}	6.3335×10^{-6}	2.0836×10^{4}
1 a.u.	3.1578×10^{5}	1	2	6.5797×10^{9}
1 Ry	1.5789×10^{5}	0.5	1	3.2898×10^{9}
1 MHz	4.7993×10^{-5}	1.5198×10^{-10}	3.0397×10^{-10}	1

Additional Useful Definitions:

$1\,\text{Hz} = 1\,\text{s}^{-1}$

$1\,\text{MHz} = 10^6\,\text{Hz}$

$1\,\text{keV} = 10^3\,\text{eV}$

$1\,\text{MeV} = 10^6\,\text{eV}$

MASS

Unit	kg	g	u
1 kg	1	1×10^{-3}	6.024×10^{26}
1 g	1×10^{3}	1	6.024×10^{23}
1 amu[a]	1.660×10^{-27}	1.660×10^{-24}	1

[a] 1 atomic mass unit $\cdot\, c^2 = 1.4924418 \times 10^{-10}\,\text{J} = 931.5016\,\text{MeV}$

LENGTH

Unit	m[a]	cm	mm	μ	Å	f
1 m (meter)	1	10^{2}	10^{3}	10^{6}	10^{10}	10^{15}
1 cm (centimeter)	10^{-2}	1	10	10^{4}	10^{8}	10^{13}
1 mm (millimeter)	10^{-3}	10^{-1}	1	10^{3}	10^{7}	10^{12}
1 μ (micron)	10^{-6}	10^{-4}	10^{-3}	1	10^{4}	10^{9}
1 Å(Ångstrom)	10^{-10}	10^{-8}	10^{-7}	10^{-4}	1	10^{5}
1 f (fermi)	10^{-15}	10^{-13}	10^{-12}	10^{-9}	10^{-5}	1

[a] 1 inch $= 2.540 \times 10^{-2}$ m

CROSS SECTION

Unit	cm^2	$Å^2$	a_0^2	πa_0^2	barn
1 cm^2	1	1×10^{16}	3.5711×10^{16}	1.1367×10^{16}	1×10^{24}
1 $Å^2$	1×10^{-16}	1	3.5711	1.367	1×10^{8}
1 a_0^2	2.8003×10^{-17}	2.8003×10^{-1}	1	3.1831×10^{-1}	2.8003×10^{7}
1 πa_0^2	8.7973×10^{-17}	8.7973×10^{-1}	3.1416	1	8.7973×10^{7}
1 barn	1×10^{-24}	1×10^{-8}	3.5711×10^{-8}	1.1367×10^{-8}	1

PLANCK'S CONSTANT IN VARIOUS UNITS

Units	h	$\hbar$
J s	6.626176×10^{-34}	1.0545887×10^{-34}
erg s	6.626176×10^{-27}	1.0545887×10^{-27}
eV s	4.135701×10^{-15}	6.5821733×10^{-16}
MeV s	4.135701×10^{-21}	6.5821733×10^{-22}

APPENDIX E

Prefix Dictionary

The names of multiples and submultiples of the units are formed with the following prefixes:

Factor by which unit is multiplied	Prefix	Symbol	Factor by which unit is multiplied	Prefix	Symbol
10^{18}	exa	E	10^{-1}	deci	d
10^{15}	pecta	P	10^{-2}	centi	c
10^{12}	tera	T	10^{-3}	milli	m
10^{9}	giga	G	10^{-6}	micro	μ
10^{6}	mega	M	10^{-9}	nano	n
10^{3}	kilo	k	10^{-12}	pico	p
10^{2}	hecto	h	10^{-15}	femto	f
10	deca	da	10^{-18}	atto	a

APPENDIX F

A User's Guide to Atomic Units

Introduction

In atomic units, the following quantities are arbitrarily set equal to 1.0: m_e (the rest mass of the electron), $\hbar$ (Planck's constant divided by 2π), e (the elementary charge), and a_0 (the first Bohr radius of the hydrogen atom). The resulting unit of energy is the Hartree E_h:

$$E_h = \frac{m_e e^4}{\hbar^2} = 27.211\,\text{eV} = 4.359748 \times 10^{-18}\,\text{J}$$

Confusingly, results are often reported in another energy unit, the Rydberg

$$1\,\text{Ry} = \frac{1}{2}E_h = 13.605804\,\text{eV}$$

Summary of Energy Conversions to Atomic Units

1 eV = $3.674931 \cdot 10^{-2}$ Hartree = $1.6021773 \cdot 10^{-19}$ J
1 Rydberg = $\frac{1}{2}$ Hartree = 13.6058 eV = $2.179907 \cdot 10^{-18}$ J
1 Hartree = 2 Rydberg = 27.2114 eV = $4.359748 \cdot 10^{-18}$ J

Conversion to Wavenumbers

$k = \frac{1}{\lambda}$, where λ is the wavelength:
1 Hartree = $2.194746 \cdot 10^5$ cm^{-1}
1 Rydberg = $1.097373 \cdot 10^5$ cm^{-1}
1 cm^{-1} = $4.556335 \cdot 10^{-6}$ Hartree = $9.112672 \cdot 10^{-6}$ Rydberg

TABLE F.1 DEFINITION OF THE ATOMIC UNITS OF FUNDAMENTAL PHYSICAL QUANTITIES AND THEIR VALUES IN SI UNITS.

Unit and Significance	Symbol and/or Definition	Value
charge (charge on electron)	e	1.60218×10^{-19} C
mass (electron rest mass)	m_e	9.10939×10^{-31} kg
angular momentum	$\hbar$	1.05457×10^{-34} J · s
length (first Bohr radius in H)	$a_0 = \hbar^2/m_e e^2$	0.52918×10^{-10}m = 0.52918 Å
energy (twice the ionization potential of H with $m_n = \infty$)	$E_h = m_e e^4/\hbar^2$	27.2116 eV = 4.35975×10^{-18} J
velocity (electron speed in first Bohr orbit)	$v_e = e^2/\hbar$	2.18769×10^6 m · s^{-1}
momentum (of electron in first Bohr orbit)	$p_e = m_e e^2/\hbar$	1.992888×10^{-24} kg · m · s^{-1}
wavenumber	$\lambda_e = a_0^{-1}$	1.88973×10^{10} m^{-1}
time (electron transit of first Bohr orbit)	$t_e = \hbar^3/m_e e^4$	2.41889×10^{-17} s
frequency	$\omega_e = v_e/a_0$	4.134136×10^{16} Hz
cross section	a_0^2	2.800×10^{-21} m^2
	πa_0^2	8.79735×10^{-21} m^2
polarizability	a_0^3	1.48187×10^{-31} m^3
current density	$j_e = m_e^3 e^9/\hbar^7$	3.62517×10^{-26} m^2 · s^{-1}
dipole moment	$D = ea_0$	8.47842×10^{-3} C · m = 2.5418 D(Debye)
magnetic moment	$\mu_e = \hbar^2/m_e e$	7.6200×10^{-20} J/T

TABLE F.2 VALUES OF OMNIPRESENT PHYSICAL QUANTITIES EXPRESSED IN ATOMIC UNITS.

Quantity	Symbol	Value in atomic units
Electron mass	m_e	1
Proton mass	m_p	1835.15152
Neutron mass	m_n	1838.6827
Speed of light	c	$1/\alpha = 137.03604$
Rydberg constant	R_∞	$\alpha/4\pi = 5.807047 \cdot 10^{-4}$
Boltzmann constant	k_B	$3.1667908 \cdot 10^{-6}$

Note: In this table, $\alpha = 7.2973531 \times 10^{-3}$ is the (dimensionless) fine-structure constant: $\alpha = e^2/\hbar c$

APPENDIX G

The Greek Alphabet

A, α	Alpha	N, ν	Nu
B, β	Beta	Ξ, ξ	Xi
Γ, γ	Gamma	O, o	Omicron
Δ, δ	Delta	Π, π	Pi
E, ϵ	Epsilon	P, ρ	Rho
Z, ζ	Zeta	Σ, σ	Sigma
H, η	Eta	T, τ	Tau
Θ, θ	Theta	Υ, υ	Upsilon
I, ι	Iota	Φ, ϕ	Phi
K, κ	Kappa	X, χ	Chi
Λ, λ	Lambda	Ψ, ψ	Psi
M, μ	Mu	Ω, ω	Omega

APPENDIX H

Handy Trigonometric Identities

$$1\,\text{rad} = \frac{180}{\pi} = 57.296^\circ$$

$$1^\circ = \frac{\pi}{180} = 0.17453\,\text{rad}$$

Values of Sine and Cosine for Multiples of π

$$\sin n\pi = 0 \qquad n = 0,\ 1,\ 2,\ 3,\ 4,\ \ldots$$

$$\cos n\pi = (-1)^n$$

$$\sin \tfrac{n\pi}{2} = \tfrac{1}{2}(i)^{n+1}[(-1)^n - 1]$$

$$\cos \tfrac{n\pi}{2} = \tfrac{1}{2}(i)^n[1 + (-1)^n]$$

$$\sin \tfrac{n\pi}{4} = 0 \qquad n = 0,\ 4,\ 8,\ 12,\ 16,\ \ldots \qquad 4m\ (m = \text{integer})$$

$$\sin \tfrac{n\pi}{4} = (-1)^{\frac{1}{4}(n-2)} \qquad n = 2,\ 6,\ 10,\ 14,\ 18,\ \ldots \qquad 4m + 2\ (m = \text{integer})$$

$$\sin \tfrac{n\pi}{4} = \tfrac{1}{\sqrt{2}}(-1)^{\frac{1}{8}(n^2+4n+11)} \qquad n = 1,\ 3,\ 5,\ 7,\ \ldots$$

Sums and Differences of Angles

$\alpha =$	$90^\circ \pm \beta$	$180^\circ \pm \beta$	$270^\circ \pm \beta$	$360^\circ \pm \beta$
$\sin\alpha =$	$+\cos\beta$	$\mp\sin\beta$	$-\cos\beta$	$\pm\sin\beta$
$\cos\alpha =$	$\mp\sin\beta$	$-\cos\beta$	$\pm\sin\beta$	$+\cos\beta$
$\tan\alpha =$	$\mp\cot\beta$	$\pm\tan\beta$	$\mp\cot\beta$	$\pm\tan\beta$
$\cot\alpha =$	$\mp\tan\beta$	$\pm\cot\beta$	$\mp\tan\beta$	$\mp\cot\beta$

Exponential Definition of Trigonometric Functions

$$\sin\theta = \frac{1}{2i}\left(e^{i\theta} - e^{-i\theta}\right) \qquad e^{i\theta} = \cos\theta + i\sin\theta$$

$$\cos\theta = \frac{1}{2}\left(e^{i\theta} + e^{-i\theta}\right) \qquad e^{-i\theta} = \cos\theta - i\sin\theta$$

Fundamental Trigonometric Relations Involving Only One Angle

$$\sin^2\alpha + \cos^2\alpha = 1 \qquad \sec^2\alpha - \tan^2\alpha = 1 \qquad \operatorname{cosec}^2\alpha - \cot^2\alpha = 1$$

$$\sin\alpha\cdot\operatorname{cosec}\alpha = 1, \qquad \cos\alpha\cdot\sec\alpha = 1, \qquad \tan\alpha\cdot\cot\alpha = 1$$

$$\sin\alpha = \sqrt{1-\cos^2\alpha} = \frac{\tan\alpha}{\sqrt{1+\tan^2\alpha}} = \frac{1}{\sqrt{1+\cot^2\alpha}}$$

$$\cos\alpha = \sqrt{1-\sin^2\alpha} = \frac{\cot\alpha}{\sqrt{1+\cot^2\alpha}} = \frac{1}{\sqrt{1+\tan^2\alpha}}$$

$$\tan\alpha = \sqrt{\sec^2\alpha - 1} = \frac{\sin\alpha}{\sqrt{1-\sin^2\alpha}} = \frac{1}{\sqrt{\operatorname{cosec}^2\alpha - 1}}$$

$$\cot\alpha = \sqrt{\operatorname{cosec}^2\alpha - 1} = \frac{\cos\alpha}{\sqrt{1-\cos^2\alpha}} = \frac{1}{\sqrt{\sec^2\alpha - 1}}$$

$$\tan\left(\alpha + \tfrac{\pi}{4}\right) = \frac{1+\tan\alpha}{1-\tan\alpha}$$

$$\cot\left(\tfrac{\pi}{4} - \beta\right) = \frac{\cot\beta + 1}{\cot\beta - 1}$$

$$\sin 2\alpha = 2\sin\alpha\cos\alpha$$

$$\cos 2\alpha = \cos^2\alpha - \sin^2\alpha$$

$$\tan 2\alpha = \frac{2\tan\alpha}{1-\tan^2\alpha}$$

$$\cot 2\alpha = \frac{\cot^2\alpha - 1}{2\cot\alpha}$$

$$\sin 3\alpha = 3\sin\alpha - 4\sin^3\alpha$$

$$\cos 3\alpha = 4\cos^3\alpha - 3\cos\alpha$$

$$\sin\frac{\alpha}{2} = \sqrt{\frac{1-\cos\alpha}{2}}$$
$$\cos\frac{\alpha}{2} = \sqrt{\frac{1+\cos\alpha}{2}}$$

$$\tan\frac{\alpha}{2} = \sqrt{\frac{1-\cos\alpha}{1+\cos\alpha}} = \frac{1-\cos\alpha}{\sin\alpha} = \frac{\sin\alpha}{1+\cos\alpha}$$

$$\cot\frac{\alpha}{2} = \sqrt{\frac{1+\cos\alpha}{1-\cos\alpha}} = \frac{1+\cos\alpha}{\sin\alpha} = \frac{\sin\alpha}{1-\cos\alpha}$$

$$\sin^2\alpha = \tfrac{1}{2}(1-\cos 2\alpha),$$
$$\cos^2\alpha = \tfrac{1}{2}(1+\cos 2\alpha)$$

$$\sin^3\alpha = \tfrac{1}{4}(3\sin\alpha - \sin 3\alpha)$$
$$\cos^3\alpha = \tfrac{1}{4}(\cos 3\alpha + 3\cos\alpha)$$

Fundamental Trigonometric Relations Involving Two or More Angles

$$\sin\alpha \pm \sin\beta = 2\sin\tfrac{1}{2}(\alpha\pm\beta)\cos\tfrac{1}{2}(\alpha\mp\beta)$$

$$\cos\alpha + \cos\beta = 2\cos\tfrac{1}{2}(\alpha+\beta)\cos\tfrac{1}{2}(\alpha-\beta)$$

$$\cos\alpha - \cos\beta = -2\sin\tfrac{1}{2}(\alpha+\beta)\sin\tfrac{1}{2}(\alpha-\beta)$$

$$\tan\alpha \pm \tan\beta = \frac{\sin(\alpha\pm\beta)}{\cos\alpha\cos\beta}$$

$$\cot\alpha \pm \cot\beta = \frac{\pm\sin(\alpha\pm\beta)}{\sin\alpha\sin\beta}$$

$$\sin^2\alpha - \sin^2\beta = \sin(\alpha+\beta)\cdot\sin(\alpha-\beta)$$

$$\cos^2\alpha - \cos^2\beta = -\sin(\alpha+\beta)\cdot\sin(\alpha-\beta)$$

$$\cos^2\alpha - \sin^2\beta = \cos(\alpha+\beta)\cdot\cos(\alpha-\beta)$$

$$\frac{\sin\alpha\pm\sin\beta}{\cos\alpha+\cos\beta} = \tan\tfrac{1}{2}(\alpha\pm\beta)$$

$$\frac{\sin\alpha+\sin\beta}{\sin\alpha-\sin\beta} = \frac{\tan\frac{1}{2}(\alpha+\beta)}{\tan\frac{1}{2}(\alpha-\beta)}$$

$$\frac{\sin\alpha \pm \sin\beta}{\cos\alpha - \cos\beta} = -\cot\tfrac{1}{2}(\alpha \pm \beta)$$

$$\sin(\alpha \pm \beta) = \sin\alpha\cos\beta \pm \cos\alpha\sin\beta$$

$$\cos(\alpha \pm \beta) = \cos\alpha\cos\beta \mp \sin\alpha\sin\beta$$

$$\tan(\alpha \pm \beta) = \frac{\tan\alpha \pm \tan\beta}{1 \mp \tan\alpha\tan\beta}$$

$$\cot(\alpha \pm \beta) = \frac{\cot\alpha\cot\beta \mp 1}{\cot\beta \pm \cot\alpha}$$

$$\begin{aligned}\sin(\alpha+\beta+\gamma) &= \sin\alpha\cos\beta\cos\gamma + \cos\alpha\sin\beta\cos\gamma \\ &\quad + \cos\alpha\cos\beta\sin\gamma - \sin\alpha\sin\beta\sin\gamma\end{aligned}$$

$$\begin{aligned}\cos(\alpha+\beta+\gamma) &= \cos\alpha\cos\beta\cos\gamma - \sin\alpha\sin\beta\cos\gamma \\ &\quad - \sin\alpha\cos\beta\sin\gamma - \cos\alpha\sin\beta\sin\gamma\end{aligned}$$

$$\sin\alpha\sin\beta = \tfrac{1}{2}[\cos(\alpha-\beta) - \cos(\alpha+\beta)]$$

$$\cos\alpha\cos\beta = \tfrac{1}{2}[\cos(\alpha-\beta) + \cos(\alpha+\beta)]$$

$$\sin\alpha\cos\beta = \tfrac{1}{2}[\sin(\alpha-\beta) + \sin(\alpha+\beta)]$$

$$\sin\alpha\sin\beta\sin\gamma = \tfrac{1}{4}[\sin(\alpha+\beta-\gamma) + \sin(\beta+\gamma-\alpha) + \sin(\gamma+\alpha-\beta) - \sin(\alpha+\beta+\gamma)]$$

$$\sin\alpha\cos\beta\cos\gamma = \tfrac{1}{4}[\sin(\alpha+\beta-\gamma) - \sin(\beta+\gamma-\alpha) + \sin(\gamma+\alpha-\beta) + \sin(\alpha+\beta+\gamma)]$$

$$\sin\alpha\sin\beta\cos\gamma = \tfrac{1}{4}[-\cos(\alpha+\beta-\gamma) + \cos(\beta+\gamma-\alpha) + \cos(\gamma+\alpha-\beta) - \cos(\alpha+\beta+\gamma)]$$

$$\cos\alpha\cos\beta\cos\gamma = \tfrac{1}{4}[\cos(\alpha+\beta-\gamma) + \cos(\beta+\gamma-\alpha) + \cos(\gamma+\alpha-\beta) + \cos(\alpha+\beta+\gamma)]$$

Hyperbolic Functions

$$\sinh u = \frac{e^u - e^{-u}}{2} \qquad \cosh u = \frac{e^u + e^{-u}}{2}$$

$$\tanh u = \frac{e^u - e^{-u}}{e^u + e^{-u}} \qquad \coth u = \frac{e^u + e^{-u}}{e^u - e^{-u}}$$

$$\operatorname{csch} u = \frac{1}{\sinh u} \qquad \operatorname{sech} u = \frac{1}{\cosh u}$$

$$\coth u = \frac{1}{\tanh u}$$

$$\sinh(-u) = -\sinh u \qquad \text{odd function}$$

$$\cosh(-u) = \cosh u \qquad \text{even function}$$

$$\tanh(-u) = -\tanh u \qquad \text{odd function}$$

$$\tanh(-u) = -\coth u \qquad \text{odd function}$$

Fundamental Relations Involving One Angle

$$\sinh u = \frac{\tanh u}{\operatorname{sech} u} \qquad \cosh u = \frac{\coth u}{\operatorname{cosech} u} \qquad \tanh u = \frac{\sinh u}{\cosh u}$$

$$\operatorname{cosech} u = \frac{\operatorname{sech} u}{\tanh u} \qquad \operatorname{sech} u = \frac{\operatorname{cosech} u}{\coth} \qquad \coth u = \frac{\cosh u}{\sinh u}$$

$$\sinh u = \tanh u \cosh u \qquad \cosh u = \coth u \sinh u$$

$$\tanh u = \sinh u \operatorname{sech} u \qquad \coth u = \cosh u \operatorname{cosech} u$$

$$\operatorname{sech} u = \operatorname{cosech} u \tanh u \qquad \operatorname{cosech} u = \operatorname{sech} u \coth u$$

$$\cosh^2 u - \sinh^2 u = 1 \qquad \tanh^2 u + \operatorname{sech}^2 u = 1$$

$$\coth^2 u - \operatorname{csch}^2 u = 1$$

$$\operatorname{csch}^2 u - \operatorname{sech}^2 u = \operatorname{csch}^2 u \operatorname{sech}^2 u$$

Fundamental Relations Involving Two or More Angles

$$\sinh(u+v) = \sinh u \cosh v + \cosh u \sinh v$$

$$\sinh(u-v) = \sinh u \cosh v - \cosh u \sinh v$$

$$\cosh(u+v) = \cosh u \cosh v + \sinh u \sinh v$$

$$\cosh(u-v) = \cosh u \cosh v - \sinh u \sinh v$$

Multiples of One Angle

$$\sinh 2u = 2 \sinh u \cosh u = \frac{2 \tanh u}{1 - \tanh^2 u}$$

$$\cosh 2u = \cosh^2 u + \sinh^2 u = 2\cosh^2 u - 1$$

$$= 1 + 2\sinh^2 u = \frac{1 + \tanh^2 u}{1 - \tanh^2 u}$$

$$\tanh 2u = \frac{2 \tanh u}{1 + \tanh^2 u} \qquad \coth 2u = \frac{\coth^2 u + 1}{2 \coth u}$$

$$\sinh u + \sinh v = 2 \sinh \frac{1}{2}(u+v) \cosh \frac{1}{2}(u-v)$$

$$\sinh u - \sinh v = 2 \cosh \frac{1}{2}(u+v) \sinh \frac{1}{2}(u-v)$$

$$\cosh u + \cosh v = 2 \cosh \frac{1}{2}(u+v) \cosh \frac{1}{2}(u-v)$$

$$\cosh u - \cosh v = 2 \sinh \frac{1}{2}(u+v) \sinh \frac{1}{2}(u-v)$$

$$\sinh u \cosh v = \frac{1}{2} \sinh(u+v) + \frac{1}{2} \sinh(u-v)$$

$$\cosh u \sinh v = \frac{1}{2} \sinh(u+v) - \frac{1}{2} \sinh(u-v)$$

$$\cosh u \cosh v = \frac{1}{2} \cosh(u+v) + \frac{1}{2} \cosh(u-v)$$

$$\sinh u \sinh v = \frac{1}{2} \cosh(u+v) - \frac{1}{2} \cosh(u-v)$$

$$\sinh^2 u = \frac{1}{2}(\cosh 2u - 1)$$

$$\cosh^2 u = \frac{1}{2}(\cosh 2u + 1)$$

$$(\cosh u + \sinh u)^n = \cosh nu + \sinh nu$$

Relations Between Hyperbolic and Trigonometric Functions

$$\sinh iu = i \sin u, \qquad \sinh u = -i \sin iu$$

$$\cosh iu = \cos u, \qquad \cosh u = \cos iu$$

$$\tanh iu = i \tan u, \qquad \tanh u = -i \tan iu$$

Every hyperbolic relation may be obtained from the corresponding trigonometric relation by replacing $\sin \alpha$ by $i \sinh u$ and $\cos \alpha$ by $\cosh u$.

APPENDIX I

Integrals You Should Know

Indefinite Integrals

$$\int x \sin bx \, dx = \frac{1}{b^2} \sin bx - \frac{x}{b} \cos bx$$

$$\int x e^{bx} \, dx = \frac{e^{bx}}{b^2}(bx - 1)$$

$$\int x^2 e^{bx} \, dx = e^{bx} \left(\frac{x^2}{b} - \frac{2x}{b^2} + \frac{2}{b^3} \right)$$

$$\int x^n e^{ax} \, dx = \frac{x^n e^{ax}}{a} - \frac{n}{a} \int x^{n-1} e^{ax} \, dx$$

$$\int \sin x \, dx = -\cos x$$

$$\int \sin \frac{x}{a} dx = -a \cos \frac{x}{a}$$

$$\int x \sin x \, dx = \sin x - x \cos x$$

$$\int x^2 \sin x \, dx = 2x \sin x - (x^2 - 2) \cos x$$

$$\int \sin^2 x \, dx = \frac{x}{2} - \frac{\sin 2x}{4} = \frac{x}{2} - \frac{\sin x \cos x}{2}$$

$$\int x \sin^2 x \, dx = \frac{x^2}{4} - \frac{x \sin 2x}{4} - \frac{\cos 2x}{8}$$

$$\int \sin^3 x\, dx = \frac{\cos^3 x}{3} - \cos x$$

$$\int x \cos x\, dx = \cos x + x \sin x$$

$$\int x^2 \cos x\, dx = 2x \cos x + (x^2 - 2) \sin x$$

$$\int \cos^2 x\, dx = \frac{x}{2} + \frac{\sin 2x}{4} = \frac{x}{2} + \frac{\sin x \cos x}{2}$$

$$\int x \cos^2 x\, dx = \frac{x^2}{4} + \frac{x \sin 2x}{4} + \frac{\cos 2x}{8}$$

$$\int x^2 \cos^2 x\, dx = \frac{x^3}{6} + \left(\frac{x^2}{4} - \frac{1}{8}\right) \sin 2x + \frac{x \cos 2x}{4}$$

$$\int \cos^3 x\, dx = \sin x - \frac{\sin^3 x}{3}$$

$$\int x \cos^3 x\, dx = \frac{x \sin 3x}{12} + \frac{\cos 3x}{36} + \frac{3}{4} x \sin x + \frac{3}{4} \cos x$$

$$\int \sin x \cos^2 x\, dx = -\frac{\cos^3 x}{3}$$

$$\int \sin^2 x \cos x\, dx = \frac{\sin^3 x}{3}$$

$$\int \tan^2 x\, dx = \tan x - x$$

$$\int \cot^2 x\, dx = -\cot x - x$$

$$\int x^2 e^{ax} dx = e^{ax} \left[\frac{x^2}{a} - \frac{2x}{a^2} + \frac{2}{a^3}\right]$$

Integrals from 0 to ∞ [for $a > 0$]

$$\int_0^\infty x^n e^{-ax}\, dx = \frac{n!}{a^{n+1}} = \frac{\Gamma(n+1)}{a^{n+1}}, \quad n > -1$$

$$\int_0^\infty e^{-ax^2}\, dx = \frac{1}{2}\sqrt{\frac{\pi}{a}}$$

$$\int_0^\infty x\, e^{-ax^2}\, dx = \frac{1}{2a}$$

$$\int_0^\infty x^2\, e^{-ax^2}\, dx = \frac{1}{4}\sqrt{\frac{\pi}{a^3}}$$

$$\int_0^\infty x^3\, e^{-ax^2}\, dx = \frac{1}{2a^2}$$

$$\int_0^\infty x^{2n}\, e^{-ax^2}\, dx = \frac{1\cdot 3\cdots(2n-1)}{2^{n+1}}\sqrt{\frac{\pi}{a^{2n+1}}} \quad [n = 1, 2, 3, \ldots]$$

$$\int_0^\infty x^{2n+1}\, e^{-ax^2}\, dx = \frac{n!}{2a^{n+1}} \quad [n = 0, 1, 2, \ldots]$$

$$\int_0^\infty \frac{dx}{(a^2+x^2)^n} = \frac{1\cdot 3\cdot 5\cdots(2n-3)}{2\cdot 4\cdot 6\cdots(2n-2)}\,\frac{\pi}{2a^{2n-1}}, \quad [a > 0; n = 2, 3, \ldots]$$

$$\int_0^\infty e^{-ax}\cos mx\, dx = \frac{a}{a^2+m^2}$$

Integrals from 1 to ∞ [for $a > 0$]

$$\int_1^\infty e^{-ax}\, dx = \frac{e^{-a}}{a}$$

$$\int_1^\infty x\, e^{-ax}\, dx = \frac{e^{-a}}{a^2}\,(1+a)$$

$$\int_1^\infty x^2\, e^{-ax}\, dx = \frac{2\, e^{-a}}{a^3}\left(1 + a + \frac{a^2}{2}\right)$$

Integrals from -1 to $+1$

$$\int_{-1}^{+1} e^{-ax}\, dx = \frac{1}{a}\left(e^a - e^{-a}\right)$$

$$\int_{-1}^{+1} x\, e^{-ax}\, dx = \frac{1}{a^2}\left[e^a - e^{-a} - a\left(e^a + e^{-a}\right)\right]$$

$$\int_{-1}^{+1} x^n\, e^{-ax}\, dx = (-1)^{n+1}A_n(-a) - A_n(a)$$

where

$$A_n(a) \equiv \int_0^{\infty} x^n\, e^{-ax}\, dx = \frac{n!\, e^{-a}}{a^{n+1}} \sum_{k=0}^{n} \frac{a^k}{k!}$$

$$\int_{-1}^{+1} x^n\, dx = \begin{cases} 0 & n = 1, 3, 5, \ldots \\ \dfrac{2}{n+1} & n = 0, 2, 4, \ldots \end{cases}$$

Integrals from 0 ***to*** 1 ***[for*** $a > 0$***]***

$$\int_0^1 x\, e^{-ax}\, dx = \frac{1}{a^2}\left[1 - e^{-a}(1+a)\right]$$

$$\int_0^1 e^{-ax}\, dx = \frac{1}{a}\left(1 - e^{-a}\right)$$

$$\int_0^1 x^2\, e^{-ax}\, dx = \frac{2}{a^3}\left[1 - e^{-a}\left(1 + a + \frac{a^2}{2}\right)\right]$$

Integrals from 0 ***to*** $\pi/2$

$$\int_0^{\pi/2} \sin^2 mx\, dx = \int_0^{\pi/2} \cos^2 mx\, dx = \frac{\pi}{4} \quad [m = 1, 2, \ldots]$$

Integrals from 0 ***to*** π ***[***M ***and*** N ***Non-zero Integers]***

$$\int_0^{\pi} \sin^2 x\, dx = \int_0^{\pi} \cos^2 x\, dx = \frac{\pi}{2}$$

$$\int_0^{\pi} \sin mx \sin nx\, dx = \begin{cases} 0 & [m \neq n] \\ \dfrac{\pi}{2} & [m = n] \end{cases}$$

$$\int_0^{\pi} \cos mx \cos nx\, dx = \begin{cases} 0 & [m \neq n] \\ \dfrac{\pi}{2} & [m = n] \end{cases}$$

$$\int_0^{\pi} \sin mx \cos nx\, dx = \begin{cases} 0 & [m = n] \\ 0 & [m \neq n; (m+n) \text{ even}] \\ \dfrac{2m}{m^2 - n^2} & [m \neq n; (m+n) \text{ odd}] \end{cases}$$

Still More Integrals [for $a > 0$]

$$\int_y^\infty x^n\, e^{-ax}\, dx = \frac{n!\, e^{-ay}}{a^{n+1}} \sum_{k=0}^{n} \frac{(ay)^k}{k!} \quad [n = 0, 1, 2, \ldots]$$

$$\int_t^\infty z^n\, e^{-az}\, dz = \frac{n!}{a^{n+1}}\, e^{-at} \left(1 + at + \frac{a^2t^2}{2!} + \cdots + \frac{a^n t^n}{n!}\right), \quad [n = 0, 1, 2, \ldots]$$

$$\int_0^{2\pi} \sin^2 mx\, dx = \int_0^{2\pi} \cos^2 mx\, dx = \pi, \quad [m = 1, 2, \ldots]$$

APPENDIX J

Integrals Involving the Gaussian Function

General Results

$$\boxed{I_n \equiv \int_{-\infty}^{\infty} x^n\, e^{-(a+ib)x^2}\, dx}$$

$$\left.\begin{aligned} I_{2m+1} &= 0 \\ I_{2m} &= (-1)^m \left(\frac{d}{da}\right)^m \sqrt{\frac{\pi}{a+ib}} \end{aligned}\right\} \qquad m = 0, 1, 2, 3, \ldots$$

$$\boxed{J_n \equiv \int_{-\infty}^{\infty} x^n\, e^{-\alpha x^2 - \beta x}\, dx}$$

$$\begin{aligned} J_{2m+1} &= (-1)^{2m+1} \left(\frac{\partial}{\partial \beta}\right)^{2m+1} \left(\sqrt{\frac{\pi}{\alpha}}\, e^{\beta^2/(4\alpha)}\right) \\ J_{2m} &= (-1)^m \left(\frac{\partial}{\partial \alpha}\right)^m \left(\sqrt{\frac{\pi}{\alpha}}\, e^{\beta^2/(4\alpha)}\right) \\ J_{m+1} &= -\frac{\partial}{\partial \beta}\, J_m \end{aligned}$$

Specific Results

$$I_0 = \int_{-\infty}^{\infty} e^{-(a+ib)x^2}\, dx = \sqrt{\frac{\pi}{a+ib}}$$

$$I_2 = \int_{-\infty}^{\infty} x^2\, e^{-(a+ib)x^2}\, dx = \frac{\sqrt{\pi}}{2}\,(a+ib)^{-3/2}$$

$$J_0 = \int_{-\infty}^{\infty} e^{-\alpha x^2-\beta x}\, dx = \sqrt{\frac{\pi}{\alpha}}\, e^{\beta^2/(4\alpha)}$$

$$J_1 = \int_{-\infty}^{\infty} x\, e^{-\alpha x^2-\beta x}\, dx = -\frac{\beta}{2\alpha}\sqrt{\frac{\pi}{\alpha}}\, e^{\beta^2/(4\alpha)}$$

$$J_2 = \int_{-\infty}^{\infty} x^2\, e^{-\alpha x^2-\beta x}\, dx = \left(\frac{1}{2\alpha}+\frac{\beta^2}{4\alpha^2}\right)\sqrt{\frac{\pi}{\alpha}}\, e^{\beta^2/(4\alpha)}$$

APPENDIX K

Review of Complex Numbers

$$i = \sqrt{-1}$$

Real and Imaginary Parts

If $z = x + iy$, then $x = \Re(z)$ and $y = \Im(z)$.

Polar Representation of a Complex Number

If we plot $x = \Re(z)$ and $y = \Im(z)$, then we can represent z by the polar coordinates r and θ, where

$$r = |z| = \text{distance of the point } z \text{ from the origin ("modulus" of } z)$$

$$\theta = \tan^{-1}\frac{y}{x} = \text{angle that the vector to } z \text{ makes with the } x \text{ axis}$$

The following relationships hold:

$$x = r\cos\theta \qquad |z| = r = \sqrt{x^2 + y^2}$$

$$y = r\sin\theta \qquad z = re^{i\theta} = r(\cos\theta + i\sin\theta)$$

All About the Complex Conjugate

$$\text{If } z = x + iy, \text{ then } z^* = x - iy$$

If $\hat{A}$ is an operator, then $(\hat{A}\Psi)^* = \hat{A}^\dagger\Psi^*$

A real number is one that is equal to its complex conjugate, *i.e.*,

$$z \text{ real} \quad \Longrightarrow \quad z = z^*$$

de Moivre's Theorem

$$[r(\cos\theta + i\sin\theta)]^n = r^n(\cos n\theta + i\sin n\theta)$$

Simple Algebra Involving Complex Numbers

If $z_1 = x_1 + iy_1$ and $z_2 = x_2 + iy_2$, then

$$z_1 + z_2 = (x_1 + x_2) + i(y_1 + y_2)$$

$$z_1 \cdot z_2 = (x_1x_2 - y_1y_2) + i\,(y_1x_2 + x_1y_2)$$

$$\frac{z_1}{z_2} = \frac{x_1x_2 + y_1y_2}{x_2^2 + y_2^2} + i\,\frac{y_1x_2 - x_1y_2}{x_2^2 + y_2^2}$$

Multiples of i

$$i^2 = -1, \quad i^3 = -i, \quad i^4 = +1, \quad i^5 = ii^4 = i, \quad i^6 = ii^5 = i^2 = -1$$

$$i^{4n} = 1, \quad i^{4n+1} = i, \quad i^{4n+2} = -1, \quad i^{4n+k} = i^k$$

$$1/i = -i, \quad 1/i^2 = i^2 = -1, \quad 1/i^3 = 1/ii^2 = -1/i = i$$

APPENDIX L

The Dirac Delta Function

Basic Properties

Definition	$\int_{-\infty}^{\infty} \delta(x - x_0) f(x)\, dx = f(x_0)$
Nature	$\delta(x - x_0) = \begin{cases} 0 & x \neq x_0 \\ \infty & x = x_0 \end{cases}$
Normalization	$\int_{-\infty}^{\infty} \delta(x - x_0)\, dx = 1$
$\delta(x)$ is real	$\delta^*(x) = \delta(x)$
$\delta(x)$ is even	$\delta(x) = \delta(-x)$

Additional Properties

$$\delta(ax) = \frac{1}{a}\delta(x) \qquad \text{for } a > 0$$

$$\int \delta'(x) f(x)\, dx = -f'(0), \qquad \text{where } \delta'(x) = \frac{d}{dx}\delta(x)$$

$$\delta'(-x) = -\delta'(x)$$

$$x\delta(x) = 0$$

$$\delta(x^2 - a^2) = \frac{1}{2a}[\delta(x - a) + \delta(x + a)] \qquad \text{for } a > 0$$

$$f(x)\delta(x-a) = f(a)\delta(x-a)$$

$$\int \delta(x-b)\delta(a-x)\,dx = \delta(a-b)$$

$$\int_{-\infty}^{\infty} \frac{d^m\delta(x)}{dx^m} f(x)\,dx = (-1)^m \left[\frac{d^m f}{dx^m}\right]_{x=0}$$

$$\delta[f(x)] = \frac{1}{|df/dx|_{x=x_0}}\delta(x-x_0), \qquad x_0 = \text{a root of } f(x)$$

Representations

$$\delta(x) = \lim_{\alpha\to\infty}\sqrt{\frac{\alpha}{\pi}}\,e^{-\alpha x^2} \qquad \text{[normalized Gaussian of width } (2\alpha)^{-1/2}\text{]}$$

$$\delta(x) = \lim_{\epsilon\to 0}\frac{1}{\pi}\,\frac{\epsilon}{x^2+\epsilon^2}$$

$$\delta(x) = \frac{1}{2\pi}\int_{-\infty}^{\infty} e^{-ikx}\,dk \qquad \text{[Fourier transform of } 1/\sqrt{2\pi}\text{]}$$

$$\delta(x) = \lim_{\alpha\to 0}\frac{1}{\pi}\,\frac{\sin(x/\alpha)}{x} \qquad \text{[diffraction amplitude of width } \propto 1/\alpha\text{]}$$

$$\delta(x) = \frac{d}{dx}\theta(x)$$

where $\theta(x)$ is the step function, defined by

$$\theta(x) = \begin{cases} 0 & x < 0 \\ 1 & x > 0 \end{cases}$$

$$\delta(x-x') = \sum_{n=0}^{\infty}\frac{1}{\sqrt{\pi}\,2^n\,n!}\exp\left[-\left(\frac{x^2+x'^2}{2}\right)\right] H_n(x)H_n(x')$$

where $H_n(x)$ is the nth-order Hermite Polynomial.

$$\delta(x-x') = \frac{1}{\pi}\int_0^{\infty}\cos k(x-x')\,dk$$

$$\delta(x-x') = \frac{1}{2\pi}\sum_{-\infty}^{\infty}\exp[in(x-x')]$$

$$\delta(x-x') = \frac{1}{2\pi}\left[1+\sum_{1}^{\infty}2\cos n(x-x')\right]$$

$$\delta(x - x') = \lim_{\epsilon \to 0} \frac{e^{-(x-x')^2/\epsilon^2}}{\epsilon\sqrt{\pi}}$$

$$\delta(x - x') = \frac{1}{\pi} \lim_{\epsilon \to 0} \frac{\epsilon}{(x - x')^2 + \epsilon^2}$$

$$\delta(x) = \frac{1}{2L} + \frac{1}{L} \sum_{n=1}^{\infty} \cos\left(\frac{n\pi x}{L}\right)$$ Fourier series for region $+L$ to $-L$

Bibliography to the Appendixes

1. Taylor, B. N., W. H. Parker, and D. N. Langenberg, *Rev. Mod. Phys.*, **41**, 375 (1969).
2. Radzig, A. A., and B. M. Smirnov, *Reference Data on Atoms, Molecules, and Ions* (New York: Springer-Verlag, 1980).
3. The Symbols Committee of the Royal Society, *Quantities, Units, and Symbols* (London: The Royal Society, 1971).
4. McGlaschan, M. L., *Physico-Chemical Quantities and Units*, 2nd ed., (London: The Royal Institute for Chemistry, 1971).
5. Saxon, David S., *Elementary Quantum Mechanics* (San Francisco: Holden-Day, 1968).
6. Dwight, H. B. *Tables of Integrals and Other Mathematical Data*, 4th ed. (New York: Macmillan, 1961).
7. Fischbeck, H. J., and K. H. Fischbeck, *Formulas, Facts, and Constants for Students and Professionals in Engineering, Chemistry, and Physics* (London: Springer-Verlag, 1982).

Bibliography

I used the following books extensively during the writing of this text; they afforded a continual source of insights, problem ideas, and guidance. All are recommended to the serious student of quantum physics. Additional, more specialized references appear in the footnotes and in the Suggested Readings lists at the end of most chapters.

Alonso, M. and H. Valk, *Quantum Mechanics: Principles and Applications* (Reading, Mass.: Addison-Wesley, 1972).

Amaldi, G., *The Nature of Matter: Physical Theory from Thales to Fermi* (Chicago: University of Chicago Press, 1966).

Anderson, E. E., *Modern Physics and Quantum Mechanics* (Philadelphia: Saunders, 1971).

Audi, M., *The Interpretation of Quantum Mechanics* (Chicago: University of Chicago Press, 1973).

Baym, G., *Lectures on Quantum Mechanics* (New York: W. A. Benjamin, 1969).

Beard, D. B. and G. B. Beard, *Quantum Mechanics With Applications* (Boston: Allyn and Bacon, 1970).

Belinfante, F. J., *Measurement and Time Reversal in Objective Quantum Theory* (New York: Pergamon, 1975).

Bethe, H. and R. W. Jackiw, *Intermeditate Quantum Mehcanics* (New York: Benjamin, 1968).

Blokhintsev, D. L., *Quantum Mechanics* (Dordrecht, West Germany: D. Reidel, 1964).

Bockhoff, F. J., *Elements of Quantum Theory*, 2nd ed. (Reading, Mass.: Addison-Wesley, 1976).

Bohm, A., *Quantum Mechanics: Foundations and Applications*, 2nd ed., (New York: Springer-Verlag, 1986).

Bohm, D., *Quantum Theory* (Englewood Cliffs, N. J.: Prentice Hall, 1953).

Bohm, D., *Causality and Chance in Modern Physics* (Philadelphia: University of Pennsylvania Press, 1957).

Borowitz, S., *Fundamentals of Quantum Mechanics* (New York: Benjamin, 1967).

Boorse, H. A. and L. Motz, eds., *The World of the Atom* (New York: Basic Books, 1963).

Brandt, S. and H. D. Dahmen, *The Picture Book of Quantum Mechanics* (New York: John Wiley, 1985).

Cartwright, N., *How The Laws of Physics Lie* (New York: Oxford University Press, 1983).

Clark, H., *A First Course In Quantum Mechanics*, rev. ed. (New York: Van Nostrand Reinhold, 1982).

Cohen-Tannoudji, C., B. Diu, and F. Laloë, *Quantum Mechanics,* Volumes I and II (New York: Wiley-Interscience, 1977).

Cropper, W. H., *The Quantum Physicists and an Introduction to Their Physics* (New York: Oxford University Press, 1970).

Das, A. and A. C. Melissinos, *Quantum Mechanics: A Modern Introduction* (New York: Gordon and Breach, 1986).

Davis, P. C. W. and J. R. Brown, eds., *The Ghost in the Atom* (Cambridge: Cambridge University Press, 1986).

Davydov, A. S., *Quantum Mechanics*, translated, edited and with additions by D. Ter Haar (Oxford: Pergamon Press, 1965).

d'Espagnat, B., *Conceptual Foundations of Quantum Mechanics*, 2nd ed. (Reading, Mass: W. A. Benjamin, 1976).

Dicke, R. H. and J. P. Wittke, *Introduction to Quantum Mechanics* (Reading, Mass.: Addison-Wesley, 1960).

Diner, S., S. Fargue, G. Lochak, and F. Selleri, eds., *The Wave-Particle Dualism: A Tribute to Louis de Broglie on his 90th Birthday* (Boston: D. Reidel, 1984).

Dirac, P. A. M., *The Principles of Quantum Mechanics*, 4th ed., (Oxford: Clarendon Press, 1958).

Eisberg, R. M., *Fundamentals of Modern Physics* (New York: Wiley, 1961).

Eisele, J. A., *Modern Quantum Mechanics with Applications to Elementary Particle Physics: An Introduction to Contemporary Physical Thinking* (New York: Wiley-Interscience, 1969).

Eisberg, R. M. and R. Resnick, *Quantum Physics of Atoms, Molecules, Solids, Nuclei and Particles*, 3rd ed. (New York: Wiley, 1985).

Eyring, H., J. Walder, and G. E. Kimball, *Quantum Chemistry* (New York: Wiley, 1944).

Fano, G., *Mathematical Methods of Quantum Mechanics* (New York: McGraw-Hill, 1971).

Fermi, E., *Notes on Quantum Mechanics* (Chicago: University of Chicago Press, 1972).

Feynman, R. P., *The Character of Physical Law* (Cambridge, Mass.: M. I. T. Press, 1965).

Feynman, R. P., R. B. Leighton, and M. Sands, *The Feynman Lectures on Physics, Vol. III: Quantum Mechanics* (Reading, Mass.: Addison-Wesley, 1965).

Flügge, S., *Practical Quantum Mechanics, I and II* (Berlin: Springer-Verlag, 1971).

Fong, P., *Elementary Quantum Mechanics* (Reading, Mass: Addison-Wesley, 1960).

French, A. P. and E. F. Taylor, *An Introduction to Quantum Physics* (New York: Norton, 1978).

Gasiorowicz, S., *Quantum Physics* (New York: Wiley, 1974).

Gillespie, D. T., *A Quantum Mechanics Primer* (London: International Textbook, 1973).

Gottfried, K., *Quantum Mechanics,* Volume I (Reading, Mass.: Benjamin/Cummings, 1966).

Green, H. S., *Matrix Mechanics* (The Netherlands: P. Noordhoff, Ltd., 1965).

Hameka, H. F., *Quantum Mechanics* (New York: Wiley-Interscience, 1981).

Harris, L. and A. L. Loeb, *Introduction to Wave Mechanics* (New York: McGraw-Hill, 1963).

Herbert, N., *Quantum Reality: Behind the New Physics* (New York: Anchor Press/Doubleday, 1985).

Hey, T., and P. Walters, *The Quantum Univese* (Cambridge: Cambridge University Press, 1987).

Houston, W. V. and G. C. Phillips, *Principles of Quantum Mechanics* (Amsterdam: North-Holland, 1973).

Ikenberry, E., *Quantum Mechanics for Scientists and Engineers* (New York: Oxford University Press, 1962).

Jackson, J. D., *Mathematics for Quantum Mechanics* (New York: W. A. Benjamin, Inc., 1962).

Jammer, M., *The Conceptual Development of Quantum Mechanics* (New York: McGraw-Hill, 1966).

Jammer, M., *The Philosophy of Quantum Mechanics* (New York: Wiley, 1974).

Jauch, J. M., *Foundations of Quantum Mechanics* (Reading, Mass.: Addison-Wesley, 1968).

Jordan, T. F., *Linear Operators for Quantum Mechanics* (New York: Wiley, 1969).

Kompaneyetz, A. S., *Basic Concepts in Quantum Mechanics* (New York: Reinhold Publishing Corp. 1966).

Kramers, H. A., *Quantum Mechanics* (Amsterdam: North-Holland, 1958).

Landau, L. D. and Lifshitz, *Quantum Mechanics, Nonrelativistic Theory* (Oxford: Pergamon Press, 1965).

Landshoff, P. and Metherell, A., *Simple Quantum Physics* (Cambridge: Cambridge University Press, 1979).

Lawden, D. F., *The Mathematical Principles of Quantum Mechanics* (London: Methuen, 1967).

Leighton, R. B., *Principles of Modern Physics* (New York: McGraw-Hill, 1959).

Levine, I. N., *Quantum Chemistry* 3rd ed. (Boston: Allyn and Bacon, 1983).

Liboff, R. L., *Introductory Quantum Mechanics* (San Fransisco: Holden-Day, 1980).

Lowe, J. P., *Quantum Chemistry* (New York: Academic Press, 1978).

Ludwig, G., ed., *Wave Mechanics* (New York: Pergamon, 1968).

McGervey, J. D., *Introduction to Modern Physics*, 2nd ed. (New York: Academic Press, 1983).

Mandel, F., *Quantum Mechanics*, 2nd ed. (London: Butterworths, 1957).

Margenau, H., *The Nature of Physical Reality: A Philosophy of Modern Physics* (Woodbridge, Conn.: Ox Bow Press, 1977).

Martin, J. L., *Basic Quantum Mechanics* (Oxford: Clarendon Press, 1983).

Massey, H. S. W., *Atomic and Molecular Collisions* (London: Taylor and Francis Ltd., 1979).

Mathews, P. T., *Introduction to Quantum Mechanics* (New York: McGraw-Hill, 1963).

Mehra, J. and H. Rechenberg, *The Historical Development of Quantum Theory*, Volumes I–III (New York: Springer-Verlag, 1982).

Merzbacher, E., *Quantum Mechanics*, 2nd ed. (New York: Wiley, 1980).

Messiah, A., *Quantum Mechanics*, Volumes I and II (New York: Wiley, 1966).

Morrison, M. A., T. L. Estle, and N. F. Lane, *Quantum States of Atoms, Molecules, and Solids*, (Englewood Cliffs, N. J.: Prentice Hall, 1976).

Mott, N. F. and I. N. Sneddon, *Wave Mechanics and Its Applications*, (New York: Dover, 1963).

Park, D., *Introduction to the Quantum Theory*, 2nd ed. (New York: McGraw-Hill, 1974).

Pauli, W., *General Principles of Quantum Mechanics* (Berlin: Springer-Verlag, 1980).

Pauling, L. and E. B. Wilson, Jr., *Introduction to Quantum Mechanics* (New York: McGraw-Hill, 1935).

Powell, J. L. and B. Crasemann, *Quantum Mechanics* (Reading, Mass.: Addison-Wesley, 1961).

Price, W. C. and S. S. Chissick, eds., *The Uncertainty Principle and the Foundations of Quantum Mechanics* (New York, Wiley: 1977).

Rae, A. I. M., *Quantum Mechanics* (New York: Halsted, 1981).

Rapp, D., *Quantum Mechanics* (New York: Holt, Rinehart and Winston, 1971).

Reif, F., *Statistical Physics* (New York: McGraw-Hill, 1965).

Richtmeyer, F. K., E. H. Kennard, and J. N. Cooper, *Introduction to Modern Physics*, 6th ed. (New York: McGraw-Hill, 1969).

Rojansky, V., *Introductory Quantum Mechanics* (Englewood Cliffs, N. J.: Prentice Hall, Inc., 1938).

Rubinowicz, A., *Quantum Mechanics* (New York: Elsevier, 1968).

Sakuri, J. J., *Modern Quantum Mechanics*, edited by S. F. Tuan (Reading, Mass.: Benjamin/Cummings: 1985).

Saxon, D. S., *Elementary Quantum Mechanics* (San Francisco: Holden-Day, 1968).

Schiff, L. I., *Quantum Mechanics*, 3rd ed. (New York: McGraw-Hill, 1969).

Schrödinger. E., *Four Lectures on Wave Mechanics* (London: Blackie and Sons, 1928).

Shankar, R., *Principles of Quantum Mechanics* (New York: Plenum, 1980).

Sposito, G., *An Introduction to Quantum Physics* (New York: Wiley, 1970).

Sproull, R. L. and W. A. Phillips, *Modern Physics: The Quantum Physics of Atoms, Solids, and Nuclei*, 3rd ed. (New York: Wiley, 1980).

Strauss, H. L., *Quantum Mechanics: An Introduction* (Englewood Cliffs, N. J.: Prentice Hall, 1968).

Sudbery, A., *Quantum Mechanics and the Particles of Nature: An Outline for Mathematicians* (Cambridge: Cambridge University Press, 1986).

ter Haar, D., ed., *Selected Problems in Quantum Mechanics* (London: Infosearch, 1964).

Tomonaga, S. I., *Quantum Mechanics,* Volumes I and II (Amsterdam: North-Holland, 1966).

von Neumann, J., *Mathematical Foundations of Quantum Mechanics* (Princeton, N. J. : Princeton University Press, 1955).

Weider, S., *The Foundations of Quantum Theory* (New York: Academic Press, 1973).

Wheaton, B. R., *The Tiger and the Shark: Empirical Roots of Wave-Particle Dualism* (New York: Cambridge University Press, 1983).

Wichamnn, E. H., *Quantum Physics*, Berkley Physics Course, Vol. 4 (New York: McGraw-Hill, 1971).

Winter, R. G., *Quantum Physics*, (Belmont, Ca.: Wadsworth, 1979).

Yariv, A., *An Introduction to Theory and Applications of Quantum Mechanics* (New York: Wiley, 1982).

Ziock, K., *Basic Quantum Mechanics* (New York: Wiley, 1969).

Index

NOTE: This index lists only major entries for each item. Page numbers followed by an r contain reference to further information about the item; those followed by an n contain important information about the item in a footnote; and those followed by a p refer to a problem concerning the item.